AF904256

Natural Catastrophes During Bronze Age Civilisations

Archaeological, geological, astronomical and cultural perspectives

Edited by

Benny J. Peiser, Trevor Palmer
and Mark E. Bailey

BAR International Series 728
1998

Published in 2016 by
BAR Publishing, Oxford

BAR International Series 728

Natural Catastrophes During Bronze Age Civilisations

ISBN 978 0 86054 916 1

© The editors and contributors severally and the Publisher 1998

COVER IMAGE
'The Destruction of Sodom and Gomorrah', from Victor Clube and Bill Napier
The Cosmic Winter *Blackwell, Oxford 1990*

The authors' moral rights under the 1988 UK Copyright,
Designs and Patents Act are hereby expressly asserted.

All rights reserved. No part of this work may be copied, reproduced, stored, sold, distributed, scanned, saved in any form of digital format or transmitted in any form digitally, without the written permission of the Publisher.

BAR Publishing is the trading name of British Archaeological Reports (Oxford) Ltd. British Archaeological Reports was first incorporated in 1974 to publish the BAR Series, International and British. In 1992 Hadrian Books Ltd became part of the BAR group. This volume was originally published by Archaeopress in conjunction with British Archaeological Reports (Oxford) Ltd / Hadrian Books Ltd, the Series principal publisher, in 1998. This present volume is published by BAR Publishing, 2016.

Printed in England

BAR
PUBLISHING

BAR titles are available from:

BAR Publishing
122 Banbury Rd, Oxford, OX2 7BP, UK
EMAIL info@barpublishing.com
PHONE +44 (0)1865 310431
FAX +44 (0)1865 316916
www.barpublishing.com

Contents

INTRODUCTION

Benny J. Peiser
School of Human Sciences, Liverpool John Moores University, Trueman Street, Liverpool, L3 3AF, UK

Trevor Palmer
Faculty of Science and Mathematics, Nottingham Trent University, Clifton Lane, Nottingham, NG11 8NS, UK

Mark E. Bailey
Armagh Observatory, College Hill, Armagh, BT61 9DG, Northern Ireland, UK

1. Background

The Second SIS (Society for Interdisciplinary Studies) Cambridge Conference, entitled *Natural Catastrophes during Bronze Age Civilisations: Archaeological, Geological, Astronomical and Cultural Perspectives*, was held at Fitzwilliam College between 11-13 July 1997. The one hundred or so participants, who came from as far afield as North America, Australasia and Japan, as well as from all corners of Europe, were a vibrant blend of enthusiastic amateurs and professionals from all the subject areas under consideration, in keeping with the traditions of the SIS. The event was dedicated to the SIS Vice-Chairman, Geoffrey Bennett, who organised the First Cambridge Conference, but was unable to attend the Second because of terminal illness.

The SIS was formed in 1975 to provide a forum for the discussion of all aspects of catastrophism and chronology. At that time, the gradualist paradigm was supremely dominant, as it had been throughout the previous hundred years, and any attempts to suggest catastrophist explanations for events in geology, evolution or ancient history were viewed with great suspicion and generally ignored [12, 28, 30, 41]. That fate certainly greeted the publication in 1948 of Claude Schaeffer's *Stratigraphie comparée et chronologie de l'Asie occidentale* [32], despite the eminençe of the author, who at various times occupied chairs at the École de Louvre and the Collège de France [10]. Schaeffer's main professional achievement was the excavation of a tell at Ras Shamra in Syria, which he was able to identify as ancient Ugarit.

On the basis of findings here and at other sites throughout the Middle East, Schaeffer claimed that there had been at least five occasions in the Bronze Age when catastrophic destructions occurred in widespread fashion, often with evidence of earthquakes and/or fire.

Two of these were in the Early Bronze Age, the first around 2300 BC, co-incident with the end of the Old Kingdom in Egypt, involving sites in Syria (Byblos, Hama and Ugarit), Palestine (Beth Shan) and Anatolia (Alaça Hüyük, Alishar, Tarsos and Troy), whilst the second occurred perhaps 200 years later, affecting many of these same locations, together with others such as Bait Mirsim, Jericho and Tell el-Ajjul in Palestine and Tell Brak in Mesopotamia. The end of the Middle Bronze Age was marked by destructions at many sites, including Ugarit in Syria, Beit Mirsim, Jericho, Bethel, Hazor and Lachish in Palestine, Alaça Hüyük, Alishar and Boghazköy in Anatolia and Tepe Gawra in Mesopotamia. This was also the time the Hyksos invaded Egypt. Schaeffer further claimed that there were two episodes of widespread catastrophic destruction in the Late Bronze Age, the first around 1365 BC, the time of the Amarna Period in Egypt, affecting locations in Syria (Alalakh and Ugarit), Palestine (Beit Mirsim, Beth Shan, Megiddo, Tell Hesi, Beth Shemesh, Lachish and Ashkelon), Anatolia (Boghasköy, Tarsos and Troy) and Mesopotamia (Chagar Bazar and Tell Brak), and the other around 1200 BC, bringing to an end some Bronze Age cultures, with destructions at most of the same sites in Syria, Palestine and Anatolia as in the previous wave [8,22,31].

Schaeffer was convinced that these catastrophic destructions were the result of natural events, rather than human activity. However, he was undecided as to the precise causes, although undoubtedly favouring the involvement of earthquakes. He did not consider the possibility of the involvement of extraterrestrial factors, a point picked up by the Belgian amateur geologist, René Gallant, in his 1964 book, *Bombarded Earth* [8].

Gallant (who was to become an SIS member, and who addressed a Society meeting in London in 1984, shortly before his death [25]) argued that the seismic activity and climate changes which, according to the evidence provided by Schaeffer, occurred at the times of the destructions, were both likely to have resulted from large meteoritic impacts. *Bombarded Earth*, however, received even less attention than Schaeffer's major work had done.

However, if the ideas of Schaeffer and Gallant made very little impression on the consciousness of others, a very different reaction, although one which was no more positive, greeted those of another catastrophist, the Russian-born psychoanalyst, Immanuel Velikovsky. Largely on the basis of myths from around the world, Velikovsky came to the conclusion that several of the planets of the Solar System had threatened the Earth in historical times. In particular, he believed that Venus had caused major catastrophes by passing close to the Earth at a time corresponding to the end of the Middle Bronze Age in the Middle East, and Mars did similarly a few hundred years later. These ideas were outlined in his 1950 book, *Worlds in Collision* [35].

Despite the mythological origin of Velikovsky's ideas, he made several successful scientific predictions in *Worlds in Collision* and at a graduate forum at Princeton University, a transcript of this talk subsequently being included as a supplement to his later book, *Earth in Upheaval* [37]. Amongst these predictions were that Jupiter would be found to emit radio waves and that, contrary to what was generally believed at the time, the surface of Venus was very hot.

Furthermore, by comparing accounts of catastrophies in different traditions, Velikovsky came to the conclusion that the currently accepted chronologies of certain civilisations were incorrect, and that the supposed "Dark Ages" between the Bronze Age and the Iron Age periods in Greece (and similar ones elsewhere) had never existed. His proposals for a revised chronology for the ancient world were given in *Ages in Chaos* and subsequent books [36, 38, 39].

The eminent physicist Albert Einstein, who from 1921-1924 had been co-editor with Velikovsky of the *Scripta*

Universitatis atque Bibliothecae Hierosolymitarum, from which the Hebrew University of Jerusalem was to grow, found his compilation of evidence for catastrophic events at the Earth's surface convincing, but not his proposed mechanism of planetary interactions. On the other hand, because of Velikovsky's correct predictions, he considered his ideas to be worthy of further study. Many other academics took a different view, however, and in America there was an attempt to suppress publication of Velikovsky's books [13, 40, 42].

For a while there was little knowledge of these events in Britain, but then, in 1973, archaeologist Euan MacKie wrote in *New Scientist* that, no matter whether Velikovsky was right or wrong, he had formulated hypotheses which should be tested in the normal way [17]. In the same year, he suggested in *Pensée* that radiocarbon dating might provide the evidence for a test of Velikovsky's theories of global catastrophes and chronological revisions [18].

A year later, on the 5th November 1974, MacKie discussed related matters with Harold Tresman, Brian Moore and Martin Sieff over a meal at the Regent Palace Hotel in Picadilly and, as a direct consequence, the SIS came into being. The inaugural meeting took place at the Library Association Building, London, in November 1975, with Tresman in the Chair, and 70 members present [34]. (Happily, three of the four founding members of SIS, the exception being Sieff, were present at the Second Cambridge Conference, with MacKie presenting a paper and Moore chairing a session.)

From 1975 onwards, regular debates have taken place at SIS meetings, and in the pages of the Society's journal, the *SIS Review*, later re-named the *Chronology and Catastrophism Review*. To avoid possible misunderstandings, it was made clear right from the start that the Society had been formed to *examine* the ideas of Velikovsky and other catastrophists, not to promote any particular point of view [34].

Of Velikovsky's several claims, the only one which has made significant progress towards widespread acceptance is his general one that the history of life has been shaped by major catastrophes to a far greater extent than his contemporaries realised. Partly that has come about because of increased knowledge of the threat from asteroids and comets in Earth-crossing orbits, together with the growing realisation that many of the craters at the Earth's surface, previously thought to be of volcanic origin, were in fact formed by impacts. Also, many previously-sceptical scientists started to become receptive to catastrophist arguments when physicist Luis Alvarez and colleagues showed, in 1980, that the end of the Cretaceous Period 65 million years ago, when the dinosaurs and many other groups of animals became extinct, is marked in rocks around the world by a high concentration of iridium. This metal is largely absent from the Earth's crust, but present in abundant amounts in extraterrestrial materials, and taken together the evidence suggested the possibility that the extinctions could have been linked to the impact of a large asteroid or comet [1, 6, 7, 23, 26, 28, 30, 33, 41].

So far as the historical record is concerned, orthodox opinion has remained unconvinced about the need to make any major revisions to the established chronologies of ancient civilisations, but challenges continue to be made. In 1978, the SIS, in collaboration with the Extra-Mural Department of Glasgow University, organised its first residential conference to discuss the issues. It was entitled *Ages in Chaos?* and held at the Jordanhill College of Education. The consensus which emerged at the conference was that there were indeed problems with the conventional chronologies but, equally, there were major difficulties with Velikovsky's proposed revisions [2, 11, 15]. Since then, several historians with SIS associations, including Gunnar Heinsohn [14], Peter James [16] and David Rohl [31], have gone on to propose revised chronologies different from those of Velikovsky, and from each other.

Velikovsky's belief that the planets Mars and Venus, now in stable orbits, could have passed sufficiently close to the Earth in historical times to have caused global catastrophes, cannot be reconciled with the known laws of physics, so, although planetary catastrophism still receives enthusiastic support in some quarters, it has been firmly rejected by professional scientists [27, 28, 33, 41].

The British astronomers, Victor Clube and Bill Napier, have acknowledged that Velikovsky may have been correct in suggesting that some myths might have been derived from objects which had been prominent in the ancient sky, and caused catastrophes on Earth, but these cosmic bodies must have been comets, not planets. By extrapolating backwards in time the orbits of Encke's Comet, the Taurid meteor stream and associated Apollo asteroids, Clube and Napier concluded that all were products of a huge comet which came into an Earth-crossing orbit around 20,000 years ago and began to break up, with particular disintegration events occurring about 7500 and 2700 BC. Fragments would have struck the Earth at intervals throughout the Bronze Age, with devastating consequences [6, 7]. Clube put these ideas before the general public for the first time at an SIS meeting in London in 1982, and developed them at another in Nottingham the following year [5, 23]. The model which he and Napier advocate, that small but frequent impacts occur as a consequence of the break-up of a giant comet, has been termed *coherent catastrophism*, in contrast to *stochastic catastrophism*, which involves larger individual impacts occurring in isolated fashion over long intervals of time [33, 40].

In an issue of *SIS Review* published in 1979, Euan MacKie followed up his earlier suggestion of using radiocarbon dating to test for possible correlations between catastrophic events in different locations by carrying out a survey of published data. He tentatively concluded that the end of the Old Kingdom in Egypt, which Schaeffer had included as part of the first wave of Early Bronze Age catastrophes in the Middle East, could also have been contemporaneous with the end of the Chalcolithic in the western Mediterranean, the fall of the Harappan civilisation in India, and the end of the Neolithic in northwestern Europe [19]. One of the sites associated with the last-mentioned event, Skara Brae in Orkney [4], was investigated by Brian Moore and Peter James, who concluded that the evidence was consistent with a catastrophic destruction around 2300 BC [24].

More generally, archaeological, geological and climatic evidence for a world-wide catastrophic event around 2300 BC was presented in the pages of the *SIS Review* by the American engineer, Moe Mandelkehr [20, 21, 22]. At this time, for example, there were global crustal deformations, sea-level discontinuities, earthquakes, volcanic activity, a geomagnetic transient and a transient in the atmospheric radiocarbon concentration [22].

The First SIS Cambridge Conference, held between 16-18 July 1993, was entitled *Evidence that the Earth has Suffered Catastrophes of Cosmic Origin in Historic Times.* At this conference, Bob Porter outlined the destructions which had occurred at various sites during the Bronze Age, and concluded that there was strong evidence of a widespread catastrophe of possible extraterrestrial origin only towards the end of the Early Bronze Age. Even here, however, there was doubt about the precise dating of events at the different sites. Catastrophic events at the end of the Middle Bronze Age, and at other times, remained a possibility but, if any had occurred, they were on a much smaller scale than had been envisaged by Velikovsky [29]. A similar conclusion was also reached by the Old Testament historian, John Bimson [3]. Both Porter and Bimson considered comets to be a far more plausible cause of Bronze Age catastrophes than planetary encounters.

All who attended the First Cambridge Conference considered it to be a great success, characterised by stimulating discussions on a wide range of topics. However, in retrospect, the SIS Council thought that, perhaps, the programme had been *too* wide ranging. Despite the title there had been papers on, for example, the age of the Earth, the age of Venus, and the identity of Job, and there had also been one (by Victor Clube) on catastrophes in the Christian Era, as well as those focusing on events in earlier times. Hence, when planning the Second Cambridge Conference, it was decided to narrow the title to include only Bronze Age catastrophes and, apart from papers concerned with present-day scientific findings which could throw light on past events, to exclude from the formal programme topics which were not clearly related to the subject of the conference.

On the other hand, the meaning of Bronze Age was interpreted loosely, partly for reasons which, since not all who read these Proceedings are likely to be particularly knowledgeable about archaeology, may need a brief explanation. The term refers, of course, to a time characterised by the use of bronze weapons and tools, but it was not an all-or-nothing situation: iron was used, albeit rarely, in the Bronze Age, and bronze continued to be used in the Iron Age.

In any case, metals (of whatever type) were far from common, so the different levels at particular locations are generally classified on some other basis, e.g. style of pottery, enabling correlations to be attempted between different sites, but not without some element of subjectivity. Also, the introduction of a new metal-working technique has to start somewhere, and it could take a long time for it to spread to a far-off region, or to be developed independently. Hence, the Bronze Age undoubtedly started and finished at different times in different places. For example, as we have already noted, the Early Bronze Age in the Middle East overlapped to a considerable extent with the Neolithic in north-western Europe. Furthermore, it is generally believed that the Iron Age in some locations did not begin for several centuries after the end of the Bronze Age, the intervening period being a Dark Age, thus complicating the picture still further. So, a broad view was taken by the organisers of the Second Cambridge Conference as to the period covered by the term Bronze Age and, in consequence, it should be understood that it was concerned with events between about 3500 BC and 500 BC.

2. The Proceedings

The first paper in the Proceedings is based on the keynote address by science journalist Robert Matthews. In this, Matthews makes two main points: (1) that observations made in the distant past may be far more accurate than we generally assume; and (2) that, because of the dangers from asteroids and comets, the Earth is not, and never has been, a safe place to live. He concludes with a quotation from George Santayana: 'Those who do not remember the past are condemned to relive it.'

Then follows a series of papers by astronomers concerned with those hazards from space. Firstly, Mark Bailey reviews recent advances in our knowledge of Near-Earth objects, some of which originated in the cometary regions of the Solar System and some in the main asteroid belt. Calculations indicate that giant comets are likely to come into the inner Solar System and break up every 0.1 to 1 million years. Bailey points out that it is now sometimes difficult to make a clear distinction between asteroids and comets. Regardless of that, they undoubtedly pose a threat, and some may have struck the Earth in the astronomically-recent past. Bill Napier then assembles data from a variety of sources to present a picture of the current interactions between the Earth and its cosmic environment. In his view, the Taurid/Encke complex of interplanetary material has been a regular and occasionally conspicuous hazard over the past 12,000 years or more. This has resulted in impacts such as that which devastated the Tunguska region of Siberia in 1908; in an occasional contamination of the stratosphere by cometary dust, leading to freezing episodes which may have lasted decades; and in small-body impacts into an ocean, causing catastrophic flooding of coastal areas.

After these two papers comes one from Duncan Steel which is more speculative, although based on the same astronomical data and interpretations. Steel makes the intriguing suggestion that the construction around 3500 BC of the Great Cursus near Stonehenge, and that around 3100 BC of the first stage of Stonehenge itself, were intended as predictors of catastrophes, since these were the approximate times when the orbit of the giant proto-Encke comet intersected that of the Earth.

Finally, for this section, Gerrit Verschuur takes both a scientific and a philosophical view of the Earth's place in space. Impacts have been the rule rather than the exception, and will be in the future. The problem of humankind is that hope prevents us from seeing that the cosmic events which have destroyed civilisations in the past will continue to do so, unless we take preventative action.

The next and largest group of papers are concerned with archaeology, geology and climatology. To start this section, Bruce Masse attempts to re-evaluate events on Earth in the light of estimates made by astronomers of the rates of impact of asteroids and comets. On the assumption that 20-30 impacts causing at least local catastrophes are likely to have occurred in the past 6000 years, he examines literary traditions, together with archaeological and palaeo-environmental data, to see if any previously unknown Bronze Age catastrophes can be identified. The most significant one appears to be a cometary impact in the ocean around 2800 BC, which released almost a million megatons of energy, causing devastation on a global scale.

After this come three papers which are concerned, at least in part, with happenings around the time of Mandelkehr's supposed 2300 BC catastrophic event, close to the end of the Early Bronze Age in the Middle East. Firstly, Marie-Agnès Courty presents new archaeological evidence of a dust layer and burnt surface horizon apparently caused by an air blast in northern Syria around 2350 BC. A previous hypothesis involving a local volcanic eruption has now been rejected, with a cosmic catastrophe appearing more consistent with the evidence, but whether such an impact event actually took place at the time has still to be established. Regardless of that, Courty stresses the importance of high temporal resolution investigations in the assessment of causal relationships between natural catastrophes and societal collapse.

Evidence for an adverse climate change in Ireland at about the same time, and on several other occasions, is then given by Mike Baillie. Narrowest-ring events in Irish oak chronologies corresponding to 2345 BC, 1628 BC and 1159 BC line up with similar events in other tree-ring chronologies and also large acidities in Greenland ice records. They also correspond to the approximate ages of the Hekla 4, Santorini and Hekla 3 volcanic eruptions, respectively. However, the narrowest-ring events are imposed on pre-existing climatic downturns, which, as with similar events around 207 BC and 540 AD, suggests a scenario of stratospheric dust loading and bombardments from space, the latter triggering or at least augmenting the volcanic eruptions.

Benny Peiser then summarises a survey he has made of around 500 reports of late 3rd millennium BC civilisation collapse and climate change, which shows a significant clustering around 2300 BC. Most sites in Europe, the Middle East, India and China where civilisation collapsed at this time show clear signs of natural disasters and/or rapid abandonment, whilst around the world there is strong evidence of water-level and vegetation changes, glacier and desert expansion, seismic activity, floods and extinctions of animal species. He concludes that only extraterrestrial bodies acting on terrestrial systems could produce the range of glaciological, geological and archaeological features reported.

The next group of papers is concerned with events which are slightly more recent, occurring around the time certain Late Bronze Age cultures came to an end. Firstly, Amos Nur argues that large earthquakes are likely to have contributed to the physical and political collapse of Late Bronze Age civilisations around the eastern Mediterranean. It is known that, every few centuries, massive earthquakes occur in bursts that sweep across about 1000 km of the eastern Mediterranean over a time-scale of approximately 50 years. In Nur's scenario, the burst at the end of the Late Bronze Age probably began between 1225-1175 BC, and made urban centres vulnerable to opportunist military attacks.

Then, Lars Franzén and Thomas Larsson present evidence from sites in Tunisia and Sweden showing that a major atmospheric cooling event, accompanied by excessive precipitation, which led to flooding, occurred around 1000 BC. Other sources indicate that the event was sudden and widespread, and the finding of small glassy spherules points to a possible impact origin. Franzén and Larsson suggest that an asteroid or comet of diameter in the range 0.5-5 km may have landed in the eastern Atlantic around 1000 BC, affecting in particular Europe, North Africa and the Middle East.

After this, Bas van Geel and colleagues show that a sharp rise in the ^{14}C content of the atmosphere towards the end of the Bronze Age in north-western Europe, around 850 BC, was accompanied by a rapid transition from a relatively warm and dry climate to one which was cooler and wetter. They suggest that a reduced sunspot activity at that time allowed more high-energy galactic cosmic rays to reach the top of the atmosphere, leading to an increased production of ^{14}C cloudiness and precipitation.

The final paper in the section on archaeology, geology and climatology is by Euan MacKie, who begins by warning that astronomers will have to produce clear evidence of comet swarms or the likelihood of large impacts at specific dates before most archaeologists will be willing to re-examine their data with this in mind. He then briefly suggests some examples of instances where such a re-examination might be productive, including two around the end of the Bronze Age in north-western Europe. One of these concerns a site ten miles west of Glasgow, where there are two phases of "cup and ring" rock carvings, the first perhaps from the latter part of the 3rd millennium BC, and the other probably from the 6th or 7th century BC. According to Victor Clube and Bill Napier, these could be representations of comets, but that suggestion is not currently being taken seriously by archaeologists. The other example concerns the vitrified forts of Scotland, dating from the period after 800 BC, whose timber-framed construction might have been intended as a protection against earthquakes.

The Proceedings are then brought to a close by five papers on the subject of history and culture. In the first paper, Gunnar Heinsohn considers the origins of kingship, priesthood and blood sacrifice in the Early Bronze Age. Nicolas-Antoine Boulanger, in the eighteenth century, believed they were reactions to major catastrophes taking place at the time, but that view has been disregarded almost ever since. However, in the light of increased knowledge about cosmic events, Heinsohn argues that Boulanger was correct after all. Re-enacting catastrophic events as rituals involving blood sacrifice would have had a therapeutic effect on traumatised survivors. Significantly, according to Heinsohn, there was a gradual abandonment of blood sacrifice in the Iron Age, when cosmic catastrophes were much rarer events than they had been in the Bronze Age.

Similarly in the next paper, David Pankenier suggests that, contrary to what has generally been supposed, legends and rituals from Bronze Age China may reflect actual events. In particular, around the time of the transition from the Xia to the Shang dynasty in the middle of the second millennium BC, there is a story of ten suns appearing in the sky and then, a few years later, of five planets criss-crossing, and stars falling like rain, after which there was an earthquake and then a drought. It would not be difficult to see this as an indication of the appearance of multiple comets in the sky, and impact-induced catastrophes. The same or a different cometary catastrophe could also form the basis for the legend of the battles between the wicked Chi You and the Yellow Emperor, which featured in ritual games.

Finally come three papers which, on the assumption that major catastrophes were indeed a feature of the Bronze Age and the first few centuries of the first millennium BC (whatever Age one wishes to call this latter period at particular locations), consider how humankind reacted when more peaceful times came along.

Firstly, William Mullen describes how the Milesian School of pre-Socratic philosophers in the sixth century BC set out to explain terrifying phenomena such as thunder, lightning, earthquakes and eclipses in terms of the same processes which it used to explain the orderly arrangement of the Earth and the heavens, thus moving away from the old view which associated them with the unpredictable activities of the Olympian gods. World-destructions could occur, but only in cycles which stretched over vast periods of time. Mullen suggests that the hidden agenda may have been a desire to reassure the population that they were now safe from the cosmic catastrophes which had occurred in the past.

In similar fashion, Irving Wolfe then argues that a cultural crisis occurred in the sixth century BC, with the appearance of new religions, new philosophies, new art forms, new types of games and new forms of social organisation, all of which were very different from what had existed previously. In many ways, these laid the foundations for the cultural characteristics of our modern age. According to Wolfe, the cost has been that, ever since the middle of the first millennium BC, humankind has been suffering from a collective form of Post-Traumatic Stress Disorder, denying not only past catastrophes, but also the possibility of future ones.

The denial of past and future cosmic catastrophes was certainly a feature of the influential philosophy of Aristotle, and has been a characteristic feature of scientific thought over the past few centuries. However, in the concluding paper of the Proceedings, Victor Clube argues that the situation in between was somewhat different. The relatively tranquil period in the middle of the first millennium BC did not last for long, and further episodes of cosmic bombardment conditioned people once again to believe that the world might come to an end in this way. Clube suggests that this provides strong support for coherent rather than stochastic catastrophism, because frequent small-scale events would keep the issues in people's minds, which would not be the case if there were vast periods of time between impacts. According to Chinese astronomical records, there have been seven peaks of fireball activity in the past 2000 years, at times which indicate an association with the Taurid/Encke complex. However, the past two centuries have been a quiet period and, because of the influence of Lyell and Darwin (who established the gradualistic paradigm, largely for philosophical reasons), and of Newton (who played down the threat from space on religious grounds), the future seemed secure. We now know otherwise but, in contrast to previous generations, who could only hope and/or pray, we may soon have the capability for defending ourselves. However, Clube warns that the prospect of safeguarding the future of civilisation is not being helped by those who cling to gradualistic, Earth-centred views, or by those who adopt what he sees as erroneous forms of catastrophism. To produce the best answer, we must fully understand the problem.

Acknowledgements

We are grateful to Alasdair Beal, Birgit Liesching, Bill Napier and David Roth for their help in preparing papers for these Proceedings.

References

1. Alvarez, L.W., Alvarez, W., Asaro, F. and Michel, H.V., 1980, Extraterrestrial cause for the Cretaceous-Tertiary extinction, *Science* **208**, 1095-1108.
2. Bimson, J.J., 1982, Can there be a revised chronology without a revised stratigraphy?, in *Ages in Chaos? Proceedings of the Glasgow Conference, 1978*, 16-26, SIS.
3. Bimson, J.J., 1994, The nature and scale of an Exodus catastrophe re-assessed, in *Proceedings of the 1993 Cambridge Conference*, 33-44, SIS.
4. Childe, V.G., 1931, *Skara Brae: A Pictish Village in Orkney*, London: Kegan Paul, Trench, Trubner & Co.
5. Clube, V., 1980/81, Cometary catastrophes and the ideas of Immanuel Velikovsky, *SIS Review* **V**, 106-111.
6. Clube, V. and Napier, W., 1982, *The Cosmic Serpent*, Faber and Faber, London.
7. Clube, V. and Napier, W., 1990, *The Cosmic Winter*, Basil Blackwell. Oxford.
8. Gallant, R., 1964, *Bombarded Earth*, John Baker, London.
9. Gammon, G.J., 1980, Bronze Age destructions in the Near East, *SIS Review* **IV**, 104-108.
10. Gammon,G.J., 1980/81, Dr Claude Schaeffer-Forrer, 1898-198 2: an appreciation, *SIS Review* **V**, 70-71.
11. Gammon, G.J., 1982, The nature of the historical record, in *Ages in Chaos? Proceedings of the Glasgow Conference, 1978*, 12-15, SIS.
12. Gould, S.J., 1987, *Time's Arrow, Time's Cycle*, Harvard University Press.
13. de Grazia, A., 1966, *The Velikovsky Affair*, Sidgwick and Jackson, London.
14. Heinsohn, G., 1988, *Ghost Empires of the Past*, SIS.
15. James, P.J., 1982, in *Ages in Chaos? Proceedings of the Glasgow Conference, 1978*, 34-52, SIS.
16. James, P.J., 1991, *Centuries of Darkness*, Jonathan Cape, London.
17. MacKie, E.W., 1973, A challenge to the integrity of science, *New Scientist* 11 January, 76-77.
18. MacKie, E.W., 1973, A quantitative test for catastrophic theories, *Pensée IVR* **III**(Winter), 6-9.
19. MacKie, E.W., 1979, Radiocarbon dates and cultural change, *SIS Review* **III**, 98-100.
19. Mandelkehr, M.M., 1983, An integrated model for an Earth-wide event at 2300 BC. Part I: The archaeological evidence, *SIS Review* **V**, 77-95.
21. Mandelkehr, M.M., 1987, An integrated model for an Earth-wide event at 2300 BC. Part II: Climatology, *Chronology and Catastrophism Review* **IX**, 34-44.
22. Mandelkehr, M.M., 1988, An integrated model for an Earth-wide event at 2300 BC. Part III: The geological evidence, *Chronology and Catastrophism Review* **X**, 11-22.
23. Moore, J.B., Abery, J. and James, P.J., 1984, Global catastrophes: new evidence from astronomy, biology and archaeology, *SIS Review* **V**, 89-91.
24. Moore, J.B. and James, P.J., 1984, Skara Brae: a time capsule of catastrophism?, *SIS Review* **VI**, 104-107.
25. Moore, J.B., Palmer, T. and James, P.J., 1985, Comets, meteorites and Earth history, *SIS Review* **VII**, 2-5.
26. Palmer, T., 1985, Catastrophism and evolution, *SIS Review* **VII**, 9-20.
27. Palmer, T., 1997, Review of *Origins* by J.E. Strickling, *Chronology and Catastrophism Review* **(2)**, 47-49.
28. Palmer, T., 1998, *Controversy - Catastrophism and Evolution: The Ongoing Debate*, Plenum Press, New York.
29. Porter, R.M., 1994, Bronze Age multi-site destructions, in *Proceedings of the 1993Cambridge Conference*, 45-50, SIS.
30. Raup, D.M., 1986, *The Nemesis Affair*, Norton, New York.
31. Rohl, D., 1995, *A Test of Time*, Century, London.
32. Schaeffer, C.F.A., 1948, *Stratigraphie comparée et chronologie de l'Asie occidentale (IIIe et IIe millénaires)*, Oxford University Press.
33. Steel, D., 1995, *Rogue Asteroids and Doomsday Comets*, Wiley, New York.
34. Tresman, H., 1994, The SIS, its history and achievements: a personal perspective, *Proceedings of the 1993 Cambridge Conference*, 2-6, SIS.
35. Velikovsky, I., 1950, *Worlds in Collision*, Doubleday, New York.
36. Velikovsky, I., 1952, *Ages in Chaos*, Doubleday, New York
37. Velikovsky, I., 1956, *Earth in Upheaval*, Gollancz, London.
38. Velikovsky, I., 1977, *Peoples of the Sea*, Sidgwick and Jackson, London.
39. Velikovsky, I., 1978, *Ramses II and his Time*, Sidgwick and Jackson, London.
40. Velikovsky,I., 1983, *Stargazers and Gravediggers*, William Morrow, New York.
41. Verschuur, G.L., 1996, *Impact*, Oxford University Press.
42. Vorhees, D., 1993, Velikovsky in America, *Aeon* **III(4)**, 32-58.

The Past is our Future

Robert A. J. Matthews
Science Correspondent, The Sunday Telegraph, London, UK
Visiting Fellow, Department of Computer Science, Aston University, Birmingham, UK

The past is another country: they do things differently there.
(L. P. Hartley, *The Go-Between,* 1953)

1. Introduction

It can be hard to resist dismissing the beliefs of our predecessors as in some way inferior to our own: less sophisticated, perhaps, or less rigorous. Nowhere is this temptation stronger than in science, the whole enterprise of which is underpinned by a conviction that we are making more or less steady - if asymptotic - progress towards The Truth. Examine the literature references at the end of the typical paper in *Nature*, and one finds an exponential decline in the numbers of citations as one goes further back in time. The vast bulk will be from the last five years, with barely any dating back more than 15 years. It would be easy to conclude that most of what was done over 15 years ago proved to be unreliable, or uninteresting - or plain wrong.

Of course, so peremptory a dismissal of past work would be a mistake. By its very nature, science tends to progress by pulling together many strands of research into a unified whole. The result is that entire tranches of research disappear into a broader synthesis. Perfectly sound research by Faraday, Oersted, Ampere and Gauss is rarely cited directly today, having been subsumed into Maxwell's equations. Newton's law of gravity has been absorbed into the field equations of general relativity, which themselves appear deducible from the quantum field-theoretic concept of mass-less spin-2 gravitation exchange.

Other scientific ideas vanish from the literature not because they are wrong, but because they are no longer 'fashionable', or lack definitive observational support. Aristarchus of Samos outlined the heliocentric theory of the solar system 1,800 years before Copernicus. Kant's concepts of the Solar System's membership of the galaxy, and of the existence of many such galaxies throughout the universe were only confirmed in the 1920s. Even in that most hard of hard sciences, elementary particle physics, the effects of fashion can be seen in the evolution of, for example, superstring theory.

The 2nd SIS Cambridge Conference brings together people from many disciplines who are less keen to dismiss 'old' beliefs as wrong beliefs. Specifically, the view that binds many of the delegates at this conference is a belief that the threat of catastrophe to present and future civilisations can be gauged by taking seriously the beliefs, observations and actions of Bronze Age people.

Such a conviction is, I suspect, not widely shared; indeed it may be regarded as somewhat eccentric by those not familiar with the field. Such sceptics have, I would argue, fallen victim to two long-standing myths. The first myth is that, as L.P. Hartley puts it, they did things differently in the past - and, more specifically, that they did things less well, less conscientiously and less rationally. Second, there is the myth that because we have not suffered any great cosmic catastrophe in our own time, our planet must *ipso facto* be immune from such catastrophe. In this keynote address, I shall highlight the dangers of believing these two modern myths.

2. The reliability of past observations

The devices wielded by contemporary scientists may bear little resemblance to those used by the ancients, yet the *care* with which the available instrumentation is used creates a bond between researchers that spans the centuries. The Hubble Space Telescope has a pointing accuracy of better than 10 milliarc seconds, an impressive figure achieved through extremely sophisticated computerised control systems. At first glance, this makes Tycho Brahe's 16th century catalogue accuracy of around 75 arc seconds seem almost comically incompetent - until one learns that it was achieved using nothing more than an arrangement of sighting sticks. Furthermore, there is nothing laughable in the achievements made possible by Brahe's skill and care in using such simple apparatus. For example, it allowed him to measure the parallax of the Comet of October 1577, showing that it lay beyond the orbit of the Moon, and thus presented a challenge to the Aristotelian concept of celestial spheres and the immutability of the heavens. His sighting sticks also revealed the complex nature of the Moon's motion, laying the observational foundations for lunar theory, whose elucidation became a key challenge to celestial mechanics for centuries afterwards. Most famously, of course, Tychos' measurements of the positions of the planets were sufficiently accurate to allow Kepler to establish the non-circular nature of their orbits, and thus provide a crucial test of Newton's subsequent law of gravitation [11].

Recent research [10] has shown that between AD 830 and 1020 Arab astronomers achieved previously unrivalled accuracy in the measurement of meridian solar altitudes (to within 0.02 degrees) and equinox times (to within 1.2 hr). The altitude determinations have an accuracy close to the limit possible with unaided human eye, and had no rival for five centuries.

Further back still, the attempts of the ancients to determine key time periods in astronomy are particularly impressive. As far back as 430 BC, the Greek astronomer Meton had determined the length of the tropical year, i.e. the interval between two successive Vernal equinoxes, as 365 + $(5/19)^{th}$ of a day - 99.43 per cent accurate. By the 2nd century BC, Hipparchos of Nicaea had pinned the value down to 365.25 - $(1/300)^{th}$ of a day: a 99.88 per cent accurate result. For the length of the synodic month - the period between two successive new moons - Hipparchos himself adopted a Babylonian figure of 29 days 12 hr 44 min 3.3 seconds - which is within half a second of the modern value, or 99.9987 per cent accurate [8].

It would of course be wrong to presume from this that the ancients are an unimpeachable source of astronomical data. Babylonian lunar eclipse contact timings have turned out to be extraordinarily unreliable, with errors of 10 per cent or more [14]. The attempts of the ancient Greeks to determine relative distances of the moon and sun were also far from impressive: Aristarchus of Samos, for example, deduced mean distances for these two objects as 9.5 and 180 earth diameters respectively; the actual values are 30 and 11,740 [8, 54-55].

Nevertheless, there can be little doubt that the ancients were generally conscientious observers who made the best use of what limited technology they had to make sense of the cosmos - and that they frequently succeeded astonishingly well.

3. The reliability of past beliefs

One might argue that the quantitative efforts of our scientific forebears would be expected to stand up to scrutiny reasonably well, as their conclusions were necessarily constrained by the mathematical underpinning of their work. This might lead one to expect that their more qualitative claims would be less constrained, more influenced by contemporary beliefs - and thus far more suspect. Yet again, this seems to say more about our own prejudices against the past than about reality.

For example, reports from the occupants of the Normandy village of Laigle of a storm of 'stones from the sky' on the night of 26 April, 1803 may seem to be a classic example of myth-making by a barely-literate population. Indeed, this appears to have been the view of the contemporary French scientific establishment, which dispatched the young Jean Biot to debunk such a wild notion. The result was the establishment of the reality of meteorites as extra-terrestrial objects [6].

In the late eighteenth century, the French philosopher and mathematician Pierre-Simon de Laplace raised the possibility that comets might have catastrophic effects on the earth, a claim that until very recently would have seemed ludicrous. Yet he was merely following the lead of Sir Edmond Halley, who in 1694 had suggested that the biblical flood may have been caused by a comet whose impact created the Caspian Sea and other great lakes of the world [12]. Halley's claims are now considered to be almost mainstream, with the Manicouagan and Clearwater Lakes, Canada, now thought to be impact-related.

A recent historical discovery by Japanese researchers [15] raises some intriguing questions about the role of historical accidents on past beliefs. Isshi Tabe, an amateur astronomer, has discovered a drawing entitled "Nouvelles d'escouvertes dans le globe de Jupiter" by Giovanni Cassini at the Paris Observatory. It records the appearance of a dark spot on the face of the giant planet on the night of 5 December 1690, and sketches its changes through the following nights until 23 December.

The drawing bears striking similarities to the aftermath of the impact of comet Shoemaker-Levy 9 observed by modern astronomers almost 300 years later. The spot does not gradually come into being like some cyclonic storm on the planet, but suddenly appears as a dark, round patch. Measurements from the drawing suggest that it was around 7,500 km across - a huge disturbance, but very similar in size to the "spots" created by some of the fragments of SL-9.

Particularly impressive is the description of what happened to the spot over the next 18 days. According to Cassini, the spot became crescent-shaped, and stretched out parallel Jupiter's equator. It then split into two, then merged, and finally turned into three narrow patches before vanishing.

Using measurements of Jovian windspeeds made by the Voyager fly-by missions, Tabe and Junichi Watanabe - who organised SL-9 impact observations at the Okayama Astrophysical Observatory - have found that Cassini's 300 year old account fits in well with modern data about conditions in Jupiter's atmosphere. His description also mirrors almost exactly the evolution of one of the spots created by the SL-9 impact. According to Tabe and Watanabe, the results are consistent with a comet around half a kilometre across impacting into Jupiter on 5 December 1690.

Cassini would, of course, have been quite unable to detect the comet with the telescopic equipment of the day. Had he been, he might have been able to avert a substantial shift away from concern about cosmic catastrophes, which flowed from the Newtonian view of the solar system as a 'majestic clockwork' driven by immutable laws. Newton's discovery of the inverse-square law of gravity famously gave astronomers an unprecedented ability to predict cosmic events. This ability reached its apogee in 1705, when Halley himself showed that even the appearance of comets - previously regarded as capricious 'portents' of doom - could be understood. Specifically, he showed that one bright comet - now bearing Halley's name - had been seen on a regular basis since the 15th century. Halley famously used Newton's newly-discovered law of gravity to predict its return in 1758. The comet came in more or less on schedule, thus providing impressive - though as we now know, misleading - support for the notion that the solar system cannot spring nasty surprises on us.

If Cassini had been able to demonstrate that a comet was responsible for the event on Jupiter, one cannot help but think that the subsequent complacency about our cosmic environment may never have descended.

4. The reliability of 'mythological' beliefs

As we reach further back in time, beliefs founded on what we would recognise as scientific principles and solid quantitative evidence become rarer, replaced with beliefs that appear positively fantastical. It is here that the temptation to reject as useless the testimony of our forebears is strongest. Yet once again one can find examples which challenge such peremptory dismissal. One of the most impressive cases centres on the following report that appears in the chronicles of the medieval monk Gervase of Canterbury [3]:

> "In this year, on the Sunday before the Feast of St John the Baptist, after sunset when the moon had first become visible a marvellous phenomenon was witnessed by some five or more men who were sitting there facing the moon, Now there was a bright new moon, and as usual in that phase, its two horns were tilted towards the east; and suddenly the upper horn split in two. From the midpoint of the division a flaming torch sprang up, spewing out, over a considerable distance, fire, hot coals and sparks. Meanwhile the body of the moon which was below writhed, as it were, in anxiety, and, to put it in the words of those who reported it to me and saw it with their own eyes, the moon throbbed like a wounded snake. Afterwards it resumed its proper state. This phenomenon was repeated a dozen times or more, the flame assuming various twisting shapes at random and then returning to normal. Then after these transformations, the moon from horn to horn, that is along its whole length, took on a blackish appearance. The present writer was given this report by men who saw it by their own eyes, and are prepared to stake their honour on an oath that they have made no addition or falsification in the above narrative."

From Gervase's account, we can date the time of these extraordinary events as the evening of 25 June 1178 in the Gregorian calendar (18 June in the Julian calendar). The florid language leads one to suspect that, *pace* Gervase's implorations to the contrary, the whole account is little more than an amusing medieval fantasy. Hartung has shown, however, that the account is consistent with the description of an impact between the moon and an object that would have left a crater at least 10km in diameter [5], and was even able to deduce rough selenographic coordinates for the site of the crater. It turns out that there is indeed a very plausible candidate for the impact site consistent with these coordinates: the 20-km wide crater Giordano Bruno.

Still more impressive evidence for the reliability of Gervase's account emerges from an analysis of the effect of the impact on lunar dynamics [2]. Observations of oscillations of the moon about its axis carried out using the Laser Ranging Retro Reflectors placed on the moon by both Soviet and American lunar missions reveal the presence of a 15-metre 'wobble' of period 3 years. Such oscillations are consistent with an impact at the site of Giordano Bruno in the recent past.

Of course, one cannot expect many such 'fantastical' accounts to be amenable to such detailed analysis and apparent confirmation. Much more representative are the examples of *De excidio et conquestu Britanniae*, written by the 6th century cleric Gildas, and the *Annales Cambriae* [3, 106-9]. While both can be dismissed as overwrought accounts of disasters wreaked by enemies, they can also be seen as descriptions of the effects and aftermath of a cosmic impact in the mid 5th century AD. While there are obvious dangers of seeing descriptions of such impacts in every ancient tale of woe, from what we have seen, we should perhaps now be more willing to treat accounts of 'heaven-sent' disasters somewhat more seriously. Among the testimony of the past that might benefit from such open-mindedness one might count the following, some of which are discussed elsewhere in this volume:

- The widespread nature of Flood and Judgement Day legends: is this simply testimony to the power of cultural importation, or is it the result of a folk-memory of genuinely widespread catastrophes ?

- The Aztec ball-game: is it a re-enactment of a cosmic 'ball-game' with catastrophic consequences ? [9]

- Glyphs such as the Panorama stone, Ilkley, UK: mere pretty decoration, or images of comets ? [3, p. 176]

- Stonehenge: a place of agrarian worship, an eclipse observatory - or a predictor of recurring comet hazard ? [13]

- Pyramids: majestic tombs of appealing geometric simplicity - or representations of the conical form of the Zodiacal Light, boosted by comet disintegration ? [12, pp. 163-166].

5. Why should we care ?

I hope I have shown that past observations, beliefs and even 'myths' should not automatically be dismissed as unreliable, muddle-headed or just crazy. Yet apart from ridding us of what one might call temporal chauvinism, what purpose is served by taking them seriously? From the Dark Age chronicles of Gildas, through the reports of Gervase of Canterbury to the observations of Cassini, we are presented by evidence that comet and asteroid impacts are not something that *might* occur: they have occurred, and will do again. As such, the evidence of the past can counter the effect of a series of accidents of history that gave birth to perhaps the greatest myth of all: that the earth is a safe haven for humanity.

I have already alluded to some of the events that contributed to this potentially catastrophic line of reasoning. Ironically, many of them stem from precisely the kind of rational, objective and quantitative evidence that we usually regard as the best defence against the emergence of myths. Prime among them is Newton's discovery of a fundamental law governing the behaviour of all celestial bodies: the inverse-square law of gravity, which formed the centrepiece of the image of the solar system as a majestic clockwork, a comfortingly cyclic metaphor for what had previously seemed incomprehensible and capricious.

Halley's successful prediction of the return of his eponymous Comet suggested that even these apparently lawless objects were part of that same clockwork, thus suggesting that all comets were no more 'portents of doom' than eclipses, whose cyclic and predictable nature had been revealed far earlier.

This of course overlooks the fact that comets could still emerge from the abyss of space without warning. Yet the terror engendered by this possibility was neutralised by another historical accident: the close approach of Lexell's Comet in June 1770, which made its closest approach to the earth only a fortnight after being discovered, but with no obvious ill-effects - despite coming within just 2.3 million km of hitting the earth, (still a record close-approach for a comet). The later failure of the disintegration of Comet Biela in 1846 to trigger any obvious calamity served to confirm the idea that comets should hold no terror for us. Once again, one cannot but reflect on how different our attitude might have been had Cassini been able to detect the comet responsible for the disfigurement of Jupiter in 1690.

6. Conclusions

The reluctance to return to the ancient view of the earth as a place vulnerable to attack from the heavens is - perhaps understandably - proving hard to overcome. However, as the amount of evidence in support of the ancient view mounts, not least from ancient sources, there can be little doubt that this possibility must be accepted eventually. The process is already underway: by the early 1980s, Luis Alvarez and his co-workers were winning supporters for their claims for a bolide-related extinction of the dinosaurs [1] which had been unsuccessfully advocated years earlier by Nininger [7] and De Laubenfels [4].

But recognition of the threat from cosmic impacts is of only restricted value: ultimately, it must be turned into action. That will require the establishment of detection programmes, to discover the size and nature of the threat, and perhaps the development and deployment of ameliorative measures. All this will require funding, and thus considerable justification. That, in turn, demands evidence - and as I hope I have shown, the historical and cultural record of our planet can provide that evidence. The need for greater recognition of this has never been more urgent - for, more than any other threat we face, the threat of cosmic impacts has the power to reveal the truth of the words of George Santayana: "Those who do not remember the past are condemned to relive it."

References

1. Alvarez, L.W., Alvarez, W., Asaro, F., Michel, H.V. 1980 *Extraterrestrial cause for the Cretaceous-Tertiary extinction* Science **208**, 1095

2. Callame, O. Mulholland, J.D. 1978 *Lunar crater Giordano Bruno: AD 1178 impact observations consistent with laser ranging results* Science **199**, 875
3. Clube, V., Napier, W.M. 1990 *The Cosmic Winter* (Oxford: Basil Blackwell) 159
4. De Laubenfels, M.W., 1956 *Dinosaur extinction:one more hypothesis* J. Paleontology **207**
5. Hartung, J.B. 1976 *Was the formation of a 20 km diameter impact crater on the moon observed on June 18, 1178 ?* Meteoritics **11**, 187
6. Milton, R., 1994 *Forbidden Science* (London: Fourth Estate), p. 3
7. Nininger, H.H., 1942 Cataclysm and evolution, *Popular Astronomy* **50**, 270
8. Pederson, O. Pihl, M. 1974 *Early Physics & Astronomy* (London: Macdonald & Janes) p 42.
9. Peiser, B.J. 1996 *Cosmic Catastrophes and the Ballgame of the Sky Gods in Mesoamerican Mythology,* Chronology & Catastrophism Review **17,** 29
10. Said, S.S., Stephenson, F.R. 1995 *Precision of medieval Islamic Measurements of Solar Altitudes and Equinox Times.* Journal for the History of Astronomy **26,** 117
11. Singer, C. 1959 *A Short History of Scientific Ideas to 1900* Oxford University Press
12. Steel, D. 1995 *Rogue Asteroids and Doomsday Comets* (Chichester: Wiley)
13. Steel, D. 1998 (see elsewhere in these proceedings)
14. Stephenson, F.R., Fatoohi, L.J. 1993 *Lunar eclipse times recorded in Babylonian history* Journal for the History of Astronomy **24,** 255
15. Tabe I., Watanabe J., Jimbo M. 1997 *Discovery of a Possible Impact Spot on Jupiter Recorded in 1690* Pub. Astron. Soc. Japan, **49**, L1-L5

Sources and Populations of Near-Earth Objects: Recent Findings and Historical Implications

Mark E. Bailey

Armagh Observatory, College Hill, Armagh, BT61 9DG, Northern Ireland, UK

Summary

Near-Earth Objects (NEOs) comprise a heterogeneous population of minor bodies, encompassing both comets and asteroids and originating from sources as diverse as the outer Oort cloud and the main asteroid belt. Their orbital evolution is chaotic, frustrating attempts to infer their origin by simple back-tracking of their motion, while the cometary NEOs can sometimes evolve into orbits indistinguishable from those of asteroids, and vice-versa. Physically, comets may also occasionally resemble outer main-belt asteroids, for example when far from the Sun and not outgassing, or close to the Sun and heavily mantled or devolatilized; and asteroids may sometimes appear cometary. However, the two classes of body have very different principal modes of evolution and decay, reflecting their fundamentally different internal structure and chemical composition. Near-Earth Objects, in frequently passing close to the Earth, provide outstanding opportunities to resolve the differences between evolved comets and asteroids, whilst occasional impacts on the Earth provide an additional powerful motivation for their scientific study. This paper reviews progress towards understanding NEOs and considers the effects of possible variations in the flux and size distribution of impacting objects. Evidence for a significant number of 'asteroids' in Halley-type cometary orbits, originating from the evolution and decay of Halley-type comets, and for impacts in the astronomically recent past is briefly mentioned.

1. Introduction

Recent years have seen an almost explosive growth in our knowledge of the origin and dynamics of small solar system bodies. In addition to focusing on standard cosmogonical questions, for example on theories of the formation of the solar system and their implications for the origin of the Earth and other planets, including extra-solar planetary systems, much attention has also been directed towards understanding the various subsets of objects which may come close to the Earth. These bodies are generally planet-crossing, and during the course of their dynamical evolution may collide with the Earth or another planet.

The global effects of the largest such collisions are well documented (e.g. craters on the Moon, biological mass-extinctions on Earth), and are known to affect the evolution of life. At the other extreme, the more frequent, small-scale impacts (which occur on the average roughly every few hundred or few thousand years) are usually regarded as of little consequence, certainly in comparison with the effects of the impact of a kilometre diameter asteroid, which although not causing a mass extinction of life might nevertheless ruin civilisation. Studies of the number of Near-Earth Asteroids (NEAs) or of the number of 10-20 km diameter craters on Earth indicate that such impacts occur roughly every 100,000 years [73, 38, 37].

The significance of the size of the impacting bodies hinges, in this context, on the characteristic dimension of the objects capable of producing global catastrophes. The boundary is usually placed - for both comets and asteroids - at a diameter of about 1 km; objects this size and larger produce craters with diameters in excess of 10-20 km. It is a fortunate coincidence that such objects, though rare in space, are fairly easy to find with modest astronomical equipment. Proposals to survey the whole sky in order to provide an accurate assessment of the impact hazard to civilisation (e.g. [61]), known collectively as Spaceguard projects, have therefore focused on how to identify a more or less complete sample of the kilometre-size objects which, it is assumed, present the greatest extraterrestrial threat.

A second issue concerns the impact frequency. Virtually everyone agrees that the greatest risk is carried by the greatest body (e.g. the 'killer' asteroid which finished the dinosaurs). It is again fortunate, for us, that the size-frequency distribution of such objects, independent of whether the objects are primarily comets or asteroids, is such that these heart-stopping moments occur only at intervals on the order of 10-100 million years (Myr). It is ironic, however, that having achieved this useful understanding of the Earth's potentially violent near-space environment, the gap between impacts is just too long to grasp. In practice it appears almost infinite, and most people routinely accept, or sometimes even invite, day-to-day individual risks of a life-threatening nature that have much greater annual probabilities of occurrence. Partly because of this, most agencies with the resources to consider funding a viable Spaceguard programme find difficulty in justifying the adoption of an appropriate long-term view.

Thus, although the primary reason for pursuing such a project is straightforward: namely, to discover the largest Earth-colliding NEO before it discovers us; it is often hard to pin down reasons why the programme should be given a high priority. The tangible and perhaps paradoxical sense of urgency, which is no doubt felt by many in the research community, may turn out to be misplaced; but in order to resolve the question whether there exists a NEO with the capacity to destroy civilisation within (say) the next 200 years, it is absolutely essential that the astronomical survey should begin sooner rather than later. Moreover, we are currently facing a period comparable at least to the orbital periods of the impacting objects before the required NEO catalogue can possibly be said to be sufficiently complete. If inert Halley-type comets provide the greatest threat (e.g. [10]), the period in question probably runs to at least several hundred years (cf. [75]).

At the other extreme of impact frequency, considering objects with diameters measured in metres or tens of metres, although these bodies run into the Earth every year or every few hundred years, such members of the NEO ensemble are very hard to detect in space. However, they are actuarially of very little consequence, and one is again tempted (perhaps justifiably in this case) to play the odds, and to *assume* that impacts of such small bodies will probably occur far from a major city or conurbation. 'Small' objects, it is argued, although arriving on time-scales of human concern, can be ignored.

Much smaller particles, of the size of dust grains, run into the Earth more or less continuously, at a current rate on the order of 4 x 10^7 kg yr^{-1} [50], albeit with substantial variations at certain times of the year or in certain years. It has often been argued (e.g. [43, 21, 85, 86, 23, 24, 63]), that although the individual dust particles are minute they nevertheless exert a global influence through effects on the transparency of the atmosphere and associated climate change.

The observed fluctuations in the dust flux brings us to the third key point, namely probable variation in the mean impact rate of even the larger bodies (e.g. [79]). This aspect of the subject also suffers from an apparent disconnection, in that the events which would appear most likely to produce substantial time-dependence in the NEO flux would seem to occur only at intervals measured in tens or hundreds of Myr; and in any case the ensuing variability would be expected to have a characteristic timescale on the order of the dynamical lifetime of the NEOs themselves, i.e. in the range 0.1-10 Myr. An example would be the passage of a star through the Oort cloud. Such close stellar encounters disturb the orbits of long-period comets in the inner core of the Oort cloud and produce short-lived, intense 'comet showers' lasting perhaps a million years during which the cometary influx may exceed the present value by a factor of 1000 or more. But sufficiently close encounters to cause this effect are rare, occurring at intervals of at least 200 Myr; a recent analysis of data from the Hipparcos satellite [36] suggests that it is highly unlikely that a substantial comet shower will occur within the next 0.5 Myr.

A similarly long timescale is associated with asteroid showers. These occur following random collisions between large asteroids in the main asteroid belt, which scatter kilometre-size fragments and smaller objects (ranging from the size of meteorites down to dust [48, 88]) onto Earth-crossing orbits. The enhanced NEO population again persists for several Myr, but the largest collisions, known as asteroid-family producing events, are rare, and have probably only occurred a handful of times in the geological record (cf. [74]).

Why, then, are a few astronomers apparently hell-bent on pursuing the idea that some aspects of the development of civilisation, even the evolution of the Earth (the past 30,000 years, say), might be explained in part as owing to the cumulative effects of comet and asteroid bombardment? Were this thesis correct then Project Spaceguard would certainly acquire a new sense of urgency; but that hardly explains the academic motivation. Instead, as has been previously noted (e.g. [20, 23, 42]), what is at stake, no less, is a coherent understanding of Man's place in the Universe and the survival of the human species. What - astronomically speaking - has been left out?

1.1 Streams

Early indications of the importance of streams of small solar system bodies were provided by arguments based on the Taurid meteor showers, the meteoroids of which comprise a substantial proportion of the interplanetary dust flux onto the Earth. The general picture, initially developed by Whipple (e.g. [84, 83]), involves the dynamical evolution and possible fragmentation of Encke's comet and/or its progenitor throughout the past ~10^4 years, and is now principally associated with the names of Clube, Napier, Steel and Asher (e.g. [20, 21, 23, 62, 78, 3, 77].

Evidence of another kind had surfaced somewhat earlier, during the 1960s and 1970s, this time involving the Kreutz Sun-grazing comet group. As described by Marsden [53, 54, 55], the Kreutz Sun-grazers probably originate from the break-up of a single 'giant' comet (possibly as large as 100 km across [66]); the recent surprise has been the continuing high rate of accretion of cometary debris by the Sun. Observations by the SOHO satellite, launched primarily to achieve a greater astrophysical understanding of the Sun, have shown that comets hit the Sun roughly once per week. Two notable examples, amongst the brightest SOHO Sun-grazers observed, namely X/1998K10 and X/1998K11 (provisionally labelled SOHO-54 and SOHO-55), followed each other into the Sun with an orbital separation of only 4 hours. The observed rate of cometary accretion by the Sun may be compared with theoretical estimates based on the random influx of comets from the Oort cloud, producing about one cometary impact per *century*.

Both the Taurids and the Kreutz group highlighted the possible importance of *streams* of debris in the inner solar system, and therefore the occurrence of multiple impacts within a relatively short timespan.

Evidence of the importance of streams of a third kind came to light in 1994, with the discovery of the object called Comet Shoemaker-Levy 9. The stream's progenitor, a cometary nucleus some 5-15 km in diameter [72], passed too close to Jupiter in 1992, was broken by the Jovian tide, and split into about a dozen pieces. These then broke into further fragments which, owing to slight differences in their positions and velocities when close to Jupiter, evolved on neighbouring orbits so as to collide with the giant planet over a period of about a week during July 1994. The impacts were witnessed world-wide, and graphically demonstrated the potential significance and planet-scale consequences of such events. As was pointed out by Steel [76], whereas the a priori probability of Jupiter being struck by ~ 20 kilometre-size comets during a particular week is tiny (on the order of 10^{-70}, the existence of the stream completely changes the picture. Thus, Comet Shoemaker-Levy 9 confirmed the significance of *streams* for assessing the potential hazard to civilisation posed by cometary impacts.

2. Populations of Near-Earth Objects

Modern cosmogonical ideas usually ascribe the origin of small solar system bodies to collisional processes in the protosolar and protoplanetary nebula, followed by the gravitational accumulation of objects into bodies the size of observed planets. The whole process was completed within the first few 100 Myr of the 4,600 Myr age of the solar system. The collisional aggregation process turns out (perhaps obviously) to be highly inefficient, and requires the removal of many of the original comet-size building blocks, which are known as 'planetesimals'. These bodies, which have sizes in the 1-10^2 km range, are conveniently associated with observed comets and asteroids or the parent bodies of asteroids in the main belt. They were mostly gravitationally ejected from the solar system or scattered into long-lived orbits of various types having dynamical lifetimes measured in billions of years. Subsequent orbital evolution of these source objects ultimately populated the various source orbits for observed NEOs, which have dynamical lifetimes in their present orbits measured in millions, rather than billions, of years (e.g. [40]).

The resulting orbits of the observed small solar system objects are conventionally divided into long-period and short-period types, the separation occurring at an orbital period P = 200 yr. Subsets of these principal dynamical classes have been introduced as our knowledge of the whole population has increased, while the historical separation into two physical types - comets and asteroids - has remained, despite the empirical distinction in some cases becoming increasingly blurred. We thus have, for example, short-period comets, Apollo, Amor and Aten asteroids, and the newly discovered object (1998DK$_{36}$) with an orbit located entirely within that of the Earth [81].

2.1 Comets

Most observed long-period comets (those with P > 200 yr) originate in the Oort cloud, which is a nearly spherical reservoir containing orbits extending up to half-way to the nearest star. The periods of revolution of Oort cloud comets, sometimes called 'new' comets, typically range up to several tens of Myr. Oort's primary motivation for introducing the comet cloud appears to have been to provide a primordial reservoir for long-period comets which was stable for the age of the solar system [65]. However, this justification has now been overtaken to the extent that the outer cloud is known to be dynamically unstable, due to the combination of stellar and giant molecular cloud perturbations (e.g. [64, 5, 7, 8]). According to current theories, the unstable outer Oort cloud (which is the principal source of observed long-period comets) has to be replenished from within: from a more massive, dense inner core of the Oort cloud [41, 6], itself a natural product of its formation by the scattering of planetesimals from the region of planet formation (e.g. [35, 25]).

Short-period comets (those with P < 200 yr) are usually divided into three categories: Halley-types (with periods in the range 20 < P < 200 yr and a wide range of inclinations); Jupiter-family (with P < 20 yr and normally low inclinations); and Near-Earth-Asteroid (NEA) or Encke-types, with exceptionally short orbital periods and aphelion distances Q close to or within the orbit of Jupiter.

A few objects are difficult to classify on this scheme. For example, 96P/Machholz 1, which has P = 5.24 yr, is probably best regarded as a Halley-type of unusually short period; while the recently discovered comets P/1994N2 (McNaught-Hartley) and P/1997B1 (Kobayashi), with periods a little over 20 years, are possibly best viewed as members of an intermediate-period low-inclination population, dynamically half-way between Halley-types and the Jupiter family.

Comets, of course, only exhibit significant activity when they come close to the Sun. A cometary nucleus which lies in an orbit that never passes close to the Sun might never show cometary activity, and might therefore be described or observed as asteroidal. However, advances in technology have meant that activity can now be detected in many objects which were formerly classed as asteroids, and from time to time also in objects that are only transiently active. The archetype for such behaviour is a body first observed in 1977 and subsequently classified as an asteroid, namely (2060) Chiron. This object moves in an orbit with period P = 50.7 yr and perihelion distance q close to the orbit of Saturn (q = 8.454 AU, where 1 AU is the distance of the Earth from the Sun). Chiron was subsequently observed to be outgassing [59], and now also has a cometary designation, namely 95P/Chiron.

Another example is provided by the asteroidal object (4015)Wilson-Harrington, which was at first thought to be a perfectly ordinary asteroid - until its orbit was shown [56] to be compatible with observations of the comet 1949III. The imprecise cometary orbit was subsequently amended, and the object is now known also as comet 107P/Wilson-Harrington. This body has the second shortest orbital period of all known comets (P = 4.30 yr); only 2P/Encke (P = 3.28 yr) has a shorter period. A third anomalous case is provided by comet 133P/Elst-Pizarro, which despite its asteroidal orbit (lying within the main belt and apparently of remarkable long-term stability [44]) is notable for showing a distinct cometary tail.

The example of Chiron provides the opportunity for an exceptionally interesting case study because its orbit, in passing close to Saturn, is unstable with a half-life on the order of 0.1 Myr for transfer onto a Jupiter-crossing orbit and hence into the Jupiter-family of short-period comets. Chiron has an overall dynamical lifetime against ejection from the solar system of 1-2 Myr [39], so statistically it has probably entered the inner solar system on several previous occasions, with a possibility of having been Earth-crossing at some stage of its evolution. (The same applies, of course, to its future evolution.) This is an important point, so far as attempts to understand the Earth's history over the past 1 Myr or so are concerned, because Chiron was the first 'giant' comet to be positively identified. Chiron is currently thought to have a diameter on the order of 200 km (e.g. [80]).

These examples show that comets sometimes masquerade as asteroids, and vice versa, and that the inferred physical properties of a given object have to be considered on the merits of each case. In borderline cases, typical of NEOs, the decision whether an object should be classified as 'cometary' or 'asteroidal', connoting an origin in the outer or inner solar system respectively (with implications for the bodys physical structure and composition, e.g. icy and fragile versus rocky and strong), is best regarded as provisional.

2.2 Asteroids

With this caveat, we turn to asteroids. Some 40,000 asteroids currently have approximately known orbits, and almost a quarter of this number have now received numerical designations (and usually a name, e.g. 2060 Chiron), indicating that the object has been well observed and has an accurately determined orbit. The others have been given preliminary designations which indicate the date of discovery, for example the outer solar system object 1997RX$_9$ was discovered in the 'R-th' half-month of 1997 (i.e. the first 15 days of September, the letter I not being used for this purpose), and was the 248th new asteroid to be identified in that period (the letters cycle as many times as necessary through RA, ... RZ, RA$_1$, ... with I again being omitted).

Most of these objects belong to the main asteroid belt, having low-eccentricity, low-inclination orbits lying between Mars and Jupiter, with semi-major axes in the approximate range 2.1-3.3 AU. While the sample of large main-belt asteroids is virtually complete (i.e. the ~682 objects with diameters d > 50 km), the fraction with smaller diameters is increasingly incomplete. Less than 10% of the estimated ~10^5 main-belt asteroids with d >10 km have currently been found, and an even smaller proportion of the ~10^8 objects with d > 1 km.

So far as the main belt is concerned, an increasingly important distinction is now drawn between the background 'field' population and those moving on orbits which, when averaged over very long timescales, are roughly similar to one another. The latter, so-called 'family' asteroids, represent the results of catastrophic collisions that occurred between asteroids possibly hundreds of millions or billions of years ago. In both cases, however, the size distribution is steep (at least for diameters less than about 40 km), with the number of objects larger than a given diameter d increasing

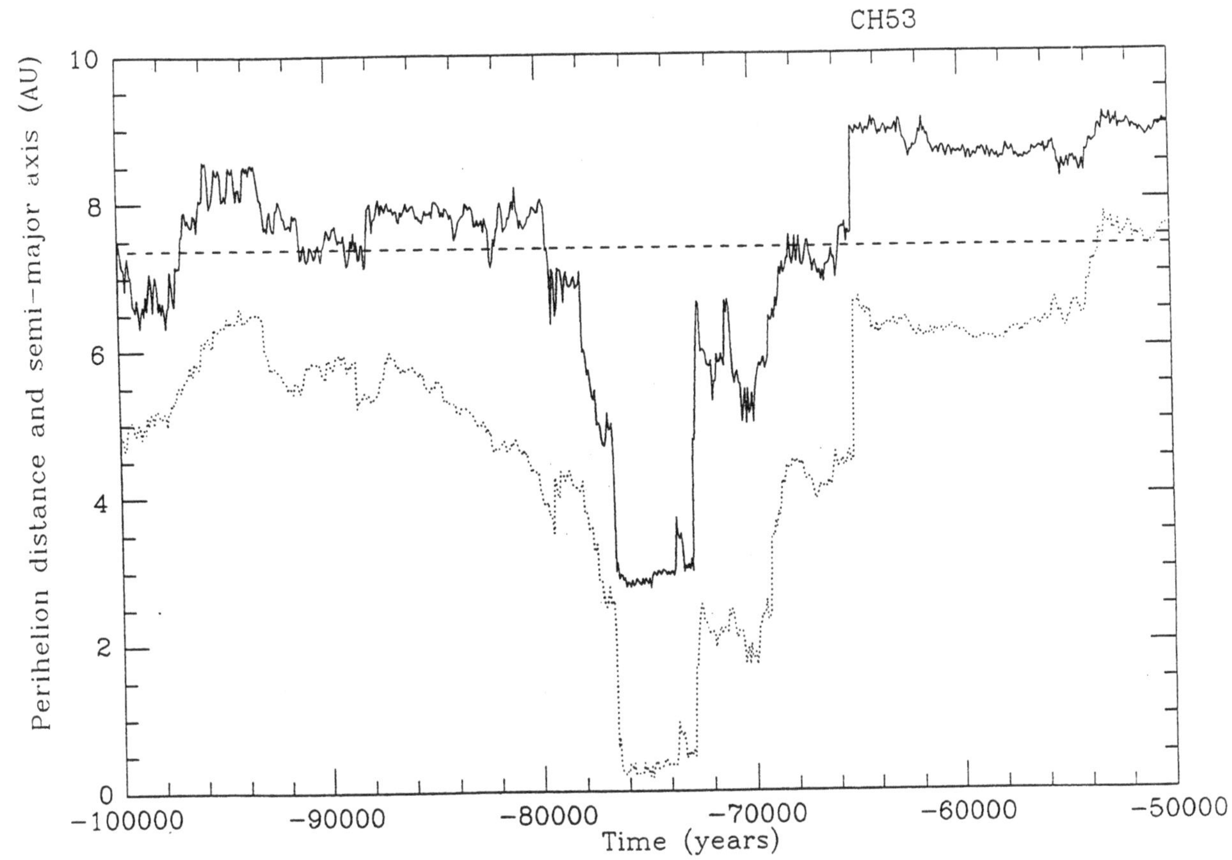

Figure 1: Dynamical evolution of a test particle labelled CH53, with initial orbital elements similar to those of Chiron, illustrating the evolution 75,000 years ago when the body temporarily became an Earth-crossing Jupiter-family comet (after [39]).

with decreasing size as a power-law approximately as d^{-3} or d^{-4} for the field and family populations respectively. This has important implications for the cometary versus asteroidal source of NEOs, since the cometary size distribution appears to be much flatter, i.e. $N(>d)$ d^{-2}, so that comets or cometary sources generally dominate for diameters larger than 1-10 km and asteroids at small diameters (cf. [63]).

The complexity of the main asteroid belt also provides many possibilities for the separation of asteroids into distinct physical types and dynamical classes, reviews of which may be found elsewhere (e.g. [16]). Here, however, whilst recognising that collisions in the main belt and the subsequent dynamical evolution of the collision fragments certainly represent a significant source of observed Near-Earth Objects (e.g. [51]), the principal focus of this review is on the physical and dynamical types that lie predominantly beyond the main belt.

Leaving the main belt aside, therefore, the first main subdivision of asteroids is into Near-Earth Asteroids (NEAs) of various dynamical types. By convention, observed NEAs (which all have possibilities to come close to the Earth), are divided into Apollo, Amor and Aten asteroid types, with a handful of other types occasionally introduced for particular physical or dynamical reasons. Apollos have semi-major axes greater than 1.0 AU and perihelion distances less than the aphelion distance of the Earth's orbit (i.e. $q < 1.017$ AU); Amors have q just outside the Earth's orbit ($1.017 < q < 1.3$ AU), but may evolve to become Earth-crossing; and Atens have semi-major axes less than 1.0 AU and aphelion distances which extend beyond the Earth's perihelion distance (i.e. $Q > 0.983$ AU). Broadly speaking, NEAs have orbital periods less than 4 or 5 years, and orbits which cross or pass close to the Earth. Their aphelia generally lie in the main belt, and this region is therefore often regarded as the principal source of observed NEAs, of which some 500 are currently known. Many of these objects, however, have very small dimensions (some are scarcely more than house-size boulders), whilst the total number of objects with diameters $d > 1$ km is estimated to be ~2000 [69, 60]. Roughly 10% of NEAs with $d > 1$ km have so far been found, again highlighting the importance of identifying the remaining 90% as soon as possible in order to reduce the unforeseen extraterrestrial impact hazard.

Moving further through the solar system brings us first to a number of small groups of outer main-belt asteroids, then to the Trojans (a populous group of objects with orbital periods close to that of Jupiter), and finally to the Centaur region.

Centaurs, of which (2060) Chiron was the first to be identified, have low-inclination orbits between Saturn and Neptune and were originally named after the Centaurs of Greek mythology. Their significance is that they are prototypes for giant comets, defined here to be icy objects with diameters greater than 100 km. Chiron, for example, which shows well-defined outgassing and hence has cometary characteristics in addition to an unstable orbit, has a diameter on the order of 200 km, while the more recently discovered $1997CU_{26}$ has an estimated diameter of about 300 km [45]. Centaurs not only have very large sizes, but also have orbits that are remarkably unstable, evolving into Jupiter-family cometary orbits on a timescale in the range 0.1-10 Myr. There are currently 7 known Centaurs with sizes ranging up to 300 km; and assuming that the presence of an object like Chiron in the population is typical, one such should enter the inner solar system about once every 100,000 years. That such a timescale is close to the mean interval between random impacts of kilometre-size NEAs with the Earth, demonstrates that giant comets may be at least as important as asteroids for the long-term extraterrestrial impact hazard.

The next significant asteroid 'belt' is the Edgeworth-Kuiper belt (EKB) or Kuiper belt, a flat disc-like distribution of comets or planetesimals beyond Neptune, whose existence was anticipated in papers by Edgeworth [27, 28], Kuiper [49], Whipple [82] and Fernández [34]; for a review of Edgeworth's contribution, see McFarland [58]. The name 'Kuiper belt' was proposed by Duncan *et al.* [26], four years before the discovery of the first comet or planetesimal in that region, namely $1992QB_1$. Subsequent work has identified 67 outer solar system objects in such orbits, up to 1 August 1998, many of which have low-eccentricity, low-inclination EKB orbits and dimensions well in excess of 100 km. The largest Edgeworth-Kuiper belt object (EKO) to date (we ignore Pluto), is $1996TO_{66}$, with a diameter on the order of 800 km assuming a conventionally low albedo. Although most EKOs have orbits lying within the EKB, at least one recent discovery ($1996TL_{66}$ [52]) suggests the presence of a more extended flattened distribution of orbits, corresponding to a trans-Neptunian disc and possibly representing a link between the EKB and the inner core of the Oort cloud.

2.3 Giant Comets

Taking these observations into account, it is clear that the cumulative frequency distribution of cometary diameters d is relatively flat and extends up to very large sizes, certainly beyond d = 300 km. Supporting evidence for a law of the form $N(>d) \propto d^{-2}$, reviewed by Bailey *et al.* [14], comes primarily from the observed cometary brightness distribution, the existence of the Kreutz Sun-grazer group, the exceptional Comet Sarabat in 1729, outer solar system bodies such as (2060) Chiron and (5145) Pholus, in Centaur orbits between Saturn and Neptune, and objects such as $1992QB_1$ and 1993FW in the Edgeworth-Kuiper belt.

Observations thus indicate the presence of giant comets, not only in the near-parabolic flux (which leads to high-inclination Halley-type comets [32, 33]), but also in the EKB and Centaur populations, which contribute to the observed Jupiter family, and hence also produce objects capable of evolving onto Earth-crossing orbits.

The two principal classes of short-period comet, Jupiter-family and Halley-types, therefore both include occasional large members. Although their precise mass distribution is uncertain, the total number of objects of either type suggests an average interval between the arrival of giant comets in short-period orbits, whether Jupiter-family or Halley-type, on the order of 0.1 Myr [14]. These 'giants' are substantial objects, 10^4 times the mass of Halley's comet, and their fragmentation in the inner solar system would produce a significant perturbation on the present distribution of interplanetary matter [22, 77]. The instantaneous number of NEOs is thus a function not only of individual collisional events in the main asteroid belt [60, 88], but also of the time since the most recent injection and break-up of a giant comet in the inner solar system.

3. Dynamics

The diverse sources of NEOs, both cometary and asteroidal, demonstrate the importance of studies of their long-term dynamical evolution. Such work shows that the orbits often become planet crossing, and as a consequence are extremely chaotic. In most cases, the exact evolution is unpredictable on time-scales longer than that required to make more than a handful of close approaches to the major planets, typically a few thousand years or so.

This means that Chiron, to take an extreme case, could have been a short-period Jupiter-family comet as recently as $\sim 10^4$ yr ago. This highlights a key point if one is interested in compiling the inventory of Earth-crossing orbits over a timescale comparable to that of human history: Chiron could have been Earth-crossing a relatively short time ago. An example illustrating this point is shown in Figure 1.

A detailed investigation of the dynamics of 'giant-comet Chiron' was carried out by Hahn & Bailey [39]. More recent studies have shown that Chiron generally undergoes significant orbital evolution after an average time-scale on the order of 0.15 Myr, the object in the first instance being scattered to the control of Jupiter. In the case of the second Centaur to be discovered, namely Pholus, the times-cale to be handed down to Jupiter is about a million years, while the ejection half-life for both objects is in the range 1-2 Myr. Chiron has an 88% probability of becoming or having been a Jupiter-family comet at some time during its dynamical evolution, whereas Pholus has about a 42% chance of suffering the same fate; the probability of Pholus becoming Earth-crossing is estimated to be on the order of 0.02 to 0.2 per million years [12, 2].

Turning to Halley-type orbits, their long-term dynamical evolution may be illustrated by that of Halley's comet itself, and by the evolution of the asteroid (5335) Damocles (formerly designated 1991DA), which has an orbital period of about 40 years and inclination close to 62 degrees. As previously discussed [14, 4, 29], an important result is that many Halley-type objects eventually evolve to a Sun-grazing end-state, and may produce compact meteoroid streams which lead to meteor storms [30]. A further important result is that the probability of a random long-period comet with perihelion distance less than 4 AU to evolve into a Halley-type orbit is on the order of 1% [33]. This implies the existence of a very large number of extinct or inert objects in Halley-type orbits, sufficient to make a substantial contribution to the terrestrial cratering rate and the overall population of NEOs [33].

Figure 2 shows a possible evolution of Halley's comet for a period plus or minus one million years from the present. One can easily identify the presence of mean-motion resonances in the semi-major axis plot, strong excursions in the inclination, and large variations in the perihelion distance as a result of secular resonances. In particular, note the future evolution into a Sun-grazing state within ~0.1 Myr. These features are typical of virtually every Halley-type orbit that has been investigated, even of the exceptional long-period comet Hale-Bopp during times when its orbit happens to become a Halley-type [15].

Figure 3 shows the long-term evolution of the Halley-type 'asteroid' Damocles, over a time interval a little longer than one million years. This object too is subject to mean-motion resonances, and to strong secular perturbations that cause its perihelion distance to evolve from near the orbit of Jupiter (with an inclination near 80 degrees) to its current orbit with present perihelion distance close to Mars. In many cases Damocles finally becomes Sun-grazing, and this test particle too would presumably end its life by falling into the Sun.

The main conclusion to be drawn from these results is that the source reservoirs of short-period comets, whether Centaurs (which become Jupiter-family objects) or long-period comets (which become Halley-types), contain significant numbers of very large members which we describe as giant comets. Objects from either source may pass close to the Sun or Jupiter, and may be tidally disrupted during the close

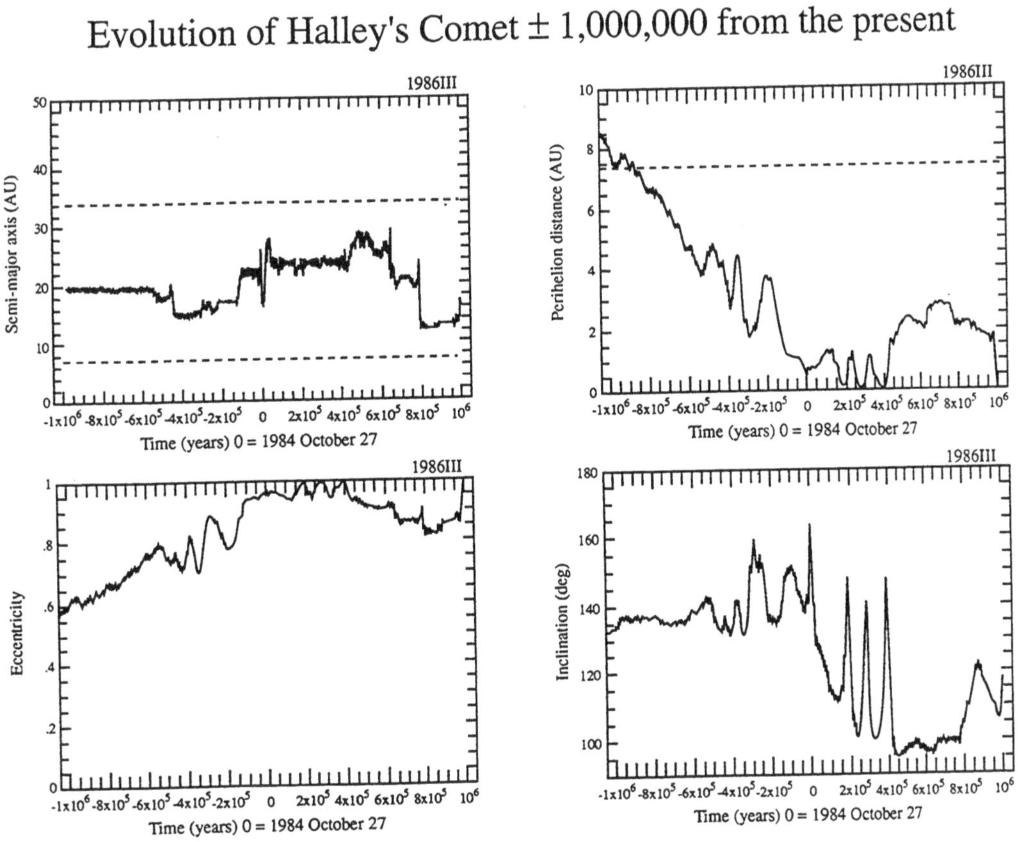

Figure 2: Long-term dynamical evolution of a test particle with initial elements similar to those of comet 1P/Halley.

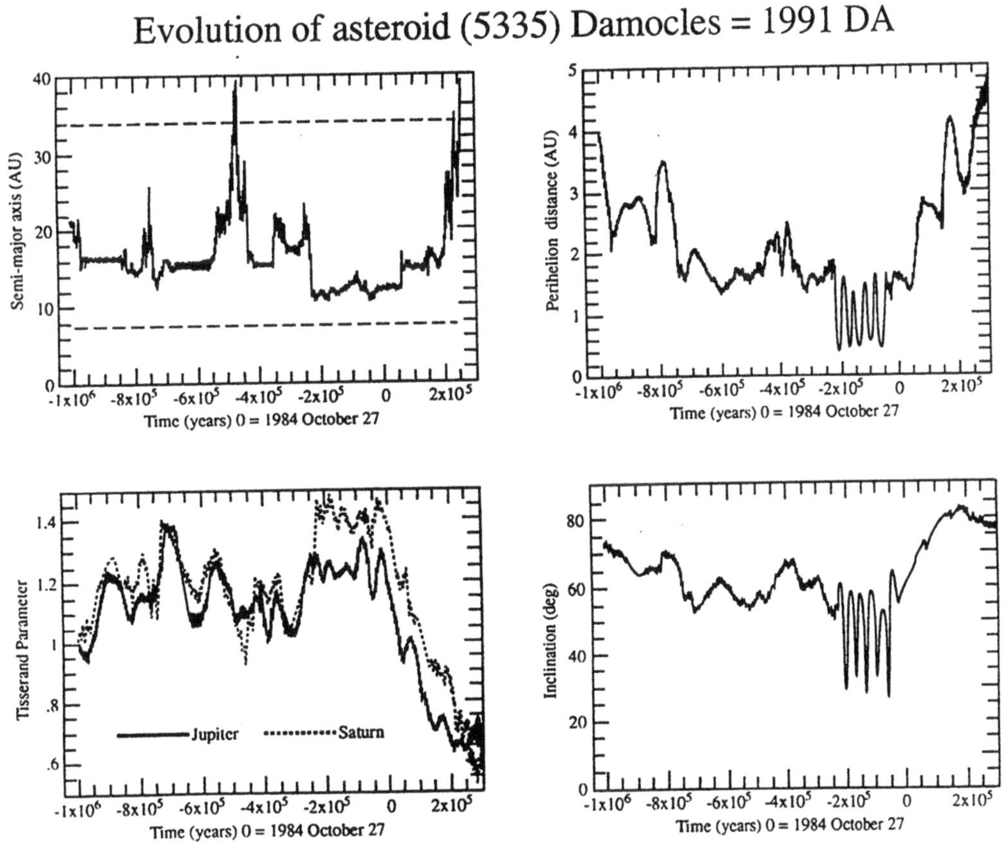

Figure 3: Long-term dynamical evolution of a test particle with initial elements similar to those of asteroid (5335) Damocles.

encounter. What, then, are the implications of this picture for our understanding of mass extinctions, on the one hand, and the current distribution of near-Earth objects and interplanetary matter on the other?

4. Implications

4.1 Long View

As pointed out by Bailey *et al.* [14] and Napier [63], the 'one-off' impact hypothesis has difficulties which are most naturally overcome by postulating an alternative giant-comet break-up scenario, involving interactions between the Earth and the giant comet debris stream. Instead of the standard picture for the mass extinction at the Cretaceous-Tertiary K/T boundary, i.e. a random ~10 km diameter asteroid running into the Earth, we may instead postulate the evolution of a giant comet in a short-period, high-inclination Halley-type orbit, which is potentially Sun-grazing. A possible analogue might be to consider the orbital evolution of Comet Machholz 1 [13]. The dynamical lifetime would be on the order of 1 Myr, or some 10,000 or 100,000 revolutions, while Sun-grazing phases might be expected to recur roughly every 10^3 revolutions.

Such evolution would lead to intermittent break-up and decay of the large body, feeding dust and larger bodies of 'Tunguska' size (i.e. 10-100 m), and possibly small comets, into longer lived Earth-crossing meteoroid streams, finally producing a structure somewhat similar to that of the Kreutz Sun-grazing group. Fortunately, this stream of cometary debris does not currently intersect the Earths orbit.

The dynamical results reviewed above suggest that a large comet, with a dynamical lifetime exceeding a million years, will occasionally evolve into a short-period Earth-crossing orbit with the capability of making tens of thousands of revolutions before finally dissipating, possibly becoming Sun-grazing every 1000 revolutions or so in the process due to secular perturbations. One would expect the giant comet to undergo intermittent break-up and decay, leading to streams of cometary fragments similar to the structure of the Kreutz Sun-grazing group. One might even expect correlated impacts on the Earth around the times when secular perturbations cause the stream bodies to become Earth-crossing. These episodes of small perihelion distance would occur at extrema of the inclination, increasing the chance that an impact may occur soon after a Sun-grazing phase [13].

Supporting evidence for such a scenario (reviewed by Bailey *et al.* [14]) includes the organic 'cometary' dust both below and above the boundary, the variable iridium concentration either side of the K/T boundary, contra-indicating a one-off random asteroid impact, and evidence for climatic change - cooling - just before the boundary, possibly associated with accretion of dust. In summary, observations of the K/T mass extinction do not strongly support a one-off random impact, but instead suggest that correlated, possibly smaller accretion events may have played a major role.

4.2 Short View

We now turn to the other class of giant comet, namely the bodies deriving from the Centaur or Chiron-types, which evolve into Jupiter-family short-period orbits. These orbits suffer strong Jovian perturbations, and the timescales for undergoing significant orbital evolution are much shorter, typically in the range 10^3-10^5 yr.

Consider the behaviour of a giant comet captured into a Jupiter-family orbit, as must happen on the average about

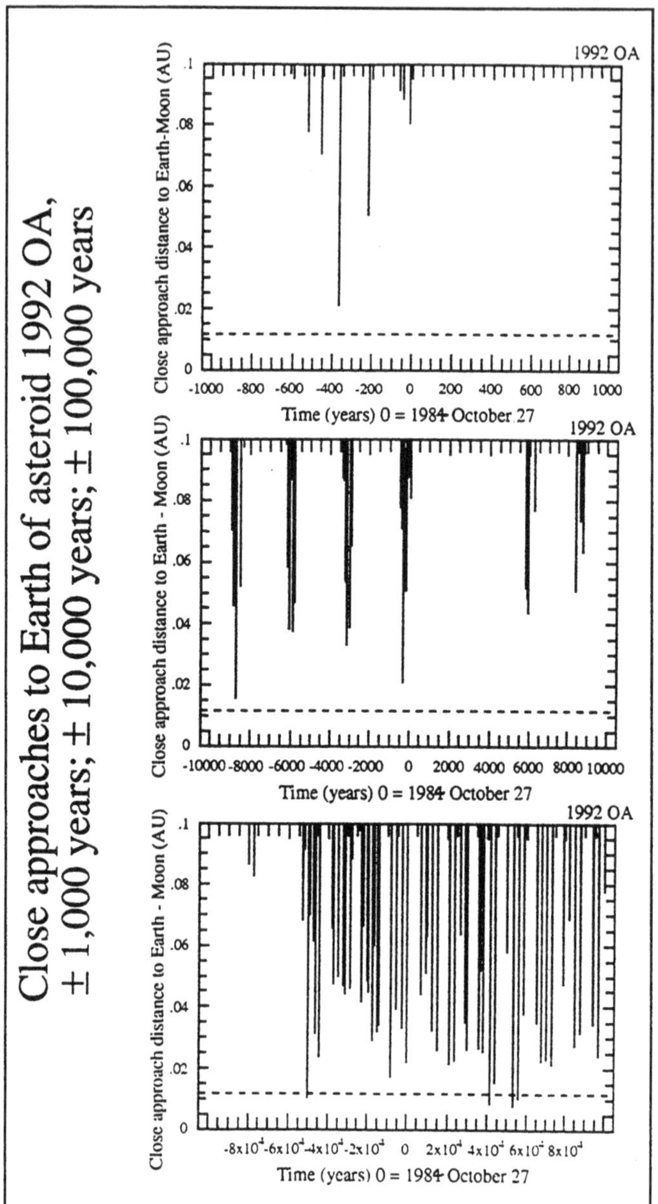

Figure 4: Structure in the frequency distribution of close approaches to Earth of the asteroid (5496) 1973NA = 1992OA to the Earth over periods 10^3, 10^4 and 10^5 yr from the present. Note the bunching in the distribution of close encounters, characteristic of close encounters of near-Earth asteroids and their debris with the Earth, demonstrating episodes of enhanced collision probability on time-scales ranging from decades to millennia.

once every 0.1 Myr. The lifetime as a potentially Earth-crossing Jupiter-family object is 10^3-10^4 yr before one of disruption, dynamical ejection or evolution back to the control of Saturn takes place (cf. Fig. 1). However, the overall dynamical lifetime, as in the case of Chiron and Pholus, is still about a million years. In principle, as shown by the results of long-term dynamical evolution studies, a single object could return to the inner solar system many times before becoming inactive or leaving the solar system as a result of the combined perturbations of Jupiter and Saturn. A corollary of this picture, supported by dynamical studies, is that Chiron has 'probably' already been a short-period comet.

It is therefore interesting to speculate that there may have been several recent episodes of deterioration of the Earth's near-space environment owing to the evolution and break-up of one or more giant comets in the Jupiter family short-period system. This would lead to streams of debris in the

Figure 5: Bronze Age engravings suggestive of a possible cometary inspiration, found on a stone in Cairn T, Loughcrew, Co. Meath, Ireland. The figure, courtesy of The National Museum of Ireland, has been taken from O'Sullivan [67].

inner solar system and episodes of multiple bombardment, in much the same way that exceptionally strong meteor showers occur intermittently, for example every 33 years in the case of the Leonid meteors. The model predicts periods of enhanced 'hazard' separated by longer intervals of relative safety.

The general picture is illustrated in Figure 4, which shows the evolution of close encounters to the Earth of the Apollo asteroid (5496) 1973NA = 1992OA. The plot (taken from Bailey *et al.* [14]) shows the distribution of close approaches of this body to the Earth on various time-scales, from plus or minus 100,000 years (bottom) to plus or minus 1,000 years (top). Viewed on the long time-scale, the close approaches occur in bunches or groups, which when seen at higher resolution break into typically half-a-dozen close approaches, the individual groups each being separated by thousands of years. The episodicity shown in this diagram is likely to be typical of the frequency distribution of impacting objects arising from the break-up and fragmentation of a single body, whether asteroid or giant comet.

4.3 Historical record

The detailed implications of this astronomical picture have still to be worked out, although if the distribution of small bodies in the inner solar system is indeed dominated by streams produced by a relatively small number of large bodies (rather than by random injections of asteroids from the main belt), the prediction of a more active 'sky' during past millennia would appear to be robust. Indirect evidence supporting such a picture, reviewed by Bailey *et al.* [11], includes aspects of the Babylonian and Chinese 'obsession' with making celestial observations, the belief that terrestrial and astronomical events were physically related (e.g. [68]), the 'confusion' of early Greek writers between the observed Milky Way and zodiacal light, and the 'astronomania' in many parts of western Europe, indicated by the megalithic monuments and stone circles showing apparently significant astronomical orientation in addition to presumed calendrical and religious uses.

So far as the Milky Way is concerned, it is interesting to note that early Greek sources describe 'stars' as lying below the Sun and Moon (Anaximander, Leucippus), call the Milky Way the *former* path of the sun (Metrodorus), and describe the Milky Way's location in the sky as 'in the Earths shadow' (Anaxagoras, Democritus). These notions are difficult to understand unless 'stars' is interpreted to mean 'shooting stars' (i.e. meteors), and the 'Milky Way' is understood to refer to an earlier, more intense zodiacal light, presumably emanating from a more massive former zodiacal cloud. Further arguments that the zodiacal light may have been much brighter in the past have been presented by Jones [46, 47].

Megalithic monuments are of direct relevance to the Bronze Age theme of these proceedings, and here it is interesting to note that a large number of stones show engravings known as cup-and-ring marks, many of which look decidedly cometary (cf. Figures 2.2 and 2.4 of Bailey *et al.* [11]). Many authors have drawn attention to the similarities, across widely different cultures, in the rock art of this period, possibly suggesting a common extraterrestrial or cometary inspiration for the rock carvings (cf. [9]). Examples of possibly influential astronomical phenomena include the appearance of exceptionally bright comets (e.g. the previous appearance of Comet Hale-Bopp around 2215 BC [57]), bright fireballs, the fall of meteorites, meteor storms, and occasional more energetic impacts of small bodies in the Tunguska class or larger, such as the ~350 MT projectile which produced the Rio Cuarto string of craters in Argentina 2000-4000 years ago [70, 71]. Such impacts are statistically certain to have occurred a number of times over the span of human history, and it is clearly a valid question to ask what would be their effects on individuals and societies (cf. [17, 18]).

The conjunction of these ideas, linking astronomy and history, therefore suggests that human societies may have been witness to a somewhat more active celestial environment during past millennia. The evidence is not yet conclusive, but the astronomical arguments are testable, in that they predict a non-uniform spatial distribution of small bodies in the inner solar system, dominated by streams.

Were this confirmed, then future periods of significantly enhanced accretion might be expected [1], with profound implications not only for our understanding of the historical record (e.g. [17,18]) but also for the development of a more complete picture of environmental change including both global warming and global cooling (e.g. [23, 9, 24]).

5. Conclusion

Arguments based on the capture of comets into Halley-type orbits from the near-parabolic flux lead to the prediction [31, 32, 33, 10] that a very large number of bodies (on the order of 1000 at least) resembling inert, devolatilized Halley-type asteroids such as (5335) Damocles should exist. So far we have found only a handful of such objects, but searches should reveal more. In particular, given our state of relative ignorance, we should ensure that any adopted 'Spaceguard' search strategy is sufficiently comprehensive to resolve the question of the existence or otherwise of large numbers of such objects. Halley-type 'asteroids', originating from the evolution and decay of Halley-type comets, probably represent a significant (though still unseen) contribution to the impact hazard.

Tidal break-up is increasingly accepted as a viable mechanism to destroy large comets. Whereas Halley-type comets (if they survive thermal stress and are not dynamically ejected) will eventually be broken up by grazing encounters with the Sun, as happened to the progenitor of the Kreutz group, objects originating in the Centaur population, such as the progenitor of Comet Shoemaker-Levy 9, must make many close approaches to Jupiter. The evolution of this object, the fragments of which finally fell onto Jupiter in July 1994, illustrates how the break-up of a giant Jupiter-family comet might in principle suddenly inject new short-period comets, or comet fragments, into the Jupiter-family system.

A further short time-scale phenomenon is that of ordinary cometary splitting. Streams of material due to split comets may contain meteoroids, meteorites and even asteroids. Viewed globally, the main consequence of streams is that the 'hazard' may be intermittent on a 100 to 1000-year time-scale - and potentially recurrent - highlighting again the importance of a correct assessment of the historical record, the climatological record and of interpreting, for example, the Chinese literature summarized by Yau *et al.* [87]. It is important to bear in mind such 'additional arguments' when considering the nature of the interplanetary complex and the possible futures that our descendants might experience.

Taking a slightly longer view, the dynamical capture of the largest, so-called giant comets, both Halley-type and Jupiter family, is crucial for understanding certain features of the long-term terrestrial record. The Halley-type comets and the Jupiter family have totally different dynamical characteristics, leading them to produce totally different impact hazards in a general sense. Whereas Halley-types might in principle persist in the circumterrestrial environment for periods ranging up to a million years, the Centaurs dip in and out on time-scales ranging from 10^4 to 10^6 yr.

These arguments indicate that there should be large fluctuations in the distribution, number and other properties of interplanetary objects. The resulting picture for the potential hazard to civilisation is qualitatively different from one based on the idea of random 'one-off' impacts, although the recommended initial response - to survey the sky - should be the same. The present picture highlights the role of correlated, multiple impacts, because streams of meteoroids certainly exist, and streams of larger, Tunguska-size objects may also exist in the inner solar system. In assessing the present, instantaneous hazard, we need to know whether the Earth has recently passed through such a stream, or is shortly about to do so. Given a more complete understanding of the small bodies in potentially Earth-crossing orbits, the important time-scales for planning and mitigation could be as short as 100 to 1000 years. A key point is that these timescales are precisely within the range of interest for a correct interpretation of the historical record. This highlights the importance of interdisciplinary studies aimed at assessing the possible extraterrestrial factors which could have driven the development and collapse of previous civilisations.

In summary, the scientific study of NEOs is timely and of strong cultural and economic value. The subject is multidisciplinary, and requires greater resources for researchers to provide a correct assessment of the full implications of the impact hazard. There is a need for more observations, improved theoretical understanding on all fronts, and eventually an international Spaceguard programme, to which scientists from many nations would be expected to contribute. The pay-off, in terms of a correct assessment of the historical and archaeological records, could be immense.

Acknowledgements

It is a pleasure to thank D.J. Asher, S.V.M. Clube, S.P. Manley, W.M. Napier and B.J. Peiser, for comments and discussions which have substantially improved this text. This work was supported by DENI.

References

1. Asher, D.J. & Clube, S.V.M., 1993, An extraterrestrial influence during the current glacial-interglacial, *Q. Jl. Roy. Astron. Soc.*, **34**, 481-511.
2. Asher, D.J. & Steel, D.I., 1993, Orbital evolution of the large outer solar system object 5145 Pholus, *Mon. Not. Roy. Astron. Soc.*, **263**, 179-190.
3. Asher, D.J., Clube, S.V.M., Napier, W.M. & Steel, D.I., 1994a, Coherent catastrophism, *Vistas in Astron.*, **38**, 1-27.
4. Asher, D.J., Bailey, M.E., Hahn, G. & Steel, D.I., 1994b, Asteroid 5335 Damocles and its implications for cometary dynamics, *Mon. Not. Roy. Astron. Soc.*, **267**, 26-42.
5. Bailey, M.E., 1983a, The structure and evolution of the solar system comet cloud, *Mon. Not. Roy. Astron. Soc.*, **204**, 603-633.
6. Bailey, M.E., 1983b, Is there a dense primordial cloud of comets just beyond Pluto? *Asteroids Comets Meteors*, eds. Lagerkvist, C.-I. & Rickman, H., pp. 383-386, Uppsala Observatory.
7. Bailey, M.E., 1986, The mean energy transfer rate to comets in the Oort cloud and implications for cometary origins, *Mon. Not. Roy. Astron. Soc.*, **218**, 1-30.
8. Bailey, M.E., 1991, Comets and molecular clouds: the sink and the source, *Molecular Clouds*, eds. James, R.A. & Millar, T.J., pp. 273-289, Cambridge University Press.
9. Bailey, M.E., 1995, Recent results in cometary astronomy: implications for the ancient sky, Vistas in Astron., **39**, 647-671.
10. Bailey, M.E. & Emel'yanenko, V.V., 1998, Cometary capture and the nature of the impactors, *Meteorites: Flux with Time and Impact Effects*, eds., Grady, M.M., Hutchison, R., McCall, G.J.H. & Rothery, D.A., *Geol. Soc. Spec. Publ.* No. **140**, 11-17.
11. Bailey, M.E., Clube, S.V.M. & Napier, W.M., 1990, *The Origin of Comets*, Pergamon Press, Oxford.
12. Bailey, M.E., Chambers, J.E., Hahn, G., Scotti, J. & Tancredi, G., 1992a, Transfer probabilities between Jupiter and Saturn-family orbits: application to 1992AD = 5145, *Observations and Physical Properties of Small Solar System Bodies*, eds. Brahic, A., Gerard, J.-C. & Surdej, J., *Proc. 30th Liège Int. Astrophys. Coll.*, pp. 285-287.
13. Bailey, M.E., Chambers, J.E. & Hahn, G., 1992b, Origin of Sungrazers: a frequent cometary end-state, *Astron. Astrophys.*, **257**, 315-322.

14. Bailey, M.E., Clube, S.V.M., Hahn, G., Napier, W.M. & Valsecchi, G.B., 1994, Hazards due to giant comets: climate and short-term catastrophism, *Hazards due to Comets and Asteroids*, ed. Gehrels, T., pp. 479-533, University of Arizona Press.
15. Bailey, M.E., Emel'yanenko, V.V., Hahn, G., Harris, N.W., Hughes, K.A., Muinonen, K. & Scotti, J.V., 1996, Orbital evolution of Comet 1995,O1 Hale-Bopp, *Mon. Not. Roy. Astron. Soc.*, **281**, 916-924
16. Binzel, R.P., Gehrels, T. & Matthews, M.S., 1989, *Asteroids II*, University of Arizona Press.
17. Clube, S.V.M., 1994a, Hazards from space: comets in history and science, *The Mass-Extinction Debates: How Science Works in a Crisis*, ed. Glen, W., pp. 159-169, University of Stanford Press.
18. Clube, S.V.M., 1994b, Revelation and catastrophe during the Christian era: a basis for historical interpolation and future extrapolation, *Chronology and Catastrophism Review Special Issue: Proc. 1993 Cambridge Conference*, 66-73.
19. Clube, S.V.M., 1998, The problem of historical catastrophism, This volume.
20. Clube, S.V.M. & Napier, W.M., 1982, *The Cosmic Serpent: a Catastrophists View of Earth History*, Faber & Faber, London.
21. Clube, S.V.M. & Napier, W.M., 1984, The microstructure of terrestrial catastrophism, *Mon. Not. Roy. Astron. Soc.*, 211, 953-968.
22. Clube, S.V.M. & Napier, W.M., 1986, Giant comets and the Galaxy: implications of the terrestrial record, *The Galaxy and the Solar System*, eds. Smoluchowski, R., Bahcall, J.N. & Matthews, M.S., pp. 260-285, University of Arizona Press.
23. Clube, S.V.M. & Napier, W.M., 1990, *The Cosmic Winter*, Basil Blackwell, Oxford.
24. Clube, S.V.M., Hoyle, F., Napier, W.M. & Wickramasinghe, N.C., 1996, Giant comets, evolution and civilization, *Astrophys. Space Sci.*, **245**, 43-60.
25. Duncan, M., Quinn, T.. & Tremaine, S.D., 1987, The formation and extent of the solar system comet cloud, *Astron. J.*, **94**, 1330-1338.
26. Duncan, M., Quinn, T. & Tremaine, S.D., 1988, The origin of short-period comets, *Astrophys. J. Lett.*, **328**, L69-L73.
27. Edgeworth, K.E., 1943, The evolution of our planetary system, *J. Brit. Astron. Assoc.*, **53**, 181-18.
28. Edgeworth, K.E., 1949, The origin and evolution of the solar system, *Mon. Not. Roy. Astron. Soc.*, **109**, 600-609.
29. Emel'yanenko, V.V. & Bailey, M.E., 1996a, Dynamical evolution of Halley-type comets, *Mon. Not. Roy. Astron. Soc.*, **278**, 1087-1110.
30. Emel'yanenko, V.V. & Bailey, M.E., 1996b, Regular and stochastic motion of meteoroid streams in Halley-type orbits, *Physics, Chemistry, and Dynamics of Interplanetary Dust*, eds. Gustafson, B.A.S. & Hanner, M.S., IAU Coll. No. 150, ASP Conf. Series, **104**, 121-124, Astron. Soc. Pacific, San Francisco.
31. Emel'yanenko, V.V. & Bailey, M.E., 1996c, Dynamical evolution of comets and the problem of cometary fading, *Earth, Moon & Planets*, **72**, 35-40.
32. Emel'yanenko, V.V. & Bailey, M.E., 1997, The capture of Halley-type and Jupiter-family comets from the near-parabolic flux, *Dynamics and Astrometry of Natural and Artificial Celestial Bodies*, eds. Wytrzyszczak, I.M., Lieske, J.H. & Feldman, R.A., IAU Coll. No. 165, pp. 159-164, Kluwer, Dordrecht, The Netherlands.
33. Emel'yanenko, V.V. & Bailey, M.E., 1998, Capture of Halley-type comets from the near-parabolic flux, *Mon. Not. Roy. Astron. Soc.*, **298**, 212-222.
34. Fernández, J.A., 1980, On the existence of a comet belt beyond Neptune, *Mon. Not. Roy. Astron. Soc.*, **192**, 481-491.
35. Fernández, J.A., 1985, The formation and dynamical survival of the Oort cloud, *Dynamics of Comets: Their Origin and Evolution*, IAU Coll. No. 83, pp. 45-70, Reidel, Dordrecht, The Netherlands.
36. Frogel, J.A.& Gould, A., 1998, No Death Star - for now, *Astrophys. J. Lett.*, **499**, L219-L222.
37. Grieve, R.A.F. & Pesonen, L.J., 1996, Terrestrial craters: their spatial and temporal distribution and impacting bodies, *Earth, Moon & Planets*, **72**, 357-376.
38. Grieve, R.A.F. & Shoemaker, E.M., 1994, The record of past impacts on Earth, *Hazards due to Comets and Asteroids*, ed. Gehrels, T., pp. 417-462, University of Arizona Press.
39. Hahn, G. & Bailey, M.E., 1990, Rapid dynamical evolution of giant comet Chiron, *Nature*, **348**, 132-136.
40. Harris, N.W. & Bailey, M.E., 1998, Dynamical evolution of cometary asteroids, Mon. Not. Roy. Astron. Soc., **297**, 1227-12 36.
41. Hills, J.G., 1981, Comet showers and the steady-state infall of comets from the Oort cloud, *Astron. J.*, **86**, 1730-1740.
42. Hoyle, F., 1993, *The Origin of the Universe and the Origin of Religion*, Moyer Bell, Rhode Island.
43. Hoyle, F. & Wickramasinghe, N.C., 1978, Comets, ice ages, and ecological catastrophes,*Astrophys. Space Sci.*, **53**,523-526.
44. Ipatov, S.I. & Hahn, G.J., 1997, Evolution of the orbits of the objects P/1996R2 (Lagerkvist) and P/1996N2 (Elst-Pizarro), *Lunar and Planet. Sci. Conf.*, 28, 619-620.
45. Jewitt, D. & Kalas, P., 1998, Thermal observations of Centaur $1997CU_{26}$. Preprint.
46. Jones, H.D., 1990, Zodiacal light and the pyramids, *J. Brit. Astron. Assoc.*, **100**, 162.
47. Jones, H.D., 1993, Zodiacal light and ancient symbolism, *J. Brit. Astron. Assoc.*, **103**, 283-284.
48. Kortenkamp, S.J. & Dermott, S.F., 1998, A 100,000-year periodicity in the accretion rate of interplanetary dust, *Science*, **280**, 874-876.
49. Kuiper, G.P., 1951, Origin of the solar system, *Astrophysics - A Topical Symposium*, ed. Hynek, J.A., pp. 357-424, McGraw-Hill, New York.
50. Love, S.G. & Brownlee, D.E., 1993, A direct measurement of the terrestrial mass accretion rate of cosmic dust, *Science*, **262**, 550-553.
51. Lupishko,D.F. & Di Martino, M., 1998, Physical properties of near-Earth asteroids, *Planet. Space Sci.*, **46**, 47-74.
52. Luu, J., Marsden, B.G., Jewitt, D.C., Trujillo, C.A., Hergenrother, C.W., Chen, J. & Offutt, W.B., 1997, A new dynamical class of object in the outer solar system, *Nature*, **387**, 573-575.
53. Marsden, B.G., 1967, The Sungrazing comet group, *Astron. J.*, **72,}** 1170-1183.}
54. Marsden, B.G., 1989, The Sungrazing comet group. II, *Astron. J.*, **98**, 2306-2321.
55. Marsden, B.G., 1990, The Sungrazing comets revisited, *Asteroids Comets Meteors III*, eds. Lagerkvist, C.-I., Rickman, H., Lindblad, B.A. & Lindgren, M., pp. 393-396, Uppsala Observatory.
56. Marsden, B.G., 1992, (4015) 1979VA = Comet Wilson-Harrington (1949III), *IAU Circular* No. 5585.
57. Marsden, B.G., 1998, Personal communication.
58. McFarland, J., 1996, Kenneth Essex Edgeworth - Victorian polymath and founder of the Kuiper belt? *Vistas in Astron.*, **40**, 343-354.
59. Meech, K.J. & Belton, M.J.S., 1989, (2060) Chiron, *IAU Circular* No. 4770.
60. Menichella, M., Paolicchi, P. & Farinella, P., 1996, The main belt as a source of near-Earth asteroids, *Earth, Moon & Planets*, **72**, 133-149.
61. Morrison, D., 1992, *The Spaceguard Survey: Report of the NASA Near-Earth-Object Detection Workshop*, NASA, Washington, DC.
62. Napier, W.M., 1983, The orbital evolution of short-period comets, *Asteroids Comets Meteors*, eds. Lagerkvist, C.-I. & Rickman, H., pp. 391-395, Uppsala Observatory.
63. Napier, W.M., 1998, Cometary catastrophes, cosmic dust and ecological disasters in historical times: the astronomical framework, This volume.
64. Napier, W.M. & Staniucha, M., 1982, Interstellar planetesimals - I. Dissipation of a primordial cloud of comets by tidal encounters with nebulae, *Mon. Not. Roy. Astron. Soc.*, **198**, 723-735.
65. Oort, J.H., 1950, The structure of the cloud of comets surrounding the solar system and a hypothesis concerning its origin, *Bull. Astron. Inst. Neth.*, **11**, 91-110.
66. Opik, E.J., 1966, Sun-grazing comets and tidal disruption, *Irish Astron. J.*, **7**, 141-161.
67. O'Sullivan, M., 1993, *Megalithic Art in Ireland*, Town House and Country House, Dublin.
68. Pankenier, D.W., 1994, Astrological origins of Chinese dynastic ideology, *Vistas in Astron.*, **39**, 503-516.
69. Rabinowitz, D., Bowell, E., Shoemaker, E. & Muinonen, K.,

1994, The population of Earth-crossing asteroids, *Hazards due to Comets and Asteroids*, ed. Gehrels, T., pp. 285-312, University of Arizona Press.

70. Schultz, P.H. & Lianza, R.E., 1992, Recent grazing impacts on the Earth recorded in the Rio Cuarto crater field, Argentina, *Nature* **355**, 234-237.
71. Schultz, P.H. & Beatty, J.K., 1992, Teardrops on the Pampas, *Sky & Telescope*, **83**, No. 4 (April), 387-392.
72. Sekanina, Z., Chodas, P.W. & Yeomans, D.K., 1998, Secondary fragmentation of comet Shoemaker-Levy 9 and the ramifications for the progenitors breakup in July 1992, *Planet. Space Sci.*, **46**, 21-45.
73. Shoemaker, E.M., 1983, Asteroid and comet bombardment of the Earth, *Annu. Rev. Earth & Planet. Sci.*, **11**, 461-494.
74. Shoemaker, E.M., 1984, *Large body impacts through geologic time, Patterns of Change in Earth Evolution*, eds, Holland, H.D. & Trendall, A.F., 15-40, Springer-Verlag, New York.
75. Shoemaker, E.M., Weissman, P.R. & Shoemaker, C.S., 1994, The flux of periodic comets near Earth, *Hazards due to Comets and Asteroids*, ed. Gehrels, T., pp. 313-335, University of Arizona Press.
76. Steel, D.I., 1995, *Rogue Asteroids and Doomsday Comets: The Search for the Million Megaton Menace that Threatens Life on Earth*, John Wiley \& Sons, New York.
77. Steel, D.I., 1996, The limits of NEO-Uniformitarianism, *Earth, Moon & Planets*, **72**, 279-292.
78. Steel, D.I., Asher, D.J. & Clube, S.V.M., 1991, The structure and evolution of the Taurid Complex, *Mon. Not. Roy. Astron. Soc.*, **251**, 632-648.
79. Steel, D.I., Asher, D.J., Napier, W.M. & Clube, S.V.M., 1994, Are impacts correlated with time? *Hazards due to Comets and Asteroids*, ed. Gehrels, T., 463-477, University of Arizona Press.
80. Stern, A. & Campins, H., 1996, Chiron and the Centaurs: escapees from the Kuiper belt, *Nature*, **382**, 507-510.
81. Tholen, D. & Whiteley, R., 1998, Astronomers find new class of asteroid, *www.ifa.hawaii.edu/info/press-releases/1998D-K36.html.*
82. Whipple, F.L., 1964, The evidence for a comet belt beyond Neptune, *Proc. Natl. Acad. Sci.* (USA), 51, 711-718.
83. Whipple, F.L., 1967, On maintaining the meteoritic complex, *The Zodiacal Light and the Interplanetary Medium*, ed. Weinberg, J., NASA SP-150, 409-426, Washington DC.
84. Whipple, F.L. & Hamid, S.E., 1952, On the origin of the Taurid meteor streams, *Helwan Obs. Bull.* No. **41**, 1-30, Royal Obs., Helwan, Cairo.
85. Wickramasinghe, N.C., Hoyle, F. & Rabilizirov, R., 1989a, Greenhouse dust, *Nature*, **341**, 28.
86. Wickramasinghe, N.C., Hoyle, F. & Rabilizirov, R., 1989b, Extraterrestrial particles and the Greenhouse effect, *Earth, Moon & Planets*, **46**, 297-300.
87. Yau, K., Weissman, P., Yeomans, D., 1994, Meteorite falls in China and some related human casualty events, *Meteoritics*, **29**, 864-871.
88. Zappalà, V., Cellino, A., Gladman, B.J., Manley, S.P. & Migliorini, F., 1998, Asteroid showers on Earth after family breakup events, *Icarus*, **134**, 176-179.

Cometary Catastrophes, Cosmic Dust and Ecological Disasters in Historical Times: the Astronomical Framework

W. M. Napier
Armagh Observatory, College Hill, Armagh BT61 9DG, Northern Ireland, UK

Summary

Available data from the current near-Earth environment yield a coherent picture of the interface between the Earth and its environment. The Taurid Complex of meteors, meteoroids, asteroids and Comet Encke has probably been a visible and recurring hazard from the last Ice Age through to the present day. These data lead to the expectation that history has been punctuated by (i) Tunguska-like impacts, singly or as part of fireball swarms; (ii) freezing events of decadal duration, due to cometary dusting of the stratosphere; and (iii) occasional disastrous inundations of coastal areas due to small-body ocean impacts. There is a reasonable expectation that one or more areas the size of a small country have been destroyed by fire and blast caused by the impact of a bolide or bolide swarm some time within recorded history, and that the Earth has been subjected to one or more significant climatic coolings over the same period. On time-scales ~100,000 years, truly global celestial hazards come in the form of (i) giant comet dusting with the potential of generating an ice age, and (ii) impacts yielding significant worldwide extinction of life. A periodic Galactic modulation of these effects may exist, and provide a unifying 'clock' for terrestrial catastrophism. There may thus be a continuity of astronomical effects from geological to human time-scales.

1. Introduction: the nature of the problem

The object of this contribution is to describe, for non-specialists, the likely hazards posed by the interplanetary environment over timescales relevant to history (see [53] for a more technical review). It is expected that repeated and spectacular celestial signs were associated with some of these hazards: modern astronomical evidence does not support the common supposition that the night sky has been unchanging for 5,000 years. Occasional disastrous inundations and the raining of fire from heaven are reasonable expectations from the current astronomical evidence. Brief but intense episodes of stratospheric dusting from comets are also expected, and these may lead to cooling episodes and consequent trauma. That being so, one may contemplate a framework for many facets of culture in antiquity which are otherwise quite baffling. The identification of specific events in the historical, cultural and palaeoclimatic records is of course a matter for experts in those areas; this paper merely describes the astronomical framework.

Consider a sphere centred on the Sun and of radius 1 AU, so that its surface passes through the Earth. All classes of object passing within the sphere have the potential to collide with the Earth (it will emerge that this applies even to many objects which enter the sphere far from the ecliptic plane). In asking what interactions we might expect between the Earth and its environment, we are drawn, as a first approach, to ask: What are the dominant contributors to the mass flux through the sphere? What are their overall residence times within it? Where do the dominant collision cross-sections lie? Is the flux steady or variable in time? What is the nature of the interactions with the Earth?

The answers to these questions are still only partly understood, but they lie at the heart of the 'catastrophism' issue on all temporal scales from the geological to the historical. Some facts appear to be fairly well established:

- The most massive entity in the near-Earth environment is a flattish system of dust and boulders, in the invariable plane of the planets, known as the zodiacal cloud. There is evidence that its mass is highly variable over short time-scales, say of order 10,000 years.
- In terms of target area, the largest structures in the current near-Earth environment are short-lived trails of dust associated with active comets. Intersection with these trails takes place at characteristic intervals of a century, lasts an hour or so, and gives rise to meteor storms of such intensity that they have often been seen as signalling 'the end of the world'. The current Earth-crossing trails are generally associated with small, insignificant comets.
- The most massive single entities capable of entering the 1 AU sphere are rare, giant comets. Their mass influx dominates over all other sources: in contemplating an active comet of this magnitude, we have to think of something equivalent to the simultaneous disintegration of 10,000 Halley comets! Debris from such a monster, breaking up in a short-period, Earth-crossing orbit, will overwhelmingly dominate all other material in the near-Earth environment, at least sporadically.
- Near-Earth space is now recognised to contain many asteroids up to ~10 kilometres across, with the potential on collision to wreak great damage on many spatial scales from local to global.

Thus in looking for dominant processes, we are led first to consider the zodiacal dust cloud, comet dust trails, exceptionally large comets and the effects of asteroid impact. It will emerge that these types of hazard are closely interrelated.

2. The evidence from astronomy

2.1 Giant comets and the zodiacal cloud

The zodiacal cloud is a disc of interplanetary dust within which the inner planets orbit. Estimates of the mass of the zodiacal cloud, considering dust particles in the shooting star range 10^{-6}-10^{2}g, range from 2.5×10^{19}g [66] to 3×10^{20}g [34]. The latter is about 1500 times the mass of Halley's comet (taken as 2×10^{17}g), or twenty times that of the near-Earth asteroid system. Various factors combine to remove dust from the cloud: collisions will erode a 1g particle away in 7000 yr; inspiralling will remove cometary particles in about 10,000 yr and so on. Without replenishment the cloud will be gone, or reduced to insignificance, in a few tens of thousands of years [30]. All current sources of replenishment, however, fail by about two orders of magnitude to replace this loss rate [40, 66]. Further evidence of disequilibrium comes from a detailed analysis of the collisional and other processes going on within the dust: it turns out that particles of about millimetre size upwards (10^{-5}g) are, through their mutual collisions, creating smaller fragments faster than these latter are removed [30]. Since the lifetime of 10^{-5}g dust particles in the zodiacal cloud is a few tens of thousands of years, this suggests again that a major injection of dust into the zodiacal cloud has taken place on this timescale or less.

Exceptionally large comets are the only known source capable of replenishing the zodiacal cloud, albeit erratically. The mass distribution of comets is a power law (index about -1.7) such that the bulk of the mass is concentrated in the rarest, largest bodies. The much more common 'ordinary' comets (such as Halley's, Hale-Bopp, Hyakutake of recent years) are minor contributors to the overall mass influx to the inner regions of the planetary system. There is no known upper limit to the size of comets. Comet Sarabat of 1729 approached on a high inclination, parabolic orbit which never took it closer than 4 AU to the Sun, well beyond the range of water outgassing, and yet was an easy naked eye object for several months. Whipple [67] estimated its mass to be 10^{21}g, about 5,000 times that of Halley's Comet. However recent years have seen an order of magnitude upwards revision in estimated comet masses [6], and Whipple's estimate for the mass of Comet Sarabat may well be a severe lower limit.

Archetypal bodies of the giant comet type are also found in low-inclination, unstable, chaotic orbits beyond Saturn, with masses again of order 10^{21}g. Chiron and Pholus, about 150-200 km in diameter, are examples of these 'Centaurs'. Their dynamical lifetimes are remarkably short, and there is a high probability that a Centaur will eventually be thrown into a short-period, Earth-crossing orbit of low inclination and small perihelion distance, where it will disintegrate under the influence of sunlight. An icy body of diameter 200 km has a mass 4×10^{21}g, 10 to 100 times greater than that of the current zodiacal cloud. The active lifetime of a very large comet is likely to be an order of magnitude or more shorter than the half-life of the zodiacal cloud, whence such bodies are more than adequate to replenish it.

Zodiacal dust particles which enter the stratosphere have been captured by high-flying U2 aircraft. About two thirds of these so-called Brownlee particles, which have microscopic dimensions, turn out to have a fluffy structure, often possessing a 'cluster of grapes' constitution; this fragile, open structure is indicative of a cometary origin. The remaining third are more solid and may (or may not) be the products of collisional grinding from the main asteroid belt. Thus a cometary provenance is again indicated for the bulk of these small particles.

However the bias of cometary mass towards the largest bodies implies that the 'refuelling' of the zodiacal cloud is sporadic. During such refuelling, the Earth first passes through an extraordinarily dense meteor stream and then, as the stream disperses, a massive zodiacal cloud; meanwhile the Earth's stratosphere acquires a significant optical depth of reflective dust. Thus one is led to ask: what happens to the Earth's climate during a major cometary disintegration?

2.2 Sub-structures: the Taurid Complex

Much information about structure within the zodiacal cloud has been obtained from a study of meteor radiants, supplemented i n recent years by observations from satellites and interplanetary probes (e.g. [46]). A broad stream of material is imbedded within it, comprising the so-called Stohl stream, which is composed of dust particles in near-equal, high eccentricity orbits. These manifest themselves, on entering the Earth's atmosphere, as shooting stars. Radar observations, which detect generally smaller particles than those which give rise to shooting stars, reveal that about 90% of them concentrate in these helion and anti-helion streams. Within the Stohl stream are the Taurid meteor streams (comprising several closely related branches). The Taurid streams are old and somewhat dispersed, and have clearly derived from a large progenitor body. Their detailed orbital structure shows them to have been created in a few highly active episodes. It turns out that the plane of the zodiacal dust cloud is inclined at about 1.5° to the ecliptic plane and in fact aligns itself closely to the symmetry plane of the Taurid complex.

Contained within the Taurid meteor streams, in turn, is a small, short-period comet (Encke's Comet) and a number of asteroids which have orbits similar to that of the comet. Comet Encke is a faint, low-inclination, short-period comet of high eccentricity (i =12°, P = 3.3 yr, e = 0.847). Its orbit is unique amongst the known active comets: it lies entirely inside that of Jupiter. Its perihelion distance 0.34 AU takes it well inside the orbit of the Earth, and the precession of its orbit (the nodes having a period of about 5400 years) ensures that it has undergone a series of close encounters with the Earth in the past (non-gravitational forces make the passage times uncertain, but 500-600 BC and 3600-3700 BC are reasonable estimates based on the current orbit). It was seen as a naked eye object in 1786 and 1796 and about a score of times in the 19th century. On four such occasions it was recorded as a 4th magnitude object, and was once seen at magnitude 3.5, making it easily detectable by eye. In the 20th century it has been recorded at 5th magnitude just three times. However, searches have failed to reveal any detection of the comet in earlier records.

Encke's Comet is only a few kilometres across and is much too small to be the source of the current Taurid meteors. Most likely, it is simply one of the co-moving cometary asteroids which has had a recent, and temporary, burst of activity. It is a characteristic feature of many short-period comets that they are dormant for periods of time of uncertain duration (perhaps centuries or millennia) and then flare into activity, again for uncertain durations.

An active periodic comet releases meteoroids, and these will spread to form a stream of material along the orbit which, in the absence of replenishment from the parent, will eventually disperse. The comet itself, if dark and rich in volatiles, and if its dust grains are fluffy, may disintegrate completely; otherwise it will become dormant or extinct, and asteroidal in appearance [23]. When the Earth intersects the meteoroid stream a meteor shower is seen, the meteors emanating from a point in the sky. Given enough time, differential precession acting on meteor orbits of different sizes will split a single shower into four branches, two daytime showers and two night-time ones. Thus for example the periodic Encke's Comet (P/Encke) is associated with the Northern and Southern Taurids as well as with the daytime ζ-Perseids and β-Taurids.

If a meteor shower is found to be associated with an asteroid, then there is a strong presumption that the asteroid is a degassed comet. The Geminid meteor stream, for example, is not associated with an active comet. Imbedded within the stream, however, is Phaethon, an Apollo asteroid about 5 km in diameter. The Geminid stream is narrow, and without replenishment it would be gone within a few millennia. The inference is that Phaethon is a comet which has, in recent millennia, given rise to the Geminids before developing a protective mantle. The Quadrantids are another example of a major stream whose origin seems to lie in an outburst, 500 years ago or less, of some body now hiding in an asteroid-like orbit [35].

A number of asteroids occur in orbits close to that of P/Encke, and given the propensity of some comets to split and develop mantles, it may be that they are the remnants of the progenitor giant. Given the incompleteness of discovery of near-Earth asteroids, there are likely to be 100-200 of them over a kilometre across, some 5% of the NEO population. Whether these 'Taurid asteroids' are a physically real grouping is currently a somewhat controversial issue. High-eccentricity asteroids will tend to be discovered with the orbital characteristics of the known Taurid asteroids, suggesting that the concentration of orbits is no more than a discovery selection effect (Valsecchi, *personal communication.*) On the other hand such an effect does not explain the coincidence that P/Encke and the Taurids are imbedded in the same high-eccentricity envelope of orbits. In a recent study, Babadzhanov [4] has discovered that 40 meteor showers may be matched with ten of the Taurid Complex asteroids, thus supporting the hypothesis that the latter are indeed degassed comets some thousands of years old.

The Taurid Complex is a massive component of the near-Earth interplanetary space (about half the dust mass of the zodiacal cloud is contained within the Stohl stream). It is best explained as the fossil remnants of an erstwhile very large progenitor comet, the Stohl stream being a bridge between the recognisable Taurid meteors and the zodiacal cloud into which the material ultimately disperses.

2.3 The origin of the Taurid complex

The source of the very large progenitor is uncertain. Comet Encke itself appears to be a small, relatively unimportant body within the stream which underwent a minor outburst in the mid-eighteenth century and which is in process of returning to its dormant state (another Taurid asteroid, Oljato, also shows signs of activity). Dynamical highways have been found which connect Encke's Comet to both the main asteroid belt and a family of short-period comets whose dynamics are closely controlled by Jupiter [65]. However the travel time along these highways is of order 100,000 years, by which time any comet would have degassed and the Taurid meteor complex would have long dispersed (the Taurid meteors, without replenishment, would disperse in less than 30,000 years [60]). However in general outgassing from an active comet generates small but cumulative non-gravitational accelerations, and these may cause a sufficiently massive comet to decouple from the influence of Jupiter and enter a relatively stable Earth-crossing orbit on a time-scale of about 30,000 years [61].

Very large comets may also arrive from the Oort cloud, however, possibly via the family of 'dark Halleys' necessary to explain the visible population of comets in orbits like that of Comet Halley (Bailey, this volume). This is evidenced not only by Comet Sarabat of 1729 but also by a large group of comets known as the Kreutz sungrazers. These are comets in very similar, retrograde orbits, grazing or falling into the Sun at high inclinations. Comets belonging to this group are currently being discovered every few weeks by the SOHO satellite and are clearly debris from the breakup, by solar tides, of an exceptionally large progenitor. There are probably tens of thousands of fragments. A bright comet of 372 BC may have been, if not the progenitor, at least one of the major fragments: Ephorus reported that this comet, 'a flaming body of exceptional size', split in two. The hierarchic disintegrations have yielded fragments which were considerable comets in their own right, such as the daytime comet of 1882 which split into four or five fragments as it skimmed past the Sun at only 200,000 kilometres altitude. The Kreutz sungrazers illustrate the operation of a secular trend (known as the Kozai cycle for short-period comets and asteroids) in which orbital elements such as inclination and perihelion distance oscillate in a correlated fashion under the gravitational influence of the giant planets [7]. Objects in retrograde orbits tend towards a sungrazing state when their inclinations are high, while prograde bodies tend towards sungrazing at low inclination.

It is interesting to consider the position had the Kreutz progenitor been in a prograde rather than retrograde orbit. In that case the progenitor of the Kreutz group would disintegrate near the ecliptic plane, in a low-inclination, Earth-crossing orbit. Twice a year the Earth would run through a stream of material including these disintegrating comets and their meteoric debris, one passage occurring during daytime and the other at night. Now further suppose that the characteristic period of the debris was not ~1000 years as for the present Kreutz Complex, but ~3.3 years as for the Taurid one. Then the fluxes of both cometary fragments and the annual meteors would be increased by a factor ~300 for the same cometary disintegration rate (which would, however, be greatly enhanced due to its much more prolonged exposure to sunlight). The celestial fireworks in the night sky, especially in the form of annual fireball storms, would then be of an intensity quite outside that of modern experience. It is likely that for periods of centuries or millennia one or two comet fragments would dominate, being bright, recurring objects in the night sky. Comet splitting and even multiple disintegration would be a common observed phenomenon. If such a disruption had happened within historical times, it is difficult to imagine pastoral societies remaining indifferent to this annual show.

In fact, the most active recent phase of the Taurid progenitor comet appears to have been about 3000 BC, although it may have a pedigree an order of magnitude longer than this. It is hardly conceivable, and certainly not in keeping with the known characteristics of comets, that such a major episode of meteor formation could have taken place in the near-Earth environment with the progenitor comet invisible (even at the present time, Comet Encke is only just below naked-eye detectability). Conservative estimates have the short-period progenitor at that time brighter than Venus. Thus the night sky around 3,000 BC, and for a period of at least centuries and probably one or two millennia after it, was disturbed, contained one or a few major comets recurring annually, coupled with epochs (set by orbital precession) when the annual meteor storm reached prodigious levels. Meteor storms are probably the most impressive spectacles the sky has to offer. At some intensity level beyond modern experience, they may become an ecological hazard. They are discussed later.

These considerations suggest that two types of celestial hazard present themselves over relatively short time-scales (say 10,000 years), namely impacts and dustings. We examine each in turn.

2.4 The small body populations

As a first approach to the impact hazard problem, we may ask, what is the largest impact we may expect on time-scales of direct human interest, say over the period of recorded history (~5000 yr); and what are the consequences of impact? Attempts to assess the hazard have involved extrapolating the size distribution of large terrestrial craters down to small ones, counting small craters on the Moon, counting bright fireballs with networks of all-sky cameras in the USA, Canada and Europe, military detection systems (satellite and acoustic), and direct observation of small bodies in space (the *Spacewatch* project).

All these approaches have problems. To go from the relatively complete sample of large, young terrestrial impact craters to the almost unknown small one [52] involves a large, uncertain extrapolation; using counts of small lunar craters [58] to deduce present-day impact rates [14, 20] involves the dubious assumption that the current small-body flux equals that averaged over the age of the lunar surface (≤3 Gyr); counting bright fireballs [13] involves small number statistics; and observing bodies in near-space has to date involved assumptions about their albedos (and hence diameters, masses and impact energies). We discuss below the approaches based on direct observation.

Fireball network surveys began around 1964 in the USA, Canada and Europe, with search areas about two million square kilometres, and several hundred photographic meteors in the size range 0.1 to 1 metre were detected. Only thirteen bodies with initial sizes over a metre were detected by these networks (three of which had orbits which made them possible Taurid Complex bodies). There are large differences in the atmospheric behaviour of incoming fireballs, reflecting differences in their density and strength. In the range up to about 2 metres diameter, the so-called Type II and Type III fireballs are about equal in number, the former being associated with stony (carbonaceous) bolides of density ~2g cm^{-3}, while the Type III bolides have low densities (~0.5g cm^{-3}) and are thought to be associated with comets. Beyond 2 metres or so, the relative flux of stony bodies drops rapidly; and at about 10 metres, virtually in the regime of small-number statistics, cometary bolides outnumber stony ones by about two to one. The numbers are then dominated in particular by Type IIIB fireballs, an extreme population type whose mean density is ~2g cm^{-3}. A 1-metre Type III bolide weighs about 10^5g and has a pre-entry impact energy of about 7.5×10^{-6} megatons; a 100-metre bolide is 7.5 megatons. Estimates of the energy of the Tunguska bolide, which devastated ~10^4 km^2 of forest in the Cental Siberian Plateau in 1908, are in the range 5-30 megatons.

Based on the fireball network data, it has been estimated that the whole-Earth annual flux of Type IIIB bolides in the diameter range 0.1 to 1 metre is 21,000, while in the range 1 to 10 metres, the annual flux is 980 [13]. If we assume a power law distribution of masses, then the flux decline of a factor ~20 per decade of diameter leads one to expect that Tunguska-type events should be annual affairs! If this were so then our perspective on the sky, not to mention human history, might be quite different from the popular one. On this basis it seems there must be a downturn in the population somewhere between ten and one hundred metres. The fireball data thus, unfortunately, do not allow us to make an accurate estimate of the impact rate in the 'Tunguska' range. However they do indicate that there exists a large interplanetary population of extremely fragile boulders, derived presumably from the disintegration of comets. Virtually nothing is known about this population. Given the fragility of these bodies, they may well be short-lived; in that case, one expects them to concentrate in thin streams along the tracks of erstwhile or active comets. A mismatch of a factor two or so between the rates inferred from the time-averaged lunar cratering and the inferred 100-metre Earth-crossing population would in principle allow for hugely enhanced Tunguska rates concentrated in short intervals, analogous to the rare meteor storms. The task of discovering such boulder swarms, however, seems to be beyond current technology.

The few-metre fireball fluxes seem to agree with those derived from the Spacewatch project, an automated sky survey capable of detecting bodies between 100 and 5 metres across. It has been found that their numbers are ~40 times those expected by extrapolation from the size distribution of the larger Earth-crossers. This is again consistent with the view that there is a population of sub-Tunguska bodies in the near-Earth environment, with a predominantly cometary constitution [15]; a subset of them appear to have low inclinations, almost circular orbits and perihelia close to that of the Earth. However, there still remains great uncertainty about absolute impact rates, in part because the albedos of the Spacewatch objects are uncertain.

Based on US Air Force acoustic detection of energetic fireballs over the period 1960 to 1972, ReVelle [57] computed the whole-Earth annual flux of fireballs of energy ≥ E kilotons to be

$$F = 7.2E^{-0.73}$$

This would yield a 10 megaton impact every century or so, and a 15 megaton Tunguska every 150 years. The formula is based on only ten events, only one of which was in the megaton class. The rate is nevertheless in line with one's terrestrial experience that Tunguska-type impacts are at least not annual events.

Satellite observations of the Earth between 1975 and 1992 have revealed that some 136 sub-Tunguska bolides impacted the Earth's upper atmosphere during this interval, yielding a rate of a few per annum for objects 10-30 m in diameter. For 100 m sized Tunguska-like bolides the current impact rate could then be one in 30-100 years [18, 20].

What cosmic inputs, then, can we reasonably expect to have taken place over, say, the last 5,000 years? Assume that an impact of at least E_o megatons takes place at mean intervals of 100 years. Then for a constant population index -1.73 characteristic of most fragmenting systems, the mean recurrence time Δt in years between impacts of energy at least E megatons is given by

$$\Delta t\ (\geq E) \sim 100(E/E_o)^{0.73}$$

E	E_o=5	10	20
10	170	100	60
50	540	320	200
100	890	540	320
1000	4800	2880	1740
10000	25700	15490	9340

Table 1. The mean interval in years between sporadic impacts of various energies ≥ E megatons as a function of centennially-expected impacts of energy E_o = (5, 10, 20) megatons respectively.

Table 1 shows the waiting time in years between impacts of energies at least (10, 50, 100, 1000, 10,000) megatons for a reasonable range of centennially-expected impacts of energy at least E_o.

These are the rates of *sporadic* impacts (the possibility of brief, highly-enhanced episodes of risk is discussed later).

The range of uncertainty, a factor of three, is probably an underestimate. Nevertheless, on the basis of Table 1, there is little reason to change significantly the earlier conclusion [18] that 'a few dozen sporadic impacts in the tens of megatons, and a few in the 100 to 1,000 megaton range, must have occurred within the past 5,000 years.' The effects of such small impacts have been intensively studied over the past few years, and it is to this issue that we now turn.

3. The effects of impact

Much uncertainty in assessing the impact hazard, whether in the near past or near future, arises from factors such as the (unknown) turndown or slope change of the distribution, and temporal surges of fireballs due either to the injection of new material into the inner planetary system, or the existence of spatial concentrations of material already in Earth-crossing orbits, or its creation by disruption. If we take the fireball (civilian and military) data as a guide, and allow for the possibility of brief high-risk periods, it may be that the largest impact expected over the last 5000 yr (and within the next 5000) lies in the energy range 1,000 to 10,000 megatons. Atmospheric protection probably holds up to the 10-megaton (50 metre) energy range, beyond which air-bursts may generate sufficient overpressure at ground level to cause severe damage [15].

Oceans cover three-quarters of the Earth's surface, a 1,000 Mt sea impact couples efficiently to the wave generated, and the specific wave energy drops only linearly with distance, since it perturbs a surface rather than a volume. At 100 kilometres from the impact site, the ocean wave so created is about 50 metres high. On approaching a shoreline, the same wave energy is carried by a progressively dwindling mass of water, and so the wave rears up. This run-up factor is in the range 10-40, and depending on the distance from the epicentre, open ocean waves may become tens or even hundreds of metres high. A series of oscillations ensures that the target coast is hit by a succession of waves, characteristically 5-10, over a period of a few hours. The inshore flood plane may extend for tens to hundreds of kilometres. A 200-metre body impacting anywhere in the North Atlantic would overrun all low-lying areas (Denmark, Holland, Manhattan etc.) bordering the ocean; while Yabushita & Hatta [69] found that the average tsunami from a 200 m Pacific impactor would be of order 15 m high at Japan, Taiwan and Shanghai, and about 20 m high in Hawaii. For comparison, they assessed that wooden houses would be destroyed by a 2 m tsunami, stone houses by a 7 m one and brick houses by a 20 m one. They estimated that a Pacific impact would destroy most of the artificial constructions around the Rim.

Land impacts are less common than ocean ones and their potential for destruction appears to be smaller (although the smaller area of land may be offset by the likely greater frequency of damaging airbursts). However, because they may occur over inhabited areas, they may have cultural implications which distant impacts over water do not, especially as they may be associated with spectacular celestial phenomena. The energy of the 1908 Tunguska impact was probably of order only 5-30 megatons, whereas most of the craters listed in Table 4 required ~100 million megaton impacts for their creation; nevertheless had the Tunguska bolide struck, not in the remote Central Siberian Plateau, but in London, Manhattan or any urban area, there would surely now be a sharpened perception amongst the general population of the celestial hazard represented by the interplanetary population, and a shift in ones historical perceptions.

A land impact will generate blast, heat, earthquake and dust:

Blast. Air pressure momentarily increases as the shock passes, and the defining pressure excess (or overpressure) for substantial blast damage is often taken to be 4 psi. Corresponding to this overpressure is a wind speed of ~70 ms^{-1} or 160 mph, and this itself creates a dynamic overpressure of 0.6 psi. At 10^3 Mt, the 4 psi contour reaches out to 150 km, encompassing an area of about 70,000 square kilometres, while at 10^4 Mt the area of blast devastation is ~200,000 km^2, about that of the United Kingdom, for example (flying debris causes extensive injury out to 2 psi, especially shards of glass in urban areas). Assuming that, say, civilizations occupied ~20% of the land area of the Earth 5000 years ago, there is a reasonable expectation that an area the size of a small country has been subjected to a Tunguska-like blast over the period of recorded history. This is for the baseline rate.

Heat. The threshold for fire ignition may be taken as ~100 J cm^{-2} but is a strong function of visibility at the time of impact. At 1,000 megatons, this corresponds to the lighting of fires over an area about 200 km across. Flash burn injuries occur at ≥ 20 J cm^{-2}. (At 5×10^5 Mt, beyond anything likely to have been encountered in history, everything combustible along the line of sight of the rising fireball will be ignited: this corresponds to a region ~2,000 km across, an area about the size of Europe.) Five minutes after impact, the plume from a thousand megaton land impact has risen to a height in excess of 200 kilometres.

Earthquake. Laboratory-scale experiments, nuclear weapons detonations, rocket casing impacts and theoretical considerations have all been used to yield estimates for the efficiency f with which an impact will couple energy to the ground, and the results have ranged widely. None of the existing calibrators corresponds closely to the real thing. The range for impacts may be straddled by airbursts on the one hand and underground explosions on the other. For example the Tunguska impact, an airburst, yielded an earthquake of magnitude 5 on the Gutenberg-Richter scale: this corresponds to a coupling efficency f of only a few times 10^{-5}. At the other extreme, underground nuclear explosions yield f up to about 0.05 [28]. A simple theoretical estimate based on the median tensile strength of rock suggests f ~0.01 or 0.02. If we adopt $f = 10^{-3}$, then an earthquake of magnitude 8 or 9 would require an impact of energy $E \geq 10^4$ Mt.

An earthquake with M = 8 or 9 will cause devastation over a region 500-1000 km across. Unless the coupling efficiency is higher than most estimates, the seismic destruction initiated by say a 10^4 Mt impact may not extend as far as that caused by fire and blast, as at most only a few hundred megatons of this energy will couple into the ground in the form of earthquake. Presumably an impact in a seismically sensitive area might trigger a more extensive release of ground energy, but to the author's knowledge there are no quantitative estimates for this. The intensity of an impact-induced earthquake may be no greater than that of natural ones in this energy range, and their frequency is much less. Earthquakes have been invoked to account for the Bronze Age destructions, and need no impact to induce them. Other evidence (air blast, fireball reports, climatic downturn or whatever) would have to be adduced before a case could be made for impact-induced earthquake as an ancillary agent of destruction in any specific case.

Dust. Nuclear explosion data [28] reveal that about 0.3 million tons of vaporized and melted dust are raised into a plume per megaton of explosion, about 8% being in sub-micron form. Broadly similar figures apply to the dust resulting from a land impact [70]. In terms of climatic perturbation, the most dangerous dust is the sub-micron, which contributes most of the optical depth and has a settling time of order months to a year. Such dust, generated by a 10^4 Mt impact, is comparable in mass and effect with that from the greatest volcanic outbursts. The latter have yielded measurable climatic coolings (a few tenths of a degree) of duration a year or more. A 10,000 megaton impact therefore represents a threshold beyond which the cooling may begin to exert agricultural and biotic effects on a global scale. However, on a 5,000-year timescale, it is unlikely that the climatic effects would outweigh those expected from great volcanic explosions.

4. Correlated impacts

The 1994 impact of Comet Shoemaker Levy 9 on to Jupiter dramatically confirmed the prediction that trouble might on occasion come in droves [19, 20, 62]. Apart from tidal disintegration of comets, 'spontaneous' disintegration is also a common phenomenon and may yield a hierarchy of comets, some becoming short-lived 'asteroids'. Small (Tunguska-sized) bodies, deriving from such hierarchic breakup, will tend to spread out along individual orbits over ~200 yr, but will also yield a fairly concentrated core-stream, which may be quite stable in particular circumstances (e.g. if they inhabit a mean-motion resonance with Jupiter: [3]). Intersections with such core-streams will recur on two characteristic timescales, of order decades (dictated by commensurability between the periods of Earth and core) and of order millennia (dictated by the orbital precession period of the core material). When the core and Earth are brought into intersection, there will be an epoch lasting from one to a few years when an annual meteor shower, occurring at the same date each year, becomes remarkably strong and the risk of impact from larger bodies is then correspondingly enhanced. In short, rather than purely sporadic bombardment by individual objects, the impact hazard is likely to be structured.

A system of N bodies of period P years passing through the 1AU sphere would, if distributed at random, strike the Earth at mean intervals $\sim P/N$ Gyr. Thus a Chiron-sized comet in an Encke-like orbit would, if broken into pieces 10 km across, yield 8000 Halley-sized Earth-crossing comets hitting the Earth at mean intervals of 400,000 years, each impact being at the level associated with a great mass extinction. If broken into pieces 50 metres across, the giant comet would generate about 6.4×10^{10} Tunguska-like fragments hitting the Earth once every five or six days! While these modes of disintegration are overly simplistic, they do illustrate that surges of strong bombardment could in principle occur in the wake of a large disintegrating comet.

The most extreme known manifestation of fine structure in the current zodiacal cloud occurs in the form of rare, intense meteor storms, which take place when the orbit of the Earth intersects a concentration of debris within a meteor stream. Table 2, taken from Kresak [38], illustrates that the flux of meteors may on rare occasions, for a period of a few hours, reach levels hundreds of times that of the sporadic background.

storm	year	hourly rate
Leonids	1799	30,000
	1833	100,000
	1966	150,000
Andromedids	1885	15,000
Dracondis	1933	20,000

Table 2. Major meteor storms of the past 200 years.

Some of these rates may be conservative. For example von Humboldt estimated that the Leonid storm of 11 November 1799 produced 300 meteors a second, or close to a million an hour.

For all their intensity, these showers are only a small addition to the integrated flux over say a 100-year period. Since the uncertainty in current impact rates (say in the 10-10,000 megaton range) is substantially more than a factor of two, it is clear that Tunguska-like showers could occur in principle pro-rata without affecting the overall rate estimated from, say, lunar crater counts.

The concentrations of dust in the wake of a comet take the form of a long ribbon of material in the orbital plane, characteristically ten times as broad as it is high and with a length up to a thousand times its breadth. There may be filamentary structure within these ribbons [39]. Dust concentrations corresponding to particle densities 400±200 times the sporadic background have been detected in the infrared with the Infrared Astronomical Satellite (IRAS). Eight such IRAS trails were associated with well-known periodic comets; others were orphans with no obvious parents. No meteor storm of the last two centuries can be identified with any of the IRAS trails.The intensities of these storms correspond to passages just outside the observed boundaries of an IRAS comet trail [39]. The IRAS trail of Comet Encke is exceptionally broad and massive even although the comet must have been dormant before its discovery in 1786 (otherwise it would have been easily visible with the naked eye). Most IRAS trails disperse rapidly. It is possible that the 'Comet Encke' IRAS trail derives in fact from some other body in a similar orbit to Comet Encke; or it may feed out from an undiscovered core of material trapped in a stable resonance with Jupiter [3].

A rule of thumb given by Kresak [39] is that the Earth runs through a cloud of density D times the sporadic background once every D years. This formula predicts meteor storms of awesome intensity at millennial intervals, but even so it is based on the present-day environment. In the epochs when the Comet Encke progenitor was undergoing one of its catastrophic disruptions, and in particular when the major phase of Taurid meteor formation was under way, annual fireball storms of huge intensity would be expected. Further high-activity epochs, unfortunately not yet predictable in time, are expected when orbital precession brought the 'IRAS trail' of the active progenitor or its subsidiaries into annual intersection with the Earth's orbit.

But what are the largest bodies which we might expect to find in swarms? Unfortunately the data to answer this question are all but non-existent. Fireball swarms, which must have involved multi-metre bodies, are explicitly recorded in Chinese annals. In the compilation of Tian-shan [63], out of 147 showers recorded over a ~2,000-year period, over a dozen seem to have contained substantial boulders, especially since five showers were seen in daylight. Thus:

> Dynasty *Han*, Reign *Yuan-yan*, Year 1, Month 4, day *Ding-you*. At the hour of *rifu* the sky was cloudless. There was a rumbling like thunder. A meteor with a head as big as a *fou*, and a length of some ten-odd *zhang* colour bright red and white, went southeastward from below the Sun. In all directions, meteors, some as large as basins, others as large as hens' eggs, brilliantly rained down. This only ceased at evening twilight.

The Leonids have likewise been reported as having fallen with noise, or great noise, in 1798, 1666, 1602, 1566, 1533 and 1002 AD.

Seismometers were left on the Moon by the Apollo astronauts to record the vibrations from impacting bodies (and any internally caused moonquakes). Unfortunately precise calibration has not been possible and so there is some uncertainty over the masses of the recorded impactors; generally they were 'boulder' sized. The seismometer signals were recorded over a five year period. Over a period of about five days, covering the period 26-30 June 1975, these lunar seismometers recorded a boulder flux about a hundred times stronger than the sporadic background. The date corresponds to the passage of the Earth through the daytime Taurid meteors (the Beta Taurid stream).

Historically, too, there have been epochs of about a century's duration when strong peaks in the incidence of fireballs were recorded in Korean and Chinese annals. The 11th century AD peak is particularly strong and is difficult to attribute to say calibration or climatic factors. It seems to have comprised mainly Taurid fireballs.

5. The influx of dust

Sufficient dust is likely to be generated during the disintegration of a giant comet to create episodes of sharp climatic cooling, and we now examine this hazard from the astronomical perspective. A comet 200 km in diameter thrown into a Taurid-like orbit ($P \sim 3.3$ yr, eccentricity ~0.85) will lose $M \sim 10^{18}$g yr^{-1} due to outgassing, more than half of it as meteoroidal dust with diameters in the range 0.01 microns to several mm [26]. A particle size distribution $n(a)da \propto a^{-3}\, da$, $a > 0.01$ mm, is indicated by the Halley data, implying a significant pile-up of size in the submicron range. Outgassing and dust production will not be uniform with time: the overall active lifetime of the comet ~3000 yr may be interspersed with dormant periods when the surfaces become temporarily crusted. During its active lifetime such a comet could generate a zodiacal cloud of mass ~300 times that of the present one.

For dust particles in bound orbits the main loss mechanism from the zodiacal cloud is due to Poynting-Robertson drag forces. A zodiacal cloud resulting from a cometary evaporation episode would have an average half-life of ~10^4 yr. Modelling this temporary zodiacal cloud as a disc of mass $5\text{x}10^{20}$g, radius 1 AU and thickness 0.2 AU, and allowing for gravitational focussing, one finds that ~10^9 tons of dust are swept up by the Earth annually over the few millennia when the comet is active. This may be compared with the observed current rate of infall of micrometeorites, about 40,000 tons/yr [43].

A porous mineral/organic interplanetary dust particle of radius 0.1-0.3 microns and density 1 g cm^{-3} has a settling time through the atmosphere of about 3-10 yr [36]. Brownlee (cometary) particles have very rough surfaces which make them efficient scatterers of radiation. Their mass extinction coefficient is of order 10^5 cm^2 g^{-1}, comparable with that of smoke [68]. The presently observed influx of comet dust (40,000 tons/yr, taking a decade to settle) thus yields an optical depth, in the stratosphere, ~0.01. An enhancement of the comet dust load by two or three powers of ten, within the range of fluctuation of the zodiacal cloud, will in effect envelope the Earth in a highly reflective dust cloud. This can hardly fail to have a dramatic effect on climate (a mean surface temperature change ~ 1° is induced by a 1% decrease in incident sunlight). At high optical depths complicating factors such as dust coagulation will come into play, but it seems clear that even a relatively modest comet thrown into an Encke-like orbit has the potential to significantly decrease the solar constant with a corresponding effect on climate.

Indeed, a disintegrating giant comet may induce climatic cooling for a time in excess of critical time constants in the Earth system (in particular the cooling time of the oceans). The Earth's albedo need only be raised from 0.4 to 0.5 for a decline in precipitation from 80 cm/yr world-wide to 50 cm/yr world-wide to occur, cooling the stratosphere to around -110°C, low enough for the formation of a permanent haze of high-altitude, highly reflective ice crystals, so locking the Earth into a cold state [32]. In that case load shifting between hydrosphere and cryosphere may occur, with the potential for irreversible geophysical change. Such an event might lead to a full-blown ice age [7, 19, 22, 32], which would presumably bring civilisation to an end over most of the globe. The injection rate of giant comets from chaotic, trans-Saturnian orbits into short-period, Earth-crossing ones is ~10 Myr^{-1} [7], comparable to the mean interval between glaciations; thus the timescale is of the right order.

It is possible, however, that there has been a continuum of lesser climatic events, significant on millennial timescales, associated with periodic immersions in the dense dust trails of active short-period comets. The Earth probably skirts the fringes of IRAS-like trails every century or so (giving us meteor storms: [39]). A reactivation of one of the remnant Taurid asteroids, however, coupled with orbital precession, could lead to periodic intersections with dense dust concentrations and a significant dumping of sub-micron dust into the stratosphere. For strong climatic effects to be expected, the 'IRAS trail' would have to be at least 1000 times denser than those now observed, and might be visible as a milky patch of light in the night sky, drifting through the zodiacal light.

To assess the consequences of such 'lesser' events, limited comparisons may be made with nuclear winter studies. The earliest nuclear winter models were one-dimensional, assuming a hemispheric distribution of smoke and dust, and neglecting feedback effects from cryosphere and oceans. Typically in such runs the initial optical depth was $\tau \sim 4$, the bulk of the absorption being due to smoke injected into the troposphere, while fine dust (≤10 μm radius) reaching the stratosphere contributed $\tau \sim 1$. The overall optical depth declined, in these models, to ~2 after three months. Dramatic temperature drops $\Delta T \sim$ -40°C were attained within about twenty days of the dust injection, and the recovery time of the climate was over a year. Second generation models incorporated snow and ice feedbacks, and took account of the heat content of the oceans, which had the effect of moderating the land temperature response, and a third generation of models has further moderated these predictions since the infrared opacity of the smoke allows for a compensatory 'greenhouse effect', and the smoke distribution is patchy, allowing sunlight to penetrate from time to time. Thus these later models point to a 'nuclear fall' rather than 'nuclear winter'.

However the dust influx from an IRAS-type trail is prolonged, and probably involves not only intense meteoric input but also the disintegration of larger bodies on atmospheric entry to ~ μm-sized dust particles. The dust veil from such encounter events could easily yield sharp cooling, sudden in onset, with measurable climatic and agricultural effects, recovery taking some years. The AD 536 dust-veil (τ ~2.5) might be of this character as it appears not to be associated with a volcanic acidity signal. If astronomical in cause, one would expect a highly disturbed sky to have been seen, with accounts of fireballs, meteor showers and perhaps a great comet. A causal relationship has been proposed between this dust veil and the severe cold and famine at this time, the latter documented throughout the Old World and presaging the arrival of the Justinian Plague [8]. Both the AD 536 event and the 17th century Little Ice Age are correlated with known surges in the Taurid meteoroid flux, the likely source of the 1908 Tunguska object. Such coolings are global in extent, sudden in onset and decadal in duration. In terms of risk, therefore, they may be at least as significant as those of small impacts. Orbital precession would lead to recurrence of such events and a global cycle of a few thousand years (Clube, this volume). The subject is currently all but unexplored.

6. The evidence from geology

6.1. Impacts and extinctions

From the astronomical perspective, modern catastrophist studies arose from a number of discoveries which were made throughout the 1970s. The Apollo landings finally confirmed that most or all lunar craters have an impact rather than volcanic origin; a series of probes clearly showed all the bodies of the solar system to have been heavily bombarded; a population of small bodies was found to be orbiting in the interplanetary regions; and increasing numbers of impact craters were found on Earth. The finding that a system of massive nebulae orbits within the disc of the Galaxy raised the prospect that the Galaxy might exert an influence on the Earth. These so-called giant molecular clouds have masses typically half a million times that of the Sun; a grazing encounter with one such may strip the outer Oort cloud of its comets and flood the inner planetary system, yielding an episode of bombardment of the planetary system. Collision rates on Earth were deduced both from terrestrial impact craters [29] and from surveys of Earth-crossing asteroids. Napier & Clube [52] used these early quantitative results, along with the new molecular cloud data, to postulate that the great mass extinctions and geological disturbances were caused by the impacts of 10 km planetesimals (yielding blast, cosmic winter through ejected dust and ozone depletion) during periodic bombardment episodes. These were supposed to occur at ~50 Myr intervals when the Sun encountered the spiral arms of the Galaxy (the hypothesis that mass extinctions correlate with spiral arm crossings has recently been revived [42]). The turning point in such studies came with the finding that iridium, a rare element supposed to have an extraterrestrial origin, was found in excess at the K/T boundary where the dinosaurs and other creatures of 65 Myr ago became extinct [2]. The discoverers likewise attributed this iridium excess to the impact of a 10 kilometer asteroid, and the associated extinctions to the cosmic winter induced by dust thrown into the stratosphere.

The interest of the big picture in the present context lies in the extent to which rare, large events in the geological record might yield insights into more frequent, smaller ones in the historical one. Do we indeed expect continuity of process from geological to historical time-scales? To what extent can we apply the lesson of *tyrannosaurus rex* to *homo sapiens*?

It appears that from time to time throughout geological history, the biosphere has been overwhelmed by catastrophe. Life has survived, but often at a greatly reduced level and sometimes, as with the end-Permian extinctions, scarcely at all, the recovery time being measured in many millions of years. These great mass extinctions were planet-wide and traumatic events. Table 3 shows the extinction percentages for the largest of them, sometimes known as the Big Five: at the genus level, extinctions are of order 50%. Since a genus usually involves several (sometimes many) species, and the extinction of a species will in turn usually involve the elimination of millions or billions of individuals, it is clear that a geologically measurable mass extinction involves a drastic reduction in the biosphere, in terms of diversity and numbers. For example the loss of 83% of marine genera at the end-Permian involved the extinction of 96% of all species [56]. The end-Triassic extinction thus approached something like the elimination of life on Earth.

date (Myr)	extinction	family	genus
438	end-Ordovician	22–27	57
367	late Devonian	19–21	50
245	end-Permian	50-57	83
225	end-Triassic	20–28	48
65	end-Cretaceous	15–17	50

Table 3. Percentage extinctions by family and genus for the Big Five.

Now, with a greatly enhanced database of terrestrial impact craters and improved dating for many geological and biological events, it is becoming possible to take a more searching look at the 'bombarded planet' thesis. Although over 200 impact craters over 5 km across have been detected on Earth, and the number is increasing every year, only about thirty have ages known to better than 10 million years. Of these thirty, about half are significantly associated with mass extinctions of life recorded in the fossil record (Table 4).

extinction Myr	crater age Myr	diameter km
11.2	–	
36.6	35.0±5	100
	35.5±0.6	85
	37.0±0.2	7.5
	38.0±0.4	28
65.0	65.0±0.1	170
91.0	88.0±3	24
	89.0±2.7	19
	95±7	25
113.0	115±10	39
144.0	142.5±0.8	22
	144.7±0.8	70
176.0	175±3	80
193.0	186±8	23
216.0	214±1	100
245.0	247.5±5.5	40
367.0	368.0±1.1	52

Table 4. Impact crater/mass extinction correlation.

A number of clear, statistically verifiable associations emerge from these events, taken over the 600 Myr of the Phanerozoic. Thus the mass extinctions are found to correlate not only with impact craters but also with a wide variety of global geological upsets [21, 50, 51, 55]. For example there is a close coincidence in date between the K/T extinctions, the Chicxulub crater and also that of the creation of the Deccan Traps. These correlations are at confidence levels in the range 96-99% which, given the intrinsic noisiness of the data, suggests a remarkably strong external forcing. The remarkable result follows that, in some way, celestial disturbances induce not only mass extinctions of life but also a wide variety of geological disturbances including world-wide vulcanisms, flood basalt outpourings and mountain-building events. These statistical correlations raise a paradox: the dynamic effects of an impact are over in less than an hour, whereas mountain building, although often geologically rapid, is still a process taking a million years or more.

Further, the relationship between impact and extinctions is not a simple one-to-one. For example the Siljan (52 km) and Araguainha (40 km) impact structures seem much too small to have been the cause of the massive Late Devonian and end Permian extinctions respectively, and there are craters of comparable size which are not associated with mass extinction (such as the 73 Myr-old Kara structure 65 km across, and the 128 Myr old Tookoonooka structure 55 km in diameter). Multiple cratering is evident at several of the Phanerozoic mass extinction boundaries (Table 4), and a few crater chains have been claimed to exist, for example at the late Triassic boundary [59].

7. Mass extinction boundaries

A further clue to the nature of the celestial forcing comes from detailed study, by palaeontologists, of the patterns of extinction. Most regard the *impact=extinction* equation as incompatible with these detailed patterns [44, 48, 49]. The situation is not so much a clash between two world systems (catastrophists and uniformitarians) as one in which two groups of people are simply talking past each other.

A key feature in the 'traditionalist' armoury is the apparently prolonged nature of some of the extinctions and markers of geophysical stress. Thus McLean [49] points out that many lineages of dinosaurs were dwindling throughout a late Cretaceous cooling, that a K/T greenhouse warming began about 10,000 years before the boundary and extended at least 200,000 years into the the early Tertiary. In a 22-authored paper amounting to a major review of the K/T boundary [44], it is argued that many groups (ammonite cephalopods, bryozoa, bivalves, etc,) were in decline before the K/T boundary was reached, while others (diatoms, diolaria, fish, amphibians, terrestrial plants, *etc.*) passed through with little or no change. What appears to have been an extinction in waves is seen in the fossil record of a type of plankton known as foraminifera. Two extinction waves have been identified and seem to have had a geographical structure. The first wave began just before the K/T boundary was reached and affected species in equatorial and middle latitudes. Those species which survived this event seem to have carried through the boundary, only to be removed *en masse* 40,000-100,000 years into the Palaeocene. There does not seem to have been a significant pulse of foraminifer extinction at the K/T boundary itself.

On the other hand, there is evidence that huge populations of plankton underwent a sudden and catastrophic collapse at the K/T boundary. Of 90 or so recognised species of calcareous plankton in the late Maastrichtian, about 80 vanished suddenly at the K/T boundary. There does not seem to have been a forewarning of the extinction, say through a progressive decline in the number of species. This spectacular loss of calcium-shelled creatures produced a word-wide volcanic boundary clay. Overall, the fine structure of the K/T extinctions has led many palaeontologists to the view that they were caused by a long period of environmental instability, perhaps driven by fluctuations in climate acting in combination with volcanic and impact-related event(s). Such considerations apply also at other extinction boundaries, and have led most palaeontologists to continue to seek internal processes rather than large impacts to account for the great mass extinctions, with climatic changes and sea-level recessions high on the list of suspects.

Here, then, is a paradox. On the one hand the dynamic effects of a large impact are over in an hour or less, although both geological processes and palaeontological data imply that complex and prolonged stresses, perhaps 100,000 years in duration, act on the Earth during these times of global crisis. On the other hand, these stresses are statistically correlated with large impact craters. The paradox is that, if stray impacts will not do, neither will purely internal processes.

Of course, the paradox is only such if one adheres to either the simple 'stray impact' picture or the 'nothing higher than the rooftops' one. The evidence, however, fits better with the more complex reality of a bombardment episode, within which the prolonged effects of cometary dusting act along with the prompt ones of multiple impacts.

There is in fact independent geochemical and palaeontological evidence to support the hypothesis of contemporaneous dusting and multiple bombardments: (i) Around the Eocene-Oligocene boundary, a series of step-wise extinctions took place, associated with sharp cooling events and a much greater range of mean temperatures over the course of a year. Several impact craters were formed around this time (Table 4), but once again detailed stratigraphy does not support a simple one-to-one relation. McGhee [47], independently of the astronomical picture described here, postulated that a series of impacts took place around that time. (ii) Pacific ocean core studies [24] reveal order-of-magnitude excesses of ^{3}He at about 37 Myr and 63 Myr BP, corresponding to the Eocene-Oligocene and K/T boundaries. The significance of ^{3}He is that it is a volatile isotope brought in by dust (rather than say, a large bolide). Its overabundance at these boundaries is therefore *prima facie* evidence that an episode of dust influx, rather than purely impact, was involved in the mass extinctions at these boundaries. (iii) At a well-studied Danish K/T site, amino acids apparently of non-terrestrial origin, are found in a layer two metres deep straddling the extinction level of the K/T extinctions [71]. Such compounds would be destroyed in the fireball of a large asteroid and, as with the ^{3}He, again indicate that a major deposition of dust was involved.

To summarise, the evidence on the ground seems to indicate that:

1. The correlations between impact craters, mass extinctions and a variety of global disturbances imply a remarkable degree of external forcing;

2. isolated stray impacts are inadequate to account for the complex and sometimes drawn-out nature of the mass extinctions;

3. likewise the sea-level changes, mountain-building events, anoxic episodes, basalt floodings and so on require the prolonged application of stress.

The paradox may be resolved when one notes the nature of the dominant perturbers. Thus the Earth is subject to episodes of bombardment which may persist for times of order a million years. During such episodes, the mass influx is dominated by very large comets. But a giant comet is a prolific dust machine. Over the millennia of its disintegration, the inner planetary system is copiously flooded by meteoric dust as well as by larger bodies, up to asteroidal dimensions. It is readily shown [7, 19, 32] that, during the peak of its disintegration, the Earth is wrapped in a reflecting dust blanket of significant optical depth. This will result in drastic climatic upset (*loc. cit.* and [22]). Thus massive comets are expected to yield global climate, geophysical and biological change through prolonged dusting, as well as directly through impacts. The latter may therefore not be the sole or even primary cause of the mass extinctions.

There is, however, one more statistical test to be applied.For if the messengers carrying information from Galaxy to Earth are comets, there is the possibility that Galactic cyclicity will be detectable in the terrestrial record, since the long-period comet system is sensitive to the regular ebb and flow of Galactic tides [12]. On the other hand, if say terrestrial control is predominantly due to the impacts of stray main belt asteroids, no periodicity is expected.

8. Galactic periodicity

There are several reservoirs from which large bodies ultimately arrive at the Earth's environment. Mutual collisions in the main asteroid belt knock asteroids into unstable regions of orbital parameter space, where planetary perturbations act to feed asteroids into Earth-crossing orbits. The influx of asteroids of diameter $\geq d$ km to the Earth-crossing regions from this source is

$$F_a(\geq d) \sim 400/d^3$$

asteroids/Myr. That of bright long-period comets into the inner solar system has been given by Bailey *et. al.* [7] as

$$F_c(\geq d) \sim (5/d)^2$$

comets/AU/yr, with injection into short-period orbits occurring every ~100,000 years [31]. The inverse cube d relation for the asteroids arises because the more massive the asteroid the more difficult it is to shift by collision, and ensures that large Earth-crossing asteroids dwindle in numbers relative to comets as d increases. It emerges that Earth-crossing bodies more than a few kilometres across are more likely to have a cometary rather than a main belt provenance, with a significant proportion ultimately deriving from the Oort cloud. This opens the door to the prospect that Galactic disturbances of the Oort comet cloud might be discernible in the terrestrial record. In particular the Sun bobs up and down as it orbits the Galaxy, subjecting the Oort cloud to a Galactic tide whose strength ebbs and flows with a periodicity somewhere in the range ~26-45 Myr.

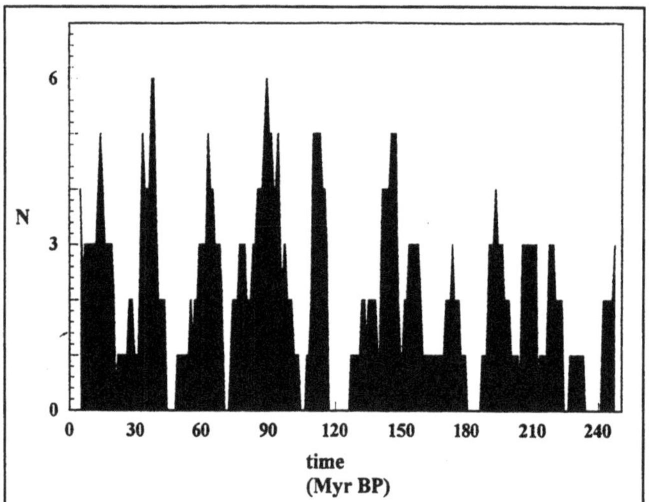

Figure 1. Global geological disturbances according to Rampino & Caldeira [54]. A rectangular smoothing 7 Myr wide has been applied to their data.

In fact, claims of a ~26-33 Myr cycle in the extinction, geological and cratering records go back over 70 years and have generated much controversy over the past decade. Recent improvements in dating of craters (especially the use of argon-argon rather than stratigraphic dating), coupled with better geological datasets [54] have justified a reanalysis of these old claims [21]. In Figure 1 is plotted the major global geological disturbances of the past 250 Myr, after Rampino & Caldeira [53]. A running mean of width 7 Myr has been used (mass extinctions are excluded from this list). Formal analysis confirms the impression that a periodicity (P = 27 Myr, phase ϕ = 7.7 Myr) is indeed present. Caution is called for, as the ages of some geological events have been arbitrarily assigned to geological stages which have a characteristic separation of 5 or 6 Myr. This might artificially quantize the data and so convert an episodic structure into a spuriously periodic one. However the data do indicate that the *whole Earth* switches on and off recurrently, on a timescale characteristic of the variation of Galactic tide acting on the Oort cloud. A bombardment episode will indeed disturb many elements of the Earth system yielding a complex interaction between mass extinction, climatic disturbance, basalt flooding, mountain building, sea-floor spreading and so on. Thus at least tentatively, the data seem to support the basic Galactic hypothesis in which the Oort comet cloud is periodically disturbed as the Sun oscillates vertically through the Galactic plane [6, 21, 45].

These celestial processes undoubtedly operate on geological time-scales. These will include stray impacts [2], impact showers due to fragmenting collisions in the main asteroid belt [72], corresponding dust generation events and even rare molecular cloud penetrations [1]. In essence, such mechanisms add sporadic noise to the Galactic signal.

To sum up, both the geological and astronomical evidence suggest that we may be moving towards a *unified* theory of terrestrial catastrophism, in which a single underlying process yields a spectrum of catastrophes, from the rare, global ones which have on occasion overwhelmed the planet, to more frequent but less drastic events which might have only regional or limited effects.

9. Concluding remarks

Estimates of the impact rate on Earth not unlike modern values are to be found scattered through the scientific literature of the last 50 years (e.g. [9]). However these had little effect on the prevailing consensus, which was that impacts are of little consequence on either human or geological time-scales. It was not until the late seventies and early eighties, with the widespread acceptance that lunar craters have an impact origin, that this common perception began to change. Kresak [37] and Napier & Clube [52] appear to have been the first astronomers in modern times to estimate the high historical impact rates which are now generally accepted. Kresak deduced, on the basis of fireball studies, that one Tunguska-like impact should take place on Earth every 50 years; while Napier & Clube pointed out that the impact rates of large terrestrial craters, extrapolated to low energies, yield a flux of 'one $10^{3\pm1}$ MT event every ~10^4 yr', with 'a 20% probability that the event lies in the upper half of the stated range', and a flux which 'also corresponds to one $10^{2\pm1}$ MT event per ~1,000 yr ...' (see also [18]). These crude early estimates were based on little more than an extrapolation of terrestrial impact crater flux [29], coupled with simple and uncertain scaling laws between crater diameters and impact energies; but they have held up surprisingly well. Probably no authority would now defend the view, current until about 1980 (e.g. [10]), that only one Tunguska-like impact is expected on the planet every 2,000 years.

The current collision rates with bodies about a kilometre or more in diameter appear to be known to within a factor of two or so and are high in geological terms. Although their destructive effects as a function of impact energy are not very well known it is expected that they would be calamitous on a global scale [64]. No such impact is likely to have occurred on historical time-scales. On a similar time-scale, the near-Earth injection and disintegration of a giant comet is expected and the climatic effects are likely to be no less calamitous, with the onset of an ice age as a distinct possibility [19, 22, 32].

At the sub-kilometre level, the inferred bombardment rates depend on uncertain extrapolations and albedos, and are not reliably known to a factor much better than three or four. At this sub-kilometre level, impacts may well be civilization-destroying on a regional scale, merging into global but relatively infrequent hazards at the top end. The effects of brief comet dusting, accompanied by fireball storms of great intensity, are probably significant on millennial time-scales, at least in the wake of a giant comet episode. Although much attention is currently paid to the hazard represented by stray impacts, our present state of ignorance does not allow us to state that they outclass climatic downturns, either in frequency or in their effects; indeed, fairly conservative modelling yields temperature downturns sufficient to collapse modern agriculture on time-scales an order of magnitude shorter than those associated with a *global* civilization-destroying impact.

There are likely too to have been epochs when the sky contained one or more visible, periodic comets, associated with annual fireball storms of huge intensity, and perhaps also with devastating impact. Such phenomena, enduring for centuries, surely had a profound effect on the minds of early peoples. At a minimum, traces of this ancient sky should still be detectable in the artefacts and belief systems of the earliest cultures [18, 20, 33].

Telescopic searches for near-Earth objects are a vital component of this continuing research; unfortunately, since much of the hazard may lie in sub-telescopic swarms or large but distant bodies, these searches may yield an incomplete or biased risk inventory. The Earth, however, may be regarded as a test particle whose past (as preserved in ice cores, peat bogs and even cultural artefacts), preserves evidence relevant to its future. And of all the lines of evidence which can be brought to bear on the celestial hazard issue, that of ground experience is surely the most direct. Thus assessments of our future security are inextricably linked to our understanding of the past.

Acknowledgements

The author is indebted to Benny Peiser for the invitation to attend the SIS Conference, and to David Asher, Mark Bailey, Victor Clube and Duncan Steel for many discussions on the above topics. He is particularly indebted to the Leverhulme Trust for support.

References

1. Allen, A. & Yabushita, S., 1997. Did an impact alone kill the dinosaurs? *Astron. Geophys.* **38**, 15-19.
2. Alvarez, L.W., Alvarez, W., Asaro, F. & Michel, H.V., 1980. Extraterrestrial cause for the Cretaceous-Tertiary extinction. *Science* **208**, 1095-1108.
3. Asher, D.J. & Clube, S.V.M., 1993. An extraterrestrial influence during the current glacial-interglacial. *Q. J. R. Astr. Soc.* **34**, 481-511.
4. Babadzhanov, P.B., 1998. Meteor showers associated with the Taurid Complex asteroids. In press.
5. Bailey, M.E., 1995. Recent results in cometary astronomy: implications for the ancient sky. *Vistas in Astron.* **39**, 647-671.
6. Bailey, M.E., Clube, S.V.M. & Napier, W.M., 1990. The Origin of Comets, Pergamon, Oxford.
7. Bailey, M.E., Clube, S.V.M., Hahn, G., Napier, W.M. Valsecchi, G., 1994. Hazards due to giant comets: climate and short-term catastrophism. *Hazards due to Comets and Asteroids* (ed. Gehrels, T. pp. 479-533, University of Arizona Press, Tucson, Arizona.
8. Baillie, M.G.L., 1994. Dendrochronology raises questions about the nature of the AD 536 dust-veil event. *The Holocene* **4**, 212-217.
9. Baldwin, R.B., 1949. The Face of the Moon. Univ. Chicago-Press.
10. Brandt, J.C. & Chapman, R.D., 1980. *An Introduction to Comets*. Cambridge University Press, Cambridge.
11. Bahcall, J.N. & Bahcall, S., 1985. Nature **316**, 706-708.
12. Byl, J., 1983. Galactic perturbations of nearly-parabolic cometary orbits. *Moon and Planets* **36**, 263-273.
13. Ceplecha, Z., 1994. Impacts of meteoroids larger than 1m into the Earths atmosphere. *Astron. Astrophys.* 286, 967-970.
14. Chapman, C.R. & Morrison, D., 1994. Impacts on the Earth by asteroids and comets: assessing the hazard. *Nature* **367**, 33-40.
15. Chyba, C.F., 1993. Explosions of small Spacewatch objetcs in the Earths atmosphere. *Nature* **363**, 701-703.
16. Clube, S.V.M., 1995. The nature of punctuational crises and the Spenglerian model of civilization. *Vistas in Astron.* **39**,673-698.
17. Clube, S.V.M. & Asher, D.J., 1993. An extraterrestrial influence during the current glacial-interglacial. *Q. J. R. Astr. Soc.* **34**, 481-511.
18. Clube, S.V.M & Napier, W.M., 1982. *The Cosmic Serpent*, Faber, London.
19. Clube, S.V.M & Napier, W.M., 1984. The microstructure of terrestrial catastrophism. *Mon. Not. R. Astr. Soc.* **211**,953-968.
20. Clube, S.V.M & Napier, W.M., 1990. *The Cosmic Winter*, Blackwells, Oxford.
21. Clube, S.V.M & Napier, W.M., 1996. Galactic dark matter and terrestrial periodicities. Q.J.R. astr. Soc. **37**, 617-642.
22. Clube, S.V.M., Hoyle, F., Napier, W.M. & Wickramasinghe, N.C., 1996. Giant comets, evolution and civilization. *Astrophys. Space Sci.* **245**, 43-60.
23. Coradini, A., Capaccioni, F., Capria, M.T. *et al*,, 1997. Transition elements between comets and asteroids. Part 1: Thermal evolution models. *Icarus* **129**, 317-336.
24. Farley, K.A., 1995. Cenozoic variations in the flux of interplanetary dust recorded by ^{3}He in a deep-sea sediment. *Nature* **376**, 153-155.
25. Fechtig, H., 1982. Cometary dust in the solar system. In *Comets*, (ed. Wilkening, L.), IAU Coll. No. 61, pp. 370-382, University of Arizona Press, Tucson.
26. Fulle, M., 1990. Meteoroids from short period comets. *Astron. Astrophys.* **230**, 220-226.
27. Genuth, S.S., 1998. *Comets, Popular Culture, and the Birth of Modern Cosmology*, Princeton University Press, Princeton.
28. Glasstone, S. & Dolan, P.J., 1977. *The Effects of Nuclear Weapons*, U.S. Dept. of Defense and U.S. Dept. of Energy.
29. Grieve, R.A.F., & Dence, M.R., 1979. The terrestrial cratering record. II. The crater production rate. *Icarus* **38**, 230-242.
30. Grün, E., Zook, H.A., Fechtig, H. & Giese, R.H., 1986. Collisional balance of the meteoritic complex. *Icarus* **62**, 244-272.
31. Hahn, G. & Bailey, M.E., 1990. Rapid dynamical evolution of giant comet Chiron. *Nature* **348**, 132-136.
32. Hoyle, F., 1984. On the causes of ice ages. Earth, Moon, and Planets **31**, 229-248.
33. Hoyle, F., 1993. The Origin of the Universe and the Origin of Religion, Rhode Island, Moyer Bell.
34. Hughes, D.W., 1996. The size, mass and evolution of the Solar System dust cloud. *Q. J. R. Astr. Soc* **37**, 565-592.
35. Jenniskens, P., Betlem, H., de Lignie, M., Langbroek, M. & vanVliet, M., 1997. Meteor stream activity. The Quadrantids.

A very young stream. *Astron. Astrophys.* **327**, 1242-1252.
36. Kasten, F., 1968. Falling speed of aerosol particles. *J. Appl. Met.* 7, 944.
37. Kresak, L., 1978. The mass distribution and sources of interplanetary boulders. *Bull. Astron. Inst. Czech.* **29**, 135.}
38. Kresak, L., 1993a. Meteor storms. In: *Meteoroids and their Parent Bodies* (eds. Stohl, J. & Williams, I.P.), pp. 147-156, Bratislava.
39. Kresak, L., 1993b. Cometary dust trails and meteor storms. *Astron. Astrophys.* **279**, 646-660.
40. Kresak, L., 1980. Sources of interplanetary dust. In: *Solid particles in the Solar System* (eds. Halliday, I. and McIntosh, B.A.), IAU Symp. No. 90, pp. 211-222, Reidel, Dordrecht.
41. Kresak, L., 1992. Meteor storms. In: *Meteoroids and their Parent Bodies*, Astr. Inst., Slovak Academy of Sciences, Bratislava.
42. Leitch, E.M. & Vasisht, G., 1998. Mass extinctions and the sun's encounter with spiral arms. *New Astron.* **3**, 51-56.
43. Love, S.G. & Brownlee, D.E., 1993. A direct measurement of the mass accretion rate of cosmic dust. *Science* **262**, 550-553.
44. Macleod, N., Rawson, P.F., Forey, P.L. *et al.*, 1997. The Cretaceous-Tertiary biotic transition. *J. Geol. Soc.* 154, 265-29 2.
45. Matese, J.J., Whitman, P.G., Innanen, K.A. & Valtonen, M.J., 1995. Periodic modulation of the oort cloud comet flux by the adiabaticaly changing tide. *Icarus* **116**, 255-268.
46. McBride, N., Taylor, A.D., Green, S.F. & McDonnell, J.A.M., 1995. Asymmetries in the natural meteoroid population as sampled by LDEF. *Planet. Space Sci.* **43**, 757-764.
47. McGhee, G.R., 1994. *The Late Devonian Mass Extinction*. Columbia University Press, New York.
48. McLean, D.M., 1991. Impact winter in the global K/T extinctions: no definitive evidences. In: *Global Biomass Burning: Atmospheric, Climatic, and Biospheric Implications* (ed. Levine, J.S.), pp. 493-508, The MIT Press.
49. McLean, D.M., 1995. K-T transition greenhouse and embryogenesis dysfunction in the dinosaurian extinctions. *J. Geol. Education* 43, 517-527.
50. Napier, W.M., 1998a. Galactic periodicity and the geological record. In: *Meteorites: Flux with Time and Impact Effects* (eds. Grady, M.M., Hutchison, R., McCall, G.J.H. & Rotherby, D.A.), Geol. Soc. London, Special Publications **140**, 19-29.
51. Napier, W.M., 1998b. NEOs and impacts: the Galactic Connection. In: *The Role of Asteroids and Comets in Earth History* (eds. Henrard, J. & Yabushita, S.), Kluwer, Reidel. In press.
52. Napier, W.M. & Clube, S.V.M., 1979. A theory of terrestrial catastrophism. *Nature* **282**, 455-459.
53. Napier, W.M. & Clube, S.V.M., 1997. Our cometary environment. *Rep. Prog. Phys.* **60**, 293-343.
54. Rampino, M.R. & Caldeira, K., 1993. Episodes of terrestrial geologic activity during the past 260 million years: a quantitative approach. *Cel. Mech. and Dynam. Astr.* **54**, 143.
55. Rampino, M.R. & Haggerty, B.M., 1994. The Shiva hypothesis: impacts, mass extinctions, and the Galaxy. *Earth, Moon, and Planets* 72, 441-460.
56. Raup, D., 1979. Size of the Permo-Triassic bottleneck and its evolutionary implications. *Science* **206**, 217-218.
57. Revelle, D.O., 1995. Historical detection of asteroid impacts by large bolides using acoustic-gravity waves. *International Conference on Near-Earth Objects*, New York, April 24-26.
58. Shoemaker, E.M., 1983. Asteroid and comet bombardment of the Earth. *Ann. Rev. Earth Planet. Sci* **11**, 461-494.
59. Spray, T.G., Kelley, S.P. & Rowley, D.B., 1998. Late Triassic Multiple Impact Event on Earth. *Nature* **392**, 171-173.
60. Steel, D.I. & Asher, D.A., 1996a. The orbital dispersions of the macroscopic Taurid objects. *Mon. Not. R. Astr. Soc.* **280**, 806-822.
61. Steel, D.I. & Asher, D.A., 1996b. On the origin of Comet Encke. *Mon. Not. R. Astr. Soc.* **281**, 937-944.
62. Steel, D.I., Asher, D.J., Napier, W.M. & Clube, S.V.M., 1994. Are impacts correlated in time? In: Hazards due to Comets and Asteroids} (ed. Gehrels, T.) pp. 479--533, University of Arizona Press, Tucson, Arizona.
63. Tian-Shan, Z., 1977. Ancient Chinese records of meteor showers. *Chinese Astronomy*, **1**, 197.
64. Toon, O.B., Turco, R.P., Covey, C. *et al*,, 1997. Environmental perturbations caused by the impacts of asteroids and comets. *Rev. Geophys.* **35**, 1-78.
65. Valsecchi, G., Morbidelli, A., Gonci, R. *et al*,, 1995. The dynamics of objects in orbits resembling that of P/Encke. *Icarus* **118**, 169-180.
66. Whipple, F.L., 1967. On maintaining the meteoritic complex. In: *The Zodiacal Light and the Interplanetary Medium*, NASA SP-150, (ed. Weinberg, J.L.) 409-426, NASA SP-150, Washington DC.
67. Whipple. F.L., 1975. Do comets play a rôle in galactic chemistry and -ray bursts? *Astron. J.* **80**, 525-531.
68. Wickramasinghe, N.C., Hoyle, F. & Rabilizirov, R., 1989. Extraterrestrial particles and the greenhouse effect. *Earth, Moon, and Planets* **46**, 297-300.
69. Yabushita, S. & Hatta, N., 1994. On the possible hazard on the major cities caused by asteroid impact in the Pacific Ocean. *Earth, Moon, and Planets* **65**, 7.
70. Zahnle, K., 1990. Atmospheric chemistry by large impacts. In: *Global Catastrophes in Earth History* (eds. Sharpton, L.V. & Ward, P.D.), Geol. Soc. America Special Paper 247, pp. 271-288, Geol. Soc. America, Boulder, Colorado.
71. Zahnle, K. & Grinspoon, D., 1990. Comet dust as a source of amino acids at the Cretaceous/Tertiary boundary. *Nature* **348**, 157.
72. Zappalà, V., Cellino, A., Gladman, B.J., Manley, S. & Migliorini, F., 1998. Asteroid showers on Earth after family breakup events. *Icarus* **134**, 176-179.

Before the Stones: Stonehenge I as a Cometary Catastrophe Predictor

Duncan Steel
Spaceguard Australia P/L, P.O. Box 3303, Rundle Mall, Adelaide, SA 5000, Australia

Summary

The original purpose of Stonehenge remains a mystery, but to identify that purpose we need to concentrate upon the original structures: Stonehenge I and the Great Cursus. It seems clear that the constructors of Stonehenge were concerned in some way with celestial events, but the astronomical interpretations in terms of a luni-solar observatory or temple pertain to the later stone structures of Stonehenge II and III, which were erected many centuries after the first simple constructions. Other than a common site, there is nothing definitive to link the early and later phases. It is suggested here that the people who built the first structures were motivated by quite different celestial concerns than the Sun, the Moon and eclipses; it is argued that in the second half of the fourth millennium BC the Earth was subject to recurrent intersections with a swarm of meteoroids released through cometary decay, and that first the Cursus and then Stonehenge I were built as part of a system to predict imminent (within a day) passages through that swarm, with attendant celestial fireworks and possibly real physical danger. This hypothesis, stemming from the recognition that the present population of Earth-crossing small bodies does not appear to be in equilibrium, leads to possible explanations for (i) The chronology of Stonehenge I and the Cursus; (ii) The subsequent hiatus in development at the site; (iii) The orientations of those structures; (iv) The apparent cyclicity of 19 years later taken up by the constructors of Stonehenge II/III; (v) The association with the Stonehenge site of numerous long barrows; (vi) The reason for the importation of the bluestones; and (vii) The design of the later megalithic constructions in terms of massive lintels bridging vertical columns. Most important, this hypothesis provides an explanation for the motivation behind these enormous undertakings: something must have really worried the early Stonehenge people for them to have invested so much in its construction, and had such an effect that many generations later the myths led to the development of Stonehenge II/III as a place of ritual, not science.

1. Introduction

The purpose of Stonehenge remains an enigma. In this paper I will present some facets of a radical alternative hypothesis for why Stonehenge was built. Some preliminary thoughts on this matter were given in [47], and the present paper represents an advance on the basic theme begun there; nevertheless there are many points which will require further research, discussion and revision. This paper is an interim presentation of work in progress. I am under no illusions that it will be received favourably by many scientists concerned with Stonehenge and megalithic monuments, either from the archaeological or the astronomical disciplines, but as Sir Fred Hoyle wrote in his autobiography, "To achieve anything really worthwhile in research it is necessary to go against the opinions of one's fellows."

In his excellent essay entitled 'Stonehenge from without', Wrigley [63] wrote the following:

> "Now that we know with some precision what was constructed and when, the next question is obviously WHY?, and this presents a challenge to the intellect and imagination of us all.
>
> All historical interpretations, all reconstructions of the past that go beyond WHAT? and WHEN?, inevitably depend on the use of analogy. They cannot therefore aspire to proof in the scientific sense but only to a greater or lesser degree of conviction. They are also necessarily idiosyncratic, since they result from an interaction between the data (with which of course they must be consistent) and the mind of the interpreter, which has its own peculiar resources of reading and experience. And it is through the aggregation of idiosyncrasies that understanding can advance."

My hypothesis stems from my own research background, and so of course it is idiosyncratic. The solution to the Stonehenge enigma cannot be simple or obvious, else it would have long since been arrived at, and added to the canon of mysteries resolved; it is not surprising, then, that my hypothesis will seem complicated to someone not familiar with its astronomical basis. There is also a wider question than the motivation behind Stonehenge: the consequences of this hypothesis being correct, with regard to our understanding of human development, are very considerable.

Considerations of space and time dictate that I should not here tarry with discussions of the well-documented features of Stonehenge (Wrigley's WHAT? and WHEN?). Three books I might cite with rather different qualities and scopes are those by Castledon [15], Chippindale [17] and North [41]. Similarly the astronomical interpretations of Stonehenge in terms of a luni-solar temple or eclipse predictor do not need any major discussion here; these are usually associated in the main with three names, and here I will cite just one publication by each: Newham [40], Hawkins [26] and Hoyle [27].

2. Astronomical context

In this section I will discuss the astronomical context for the interpretation of Stonehenge I and the Cursus which I present later. But I will begin by stating where I believe that the luni-solar observatory/temple interpretation fits in.

2.1 The luni-solar observatory/temple theory

In essence I believe that the midsummer sunrise apparent alignment recognized some centuries ago, and then taken up by (for example) Lockyer about a century ago, with later embellishments, and then expanded upon so as to include the motions of the Moon and the possibility that Stonehenge was an eclipse predictor by Newham, Hawkins and Hoyle in the 1960's, is basically a repeat of a mistaken interpretation first made by the Beaker people in about 2200 BC. What happened, I suggest, is that the Beaker people knew from their oral traditions, and the remnant, overgrown constructions, that Stonehenge was an important location, to do with something celestial which occurred near mid-summer and which appeared in the north-east close to the sunrise point, had awful consequences, and recurred every 19 years. Having stood and watched from Stonehenge for long enough, they may have discovered successively the azimuth of midsummer sunrise near 50°, the fact that the number of days between such events gives the length of the year, the remarkable circumstance of the furthest northern rising of the Sun being at 90° to the furthest south rising of the Moon as viewed from the latitude of Stonehenge, the 19-year Metonic cycle, the 18.61 year cycle of the regression of the lunar nodes, and how to predict eclipses through the 18.03 year Saros (*e.g.*, see [27] for a description of these matters). These discoveries resulted in the motivation - ritual and worship based upon the Sun and Moon - for building Stonehenge II and III between *ca.* 2200 and 1100 BC; I suggest that they are the result of set of flukes, the full extent of which has hitherto been unrecognized, although certain facets are certainly clear (see [27] again). There is no basis to any argument that the set of flukes (including the various features described later) are unlikely, because if they had *not* occurred then Stonehenge II and III would not have been built, and we would not be puzzled by their purpose.

I reject the notion that Stonehenge I was built as a luni-solar observatory because I can see no reason to suppose that it was used in this way. Whilst it is true that the discovery [43] of unoccupied pit 97 close to the Heel Stone indicates the earlier existence of a twin to that stone (or perhaps a previous positioning of it), the two producing a portal through which the midsummer sunrise might be viewed around 3100 BC (or a stone in that pit alone being the marker), the eclipse-prediction theory largely hangs on the Station Stone locations, and these were not emplaced until at least a few centuries later (as shown by the fact that they cut across the Aubrey Holes, which date from 2900-2800 BC). It is true that the 'A post' could have been used in the Stonehenge I phase to observe or chart moonrises at azimuths north of the midsummer sunrise, but in my hypothesis these had an alternative usage.

In summary, then, my suggestion is that the early developments at Stonehenge (Stonehenge I and the Cursus) were provoked by celestial events of a temporary nature unrelated to the Sun and the Moon, and that the cessation of these events led to the abandonment of the site for some time after *ca.* 2700 BC. The handing on of stories concerning the usage and importance of the Stonehenge site as cultural changes occurred (the Windmill Hill culture being superseded by the Beaker, perhaps with an influx of people from elsewhere) led to systematic observations of the Sun and the Moon being made later in the third millennium BC, and the eventual discovery of the solar and lunar phenomena mentioned above. These making a strong impression on the Beaker people, Stonehenge II and eventually III were erected in a form of temple, a place of ritual. Thus I believe that the interpretation of Newham, Hawkins and Hoyle is broadly correct, but that it applies only to the later developments, with Stonehenge I and the Cursus having a quite different motivation and usage.

In this context I suggest that Hoyle [27] was almost prescient in writing the following:

> "Stonehenge I is essentially very simple, a set of marked positions and a few naturally occurring boulders - there may even have been wooden posts instead of boulders in the beginning. This simple structure was sufficient for the astronomical needs. Once simplicity became replaced by complexity, as in Stonehenge III, one can be virtually certain that science had been displaced by ritual. In seeking the forerunners of Stonehenge I, we must look for something simple, we must look around the year 3000 BC, not around 2000 BC, and we must look in the right place, which is where?"

My answer to that question is that one must look for the most spectacular and worrying phenomena (social effects) which can appear in the sky, and which *may* have a direct physical effect upon humans.

2.2 Meteor storms

Barring being so unfortunate as to witness the entry of a large asteroid or comet into the atmosphere, the most spectacular celestial events one might wish to see may be argued to be meteor storms. These are experienced when the Earth happens to pass through the high spatial density of meteoroids in a stream spawned by a comet, and in particular passage close to the cometary nucleus. Occasionally an outburst occurs with no known parent object, but in the past couple of centuries most of the storms have been linked to three specific comets: the Andromedid storms in the latter half of the nineteenth century associated with the break-up of 3D/Biela earlier in that century [8], the October Draconid storms associated with 21P/Giacobini-Zinner with a 13 year periodicity (twice the comet's orbital period: [32]), and the great Leonid storms produced about every 33 years by 55P/Tempel-Tuttle, the next epoch being imminent [32, 35, 64, 65, 66].

In order for a meteor shower or storm to be produced, the meteoroid stream must intersect the Earth's orbit, and so at some stage in its orbital evolution the comet's orbit must have a node at 1 AU. This is/was the case for the above comets, but it must be recognized that this circumstance is only temporary: orbital evolution will take the comet's path away from that of the Earth such that our planet can no longer pass close by the comet itself (the storms are produced mainly by meteoroids released during the past few perihelion passages), so that storms will cease although weaker shower activity may persist due to meteoroids which have undergone slightly different perturbations than those afflicting the comet. Thus, for example, the Leonid meteor storms did not occur in the epoch prior to about a millennium ago (but may have done so rather further back in time), and similarly the non-observation of the Geminid meteor shower prior to the early nineteenth century is explicable in terms of orbital evolution (see [46] for a background discussion).

In the present context the features of meteor storms which are significant are: (i) Their cyclicity, with a shower being seen at the same time every year but with major enhancements (storms) occurring every N years where N is an integer multiple of the parent comet's orbital period; (ii) Their temporary nature, occurring only for some centuries whilst the comet happens to have a node near 1 AU; and (iii) Their spectacular nature, with apparitions often having a severe psychological effect upon many of the people who witness them. For example, the Leonid meteor storm in 1833 had an immediate terrorizing effect upon many viewers, and longer-term repercussions for various religious groups who interpreted the events in terms of biblical prophecies (for example, see [28, 35, 45]).

The above description is based on meteor storms which have occurred in the present epoch (the past century or so). Clearly the meteor showers/storms active in some other epoch will be different, and the archival record contains reports of many such events in historical times (*e.g.*, see [25]). In many cases the identification of the parent comets responsible for specific recorded storms even within the past millennium is difficult, if not impossible, due to the effects of non-gravitational forces on active comets and the rapid chaotic orbital evolution of most Jupiter-crossing objects. The association of specific meteor shower/storm events (and indeed the relative stability of the meteoroid streams) for 1P/Halley, 55P/Tempel-Tuttle and 109P/Swift-Tuttle for the past millennium or so has been made possible by the fact that all three comets are highly-retrograde and of intermediate period, reducing the perturbations imposed by Jupiter during relatively close approaches as compared to the effects visited upon prograde short-period comets (apparently the main sources of presently-observed meteor showers, but see [49] for a more extensive discussion and references) as they undergo jovian approaches.

After consideration of the above the reader might think it impossible that we might be able to retrospectively calculate the epochs of occurrence of meteor storms some five millennia ago. This is certainly the case for the majority of showers observed now, associated with Jupiter-crossing comets, especially since we can know little of their physical history (*i.e.*, when and to what degree they have been actively producing meteoroids). However, as will be briefly discussed later, it seems likely that the majority of the interplanetary complex of small (mm-cm) meteoroids has been spawned by just one comet - 2P/Encke - which happens (not by chance) to have a cis-jovian orbit largely immune to the gross chaos-inducing effects of close approaches to that planet. It will further be shown that - up to a point - the epochs and conditions of meteor storm production by 2P/Encke over the past several millennia can be deduced on the basis of celestial mechanics. These matters combined allow an interpretation of the early constructions at Stonehenge in terms of a warning system that meteor storms were imminent.

2.3 The meteoroid/dust trail of 2P/Encke

Before leaving the topic of meteor storms *per se*, a feature of 2P/Encke deserves special mention. Kresák [32] has suggested that one might identify the structures which result in meteor storms on Earth-transit with the meteoroid *trails* (not *tails*) detected with the Infra-Red Astronomy Satellite (IRAS) in 1983 [59]. The particle sizes responsible for the detected thermal emission are estimated to be typically of mm-cm sizes. The observation program and the timing (*i.e.*, when 55P/Tempel-Tuttle and 21P/Giacobini-Zinner were both near aphelion) meant that the comets producing meteor storms in the present epoch generally were not detectable with IRAS, but of the prominent trails detected one is that associated with 2P/Encke; indeed that trail was the only strong one identified for any known Earth-crossing comet [59]. 2P/Encke began its current active phase in AD 1786 (see [44] for a historical account), and it is generally assumed that the trail now observed (of length about 95° in mean anomaly M and sky-plane thickness inferred to be about 680,000 km) has been produced during the intervening two centuries. 2P/Encke does not cause meteor *storms* now because in the present phase of its orbital evolution its nodes occur far from 1 AU; the annual meteor *showers* which are associated with this comet are the result of meteoroid release over many millennia (see below). My point here is that if, by chance, 2P/Encke happened to be in a different phase of its precession cycle now - and the phase depends only on the dynamics, not the physical activity of the comet - then we would experience frequent meteor storms due to terrestrial passage through the IRAS-detected trail of meteoroids.

In any particular epoch, lasting some centuries, the storms would occur in only one of the four possible Earth-intersection branches (*i.e.*, ascending and descending node, pre- and post-perihelion leg combinations) before moving on eventually to another branch, as described in connection with the backwards integrations of the cometary orbit presented later. With the present period of 2P/Encke (about 3.28 years) and the observed trail length (95° in mean anomaly), if one assumes that a storm occurs for any transit through that trail then such storms would be spaced by 3, 4 or 7 years with the frequencies of occurrence of such gaps in ratios of about 4:1:1 respectively. I believe that not only would such events not escape the notice of unsophisticated (scientifically-speaking) societies today, but in fact such peoples would be: (a) Terrified, as were many by the 1833 and other Leonid displays; (b) Driven to work out when such events were to recur, after a few such experiences, so that ameliorative action might be taken (even if this could only involve hiding under the bed, figuratively-speaking). One could imagine that the ability to make predictions of such terrifying events would put great power in the hands of a few people (*cf.* the discussions of an 'astronomer-priest' class by Hoyle and others). In order to accomplish (b) they would be required to determine the length of the year (because the storm recurs at the same time of year), and also recognize a pattern stretching over many years. Note that one would expect significant showers, but not storms, to occur at the same time *every* year, due to the lesser numbers of meteoroids spread right around the cometary orbit; and that the larger meteoroids producing fireballs, audible events and possibly blast damage would likely be concentrated towards the middle of the trail (see below).

Another feature of the meteoroid trail detected with IRAS and apparently associated with 2P/Encke which should be mentioned here is the suggestion that, rather than being merely a string of debris randomly moving away from the comet which spawned it, as appears to be the case for the meteoroids producing the Leonid storms for example, instead this trail might be evidence for a source of material located in the jovian 7:2 mean motion resonance as described by Asher and Clube [1, 2, 20], and supporting evidence from recent observations of the Taurid meteor showers has been presented [4]. These authors have noted that the strong IRAS trail which has been assumed to originate from 2P/Encke in fact is at some distance from the comet, and might therefore have originated from some other source body which, like 2P/Encke itself, is actually a major fragment of the original parent (hereafter *proto-Encke*); in this interpretation, which the present author favours, 2P/Encke happens to be merely the fragment which is active in the present epoch, and is not of any greater significance than the apparently-asteroidal fragments we associate with the complex [3, 52]. Under Asher and Clube's scenario there are two centres of relative stability in the 7:2 resonance, separated by 180° in mean anomaly M, and only one of those (at present, at least) is occupied by a disintegrating source. Such resonances act as attractors, with objects occupying them having semi-major axes a which oscillate about the mean value (that precisely on the resonance; for example see the numerical integration results of Steel and Asher [52, 53]), and as the values of a change a particular body will vary between being ahead of, or behind, the mean value of M. The amplitude of this libration in M depends upon the amplitude of the variation in a, and a trail of material constrained between extreme values of M results. Since smaller (mm-cm) meteoroids are subject to stronger initial perturbations upon their semi-major axes (due to ejection speeds and radiative forces) than larger bodies, it is to be anticipated that the central region of the complex (values of a very near the resonance, values of M near the mean) will be the region where larger objects are mostly found, the smaller particles being spread on a longer trail (as observed with IRAS). It will be seen later that this phenomenon - a librating grouping of objects around a jovian mean-motion resonance producing a long trail segment of small meteoroids and a tighter concentration of larger meteoroids - may be of significance in producing a cyclicity in terrestrial intersections in accord with my hypothesis for the early usage of Stonehenge. The possible effects upon civilizations in the past 2,000 years of intersections with such a trail, producing meteor storms and/or enhancements in the fireball rates, have been discussed by Clube [19]; whilst I am in agreement with his general ideas, I note that terrestrial intersections in the epochs he describes would require multiple trails and/or faster precession in the 7:2 resonance (as discussed in [1, 2]).

2.4 Calendrical significance of meteor showers

In passing I would note that the significance of meteor showers for calendrical-definition purposes seems to have been largely neglected by those investigating both the purpose of Stonehenge and also the origin of time-keeping systems. In the case of Stonehenge and other devices apparently aligned so as to provide horizon markers, much has been made of the idea that counting the days between (say) the Sun reaching its northernmost or southernmost rising azimuths would determine the length of the year, even though that technique would be fraught with difficulties (for example: cloudiness, the fact that you only get one instantaneous chance per day to note the azimuth of rising, and the slow rate of azimuthal motion near the extremes). Hoyle [27] quite sensibly describes an alternative method which might be used, deploying horizon markers below the extremes and bisecting the dates at which the Sun passes them to determine midsummer/midwinter, but again there are various drawbacks.

On the more general question of the determination of the length of the year, the day not being an aliquot part thereof, different techniques have certainly been used by various societies, such as the ancient Egyptians using the rising of the Nile as one annual marker, whilst also watching for the heliacal rising of Sothis/Sirius [39]: again with only one opportunity per day to make an observation, the potential accuracy is limited. An alternative technique which I feel should be kept in mind is the observation of meteor showers. The peak of any specific shower occurs at a particular time (a particular solar longitude) which may be well - or poorly - defined depending on the shower in question. For example the Geminids build up over about a week to a maximum on one particular night, but without the peak being distinguishable at an hourly level, say. The Quadrantids build to a more distinct peak in a shorter activity period, and the Perseids are broadly similar, a more pronounced second peak having appeared with the recent perihelion passage of the parent comet (109P/Swift-Tuttle). The Leonids in the 1990's provide another distinct marker (modern observers working in networks define the peak to within about a quarter of an hour, meaning that *in principle* the length of the year may be determined in this way to within 0.01 day from observations spread over just one year). There are a few other candidate showers which might be mentioned: the April Lyrids, and the Delta Aquarids, for example. In any other epoch some other set of about half-a-dozen characteristic showers might be anticipated, and counting the days between their occurrences would quite quickly produce an evaluation of the year. In the case of very strong showers such as those I suggest proto-Encke was producing at the time Stonehenge I was begun, counting the days would not only be of *interest* in terms of defining the year, but it would also be *essential* for the psychological well-being of the people concerned: they would need to be able to predict when the terror in the sky was due to recur.

In my opinion the above provides a strong reason for wanting to determine the length of the year, rather than the host of other reasons which have been suggested previously. Notwithstanding the fact that some cultures (*e.g.*, the Hopi of the southwestern part of North America) seem to have used sunrise horizon markers to indicate general times for sowing and reaping *etc.*, the idea that the ancients wanted to know the length of the year *accurately* for agricultural purposes seems risible to me: farmers plant seeds when it's warm enough, and year-to-year variations in when the appropriate weather is attained may vary by many days, even weeks. For example, the pharaoic Egyptians apparently used a 'Nile year' rather than an astronomically-defined period, and yet the time between floods could vary by months, with a standard deviation of about half a month [39]. On the other hand, if one wanted to know when an abrupt meteor storm was due, then in effect the length of the year must be determined to an accuracy of better than a single day.

2.5 Meteoroid production by proto-Encke

I now turn to the gross meteoroid production by proto-Encke over the past ten or so millennia. Whilst one recent author [29] has given a fairly low evaluation for the present meteoroid production rate of the comet, others have in the past suggested that the progenitor of 2P/Encke has in fact been the dominant source of almost the entire contemporary interplanetary complex of meteoroids and zodiacal dust. Before passing on to some mention of other work, I note with interest Hughes' statement regarding his method for deriving the total dust flux from the presently-observed short-period comets [29]:

> "This value tacitly assumes that the comets that we see now are responsible for the present dust cloud, a statement that leans on the uniformitarian view of nature. Those who favour a more catastrophist approach [...] might insist that the natural process of feeding a dust cloud is dominated by the infrequent source that we are unlikely to witness."

In this I agree with Hughes entirely, as may be seen from [50]. Indeed this is the whole crux of what I perceive as being the limitation in previous interpretational work in archaeoastronomy: the assumption that the phenomena seen in the sky by the ancients were the same as those which we see now (after allowance for the precession of the equinoxes, *etc.*). To the contrary, in my opinion their execution of exceptional feats of engineering or other endeavour might rather be viewed as an indication that exceptional phenomena were being experienced.

One might first ask the question "is it *likely* that a recent large enhancement in the interplanetary complex of meteoroids and dust has occurred?" By 'recent' I mean within the last 10,000 or 20,000 years (basically, during the Holocene); there is certainly evidence suggestive of substantial short-term (meaning on a similar time-scale: 'short' in the astronomical and geological sense) variations in the interplanetary dust influx to the Earth over geological time-scales (*e.g.*, [23]). Mention of the present (Holocene) era allows an immediate objection to the basis of the question posed above: it is possible that there is a causal relationship, with the current interglacial being the *result* of a recent enhancement in the interplanetary dust and meteoroid population [1, 2, 50]. If that is the situation then we would not be here to ask the question if the enhancement had not occurred: there would have been no rise in civilization if it were not for the current interglacial conditions.

Leaving that possibility aside, the question becomes one of how often giant comets enter and are trapped in the inner solar system, producing a spike in the supply of inert Earth-crossing bodies of all sizes (dust, meteoroids, asteroids). Some of the astronomical evidence for the recent arrival of a giant comet in a cis-jovian orbit has been discussed by, for example, Asher *et al.* [3], Steel and Asher

[52, 53] and Napier and Clube [38], whilst Bailey *et al.* [10] discuss in general terms the hazards posed by giant comets, and Steel [48] outlined the evidence for the presence of asteroids in meteoroid streams. The bottom line is that one expects giant comets to arrive about every 100,000 years, and to dominate the supply of interplanetary material as averaged over such time-spans, as was suggested earlier by Clube and Napier [21]. This is of the same order as, or longer than, the current collisional lifetimes of small meteoroids in streams similar to the orbits of Jupiter-family comets [24, 42, 56], and in any case if there were a gross enhancement of the dust population then the feedback effect would further limit the lifetimes. The random arrival of a giant comet within the last 10-20,000 years would not, therefore, be an *unlikely* event; it would, however, be an *exceptional/catastrophic* event in that the dust and meteoroids released would dominate the overall supply in that same period (*cf.* comments by Hughes quoted above).

The next question is that of the evidence for such a major enhancement in the meteoroid supply. This is a matter I have dealt with elsewhere [50], so only a brief set of comments will be given here. There are various reasons for believing that the supply and loss of meteoroids and dust are out of balance, a result arrived at by Grün *et al.* [24] and others. Purely on the basis of the observations of comets and hence their dust production rate over the past few centuries, Kresák [31, 33] and Štohl [58] came to the conclusion that proto-Encke was the dominant source of interplanetary dust, as had been suggested much earlier by Whipple [60]. Kresák and Kresáková [33] concluded that the observed mass loss rates of comets are much too low to replenish the interplanetary dust complex, and the discord might be explained in one of two ways: either the major sources of dust and meteoroids now are not active comets but instead are dark (*i.e.*, numerous Earth-crossing asteroids gradually decaying away through regolith loss), or "there was an extremely active comet in a not too distant past, which is responsible for most of the present dust population." In fact I favour a combination of those explanations: that a giant comet arrived in the relatively near past and injected not only a large quantity of meteoroidal material directly into the complex (*e.g.*, [3]), but also produced many Earth-crossing asteroids which continue to act as secondary sources [48]. In his various papers Kresák identified proto-Encke as the major present source, as did Štohl ([58], and elsewhere), who, working from the evidence of the broad 'sporadic stream' of meteoroids which dominates the terrestrial influx of mm-cm sized particles, concluded that not only was proto-Encke the major source, but the present mass-loss rate of the comet is insufficient so that there must have been "a few very productive periods of meteoric material, connected presumably with some kind of catastrophes [*i.e.*, violent outbursts]." This is consistent with the general scenarios envisioned some decades ago by Whipple and Hamid [61] for the origin of the Taurid complex meteors from comet Encke.

Having summarized the above background material, which I have discussed in more detail elsewhere [50] along with other evidence for a recent (past 10-20,000 years) enhancement of the meteoroid and dust population, it is appropriate that I mention another recent publication [51] in which I show that the majority of the terrestrial influx of small meteoroids is in orbits similar to 2P/Encke and the asteroids which may have originated from a common progenitor [3,52]. Steel [51] presents an analysis of the small meteoroids for which orbits were determined in the Harvard Radar Meteor Project. This analysis results in a demonstration that a fraction of order 90 percent of the total meteoroidal influx is contained in the Helion and Anti-Helion sources, having orbits broadly similar to 2P/Encke and the Taurid meteoroid streams. The time-scale (10-20,000 years) over which proto-Encke and its daughter products have been producing meteoroids is defined by a number of considerations, amongst them being the production of multiple groups of four showers from separate precession cycles, with the main set of four showers (from the last precession cycle) necessitating about 5,000 years [6,7,52,53,55]. As noted above, there are reasons to believe that the major releases of dust and meteoroids by proto-Encke have not been continuous, but rather have occurred in catastrophic episodes. The way in which this comet could have entered its cis-jovian orbit on this time-scale, stronger non-gravitational forces than those now acting being required, has been demonstrated, whilst the orbital dispersal of asteroids apparently associated with it may be explained in the same way [52,53]. Overall, then, a self-consistent explanation exists for the phenomena observed in the present epoch; the question is what that explanation implies for the things seen in the sky, and perhaps experienced on the Earth's surface, over the past several millennia.

2.6 Summary

There is ample evidence to suggest that 2P/Encke or at least its progenitor (proto-Encke) has been the major source of meteoroids and dust in the inner solar system in recent times (the past 10,000 years or longer), and that its behaviour during that time has been characterized by catastrophic outbursts rather than a gradual decay, with one outburst *ca.* 5,000 years ago. This would have resulted in episodes, lasting for considerable periods even after the cessation of cometary activity, when the nucleus (or nuclei, if one admits the possibility of gross fragmentation with the large daughter bodies producing dormant or extinct cores now recognizable as asteroids) was accompanied by phenomenal concentrations of meteoroids and other disintegration products of different sizes.

The arrangement of these concentrations may have been simply like those groupings which follow several contemporary comets (such as 55P/Tempel-Tuttle), gradually dispersing around the comet/stream orbit, or (in view of the special orbit of 2P/Encke, and the observation of the IRAS meteoroid trail which may or may not be directly associated with that comet) the concentration may have been quasi-stable, forming a librating complex about the source with meteoroids preceding and trailing that source in its orbit. No matter what the form of the concentration, it is inevitable that the precession of the orbit would bring it around to an Earth-intercept condition, and at such times phenomenal meteor storms would occur, perhaps accompanied by large (10-100 metre) meteoroids capable of causing physical damage to objects on the surface through airbursts. The next question to address is *when* such events may have taken place.

3. What needs to be explained?

In order to focus the mind, I believe that it would be useful here to make a separate section in which the basic features of the early developments at Stonehenge requiring explanation are succinctly summarized. These may be divided into three broad subjects:

(1) *The Chronology.*

Here I will be assuming approximate dates to be:

The (Great) Cursus: *ca.* 3500-3300 BC
Stonehenge I: *ca.* 3200-2800 BC

The fact that the site then seems to have been neglected until the Stonehenge II phase after 2200 BC also provides potentially-useful information. Recent measurements (see

[18]) provides a date for the ditch in Stonehenge I as 3020-2910 BC, whereas I have used ~3100 BC as the epoch for the beginning of that stage; in the context of my hypothesis, the difference is not of significance.

(2) *The Orientations.*

The two orientations I take to be of significance, in terms of azimuths (*z*) measured from true north, are:

The (Great) Cursus: ca. 85°
Stonehenge I: ca. 50°

Although the luni-solar astronomical hypotheses for the purpose(s) of Stonehenge have in some cases been dependent upon precise evaluations of azimuthal angles, for the purposes of my hypothesis for the motivation of the early Stonehenge developments, angles precise only to the nearest degree or so are adequate, and throughout I will be giving these only to this order.

(3) The Cyclicity.

The luni-solar temple interpretation for Stonehenge II/III indicates a possible cycle of 19 years being involved. In my hypothesis I consider whether the earlier developments could have involved some cycle with a totally different astronomical stimulus, but nevertheless either the same (or perhaps a similar) periodicity which influenced the later people who studied the sky from Stonehenge, eventually bringing in the sarcens and bluestones and constructing the final megalithic monument.

(4) Other features of Stonehenge and the environs.

Apart from the above three specific features, I will also consider whether there are other features of both the actual constructions and the surroundings which might be explicable on the basis of my hypothesis.

Finally in this section I note that my suggestion that at least some megalithic monuments might be connected with cometary or meteoric events is not a new one. In particular Bailey *et al.* [11, p.13] suggested that some precise alignments not associated with any apparent solar, lunar or stellar declinations might have been oriented towards the rising points of strong meteor shower radiants, and also drew attention to comet-like carvings on various megaliths; Bailey [9] has also discussed this, and warned against the presumption that the activity seen in our skies in the present epoch is indicative of the situation some millennia ago.

4. The epochs of Earth-intersection

From the perspective of this paper, the fundamental question to ask here is "when did proto-Encke have an orbit intersecting that of the Earth (*i.e.*, when did it have a node at 1 AU)?" Due to the chaotic evolution of orbits such as this, having frequent encounters with the terrestrial planets, and also the influence of non-gravitational forces, it is impossible to answer that question on the basis of backwards integrations. All is not lost, however, as will be seen below.

The first point to note is that we are looking for terrestrial orbit intersections in *ca.* 3500 BC (when the Great Cursus was excavated) and in *ca.* 3100 BC (about when Stonehenge I was commenced; *cf.* note on the chronology in section 3 above). Both of these dates are uncertain by a century or so, but it will be seen below that the fact that the Cursus pre-dates Stonehenge I by several centuries (and not a millennium, say) is an essential feature explicable on the basis of the hypothesis I am presenting for the original motivation of the early Stonehenge developments.

A preliminary discussion of this problem was given by Steel and Asher [54] who performed backwards integrations of test particles based upon the presently-observed orbit of 2P/Encke. The basic methods employed in these integrations have been described in several earlier publications by those authors, so that it is sufficient here to state that the integrations were performed as given in section 2 of [52]. Note that that section was restricted to integrations using gravitational effects only (the application of random cometary non-gravitational effects was discussed in section 3 of [52], but was not used here). On the other hand, the start orbits used here were variational orbits with identical parameters except that the initial semi-major axes *a* were varied by up to ±0.06 AU, amounts representing the possible accumulated change in *a* over some millennia under non-gravitational forces (*cf.* [52, 53]). Here, then, all integrations were performed using purely gravitational models but with start elements varied so as to represent to some degree our uncertainty over the orbital history of proto-Encke in the past several millennia. For initial values of *a* varied by 0 and ±0.02 AU the systematic effects upon the orbital evolution (backwards in time over 20,000 years) for single test particles are shown in Figures 3 and 4 of [52], whilst for 18 test particles each spaced by 20° in *M* the effects of chaotic orbital evolution are shown in Figures 5 and 6 of the same paper (forwards integrations from 30,000 years ago with identical start elements, except for *M*). As might be anticipated, purely on the basis of random planetary encounters and without any random, unknown non-gravitational effects acting, it is clear that the orbital history of 2P/Encke or any test particle based upon it may not be determined, in the absence of any definitive observations prior to AD 1786.

Returning to the specific aim of Steel and Asher [54] - the identification of epochs in which meteor storms produced by proto-Encke would have been experienced on Earth - the point which is of interest is the time when a node occurs near a heliocentric distance of 1 AU. We found that for a start *a* which was 0.02 AU smaller than the nominal (present) value, a test particle would have nodes near 1 AU in about the epoch of interest in the Stonehenge context (second half of the fourth millennium BC). However, due to the effects of chaos and contingent close planetary encounters, it is necessary to study more examples, as is done here.

In Figure 1 are presented the results of backwards integrations over 7,000 years for test particles integrated as above but with initial values of *a* reduced by -0.01, -0.02, -0.03, -0.04, -0.05 and -0.06 AU (those figures will be used as appropriate labels for each). 'Delta' is the heliocentric distance of the node, the dashed curve being for the descending node, the solid curve for the ascending node. The dotted horizontal line (at 1 AU) shows the approximate position of the Earth. Clearly, meteor storms may occur when one of these curves intersects the dotted line. In reality those curves could be plotted as being of substantial thickness, to represent in some way the dimensions of the meteoroid stream/swarm in question. For example, the thickness on the sky plane of the IRAS-detected Encke trail was about 680,000 km [59]. We have no obvious way of knowing the lateral dimensions of the putative proto-Encke trail *ca.* 5000-6000 years ago (the dimensions along the orbit are discussed in section 6 in connection with the storm cyclicity), but by analogy with its expected narrow structure, the rate of precession and the persistence of contemporary meteor storms, we might expect proto-Encke-derived storms to be active over an epoch lasting for of order a few centuries around the times of intersections shown in Figure 1.

Some general points may be made in connection with these plots. Firstly, storm epochs occur in pairs separated by a period of *ca.* 250-400 years, and it was foreshadowed above

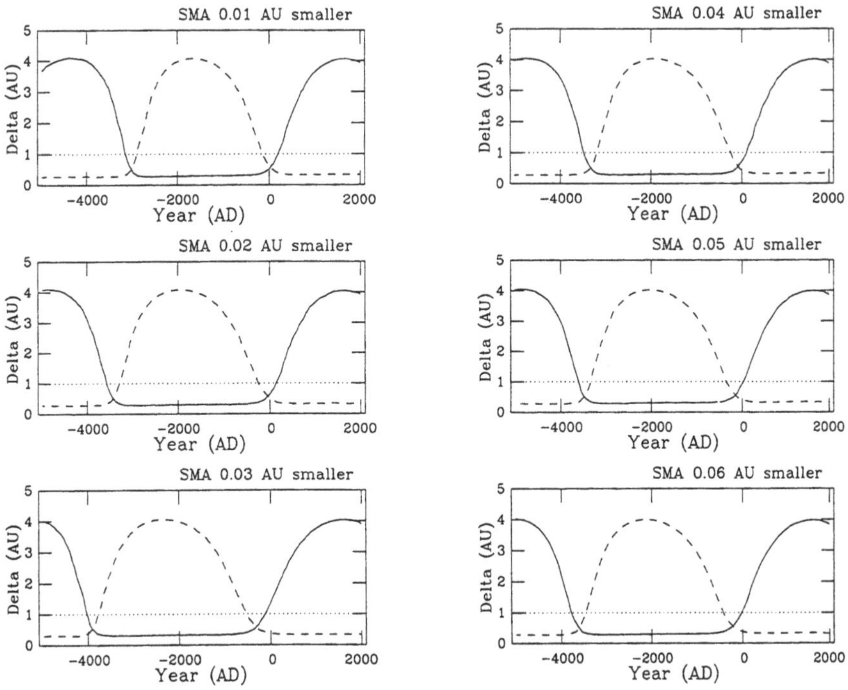

Figure 1: Results of backwards integrations of particles based upon the present orbit of 2P/Encke but with the semi-major axis (SMA) reduced by the indicated amounts. In each plot the solid line shows the heliocentric distance (Delta, in astronomical units) to the ascending node, whilst the dashed line shows the heliocentric distance to the descending node. The horizontal dotted line shows the approximate position of the Earth. In the epochs when the curves cut that dotted line, a meteor shower/storm would be expected. Going backwards in time, the first intersection (near time zero) is for the preperihelion ascending node shower (equivalent to the contemporary Southern Taurids); a few centuries earlier occurs the postperihelion descending node shower (equivalent to the daytime Zeta Perseids). About three millennia before that, another pair of intersections occur: the preperihelion descending node shower (equivalent to the Northern Taurids), and a few centuries before that the postperihelion ascending node shower (equivalent to the daytime Beta Taurids).

that the relative chronology of the Cursus and Stonehenge I (300-400 years apart in construction) might be explicable in these terms. This point will be returned to later.

Secondly, the general pattern of the intersections is consistent. Going backwards in time, the last set of intersections occurred at around the time of the start of the Christian era, or a few centuries earlier. This feature has previously been invoked as being linked to the eschatological response of people at the time and the rise of various religious movements (*e.g.*, [11, 18, 22]), but it is not of major interest here. Again, going backwards in time the first intersection (around year zero AD) is the preperihelion ascending node shower (*i.e.*, equivalent to the night-time Southern Taurids nowadays); the intersection a few centuries earlier than that is the postperihelion descending node shower (*i.e.*, equivalent to the daytime Zeta Perseids). There is then an intervening hiatus of 3000-3500 years to the preceding pair of intersections in the fourth millennium BC: the preperihelion descending node shower (*i.e.*, equivalent to the night-time Northern Taurids), preceded a few centuries earlier by the postperihelion ascending node shower (*i.e.*, equivalent to the daytime Beta Taurids).

Thirdly, the effects of chaos in the orbital evolution is clear. Looking at the dates of the intersections in the fourth millennium BC (the epoch of interest here), these push gradually backwards from the -0.01 to the -0.02 to the -0.03 set, as is to be expected (smaller orbits produce slower precession rates; *cf.* [1] for precession rates for specific orbits like these), but the corresponding intersections for -0.04, -0.05 and -0.06 occur later in time than the -0.03 pair. This is due to the contingent history of single integrations. If plausible non-gravitational forces were also imposed then the uncertainty would be even greater.

In fact making the initial orbit used in the integrations even larger than that of 2P/Encke can also produce intersections around the epoch of interest. In Figure 2 are shown the results of similar integrations with the start a incremented by +0.01, +0.02, +0.03, +0.04, +0.05 and +0.06 AU over that of 2P/Encke now. The behaviour is seen to be erratic. For +0.03 the intersections occur around 3200 BC; for +0.04 and +0.05 they occur around 5000 BC. As seen below these dates rather further back in time than might be initially anticipated are due not to chaotic effects, but rather a systematic effect: proximity to a jovian mean motion commensurability (which will be termed simply a jovian 'resonance' from here on).

In Figure 3 are plotted results pertaining to the Earth-intersection epochs for a number of test orbits. From top to bottom the rows correspond to backwards integrations using the present orbit of 2P/Encke but with initial semi-major axes stepped in 0.01 AU divisions from the present a = 2.2095 AU. To show the effects of chaos, in each row (each

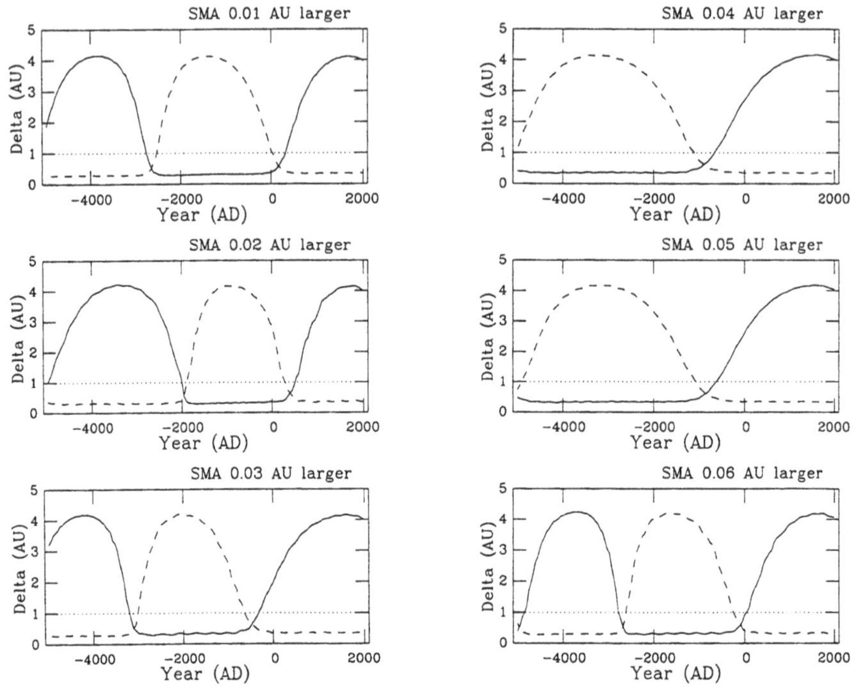

Figure 2: As Figure 1 except for start orbits slightly larger than the nominal orbit of 2P/Encke. The epochs in which intersections occur are seen to be severely affected by (i) The chance effects of close encounters; and (ii) The precession rate being altered from the general trend by proximity to the 7:2 jovian mean motion resonance (for the SMA being 0.04 and 0.05 AU larger), as is discussed in connection with Figure 3.

basic initial a) a further variation of 0, ±0.0001 AU was imposed, different symbols being used for each of the three integrations. As may be seen by comparing rows, the ordering of the intersections between the three is unpredictable, as is expected for chaotic orbital evolution and small differences in start parameters, to which extent it is not even necessary to identify which is which.

Starting at the bottom of Figure 3 and working upwards, there is a gradual progression of later intersections, reflecting the faster precession of larger orbits, until at a = 2.24-2.27 AU there is an abrupt change to slower precession. The cause of this seems to be proximity to the 7:2 jovian resonance. At the right-hand edge of Figure 3 are given the jovian and saturnian resonances for $2.15 < a < 2.27$ AU, indicating the location of the 7:2 resonance at $a = 2.257$ AU. In retrospect one could now suggest that the behaviour at a = 2.18-2.19 AU might be related to proximity to the 11:3 resonance, the precession rate trend being slightly disrupted there.

The basic question being pondered in this section is: 'When would a trail of debris associated with proto-Encke have intersected the Earth?' Figure 3 gives an indication of an answer to this question, although not directly: even ignoring chaos, non-gravitational effects *etc.* one cannot simply say that a value of a around 2.17 AU is needed, because 2P/Encke does *not* have that value now, and the precession rate would have varied as the semi-major axis (and other orbital parameters) altered until reaching their present value. On that basis one might favour a smaller semi-major axis *ca.* 5,000 years ago, say a = 2.14-2.16 AU, which has since gradually increased. Indeed if one accepts the notion that 2P/Encke is just the active macroscopic object amongst a complex of interrelated bodies, then one might equally well base the integrations upon one of the hypothesized Taurid complex asteroids listed by Steel and Asher [52], or perhaps a member yet to be discovered and thus of indeterminate orbit; a possibility is the undiscovered parent of the IRAS trail suggested by Clube and Asher [20]. If this object were librating about the 7:2 resonance as they suggest [1,2] then again it would certainly be possible to arrive at a pair of intersections around 3500-3000 BC.

The bottom line from the results presented in this section is that it is possible to choose some orbital elements which would produce Earth-intersections and hence meteor storms in the epoch of interest, when the first major developments at Stonehenge were under way in 3500-3000 BC, but it is simply not possible to define the epoch to within a thousand years due to the effects of close planetary encounters, non-gravitational forces, and also uncertainty as to whether the parent object of major interest is 2P/Encke itself or perhaps some other object either awaiting identification or totally disintegrated. Of course another possibility from that perspective is that there might have been multiple trails of debris derived from different daughter bodies spawned by proto-Encke, and these could produce storms in different epochs (*e.g.*, perhaps around 2300 BC when there appears to have been a widespread collapse of Bronze Age civilizations, and the later developments at Stonehenge were begun). On the other hand, although the absolute chronology of the intersections is unknowable on the basis of such integrations, the relative chronology is clear: two epochs of meteor storm activity, each lasting for perhaps a few centuries, would be expected to be produced by a trail, with centre separations of about 250-400 years, and that is in accord with the separation in time of the Cursus and Stonehenge I developments.

5. Orientations and observed phenomena

5.1 Recap

Outlined above has been some of the evidence for proto-Encke having arrived in a cis-jovian orbit within the past 20,000 years, and undergone major fragmentation(s) about 5-6,000 years ago, resulting in a large amount of meteoroidal material being released into a stream which has since

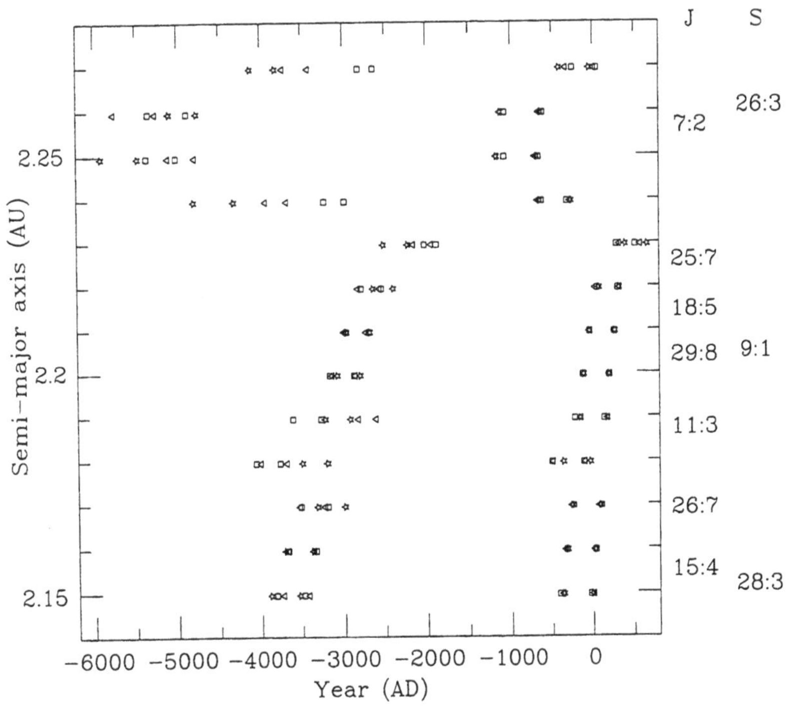

Figure 3: The epochs in which Earth-orbit intersections occur for a variety of initial orbits used in backwards integrations from the present. The initial semi-major axis (which will of course have varied during each integration) is shown on the ordinate axis, producing a series of rows. Each row has a value varied by precise multiples of 0.01 AU from that of 2P/Encke, and also two others varied by ±0.0001 AU from that main value (producing three orbits in each row, shown by the star, triangle and square symbols) in order to show the effects of chaos induced by contingent close encounters with the planets. At right are listed the locations of the various mean-motion resonances with Jupiter and Saturn. There is a general trend of faster precession (and thus intersections more recently, going back in time) for larger values of the semi-major axis, until that parameter approaches the 7:2 jovian resonance.

been dispersed to a large extent, although its influence and structure may still be perceived in the current epoch through disparate lines of investigation. It has been suggested that this general source - a single disintegrated giant comet - has been the major supplier of material to the interplanetary complex over the past 10-20,000 years. The Earth will have passed through the more densely-populated regions of this stream complex in different epochs in the past five or six millennia, and at those times phenomenal meteor storms must have occurred. The exact epochs in which such storms were experienced cannot be determined on the basis of just the modern astronomical evidence, but pairs of intersection epochs separated by a few centuries are expected, one of each pair being on the dayside, the other on the nightside. I now turn attention to what might be expected to have been seen at such times.

5.2 Giant comet debris trail

The form of giant comet break-up envisioned here is simply an extrapolation from what has been witnessed for smaller comets over the past two centuries, with frequent splitting occurring and occasionally fragmentation into many smaller daughter comets (*e.g.*, 3D/Biela, or the Kreutz group parent, or more recently P/Machholz 2, say: [5] and references therein). Such splitting has a number of causes, such as tidal disruption, although the majority of splits seem to occur when far from any planet, implicating thermal stresses as the root cause; it is clear that we must know much more about the nature of comets before fragmentation phenomena are fully understood. In the present context it is sufficient to note that any large comet in a cis-jovian orbit would be a vivid sight appearing bright in the sky (both at night and during the day when close enough to Earth) with a visibility cycle of a few years. A fragmented proto-Encke would initially form a structure similar to that of D/Shoemaker-Levy 9 as observed in the 15 months or so prior to striking Jupiter in 1994: a string of active cometary fragments forming a chain across the sky. This chain would appear to elongate and brighten as the Earth neared it, producing a vivid celestial band in the months before closest approach, with many bright condensations. One might imagine that the debris trail would be purely bimodal, with very large fragments (new daughter comets) spewing out many small meteoroids and dust grains, but no large lumps, such that the particles striking the Earth would mostly be just of the size producing visual meteors which totally ablate at high altitude. On the other hand, though, a hierarchical fragmentation of the kind envisioned here might be anticipated to produce debris of all sizes (like the 10-100 metre sizes of the recently-detected Kreutz group comets; see also the comments of Ceplecha [16] and Brandt *et al.* [13] on the likely existence of a large population of comets or cometary fragments of such dimensions), and in that circumstance transit through the trail would produce not only a meteor storm but also many atmospheric explosions of the kiloton-megaton class, with a lesser number of the Tunguska (10-100 megaton) class.

The phenomena which one might expect, therefore, are:

(a) A luminous trail of comets and debris, typically about ten degrees long but appearing to greatly elongate at such times as the trail and the Earth approach each other;

(b) A meteor storm as the Earth passes through the trail; and

(c) Possibly some real physical damage to the Earth's surface and its inhabitants if the trail contains a substantial population of small cometary fragments.

How long would trail passage take? The presently-observed IRAS trail has a sky-plane thickness of about 680,000 km. If this were characteristic of the path through the stream that the Earth had to traverse then the passage would have taken 6-7 hours. In fact the path length would be expected to be longer than this due to the relative trajectories and the trail gradually dispersing, with a transit time perhaps of order 24-48 hours, but quite possibly longer. This duration is significant, since it implies that if it were possible to get a warning that trail passage were imminent (within the next 24 hours) then the attendant effects would continue throughout the 12 hours or so during which a hemisphere is exposed to the apparent radiant of the meteors from the stream. For example, if in the case of a nighttime storm (stream intersection on its preperihelion leg) an indication were available from one night that stream entry were due, then it would be known that the storm would occur throughout the following night, with a daytime hiatus whilst the observer's location was on the opposite side of the Earth, and a continuation of the storm on the next night if the stream were especially broad.

5.3 Orientations in ca. 3100 BC

Here I describe what may have been observed if the putative stream were intersecting the Earth in about 3100 BC, at the preperihelion descending node (equivalent to the nighttime Northern Taurids; *cf.* section 3). For the sake of argument I will assume a trail of comets and debris with each component particle having essentially the same orbit. The orbit I happened to use (derived from a backwards integration) had semi-major axis a = 2.214 AU, eccentricity e = 0.8773, inclination i = 4.38°, argument of perihelion ω = 302-304° and longitude of the ascending node Ω = 113-114° at the time of intersection, but somewhat different elements are possible (a slightly smaller semi-major axis might be anticipated in most cases rendering intersections in this epoch, but the contingent history of backwards integrations makes a range of values possible). A spread of about three months in the times of perihelion passage (making the trail about 30° long in mean anomaly M) was assumed. In Figure 4 is shown a simple sketch plan with four cometary fragments spaced by three months along the trail, 15 days before the Earth is due to intercept it. If we imagine that a storm will occur for passage by the Earth through any part of the arc of the orbit defined by those four comets, then such events will occur about every decade or two on average (but see the discussion of cyclicities for different - perhaps real - trail lengths in sections 2.3 and 6). A strong meteor shower would be expected at this time *every* year, however, because small meteoroids would be expected to be quickly dispersed all around the orbit: an ejection speed of 0.1 km/sec near perihelion leads to a dispersal in orbital periods of about 10 percent and thus the orbit being filled within a century (*e.g.*, see [62]). Larger meteoroids have smaller ejection or relative velocities and therefore remain close to the parent object(s). It is passage through this concentration of larger debris which would have been of particular concern to the people of the time, requiring a prediction of imminence.

The date (or solar longitude) at which trail passage occurs in my simulations is dependent upon various parameters of the backwards integrations, but the date can be ascertained to within a few days, and that is sufficient with regard to the subject of interest. My adopted simulation showed a trail transit in around 3100 BC on August 11 or 12 on the Julian proleptic calendar. This corresponds to a Gregorian calendar date of about July 16-17, or a solar longitude of 113-114° (*i.e.*, the value of Ω given above). This implies a sunrise azimuth (z) at 53-54°, a few degrees to the south of the apparent alignment (z = 50°) of Stonehenge I. Note also that this date in mid-July is close to the time when the pharaoic Egyptians apparently awaited the heliacal rising of Sothis and the flooding of the Nile.

The essential point with regard to being able to forecast that passage through the trail (with accompanying mayhem) is about to occur is the azimuth at which different parts of the trail rise. As is the situation in Figure 4, 15 days before passage (*i.e.*, in the first few days of July on the Gregorian proleptic calendar, when the Sun is rising at an azimuth of 50-51° at about 03:50 local time) the nominal trail occupies 150° along the ecliptic across much of the night sky, and thus it has a wide range of rise times, each at a different azimuth. Given the uncertainty in the precise form of the debris trail, and epoch, there is no way to be precise about possible azimuths/times of rising, although it is easy to use variations on the basic orbital parameters given above in conjunction with one of the many commercial planetarium computer programmes in order to see how the trail would have appeared to pass by/around the Earth. It is adequate here to note that Figure 4 corresponds to a date about 10 days after the Summer Solstice, with trail passage due about 25 days after the solstice, and the reader may therefore visualize the tilt of the Earth's spin axis at these times and thus the general phenomena which occur.

As the debris trail is approached, the part in the night sky would appear to stretch in a band arrayed a few degrees north of the ecliptic (since this is just before it passes through its descending node), elongating both east and west as time progresses, assuming that the Earth is to pass through a central part of the trail. That specific part of the trail appears in the sky about 15° east of the opposition point (*cf.* my discussion [51] of the apparent and true radiants of the anti-helion meteor source), and rises at 21:00-21:15, an hour after the opposition point rises (and the Sun sets). As viewed from the Earth, parts of the trail each side of the passage point would move swiftly across the sky in ecliptic longitude, but that point itself would appear stationary in ecliptic longitude, slowly dropping in ecliptic latitude as the Earth and the trail come closer. For the above trail, 30° long in M, its apparent length as viewed from the Earth is about 40° at 75 days before trail passage; 80° at 45 days before passage; and 150° at 15 days before passage (as in Figure 4), at which time it is stretching across much of the night sky. Considering the situation at that time: (i) The back end of the trail will be rising several hours before sunset, at azimuths of 120-125°, and since these have yet to pass the node they are still north of the ecliptic; (ii) The central parts of the trail will be rising throughout the night at progressively more northern azimuths; (iii) The leading parts of the trail will be rising a couple of hours before sunrise, forming a bright column above the horizon due to forward-scatter of sunlight, at azimuths approaching the ecliptic but always south of it (*i.e.*, at azimuths typically 55-70°, larger azimuths rising earlier), these parts having already passed the node. The important point is this: 15 days before trail passage, no part of the trail is rising to the *north* of the azimuth of sunrise (z = 50-51° at that time).

We now move on to the situation just one day before trail passage: (i) The rear end of the trail has moved even further west and is rising mid-afternoon; (ii) The central parts are rising earlier than before, mostly before midnight in azimuths straddling due east; (iii) The leading parts of the trail are still rising one to two hours before sunrise, but they are now one or two degrees *north of the sunrise azimuth*. This contrast with the situation at 15 days before trail passage is due to the fact that in the earlier situation the Sun was still

Figure 4: Four comets spaced by one month each delineating a trail three months long, pictured 15 days before the Earth (on its near-circular orbit, shown by its usual symbol) is due to pass through the trail. The Sun is the black solid circle at centre. The orbit of the trail, which here is approaching perihelion, is as given in the text. This intersection epoch, in about 3100 BC, is at the descending node of the trail, producing nighttime meteor storms with a radiant just north of the ecliptic.

so close to its solstitial halt in the motion of its rising azimuth (as the word 'solstice' reflects), and the few degrees that the trail components lay south of the ecliptic were sufficient to keep them at a rising azimuth south of that of the Sun; but two weeks later, a day or so prior to trail passage (*i.e.*, a Gregorian proleptic calendar date about July 15-16), the trail is rising near the standstill point of the ecliptic whilst the Sun is rising further south.

In fact for the week or so prior to trail passage the leading components of the trail will have been rising very close to the azimuth of sunrise, sometimes very slightly to the north, but it is not until just before (meaning the day or so before) the debris trail is entered that the trail rising an hour or more before the Sun is seen to do so at as much as a degree or two north of the sunrise azimuth. One may imagine the people at Stonehenge observing during the night with trepidation, the azimuth of the trail on the horizon gradually moving north until it reaches its most northerly position and then drops to the south again. It is this attainment of the extreme point reached by the trail which indicates that trail passage is imminent (will occur that night), allowing that daytime for preparations.

The above was based upon visualizations of the trail based upon only the orbit cited. In reality one might expect a variety of orbits to occur quite quickly in the trail, due to differential perturbations, and in any case there are other orbits which could give intersections in the desired epoch (*cf.* Figure 3). Although the differences between the orbits may be small, their influence upon the precise nature of the encounter between the Earth and the trail will be critical. The above description, based upon one baseline backwards integration, indicates that imminent passage through the debris trail could be indicated by observing part of the trail rising an hour or two before the Sun and at an azimuth of 52-53°, at such time as the Sun was to rise at 53-54° (*i.e.*, trail rising a degree or two north of the Sun).

These azimuths do not, however, quite match the azimuth of about 50° for the axis of Stonehenge I. For the present hypothesis this does not present an insurmountable problem. On the one hand a small spread in orbits in the trail could lead to trail components rising slightly further north (*e.g.*, orbits with nodes inside 1 AU will still be north of the ecliptic after passing sunward of the Earth); on the other hand one could choose other integration results (*cf.* Figure 3 again) which give Earth intersections in the desired epoch (*ca.* 3100 BC) but at different longitudes (closer to the Summer Solstice such that both sunrise and trail rise may occur further north). The present paper is very much a first cut at this problem, and many more numerical experiments and simulations will be required in order to zero in on the solution. The important point here is that it does seem possible for a debris trail produced by a disintegrating proto-Encke and encountering the Earth around 3100 BC to rise with an orientation consistent with the alignment of Stonehenge I. A solution such as that suggested here is attractive in that it may lead to an explanation for the A-posts which lie at azimuths north (but not south) of the portal represented by Holes 96 and 97 (*i.e.*, the Heel Stone and its surmised twin).

5.4 Orientations in ca. 3500 BC

If there were an intersection of the debris trail with the Earth at the preperihelion descending node (equivalent to the nighttime Northern Taurids) in 3200-3100 BC, then there would also be expected to have been an intersection with the postperihelion ascending node of the trail (equivalent to the daytime Beta Taurids) in about 3500-3400 BC (see Figures 1-3). The basic scenario is as shown in Figure 5, the Earth there again being positioned 15 days before trail passage. The trail orbit at the intersection condition is similar to that for the nighttime intersection section discussed previously; my baseline simulation had elements a = 2.216 AU, e = 0.8788, i = 4.16°, ω = 234-236°, Ω = 177-178°. Since this is at the ascending node the solar longitude is 180° more than Ω, and thus it is apparent that the intersection occurs 2 or 3 days before the Vernal Equinox (about March 18-19 on the Gregorian proleptic calendar, April 17-18 on the Julian). Only a brief description of the observable phenomena seems necessary.

Firstly, it is obvious that partial obscuration of the Sun could occur (*cf.* comments on darkened and reddened Sun in section 5.5) and this would become more severe as the Earth

Figure 5: As Figure 4 except for an epoch near 3500 BC, with the trail having an ascending node at 1 AU on the post-perihelion of its orbit, producing daytime meteor storms with a radiant just south of the ecliptic.

neared the trail. Referring to Figure 5 (15 days before trail passage), we move the Earth on to a point closer to the trail, about a day prior to trail passage. The part of the trail outside of the Earth's orbit would rise in the evening sky to the south of due east, but here it is the trail sunward of the Earth which is of most interest. Being close to the equinox, the Sun would rise at an azimuth z = 90-91°, soon after 06:00. The trail sunward of the Earth would rise in the following 90 minutes in azimuths up to 15° north of the Sun (*i.e.*, at z = 75-90°). When very close to the trail one would be looking down its length, and it would form a bright bar on the sky due to forward-scattering of sunlight. The material closest to the Earth and due to pass nearby over the next 24 hours is concentrated around a rising azimuth z = 85°. This is close to the apparent alignment of the Great Cursus.

5.5 Obscuration of the Sun

Various cultures have recorded stories of the Sun (and Moon) turning dark and/or red at the time that catastrophes of some form have visited themselves upon the Earth. This apparent behaviour is of course displayed by the Moon when it is eclipsed, but the Sun is more problematical. Here I suggest that cometary dust tails produced by active cometary fragments in the proto-Encke trail might be responsible.

As the Earth passes through the debris trail discussed above, this being close to the ecliptic but of finite extent, it is clear (*e.g.*, see Figure 4 and extrapolate forward) that the trail will intervene between the Sun and the Earth, perhaps providing an explanation for such an effect. If the trail contains active cometary fragments then, since these have long gas/ion tails pointing near-radially away from the Sun and dust tails fanning between that radial direction and and the orbit vector, the Earth will be enveloped in smaller cometary products (dust, as opposed to meteoroids) for some time. It would be appropriate to make a few simple calculations here to examine whether this is a viable mechanism.

Whilst most of the mass in the terrestrial influx of meteoroids and interplanetary dust is held in small meteoroids at sizes of 100-500 μm [34], smaller particles dominate the scattering of sunlight by the interplanetary complex (*i.e.*, the zodiacal light and cometary dust tail optical phenomena are dominated by 10-100 μm particles). The *absorption* of sunlight by fresh cometary dust will be dominated, however, by the particle sizes dominating the cross-sectional area. Measurements of the dust environment very close to the nucleus of 1P/Halley using the *Giotto* spaceprobe indicated that for that comet the *number density* of particles maximized at sizes below 0.1 μm, the cross-sectional *area* peaked at sizes near 1 μm, and the *mass* at sizes near 1 mm [37]. Here I will consider the effect of ~1 μm particles in the dust tail as they pass close by the Earth through the line of sight to the Sun. Apart from absorption considerations, such small particles would be required to render any appreciable reddening in the transmitted (non-scattered) light through the preferential scattering of shorter wavelengths (*i.e.*, the scatterers need to be of comparable size to the wavelengths of visible light).

Contrary to many simple representations which show meteoroid streams being circular in cross-section, in fact these have structures which are more similar to warped sheets which may be considered to be locally flat (*e.g.*, [7]). Optical observations of comets indicate that their dust tails are similarly thin, being constrained near to the orbital plane of the comet. The proto-Encke dust tail under consideration here would be expected to be similar, but viewed very close to edge-on since the comet's orbital plane was close to the ecliptic in the epochs of Earth intersection. I will adopt in this simple treatment some extremal dimensions for this dust tail so as to derive a large volume; if the transverse thickness of the tail were ~10^5 km, the width radially outwards from the Sun ~10^7km, and the length along the orbit also ~10^7 km (an overestimate: this would take the Earth over 10 days to transit), then the volume would be ~10^{19}km^3. At the time of passage near the Earth the radial speed of the cometary trail/tail is 0.017-0.018 AU/day, so that if one imagines the Sun darkening and reddening say 3 days before passage through the core of the trail, the projected depth through the tail which the sunlight is penetrating is about 0.05 AU (7.5×10^6 km: greater than the geometrical thickness of the trail because of the viewing aspect). At any instant the obscuration is due to the dust present within a cone having a height of ~0.05 AU, with the Earth at the apex and an apical angle equivalent to the angular size of the Sun; this renders a base radius $R = 3.3 \times 10^4$ km and a volume of ~10^{16}km^3. If we require an optical thickness $\tau = 1$ then the number n of r = 1 μm particles may be derived from $\tau = n\,(r/R)^2$, assuming no shadowing effects. Thus $n \sim 10^{27}$, and the spatial density within the cone is ~10^{11} km^{-3}. If the dust particles have a mean mass density of 0.5 gm cm^{-3}, the particle masses are about 2×10^{-12} gm, and the mass of the 10^{19} km^3 dust sheet would be ~2×10^{18} gm. If instead the absorption were considered to be due to ~0.1 μm particles, the dust sheet mass derived is ~$2 \times 10_{17}$ gm.

The question is whether ~10^{18} gm is a viable mass. Firstly, it is to be expected that small dust grains of the size under consideration here would be quickly lost from the solar system through radiation pressure, so that no remnants would be expected 5,000 years later. Secondly, this mass is equivalent to the total refractory/dust content of a comet of radius around 8-10 km, meaning that it is not an unreasonable figure, in view of the fragmenting giant comet model under discussion, and in any case the assumed input parameters above might easily be reduced such that notable solar darkening and reddening would still result from total dust masses ~10^{16} gm. For example, $\tau \sim 0.25$ would be adequate to produce a marked effect, and the dust sheet volume could be rather smaller, in particular in view of the fact that observed comets do not have dust tails stretching far *along* their orbits: an along-orbit length of $\sim 10^6$ km for the dense dust region, ten times less than that assumed above, would take more than a day for the Earth to transit. Thirdly, the total mass of the present interplanetary dust and meteoroid cloud has been variously estimated at values between 2.5×10^{19} gm [60] and 3×10^{20} gm [29], with individual meteoroid streams having masses of order 10^{16} - 10^{18} gm [29]. The *Giotto* results [37] indicate that the mass held in ~1mm meteoroids is between one and two orders of magnitude higher than that in ~1 μm dust grains, so that the cometary mass release required to produce such observable effects is not unreasonable. It seems, then, that darkening and reddening of the Sun could indeed have been witnessed in the few days prior to trail passage, as a result of a dense cometary dust tail comprised largely of ~micron-sized grains being interposed between the Earth and the Sun.

6. Cyclicity

There is an apparent cyclicity set up in the later Stonehenge developments which has been extensively discussed and debated by various authors, as mentioned in section 2.1. The cyclical period in question is 19 years: either 19 tropical years (the Metonic cycle), or perhaps 19 eclipse years (the 18.03 tropical year period in which eclipse sequences repeat). In this section I show how a similar periodicity in intersection events could have been produced by a trail of debris such as that hypothesized above. It is not absolutely necessary that the trail intersection periodicity be 19 (tropical) years: in my interpretation the later luni-solar temple usage of Stonehenge, resulting from the Beaker people making observations of the Sun and the Moon over an extended period and discovering the cycle in the positions of those two orbs, could well have resulted from a transfer from a similar but distinct periodicity (*e.g.*, a 17-year trail intersection periodicity being passed into myth and legend after its cessation, with the 19-year Metonic cycle being happened upon by later observations of the movements of the Sun and Moon).

It is easy to obtain such periodicities given the approximate size of the orbits involved and trail lengths a few tens of degrees long in mean anomaly M. Let the length of the trail be measured as ΔM. Obviously if I set the orbital period to 19/6 years (implying $a = 2.15643$ AU) then there will be an intersection every 19 years (every six orbits), but only so long as the trail is not too long: for $\Delta M > 35°$ a gap of 13 years is followed by two of 3 years, and then 13 again (giving a grand cycle of 19 years). The choice of that period is arbitrary, but reference to Figure 3 will indicate that it happens to be close to the 15:4 jovian mean motion resonance (at $a = 2.15559$ AU), and there are reasons to expect prolonged proximity to such resonances, as discussed in section 2.3. In fact if the resonant period is used, the cyclicity in Earth intersections has a grand cycle of 19 years, that being the sole interval for ΔM below ~20° (but much shorter trails may produce long intervals with no intersections), the intervals becoming 3, 16 and 19 years for $\Delta M > 20°$ and 3, 13 and 16 years for $\Delta M > 40°$. An additional feature in favour of such a semi-major axis is that an average value in that region over the past 5-6,000 years would agree with the intersections occurring in the fourth millennium BC (Figure 3), notwithstanding the particular values near $a = 2.21$ AU used in sections 5.3 and 5.4 (see the comments made there).

On the other hand, as discussed in section 2.3 there has been a suggestion made that the IRAS-detected Encke trail may inhabit the 7:2 jovian resonance, corresponding to $a = 2.25705$ AU or a period near 3.39 years. Since five of those orbital periods correspond to 16.95 years, every 17 years the Earth passes through the trail within a distance equivalent to $M = 5$-$6°$ of the same point. Simple numerical experiments show that for $\Delta M < 20°$ various intervals of 17, 27 and 44 years occur, but for $20° < \Delta M < 40°$ the grand cycle is 17 years with gaps of 7 and 10 years occurring to different extents dependent upon the trail length.

Note that the length of the trail is restricted if it is in a resonance. For the p:q resonance where p and q have no common divisors there are q separate resonance centres about which librations occur. Thus for the 7:2 resonance these centres are 180° apart, whereas for the 15:4 resonance they are 90° apart, and the length of the trail ΔM must be rather less than this, leading to an expectation of a cyclicity [4].

Leaving aside the question of resonances, similar numerical experiments show that regular intervals between trail intersections, for trails of moderate length, are to be expected for any sized orbit around the region of interest (*i.e.*, the range of semi-major axes show in Figure 3). In view of the above it is clear that it is possible to understand that the people of the time would have been interested in some obvious cyclic event: the annual cycle of ordinary shower occurrence leads to a need to determine the length of the year (*cf.* section 2.4) whilst the periodicity of the major meteor storms (17 or 19 years, or perhaps some other value) leads to a need for longer-term observations and records in some form. I suggest that, long after the epoch of trail intersections had passed, the memory of this multi-year cycle led to the discovery of the Metonic and eclipse cycles, with great importance being vested in them: hence the building of the Stonehenge II/III temple. Several authors, amongst them Aubrey Burl [14], have suggested that at least part of the motivation behind megalith building may have been fear. To us, fear of the Moon seems strange; here I suggest the possibility that the darkening (and indeed reddening) of the Moon in eclipse being interpreted as a portent of forthcoming disasters, and hence a fear of the Moon, may have been a transferal from the distinct form of obscuration discussed in section 5.5. A common cyclicity (between trail intersection/ dust obscuration repeat time and the eclipse or Metonic cycles) would surely have made a strong impression upon the observers.

7. Other features

The speculative hypothesis for the original usage of the Great Cursus and Stonehenge I which has been presented above is based upon an astronomical model supported by various sources of data, and an apparent mis-balance amongst the small bodies in the inner solar system. This mis-balance seems exceptional, but may be connected with why we are here and able to ask the questions which we do ask of Stonehenge, and thus may be another aspect of the general anthropic cosmological principle. To me, the original motivation for Stonehenge must have been truly exceptional, and so I look for what exceptional events could have occurred in the sky, and on the Earth as a result.

In that vein I make here a few other comments which may be judged to be even more speculative in nature. My motivation is that by airing these ideas and opening them to disagreement, we will have moved forward.

Firstly, the stone-chambered long barrows associated with the Stonehenge site and spread widely over other areas in southern Britain. I am not an archaeologist, but in reading around the topic I have gained the impression that: (i) These date mainly from the epoch of the Cursus and Stonehenge I (3500-3000 BC); (ii) They are generally assumed to have been burial mounds for a small fraction of the population; and (iii) The dating of the bones found within the long barrows tend to indicate younger ages than the dating of the barrows themselves. To me this suggests the possibility that the long barrows were *not* constructed for funerary reasons, and the burials represent a later usage after the purpose for their erection had passed. In biology one sees many examples of 'convergent evolution,' resulting from the fact that there are only so many physical solutions to a particular problem; for example, the eyes of the octopus and the human are similar (except that our retina is back-to-front), and ichthyosaurs, sharks and dolphins (representatives of reptiles, fish and mammals) all assumed the same basic form, as set by hydrodynamical constraints. One may therefore look at a long barrow and ask 'What does this look like?' The answer to that question is 'An air-raid shelter.' One might therefore consider the possibility that long barrows were indeed built essentially as air-raid shelters, figuratively-speaking.

Even meteoroids a few metres in size entering the atmosphere, and releasing kilotons of energy at altitudes of 10-20 km, can cause some surface damage through their supersonic shocks. In the scenario envisioned here, the Earth passing through a swarm of 10-100 metre cometary fragments, accompanied by a multitude of smaller meteoroids, could cause extensive surface damage which wooden shelters would not survive. One Tunguska-type blast experienced by the south of England might be enough to induce the survivors to plan for a recurrence. Given the local supply of materials, a long barrow seems to be an excellent solution to a perceived (and perhaps real) hazard.

It might be productive to let the imagination have a loose rein in considering possibilities here. For example, if it is usually assumed that the transition from hunter-gatherers to farming societies occurred in southern Britain with humans clearing the trees at around this time, then one might wonder whether instead a Tunguska-type event might have cleared the trees (and perhaps the incumbent humans) to some extent. On the other hand, could the transition from a near-nomadic state to a geographically-fixed society led by a scientific hierarchy (which evolved into the power-holding astronomer-priests described by many authors in connection with Stonehenge II/III) with organized major works (the construction of the Cursus and Stonehenge I) for the common good have been mediated by the perceived need for predictions of the behaviour of the sky, and protection from its whims?

Returning to the long barrows, my suggestion is that their original purpose was as shelters from meteor storms; whether they were shelters from real physical danger or just used in the same way as people hid under their beds during the 1833 Leonids storm is a moot point. Being associated with death, their later usage as burial chambers may have resulted through a transfer from their original (redundant) purpose. I might also suggest that the late Iron Age fogous (souterrains with built-up walls capped with flat slab roofs) of southern Britain, and in particular western Cornwall, represent similar 'air-raid shelters' built in the time of the next intersections of the proto-Encke trail with the Earth (Figures 1-3); these seem to me to have been much too elaborate and over-engineered for the purpose usually ascribed to them (food storage).

Secondly, in my interpretation the later Stonehenge developments stem from a persisting belief of important things occurring in the sky which remained for centuries after the cometary trail encounters ceased in *ca.* 2800 BC, providing for a hiatus in developments. The trail would still be visible, however, as it slowly decayed and its orbital evolution took it away from having a node at 1 AU. Gradually the focus of the observations was switched to the Sun and the Moon, but the importance of stone-capped structures remained. That is, the trilithons of Stonehenge III, massive lintels bridging sarcen columns, and similar bluestone designs, are in mimicry of the construction of the long barrows, with vertical slabs capped by a strong horizontal stone. The story that such constructions saved the ancients from the wrath of a celestial god may therefore have been taken up and represented on a grand scale in Stonehenge III.

Thirdly, the importation of the bluestones from the southwest of Wales has long been a puzzle. My interpretation of this would be that stones of unusual appearance were taken to be of special significance because the earlier meteor storms had dropped a few meteorites, which were revered, and to a much later generation the bluestones had an appearance similar to chondrites. I can point to a similar belief arising more recently in a far-removed society. The aboriginal people of the area around Wilcannia in New South Wales have a myth of a foretold falling star, which killed many people, caused floods, and left behind strange stones. The people there nowadays present pieces of marble as being these stones, and these are clearly not meteoritic [57]. The point, however, is that over time the identification of the 'strange stones' has been transferred. It appears to me that a working hypothesis could be that a similar transfer occurred amongst the Stonehenge people. Bluestones and chips thereof have been found elsewhere in barrows (*e.g.*, Boles Barrow), and were used to make axeheads (*cf.* the Egyptians used meteoritic iron to make axeheads, but note that I am *not* suggesting that nickel-iron meteorites were dropped by the cometary trail, only that some chondritic stones might have been delivered in this way). I might therefore suggest the possibility that, if meteorites were venerated by the people, then these might also have been placed in barrows, with the earliest barrows being the most likely venues.

My final point concerns the purpose of the Aubrey Holes, which appear to be a late - but temporary - feature of Stonehenge I. The number of holes, and their usage, has long been a matter of debate (see [30] and references therein). There have been many suggestions for their employment, most of which I find bizarre (perhaps even as bizarre as many will find my own suggestions). Here I will employ again the concept of convergent evolution. I note that the Aubrey Holes were of a suitable dimension for a neolithic man (or woman) to sit in, thus providing a form of shelter in a set of fixed positions. I then ask where I have seen a similar arrangement, and the answer is in teams of meteor observers; for example, see [12] and [36]. In these examples, teams of six or eight observers are reclined in covered chairs or 'coffins' so as each to be able to monitor an assigned segment of the sky, reporting to a central recorder. This circle of observers is provided with a false horizon: a barrier within which they sit, preventing light from cars, houses *etc.* from disturbing their attention and dark adaption. In Stonehenge I the circular bank, which was about two metres high, could have fulfilled a similar function. Why so many observers (56)? A good question. It is possible that, as others have previously suggested, the large number of Aubrey Holes in effect served as a giant protractor, splitting up the horizon. Perhaps Stonehenge meteor observers looked inwards, and had their segments of the sky defined by towering wooden posts arrayed towards the centre of the circle; there is evidence for such early constructions, although they are usually interpreted as comprising a covered hut. Here I merely note that the Aubrey Holes *may* be interpreted in line with my overall hypothesis for the purpose of Stonehenge I.

8. Final comments

I am under no illusion that my suggestions will be viewed other than askance by the majority of readers, of whatever discipline. Despite the fact that I might be described as being an 'astronomer,' archaeologists should not think that my ideas will find acceptance amongst even those astronomers who work on small solar system bodies; as I have discussed elsewhere [50], despite it being clear that large departures in the population of comets and meteoroids in the inner solar system from any long-term (>10_6 year) average are to be expected on time-scales of 10^3-10^4 years, the belief persists that uniformitarian methods may be applied in this branch of science. An interesting commentary by Bailey [9] should be mentioned here, where he drew attention to the fact that recent results pertaining to our understanding of comets lead to an expectation that over the past millennia the sky may have appeared rather different to ancient peoples, and in particular intersections of the Earth with streams of cometary debris may have had some profound effects.

A criticism may be made that I have concentrated solely upon Stonehenge, and not considered megalithic structures elsewhere. In reply to that I would mention the comment of Hoyle [27], that it was *Stonehenge* which captured his imagination. Much the same applies to me, although I would also mention an interest in Stanton Drew, which is near where I was born, and I note with interest the recent discovery of evidence for a large structure involving wooden circles and a bank there, apparently contemporary with Stonehenge I. I would also note that by *not* investigating here other structures, some measure of predictive ability is afforded my hypothesis: the prediction is that for known and yet to be identified alignments at latitudes near that of southern England, the concentration will be on directions a little way north of due east for epochs around 3500 BC, and a little way south of north-east for epochs around 3100 BC. For later sites, post-dating the great meteor storms from the proto-Encke trail intersections but pre-dating the transfer of interest to the Sun and Moon (and thus occurring in the epoch 2800-2200 BC), I would expect alignments to be scattered mainly around the range of rising azimuths of the ecliptic, corresponding to the rising points of near-ecliptic meteor shower radiants associated with low-inclination Jupiter-family comets.

In astro-archaeological investigations I believe that it should be kept firmly in mind that the celestial phenomena which ancient man would have been most concerned with - objects which moved around the sky relative to the background of stars - may have been quite different to those observed now. Lest some Velikovskian or similar should attempt to use that statement to bolster belief in an absurd cosmology, let me be clear that I am not here concerned with the seven ‘planets’ or ‘wanderers’ of ancient astrology (Sun, Moon, Mercury, Venus, Mars, Jupiter and Saturn), whose recent orbital history is indisputably known; rather, I am concerned with a singular large (perhaps 50-100 km) comet trapped in a cis-jovian orbit, whose disintegration provided a persistent but variable spectacle in the sky, and real concern to the peoples of the time.

The sky of the Stonehenge people may have been quite different from that we observe today: a much more active sky, which could have had a real (not just imaginary) effect upon the lives of those people. To my mind, such a situation could have provided the motivation which would have been necessary for the building of such superlative structures as the Great Cursus and Stonehenge I. As such, I believe that my suggestion that these were built to assist in predicting imminent passages through a trail of cometary debris is worth presenting. Make of it what you will.

Acknowledgements

I thank David Asher for his assistance with the numerical integrations, Mark Bailey, Victor Clube and Bill Napier for various discussions, and Benny Peiser for his encouragement.

References

1. Asher, D.J. and Clube, S.V.M., 1993, An extraterrestrial influence during the current glacial-interglacial, *Quart. J. Roy. Astron. Soc.*, **34**, 481-511.
2. Asher, D.J. and Clube, S.V.M., 1998, Towards a dynamical history of proto-Encke, *Celest. Mech. & Dyn. Astron.*, submitted.
3. Asher, D.J., Clube, S.V.M., Napier, W.M. and Steel, D.I., 1994, Coherent catastrophism, *Vistas Astron.*, **38**, 1-27.
4. Asher, D.J. and Izumi, K., 1998, Meteor observations in Japan: new implications for a Taurid meteoroid swarm, *Mon. Not. Roy. Astron. Soc.*, in press.
5. Asher, D.J. and Steel, D.I., 1996, The orbital evolution of P/Machholz 2 and its debris, *Mon. Not. Roy. Astron. Soc.*, **280**, 1201-1209.
6. Babadzhanov, P.B., Obrubov, Yu.V. and Makhmudov, N., 1990, Meteor Streams of Comet Encke, *Solar System Res.*, **24**, 12-19.
7. Babadzhanov, P.B. and Obrubov, Yu.V., 1992, Evolution of short-period meteoroid streams, *Celest. Mech. & Dyn. Astron.*, **54**, 111-127.
8. Babadzhanov, P.B., Wu, Z., Williams, I.P. and Hughes, D.W., 1991, The Leonids, Comet Biela and Bielas associated meteoroid stream, *Mon. Not. Roy. Astron. Soc.*, **253**, 69-74.
9. Bailey, M.E., 1995, Recent Results in Cometary Astronomy: Implications for the Ancient Sky, *Vistas Astron.*, **39**, 647-671.
10. Bailey, M.E., Clube, S.V.M., Hahn, G., Napier, W.M. and Valsecchi, G.B., 1994, Hazards due to giant comets: Climate and short-term catastrophism, in *Hazards Due to Comets and Asteroids*, 479-533, T. Gehrels (ed.), University of Arizona Press, Tucson.
11. Bailey, M.E., Clube, S.V.M. and Napier, W.M., 1990, *The Origin of Comets*, Pergamon, Oxford.
12. Black, W.H., 1983, *Sky & Telescope*, **66**, 61-62.
13. Brandt, J.C., A'Hearn, M.F., Randall, C.E., Schleicher, D.G., Shoemaker, E.M. and Stewart, A.I.F., 1996, On the existence of small comets and their interactions with planets, *Earth, Moon & Planets*, **72**, 243-249.
14. Burl, A., 1987, *The Stonehenge People*, Dent, London.
15. Castledon, R., 1993, *The Making of Stonehenge*, Routledge, London.
16. Ceplecha, C., 1994, Impacts of meteoroids larger than 1 m into the Earths atmosphere, *Astron. Astrophys.*, **286**, 967-970.
17. Chippindale, C., 1994, *Stonehenge Complete*, 2nd edition, Thames and Hudson, London.
18. Cleal, R.M.J., Walker, K.E. and Montague, R., 1995, *Stonehenge in its Landscape: The Twentieth-centuryExcavations, English Heritage Arch. Rep.*, **10**, London.
19. Clube, S.V.M., 1994, Hazards from Space: Comets in History and Science, in *The Mass-Extinction Debates: How Science Works in a Crisis*, 152-169, W. Glen (ed.), Stanford University Press, Stanford, California.
20. Clube, S.V.M. and Asher, D.J., 1990, The evolution of proto-Encke: Dust bands, close encounters and climatic modulations, in *Asteroids, Comets, Meteors III*, 275-280, C.-I. Lagerkvist, H. Rickman, B.A. Lindblad and M. Lindgren (eds.), University of Uppsala.
21. Clube, S.V.M. and Napier, W.M., 1984, The microstructure of terrestrial catastrophism, *Mon. Not. Roy. Astron. Soc.*, **211**, 953-968.
22. Clube, S.V.M. and Napier, W.M., 1990, *The Cosmic Winter*, Blackwells, Oxford.
23. Farley, K.A., 1995, Cenozoic variations in the flux of interplanetary dust recorded by ^{3}He in a deep-sea sediment, *Nature*, **376**, 153-155.
24. Grün, E., Zook, H.A., Fechtig, H. and Giese, R.H., 1985, Collisional balance of the meteoritic complex, *Icarus*, **62**, 244-272.
25. Hasegawa, I., 1993, Historical records of meteor showers, in *Meteoroids and Their Parent Bodies*, 209-223, J. Štohl and I.P. Williams (eds.), Slovak Academy of Sciences, Bratislava.
26. Hawkins, G., 1973, *Beyond Stonehenge*, Harper and Row, New York.
27. Hoyle, F., 1977, *On Stonehenge*, W.H. Freeman, San Francisco.

28. Hughes, D.W., 1995, The world's most famous meteor shower picture, *Earth, Moon & Planets*, **68**, 311-322.
29. Hughes, D.W., 1996, The size, mass and evolution of the solar system dust cloud, *Quart. J. Roy. Astron. Soc.*, **37**, 593-604.
30. Kirk, T.H., 1995, A New Look at the Aubrey Circle, *Quart. J. Roy. Astron. Soc.*, **36**, 275-277.
31. Kresák, L., 1980, Sources of interplanetary dust, in *IAU Symp. 90: Solid Particles in the Solar System*, 211-222, I. Halliday and B.A. McIntosh (eds.), Reidel, Dordrecht.
32. Kresák, L., 1993, Cometary dust trails and meteor storms, *Astron. Astrophys.*, **279**, 646-660.
33. Kresák, L. and Kresáková, M., 1987, The mass loss rate of periodic comets, in *ESA SP-278, Symposium on the Diversity and Similarity of Comets*, 739-744, E.J. Rolfe and B. Battrick (eds.), ESA, Paris.
34. Love, S.G. and Brownlee, D.E., 1993, A Direct Measurement of the Terrestrial Mass Accretion Rate of Cosmic Dust, *Science*, **262**, 550-553.
35. Mason, J.W., 1995, The Leonid meteors and comet 55P/ Tempel-Tuttle, *J. Brit. Astron. Assoc.*, **105**, 219-235.
36. McKinley, D.W.R., 1961, *Meteor Science and Engineering*, McGraw-Hill, New York, p.9.
37. McDonnell, J.A.M., Lamy, P.L. and Pankiewicz, G.S., 1991, Physical Properties of Cometary Dust, in *Comets in the Post-Halley Era*, 1043-1073, R.L. Newburn, Jr., M. Neugebauer and J. Rahe (eds.), Kluwer, Dordrecht.
38. Napier, W.M. and Clube, S.V.M., 1997, Our cometary environment, *Rep. Progress Phys.*, **60**, 293-343.
39. Neugebauer, O., 1938, Sothisperiode, *Acta Orientale*, **17**, 169-195.
40. Newham, C.A., 1972, *The Astronomical Significance of Stonehenge*, J. Blackburn, Leeds.
41. North, J.D., 1996, *Stonehenge: Neolithic Man and the Cosmos*, HarperCollins, London.
42. Olsson-Steel, D., 1986, The origin of the sporadic meteoroid component, *Mon. Not. Roy. Astron. Soc.*, **219**, 47-73.
43. Pitts, M.W., 1981, Stones, pits and Stonehenge, *Nature*, **290**, 46-47.
44. Sekanina, Z., 1991, Encke, the comet, *J. Roy. Astron. Soc. Canada*, **85**, 324-376.
45. Smith, U., 1881, *Thoughts, Critical and Practical, on the Books of Daniel and the Revelation*, Pacific Press, London (and elsewhere).
46. Steel, D.I., 1994, Meteoroid Streams, in *IAU Symp. 160: Asteroids, Comets, Meteors 1993*, 111-126, A. Milani, M. DiMartino and A. Cellini (eds.), Kluwer Publishing, Dordrecht.
47. Steel, D., 1995, *Rogue Asteroids and Doomsday Comets*, J. Wiley, New York and Chichester.
48. Steel, D.I., 1995, The association of Earth-crossing asteroids with meteoroid streams, *Earth, Moon & Planets*, **68**, 13-30.
49. Steel, D., 1996, Meteoroid orbits, *Space Sci.Rev.*, **78**, 507-553.
50. Steel, D.I., 1996, The limitations of NEO-uniformitarianism, *Earth, Moon & Planets*, **72**, 279-292.
51. Steel, D.I., 1997, Meteoroid orbits: Implications for near-Earth object search programs, *Ann. New York Acad. Sci.*, **822**, 31-51.
52. Steel, D.I. and Asher, D.J., 1996, The orbital dispersion of the macroscopic Taurid objects, *Mon. Not.Roy. Astron. Soc.*, **280**, 806-822.
53. Steel, D.I. and Asher, D.J., 1996, On the origin of Comet Encke, *Mon. Not. Roy. Astron. Soc.*, **281**, 937-944.
54. Steel, D.I. and Asher, D.J., 1996, When might 2P/Encke have produced meteor storms? in *IAU Colloq. 150: Physics, Chemistry and Dynamics of Interplanetary Dust*, B.Å.S. Gustafson and M.S. Hanner (eds.), *Astron. Soc. Pacific Conf. Series*, **104**, 125-128.
55. Steel, D.I., Asher, D.J. and Clube, S.V.M., 1991, The structure and evolution of the Taurid Complex, *Mon. Not. Roy. Astron. Soc.*, **251**, 632-648.
56. Steel, D.I. and Elford, W.G., 1986, Collisions in the solar system - III. Meteoroid survival times, *Mon. Not. Roy. Astron. Soc.*, **218**, 185-199.
57. Steel, D. and Snow, P., 1992, The Tapanui region of New Zealand: A Tunguska of 800 years ago? in *Asteroids, Comets, Meteors 1991*, 569-572, A. Harris and E. Bowell (eds.), Lunar and Planetary Institute, Houston.
58. Štohl, J., 1987, Meteor contribution by short-period comets, *Astron. Astrophys.*, **187**, 933-934.
59. Sykes, M.V. and Walker, R.G., 1992, Cometary Dust Trails - I. Survey, *Icarus*, **95**, 180-210.
60. Whipple, F.L., 1967, On maintaining the meteoritic complex, in *The Zodiacal Light and the Interplanetary Medium, NASA SP-150*, 409-426, J.L.Weinberg (ed.), NASA, Washington, DC.
61. Whipple, F.L. and Hamid, S.E., 1952, On the origin of the Taurid meteor streams, *Helwan Obs. Bull.*, **41**, 1-30.
62. Williams, I.P., 1996, What can meteoroid streams tell us about the ejection velocities of dust from comets, *Earth, Moon & Planets*, **72**, 321-326.
63. Wrigley, C.C., 1989, Stonehenge from without, *Antiquity*, **63**, 746-752.
64. Wu, Z. and Williams, I.P., 1996, Leonid meteor storms, *Mon. Not. Roy. Astron. Soc.*, **280**, 1210-1218.
65. Yeomans, D.K., 1981, Comet Tempel-Tuttle and the Leonid Meteors, *Icarus*, **47**, 492-499.
66. Yeomans, D.K., Yau, K.K. and Weissman, P.R., 1996, The Impending Appearance of Comet Tempel-Tuttle and the Leonid Meteors, *Icarus*, **124**, 407-413.

Our Place in Space

Gerrit L. Verschuur
Physics Department, University of Memphis, Memphis, TN 38154, USA

Summary

Viewed from the perspective of our place in space, planet Earth is still accreting mass as part of its continuing formation process. Continuing impact events may have shaped the emergence and evolution of life during the last 4 billion years and even today the threat of impacts remains real. Asteroid and comet strikes are the rule not the exception and when an impact is violent enough the consequences can extend to the destruction of species. As regards our future in the face of the continued danger, hope stands in the way of our taking seriously the notion that past civilisations may have been wiped out by impact, or that the threat still looms over us all. From this perspective we must confront the possibility that the story of evolution is highlighted by the survival of the luckiest.

1. Astronomical perspective

To create an astronomical perspective from which to view cosmic impacts, let us take a brief tour through time so as to describe our place in space. During our journey try not to picture the universe as viewed from the surface of the Earth, nor from the time scale of a human life. Instead, imagine looking down on a tiny speck of a planet orbiting a tiny star in turn orbiting the centre of the vast Milky Way galaxy, one of billions like it throughout the universe.

Once upon a time there was no universe, at least in our corner of nothingness, but then a singularity developed which gave rise to what one species on one planet in one galaxy would some day call the Big Bang, a description of a moment in the timeless void when a stupendous amount of energy emerged as a creation event that defined the beginnings of space and time.

Space expanded and the newly born universe cooled from a pinpoint of unimaginable energy so that after about 300,000 years, perhaps a million of our time, its temperature had fallen to around 3,000 K. It was then that light escaped from between the churning masses of cooling gas and streamed out in all directions, keeping up with the expansion of space and traversing the entire ballooning universe.

Today we observe this relic radiation as redshifted radio and infrared waves whose intensity indicates that the signals are apparently produced by a warm body with a temperature of 2.7 K, an illusion, since these signals represent what is left after light generated in the 3,000 K universe has been stretched (due to the Doppler Effect) and dimmed on its enormously long journey to reach us, 12 or 15 billion years later.

Had the universe cooled more slowly than it did, much of the original hydrogen, created in the years when the temperature dropped from hundreds of millions to a few million degrees, would have been cooked up in nuclear processes to make helium and there would have been little hydrogen left to create stars like the Sun. Lucky for us, the universe did not consume all its hydrogen in the aftermath of the initial fireball.

Then, in the expanding universe, vast swaths of matter gathered to form proto-galaxies which swirled about one another while within them giant stars were born, only to die in violent supernova explosions. Lucky for us. It was inside the cores of those stars that the heavy elements were cooked up to make a rich soup that was then fed back into interstellar space to be mixed with primordial gases to form further generations of stars. Those, in turn, were made of a new mix of stuff which included the materials of which solid planets like the Earth could be built. Lucky for us.

Around many stars that faced their demise in more gentle ways, shells of gas were wafted out until they hung suspended about their parents, specifically the red super-giant stars. Those gaseous envelopes grew denser and cooler and molecular changes, driven by the laws of chemistry and physics, caused microscopic agglomerations of wondrous complexity to be formed. For example, biomolecules, life's building blocks, as well as abundant water, are even today being created in the shells surrounding super-giant stars.

It was in similar regions of space that the complex organic molecules that could seed a planet with the stuff of life were mixed with water and coagulated into comet-like objects that would someday help build the solar system and the Earth. Lucky for us.

Meanwhile, galaxies interacted and evolved and density waves wrought wondrous spiral patterns of star formation wherein more massive stars were spawned. These initially exploded to push surrounding gas and dust hither and thither. One such supernova nudged an incipient cloud of interstellar gas and dust into allowing gravity to take over and bring things under control in our small volume of space. Lucky for us.

The solar system was formed in this manner and now wanders about in the outskirts of the Milky Way galaxy, circling the galactic centre once every 220 million years, give or take a few million. Inside the orbit of the Sun about the galactic centre things are very different from what we are used to. There the frequency of supernovae is very much higher than it is out here. There you find intense X-rays and gamma rays permeating space and far more cosmic rays raining down on planets. It would probably be impossible for life as we know it to survive the radiations that permeate the inner regions of the Milky Way. Lucky for us - we live in the Outback of the Galaxy.

So here we are. In only a few million years, the Sun and its family of planets formed out of a swirling mess in space. Within the solar nebula, large masses of material called planetesimals were built through accretion and then even larger ones formed so that through this process the planets we know of today began to take shape. Lucky for us.

Those were wild times back then, about 4.6 billion years ago, when planetoids collided with monotonous regularity, and so violently that they altered the course of history of every object that was able to manifest planet-like properties in the first place.

Thus Mercury was struck and its outer layers torn off to be deposited in the Sun. Venus was struck head-on and sent spinning in the opposite direction, much of its mass no doubt lost to space. Yet it retained enough matter to form a gravitational well in which an atmosphere was trapped, an atmosphere of carbon dioxide that became so thick that it, in turn, trapped heat and raised the surface temperature to a fiery 900 F (700 K). Unlucky for the Venusians.

The third planet from the Sun was struck a healthy blow by an object somewhat smaller than Mars. That catastrophic collision stripped off the outer layers of the earth and spewed them into space. Some of the debris was trapped in orbit where it gathered, accreted, and formed the moon. In the direct aftermath of the impact, the Earth's axis was tilted which set the scene for seasons and the marvellously complex climate patterns we experience today. Recently it has been suggested that the moon helps keep the Earth's axis locked into its tilt to the plane of its orbit without which climate would be severe over most of earth. Instead, temperatures are moderate and no part of the world is intolerably hot, and only a tiny fraction is intolerably cold. Lucky for us.

In those ancient times, 4.6 billion years ago, and for a billion years thereafter, comets born in the far reaches of space carrying vast quantities of water ice mixed with biomolecules tumbled onto the newly formed Earth and brought the waters of the oceans and the raw materials for living organisms. Lucky for us.

Just like Earth, the other planets in the solar system were also struck time and time again, a process that defines accretion. Today every planet is still accreting material from space. Its just that we tend not to notice. However, we did pay attention when Jupiter dramatically accreted fragments of a comet in 1994.

Between Mars and Jupiter there was surely a planet, but it may have been struck such a violent blow that it shattered into countless fragments now called asteroids. Or did the planetesimals out there never form a planet? I've always found this idea difficult to believe, but either way it may have been lucky for us that there are asteroids wandering about all over the solar system because of the role they have played, and will continue to play, in our evolutionary history.

So the earth formed and was bombarded by comets and asteroids and the oceans churned and as the frequency of impacts fell simple organisms, perhaps some of them even brought by comets, established a foothold on Earth. Occasionally a huge comet slammed into the planet with such fury that the heat of impact evaporated the oceans and wiped out all incipient life so that the story of our origins had to start all over again with a revised opening chapter.

Much, much later, a billion or so years after the initial impact chaos had died away, during times when simple life forms were seemingly locked into a comfortable routine, large objects from space continued to sporadically inject vast quantities of energy into the planets atmosphere forcing sudden catastrophic climate change.

On each occasion, species were wiped out in large numbers and new ones emerged from the ashes. Plant life flourished and then animal life emerged that would someday lead to the emergence and subsequent evolution of *Homo Sapiens*. To set the scene for our species to strut onto the stage, the dominant creatures that once roamed earth 65 million years ago, the dinosaurs, had to be swept aside. If that had not occurred mammals would still be scurrying about under the feet of those monsters. Lucky for us something large, a comet or an asteroid, slammed into the planet and removed all vestiges of the creatures that had dominated life for over 150 million years. Had the impacting object arrived a half hour earlier or later there would have been no collision and the dinosaurs would not have been ushered off the world stage. In that case, we would not be here now.

Herein lies a sobering truth. The evolution of complex organisms may only take place on a planet where cosmic catastrophes occasionally wipe the slate clean, or at least very nearly so, so that ecological niches are re-defined. These are then available for a host of new species to burst into existence in the long-term aftermath of the collision catastrophe.

2. Historical impacts

There are so many details to this picture of our place in space, but for now I must focus on the immediate issue at hand. Victor Clube and Bill Napier in Britain and Alexander and Edith Tollmann in Austria have drawn attention to the likelihood that objects from space occasionally rained down in historic times. It is certain that some of those impactors splashed into the oceans to trigger tsunamis and flooding along coast lines and instilling terror in the minds of survivors, a terror so powerful that legends about such floods persist to this day. This possible link between flood legends and impact events is now beginning to fascinate more than just the pseudo-scientist. The issue is entering the mainstream of thought, because facts are coming to light to suggest more than an anecdotal connection.

For example, evidence has accumulated over the last decades that many early civilizations were wiped out rather abruptly. Why did they flourish and then suddenly disappear? As we have heared at this symposium, the signature of abrupt climate change has been found to be associated with many such disappearances, but what triggered the change? Was it due to volcanic activity injecting clouds of ash into the atmosphere, thus cooling the planet, or did cosmic impacts blast debris from shattered asteroids or comets into the heavens to darken the skies and cool the planet for months at a time?

The astronomer must vote for those inevitable collisions with cosmic debris that have punctuated the evolutionary saga. If there is one thing we are certain of it is that Earth is repeatedly struck by massive objects from space. This happens with surprising frequency and such impacts are expected to trigger severe climatic disruptions.

Our place in space is precarious. From down here we usually give this little consideration. The real danger recurs on time scales that we tend not to relate to, every few centuries or millennia for significant local disasters to many millions of years for minor extinction events to a 100 million years for a mass extinction event capable of bringing the planet to the edge of complete sterilisation.

Once again, consider the planetary stage from above. We observe the Earth meandering through interstellar dust clouds, into and out of spiral arms, brushing supernova explosions, mostly at a safe distance, but sometimes wandering into the remnants of devastated stars to the detriment of all life. We see space littered with swarms of debris, trillions of cometary objects in the Oort cloud half-way to the nearest stars, billions of objects in the Kuiper belt, just beyond the edge of the planetary zone, millions in the asteroid belt between Mars and Jupiter, tens of thousands in near Earth space, much of this debris left over from the solar system's formation.

Today we have a clear picture of comets and asteroids hurtling toward and about the sun. We see a blue planet orbiting amongst this mess and watch as it is occasionally struck by an object large enough so that in the fury of impact and its aftermath, sudden and catastrophic climate change is triggered to the detriment of species, and sometimes merely of civilizations.

3. Something to think about

To set this apocalyptic view in perspective, Table 1 offers a list of impact events in historic times which I have come across in my readings. It is an uncritical compilation, no doubt incomplete, but I offer it to make a point. Don't fret if your favourite event is not here, or if you would argue vehemently that I must excise another. Instead, bear in mind that 20 years ago such a table would have had few if any entries. However, now that the impact connection has been firmly established following the recognition of the K/T boundary collision of 65 million years ago, scholars are taking a fresh look at a host of astronomical, geological, archaeological and historical data that suggest that the connection may be found in recent times as well.

Table 1: Impacts in historical times?

c.3000BC	Henbury crater; 160 m diameter
c.2350BC	Early Bronze Age civilizations disappear - climate change data
c.2100BC	More Bronze Age cities disappear
c.2000BC	Campo Cielo, Argentina - bouncing asteroid
c.1800BC	Major dust event
c.1650BC	Destruction of Middle Bronze Age cities
c.1365BC	Destruction by fire in legends
c.1200BC	Destruction of Late Bronze Age cities
c.1159-1140BC	Decline in annual growth rings in Irish bog oaks
c.1000BC	Nebraska - Broken Bow crater
c.800BC	Bronze Age ends
c.850-760BC	Climate change data from the Netherlands
c.200BC	Fireball peak
c.400-600AD	A blitz of fireballs!
c.500AD	Wabar craters, Saudi Arabia
536AD	Dark Age triggered - dust veil event
c.580AD	Gregory tells of climate chaos
679AD	Gregory tells of climate chaos
c.800AD	Vikings find west coast of Europe uninhabited
c.1000	Comet and fireball peak
1176	South Island of New Zealand burned (fire from space)
1490	10,000 killed by meteorites in China
1700	Tsunamis in Japan - no earthquake cause known
1800	April 5, event in North America
1819	November 9 or 19 event, Canada and US
1885	February 24, event in Pacific
1892	May 3, event in Scandinavia
1908	June 30, event over Tunguska, Siberia
1930	August 13, event in Brazil - 800 square miles of jungle destroyed
1935	December 11, event in British Guyana
1947	Sikhote - Alin impact in USSR

The table suggests that during the last 5,000 years there may have been at least 30 events which caused significant regional destruction. Were such an impact to occur today it would create planet-wide headlines and, depending on the point of impact, would prevent many a newspaper from publishing at all.

If you fiddle the numbers to include phantom atmospheric explosions such as seen at Tunguska in 1908, and ocean impacts that produced tsunamis over the last 5000 years, it is not difficult to conclude that there may have been as many as 100 events, or about one every 50 years.

Table 1 suggested that a selection effect may be operating. In the last 200 years 10 events are listed. Is this because traces of more recent events are easier to find? Whatever the cause, the numbers suggest a significant impact event every few decades!.

In the next few decades someone may formalise Table 1 to add the stamp of scientific respectability, but I expect that the overall statistics will not change by more than a factor of two or so.

4. Will hope spell our demise?

As we focus our attention on the threat of impact, we find more and more evidence to suggest that significant collisions have occurred in historic times. From the astronomer's point of view, frequent albeit low energy impacts, those capable of producing a significant regional disaster such as occurred at Tunguska are to be expected at least once every century. But so what? Can we do anything about this? To be realistic, do we really care? Isn't this just a scientific curiosity were talking about?

To illustrate what I think will be the most likely course of action, I wish to quote from an essay I read in *Harper's Magazine*, in turn excerpted from a special issue of Social Research magazine devoted to "truth-telling, lying and deception." The quotes are from "My Grandfathers Walking Stick, or the Pink Lie by Lore Segal."

> 'When Pandora upended her box of calamities over the earth, there fell out, so says the story, a last straggler: hope. Hope pities us and lies. It pities our terrors and invites us to tell ourselves that the things we fear happen only to other people. We are a special case. When it comes to us, hope says, calamity will turn aside.'

So it is as we stare the near-earth objects in their cratered faces.

> 'Hope ignores the evidence of history and experience; it lies in order to con us out of knowing what we know and into thinking what we wish.'

This sentence haunts me with its truth. If we open our eyes to the astronomical facts we find not a shred of evidence to suggest that the planetary environment is benign, certainly not when viewed from the cosmic perspective.

We, the inhabitants of this small planet orbiting a small star in the outskirts of the Galaxy, find ourselves staring into space, overwhelmed by the view, and growing increasingly aware that we have been very lucky to have gotten this far. What do we see? We see the lurking threat of cosmic impacts and begin to realise that it may be time to prepare to avoid the end of civilization or, worse, the fate of extinction. The problem is that hope gets in the way. We hope that such warnings represent no more than a passing fad, a millennial scare perhaps. We trust that the next cosmic missile will brush by in a close call so that our lives will not be unduly disturbed.

Now consider this perspective: human beings have been lucky enough to develop a highly sophisticated, technological society *between* impact events capable of terminating civilization. There clearly has not been a significant impact disaster that we know about since, say, Greek times. Lucky for us, cultural progress has proceeded relatively unaffected by cosmic events for 3000 years. The question that now confronts us is whether we are going to do anything about making sure that this happy state of affairs continues indefinitely into the future.

Deep down we all hope we will live forever. Then, at some point in our lives, we accept the contrary fact that we are mortal. However, we continue to assume that civilization, and certainly our species, will go on forever. Such assumptions are now being shown to be without foundation.

When you take a hard look at what scientific research has taught us about the universe and its contents, about our place in space, you find not a trace of evidence to suggest that we are special on the cosmic stage. The universe we have come to discover out there does not care about our planet. We, like life on any planet anywhere, run the gauntlet of the dangers that lurk throughout space, the dangers were only just beginning to identify. Lucky for us.

I suspect that the recognition of the role of impacts in directing the course of evolution may take a long time to trickle into society's collective consciousness. We are discussing phenomena that most people would rather not consider, except when they read about them in science fiction stories. For example, it is human to deal with disturbing truths, such as the threat of impact, by gaping at a comfortable Hollywood movie on the theme. That's one of our favourite ways for coping with fear and horror; we fictionalise it. Need I remind you of that travesty called *Asteroid* shown recently on television. Its reassuring script would have had us believe that when a city is destroyed by an asteroid only three people die in the aftermath. I counted. Everyone else was merely injured and the heroes of the story clambered into and out of large craters surrounded by buildings still left standing. Impossible!

Thus, guided by hope, we deal with reality and lull ourselves to sleep.

I've suggested that from the astronomical perspective there is nothing special about this planet. At best, Earth is merely another target for the continued accretion of cosmic debris. From our point of view, however, the thing that is special about life on earth is that we have the ability to think about these issues. Once we acknowledge that we are vulnerable to cosmic catastrophes, in particular the threat of a comet or asteroid collision capable of triggering another extinction event, we may begin to consider whether our place in space is worth defending and preserving at all.

Herein lies a paradox and a tragedy; a paradox because we are able to recognise, through our unique abilities, that we are not unique, and that we are here now as the result of 12 billion years of circumstance. It is a tragedy because, lulled by hope, we will probably do nothing to assure the continued existence of our civilization, and certainly not of our species.

From the cosmic perspective, if I may again be so bold as to claim this view, we just happen to be lucky to be here, now. It was no other way. This is it, but it will not always be so. Astronomical facts tell us that planets suffer repeated humiliations in the form of impact catastrophes. But can we really believe this? Will hope ever allow us to face this truth?

In a Greek tragedy the end of the story is evident at the beginning and no amount of hope that it will be otherwise will change the plot. Yet we still hope that on the occasion we see the play something good will happen in the end. In the same vein, discovering that the sky is filled with cosmic debris should cause us to take stock of where we are and how lucky we are to be here, now. Instead, we appear to be compounding the problem by damaging the world through acid rain, ozone destruction and global warming, all of which phenomena will inevitably be triggered by an impact event as well. It is just that we struggle to do unto ourselves over a century or two what an asteroid impact could accomplish in mere moments.

5. Survival of the luckiest

I am obviously a neo-catastrophist because I can no longer think of evolution as a case of survival of the fittest. From the cosmic perspective it is survival of the luckiest that defines our existence. Of course, you could argue that the fittest would make preparations to survive an impact. That may yet be the case, but, in general, nature takes its toll independent of who we are or how prepared we might be.

In summary, astronomical discovery forces us to acknowledge our luck in getting to be here at all. The issue we must surely confront now is what we will do with our new-found knowledge, now that we've begun to see how we fit into the scheme of things we call the Universe. Will human nature cause us to shrug helplessly at the facts of existence? Will hope lure us into believing that there is little chance that a cosmic impact will threaten our civilization in the near future? I hope not.

Earth, Air, Fire, and Water: The Archaeology of Bronze Age Cosmic Catastrophes

W. Bruce Masse
U.S. Department of the Air Force
and
Affiliate Graduate Faculty, University of Hawaii

Summary

Planetary scientists and astrophysicists recently have begun to model the potential hazards on Earth from impact by asteroids and comets. These models suggest that some 25 at least locally catastrophic impacts likely occurred in various portions of the world during the past 5,000 years, during which time occurred the major developments of modern human civilization. This paper uses these cosmic impact models, coupled with data from archaeology, palaeo-environmental studies, and the systematic analysis of cosmogonic mythology and other literary traditions, in order to identify previously unknown catastrophic Bronze Age cosmic catastrophes, the most significant being a globally catastrophic oceanic comet impact estimated at between 10^5 and 10^6 megatons that occurred in 2807 BC. These data suggest that the threat of cosmic impact is very real, that the risk posed by cosmic impact is potentially much greater than currently modeled, and that such impacts have played a critical role in the development of human civilization.

1. Introduction

Near the end of his illustrious career Plato wrote a dialogue unlike any of the others he had written, one composed of a curious blend of cosmology, physics, and biology. Plato recounts an oral tradition about the Athenian lawgiver and poet, Solon, who two centuries earlier in the 7th century BC had traveled to the city of Sais on the Nile delta in order to question Egyptian priests about their knowledge of antiquity. One aged and venerable priest made the following remarkable statement [108: 451-452]:

> "O Solon, Solon, you Hellenes are never anything but children, and there is not an old man among you....[T]here is no old opinion handed down among you by ancient tradition, nor any science which is hoary with age. And I will tell you why. There have been, and will be again, many destructions of mankind arising out of many causes; the greatest have been brought about by the agencies of fire and water, and other lesser ones by innumerable other causes. There is a story, which even you have preserved, that once upon a time Phaëthon, the son of Helios, having yoked the steeds in his father's chariot, because he was not able to drive them in the path of his father, burnt up all that was upon the earth, and was himself destroyed by a thunderbolt. Now this has the form of a myth, but really signifies a declination of the bodies moving in the heavens around the earth, which recurs after long intervals; at such times those who live upon the mountains and in dry and lofty places are more liable to destruction than those who dwell by rivers or on the sea-shore. When, on the other hand, the gods purge the earth with a deluge of water, the survivors in your country are herdsmen and shepherds who dwell on the mountains, but those who, like you, live in cities are carried by the rivers into the sea. Whereas just when you and the other nations are beginning to be provided with letters and the other requisites of civilized life, after the usual interval, the stream from heaven, like a pestilence, comes pouring down, and leaves only those of you who are destitute of letters and education; and so you have to begin all over again like children, and know nothing of what happened in ancient times, either among us or among yourselves [...] In the first place you remember a single deluge only, but there were many previous ones."

Classical scholars should recognize this as part of the prologue to *Timaeus*, the dialogue in which Plato introduces the fabled lost city of Atlantis, described further in his subsequent dialogue *Critias*. Obviously Plato was fascinated by these foreign eschatological concepts, but only a generation later such catastrophic thinking no longer had an active place in the beguiling symmetry and abstract perfectionism of the Aristotelian celestial heavens. With the coming of Newtonian physics, the Age of Enlightenment, and modern social science, eschatology has been pounded and stuffed into the role of being but a minor component of studies of mythology and religion.

Scholars have paid much attention to the cosmology and biology of *Timaeus*, but have set aside the eschatology as a mere curiosity. The connection of this eschatology to something so potentially "absurd" as the sunken capitol of a lost idyllic civilization has contributed to the purposeful ignoring by modern Western science of both the message and context of Plato's Egyptian eschatology [63]. Equally damning was the fact that the eschatological message was delivered within the "suspect" context of mythology and oral history. Unfortunately, by ignoring the wisdom of the venerable Sais priest, Western science has discarded the one element of *Timaeus* likely most heavily grounded in actual historical observation as opposed to generalized philosophical speculation.

I wish to breathe new life into the concept of cyclical (or at least periodic) destructions of the world, especially as they relate to fire and water. It is my purpose in this paper to explore the likelihood that terrestrial and oceanic impacts by comets, asteroids, and meteorites are largely responsible for the powerful eschatological visions that have been handed down to us from past civilizations.

These impacts include a globally catastrophic oceanic comet impact in 2807 BC near the beginning of the Bronze Age (the last such catastrophe of truly global proportions, as will be defined below), along with several regionally and locally catastrophic impacts. A detailed study of all potential Bronze Age cosmic impacts lies beyond the scope of this paper, which is intended only as a preliminary examination of the issue. Also, the data are not yet sufficiently refined to reliably test whether or not these cosmic impacts are purely stochastic or instead have a non-random temporal distribution. However, if we assume that the evidence from the past 5,000 years is representative of longer term trends, these data suggest that current models substantially underestimate the risk of catastrophic cosmic impact.

2. Cosmography and epistemology

Although I am an environmental archaeologist by training, a term better suited for my research is "cosmography". The term cosmography has been in existence since the 14th century (*Webster's Ninth New Collegiate Dictionary* 1987), but is currently little used in the sciences. Cosmography is defined as "a general description of the world or of the universe". It is also "the science that deals with the constitution of the whole order of nature or the figure, disposition, and relation of all its various parts" (*Webster's Third New International Dictionary* 1993). As such, cosmography attempts to counter the tendency in modern Western science to segment the natural world into mutually exclusive knowledge arenas.

I am keenly interested in how a given society or cultural group uses and manipulates environmental events and processes as a means to guide or influence governmental policy, social action, ritual behavior, and religious and/or scientific thought. This may appear to smack of environmental determinism, a particularly unpopular topic in the social sciences during the past forty years. However, it does not require being an environmental determinist to accept that the environment pervades our lives today just as much as it has in past centuries.

Within the 20th century millions of lives have been lost to typhoons, hurricanes, cyclones, tornadoes, monsoons, floods, volcanic eruptions, droughts, and associated disease and pestilence. Hundreds of millions of people during this same period have periodically suffered due to starvation or by being left homeless due to environmental events. Because of the prevalence of wars in the 20th century, until recently we have tended to minimize the effects that the environment has had on human misery, even though at least some of the local and regional conflicts were largely the product of environmental perturbations. And when we did focus on the environment, it was primarily anthropogenic processes and events that captured our attention, such as pollution, the destruction of rain forests, and the reality of global warming. However, Hurricane Andrew, the recent eruptions of Mount St. Helens and Mount Pinatubo, and the effects of the El Niño/Southern Oscillation climate cycle have served to refocus scientific and governmental attention, especially in the United States, on the natural environment.

Cosmography crosscuts and blends a number of scientific disciplines and sub-disciplines including anthropology, archaeology, astronomy, epistemology, ecology, geography, geology, history, psychology, sociology, and theology. Cosmography likewise focuses attention on knowledge arenas not commonly used by archaeologists and the more traditional environmental sciences, including astrology, cognition, cosmogony, cosmology, mythology, and religion. Because past cultures were holistic, each of these knowledge arenas has the potential to shed a bit of light on past natural environmental events and processes, and when joined together can be most illuminating.

In particular, mythology is singled out as being a rich source of information on past major environmental events. Mythology to date has been largely under exploited as a resource because of our failure to understand its meaning and logic, and our failure to realize that the data contained in mythologies can be retrieved by systematic scientific methodology. Mythology, rather than being fanciful as is commonly believed in Western science, is actually a large multifaceted window on the major natural environmental events and processes that have shaped human history.

A distinction needs to be made between cosmography and the blossoming fields of archaeastronomy and ethnoastronomy. Beginning with the studies by Thom [188] and Hawkins [81] of Stonehenge and other megalithic sites in the 1960's, and subsequent work by scholars elsewhere such as Santillana and Dechend [166], Hadingham [78, 79], Aveni [6, 7], and Urton [190], archaeoastronomy and ethnoastronomy have steadily moved from the distant fringes of science to their current position as viable and exciting research arenas near the mainstream of the history of astronomy and of the historical social sciences [8, 9, 67, 101, 102, 185, 206]. Archaeoastronomy can be considered the science which systematically gleans the archaeological record in order to document the awareness and development of astronomy by past societies. Most work in this field has focused on the manner in which past societies viewed and measured the regularized movements of the fixed celestial heavens (e.g., Sun, Moon, planets, stars, and constellations), and on the creation of seasonal or yearly calendars. Ethnoastronomy focuses on the astronomies and cosmologies of traditional, largely Third World historic and modern cultures.

In contrast, the primary goal of my use of cosmography is to explain in behavioral and evolutionary perspective how individuals and cultural groups respond to environmental events and processes. These include not only earthly events, such as volcanic eruptions and typhoons, but also those naked eye temporary events which take place in the celestial heavens such as novae, supernovae, meteor storms, comets, fireballs, bolides, eclipses, and major planetary conjunctions. Indeed, celestial events and processes in general are a key component of the natural environment which until now have been largely ignored by science. An equally important goal of cosmography is that of the reconstruction of past environmental events and processes not presently known, or at least poorly known to science as determined from patterns elicited from the archaeological, documentary, oral historical, and paleoenvironmental record. Chief among these are cosmic impacts.

The path which led me to cosmography was circuitous and improbable for someone who for the first 15 years of my professional career was a conservative, data-oriented mainstream American Southwestern archaeologist. Like most archaeologists trained a generation ago, mythology, astronomy, and oral history were simply not part of my early professional venue. I had been taught that myth was largely psychological, that only a few of the more "advanced" past cultures had any significant notions of astronomy, and that oral traditions older than six or seven generations were likely to be at least partly distorted, with the distortion significantly increasing each successive generation. The thought of meaningful history being transmitted orally more than 200 or 300 years frankly was quite unthinkable. I did, however, differ somewhat from my Southwestern colleagues in that my fieldwork experiences led me to believe that the physical environment played a critical role in shaping the trajectory of individual cultures [119, 122].

My initial substantive foray away from normative archaeology occurred while conducting doctoral research in the Belau (Palau) islands of western Micronesia. Archaeological data associated with 13-14th century AD villages on long-uninhabited coralline limestone islets (*Rock Islands*) in Micronesia's Belau archipelago [121, 129], well matched the late 19th and 20th century oral traditions about warfare and settlement patterns surrounding these villages.

For example, among the village architectural features in the Rock Islands were massive defensive walls enclosing sandy coves, wall-in passes in inland areas, paired guard platforms adjacent to inland trails, and house platforms which were defensively situated. Even more poignant was the fact that our analysis of faunal remains [30, 121] demonstrated significant declines in the sizes of adult fish and shellfish food stocks along with the remarkable extirpation of pig in Belau, thus matching stories about starvation during this warfare period. I also was impressed by the finding of offertory deposits of branch coral hidden in the footings of stone house structures. This suggested that architectural remains provide insights into religious and symbolic aspects of culture in addition to the more mundane material, economic, and settlement pattern data. At this same time I perceived that historic and modern Belauan fishermen, untrained in Western science, were acutely aware of many significant aspects of animal behavior of which even experienced marine biologists were unknowledgeable [93, 120, 121].

The major breakthrough in the development of my ideas about cosmography resulted from the unique opportunity in 1989 at Hawaii Volcanoes National Park to participate in seventeen days of emergency rescue archaeology at the site of Waha'ula *heiau*. This ancient legend-laden archaeological site, purportedly the first human sacrificial temple in the Hawaiian Islands, was rapidly being overrun by lava from Kilauea Volcano.

As with the archaeology in Belau, the Waha'ula temple complex contained intriguing architectural elements that spoke to the symbolic and religious aspects of culture. Temple enclosure walls were oriented towards the cardinal directions; large blocks of bright red cinder gathered several miles from the temple were strategically incorporated into otherwise black basaltic sacred enclosure walls; paired waterworn or otherwise distinctive cobbles likewise were purposefully incorporated into the walls of enclosures and the houses of priests; and low walls marking gender and social class boundaries were used to segregate the overall temple precinct [130]. And as with Belau, I was surprised to discover how well oral histories of Waha'ula seemed to match archaeological reality.

During these seventeen frenetic days I observed close-hand the erratic and powerful behavior of living lava, complete with methane explosions, acid rain, and the destruction of a rich dryland forest. I also experienced an earthquake measuring 6.1 on the Richter scale. The earthquake collapsed the main tube system and briefly slowed the flow of lava, thus providentially allowing several extra days in which to conduct our studies.

Volcanologists, in studying the history of Kilauea volcano, had determined through radiocarbon dating of burned vegetation under lava flows of various ages, that approximately 70% of the surface area of Kilauea volcano is covered by lava younger than 500 years, while about 90% of the surface is younger than 1100 years [87]. I and my archaeological colleagues realized that such high levels of volcanic activity must have had a profound influence on traditional Hawaiian cognition and settlement patterns [29, 104, 130, 175].

In order to better understand both the significance of Waha'ula and the effects of volcanic eruption in traditional Hawaiian culture, I began to research the oral history and mythology surrounding the volcano goddess Pele. This in turn led to the realization that the majority of Hawaiian myths were firmly attached to Hawaiian chiefly genealogies, some of which were purported to go back in unbroken lineage for more than 95 generations before Kamehameha I, the monarch who unified the Hawaiian archipelago in the late 18th century. Among the myths attached to the genealogies were several Pele myths in which the volcano goddess had battles during the reigns of named genealogical chiefs (sometimes with the chiefs themselves), battles that resulted in the production of discrete named lava flows whose locations were known historically.

Much to my surprise, the radiocarbon dates for these lava flows matched not only the relative order of the named chiefs in the genealogical record, but closely matched absolute dates as well if a 22-year period were used for the length of each chiefly generation [130, 175]. Thus as with Belauan oral history, some Hawaiian myths appeared to record real environmental events which occurred up to at least 700 years in the past.

But my biggest surprise was the realization that the majority of the genealogically-fixed Hawaiian myths about powerful supernatural beings and culture heroes had little to do with earthly events such as volcanoes; rather they seemed fixated with the sky. As it turned out, an unlikely asset in my studies proved to be my unfamiliarity with archaeoastronomy and its preoccupation with the perceived more science-oriented "fixed celestial heavens". Instead, I was struck by the resemblance of the imagery of Hawaiian legendary voyages, cosmic battles between supernatural beings, and other action-filled myths with that of the nature and the occurrence of temporary celestial events such as the passage of comets and the sudden appearance of novae.

In order to refine my chronological control of mythology and historical traditions, as well as to observe and understand how various cultures respond to temporary celestial events, I created a master list of more than 1124 naked eye events recorded by Chinese, Korean, Japanese, Arab, and European observers between 200 BC and AD 1800 [123]. An updated version of this list is to appear in a forthcoming book [131]. This list of temporary celestial events is derived from data previously collected and scrutinized by astronomers and historians of astronomy [39, 89, 90, 145, 210]. These records often describe specific aspects of a particular event such as color, size, duration, and location in the sky against the backdrop of named stars, constellations, planets, Sun and Moon. Because of the largely stochastic nature of temporary celestial events, the record is marked by variation not unlike the patterns of drought and high effective precipitation found in tree rings. Some of these temporary celestial events were individually remarkable or formed unique contemporaneous clusters, while other portions of the chronological record exhibit few if any observations of temporary celestial events.

My cosmographic researches with Hawaiian, and more recently with American Southwestern oral traditions, archaeology, and ethnography revealed tantalizing patterns about how some cultures reacted to these celestial events, including as catalysts for profound culture change [123, 124, 127, 128]. It is beyond my ability in this paper to make a detailed presentation demonstrating the exact linkage between specific celestial events and specific myths; such linkage is provided elsewhere, at least for Hawaii [123, 131]. Later in this paper, I highlight a few salient aspects of this research.

By early 1991, I began to look at myths from other geographic regions besides the Pacific, and realized they, too, were filled with celestial imagery. Of note were myths describing catastrophic devastation caused by flood and firestorm. Realizing that myths in Hawaii and Micronesia encoded real environmental events, I made the assumption that these myths reflected one or more real disasters. Due to my familiarity with the literature on volcanic eruptions, I also realized that many myths did not well reflect volcanic eruptions or other known physical processes on Earth, but rather seemed to reflect disasters of cosmic origin. Particularly interesting were a series of myths (discussed later) that appeared to date to around 2815 BC, which seemed to correlate with substantive changes in the palaeo-environmental and archaeological record.

Unfortunately, being new to the finer details of astronomy and especially to the topic of cosmic catastrophe, and given the recent hoopla surrounding Supernova 1987A in the neighboring galaxy of the Lesser Magellanic Cloud, and especially given the fact that Hawaiian myths obviously encoded a number of novae and supernovae [123], I mistakingly first attempted to model the ca. 2815 BC event as the product of a near-Earth nova or supernova. Part of my model was based on perceived correlation between known historic nova/supernova events and pulses in the production of radiocarbon, which I attributed to increases in non-solar cosmic ray production from nearby exploding stars.

Armed with the ardent fervor that commonly adorns the naive, in the spring and summer of 1992 I put together a proposal for the Santa Fe Institute (SFI) to conduct a workshop on this hypothesized near-Earth nova-like event. While my selection of SFI may seem a bit odd to those who are aware of the long-term commitment made by SFI to the study of complex adaptive systems along with topics such as criticality and phase transitions at the edge of chaos, this was not as capricious a decision as it might at first seem. SFI is actually interested in all facets of system change, and they were somewhat intrigued by the possibility that the external natural environment was a bigger player in aspects of this change than previously envisioned.

Admittedly, however, the more compelling reason for this choice was the fact that my former dissertation chair sits on the SFI research board. I knew that he knew I was a solid scientist, and while he might think my recent ideas were a bit fringe and a real "shot in the dark", I had proved myself a competent professional archaeologist through many years of his observation, and under many, often difficult and interesting circumstances. For anyone who has attempted to introduce a new concept (or in this particular case, an older much abused concept) in business, politics, or science, the importance of networking and patronage cannot be overstated. I did not want to impose on our friendship, but I realized that without a lengthy record of publication on this topic, my chances of successfully convincing any other appropriate scientific organization to host the workshop was virtually nil. After more than a year of his deliberation as to the possibility that I was temporarily insane or was having chemical flashbacks from my days as an undergraduate in the San Francisco Bay area in the late 1960's, my former dissertation chairman encouraged me to submit my proposal. At the time I was Pacific Area Archaeologist for the Pacific Naval Facilities Engineering Command, and I used similar networking tactics to solicit the potential support of the U.S. Department of the Navy to help fund the workshop.

SFI took my proposal seriously and sent it to several knowledgeable reviewers, including noted cosmologist David Schramm at the University of Chicago. Schramm's compassionate but thorough review made it clear that nova events in particular, but supernovae as well, were highly unlikely candidates to produce the catastrophic effects that I had modeled from the archaeological, documentary, and oral historical record for the period of around 2815 BC.

The collapse of a pet theory, especially one representing many thousands of hours of personal research, is not a pretty sight. During several months of despondent denial, I checked and rechecked my data and wondered what went wrong. Then one day in March 1993, on a lengthy return flight to Honolulu from Navy duties on Guam and Tinian Islands, my ideas recrystallized. Armed with copies of Roger Penrose's *The Emperor's New Mind*, David Raups *Extinction: Bad Luck or Bad Genes*, excerpts from about 25 myths worldwide on the Great Flood, and especially the November 1992 issue of *Newsweek* magazine on "Doomsday Science", it suddenly dawned on me that my ca. 2815 BC catastrophe indeed was real, but as noted by Schramm and the other reviewers the event was not caused by a near-Earth supernova. Rather, this particular catastrophe was the product of a post-perihelion oceanic comet impact [124, 125].

Upon reaching home, a thorough review of my data and a brief check of the literature on cosmic impacts convinced me I was finally on the right track. I revised and resubmitted my proposal to SFI in September 1993, this time modeling the ca. 2815 BC event as a cosmic impact. It was also during the summer of 1993 that the name Velikovsky first came to my attention. After listening to my cosmic spiel, a close friend and fellow Southwestern archaeologist told me that my ideas distinctly reminded him of the work of Immanuel Velikovsky (discussed below), which he had read as a high school student. When told about Velikovsky, I was uncertain whether to be offended or amused at the comparison. Velikovsky sounded distinctly 'fringe' to me. This suspicion was reinforced by discussions with individuals at SFI who were openly scornful of his work. However, I made a mental note to someday get hold of Velikovsky's books, especially since several people had suggested that my work seemed perilously similar.

The end result of my second submittal to SFI was the same as the first. Of the roughly 25 scholars and scientists to whom the proposal had been sent either by myself or SFI, only a handful bothered to respond, with those few responses being equally mixed between those who were skeptical but thought the proposal had merit, and those who thought it uninteresting or even bad science.

My communication with SFI did have the positive result of networking with the archaeoastronomy community, and resulted in an invitation for me to give a paper at the conference "Inspiration of Astronomical Phenomena" held at the Vatican Observatory in 1994. It was at this time I discovered the work of Victor Clube and Bill Napier [42, 43], and subsequently the general field that is commonly now referred to as *neo-catastrophism* [147].

Although the present paper is incorporated into a volume resulting from a conference focusing on catastrophe, I am uncomfortable with the term neo-catastrophism. However, I recognize that several of my findings do, in fact, arguably support the basic tenets of neo-catastrophism. Indeed, much that I accomplished between 1989 and 1994 could be conceived as "reinventing the neocatastrophist wheel", as suggested by a valued colleague.

Nevertheless, I prefer instead to consider myself simply an environmental archaeologist who uses a cosmographic approach to science and history. This does not in any way diminish my respect for admiration of the work of others who have come to embrace the importance of mythology in the pursuit of historic reconstruction, and who have come to recognize the fact that environmental catastrophe is not an uncommon event in the history of humankind. The major difference between my fellow archaeological colleagues and me is that I subscribe to the notion that past societies were totally holistic in their treatment of science, religion, economics, and politics, which allows me to adopt the stance that studies of the physical environment must include the celestial heavens as well as more ground-based events and processes. And the major difference between myself and neo-catastrophists is that my cosmographic approach potentially allows for somewhat better control of the nature, context, and timing of temporary celestial events, including cosmic impacts, and thus permits at least the tentative coupling of oral historical traditions and archaeological manifestations with specific celestial and earthly events.

It is necessary to distinguish both cosmography and neo-catastrophism from the work of Erich von Däniken and other pseudo-scientists. These authors have deluged the popular press with a curious mixture of archaeological, astronomical, and historical data purposely torn from their original contexts in order to create logically flawed and scientifically untenable claims about visits from space aliens, psychic archaeology, and similar drivel [63, 182]. Unfortunately, this has given a bad reputation to virtually all serious and worthwhile studies that attempt to combine mythology and religious-ritual systems with archaeology and astronomy.

This situation was not helped by the writings of Immanuel Velikovsky in the 1950's through the 1970's, or more strictly speaking, by the scientific reaction to his work [76, 164]. Velikovsky was a European-trained psychiatrist who in preparing a treatise on Freuds own dreams and an associated comparative analysis of the lives of Oedipus/Ahkenaton/Moses, voluminously studied the eschatological myths of many cultures.

Velikovsky was struck by the possibility that the catastrophes described in these myths reflected cosmic catastrophes, a theory he thought helped to shed light on then inexplicable astronomical phenomena, such as the distinctive retrograde rotation of Venus and its likely high surface temperature, the later which Velikovsky predicted based on his model. Unfortunately, in his attempts to analyze and to explain the various data sets, he felt it necessary to manipulate aspects of archaeological and historical chronologies, and he often uncritically lumped together eschatological traditions from various parts of the world. But most damning in the eyes of astronomers and physicists, Velikovsky attempted to create a new and fatally flawed physics by which to explain objects and processes in the Solar System, including his curious notion that at some point in historical times the Earth somehow briefly stopped or significantly slowed down its rotation.

Despite my distaste for von Däniken and others of his ilk, I am willing to acknowledge that aspects of my cosmographic research were indeed thoughtfully presaged by Velikovsky. While in this vein I should also acknowledge William Whiston (A *New Theory of the Earth*, published in 1696), who succeeded Sir Isaac Newton in the Lucasian Chair of Mathematics at Cambridge, and whose work may have at least partly inspired Velikovsky's curious notions about the effects of large Earth-grazing comets.

I likewise acknowledge the earnest interdisciplinary and diverse approaches of scholars such as physicist Charles Pellegrino, the geologists Dorothy Vitaliano and Donald Patten, and the historian Vine Deloria, Jr., all of whom consider mythology to be a valuable adjunct to historical analysis. Pellegrino [154] links various environmental events, especially the eruption of Santorin volcano, with mythologies and historical texts throughout the Old World. Vitaliano [192] is particularly ambitious in creating a disciplined approach (not unlike aspects of cosmography) that she terms "geomythology". Although sometimes confusing and homogenizing the very distinct terms "mythology", "legends", and "folklore", Vitaliano nevertheless advocates that a careful study of myth, history, and geology demonstrates the keen appreciation that past cultures had for their physical world, and helps to explain the origins of specific historical geological events (e.g., volcanoes, earthquakes, floods) and specific myths or classes of myth.

The work of Patten [151] and Deloria [51] is in direct conceptual lineage with that of Velikovsky. They question the explanations provided by normative geology, particularly in their use of catastrophism as an active agent in geological change. Using the Velikovskian-like near-Earth approach of an object larger than the Moon, Patten modeled the effects of this hypothesized object as creating the Biblical Great Flood at around 2800 BC. Pellegrino [154], likewise, focuses attention on the 2800 BC date for the Biblical Flood, although he views it as a regional rather than a global catastrophe. Deloria, on the other hand, focuses attention on the the boundary between the Pleistocene and Holocene periods. He argues that the profound changes at that time, including the extinction of a number of large mammals, such as the mastodon and mammoth, were the product of catastrophe and not gradualistic factors or senseless slaughter as modeled by Paul Martin and his colleagues [118]. Deloria also effectively deals with some of the subtle ethnic biases that modern Western science has perpetuated against traditional non-Western knowledge systems. While I differ considerably from some of their modeling and especially their interpretive frameworks, I am struck by the fact that Patten comes up with a date for the Flood remarkably close to my own work on the topic [125], and am likewise intrigued by the fact that Deloria's view of the Pleistocene-Holocene boundary anticipates my belief that one or more cosmic impacts played a large role in this transition. Each author provides a number of useful observations which deserve consideration.

I especially applaud the work of geologist Russell Blong [20], whose careful research methods provide an exquisite demonstration of the value of the cosmographic approach (not labeled as such by him) to the social and physical sciences. Blong systematically studied the physical evidence and the mythology surrounding a previously unknown 17th century volcanic eruption which impacted an area larger than 100,000 km^2 along the northeastern coast and interior of Papaua New Guinea. Blong's work, which deserves a wider audience than it has apparently thus far received, provides many insights into the manner in which mythology and oral history encode environmental catastrophe. Blong's work unquestionably demonstrates the value of mythologies for providing many useful environmental details that can be systematically retrieved from oral histories and correlated with physical evidence gathered by other scientific means.

Finally, I greatly admire the work of astronomers Victor Clube and Bill Napier [42, 43, this volume] who have demonstrated an excellent understanding of much of the astronomical content of mythology. Clube and Napier, in collaboration with a handful of other colleagues, have introduced the interesting concept of cycles of impacts involving the intersection of the orbit of the Earth with debris streams from the breakup of giant comets, in particular the object responsible for the Taurid meteor streams. Unfortunately, their work has suffered from the anti-mythology and anti-Velikovsky sentiment of the hard sciences [33], but especially from their failure to identify specific matches between mythology and known or reconstructible cosmic impacts, with the possible exception of recent impacts such as that in the Tunguska region of Siberia in 1908. I will return to Clube and Napier's intriguing concepts of a Taurid 'progenitor giant comet' and 'coherent catastrophism' later in this paper.

These are but some of the individuals whose work I have discovered and have come to appreciate these past few years, including of course the fellow contributors to the present volume. At least some such as Clube and Napier have persisted with their work despite more than two decades of not just healthy skepticism but often rancorous scorn from practitioners of normative science. Their example inspires those of us who have more recently joined in this battle between paradigms.

But regardless of whether my own particular approach to history and to biological and cultural evolution be termed cosmography, catastrophism, or even Velikovskyism, I do have some tenets that may differ, at least in emphasis, from those of my colleagues. First, I rather strictly adhere to the basic principles of my archaeological heritage: sound chronology and context are fundamental components of my research, including that dealing with mythology. In fact, my introduction to cosmography and the value of mythology came about through the attempt to use genealogically-based mythology to refine the dating of Hawaiian culture history. Second, I bring together data from a large number of different scientific fields to balance and weigh my ideas, and which eventually may be used to test my ideas. Third, I have developed a logical and compelling theoretical framework for the origin and structure of mythology in which temporary celestial events, including cosmic impacts, play a significant part [123, 124, 125, 127, 131]. The mythologies included in my research are from cultures worldwide, and are not restricted to the major Old World civilizations. Finally, in contrast especially with Velikovsky, verification of the hypotheses and results generated by my research does not require the creation of a new physics nor does it violate any standard known scientific laws or principles. Likewise, my identification of previously unknown Holocene period cosmic impacts is completely amenable to scientific testing and falsification.

In fairness to Velikovsky, I have the benefit of greatly improved contextual and chronological control over mythology and archaeology, along with the benefit of the past decade of modeling and research on cosmic impacts [68] upon which to ground my own particular vision of cosmic catastrophe. But unlike nearly all other scholars noted above, I also have the benefit of a detailed working knowledge of the manner in which historic cultures viewed both the fixed celestial heavens and especially the occasional appearance within those heavens of often dramatic temporary celestial events [123]. The importance of the full suite of these temporary celestial events to past cultures and to the very fabric of mythology cannot possibly be overstated. This knowledge facilitates the identification and reconstruction of specific dated Holocene period cosmic impact events previously unknown to science, and brings into better focus the presence of other potential impacts where data sets have not as yet been fully developed.

3. Cosmic impacts

At present, the only definitive major catastrophic environmental event known to have occurred during Bronze Age times is the cataclysmic eruption of Santorin (Thera) volcano in the Aegean sea. [*Santorin* and *Thera* are used interchangeably in the historic literature; as a matter of convention I use "Thera" to refer to the island and Santorin to refer to the volcano itself.] The absolute dating of this event is still debatable [53] but seemingly took place in the late 17th or 16th centuries B.C. It likely directly impacted (largely by tsunami and ash fall) an area much greater than 500,000 km^2. Various authors in the present volume have pointed to other time periods during the Bronze Age when seemingly catastrophic events took place at least on a *local* level (i.e., a few thousand to a few hundred thousand square kilometers), and perhaps on a regional level as well. In particular, the dates of around 2350 BC, 2200 BC, 1200 BC and 1000 BC have been singled out as worthy of detailed scrutiny. These shall be returned to below.

Apart from the Santorin eruption itself, there is no clear consensus among the authors and other Bronze Age scholars as to what specific events may have triggered the various destructive layers apparent in the archaeological and geophysical record. Invading armies, earthquakes, and internecine warfare caused by the economic and social consequences of climate change have all been postulated [186]. Cosmic impact is also occasionally mentioned as a cause for at least some destructive layers (Peiser, this volume), but we are hampered in our attempts to examine this possibility because of our poor understanding of the physical, social, and archaeological signatures of cosmic impact.

Our knowledge of the effects of cosmic impact is largely based on the evidence of a single relatively small terrestrial impact of a probable stony asteroid that occurred in the Tunguska region of Siberia in 1908 [38, 82, 171]. This event was estimated at an energy of about 15 megatons (15 MT), the equivalent of about 750 times the force of the atomic bomb dropped on Hiroshima. The Tunguska impact leveled more than 2,000 km^2 of forest largely due to the fact that the asteroid exploded some 8 km above the surface of the ground. In addition to a blindingly bright atmospheric impact fireball, an air pressure blast wave which leveled the forest, and a particulate debris cloud extending many kilometers into the upper atmosphere and which was redeposited as "black rain", local effects included multiple sonic booms, earthquakes, and the generation of thermal radiation which blistered exposed portions of trees. Strong seismic waves were felt up to 600 km from the blast, while near-hurricane gusts of wind created during the passage of the shock wave traveled nearly as far. Many people were seized with terror and thought the end of the world had come. And for several nights following the impact, midnight skies over Europe were brightened to such an extent (presumably by high altitude particulates) that stars disappeared and the sky took on pinkish cast like a bright dawn. The sky was so bright that one scientist in England could actually read a book by its light at 1:45 a.m. [10, 181]. Smaller impacts, such as the fireball shower in the Russian town of Novyre Ergi in 1662 [181: 42-43) and the as yet poorly studied 1930 impact in the Brazilian rain forest [12] produced visually spectacular but less energetic effects than Tunguska.

Until the intensely watched impact of Comet Shoemaker-Levy 9 on Jupiter during the summer of 1994, our understanding of terrestrial impacts larger than Tunguska has been based primarily on theoretical modeling. This is augmented by a scattering of studies surrounding the physical evidence for the presumed dinosaur-ending Chicxulub impact on the Yucatan peninsula some 65 million years ago at the boundary between the Tertiary and Cretaceous periods (K-T boundary), along with a few other impacts which occurred prior to the Pleistocene geological period which began some 2 million years ago.

Water covers some 70% of the Earth's surface and thus oceanic or lake impacts are more than twice as likely to occur as are terrestrial impacts. Unfortunately, we are hampered in our understanding by the very lack of documented historic oceanic cosmic impacts. Our knowledge of oceanic impacts is solely limited to computer simulation and theoretical modeling, which is still largely in its infancy. Work by Jack Hills and his colleagues [82, 83] has brought much needed attention to the devastating tsunamis that can be generated by oceanic impact from even relatively small impactors. There also has been some attention paid to the effects that oceanic impacts can have on other aspects of the environment, including the atmosphere [189]. And in 1997 the Sandia National Laboratory performed a high-resolution supercomputer simulation of the effects of a 1-km wide, 60-km per second comet impact in the middle of the Atlantic Ocean, whose energy yield was close to the 300,000 MT (300 gigatons) threshold thought necessary for global catastrophe.

Despite these encouraging starts at understanding the environmental effects of oceanic impact, we are hampered by our still largely rudimentary understanding of linkages within oceanic-atmospheric circulation systems. For example, to my knowledge no one has satisfactorily modeled whether a small oceanic impactor has the capability of triggering an El Niño event or similar widespread climatic disturbance.

Currently accepted models of the risk of cosmic impact (detailed at the end of this paper) suggest that a locally catastrophic Tunguska-like impact can occur somewhere on Earth on the average roughly once every 200 years. A regionally catastrophic impact, as defined here, is one which directly impacts at least 500,000 km^2 (either land or water), or about 0.1% of the surface of the Earth. This is an area considerably greater than the size of the State of California (411,000 km^2) and is nearly the size of France (551,000 km^2). The threshold for such an impact likely is between 750 MT and 1 gigaton, and would be expected to occur by present models on the average of roughly once every 10,000 years. A globally catastrophic impact, one capable of directly or indirectly causing the death of at least one quarter of the Earth's human population is modeled to occur roughly on the average of once every 300,000 years; the energy threshold for such an impact has been put at around 300 gigatons (300,000 MT), or roughly the equivalent of 1,500,000 Hiroshima atomic bombs.

As will be discussed below, these risk models seemingly are at considerable variance with the data from archaeology, palaeo-environmental studies, and mythology. But before noting the known or hypothesized cosmic impacts and their perceived risks, the critical role played by mythology in culture history requires some discussion.

4. The structure, content, and context of mythology

> "Myth [...] is not merely a story told but a reality lived. It is not of the nature of fiction, such as we read today in a novel, but it is a living reality, believed to have once happened in primeval times, and continuing ever since to influence the world and human destinies." [115: 21]

As noted above, one of my most extraordinary findings is that individual cultural mythologies concerning the actions of gods, demigods, culture heroes, and other supernatural beings are largely cosmographic observations of actual major environmental events, in particular temporary celestial events, witnessed by the members of that culture. Because Hawaiian myths are closely tied to chiefly genealogies that extend back in time more than 95 generations, it is possible by direct matching to demonstrate that famous myths and myth cycles are actually birthing and circumcision stories which encode temporary celestial events as part of the chiefly reification process and the perpetuation of royal lineages [123, 128, 131]. As part of the reification process, the literal names of deities, culture heroes, and even the symbols and signs of royal power encode specific celestial events or classes of celestial phenomena.

For example, the famous Hawaiian myth in which Kamapua'a (*ka-ma-pu-a'a:* "sparkling bundle of eyes"), a half-hog, half-man shape-changing demigod seduces and then battles the volcano goddess, Pele, is a chiefly birthing genital chant encoding the coincidence in AD 1301 of Halley's Comet with the largest rift eruption in the historical record of Kilauea volcano. Similarly, the Polynesian myth of the demigod chief Maui (*ma-ui*: "Beautiful eye") capturing fire and latter snaring the sun and holding it still so that his mother could dry her bark cloth is a chiefly birthing story which encodes the coincidence at around winter solstice in AD 684 of Halley's Comet with a bright nova event in the Pleiades.

There is a strong correlation between the degree of fame and the magnitude of personal feats ascribed to any given legendary chief in the Hawaiian genealogies and the nature and numbers of spectacular celestial events associated with that chief. Few chiefs who reigned during "quiet times" in the celestial event record became famous in Hawaiian oral history. Indeed, the chief who reigned during the middle of the longest period of the least numbers of temporary celestial events (AD 776-813) during the past two thousand years was called *Helei-pa-wa*, "Straddling the Barren Period", and it appears that the first royal birthing temple was constructed in Hawaii at this time in order to provide an alternative chiefly reification process for the failing celestial heavens. Conversely, those chiefs who reigned at the time of known visually spectacular celestial events, or especially at the time of a series of spectacular events occurring during a short period of time, almost invariably developed a rich oral history. This is not a surprising finding since royal genital and birthing chants were used as the building blocks for the creation of orally recited genealogical name chants for the royal lineages.

The *'aha* birthing and circumcision signs of royal power were recorded in the oral histories of all chiefs beginning in the 11th century AD [95], and can be directly tied to specific celestial events. Also compelling is the association of Hawaiian gods with various types of celestial phenomena. For example, war god Ku was described by Hawaiian priests as often flying through the sky trailing a long flame-like tail [61: 75). The ornately tapering hair and highly stylized scowling mouths of the few surviving wooden and feathered images of Ku [88, 191] encodes the observable features of the nucleus and coma of near-Earth comets such as illustrated by drawings of Comet 1858 VI Donati [210: 241].

Hawaiian temples dedicated to Ku were generally filled with comet-like wooden images of a variety of shapes and abstractions [61, 191]. Many of these scowling-mouthed images have elaborate, pointed headdresses of various lengths which obviously encode the tails of comets, inspired no doubt by the position and appearance of most post-perihelion comet tails which are pushed in front of the cometary head by solar winds. And in several cases the images are represent by single small, abstract mouths variously carved into the middle of lengthy, smoothly curved logs. These images are haungtingly remniscent of long-tailed small-headed comets such as that of comet 1858 VI Donati [210: 215]. Based on a literal translation of their names and on descriptions of their physical attributes, all major Hawaiian gods were associated, in part, with major naked-eye visible specific celestial objects and phenomena.

While the mythologies of Native American cultures in the American Southwest lack the detailed genealogical ordering of Polynesia, it is still possible to perform cosmographic studies for certain aspects of Southwestern culture history. Many people are likely familiar with Katsina societies in the American Southwest in which dolls and masked dancers provide a symbiotic link with deities of the earth and sky to ensure the perpetuation of the annual seasonal round and to promote general harmony with nature [168].

Due to the exquisite chronological control that blesses Southwestern archaeology, it can be demonstrated that Katsina-like imagery suddenly appearing in the Puebloan

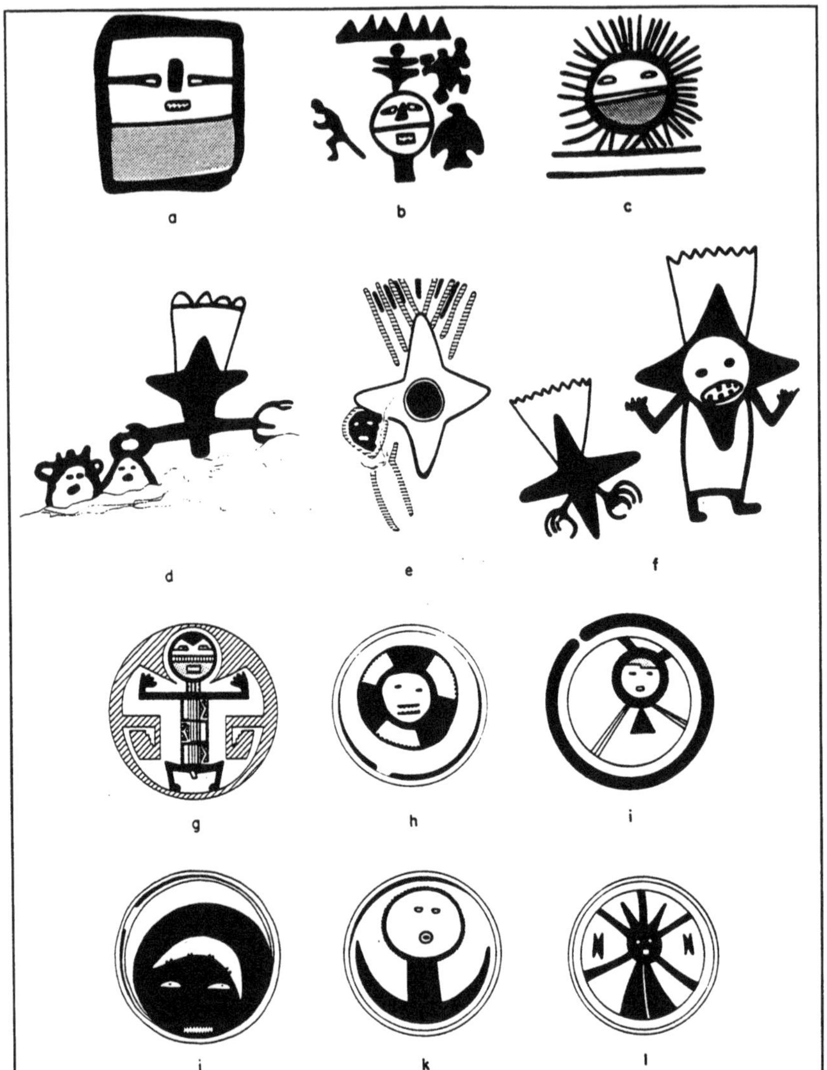

Fig. 1: Cosmographic iconography from the 14th century AD Puebloan Southwest: Figures a-c,g, possible Day-Night deity, with constellation images surrounding b. Figures d-f, eclipse star demons. Figures h-l, solar eclipse images.

archaeological record during the 14th century AD is directly correlated with the paths of three total solar eclipses [127]. The pronounced impact of eclipses on Puebloan culture may be related to the belief shared by many Native American cultures change in cosmological eras were accompanied by catastrophe and the "death" of the Sun. Many of the artistic images produced on pottery and rock art distinctly portray the actual eclipses themselves or portray "star demons", the latter which are taloned and feathered believed to appear at the time of solar eclipses (Figure 1). One of these eclipses, an event in AD 1379, is also prominently mentioned in myths surrounding the demise of the Hohokam, a major archaeological culture in southern Arizona. In fact, total solar eclipses were associated with profound social changes throughout the history of the Hohokam [124].

Other classes of celestial events are also portrayed in prehistoric Southwestern art. For example, the rapid cult-like appearance and spread of Salado polychrome pottery during the late 13th through early 15th centuries [47: Fig. 9.1) correlates well with the unique occurrence of large numbers of unusual comets. These comets are likely represented by ceramic motifs depicting feathered serpents, birds, and related celestial imagery, while other categories of temporary celestial events such as novae are also likely being depicted in Salado art. Likewise, novae and comets figure prominently in the migration legends of several Native American tribes [142, 195]. The animation and anthropomorphizing of the environment, including the celestial environment, is characteristic of early traditional societies who viewed their world holistically.

At least six compelling lessons are revealed by these various cosmographic studies. First, mythologies, at least in part, represent cosmographic records of real environmental events, especially temporary celestial events. Second, some iconographic images of gods, demigods, supernatural beings, and legendary rulers portray specific celestial phenomena and events. Third, the names of gods, demigods, supernatural beings, and legendary rulers often provide literal descriptions of specific celestial phenomena and events. Fourth, environmental events such as floods, earthquakes, volcanic eruptions, and droughts, are often cognitively linked with unusual celestial events (e.g., comets, meteor storms, supernovae, planetary conjunctions, eclipses, cosmic impacts) that may have occurred within a few years of the earthly environmental event. Fifth, oral historical records can faithfully transmit information about these events for many hundreds and seemingly even thousands of years. And sixth, even in the absence of genealogies or other means by which to chronologically fix the initial appearance of specific myths (such as provided by iconography in archaeological contexts), cosmogonic mythology is often given at least a relative ordering within each set of cultural traditions.

Thus, for example, one Cheyenne historian, in discussing the epic history of his people, provides the following remarkable sequence of events [60: 111-114]. There originally were created three types of humans - the Indian, a hairy solitary people, and fair-skinned people with long beards. They lived together in a lush and beautiful area far to the north of Montana. The hairy people moved south and were shortly followed by the Indians. The hairy people were shy and secretive, living in caves high in the mountains; eventually their population dwindled and became extirpated for a reason unknown to the Indians. At the time of a devastating flood in the south lands, the Indians moved back north and found the long-bearded white men and some wild animals were gone. The Indians increased in number and after a long period of time again moved south. While living in the south, a second devastating flood destroyed almost everything and left the Indians starving. They tried to go back to the North but discovered it to be a barren wasteland, so they returned to the south. After many hundreds of years the earth shook and fire and smoke suddenly appeared in the badlands area and destroyed the trees in the badlands area, causing a short period of starvation. Things got better and have been so up to the present.

The ethnographer who collected this epic story assumed that it represented the past 400 years of history for this particular group of Cheyenne [60]. However, as will be noted below, the earthquake and fire in the badlands likely represents a locally catastrophic terrestrial impact around 1000 BC, and the second of the two floods almost certainly is the result of a globally catastrophic oceanic comet impact in 2807 BC. And based on my present understanding of mythology and on my knowledge of Holocene period archaeology and paleoenvironment, it is my strong suspicion that the first flood event is that of another oceanic cosmic impact that I hypothesize occurred around 9300 BC. Thus this curious epic tale may in fact provide real glimpses of the past 12,000 years of human history.

5. Temporary celestial events and Bronze Age chronology

Few scholars would dispute the fact that the greatest obstacle standing in the way of a better understanding of Bronze Age events and processes is the lack of an accurate chronology that encompasses and unequivocally ties together the major Old World civilizations into a single unified system. Much of the passion and rhetoric of Bronze Age studies has been and continues to be focused on refinements of regional chronologies and the implications that these refinements have for explaining significant aspects of the past [92, 162]. Chinese historical studies, likewise, have been subject to intense chronological debate, in part stemming from discrepancies between the standard chronology developed during the Han Dynasty and that associated with the *Chu-shu chi-nien* (*Bamboo Annals*), the latter which considerably shortens the standard chronology [106, 148, 149]. Immanuel Velikovsky was no stranger to these chronological debates, and his adjustments to chronologies played a major role in his vision of a world filled with cosmic catastrophes.

A brief decade of archival study by no means qualifies me as an expert on Bronze Age archaeology or culture history. Nevertheless, my studies in cosmography suggest that the relationship between temporary celestial events (and earthly environmental events such as the destruction of Thera island by the Santorin eruption) and Bronze Age mythologies and religious iconography provides a potentially powerful new perspective for evaluating and testing models of Bronze Age chronology. The Hawaiian and American Southwestern systems of encoding celestial phenomena and events offer insights into how such phenomena and events may have been recorded in Bronze Age Mesopotamia, the Near East, Egypt, Europe, India, and China. Regional and globally catastrophic cosmographic events, such as the Santorin eruption and large cosmic impacts that may have occurred during the Bronze Age, are certain to appear in the cosmographic record of each affected region. The key then is to study and to identify cosmographic events within the context of each regional sequence, and to then match these against events recorded in other regional sequences.

The epigraphy, mythology, and archaeology of Bronze Age cultures are filled with images of frightful deities and beasts (Figure 2), many with destructive powers. Some of these beasts are associated with specific gods or themselves represent a specific demigod or deity. Notable is the fact that the worship or mention of these deities/beasts in documentary texts reveal that their popularity was not constant but rather waxed and waned over the centuries. In a similar vein dynastic rulers often utilized the names of certain gods as part of their royal titles; these, too, waxed and waned over the centuries.

Fig. 2: Cosmographic iconography from Mesopotamia (redrawn from Dalley 1989). Images likely represent celestial phenomena, including: Sun (a), comets (b, f), supernova or nova (e), constellations (b, d, g), and possible meteorite (h).

That many of these Old World beasts/ deities are cosmographic images emblematic of celestial processes and phenomena has been argued for years [3, 70], a concept which recently has been elegantly demonstrated for the previously mysterious Indus script [150]. Unfortunately, the tendency of modern science has been to has been to view the astronomical content of myth either as remnants of early observational attempts to classify aspects of the fixed celestial heavens or even as the hard-wired psychological product of the human genome. The influential mythologist, Joseph Campbell, expressed the latter position shortly before his death when attempting to explain the striking similarities between Old World and New World myths and iconography [28: 92]:

> "One can only wonder considering, as I have now been considering for some forty-odd years, the likenesses, both in breadth and in depth, of the two constellations of metaphorical images of the Old World and the New, whether the human psyche can possibly be so thoroughgoingly programmed that these two all but identical constellations might indeed have arisen independently in the separated hemispheres of our planet."

The structural anthropologist Claude Lévi-Strauss, provides a somewhat similar view which has had a profound influence on modern cultural anthropology and the thinking of Jungian psychologists [109: 240]:

> "In granting that myths have an astronomical significance, I do not propose to revert in any way to the mistaken

ideas characteristic of the solar mythography of the nineteenth century. In my view, the astronomical context does not provide any absolute point of reference; we cannot claim to have interpreted the myths simply by relating them to this context. The truth of the myth does not lie in any special content. It consists in logical relations which are devoid of content."

Even the late Carl Sagan, who perhaps more than anyone this century has helped to define the role and importance of science, particularly astronomy, in our modern Western world, has perpetuated the notion that mythology has little explanatory value for understanding the either astronomy or the universe. A few years before his death he made the following curious observation [164: 251]:

"The myths and folklore of many premodern cultures have explanatory or at least mnemonic value. In stories that everyone can appreciate and even witness, they encode the environment. Which constellations are rising or the orientation of the Milky Way on a given day of the year can be remembered by a story about lovers reunited or a canoe negotiating the sacred river. Since recognizing the sky is essential for planting and reaping and following game, such stories have important practical value. They can also be helpful as psychological projective tests or as reassurances of humanity's place in the Universe. But that doesnt mean that the Milky Way really is a river or that a canoe really is traversing it before our eyes. / Certain kinds of folk knowledge are valid and priceless. Others are at best metaphors and codifiers. Ethnomedicine, yes; astrophysics, no. It is certainly true that all beliefs and all myths are worthy of a respectful hearing. It is not true that all folk beliefs are equally valid - if were talking not about an internal mindset, but about understanding the external reality."

Sagan, in his notable attempt to purge the "demons" of pseudo-science and superstition from modern knowledge systems, made the proverbial mistake of throwing out the baby with the bath water. This apparently is an easy thing to do when the "baby" (i.e., the very real scientific observational core of mythology) is hidden by the suds of metaphor and religious symbolism. In fact, Sagan and modern archaeoastronomers (from whom Sagan's ideas were apparently developed) largely miss the point when it comes to the nature of mythology, as did Lévi-Strauss and Campbell.

Traditional cultures from their holistic perspective indeed viewed the heavens as a real cosmic ocean. This was an otherworld filled not only with static astral images of gods, rivers, boats, and the normal sacred pathways of Sun, Moon, planets, and stars, but was alive with sentient beings. The real power and vitality of the celestial pantheon does not come simply from observations of the largely static slow processes of the fixed celestial heavens (i.e., movement of constellations, stars, planets, Sun and Moon). Rather, these deities and beasts are energized and transformed by the observed actions of temporary celestial events, including the physical sounds and fury of cosmic impacts. It is the interplay between the temporary celestial events and the fixed celestial heavens which animated the sky for all past civilizations and which created the primary backdrop for the development of mythology, religion, and science [123]. Admittedly this relationship has been somewhat obscured because in a holistic world these deities also had earthly as well as celestial manifestations, such as storms, lightning and volcanic eruption.

A brief examination of two Old World deities, Set and Marduk illustrates this interaction between temporary celestial events, the fixed celestial heavens, and earthly physical processes. Marduk, the patron god of Babylon [19, 48, 73, 144], is one of many similar examples of Bronze Age deities who suddenly appear or become dominant in the archaeological and documentary record and who have variable popularity through time. Marduk is usually portrayed as an agricultural god, although in both myths and in titular symbols he is represented as a powerful being, sometimes as a storm god. In the striking *Epic of Creation* [48, 74, 176], Marduk battles the evil ocean goddess Tiamat, who personifies chaos. Marduk is described as having four eyes; he breathed fire and his body was filled with ever-blazing flame; "terrible" celestial radiance emanated above him in the form of five bright rays; and his weapons included great winds (tornado and unfaceable facing winds) and the "flood-weapon", and he rides about in a "storm chariot". As a prelude to his battle with Tiamat, he makes an unnamed constellation disappear and then reappear. Tiamat creates 11 zodiacal beasts in addition to herself to take part in the battle with Marduk.

Marduk slays Tiamat with an arrow and then sets about to create a new cosmos, and indeed a new astronomy for the people of Babylon. Marduk then is elevated to the station of king of the gods and he adopts the *mušhuššu* dragon (originally created by Tiamat) as his animal emblem. The *mušhuššu* ("furious snake"), is a dragon-like creature of tremendous power (see Figure 2). The name and the actual depiction of this horned snake-dragon, is strongly suggestive of a post-perihelion comet. In fact, the whole pantheon of Mesopotamian deities (Figure 2) is richly interwoven with astral imagery.

In a similar vein for Egyptian mythology, it is no accident that the Egyptian god Set (Seth) became manifest in the Pharaonic iconography and mythology of the Second Dynasty, immediately following a catastrophic oceanic comet impact (discussed below). Set, the enigmatic and complex god of chaos, has fascinated scholars over the years as to what animal, if any, that his form may symbolize, and as to his role(s) in Egyptian cosmology and religion [24, 25, 62, 91, 113, 161, 205].

Set had a dualistic and largely antipathetic relationship with Osiris (god of agriculture, the Nile, and the underworld) and Horus (god of the Sky and of the morning Sun). Set was commonly viewed in political terms as the major deity of Upper Egypt, while Osiris and Horus were aligned with Lower Egypt. In addition to representing chaos, Set was variously considered the god of the desert; the god of the "cruel sea"; the god of storm; and the god of metal (the "bones" of Set were considered to be made of iron ore). As god of metal and with bones of iron ore, it is probable that Set had a direct association with meteorites. And like the Chinese god of water, Kung Kung, Set was envisioned as having red hair and red eyes, the very personification of evil. Animals associated with Set were dangerous creatures such as the crocodile and the hippopotamus, but also swift animals such as the ass, antelope, pig, and fish.

Returning, once again, to the issue of chronology, it should be obvious that if mythological beasts/deities do represent (at least in part) naked eye visible temporary celestial events, and if certain spectacular events (e.g., bright supernovae) can be viewed over much of the Earth, then this creates a potential for establishing concordances between mythologies and archaeological materials, and thus between the culture histories of various geographic regions.

Despite some compelling arguments put forward by the chronological revisionists, I largely employ the traditional standard chronologies unless otherwise noted. This choice is by no means arbitrary and reflects what I perceive to be strong patterns apparent in my cosmographic data sets across widely disparate regions worldwide. For example, my data support an approximate 2920 BC date for the start of

Egypts First Dynasty, a mid 27th century BC date for the reign of Gilgamesh in Uruk; and approximately 2700 BC for the beginning of the reign of Huang Di in China. Hopefully, the discussions that follow will make clear the reasons for keeping the main tenets of the standard chronologies for Bronze Age cultures. I by no means argue that the chronologies suggested in this paper are absolutely definitive. They will require much future testing and refinement as more cosmographic data sets become established and are subjected to detailed scrutiny.

In order to set the tone for further discussion of Bronze Age chronology and cosmic impacts, Table 1 highlights those physical effects of environmental catastrophe, including cosmic impact and volcanic eruption, which are most likely to leave a lasting impression on eyewitness observers. These physical effects represent the building blocks of mythology and are among the structural determinants for the science of cosmography.

in virtually all cultures. In Western society the most common version of the story is that of the Biblical Great Flood of Noah, but it can be demonstrated that the Flood myth occurs completely independent of Biblical tradition in most non-Western cultures [56, 65].

In many sets of cultural traditions, the Flood is identified as the last in a sequence of global catastrophes. Although other more recent catastrophes are noted in some sets of cultural traditions, these are usually defined as being of less magnitude than the Flood. Likewise, although some sets of cultural traditions discuss worldwide cataclysms prior to the Flood, the Flood is generally viewed as the last such cataclysm and therefore initiated the present age or era in which we now live.

For more than 2,500 years the Flood has been discussed and debated by Western scholars. Ironically, for the past 150 years the Flood has become the linchpin for both fundamentalist religion and for science. Christian fundamentalists,

Table 1: Likely major physical effects with largescale catastrophic environmental events

EARTHQUAKE	MONSOON FLOODS	VOLCANIC ERUPTION	TERRESTRIAL COSMIC IMPACT	OCEANIC COSMIC IMPACT	HUMAN WARFARE
building fires by upset lamps and hearths		house fires by pyroclastic flows, incendiaries, lava	widespread structure fires by thermal radiation	structure fires by thermal radiation, ejecta re-entry	selected structure fires, occasional town burnings
		forest fires by pyroclastic flows, incendiaries, lava	widespread forest fires by thermal radition/ejecta	possible forest fires by thermal radiation/ejecta	selected burning of fields and arboreal crops
tusnami possible along coasts and lake margins	possible coastal storm surges	tsunami with underwater eruption and earthquakes	tsunami may be produced by coastal earthquakes	huge tsunami with inland surges; splash re-entry	
	intermittent torrential rainfall over several days		"black rain" from debris particulates for few hours	steady torrential rainfall for several days	
occasional low dense smoke from fires		high altitude ash plumes	high altitude clouds of debris particulate ejecta	high altitude clouds of debris and steam ejecta	occasional low dense smoke from fires
	thunder and lightning	massive electrical discharges and thunder	massive electrical discharges and thunder	massive electrical discharges and thunder	
earthquakes between ca. 7.0-8.5 on Richter scale		earthquakes possible to ca. 8.5 on Richter scale	earthquakes between ca. 8.0-10.0 on Richter scale	earthquakes between ca. 8.0-10.0 on Richter scale	
	intermittent hurricane force winds	blast wave creates local hurricane force winds	air pressure blast wave and large storm fronts	major cyclonic storms of super hurricane force	
		brilliant sunrises/sunsets	brilliant atmospheric displays day and night	brilliant atmospheric displays day and night	
		low altitude pyroclastic glows with incediaries	high altitude fireball before and after impact	high altitude fireball before and after impact	glow associated with largescale burning
booms and structure collapse noises	roar of wind and rain	claps, booms, whistles, hisses, roars	rapid sonic booms, roars, wind shrieks	rapid sonic booms, roars, wind shrieks	noise of battle
	darkness for couple of days during storm	darkness for days/months from periodic ash plumes	darkness for days/weeks by debris ejecta plume	darkness for days/weeks by debris/cyclonic storms	darkened battle smoke lasting a day or two

6. Oceanic cosmic impacts

As noted earlier, oceanic and large lake cosmic impacts should be at least twice as common as terrestrial impacts given the watery composition of much of the surface of the earth. The fact science has not yet been able to provide definitive evidence for a single oceanic or lake impact during the past 5,000 years is most curious, especially given the prevalence of flood myths worldwide [56, 65] . I suggest that this situation tells us something both about science and about the physical signatures of oceanic impacts.

The 'Flood Comet' Impact of 2807 BC

The myth of a single catastrophic Flood which devastated the whole Earth stands out above all other myths for several reasons. First, it is a truly universal myth, one which occurs

such as creation scientists [143, 208] not only believe that the Flood was a real event, but argue that it occurred exactly as related in the Bible. Indeed, a strangely convoluted mix of geology and fundamentalism has emerged in the attempt by fundamentalists provide a scientific basis for the Flood and for Biblical history in general [202]. Science, on the other hand, has used the presumed absence of geological deposits indicative of a universal Flood event to demonstrate the power and logic of scientific method and theory over that of religious fundamentalism. Equally ironic is the fact that modern social science, historical geology, and even aspects of comet research all at least partly had their roots in the refutation of the Flood myth.

There have been many studies of Flood mythology during the past 100 years. I single out especially that in 1919 by Sir James George Frazer, an eminent graduate of Trinity College and long-time professor of anthropology at Cambridge University [65]. Frazer prepared a detailed anthology of Flood myths from approximately 127 different cultures worldwide, along with at least 41 variants of those myths. Frazer's anthology constitutes the largest single compendium of Flood myths in the English language. This work, along with more recent studies [56], has established beyond doubt that Flood myths are indeed indigenous to most cultures and are not simply the result of diffusion spread by Christian missionaries.

I have elsewhere performed a detailed environmental and cosmographic analysis of the Frazer Flood myths, along with a sizable number of additional myths not available or known to Frazer [125, 126]. The analysis focused on the nature of a number of environmental variables in the Flood myths. These included the intensity and duration of rainfall and hurricane force winds; the obscuration of the sun by darkness; the time, direction and nature of the initiation of the Flood storm; the nature of probable tsunami events; the presence of fire and/or hot rainfall as part of the initiation of the Flood storm; the description and physical nature of supernatural beings associated with the Flood; the season and lunar phase during which the Flood took place; the date or genealogical position of the Flood event in oral traditions and documentary records; and the presence of temporary astronomical events associated with the Flood storm such as eclipses, planetary conjunctions, and meteor storms.

As an example of my cosmographic analyses, I briefly note some of the findings of one category, that of descriptions of supernatural beings or deities associated with the Flood event. The Flood was most commonly attributed to the actions of "giants" or large supernatural horned beings with snakelike bodies. Thus Kung Kung, a Chinese god of water who knocked over one of the pillars of heaven, was described as having "a snake's body with a human head and bright scarlet hair" [212: 9], and is elsewhere described as having a prominent horn on his head [36: 54]. The Vedic Hindu creator god, Prajapati, who saved Manu (humankind's progenitor) from the Flood, took the form of a rapidly growing giant horned fish with a golden body and lotus eyes [65: 183-193]. Sehkmet, the Egyptian lion-headed arrow-shooting goddess of the Memphite triad, was described as having a fiery glow that emanated from her body, and was also associated with the fire-spitting uraeus headdress worn by kings [113: 106].

These and dozens of similar images from a variety of cultures (such as the horned water serpent in North America) are associated with the beginning of the Flood. Each of these images is consistent with the character and general appearance of post-perihelion comets whose tails are thrust in front of the nucleus due to the force of solar winds, thus looking very much like a horn (or a headdress in the case of Hawaii).

The consistency between myths from widely separate geographic regions is striking in other environmental information categories as well. In fact, a precise date for the hypothesized comet impact, 10-12 May 2807 BC, was ascertained [125] from the concordance within the mythologies of calendrical, seasonal, and verifiable astronomical indicators (i.e., lunar phase, the presence of solar and lunar eclipses, and the presence of planetary conjunctions within ascribed constellations). It is of some interest to note that Chinese myths indicate that the Flood event occurred at the end of the life of Nu Wa, consort of first Chinese emperor Fsu Hsi (2953-2838 BC according to standard chronologies), and who herself allegedly lived until about the year 2810 BC. Thus there is only three years difference between the actual impact date and the date generally assigned to the event by at least some Han Dynasty chronologists. The date of 2807 BC likewise fits well with archaeological data from Egypt, as is discussed below.

Even the probable location (Atlantic-Indian basin near Antarctica) and general magnitude of the impact (between 10^5 and 10^6 MT, or 100 and 1000 gigatons) could be roughly estimated from the patterning within the mythology. The Flood Comet impact apparently produced massive tsunami and several days of torrential rainfall worldwide through the injection of water vapor into and through the upper atmosphere, and likewise created vast cyclonic storms which persisted for between five to seven days [125, 126]. The Flood mythology in aggregate (based on the relative numbers of deaths described in each set of myths) suggests approximately 80% of humankind perished as a direct or indirect (i.e., starvation, disease) result of the Flood comet impact.

I fully discuss elsewhere [125, 126] the Flood mythology, the nature of my analyses of these data, cosmic impact modeling, and the nature of supporting palaeo-climatic data, such as that palynology, geomorphology, and from absolute chronological referents such as Bristlecone pine tree-rings. I do not wish to repeat this information and instead refer the interested reader to the two cited papers, which are currently being revised for publication. Of particular note, however, is the fact that the date of the Flood event, 2807 BC, roughly coincides with the boundary of the last significant change in worldwide climate, that between the middle and the late Holocene periods. The transition to late Holocene may have begun as early as 3000 BC, and virtually all geologists agree that modern climate conditions were fully in place by 4000 radiocarbon years ago, which calibrates to about 2500 BC. This date range also encompasses the hypothesized beginning date for the El Niño/Southern Oscillation [165]. I believe that the Flood comet impact is largely responsible for both the change to modern climate and the beginning of the El Niño/Southern Oscillation cycle [126].

In the remainder of this section I focus on the archaeology of the Flood Comet impact. Of immediate concern is that despite the suggestion based on the mythology that four out of every five persons died at the time of the Flood Comet impact, none of the other authors in this volume identified a "catastrophe" of any kind as having occurred in the archaeological record at around 2800 BC. Likewise, although a 10^5-10^6 MT cosmic impact should generate long-term climate effects, none of the other authors in this volume identified climate change at around 2800 BC, although several authors suggested climatic disruptions at other times during the Bronze Age. Logic therefore dictates that either I have misdated the Flood Comet impact or that there was no such single "Flood Comet impact" at any time.

Three major circumstances together form a rationale for the apparent "invisibility" of the 2807 BC Flood Comet impact in the archaeological and geological record. The first is the simple fact that the 2807 BC impact date falls into a period of time during which there was an unusual rapid rise in natural radiocarbon production [152, 183, 184]. Carbon samples formed during the year of the comet impact could be represented by radiocarbon ages anywhere between 4300 and 4080 BP. In this particular case, even fine-tuned high precision calibration of the radiocarbon samples from deposits containing impact-related charcoal would likely yield a wide distribution of radiocarbon ages. This would make it difficult to directly compare and link archaeological strata from different sites and different regions. The impact may even have somehow contributed to the increase in radiocarbon production by injecting large amounts of nitrogen into the atmosphere thus forming a larger than normal pool of nitrogen atoms subject to conversion by cosmic rays into carbon-14.

A second confounding situation is that humans, like most mammals, have the capacity to rapidly rebound after catastrophic population loss. If 20th century population growth rates of 2.0% per year [211: 317-351] were applied to the presumed post-comet impact period, it would take less than 85 years, only four human generations, to fully recover the pre-impact population levels. In addition, both fertility and fecundity can increase under certain conditions, such as during periods of improvements in health and increases in food supply, thus the growth rate can be considerably higher than 2.0%. Likewise, population growth rates can increase when cultural behaviors are modified. In this regard, a large number of Flood myths indicate that incest prohibitions were removed during the generations immediately following the Flood [125]. When these possibilities are coupled with the fact that the dating of most archaeological sites from this general time period is quite course-grained, the deaths of 80% of the human population at 2807 BC may be difficult to observe in the palaeo-demographic record.

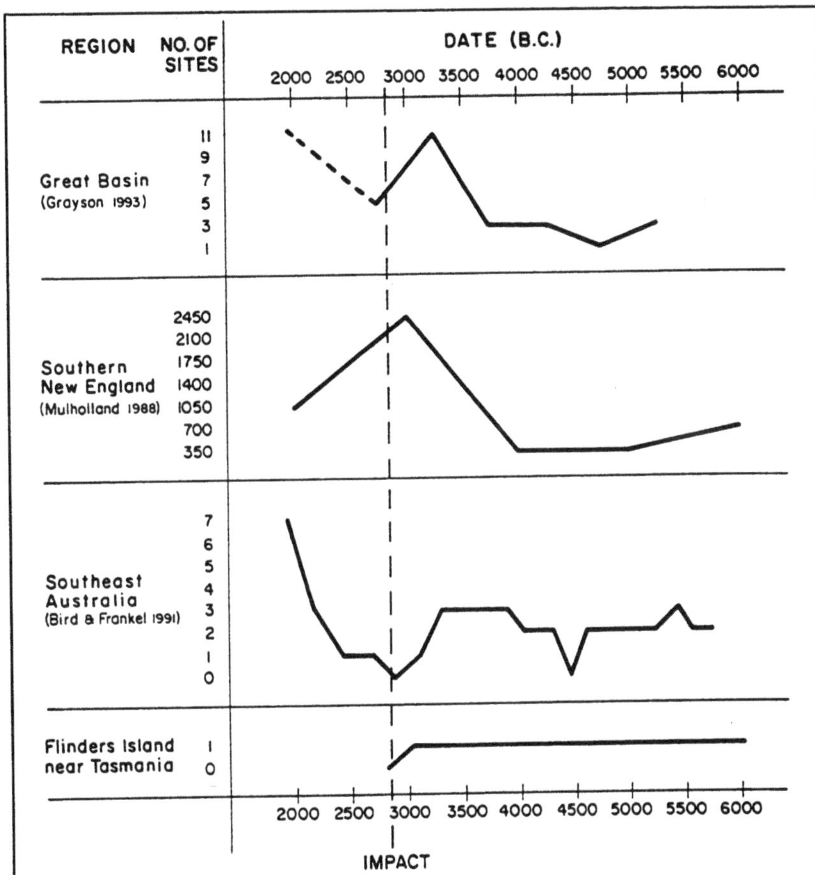

Fig. 3: Rough population estimates for selected geographical regions before and after the hypothesized·2807 BC oceanic comet impact. Based on total numbers of archaeological sites assigned to a given period (by the authors of the cited work). Flinders Island (J. Diamond, *Discover Magazine*, 1993) based solely on presence/absence of cultural remains.

For example, using a sigma (variance) of only 50 years for pertinent radiocarbon ages, good by radiocarbon dating standards, hypothetical population profiles divided into 100-year increments and based on an 80% die-off in 2807 BC would not dip below 63%. Such a signature would be difficult to distinguish from more "normal" population shifts caused by changes in settlement and subsistence patterns or by warfare, the most common non-catastrophic explanations invoked to explain demographic change in the archaeological record. Despite these difficulties, crude regional demographic studies have been undertaken in the archaeological record of North America and Australia that show an apparent dip in population (usually expressed as numbers of archaeological sites) at around 2800 BC (Figure 3). These results at the very least do not preclude the possibility that human population did indeed suffer a sizable die-off at about this time, and that 80% extirpation is not completely unreasonable. And of course, if the human mortality of the hypothesized Flood Comet impact was somewhat lower, say around 50% rather than 80%, this halving of humanity would become virtually impossible to recognize and measure in the archaeological record without greater precision and control over stratigraphy and dating than currently practiced.

The third circumstance adding to the present invisibility of the Flood Comet impact is the previously mentioned fact that a regionally or globally catastrophic oceanic impact will likely leave a very different palaeo-environmental and archaeological signature than would a terrestrial impact. Except for the actual impact crater itself a terrestrial cosmic impact would tend to primarily affect the surface of the ground and objects on that surface (i.e., buildings, trees, people) rather than to alter the subsurface. Despite the severity of a terrestrial impact, surface cultural materials will largely remain *in situ*, although these materials may show the disruptive effects of atmospheric pressure waves, thermal radiation, and earthquakes associated with the impact.

Oceanic or large lake impacts, on the other hand, have the ability to produce vast columns of water through coastal tsunami, by torrential rainfall from the injection of water vapor into and the supersaturation of the upper atmosphere, and by the re-entry of splash ejecta into the lower atmosphere. The injection of such large volumes of water vapor into and occasionally through the atmosphere conceivably may serve to purge the atmosphere of certain types of particulates (e.g., dust) that might be deleterious to the growth of surviving plants, although the formation of upper atmospheric ice crystals may partly obviate this benefit. Floods are notorious for their ability to alter landscapes by cutting, filling, and slumping, and no two landscapes will receive identical treatment. In addition, oceanic impact craters are difficult to detect. As of 1993, occanic craters made up only about 2% of the approximately 140 Earth impact craters known by that date [77].

Although circumstances conspire to make it difficult to perceive the location or the physical signature of an oceanic impact, there are indeed several aspects of the Bronze Age archaeological record which appear to support a 2807 BC oceanic impact. An example is the massive flood deposit discovered earlier this century at the site of Shurrupak (Fara), home of the legendary Mesopotamian Flood hero Atrahasis [105]. This deposit is placed in the middle or latter part of the Early Dynasty I period [209], which scholars variously date between 3000-2750 [157] and 2900-2700 BC [103]. It is possible, although not certain that similar flood deposits date to this time period at Tell Oheimir, the site that many scholars believe to be legendary Kish [156: 110]. This possibility is given added poignancy in that the *Sumerian King List* states [146: 265]: "After the Flood had swept over (the earth) (and) when kinship was lowered (again) from heaven, kingship was (first) in Kish". The present estimated dating for the earliest substantive deposits and construction at the archaeological site believed to be Kish is indeed at around 2800-2700 BC [156].

However, not all flood-related deposits in Mesopotamia date to this time, such as the massive deposits at Ur [209] dating to the Ubaid period (ca. 4000 BC), and at least some deposits at Kish date to around 2650 BC [156]. I would expect that most alluvial flood deposits in the archaeological record are related to factors other than oceanic or large lake cosmic impacts. Nevertheless, based on our current understanding of cosmic impact, it can be assumed that at least some of the deposits in Mesopotamia and elsewhere have a basis in cosmic impact events, including impacts other than that of the Flood Comet itself. Due to the biases prevalent among archaeologists and geomorphologists (and among scientists in general) concerning the historicity and basic reality of the Flood, much remains to be accomplished before alluvial deposits in Mesopotamia and elsewhere will be thoroughly studied and satisfactorily evaluated against the cosmic impact paradigm.

A date of 2807 BC would fall near the end of the First Dynasty of Egypt, which is currently dated by many Egyptologists to between about 2950-2920 and 2800-2770 BC [172, 205]. If we were to split the difference (2935-2785 BC) and to assign an arbitrarily equal reign to each of the eight kings of this dynasty, then the seventh king, Semerkhet, would have ruled between approximately 2821-280 3 BC. The 3rd century BC historian, Manetho (according to Eusebius), provides us with this brief but notable chronicle for Semerkhet [193: 33]: "In his reign there were many portents and a very great calamity."

Nowhere is this "calamity" precisely defined in historical documents, but several lines of evidence suggest that the Flood Comet impact terminated the reign of Semerkhet. Among the indicators of a untimely end for Semerkhet is the fact that of all of the First Dynasty Kings and the first two kings of the Second Dynasty, he is the only one not to be represented by a mastaba tomb at Saqqara [84]. Another more subtle difference is based on Paul Reisner's detailed examination of servant or retainer burials at Abydos associated with First Dynasty royal tombs [86: Table 12]. Reisner noted that while only a portion of the retainers associated with the second through sixth Kings exhibited evidence of *sati* or human sacrificial burial, this percentage is increased to 100% with that of Semerkhet and that of the eighth and final King, Qa'a. In addition, the tomb of Qa'a exhibits an unusual episode of remodeling during its construction. These and other bits of information noted below suggest that the Abydos tombs, which likely housed the actual bodies of the kings, were built early during the reign of each king, and were then consecrated on the death of each king by the act of sacrificing selected retainers to accompany the king into the afterlife. The Saqqara tombs, apparently built to honor royal officials upon the death of each king, were seemingly not constructed until later in the reign or upon the death of the pertinent king.

If these scenarios are correct, the fear created by the Flood Comet impact may have enhanced the need for retainer sacrifice at Abydos, and would have obviated the need to erect a post-death tomb for his officials at Saqqara. Also, the remodeling of the tomb of Qa'a may have been necessitated by actual damage sustained from the torrential rainfall or other effect of the Flood Comet impact. Also, since we know that Egyptian kings were closely associated with celestial phenomena, the differential mortuary treatment given Semerkhet may simply be indicative of being too closely associated with the Flood Comet impact and having fallen out of favor in the eyes of his priests and by the general populace. Curiously, the term *merkhet* means 'astronomer's staff' [70: 190-191], but it is unclear if this term can be logically derived from the name *semerkhet.* More significant from an etymological standpoint is the possibility that the title of the succeeding king, Qa'a, translates as meaning 'abundance' in the sense of 'flood' [198].

Of considerable interest is the pattern of destruction visible for First Dynasty tombs at both Abydos and Saqarra, as well as for other First Dynasty structures elsewhere, such as the upper class graves at Helwan [161, 176]. First Dynasty structures consistently show the effects of *intentional* burning, and most, especially those at Helwan, appear to have been razed. This strikingly differs from the Second Dynasty tombs which do not show the effects of fire and the walls of which still stand substantially above the ground surface. Also of interest is the fact that the fragmentary Palermo stone [198] records both a break in the recording of Nile River water levels after the reign of Semerkhet as well as a lowering of the water level of the Nile by approximately 1.0 meter between the reign of Semerkhet and the middle of the Second Dynasty [22: Fig. 4]. While by no means conclusive, these patterns are logically consistent with thermal radiation, flooding, earthquakes, and climate change associated with a cosmic impact at the end of the First Dynasty.

In the Palestine archaeological record, the Flood Comet impact would have occurred during the middle of the Early Bronze II period, which has been assigned a date of approximately 3050-2700 BC [132, 178]. The end of this period is notable for the unique large-scale movement of people (50% of all regional population) to the hill country of Galilee, Samaria, and Judah [132: 112]. In fact, the period of 2800-2700 BC witnessed substantive population movement and changes in settlement systems worldwide. For example, settlements in the southern Sinai and Negev contain extensive evidence of contact with Egypt during the First Dynasty, but this dramatically disappears in the Second Dynasty [159: 33-37, 132: 108-117]. In Central Europe, the abrupt appearance and rapid spread of the Corded Ware/Single Grave complex has been dendrochonologically dated to between 2800-2750 BC [203]: 284-288].

Indeed, the Flood Comet impact conceivably is responsible for the widespread and rapid dispersal of at least four large language groups worldwide, along with the material and socio-cultural trappings represented by these groups. In each case the initial spread of these languages has been independently judged to have occurred during the period of 3000-250 0 BC. These language movements include the spread of Indo-European language speakers throughout Europe [116, 117, 160], the spread of Austronesian speakers throughout the western Pacific [14: 121-132], the spread of Bantu speakers throughout Africa [155], and the spread of Uto-Aztecan speakers in Mesoamerica and the Southwest (Jane Hill, personal communication 1996).

Languages dispersals are not be surprising given the potential for "bottlenecking" of language and population created by the hypothesized globally catastrophic oceanic comet impact at 2807 BC. In fact, many of the Flood myths themselves describe a proliferation of languages and dialects once populations had stabilized and started to grow again after the Flood [125].

Another indication of population/language flux is attested by the sudden appearance of thick walls and fortifications. Such features are often seen as a response to the threat or reality of raids and other unpleasant incursive encounters by peoples who have been displaced from their homelands or who otherwise covet the resources of their perhaps more centralized and sedentary neighbours. Hadrian's Wall in Britain, the Great Wall of China, and the so-called "Amorite Wall" in Mesopotamia [157: 42-43] are examples of such defenses.

Between Early Dynastic I and II and in the latter portion of Early Bronze II, the archaeological record exhibits the first wave of massively walled cities and other substantive fortifications springing up throughout the Near East and Mesopotamia [2, 132, 159] and perhaps southern Europe and the Mediterranean as well [53, 203], including apparently at the Egyptian city of Hierakonpolis [173: 26]. It is likely that the first massive community walls in China were erected at this time [32: 248-249]. This is not to say that walls and fortifications did not exist prior to 2800 BC, especially in Mesopotamia, but rather that the period of 2800-2600 BC witnessed an unprecedented proliferation and strengthening of such features.

Other possible Oceanic Cosmic Impacts

The Flood Comet impact is not the only "flood" event described in mythology and historic traditions, although usually it is considered the greatest of flood events if multiple floods are recorded within a single tradition, or the last of the worldwide flood catastrophes. Native American traditions, especially those from Mesoamerica and North America, typically encode three and occasionally four creations prior to our modern era, the latter which arguably began with the Flood of the Flood Comet impact [125]. The prior creations are associated with various combinations of the elements fire, flood, and hurricane force winds, all of which are components of cosmic impact. Hindu traditions likewise include previous sets of creations, with four different ages or eras belonging to each set. The third age was destroyed by water, and we are now in the fourth or last age of the current set of creations. And as noted in the quote at the beginning of this paper, Egyptian traditions include recognition of multiple catastrophic floods and conflagrations, but contend that their effects have not been as severe in Egypt as in surrounding regions.

Chinese traditions appear to record multiple flood episodes which are a bit confused [36: 84-90], but may represent four somewhat distinct flood events [18: 146]. These seemingly begin with the previously discussed Flood Comet impact of 2807 BC, associated with the knocking over the pillar of heaven by the god of water, Kung Kung, leading to the collapse of the sky and its eventual repair by empress Nu Wa at the end of her reign. The second, also attributed to Kung Kung, may have occurred during the reign of Chuan Hsu (2514-2436 BC according to standard chronology). The third and fourth floods are associated with the successive reigns of legendary emperors Yao and Shun [212: 92]:

> "The whole world was submerged and all the world was an endless ocean. People floated on the treacherous waters, searching out caves and trees on high mountains. The crops were ruined and survivors vied with fierce birds and beasts for places to live. Thousands died each day".

The myths are associated with the heroic efforts to stop the floods by Kun, and especially by his son Yu, the later who succeeded Shun as emperor of China. In some versions of the myth, Kung Kung is again seen as the villain who created or enhanced the flood. The myths seem to indicate two or three peaks to the flooding, which may represent point events (such as would occur from cosmic impact), but in general the flooding is described as occurring off and on over several decades of time as opposed to being a single flood occurring during a single given year. The dating of these flood "peaks" and that of the general period of flooding itself seemingly includes the last portion the reign of emperor Yao and the first 22 years of the co-regency of Yao with his successor Shun [195: 96; 212: 91-97]. This translates to around 2297-2265 BC according to standard chronology, and is potentially significant because it falls within a general time period (2350-2000 BC) corresponding to several defined impacts in Mesopotamia and the Near East, discussed below in the section on terrestrial cosmic impacts.

Greek mythology and early Christian tradition each recognizes three major flood events [65: 157-174], although it is not entirely certain that these represent different events within each tradition. The earliest Greek flood event is that during the reign of Ogyges (Ogygus), said to be the first king of Thebes in Boetia. The second and most familiar of the Greek flood legends is that of Deucalion, a legendary king of Phthia. Both Frazer [65: 146-157] and Graves [72: 138-143] suggest that the flood of Deucalion is the same as the Noachian Flood, thus this event would be that of the Flood Comet impact of 2807 BC. The third flood event is attributed to the time of Dardanus, a legendary king of Arcadia who was driven from his homeland by a flood which submerged the lowlands [65: 163], and who later helped found the city of Troy [72: 621-630]. The current view of the infamous destruction of Troy, immortalized by the Greek poet Homer, is that it occurred sometime around the 12th century BC. This then should put the flood event of Dardanus in the century or two before that time.

Roman and early Christian scholars recognized the floods of Ogyges and Deucalion as being separate from the Noahs Flood, but seemingly did not recognize the flood of Dardanus [65]. The Roman scholar Varro, writing around 36 BC, indicates that the flood of Ogyges occurred some 2100 years before his time, or around 2136 BC. It is a curious coincidence that shortly before the destruction of Sodom and Gomorrah (an apparent terrestrial impact described below) at around this same general time period, Abraham lived "near the Oak called Ogyges" [94, I: 35]. However, other authors, such as the Christian scholar Julius Africanus writing in the third century AD, provide other dates for the Ogygian flood, which according to Africanus occurred 1020 years before the first Olympiad, or at about 1796 BC [65: 158]. Isidore, bishop of Seville at the beginning of the 7th century AD stated that the first flood was that of Noah, followed by the flood of Ogyges at the time of the patriarch Jacob, and with the last flood being that of Deucalion at about the time of Moses. The Church historian Eusebius, writing in the fourth century AD, believed the flood of Ogyges to have occurred 2200 years after the flood of Noah, and was followed some 250 years later by the flood of Deucalion.

It is clear that the dating of these several flood traditions is confused and is not to be trusted without a much more detailed and exhaustive cosmographic study. However, what does seem to be clear is the fact that early scholars recognized several catastrophic flood events separate from that of the Flood Comet impact (i.e., the flood of Noah), which for the most part were not quite as devastating as the Flood Comet impact.

There is much still to learn about oceanic and large lake impacts. The Flood Comet impact demonstrates that the very recognition of such impacts in the palaeo-environmental and archaeological record is fraught with difficulty. Of course it is also important to remember that not all flood events in mythology and oral history necessarily relate to cosmic impact. The Mediterranean and Near East are crossed by major faults and the boundaries of crustal plates (see Nur, this volume) which have the potential to create horrific tsunami, through earthquakes and volcanic eruption. It will undoubtedly take years of detailed study of coastal (including lakeshore) margins and other appropriate geological deposits to begin the process of identifying and separating out tsunami caused by cosmic impact from tsunami caused instead by earthquake, landslips, and volcanic eruptions.

7. Terrestrial cosmic impacts

Before exploring the possibility of Old World Bronze Age terrestrial cosmic impacts, I discuss two late Holocene period terrestrial impacts in the Americas which are instructive regarding the nature and consequences of terrestrial impacts as well as illustrating the richness of details preserved in oral traditions. The first is a substantive regionally catastrophic impact in northern Argentina perhaps dating to around 2000 BC, with the second being a locally catastrophic impact in Nebraska likely dating to the beginning of the first millennium BC.

The Rio Cuarto, Argentina Impact

The cover of the 16 January 1992 issue of the journal *Nature* was graced by a remarkable aerial photograph of a portion of a meteoritic crater field in the Rio Cuarto watershed on the Pampas of northern Argentina [170]. This crater field consists of a linear alignment of more than 11 distinctly elongated craters covering an area roughly 50-km long and 10-km wide, with the largest individual crater being 4.5 km long by 1.1 km in width. Based on the nature of the craters and on experimental evidence, it was deduced by the authors that the crater field was formed by a disintegrating object with an angle of impact of less than 7° from the horizontal, and which was traveling from the northeast to the southwest. The energy released during the impact was very conservatively estimated at 350 MT, or about 23 times the magnitude of the 1908 Tunguska impact or 1150 times the magnitude of the Hiroshima atomic bomb. In terms of geological dating, while the authors of the article did not attempt to precisely define the date of impact, it was clear that it took place during the last 12,000 years of the Holocene period.

In the mythology of South America are pan-regional myths about natural disasters which destroyed most or all of human life at some time in the remote past. The most prevalently mentioned disasters are the Great Flood and the "Great Fire" (or "World Fire"), with other disaster motifs including the "Great Darkness" (or "Long Night") and the "Great Cold" [16]. Each disaster is viewed as a distinct event separated by lengthy periods of time, although a few of the Flood myths indicate the Flood was immediately preceded (1-3 days?) by a conflagration which appears to be different from the Great Fire.

The myths regarding the Great Fire clearly encode a cosmic impact, as is evident from the Toba and Toba-Pilagá cultural groups in the Gran Chaco region of western Paraguay, southern Bolivia, and northern Argentina [204]:

> "The people were all sound asleep. It was midnight when [...] the moon [took] on a reddish hue [...] Fragments of the moon fell down upon the earth and started a big fire. From these fragments the entire earth caught on fire. The fire was so large that the people could not escape. Men and women ran to the lagoons covered with bulrushes. Those who were late were overtaken by the fire. The water was boiling, but not where the bulrushes grew. Those who were in places not covered with bulrushes died and there most of the people were burned alive. After everything had been destroyed the fire stopped. Decayed corpses of children floated on the water. A big wind began to blow and a rainstorm broke out. The dead were changed into birds. The large birds came out from corpses of adults, and small one from the bodies of children. / Norkaló is the great fire, Netatlta is the great flood./ The great fire was coming, and in order to escape it the people crawled into a very deep hole in the ground. The entire earth caught fire and was burned; everything burned, even the fish. With everything on fire, the intense heat brought the water in the lagoons and rivers to a boil; the fish were burned and died, and the boiling water turned into steam./ When the great fire was over the people had to wait before getting out of the hole, for everything was still hot. [...] / The whole earth lay in ashes; nothing was left but a tiny tree, a little algarrobo tree. / When the earth was full of people a dog appeared. It was mangy and dirty, but it had a beautiful face, with a beard like a monkey's. It was covered with mud and fleas as well as being mangy. Nobody liked the dog; people feared that it might be carrying some disease. / A man called the dog. He had been notified by his daughter who seemed to know who the animal really was. Affectionately he covered the dog with his mantle and invited it to stay in his house. As it was already night, the man went to sleep. / That same night a handsome and well-dressed man appeared from the dog, and he said to the kind man Get up, my son, for tomorrow all those people will be burned, and no mercy will be shown them! That dog was God. / Those people will be burned, he said, but you are going to dig a well for yourself, and you will stay inside it with whole family until the fire is over. Put a gourd outside, and when it bursts the fire will be burned out. Still, you mustn't go out right away, for if you're in a hurry you will emerge in the form of an animal. You will find it easy to stay inside for one day, but you must wait for three days. During that time you must not go out! [...] / When the remaining people emerged after the three days had passed the earth was completely bare, as it had been after the flood, except that now there were ashes everywhere. These people began to multiply; every year another child was born, until once more there was a large population." [204: 64-79]

As noted by Metraux [146: 36]:

> "Outside the Chaco the story of the World Fire was found among the following tribes: Tupinamba, Apapocuva-Guaraní, Tembé, Shipaya, Caraja, Mura, Cashinaua, Witoto, Paulipang, Arawak, Yuacare [...]. Various causes are given for the disaster. According to some stories it was sent by the culture hero or some supernatural being (Yuracare, Tupinamba, Arawak). In the opinion of the Cashinaua it was started by a spark fallen from the sky. The Tembé say that a child set a river aflame, and then the earth, with a 'candle' given to him by a mysterious stranger. The Witoto provoked a general conflagration when they attempted to burn the sky god Huisiniamui. The Apapocuva-Guaraní believe that a Big Fire will sweep the world soon after 'Our Father Great' removes one of the props which support the earth. [....] / The only survivors of the cataclysm were the individuals who hid in some underground cave or dwelling (Arawak, Mura, Yuracare). In the Tupinamba version, the culture hero saves only one man, by taking him to the sky. The Tembé say that a pregnant woman and a boy escaped death by hiding in a banana grove. Later they found untouched manioc cuttings which they planted anew. The Mura family was helped in their distress by a spirit who gave them water and plants."

John Bierhorst [16: 19] has defined a number of broad cultural areas in South America based on the sharing of myth motifs and on the general coherence of mythological tale types. These "mythological regions" are depicted in Figure 4. A striking aspect of this attempt to create cultural groupings based on myths is the fact that the area in Chile and Argentina immediately south of the Gran Chaco lacked coherence and therefore could not be placed into a defined

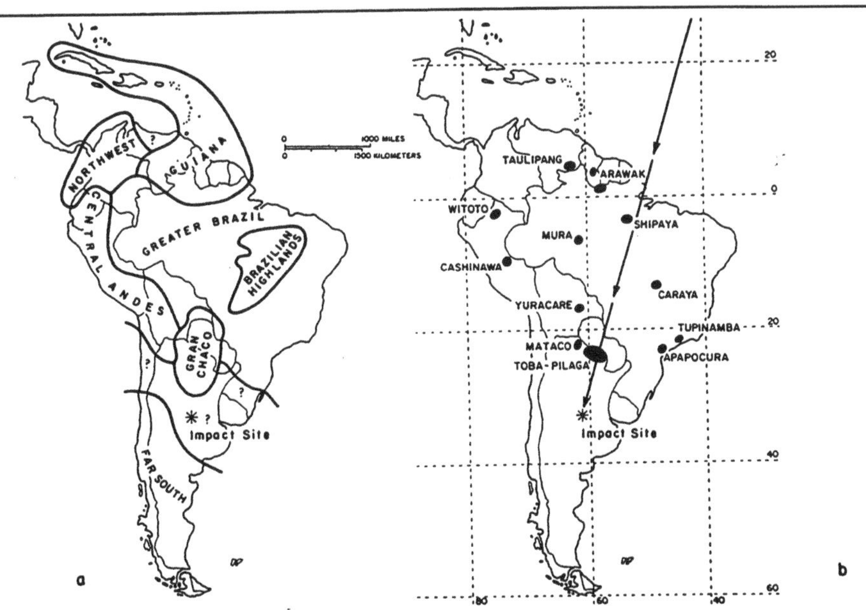

Fig. 4: Comparison of South American "mythological regions" with the distribution of "Great Fire" myths and the location of the Rio Cuarto impact crater field. Figure a depicts the boundaries of mythological regions (adapted from Bierhorst 1988: 19), Figure b depicts the historic locations of South American tribal groups with "Great Fire" myths in relation to the reconstructed path of the Rio Cuarto impactor (derived from Schultz and Lianza 1992).

mythological region. Linguists have likewise met with little success in their attempts to define historical language groups in this region [180: 23]. This mythological and linguistic "no man's land" takes on added significance when it is realized that the Rio Cuarto impact crater field lies at its very heart (Figure 4).

It is virtually certain from the reconstructed path of the Rio Cuarto impactor and from the distribution of Great Fire myths that the Rio Cuarto impact is the source of the Great Fire myths. Since these impact myths are part and parcel of Bierhorsts mythological regions, the failure by Bierhorst to place northern Argentina and northern Chile into one or more definable regions is undoubtedly related to the level of direct disruption of cultural systems caused by the Rio Cuarto impact event.

The sample of Great Fire myths noted here come from ral groups representing at least nine completely separate linguistic families: Arawakan (Arawak); Bora-Huitotoan (Witoto); Cariban (Taulipang); Guaycurú-Charruan (Toba-Pilaga); Macro-Ge (Caraya); Macro-Pano-Tacanan (Cashinawa, Yuracare); Mataco-Maccá (Mataco); Mura-Matanawí (Mura); and Tupian (Apapocura, Shipaya, Tembe, Tupinamba). Therefore, one cannot assume a simple diffusionist model within one or two language groups by which to explain this distribution.

Indeed, the widespread distribution of so many Great Fire myths across such a large and linguistically complex region suggests that the direct physical effects from the impact (e.g., thermal radiation, air pressure blast waves; ejecta) may have been felt over an area of between 1,000,000 km^2, the approximate size of the disrupted mythological and linguistic region, to perhaps more than 4,500,000 km^2, the latter representing the area in which myths relate a great loss of life, but which is only roughly half of the maximum extent of Great Fire myth distribution. Assuming that 4,500,000 km^2 represents the general extent of direct impact destruction, this is an area 6.5 times the size of the State of Texas or more than 14 times the combined size of the United Kingdom and Ireland. This suggests that the 350 MT figure derived for the impact by Schultz and Lianza [170] is substantively underestimated. It is possible that the size of the affected Rio Cuarto impact area may have been exaggerated by the extremely low flight angle of the impactor, or perhaps additional as yet undefined hits from the fragmenting impactor occurred in Brazil, therefore the original object was much larger than currently hypothesized.

As for dating the Rio Cuarto impact, the South American myths that I have scrutinized are somewhat equivocal regarding the order in which the various types of "world-wide" disasters occurred, although there is a tendency to place the Flood prior to the *Great Fire*, and to put both of these events prior to the *Great Darkness*. The Flood Comet impact is securely dated to 2807 BC, while the Great Darkness possibly correlates with a year-long diminishing of sunlight observed throughout Europe and the Middle East in AD 536-537 [169, see also Baillie, this volume].

More recent analysis of the impact site by Peter Schultz apparently places the dating tentatively at around 4000 BP (Peiser, this volume), which fits comfortably within the date range provided above. An absolute date for this event may be possible through the examination of ice-cores and tree-rings, and from a more exhaustive of study of the mythology. For example, one step toward potentially more precise dating is the realization that the impact coincided with a partial lunar eclipse [204: 69]:

> "Long ago Moon was attacked and wounded, and thus the great fire originated. As soon as people noticed blood on Moon [i.e., the reddish cast of a lunar eclipse], they started to chant and to shout and they struck their dogs to make them bark. [...] A fragment of Moon fell down and caused a fire."

The Broken Bow, Nebraska Impact

A 1.6-km wide depression west of Broken Bow, Nebraska, and some 300 km southwest of the badlands of South Dakota, has recently been pointed out as a candidate Holocene period impact crater [140]. This suggestion has met with skepticism, including that by the late Eugene Shoemaker. The estimated date of the geological formation is about 3,000 years old (1000 BC), and the energy released from an impact of this size would have been more than one hundred times that of the Tunguska impact.

Despite the current aura of skepticism by some specialists, there is, in fact, evidence from Native American oral traditions to support the impact crater claim. The following is from the 1937 publication of a tale collected earlier from then very old Sioux oral historian, Red Bird [40: 309-310]:

> "Long ago the Badlands did not exist. Instead there was a high plain covered with grass or with trees. Much game could be found there - deer and elk and buffalo. So each autumn many tribes camped there to hunt and to make ready their winter supply of meat. Tribes that were at war with each other at other times of the year met peacefully on the plains in the autumn and enjoyed friendly powwows. Quarrels were forbidden. / But after many snows, a fierce tribe from the mountains toward the setting sun came and disturbed the peace of the plains. Wanting the high tableland for themselves only, they drove off the other people. They claimed all the grass for their own horses [sic], killed the best of the game, and refused to allow other tribes to set up camp as in the old days. / Again and again the Sioux and their friends from the north and the east fought for their ancient hunting ground, but the mountain tribe from the west defeated them. Many of their warriors died in battle. At last the men of the plains held a great council of all the tribes. To appease the spirits, people fasted for many days. Some tortured themselves. The medicine men did everything they could to win the favor of the spirits, but they received no answer. Their people were in despair. / At last the spirits sent their answer. The midday sky became as black as a midnight sky. Lightning flashed, seeming to come from the ground, and thunder rolled. Strange fires lighted the entire country with their flames. The earth shook. Where the western tribe had their camp, waves of land like waves of a great water rolled back and forth. In the troughs of these gigantic waves, the whole warring tribe with their tipis and their horses [sic] sank and were engulfed by the next waves. With them went the grass, the trees, the streams, the game animals - everything that had been on the high plain. / As the frightened people from the north and the east watched from a distance, the great commotion suddenly stopped. The fires burned themselves out. When dawn came, the peace loving tribes saw that the rolling waves of earth had become fixed, bare rocks. Their beautiful plain was now a wasteland on which nothing could ever grow."

As noted earlier, a similar legend existed among the Cheyenne, as part of a long oral history of the Cheyenne people published by George Dorsey in 1905 [60: 113]:

> "The people returned to the south and lived as well as they could, in some years better, in others worse. After many hundreds of years, just before the winter season came, the earth shook, and the high hills sent forth fire and smoke. During that winter there were great floods. The people had to dress in furs and live in caves, for the winter was long and cold. It destroyed all the trees, though when spring came there was a new growth....The people continued to live in the south, and they grew and increased. There were many different bands with different languages, for the red men were never united after the second flood."

The hypothesized Broken Bow impact crater is admittedly a substantive distance (300 km) from the "Badlands" presumably being described in these epic tales. However, an impact near to the Badlands certainly could be easily linked to the badlands themselves by the very fact that the area of devastation around the impact crater would appear similar to the desolate Badlands. Alternatively, the Badlands may have themselves been directly affected by fragments of or ejecta from the Broken Bow impact, or by an as yet unidentified separate impact.

Armed with knowledge of the cosmographic parameters of the Flood Comet impact of 2807 BC and the terrestrial impacts in Argentina and Nebraska, I now turn to the evidence for other impacts during Bronze Age times. The emphasis of this exercise is to compare roughly contemporaneous myths from widely separate regions which treat cosmic battles or natural disasters, and to model the possibility that they may encode the same events. Another goal of this exercise is the attempt to refine Old World Bronze Age chronologies.

8. Celestial events of circa 2700-2650 BC

Several distinct sets of epic tales from widely separated cultures and regions are seemingly reliably dated to the 27th century BC. In each case the main plot of the story is said to have taken place within a few generations after a great flood had swept the Earth. These tales are remarkable not only because they are among the earliest known epic literature, but also because of the sobering fact that there are several general consistencies in their story lines. These parallels are numbered below for sake of discussion. Although the evidence for actual cosmic impacts based on these particular stories is somewhat weak, the analysis performed here demonstrates the utility of the cosmographic approach to elicit details of celestial events encoded in worldwide mythology.

China: Yellow Emperor, Huang Di

I begin with myths surrounding Huang Di, the legendary "Yellow Emperor" of China [212]. The standard chronology places his reign at about 2697-2597 BC [194, 200]. My choice of Huang Di as the first epic for discussion is due to the simple fact that the astronomical connections between his exploits and likely contemporary celestial events are made surprisingly clear in the *Bamboo Annals* [106]. The Bamboo Annals are the 3rd century BC official history of the State of Wei, which escaped the notorious burning of historical texts by emperor Ch'in Shih-huang-ti in 213 BC [45] by virtue of being buried in a tomb prior to that date. Therefore, the Bamboo Annals story line of the Huang Di epic serves as a springboard for discussion of the other geographically separated epics [106: 108-109]:

(1) *Ursa Major Supernova or Nova-like Event* (ca. 2697 BC). Huang Di, as is the case for virtually all demigods and culture heroes, had a miraculous birth. His mother became pregnant after witnessing "a great flash of lightning" around the star, Dubhe (part of the bucket of the Big Dipper asterisim in Ursa Major). It is clear from other stories that this event was an exceptionally bright nova or supernova, which I suggest may have been the event which led to the creation of the modern Owl Nebula (see discussion below). This event occurred at around 2697 BC based on the assumption that Huang Di began his "reign" at birth. The presumed nova event may have been visible for at least 25 months, the length of time given for Huang Di's gestation period. Huang Di could speak at birth, his countenance was "dragon-like", and he had four faces.

(2) Simultaneous Supernovae or Nova Events (ca. 2677 BC). In the 20th year of Huang Di's reign, "brilliant clouds" appeared in the heavens. The Bamboo Annals note:

> "The auspicious omen of brilliant clouds was in this way: The vapours of the red quarter [the south] extended

so as to join those of the green [east]. In the red quarter were two stars, and in the green, one; - all of a yellow colour, which appeared on a day when the heavens were clear and bright, in Shê't'e [likely *shê-thi-ko*, the first year in the 11-year Jupiter cycle; see 141: 251-252], and were named the brilliant stars. The emperor in yellow robes fasted in the Middle palace" [circumpolar stars immediately around the pole star, then in Draco].

This passage would appear to encode three simultaneous nova or supernova events. Based on the names and nature of the Gods associated with these three stars, I elsewhere have suggested that the events took place in the constellations Sagittarius, Draco, and Leo [123: 474].

(3) Taurid Meteor Storm (ca. 2647 BC). The Bamboo Annals state that in the autumn of the 50th year of Huang Di's reign, the heavens were "wrapped in mist" for three days and three nights. Along with the mist there appeared a number of male and female phoenixes (*Feng* bird), unicorns (*Ch'i-lin*), unidentified creatures with four horns termed *Lo*, and creatures termed *yin* worms which appeared like rainbows.

The nature of this "mist" is not clear, but there is reason to believe that it encodes a meteor shower, as suggested by the following imagery of part of a celestial battle between Huang Di and Chi You [212: 43]:

"These wars were full of tricks and subtle maneuver. One time Chi You [Chhih-Yu] covered the whole battlefield with thick, white fog until the Yellow Emperor's troops could see nothing at all. The horned oxen troops appeared and disappeared in the fog, slashing, slicing, stabbing into thin air. Unearthly shrieks and groans of battle filled the fog [...]."

In order to escape from the fog, the "South-pointing chariot" was invented. This had a mechanical finger which always pointed South regardless of the direction the chariot was pointed. The imagery of the "horned oxen troops" with their bronze ox heads conjures up the image of Shen Nung, Chhih-Yu's father who is likely related to the constellation Shen (Orion). The South-pointing chariot possibly refers to an alignment of circumpolar stars, such as the "Chariot Pole" [114: 80-81], a grouping of stars closely identified with the Yellow Emperor ("Yellow Thearch"). This fits comfortably with the fact that Huang Di is credited with inventing the first wheeled vehicle.

(4) Oceanic or Lake Cosmic Impact (ca. 2650 BC). After the mists ended, Huang Di went to the Lo River and saw a great fish. He sacrificed five victims to the fish, upon which torrents of rain came down for seven days and seven nights and the fish floated off to sea.

In other versions of the mist and torrential rain stories [18, 212], we learn that the events during the 50th year of the Yellow Emperor's reign were actually part of a cosmic battle between Huang Di and Chhih-Yu. Chhih-Yu is said to be the name of either a tribe of giants, or that of a ferocious god in heaven, the son of ox-headed second sovereign and Fiery Emperor, Shen Nung (and thus likely associated with the ox constellation Orion). He is described as having one or more horns on a bronze head, and a skull consisting of iron. This particular demigod likely represents the apparition of a particular form of comet, also called *Chhih-Yu*, referred to as "banner" comets. Ho Peng Yoke [89: 136] has reported this cometary form as a tailed comet, with its tail being "bent like a flag".

Chhih-Yu had as supporters the gods of Wind and Rain, while Huang Di had the Ying Long dragon, who was a water-gatherer and could cause heavy rains. After fighting for several days, Huang Di finally won by using the skin of the *kui* beast as a large drum. Chhih-Yu's army was paralyzed by the sound of the drum striking nine times, which caused the valleys and the mountains to quake and the sky to change color. Yuan Ke [212: 45] provides us with this most interesting description of the *kui*:

"It was grey and looked like an ox without horns. Although it had only one leg, it sometimes rushed out of the sea leaving violent storms in its wake. It would open its mouth to roar like thunder, and its body shone like the sun and the moon".

Mesoamerica: Seven Macaw and the Hero Twins

The *Popul Vuh* is the Quiche Mayan creation epic copied and translated by a Dominican friar, Francisco Ximénez, between 1701-1703 [187: 28-30]. Of interest is in the story line of what happened immediately following a worldwide flood which destroyed the wooden manikin people of the Earth.

(1) Ursa Major Supernova or Nova-like Event (ca. 2697 BC). A deity by the name of Seven Macaw (Big Dipper), who is married to Chimalmat (Little Dipper) decides to become the new Sun and Moon [187: 86]. The imagery quoted below is an apt description of the bright Ursa Major nova-like event:

"I am great. My place is now higher than that of the human work, the human design. I am their sun and I am their light, and I am also their months. / So be it: my light is great. I am the walkway and I am the foothold of the people, because my eyes are of metal. My teeth just glitter with jewels, and turquoise as well; they stand out blue with stone like the face of the sky./And this nose of mine shines white into the distance like the moon. Since my nest is metal, it lights up the face of the earth. When I come forth before my nest, I am like the sun and moon for those who are born in the light, begotten in the light. It must be so, because my face reaches into the distance, say Seven Macaw./ It is not true that he is the sun, this Seven Macaw, yet he magnifies himself, his wings, his metal. But the scope of his face lies right around his own perch; his face does not reach everywhere beneath the sky".

Seven Macaw is wounded in the jaw by a blowgun dart from Hunahpu, one of a pair of twin boys (warrior gods) who want to restore things to their natural order. Hunahpu and his brother Xbalanque remove his precious shining teeth and eyes and Seven Macaw fades away. The location of where Seven Macaw is wounded is on top of his giant nance tree [187: 90-91], which Freidel and his colleagues convincingly argue is likely a metaphor for the World Tree composed of the Big Dipper, the Milky Way, and the Ecliptic [67: Fig. 2.13]. As for the Hero Twins themselves, they too are celestial in that they eventually become the Sun and the Moon. Whether or not they may also represent other types of celestial phenomena, such as novae and comets (including perhaps paired comets), is unknown.

An intriguing possible additional confirmation of the association of the Seven Macaw with the Big Dipper is that of a curious scorpion motif that accompanies Classic Mayan depictions of this myth [187: 90-91]. Although Freidel, Schele, and Parker [67: 75-76] build a convincing argument for the Scorpion image to represent the constellation Scorpius, I am struck by the fact that early historic period Nahuatl speakers of Mexico seemingly interpreted Ursa

Major as representing a scorpion [4: 66].

(2) Simultaneous Supernovae or Nova Events (ca. 2680 BC). There are no unmistakable equivalents in the Popul Vuh for the hypothesized three simultaneous novae in the 20th year of the reign of Huang Di. On the other hand, the two sons of Seven Macaw, Zipacna and Earthquake, were said to have "magnified" themselves to some degree, perhaps not unlike Seven Macaw himself (Tedlock 1986: 99). This suggests that the sons of Seven Macaw constituted two of the three simultaneous novae, a possibility further supported by the fact that a typical human generation (between father and first-born son) is approximately 20 years, roughly the amount of time between the Ursa Major event and the three simultaneous nova events.

(3) Taurid Meteor Storm (ca. 2650 BC). The elder son of Seven Macaw is Zipacna, maker of mountains. This is a creature likely similar to the cipactli, a Nahua mythological creature similar to a crocodile or a shark, and which possibly represents an unidentified constellation or perhaps is representative of comets. Sometime after the death of his father, Zipacna is encountered by the Four Hundred Boys 'bathing on the shore' (of the Milky Way?). The Four Hundred Boys try to kill Zipacna by shoving a tree onto Zipacna who has dug a deep pit. After three days, the boys think that Zipacna is dead and celebrate by getting drunk. Zipacna kills them by shoving their hut down on top of them and crushing them. The boys then enter the constellation, Hundreth or Motz, which is the Pleiades. Hunahpu and Xbalanque then kill Zipacna by building a giant artificial crab (a favorite food of Zipacna) and placing it under a mountain. Zipacna tries to catch the crab and is instead himself trapped under the mountain, at which point he turns to stone.

This sequence appears to encode a three-day long great meteor storm associated with the two Taurid streams near the Pleiades, and closely matches the event in the Gilgamesh epic described below.

(4) There are no known correspondences between the Bamboo Annals and the Popul Vuh regarding the 'battle' between Chhih-Yu and the Yellow Emperor, unless it is with the actions and adventures surrounding the Hero Twins in the land of the underworld, Xibalba.

Mesopotamia: Gilgamesh and Enkidu

The epic of Gilgamesh has fascinated Western civilization since the discovery of the original clay tablets at Nineveh by Hormuzd Rassam in 1864 and their initial translation eight years later by George Smith [31: 264-278]. The Gilgamesh epic [48, 97], in much the same manner as was the case for the Mauiakalana epic in Polynesia, is a story of multiple spectacular temporary celestial events. The epic was created both to explain these events and to honor the King in whose reign these events occurred. As history tells us, the story takes place during the reign of Gilgamesh, a legendary but still historical ruler of the Mesopotamian city of Uruk during the 27th century BC, and who several centuries later was worshiped as a god [57: 53].

Gilgamesh had a miraculous birth. In the Sumerian King List [146: 265-266], the father of Gilgamesh was identified as a *lillû* spirit/demon (unidentified as to meaning). Another myth states that the grandfather of Gilgamesh, a king of the Babylonians, was warned by his magicians that a son born to his daughter would one day usurp the throne. The newborn Gilgamesh was thrown from the top of the acropolis, but was saved by an eagle. In addition, the adult Gilgamesh is described as having been a giant, nearly 5 m tall.

The epic begins with the hero having two dreams. In the first dream, the stars suddenly appeared in the sky and a meteorite fell from the sky and landed next to Gilgamesh. In the second dream an axe appeared next to Gilgamesh. (Early civilizations often confused or equated meteorites with polished stone axes of Neolithic cultures [141: 435]). Both of these were seen by the mother of Gilgamesh as symbols of the wild man Enkidu, who was to become the beloved friend of Gilgamesh. The pairing of Gilgamesh and Enkidu may be analogous to the pairing of the Hero Twins in the Popul Vuh, presumably with similar celestial overtones.

(1) Ursa Major Supernova or Nova-like Event (ca. 2700 BC). The first adventures of Gilgamesh and Enkidu is that of the killing of mighty Humbaba (Huwawa, Humbala), giant guardian of the cedar forest. Humbaba was said to have been clothed in seven layers of "terrifying radiance", spouted fire when he talked, and the "flood-weapon" was employed when he shouted. On the journey to kill Humbaba, Gilgamesh had three dreams, two of which are poorly preserved, but the third of which is of some interest [97: 34]:

> "The heavens roared and the earth rumbled; (then) it became deathly still, and darkness loomed. A bolt of lightning cracked and a fire broke out, and where(?) it kept thickening, there rained death. Then the white-hot flame dimmed, and the fire went out, and everything that had been falling around turned to ash."

A Hittite version of the killing of Humbaba [97] suggests that this took place in the celestial heavens in that Humbaba, standing on his "mountain" told Gilgamesh and Enkidu: "I shall carry you off and cast you down from the sky, smash you on the head (?) and drive you down into the dark earth." Despite this threat, Enkidu had no problem cutting off Humbaba's head with his sword.

The description of Humbaba as guardian of the cedar forest with seven bright rays is analogous to the description of Seven Macaw in the Popul Vuh. In fact, a cylinder seal depicting the slaying of Humbaba [97: Fig. 3] contains a motif of two rows of dots totaling seven dots, which parallels images that Parpola [150: 194-195] has argued represents Ursa Major in both Mesopotamian and Harrapan script.

The face of Humbaba, as illustrated in Figure 2 (and in other representations) is striking in that it seems to be alive and to pulsate with energy. This vibrancy is echoed in a surviving fragment of Tablet V of the Gilgamesh epic [97: 42]: "Gilgamesh spoke to Enkidu, saying: My Friend, Humbaba's face keeps changing!" Unfortunately the remainder of the description is lost.

(2) There is no apparent equivalent in the Gilgamesh epic for the hypothesized three simultaneous novae in the 20th year of the reign of Huang Di.

(3) Taurid Meteor Storm (ca. 2650 BC). The next adventure of Gilgamesh and Enkidu is that of killing the Bull of Heaven, sent down to Earth by the goddess Ishtar (Inana), goddess of love, because Gilgamesh has spurned her advances. The sky god Anu gives the reigns of the *Bull of Heaven* to Ishtar to lead down to earth. The Bull of Heaven snorts three times, the first two which creates a huge pit into which 100 and then 200 young men of Uruk, respectively, plunge into each time. At the third snort Enkidu falls in the pit. Enkidu then climbs out of the pit, grabs the tail of the Bull of Heaven in both hands, and then Gilgamesh kills it with a sword between the shoulders.

The similarity between this story and the death of the 400 boys in the Popul Vuh is striking. Thus it should be no surprise that Ishtar ("Queen of the Stars" according to Allen [3: 463], is associated with the planet Venus and the Bull of

Heaven) is the term commonly ascribed to the constellation Taurus. The story undoubtedly encodes an early observation of a dramatic Taurid meteor storm.

(4) Oceanic or Lake Cosmic Impact (ca. 2650 BC). Gilgamesh has additional exploits which do not seem to correlate well with the above defined actions of Huang Di. These include his attempted seduction by the goddess Ishtar (Venus); the death of his friend Enkidu; a long and treacherous journey to visit with Utanapishtim, the survivor of the Flood; and his recovery of the "plant of rejuvenation" from the waters of death, only to have it snatched from him by a snake. It is possible the the purpose of retelling the story of the Flood in the Gilgamesh epic, a lengthy sub-story not directly involving either Gilgamesh or Enkidu, is due to the occurrence of a moderate (large local or regional) oceanic or lake cosmic impact at around 2650 BC. In fact, fear of a repeat of the 2807 BC. Flood Comet impact itself may have played a substantive role in the creation of the Gilgamesh epic. This would correspond to the copious rainfall noted in the Huang Di myth as well as the previously mentioned alluvial flood deposits at Kish dating to around 2650 BC.

Egypt: Horus and Set

While Egyptian mythology regarding the 27th century cosmic events is perhaps more subtle than the Huang Di, Gilgamesh, and Popul Vuh epics, the cosmographic relation between these cosmic events, Egyptian religion, and Egyptian culture is extraordinary and most compelling. The mythology concerning these cosmic events is enshrined as a series of stories about the sky-god Horus and the great cosmic disturber, Set.

(1) Ursa Major Supernova or Nova-like Event (ca. 2700 BC). I begin this analysis with the only event chronicled by the Egyptian historian, Manetho (Eusebius version), for the first king of Egypt's Third Dynasty, Sanakhte [193: 43] "[...] in whose reign the Libyans revolted against Egypt, and when the moon waxed beyond reckoning, they surrendered in terror." This unusual "waxing" of the Moon is almost certainly the Ursa Major nova-like event described above. Current dating puts the beginning of the Third Dynasty and Sanakhte's reign at about 2700 BC [96], which well matches the Chinese date for the Ursa Major nova-like event. It is likely that Sanakhte's brother, Djoser, second king of the Third Dynasty, would have witnessed this celestial event as well.

The Manetho reference may be paralleled in the myth of Horus Behdety [26: 56-95, 91: 67-68], which has been dated to the Third Dynasty [64], presumably to the time of King Djoser since the myth is directly linked to his minister and chief architect, Imhotep [161: 143-144]. In this myth there is a revolt against the King of Egypt, Ra, while visiting Nubia. The King sails down to Edfu, where he orders his son, Horus to fight the enemy. In some versions of the myth the god Set is considered the leader of the enemy, with his army being composed of demon-like creatures. Set turned himself into a serpent, while his supporters became crocodiles and hippopotami and attacked the solar boat of Ra. Horus takes the form of a winged sun-disk, and chases the enemy through Upper and Lower Egypt, and all the way to the frontiers of Asia. Eventually, the insurrection is put down, Set is beheaded, and Ra orders that the winged sun-disk be placed in the temples and on the shrines of all deities to ward off enemies.

In terms of a link between the Ursa Major nova-like event and the Horus Behdety myth is the possibility that the name Horus is derived from the word, *hor*, meaning "face" [91: 67]. In fact, Djoser was the first Egyptian king to adopt the curious title of *Golden Horus*. Also of interest is the description of the winged disk as "shining with very many colours" [26: 59]. These images are suggestive of a near-Earth nova-like event. Also, it is notable that Imhotep, the minister to Djoser (and who later himself became deified), was also a noted astronomer and "magician."

But more intriguing is the fact that the Ursa Major nova-like event is seemingly contemporary with the construction of Djoser's temple complex, that of the famous so-called *Step Pyramid* [84: 64-67, 161: 171-182, 177: 98-105, 174: 102-113]. This assumes that the Third Dynasty begins at around 2700 BC, as suggested by Kitchen [96], and likewise assumes that the initial construction of Djoser's original mastaba tomb had begun before the appearance of the Ursa Major nova-like event.

By all accounts the Step Pyramid, the first pyramid-like structure in Egypt, signals a major shift in cosmology and world view. Prior to the Step Pyramid, Egyptian rulers and high officials were buried in massive mud-brick tombs, the largest of which was some 67 m in length, 37 m in width, and perhaps 4-5 m in height. Beginning with the Step Pyramid, for the first time mortuary edifices were constructed entirely of stone, and at a scale far greater than anything done previously in the history of civilization. The Step Pyramid complex is situated in a massive walled enclosure 545 m in length and 270 m in width, which enclosure an area of nearly 16 hectares. The Step Pyramid itself rises in six levels to a height of more than 62 m. As such, it has been aptly described as "a gigantic stairway for the monarch's ascent into the sky" [80: 62].

The relationship between the king and the sky is particularly interesting, although there is a debate as to whether the emphasis was in connection with the Sun or with the stars, or both [13, 58, 158, 161]. As noted by Rice [161: 195-196]:

> "Some authorities have proposed that the early royal religion in Egypt up to and including the Third Dynasty, was linked with the stars. The evidence, either way, is slender but certainly the Pyramid Texts, assuming these to be more ancient than the time in which they were first inscribed on the walls of the Sixth Dynasty pyramids, seem to identify the divine King as a star and it is amongst the stars, or even beyond them, that he seeks his eternal habitation. The stars are valuable instruments for measurement and the Egyptian engineers and architects of even the earliest periods seem to have been capable of making complex and sophisticated empirical observations which they used to align their buildings. The precision with which the monumental buildings of the Archaic period and the early Old Kingdom are aligned is legendary; that precision was achieved by careful sitings on selected stars and the skilful [sic] use of water channels, the consequence of the careful observation of the behavior and properties of water which large-scale irrigation projects and techniques had made familiar. With the advent of the Fourth Dynasty, however, the sun cult, the perogative [sic] of the hierophants of Heliopolis, began to rise above the other cultures of national or royal status."

Contrary to Rice and other scholars, I suspect that the Pyramid Texts, beginning with the tomb of King Unas at the end of the Fifth Dynasty, are indeed original to that period (ca. 2350-2200 BC) and for the most part do not represent earlier texts.

A simple test of the emphasis on Sun versus stars would be that of checking the symbolism of location, that is, whether or not critical aspects of the complex are oriented sensitive to the path of the Sun (generally east-west, and specifically toward solstitial and equinoctial points), or are instead oriented north toward the circumpolar stars. As noted by Wilkinson [205: 60-73], both strategies are utilized variously throughout Egyptian culture history. The answer to this question may well reside with the nature of the architectural features of the Step Pyramid complex [174: 102-113].

The first item of note for the Djoser Step Pyramid complex is that the single entry is indeed from the east. However, the long axis of the overall complex runs north-south. In fact, the majority of features appear to be in a north-south orientation. This includes the extraordinary *heb-sed* courtyard, a full-scale replica for the king to perform his ritual festival for renewal of his rulership rights in the afterlife; the doors of this replica were left permanently ajar [64: 79-88]. The *sed* festival began around the middle of September and may have coincided with the autumnal equinox. More important, however, is the *saite* or funerary temple on the north side of the pyramid, and the north orientation of the entryway tunnel(s) from the kings burial chamber beneath the Step Pyramid. While later entryway tunnels almost always were placed on the north side of pyramids, beginning in the Fourth Dynasty the funerary temple uniformly shifted to the east side.

Of special interest is Djoser's *serdab*. The *serdab* is a small, completely enclosed tiny chamber, attached to the east side of the funerary temple and against the northern base of the pyramid itself. A life-size seated statue of Djoser next to and facing the north wall of the *serdab*; two small holes were drilled through the inward slanted wall in direct orientation with the eyes of the statue. Clayton [41: 35] suggests that the holes were drilled to enable the *ka* or spirit of Djoser (as manifested through the statue) to partake of food offerings and incense placed on an alter outside the *serdab* by mortuary priests. While this may indeed have been part of the ritual, it is also true that the crystal eyes of the statue would have been oriented upwards toward the circumpolar stars.

It is also of some interest that relatively rare depictions of the baboon god Hedjwer ("Great White"), suddenly appear on Third Dynasty reliefs of King Djoser's pyramid [177: 73]. Hedjwer, like the more commonly known baboon deity, Thoth, is believed to be a personification of the Moon. Also intriguing is the fact that Hedjwer was depicted in the royal title of Semerkhet, the seventh king of First Dynasty, and the unfortunate recipient of the Flood Comet impact described above.

But perhaps the most intriguing architectural detail about the Step Pyramid is its construction history. The Step Pyramid began as a large but otherwise typical mastaba tomb not unlike all others of the First and Second Dynasties, and that of Djoser's brother, Sanakhte, first king of the Third Dynasty. As this original mastaba was nearing completion, it was remodeled twice and then suddenly and hugely transformed into the first ever pyramid complex which itself went through a couple remodelings. While some of the minor architectural elements of the Djoser Step Pyramid complex are presaged in the first two dynasties and in the mastaba tomb of his brother, Sanakhte, the major elements are all new, including the *saite* funerary temple, the replica of the *heb-sed* court, the *serdab* chamber, the enormous enclosure wall, the complete use of stone for all construction, and of course the pyramid itself.

In the words of Michael Rice [161]:

> "But it was not until the beginning of the Third Dynasty, around 2680 BC, that the titanic complex that was to preserve for all eternity the body of the second King of the dynasty, the Golden Horus, Djoser Neterikhet, was suddenly to appear on the Saqqara escarpment, built entirely of stone, on a scale never before contemplated on the face of the earth. [...] The Djoser complex is unique. Once again, it is totally without precedent, not merely in Egypt but in the entire world."

It is not very a satisfactory explanation to attribute this construction endeavor simply to the coincidence of a wealthy and powerful monarch with the genius of his principal architect, as has been commonly done for Djoser and Imhotep. What is the source of their inspiration, but even more critical is how did they willingly convince the populace of Egypt to support them in this ambitious endeavor?

(2) Simultaneous Supernovae or Novae Events (ca. 2700 BC). There is no definitive record in the myths or historical literature of Egypt of the three simultaneous supernova or nova-like events reported in the Bamboo Annals. However, one or more of these may have been observed in Egypt. For example, the hypothesized supernova or nova event in Draco [123] would have been circumpolar and perhaps was not too far from the circa 2697 BC. Ursa Major nova-like event witnessed as scant generation earlier. This event, had it actually happened, would surely have further encouraged the architectural and religious expression begun with the Step Pyramid complex. It may even be possible that the hypothesized simultaneous events of circa 2677 BC were the penultimate trigger for the final construction phases of the Step Pyramid complex.

(3) Taurid Meteor Storm (ca. 2650 BC). Clear and unequivocal descriptions of the Taurid meteor storm of circa 2647 BC are lacking in Egyptian mythology. However, there is reason to believe that the greatest Egyptian myth of them all, that of the murder of Osiris by Set, is a cosmographic encoding of this cosmic event.

There are a number of versions of this popular story scattered throughout the history of Egyptian literature, beginning with the Pyramid Texts at the end of the Fifth Dynasty [25, 80, 91]. In brief, Osiris was an enlightened king of Egypt, who was married to the goddess Isis, and who taught his subjects how to become civilized. After a period of travel abroad to teach the rest of the world, Osiris returned to Egypt where his brother Set was planning to overthrow him and seize the throne. On the seventeenth day (of the lunar cycle) of the month Hathor (believed to be October), Set and his fellow conspirators challenged Osiris to see if he could fit into a coffin-like chest. Once inside the chest, the conspirators nailed it shut, weighted it down with lead, and threw it into the Nile. Isis spent a long time searching for Osiris, and discovered the chest in Byblos. Through magic Isis managed to conceive from her dead husband. Isis then took refuge in the Egyptian Delta to give birth to and to raise her son Horus because she knew that Set would kill Horus if he got the chance. Set found the chest and tore Osiris's body into a large number of pieces, scattering them throughout the kingdom. Isis recovered the pieces of Osiris's body, except for his phallus, and was able to resurrect him. Set discovered the baby Horus and, taking the form of a viper, bit him. Isis called upon Ra, who stopped the sun and it became dark in daytime until he was able to cure Horus. Isis continued to protect Horus. Isis finally ends up poisoning Ra so that she could obtain his powers. From his spittle and dust she fashioned a viper which bit Ra. The fire departed from him until he started shivering and his eyes trembled so much that he could not see the sky. After obtaining the secret words of Ra, Isis counteracted the poison and then restored Ra to his full strength.

This story is laden with veiled cosmographic meaning. First, Osiris is identified with the constellation Orion, Isis is identified with Sirius, and the Venus is identified with the *bennu*, a heron-like bird related to the fabled Phoenix [70, 110]. Previously we have learned that Set, himself, is associated with meteorites. Another curious fact is that the sacred *benben*, makes its first appearance about the time of the Third Dynasty. These are cone-shaped capstones for pyramids, and a critical component of Fifth Dynasty solar temples and other Old Kingdom special structures [13, 64, 205]. It is clear from a number of sources that *benben* are meteorites shaped by their passage through the atmosphere. This relationship is enhanced by the likelihood that the term *benben* itself is derived from *weben*, which means "to shine" [158: 27].

The key to understanding the origin of the Osiris-Set conflict is contained in the following passage from the Pyramid Texts and associated discussion by Quirke [158: 54):

> "Egyptian texts tend to reticent on the catastrophic moment of murder [of Osiris], but the ritual for the 'end of (mummification) operations' gives a brusque summary, from the moment that Osiris first arrived at Abydos: *Osiris then said how great is this land, which is called Tawer (great land = the province of Abydos) to this day; then Osiris was perfect within it exceedingly greatly; then Seth heard him; Seth came in haste, and arrived against Osiris within Nedyt in Hatdjefau (two places in the area of Abydos) under a tree called* aru *in the first month of inundation, day 17, and committed an act of great violence against him, and had him sunk in the waters; then Nun rose to cover him exceedingly greatly, and he was borne away to hide his mysteries.../* The text goes on to describe the cosmos in chaos at the death of Osiris, and the tears of the gods turning into materials used in mummification such as honey, resins and incense; the rest of the ritual prescribes the fashioning of statues of enemies, to be destroyed, and of Osiris, to be mummified and buried./ The date in the ritual does not agree with the usual date given for the death of Osiris, presumably because the ritual tailored the myth to suit the time of year that it was due to performed."

This date, the 17th day of the first month of inundation, is actually most likely the correct date for the "death" of Osiris. I suspect that scholars have likely confused the month of Hathor (October) with the "first month of the inundation", a very special time for the sky-goddess Hathor in her role as fertility goddess. Hathor, when associated with the rise of the Nile and the Dog-star Sothis (Sirius), is depicted lying in a ship as a cow with a star between her horns.

The rise of the Nile typically occurs in late June or July [27, p. 447]. This also coincides with the date of the very important New Year's Festival [64]. However, most extraordinarily, around 2650 B.C., the 17th day (two days after Full Moon) of the first month of inundation would have also coincided with one of the greatest fireworks displays ever witnessed on Earth. By figuring in approximately 102 days difference for precession, the Taurid meteor stream, whose modern maximum occurs around 3 November, and whose modern range encompasses the period of 15 October to 25 November [21, Table 5.1], would have been in full bloom in late July of circa 2650 B.C. It is small wonder that Osiris, as the constellation Orion, was virtually annihilated by Set, the latter who in addition of being the god of meteorites may possibly have been envisioned as a neighboring constellation incoporating portions of Taurus, including the Pleiades and the Hyades. Given its seeming source for the Taurids, it is little wonder that the Egyptians gave the name 'thousands' to the Pleiades asterism [70, Table 21].

(4) Oceanic or Lake Cosmic Impact (ca. 2650 BC). It is not so clear, however, from the Egyptian myths as to whether or not the Taurid storm was accompanied by one or more actual cosmic impacts on Earth. However, this is certainly a reasonable supposition and would help to explain the association of the Taurids and the 'fire-drill' constellation (Pleiades) with actual meteorities. The fact that the step pyramid of Djoser's successor, Sekhemkhet, was never finished, and the substantive differences in the pyramid construction techniques of the last two kings of the Third Dynasty, Khaba and Huni [41], may have been influenced by the Taurid storm and possible associated impacts.

India: Rahu

As with the Egyptian traditions, I have not yet had the opportunity to fully analyze Hindu traditions in relationship with the celestial events of the 27th century BC. However, I am particularly struck by the legends surrounding Rahu, an eclipse demon who is considered of importance equal to the five naked eye visible planets, the Sun, the Moon, and Ketu, a deity representing comets [150: 199-201].

(1) Ursa Major Supernova or Nova-like Event (ca. 2700 BC). Rahu appears in association with the tortoise Kurma, the second of the twelve avatars or incarnations of the Vishnu, and immediately following Matsya, the fish avatar associated with the Flood [54: 67-75]. Rahu is associated with the illicit drinking of Soma, the nectar of the Moon whose loss causes the Moon to wane or eclipse [49: 316]:

> "While the gods were sharing the nectar, a *danava* [genii] named Rahu, disguised as a god, was drinking his share; the Sun and the Moon discovered him when the nectar had only reached his throat, and they informed the other gods. Vis[hnu] instantly cut off with his discus the *danavas* head, which was thrown into the sky and began to utter piteous cries, while the headless trunk fell upon the ground and, rolling thereon, made the earth tremble, with her mountain, forests, and islands. From that time a quarrel has existed between Rahu's head and the sun and the moon. And to this day it always attempts to swallow the sun and the moon." (*Bhagavata Purana* 10.8.9.)

The tail of Rahu became Ketu, the monster which gives birth to comets and meteors.

Presumably when Rahu drank the Soma, his countenance was much like that described for the nectar itself [197: 4]:

> "The poets never tire of stressing Soma's sensuous appeal. In appearance it was brilliant, reminding them of the Sun, of Fire, of the rays of the Sun, of the round bowl of the heavenly firmament, the 'back of the sky'".

This description, a head-shaped brilliant celestial object sounds very suspiciously like the nova-like event in Ursa Major, although I have not yet been able to establish a clear link in Hindu mythology between Rahu and this constellation.

Evaluation of 2700-2600 BC celestial events: Taurids, Owl Nebula, and oceanic impact?

The myths detailed above for the period of 2700-2650 BC encode nova-like events and meteor storms, and possibly the passage of paired or otherwise unusual comets. The evidence is compelling that the various sets of myths are indeed describing the same events, and thus can serve at least as indicators for the construction of regional chronologies. The several descriptions of an early episode of the Taurid meteor showers at around 2650 BC is notable.

Especially impressive are the several myths which appear to identify a nova-like event in Ursa Major, in particular the "bucket" area of the Big Dipper asterism. It is significant that only a single notable nebula is known from the region of the bucket of the Big Dipper. This object is the peculiar Owl Nebula (Messier 97; NGC 3587), located about 2.5° southeast of the star Merak (*beta ursa majoris*) and approximately 11° south of Dubhe (*alpha ursa majoris*). The Owl Nebula, situated some 1600 light years from Earth, is remarkable in three respects [1, 37, 138]. First, it is one of the largest known planetary nebula, being nearly 3.5 light years in diameter (170"). Second, the multiple shell structure of the still expanding object is complicated and turbulent. Third, the shape of the nebula, nearly circular with two dark eye-like spots, is remarkably like that of a face (thus the appellation "owl"). It is believed that the nebula was initially formed about 6,000 years ago (ca. 4000 BC).

Leaving aside, form the moment, the puzzling issue of the mechanics of shell formation, the similarity of the images of Humbaba (Fig. 2) to that of telescopic photographs of the Owl Nebula [134: Fig. 47, 138: Fig. 26], is remarkable. The similarity is further illustrated by the striking similarity between Humbaba and a drawing of the Owl Nebula sketched in 1848 by William Parson, the 3rd Earl of Rosse. Given the strong documentary evidence for the appearance of a face-like, nova-like event in Ursa Major at around 2700 BC, and given the lack of other notable nebula, including supernova remnants, around the vicinity of the bucket of the Big Dipper, I suggest that the formation of the Owl Nebula is indeed the Chinese birth event of Huang Di, and therefore the event which inspired the creation of Seven Macaw in the Popul Vuh and Humbaba in the Gilgamesh epic.

Acceptance of this relationship, however, requires a satisfactory explanation for the rapid production of a naked eye visible gas shell of sufficient size, brightness, and duration (perhaps visible for 25 months) for people to see actual structures (e.g., dark spots and filaments) that might create the impression of a human face. The myths strongly suggest that the naked eye visible temporary object was comparable to or even larger than the size of the full Moon, and thus much larger than the image now seen through optical telescopes. While it is possible that these myths encode a nova-like event completely distinct from the Owl Nebula, the possibility that the myths encode an early eruptive episode in the evolution of the Owl Nebula itself is deserving of further investigation. I leave this puzzling but most intriguing mystery to my colleagues in astronomy and astrophysics.

Much less secure is the evidence for encodings of cosmic impacts in the myths dating to the 27th century BC. The vivid description of "unearthly shrieks and groans" during the cosmic battle between Chhih-Yu and Huang Di suggests the presence of bolides in conjunction with the presumed Taurid meteor storm(s), but clear evidence of actual ground-disturbing impacts is lacking in the myths used in my analysis. However, as previously noted, there is some suggestion of an oceanic or large lake cosmic impact at around 2650 BC, of substantially less magnitude than the Flood Comet impact of 2807 BC.

9. Celestial events of circa 2350-2000 BC

The period between 2650 BC and 2350 BC seems to lack definitive supernovae/bright novae and cosmic impacts, which may be one reason for the fact that there are few myths about rulers during this period, with the possible exception of Khufu and Khafre (the builders of the great pyramids of Giza). However, as with the myths discussed earlier for the period of 2700-2650 BC, the period of circa 2350-2000 BC witnessed spectacular temporary celestial events amenable to cosmographic study and comparison. I once again begin with descriptions in the Chinese Bamboo Annals [106].

China: Yao, Shun, and Yu

Yao, Shun, and Yu are among the most famous legendary rulers of ancient China, with the majority of extant myths relating to Shun and especially Yu rather than to Yao [18: 74]. According to standard chronology, Yao's reign began in 2357 BC, and lasted a full one hundred years. Yao was followed by Shun, who began his reign in 2255 BC, who in turn was followed by Yu in 2205 BC, the later who died in 2197 BC.

Yao, in typical fashion, had a miraculous birth, his mother purporting to have been impregnated by a red dragon. He was said to be of giant stature, and had the virtue of a sage. Among other things, he is credited for astronomical observations and for an elaborate system of canal works. Shun is likewise credited with a miraculous birth, being born with double eyes, and the countenance of a dragon with a large mouth and black body. Shun was chosen by Yao for succession over Yao's own son, and became famous for his good and humane governance. Yu's mother became pregnant after seeing a falling star go through the Pleiades. Yu was born from his mother's back, and he had a tigers nose and a large mouth. His head bore the resemblance of the stars *kou* (Sagittarius) and *ke'en* (cancer), while his breast resembled Ursa Major. Yu is best known for his success in stopping floods that were threatening all of China.

Notable events during the reigns of these three rulers include the following:

(1) Possible Terrestrial Cosmic Impact (ca. 2345 BC). In the 12th year of his reign (ca. 2345 BC), Yao meets the Divine Archer, Yi (*Shên-I*). According to myths assembled by Werner [199: 180-182]:

> "At this time terrible calamities began to lay waste the land. Ten suns appeared in the sky, the heat of which burnt up all the crops; dreadful storms uprooted trees and overturned houses; floods overspread the country. Near the Tung-t'ing Lake a serpent, a thousand feet longs, devoured human beings, and wild boars of enormous size did great damage in the eastern part of the kingdom. Yao order Shên I to go and slay the devils and monsters who were causing all this mischief, placing three hundred men at his service for that purpose. / Shên I took up his post on Mount Ch'ing Ch'iu to study the cause of the devastating storms, and found that these tempests were released by Fei Lien, the spirit of the Wind, who blew them out of a sack. As we shall see when considering the thunder myths, the ensuing conflict ended in Fei Zlien suing for mercy and swearing friendship to his victor, whereupon the storms ceased. / After this first victory Shên I led his troops to the banks of the Hsi Ho, West River, at Lin Shan. Here he discovered that on three neighboring peaks nine extraordinary birds were blowing out fire and thus forming nine new suns in the sky. Shên I shot nine arrows in succession, pierced the birds, and immediately the nine false suns resolved themselves into red clouds and melted away."

(2) Crater Supernova (ca. 2314 BC). A likely supernova appears in the constellation, *"I"* (Crater), perhaps lasting 14 months. [This event seems very similar to that encoded in point 5 below].

(3) Possible Oceanic or Lake Cosmic Impact(s) (ca. 2297-2265 BC). Kwan (Gun) attempts to regulate the Ho river

during a period of disastrous flooding (ca. 2295-2287 BC), but does not succeed. Kwan is then sacrificed. His son Yu, is born from Kwan in the shape of a dragon. Yu continues to battle the flooding and finally succeeds around 2265 BC.

(4) Crater Supernova or Nova (ca. 2286 BC). A supernova or nova appears in the constellation Crater. This event occurs during the first month of spring, at a time when the five planets look like "threaded pearls". A check with a software astronomy program (Redshift Multimedia Astronomy, version 1.2°, by Maris) indicates that at dawn on 25 March 2284 BC, four of the five naked eye visible planets (Mercury, Jupiter, Venus, Mars) were evenly spread out over a space of about 10° in the constellation Pisces, like threaded pearls; only Saturn was missing from this alignment.

(5) Scorpius Supernova or Nova (ca. 2254 BC). A "brilliant star" (presumably a nova or supernova) appears in the constellation Scorpius at the beginning of Shun's reign.

(6) Terrestrial Cosmic Impact (ca. 2240 BC). "Auspicious clouds" appeared during the reign of Shun. These auspicious clouds are described by the following note in the Bamboo Annals [106: 115-116]:

> "In the 14th year of Shun's reign, at a grand performance with bells, musical stones, organs, and flutes, before the service was concluded, there came a great storm of thunder and rain. A violent wind overthrew houses and tore up trees. The drumsticks and drums were scattered on the ground, and the bells and stones dashed about confusedly. The dancers fell prostrate, and the director of the music ran wildly away; but Shun keeping hold of the frames from which the bells and stones were suspended, laughed and said, 'How clear it is that the empire is not one mans empire! It is signified by these bells, stones, organs, and flutes.' On this he presented Yu to Heaven, and made him perform actions proper to the emperor; whereupon harmonious vapors responded on all sides and felicitous clouds were seen. They were like smoke, and yet not smoke; like clouds, and yet not clouds; brilliantly confused; twisting and whirling....When the essential brightness was exhausted, the clouds shriveled up and disappeared. Thereupon the eight winds all blew genially, and other felicitous clouds collected in masses [...] When the day declined, there came a fine and glorious light."

Mesoamerica: The Corn People

After the earlier described stories of the hero twins (the period of circa 2700-2650 BC), the Popul Vuh next describes a sequence of events relating to the early years of the people created from cornmeal, that is, modern humans [187]. These events appear to follow the same basic sequence identified in the Bamboo Annals. While direct allusions to cosmic impacts are difficult to interpret from this sequence and some of the matches with the Chinese myths are a bit tenuous, it at least suggests that myth cycles based on cosmic events witnessed in different parts of the world can be usefully compared.

(1) Crater Supernova (ca. 2314 BC) [187: 181]:

> "And here is the dawning and showing of the sun, moon, and stars. And Jaguar Quitze, Jaguar Night, Machucutah, and True Jaguar [the first of the modern humans] were overjoyed when they saw the daybringer. It came up first. It looked brilliant when it came up, since it was ahead of the sun."

(2) Possible Terrestrial Cosmic Impact (ca. 2345 BC) [187: 182]:

> "And then the face of the earth was dried out by the sun. The sun was like a person when he revealed himself. His face was hot, so he dried out the face of the earth. Before the sun came up it was soggy, and the face of the earth was muddy before the sun came up. And when the sun had risen just a short distance he was like a person, and his heat was unbearable. Since he revealed himself only when he was born, it was only his reflection that now remains. As they put it in their own words: 'The sun that shows itself is not the real sun.'"

(3) Possible Oceanic or Lake Cosmic Impact(s) (ca. 2297-2265 BC) [187: 187-188]:

> "And then comes the killing of the tribes. This is how they died: when there was just one person out walking, or just two wer out walking, it wasnt obvious when they took them away [...] The way was not clear [for people attempting to track and to kill the four heroes): It would get cloudy. It would get dark and rainy. It would get muddy, too. It would get misty and drizzly."

(4) Crater Supernova or Nova (ca. 2286 BC) [187: 189]:

> "First the tribes tried to plan the defeat of Tohil, Auilix, and Hacauitz. All the penitents and sacrificers of the tribes spoke to the others. They roused and summoned one another, all of them [...]. Let it be this way: if the fiery splendor of Tohil, Auilix, and Hacauitz is so great, then let this Tohil become our god! Let him be captured! Dont let them defeat us completely!"

(5) Terrestrial Cosmic Impact (ca. 2240 BC) [187: 197-198]:

> "We are going back to our own tribal place. Again it is the time of our Lord Deer, as is reflected in the sky. We have only to make our return. Our work has been done, our day has been completed. Since you know this, neither forget us nor put us aside. You have yet to see your own home and mountain, the place of your beginning. Let it be this way: you must go. Go see the place where we came from, were the words they spoke when they gave their advice. And then Jaguar Quitze left a sign of his being: This is for making request of me. I shall leave it with you. Here is your fiery splendor. I have completed my instructions, my counsel, he said when he left the sign of his being, the Bundle of Flames, as it is called. It wasnt clear just what it was; it was wound about with coverings. It was never unwrapped. Its sewing wasnt clear because no one looked on while it was being wrapped. In this way they left instructions, and then they disappeared from there on the mountain of Hacauitz. Their wives and children never saw them again. The nature of their disappearance was not clear. But whatever the case with their disappearance, their instructions were clear, and the bundle became precious to those who remained."

Mesopotamia: Destructions of Ur

The epic stories gathered thus far from Mesopotamia, the Near East, and Egypt do not exhibit meaningful chronological patterning, unlike those from China and Mesoamerica. This situation is likely simply the result of the archaeological record rather than the failure of Near Eastern cultures to recite epics in purposeful chronological order. Because of the lack of well defined chronological order, each story is here treated without reference to the numbering used for the Chinese myths.

Standard chronology places the rise and height of Mesopotamias Agade empire at around 2340-2159 BC, followed by the Third Dynasty of Ur at around 2112-2004 BC [103: 44-56]. There is a seeming interruption in normal activities and polities both before and after the Third Dynasty of Ur, and perhaps just before the beginning of the Agade empire. The Agade or Akkadian empire included the reigns of Sargon and his two sons, Rimush and Naram-Sin. There is a sizable mythology surrounding Sargon and especially Naram-Sin, the later who was deified in his own lifetime. In several respects, the stories about Sargon and Naram-Sin parallel those of Yao, Shun, and Yu, suggesting contemporaneity between the two sets of rulers.

Of special note in the literature surrounding the Third Dynasty of Ur is a long lament of some 436 lines, which has been entitled by it translator Samuel Kramer as *Lamentation Over the Destruction of Ur*. The tablets upon which this work was imprinted come primarily from Nippur, and have been dated to the period between the fall of the Third Dynasty of Ur and the beginning of the Kassite period in Mesopotamia (ca. 2000-1600 BC). Although the tablets do not refer to a specific ruler by which to tie the events being depicted, it does refer to depredations by Elamites and Subarians (Assyrians), whom scholars have mistakingly believed were the primary causes of the destruction to Ur [98: 455-463]:

> "Enlil [powerful Sumerian deity sometimes referred to as raging storm and wild bull] called the storm; the people groan. The storm of overflow he carried off from the land; the people groan. The good storm he carried off from Sumer; the people groan. To the evil storm he issued directions; the people groan. To Kingaluda, the tender of the storm, he entrusted it. The storm that annihilates the land he called; the people groan. The evil winds he called; the people groan. Enlil brings Gibil [god of fire] to his aid. The great storm of heaven he called; the people groan. The land-annihilating storm roars below; the people groan. The evil wind, like the rushing torrent, cannot be restrained; the boats of the city it attacks (and) devours, At the base of heaven *it made the...whirl*; the people groan. In front of the storm *fires burned*; the people groan. To the battling storms was joined the scorching heat. [...] fires burned. The day *was deprived* of the rising of the bright sun, *of* the good *light*, in the land the bright sun rose not, like the evening star it shone. The night *was deprived by* the South Wind of *its customary feasts and banquets; at the side of their cups* dust was piled high; the people groan. [...] The destructive storm makes the land tremble and quake; Like the flood storm it destroys the cities. The storm ordered by Enlil in hate, the storm which wears away the land, covered Ur like a garment, *enveloped* it like linen.[...] Its people, not potsherds, filled its sides; its walls were breached; the people groan. In its lofty gates, where they were wont to promenade, dead bodies were lying about; in its boulevards, where the feasts were celebrated, *scattered they lay*. In all its streets, where they were wont to promenade, dead bodies were lying about; in its places, where the festivities of the land took place, the people lay in heaps. The blood of the land, like bronze and lead...its dead bodies, like fat placed in the sun, of themselves melted away....My houses of the outer city have been destroyed - 'alas for my city', I will say; my houses of the inner city verily have been destroyed - 'alas for my house', I will say....In the rivers of my city dust has gathered, into *fox-dens* verily they have been made; in their midst no sparkling waters flow, gone is its *riverworker*. In the fields of the city there is no grain, gone is its fieldworker, my fields verily like fields torn up by the pickaxe have brought forth....My palm groves and vineyards that abounded with honey and wine verily have brought forth the mountain thorn. My plain where the *kazallu* and *strong drink were prepared* verily like an oven has become parched. My possessions like *heavy locusts* on the move verily...*have been carried* off - 'O my possessions', I will say....Verily my (precious) metal, stone, and lapis lazuli have been scattered about - 'O my possessions', I will say....*Woe is me*, the city has been destroyed, the house too has been destroyed; O Nanna [Moon god and patron of Ur], the shrine of Ur has been destroyed, its people are dead."

Most aspects of this lament refer to the "storm" as originating in Heaven, separate from human influence. The fact that the spoils of the city, such as precious metal, stone, and lapus lazuli are described as laying scattered over the ground likewise appears to support a natural versus human agency for the cause of destruction. The lament likely describes a terrestrial cosmic impact and its devastating effects, as is evidenced by fires, hurricane-force winds, dust obscuration, and the destruction of the cities themselves. Several specific cities are mentioned in the lament as having been destroyed and/or abandoned (e.g., Nippur; Kish; Isin; Uruk; Ur, Eridu; Larak; Umma, Lagash). These cities occupy an area of the Sumerian plain covering about 17,000 km^2, indicating that the impact was of at least of substantive local nature.

The Lamentation Over the Destruction of Ur bears at least superficial resemblance to another lengthy text, similarly entitled *Lamentation Over the Destruction of Sumer and Ur* [99], suggesting that the two laments describe the same event or series of events. However, the second lament omits any discussion of heavenly fire or scorching winds as part of the destruction, and makes the curious implication that Ur was being destroyed for a second time by the Elamites. (The "first" time may coincide with the lamentation, *The Curse of Agade*, discussed below). This second Ur lamentation is also of considerable interest in that it strongly suggests that a human hand, the Gutians, the Elamites, and other "men of the mountain", played a major role in the destruction, although perhaps as scavengers rather than as the originators of the destructive storm itself.

The *Lamentation Over the Destruction of Sumer and Ur* refers specifically to events which took place during or immediately after the last ruler of the Third Dynasty of Ur, Ibbi-Sin, at or around 2000 BC [99]:

> "An [Sumerian sky god] has made the Sumerians tremble in their dwelling places, the people are terrified. Enlil has made the day break bitter, has struck the city dumb [...] Enki has deprived the Tigris (and) the Euphrates of water [...] on the Land fell a calamity, one unknown to man, one that had never been seen (before and) for which there were no words, one could not be withstood. On all the lands, the terrified, a disruptive hand was placed, in their cities their city-gods stood aside, the people, the terrified, could hardly breathe. The storm *fettered* them, it returns not the day to them, the 'returned day' that it obtained for them *came* not *as* a [...] day. Enlil, the shepherd of the blackheads [Sumerians], this is what he did - Enlil, in order to destroy the righteous houses, to decimate the righteous, to set an evil eye on the sons of the righteous, the noble - On that day Enlil brought down the Guti from the mountain; and, whose coming is the Flood of Enlil, that none can withstand. They filled the steppe with great winds of the steppe [...] They *laid waste* to the [steppe] (and) whatever *flourished* in it, none could travel there. The [...] in the dark, dark days, They *overwhelmed* readily the bright day with *tumult* [...] On that day Heaven was crushed, Earth was smitten, the face was *blinded by the*

> *storm*. Heaven was darkened, was overcast in shadow, it was turned into the nether world. Utu [Sun god] lay (motionless) on the horizon [...] Nanna [Moon goddess] lay in [...] was terrified. [Missing lines 87-93]. The...w-ere piled up in heaps. The [...] were spread in heaps [...] In the Euphrates there were cadavers...are massacred [...] On Kazallu, the *awe-inspiring* city, a disruptive hand was placed [...] Its river has become empty, it poured no water. Like a river cursed by Enki, it came to an end at its source. In the field there was (neither) grain (nor) vegetation, the people had nothing to eat. Its orchards (and) gardens are parched like an oven, their *produce perished* [...] On that day the word - who knows its meaning? - attacked like a storm. The word of Enlil that winds to the right, *knows* the left. Enlil who decrees the fates, this is what he did: Enlil brought down Elam, the foe, from the mountain [...] On the bank of Idnun of Nanna a heavy arm was placed. The settlements of the Edanna of Nanna were destroyed like a distended stall. Those who fled from it were devoured by the wild beasts like fleeing kids [...] In Ur, no one took charge of food, no one took charge of water. Who was (formerly) in charge of food, stood away from the food, pays no *heed* to it. Who was (formerly) in charge of water, stood away from the water, pays no *heed* to it. Below, the Elamites are in charge, slaughter follows in their wake. Above, the Halma-people , the men of the mountains, took captives. The Tidnumites daily fastened the mace to their loins. Below, the Elamites, like those who bring forth woe, *brandish* their *weapons*. Above, like chaff blown about by the wind, the steppe [...] Zur, the great wild ox that (formerly) stepped forth confidently (in combat), has been made prostrate. Enlil, he who decrees the fates, this what [he did]: For a second time he brought down the Elamites, the men of the mountain from the mountain."

This second Ur lament, in its suggestion of strong winds and "dark, dark days" evokes the effects of a distant terrestrial impact, perhaps that of the Rio Cuarto impact in far away Argentina. If this is indeed the case, then the devastating drought noted in the text may be due to the disruptive effects on climate of the Rio Cuarto impact.

There is yet a third text which invokes the desstruction of cities in Mesopotamia. This is *The Curse of Agade* whose tablets date to the 18th century BC [100: 646-651]. This text is of interest for several reasons. First, it specifically links one period of widespread destruction in Mesopotamia to sometime during the reign of Naram-Sin, grandson of Sargon and fourth king of the Agade Dynasty. Standard dating [144: 199] places his reign at around 2254-2218 BC. Second, *The Curse of Agade* invokes the Gutians as at least partial causative agents for the destruction of Agade, thus lending a new dimension to the second Ur lament which states that the Gutians had ransacked Mesopotamian cities at least once prior to the destruction of Ur at around 2000 BC.

Yet another interesting aspect to *The Curse of Agade* is that the first nine lines of text provide some context for Sargons rise to power [100: 647]:

> "After the frowning forehead of Enlil had killed (the people of) Kish like the 'Bull of Heaven', after he had ground the house of Erech into dust, like a giant bull, after in due time, to Sargon the king of Agade, from below to above, Enlil had given him lordship and kingship, then did holy Ianna, the shrine of Agade, erect as her nobel chamber, in Ulmash did she set up a throne."

The imagery of this passage lends some support for the reality of a cosmic impact at around 2345 BC. Given the association of Enlil with the snake-dragon, given his attribute of being the "raging storm", and coupling these with the "other gods not being able to look upon his splendor" [19: 166], the "frowning forehead of Enlil" invokes an image possibly that of an impact fireball. Also of significance is the phrase "Bull of Heaven" which suggests an association with the Taurid meteor showers, or at least with a meteor shower not unlike the Taurids.

The remainder of the text of the *The Curse of Agade* conveys little imagery suggestive of cosmic impact. Therefore although the reign of Naram-sin presumably brackets the Chinese impact of circa 2240 BC., *The Curse of Agade* seemingly cannot be used to lend support to the Chinese data.

Ancient Palestine: The Destruction of Sodom and Gomorrah

One of the most poignant passages of the Bible (Revised Standard Version) is that in Genesis 19 which describes the fiery destruction of the cities of the Vale of Siddim: Sodom, Gomorrah, Admah, and Zeboiim; a fifth city in the vale, Zoar, was spared so as to provide sanctuary for Lot and his daughters:

> "Then the Lord rained on Sodom and Gomorrah brimstone and fire from the Lord out of heaven; and he overthrew those cities, and all the valley, and all the inhabitants of the cities, and what grew on the ground [...] And Abraham went early in the morning (24 hours later according to the JPS Torah Commentary [167: 139]) to the place where he had stood before the Lord; and he looked down toward Sodom and Gomorrah and toward all the land of the valley, and beheld, and lo, the smoke of the land went up like the smoke of a furnace."

The Jewish historian, Flavius Josephus, writing in the 1st century AD, describes this event as follows [94, I: 38-39]:

> "God then cast a thunderbolt upon the city, and set it on fire, with its inhabitants; and laid waste the country with the like burning [...] Now he and his daughters fled to a certain small place, encompassed with the fire, and settled in it: it is to this day called *Zoar*, for that is the word which the Hebrews use for a *small thing*. There it was that he lived a miserable life, on account of his having no company, and his want of provisions."

Elsewhere, Josephus makes the following comments based on his own visit to the area [94, IV: 130]:

> "The country of Sodom [...] was of old a most happy land, both for the fruits it bore and the riches of its cities, although it be now all burnt up. It is related how, for the impiety of its inhabitants, it was burnt by lightning; in consequence of which there are still the remainder of that divine fire, and the traces [or shadows] of the five cities are still to be seen."

Josephus also refers to the destructions as having been caused by thunder [94, IV: 269].

It appears from these passages that the Sodom and Gomorrah story encodes the observation of a terrestrial cosmic impact event not unlike that recorded for the Rio Cuarto and Broken Bow impacts and for the 1908 Tunguska impact. The gross magnitude of the event can be estimated from the story line. The impact was seemingly more energetic than

Tunguska, perhaps similar to that of Broken Bow, and clearly less substantial than Rio Cuarto. This is based on the presumption of the complete destruction of four or five cities and surrounding fields and other plant life in a single valley; burning lasting at least into the next day; the hiding in a cave; and the failure of regional populations to reoccupy the impact region, the location of which could still be identified two thousand years later.

Although the cosmic impact signature is manifest, what is not clear is the date of the impact event and the exact location of the impact. The impact occurred during the life of Abraham, which has been variously dated to between about 2300 and 1800 BC. Standard biblical chronology based on Bishop Ussher's genealogical date of 2348 BC for Noah's Flood, would place the birth of Abraham in the year 2056 BC, and would place the actual destruction of Sodom and Gomorrah in 1957 BC. And while most scholars believe the Vale of Siddim and the cities of Sodom and Gomorrah included the region around the southeastern portion of the Dead Sea, Pellegrino [154] and presumably others have suggested instead that the fertile plains of the Tigris and Euphrates rivers were the site of the doomed cities.

Additional details and potential resolution to these questions are provided in the remarkable collection of Jewish oral history gathered by Louis Ginzberg [69] at the beginning of this century. The first clue is the association of the birth of Abraham (Abram) with the sudden appearance of a bright new star while his father served in the court of Nimrod. If we accept the logic of a number of scholars that Nimrod was in fact the famous historical Naram-Sin, fourth king of the dynasty of Akkad in Mesopotamia, this would place Abraham in the middle of the 23rd century BC according to standard chronology [144].

We know from the Chinese Bamboo Annals that in the 42nd year of the reign of emperor Yao (ca. 2315 BC according to standard chronology), a likely brilliant supernova occurred in the constellation Crater; likewise, another supernova or nova event occurred in the constellation Scorpius in the 1st year of the reign of emperor Shun (ca. 2255 BC according to standard chronology). These are the only new stars unambiguously mentioned in the extant Chinese records between the reign of Huang Di (ca. 2357 BC) and that toward the end of Zhou Dynasty in approximately 532 BC, with the exception of a possible nova in Scorpius recorded on oracle bones from the Shang Dynasty in the 14th or 13th century BC [39, 141]. While there were surely other significant naked eye novae and supernovae between 2650 BC and 532 BC, we can assume that the three noted in the 24th and 23rd century BC were of special character and are likely candidates for the birth star of Abraham. This would then suggest that the destruction of Sodom and Gomorrah, which allegedly occurred in the 99th year of Abrahams life, took place during the period of around 2216 BC to 2156 BC. Given the emphasis by the Chinese on the supernova or nova-like event of circa 2287 BC, which occurred at the time of the ascension of Emperor Shun to power, the most likely date for the Sodom and Gomorrah impact would be that of circa 2188 BC.

Egypt: End of the Old Kingdom

I have not yet been able to perform a detailed analysis of Egyptian myth and archaeology during the period of 2350 BC to 2000 BC. However, it is noteworthy that according to standard chronology [96] the beginning of this period of time, 2350 BC, marks the transition between the Fifth and Sixth Dynasties, and that the weak Seventh and Eighth Dynasties and the turbulent First Intermediate Period follow shortly after 2200 BC and reach their conclusion at around 2000 B.C. Thus the dates of 2350 BC, 2200 BC, and 2000 BC represent times of substantive change in Egypt.

It is also of interest that after the steller orientation of Third dynasty architecture and cosmology and the subsequent change to more solar emphasis with the Fourth and Fifth Dynasties, with the development of the Pyramid Texts at the time of King Unas at the end of the Fifth Dynasty there is again a return to stellar imagery. This steller imagery persists throughout the Sixth Dynasty. It is my belief that this reorientation towards stars reflects first the hypothesized cosmic impact at around 2345 BC, and then the three subsequent supernovae or novae-like events in Crater and Scorpius noted in the Bamboo Annals between 2314-2254 BC.

One of the utterances in the tomb of Unas may even reflect the cosmic impact of circa 2345 BC. The beginning of utterances 273-274 ("The King Feeds on the Gods") is very suggestive in its imagery [110: 36], and invokes aspects of the earlier noted tale of Yi, the Divine Archer:

> "Sky rains, stars darken, the vault s quiver, earth's bones tremble, the planets stand still at seeing Unas rise as power, a god who lives on his fathers, who feeds on his mothers! Unas is master of cunning whose mother knows not his name; Unas's glory is in heave, his power is in lightland; Like Atum, his father, his begetter, though his son, he is stronger than he! The forces of Unas are behind him, his helpers are under his feet, his gods on his head, his serpents on his brow, Unas's lead-serpent is on his brow, soul-searcher whose flame consumes, Unas's neck is in its place. Unas is the bull of heaven who rages in his heart, who lives on the being of every god, who eats their entrails when they come, their bodies full of magic from the Isle of Flame [...] Unas eats their magic, swallows their spirits: Their big ones are for his morning meal, their middle ones are for his evening meal, their little ones for his night meal, and the oldest males and females for his fuel. The Great Ones in the northern sky light him fire for the kettles' contents with the old ones' thighs, for the sky-dweller serve Unas, and the pots are scraped for him with their womens legs [...] Unas has risen again in heaven, he is crowned as lord of lightland [...]."

There is one other interesting Egyptian text that although- written in the Nineteenth or Twentieth Dynasty (1295-1069 BC, according to Kitchen [96], may reflect the period of 2350-2000 BC. I suspect, however, that the text can also equally as well reflect the Second Intermediate Period at the time of the Hyksos This is the text referred to as *The Admonitions of Ipu-wer* [207: 441-44, 110: 149-163]. This text is in the same genre as the second Ur lamentation and especially *The Curse of Agade*. While there is nothing that specifically encodes a cosmic impact, it is clear that Egypt is undergoing at least some form of social disruption. Unlike the views of Lichtheim [110: 149-150], I see no reason to assume that the text is literary propaganda or ahistorical in content. Based on my cosmographic researches, the text represents an individual's (Ipu-wer) view of the state of his society at some point between 2350-1550 BC.

Evaluation of the 2350-2000 BC cosmic impacts

In each case presented above, that of the destruction of Ur (especially the first lament), the destruction of Sodom and Gomorrah, and the great windstorm witnessed by Chinese emperor Shun, the mythic story line is indicative of a cosmic impact. The question then becomes: How many impacts? - when? - where? - and of what magnitude?

The impact witnessed by Shun (ca. 2240 BC) appears most

similar in magnitude and effect to the Tunguska impact described above, a distant impact as witnessed several hundred km from the impact site itself. It does not seem likely that this impact is directly connected with the Ur or the Sodom and Gomorrah impacts (which would imply a regional impact significantly larger than that of Rio Cuarto, for which there is no good evidence based on the lack of recognizable craters), unless as part of a single swarm of Tunguska-like meteorites with multiple impacts. The lack of definable cosmic impact language (other than that for Sargon) in *The Curse of Agade* reinforces the limited nature of the circa 2240 BC impact.

The Ur impact (of the first lament) and the Sodom and Gomorrah impact describe the direct effects of terrestrial impacts, but there is ambiguity about both the dating and the magnitude of the events. If we assume that the birth of Abraham did coincide with one of the bright nova/supernova events in the Bamboo Annals, and if we assume that Abraham was indeed 99 years old at the time of the destruction of Sodom and Gomorrah, then this event took place at around 2215 BC, 2188 BC, or 2155 BC, with the 2188 BC date being most likely given the great significance given by the Chinese to the supernova or nova-like event of circa 2287 BC in the 70th year of the reign of Emperor Yao.

Using the standard "middle" chronology for Mesopotamia, the date of the impact of the second Sumer and Ur lament would have occurred at around 2000 BC, at the end of the reign of Ibbi-Sin. As previously noted, there are enough differences between the two laments to suggest that two different impacts are represented. The first impact may be the same as that described for Sodom and Gomorrah, which in keeping with the logic of Pellegrino [154] who has argued that Sodom and Gomorrah, in actuality, are referents to cities on the Mesopotamian plain. The archaeological evidence for widespread destruction layers at around 2300 BC discussed by Peiser (this volume), suggests that a cosmic impact may have been a factor, a date which fits well with the estimate of 2188 BC for the Sodom and Gomorrah impact and possibly with the impact described in the first Ur lament.

However, a date of around 2345 BC is also possible for the first Ur lament, since it cannot be directly ascertained from the list of destroyed cities that what was destroyed represented either the Agade Empire or that of Dynasty III at Ur. This dating then raises the specter that the reference in the Bamboo Annals (the story of Archer Yi killing the nine suns) to a brief period of "intense heat" and abnormal solar activity at around 2345 BC, is the product of a cosmic impact rather than some kind of unusual solar activity. A clay plaque from the Mesopotamian city of Khafajah depicts a god armed with a bow and a knife killing a fiery cyclops [144: Fig. 36], possibly alludes to this same story.

In sum, the period of 2350-2000 BC witnessed at least four cosmic impacts (ca. BC 2345, 2240, 2188, 2000), and perhaps a fifth (ca. BC 2297-2265). The circa 2000 BC event is linked with that of the large regional terrestrial cosmic impact at Rio Cuarto, Argentia.

The circa 2345 BC destructive layers throughout much of the Near East and Mesopotamia are arguably the result of a large local or regional cosmic impact. The lack of a known crater associated with this hypothesized regional impact is admittedly disturbing, with one possible explanation being that the impact occurred in or next to a body of water, such as a large lake (e.g., the Dead Sea, Lake Baikal), or the edge of an ocean (e.g., Red Sea, Persian Gulf, the Mediterranean Sea) which has helped to hide the crater. Alternatively, a crater could be hidden by sand, such as was the case for the Wabar crater in Saudia Arabia [85: 118-119]. Another possibility is that rather than a single impact, the circa 2345 BC event was the product of a swarm of Tunguska-like impacts such as modeled by Clube and Napier [43] and other advocates of coherent catastrophism.

The event of circa 2240 BC is likely a local terrestrial cosmic impact restricted to Asia, while the circa 2188 BC event is seemingly a local terrestrial cosmic impact near the southeastern shores of the Dead Sea.

The circa 2297-2265 BC event is the least well documented of the five hypothesized cosmic events and may represent an oceanic (or large lake) cosmic impact in the vicinity of Southeast Asia. An example of the mythology associated with this lengthy period of "flooding" is as follows [212: 98]:

> "The regulation of the rivers was fraught with dangers and setbacks. One year Yu and his people were working in Ba Zhu at the foot of Wushan Mountain (now in Sichuan Province). They were cutting trees and digging the earth when a sudden violent strom blew up. Black clouds darkened the sky, rocks flew through the air on heavy winds, the very cliffs and mountains trembled. Rough waves blew up like hills, and Yu's people scattered out of the way of disaster like autumn leaves. Powerful as he was, Yu could not control the storm."

Perhaps rather than flooding caused by cosmic impact, these stories simply reflect a lengthy period of unusual weather events and conditions caused by factors similar to those producing El Niño and similar climatic perturbations. However, there are enough tantalizing images so as to not be able to entirely rule out one or more oceanic/large lake cosmic impacts, such as the presence in some stories of Kung Kung (who represented the Flood Comet impactor of 2807 BC), and the fact that Yu had a miraculous birth in which he was born in the shape of a dragon.

10. Cosmic impacts of circa 1800 BC

The *Bamboo Annals* and other Chinese sources suggest that several striking celestial and environmental events occurred during the reign of Jie, the last emperor of the Xia Dynasty. Standard chronology indicates that this reign occurred between 1818-1766 BC, although the Bamboo Annals indicates a reign of 31 years for Jie rather than the 52 years of the standard chronology. Of interest are three passages in the *Bamboo Annals* [106: 125-127]. The first reads:

> "In his 10th year, the five planets went out of their courses. In the night, stars fell like rain. The earth shook. The E [I river] and Loh [Lo river] became dry."

"Stars falling like rain" is an apt description of a great meteor storm. Unfortunately, there is not enough detail (e.g., month, constellation, season) provided to suggest a particular meteor shower. The earthquake associated with the meteor storm may be indiciative of meteorite or even larger cosmic impact. The description of the five planets going out of their courses is odd. We do know that the period of 1794-1793 BC witnessed an unusual triple conjunction of Jupiter and Saturn in the constellation Leo, near Regulus, and that four of the five planets were in Leo just prior to summer solstice in 1793 BC (Multimedia Astronomy Redshift version 1.2°, by Maris). Since the constellation Leo is representative of a lion in Mesopotamian astronomy [3, 70], the presence of conjunctions or other unusual celestial events in and around Leo may have played a role in the prevalence of lions and griffins displayed on cylinder seals [44].

The second passage reads: "In his 29th year, the forces of Shang took Koo. Three Suns appeared together". Yuan Ke

[212: 136] describes this event as follows:

"One of King Jie's trusted followers was a man called Fei Chang. One day as he was strolling on the bank of the Yellow River, he noticed two suns in the sky, one rising brilliantly in the east amid coloured clouds, the other setting into a grey mist in the west. Terrible thunder echoed from the heavens, and Fei Chang was deeply troubled by the strange power of these signs. He thought to himself, 'There cannot be two suns in the sky; the people cannot be governed by two monarchs.'"

The third passage reads:

"In his 31st year, Shang proceeded by way of Orh against the capital of [Xia]; and overcame [King Jie]. Amid great thunder and rain a battle was fought in Ming t'eaou, when the army of [Xia] was defeated."

The speech of King Cheng Tang, who ousted King Jie and founded the Shang Dynasty , is recorded in the *Shu Ching* or Book of History [107: 186-187]:

"The way of Heaven is to bless the good and to punish the bad. It sent down calamities on the House of [Xia], to make manifest its crimes. Therefore, I, the little child, charged with the decree of Heaven and its bright terrors, did not dare to forgive the criminal."

While the above descriptions are not nearly so convincing as those for the 2350-2000 BC period, it is possible that "Heavens calamities and bright terrors" may describe, among other things, a Tunguska-like impact. Unfortunately, the Egyptian and Mesopotamian literature appear silent in this regard. However, it is of interest that the period of ca. 1800-1750 BC is a time of considerable flux in Egyptian history (the end of the Twelfth Dynasty and the start of the Second Intermediate period), and in the Near East (the rise of the Hittite kingdom), as is also the case for China.

The one possible allusion to this hypothetical impact event outside of China is that of a fragmentary Hittite myth [71: 120], which is suggestive of an oceanic or large lake impact:

"The Moon-god [...] fell down from heaven. He fell upon the *kilamma*. But no one saw him. The Storm-god [...] sent rain after him, he sent rainstorms after him so that fear seized him. [...] The storm-god [...] saw him and he sent rain after him, he sent rainstorms after him, he sent the winds after him so that fear seized him (and) fright seized him."

Unfortunately, the text does not specify a particular ruler or other chronological referent by which to approximate a date for the presumed impact, although it is from an archaeological context suggesting sometime between 1800 and 1200 BC Also in possible support of an oceanic or large lake impact at around 1800 BC is the fact that once impressive cities in the Indo-Iranian borderlands, such as Mohenjo-daro, suffered one or more devastating floods at or shortly before this time. These flood events apparently led to the dissolution of the once fluorescing Harrapan culture [150: 24].

11. Other potential cosmic impacts and volcanic eruptions

With the exception of the mythology of the Flood Comet impact, which is detailed elsewhere [125], the texts discussed above provide the clearest depictions of cosmic impact events. Nevertheless, it is possible that even rather innocuous-seeming statements about unusual weather or atmospheric conditions in traditional histories may relate to cosmic impact effects.

For example, several passages in the *Chinese Spring and Autumn Annals* (*Ch'un ch'iu*) and the *Bamboo Annals* (*Chu-shu chi-nien*) describe the Milky Way and other celestial objects as being invisible at midnight on cloudless nights. I strongly suspect that this statement refers to the curious atmospheric effect describe above for the Tunguska impact. In a similar vein, it is stated for Shang Dynasty King, Wu Yi, in the Bamboo Annals that around 1200 B.C. (According to standard chronology) that Wu Yi [106: 138]: " [...] was hunting between the Ho and the Wei [rivers], when he was frightened to death by a great thunderstorm." Wu Yi's curious death could be attributable to the effects of a distant Tunguska-like impact event.

It is important to emphasize that not every description of spectacular temporary celestial events in mythology is necessarily an encoding of a cosmic impact event. For example, several of the authors in this volume have highlighted the period of 1200-1000 BC as having witnessed catastrophic destruction throughout much of the Near East. Because of striking celestial phenomena described in Chinese and Biblical sources for this time period, I assumed that at least some of the destruction observed during this period was due to cosmic impact.

I early on thought that the famous battle (ca. 1122 BC.) between Chou Wang (Chou Hsin), last emperor of the Shang Dynasty, and Wu Wang, founder of the Zhou (Chou) Dynasty, wold contain evidence for impact events (still a remote possibility). During this epic battle, the forces of the tyrannical Chou Wang were routed at the Huang Ho river. Allegedly Chou Wang returned to his pleasure palace, built a huge funeral pyre on one of its terraces, and immolated himself.

The main reason for my interest in this battle was the fact that there is a rich mythology about a mighty cosmic battle taking place at the same time as the Earthly battle. Wu Wang and Chou Hsian employed the aid of legions of supernatural and celestial spirits [194: 123-125]:

"Li No-cha overcame the soldier-immortal Feng-lin by means of a Heaven-Earth bracelet. The soldier called on his superior, Chang Kuei-fang to assist him, but Li No-cha mounted his Wind-Fire wheel, on which he could travel about at rapid speed. Although Li No-cha heard his name called out three times, he continued to fight, and eventually broke Chang Kuei-fang's arm with his Heaven-Earth bracelet. [...] / On arriving at K'un Lun, he [Chian Tzu-ya] made his way to the Jade Palace, and was presented to Yüan-shih. The Old Man of the South Pole [ruling deity of the Star of Longevity] gave him a list of Promotions to Immortality, and was told to erect a Spirit Terrace to display the list of honored names....Shen Kung-pao also tried to coax Chiang Tzu-ya into giving up the list of Promotions to Immortality. When Chiang Tzu-ya resisted, Shen Lung-pao offered to wager his magical skills against those of Chiang Tzu-y, telling the general that he could take off his head and cause it to float through space [...] Shen Kung-pao took his sword and cut off his own head, and threw it into the air. Fortunately, the Old Man of the South Pole had been keeping a careful watch, and seeing that Chiang Tzu-ya was about to lose the vital document, sent the White Crane Youth, Pai-ho T'ung-tzu to carry off the head....[T]he Old Man of the South Pole challenged Chang Shao, who tried to destroy him with hot sand. The venerable immortal easily encountered the attack with his seven-feathered fan, and was unharmed, while the attacker was swiftly dispatched by the White

Crane Youth's jade scepter."

It is likely that large near-Earth comets and novae/supernovae, are represented in this story, but there are no clear indications of actual impact events. Likewise, the documentary record strongly implicates the hand of real armies in the 1200-1000 B.C. destruction in the Near East, rather than supernatural (i.e., cosmic) forces [55, 103].

Nevertheless, it is curious that the hypothetical cosmic battle around 1122 BC would create such a large body of myths in Chinese literature, as well as being the presumed subject matter of other great epics dating from this approximate time period, including the Sanskrit poem *Mahabhaata*, and the Homerian epics of the Iliad and the Odyssey. Because there is such a great literature on presumably this same set of celestial events, detailed cosmographic analysis should provide at least partial reconstruction of the temporary celestial events themselves. It may well prove once this analysis is done that bolides and/or Tunguska-like events may have played a role in the seemingly spectacular nature of these celestial events.

Future cosmographic analyses of many basic mythological motifs, such as the battles or adventures of supernatural hero twins or pairs, should tell us much about spectacular celestial phenomena and events that have been observed for countless millennia. In the specific case of twins or pairs, I suspect that at least some of these epics may shed light on the history of our encounters with giant and/or fragmenting comets. Biblical statements such as "Enoch walked with God; and he was not for God took him" (Genesis 5:24, RSV) and related texts such as the Enochic Book of the Giants [59], the pseudepigraphal Books of Enoch [35], and the legends compiled by Ginzberg [69] are replete with imagery suggestive of the observation of giant comets and other celestial objects and phenomena.

Once we have completed such analysis, we should have an improved understanding about much of what has been mysterious in the past. For example, curious images of the late 4th millennium BC from the Temple of the Eyes at Tell Brak (Figure 5), are obvious representations of tailed comets, and are strikingly similar to Hawaiian wooden image depictions of comets. Some of these images have multiple or paired eyes suggesting paired or fragmenting comets. This may also be the inspiration for multiple-eyed or multiple-faced deities (e.g, the Roman god of gates, Janus), demigods, and culture heroes. For example, the *Bamboo Annals* describe several of the legendary early rulers as having multiple eyes, pupils, and faces. However, at least some of this symbolism, such as that with Janus, may simply reflect cardinal directions or the path of the Sun.

The volcanic destruction of Thera Island

Returning to the topic of the nature of environmental events encoded in mythology, it is important to emphasize that not every description of a major environmental catastrophe in mythology is necessarily an encoding of a cosmic event. For example, those familiar with the descriptions and photographs of the 1980 eruption of Mount St. Helens and the 1991 eruption of Mount Pinatubo should appreciate the eruption encoded in the following passage from Hesiod's *Theogony* [23: 76-77]. Contained in this passage are all the elements of a massive volcanic eruption on an oceanic island: The hiss and roar of billowing clouds of steam; the roar of snake-like jets of dark ash; the whistling of incendiary volcanic bombs; the terrifying deadly rush of pyroclastic flows (nuée ardente); earthquakes; violent tsunami; dazzling electrical storms triggered by the uplifting and wide spreading ash column; the eerie glow of magma spilling out the bowels of the Earth; forest fires sparked by magma, pyroclastic flows, and incendiary bombs; and finally, the eventual destruction and subsidence of much of the island itself. This passage is given in its entirety so as to better distinguish the mythology of violent eruption from that encoding the effects of cosmic impact catastrophe:

> "After Zeus had driven the Titans from the sky, monstrous Earth gave birth to her youngest child Typhoeus [Typhon], after being united in love by golden Aphrodite with Tartarus [a Hell-like region of the underworld]. Typhoeus is a god of strength: there is force in his active hands and his feet never tire. A hundred snake heads grew from the shoulders of this terrible dragon, with black tongues flickering and fire flashing from the eyes under the brows of the prodigious heads. And in each of those terrible heads there were voices beyond description: they uttered every kind of sound; sometimes they spoke the language of the gods; sometime they made the bellowing noise of a proud and raging bull, or the noise of a lion relentless and strong, or strange noises like dogs; sometimes there was a hiss, and the high mountains re-echoed. The day of his birth would have seen the disaster of his becoming the ruler of men and gods, if their great father [Zeus] had not been quick to perceive the danger. He thundered hard and strong, so that Earth and broad Sky above, Sea and Ocean-streams, and the Tartarus region below the earth, all rumbled with the awful sound. Great Olympus quaked under the divine feet of its royal master as he rose up, and the earth groaned also. The heat from both sides, from the thunder and lightning of Zeus and from the fiery monster, penetrated the violet deep and made the whole earth and sky and sea boil. The clash of those immortal being made the long waves rage round the shores, round and about, starting a convulsion that would not stop. Trembling seized Hades, king of the dead in the underworld, and the Titans who stood by the Cronus and who live at the bottom of Tartarus. When Zeus' energy had risen to the peak, he took his weapons, thunder and lightning and the smoking thunderbolt, and jumped on his antagonist from Olympus and struck. He blasted all those prodigious heads of the terrible monster and dealt him a flogging until he was tamed. Typhoeus fell down crippled, and the monstrous earth groaned underneath. Flame streamed from the once powerful potentate, now struck by lightning, in the dim clefts of the rocky mountain where he fell. Large tracts of the monstrous earth were set on fire by the prodigious heat and melted like tin heated in moulded crucibles by skillful workmen, or like iron, the strongest metal, softened by the heat of fire in some mountain-cleft; even so did the earth melt in the flame of the fire then kindled. Zeus, in the bitterness of his anger, threw him into the abyss of Tartarus."

That this powerful myth encodes a cataclysmic volcanic eruption is nothing new and has been deduced by a number of other scholars [72], some of whom even link the tale specifically to the regionally catastrophic Bronze Age eruption of the Santorin stratovolcano on Thera island in the Aegean Sea [112, 154, 192]. There are several reasons to associate Typhon with the destruction of Thera.

The imagery of the Typhon myth closely reflects what we know and can reconstruct about the Santorin eruption, including the devastating nature of tsunami associated with the eruption [154: 207-231]. Second, the relative placement of this myth in alleged chronologically ordered cosmographic compendia such as Hesiod's *Theogony* and Ovid's *Metamorphoses* is between the Flood (2807 B.C.) and the fall of Troy (12th century BC. This can be further refined by linking the myth of Yi the Divine Archer shooting nine of ten suns from the sky in 2345 BC, with that of the disastrous sun-chariot ride and fall from the sky by Phaeton, an event which occurred prior to the appearance of Typhon in *Metamorphoses* [133]. The destruction of Thera in the 17th or 16th centuries BC seemingly represents the largest eruptive event in the Mediterranean during the Middle and Late Bronze Age, matching the image of Typhon as "the largest monster ever born" [72]. Also of interest in later versions of the Typhon story [72: 134] is reference to Zeus placing northeastern Sicily's Mount Etna (Aetna) on top of Typhon so that he cannot escape. This allusion may take on meaning in that evidence suggests that while Mount Etna first erupted some 2.5 million years ago, the present eruptive cycle for Mount Etna began around 1500 BC, with some 150 recorded eruptions since then to the present [50: 258]. I suspect that part of the purpose of the Mount Etna portion of the Typhon story is to formally transfer the role of "mightiest volcano" from Santorin to Mount Etna (the highest active volcano in Europe) after the destruction of Thera island and the beginning of the most recent eruptive cycle of Mount Etna around 1500 BC.

Also, it is of some note that in some versions of the myth [133: 109] the Olympian gods flee into Egypt to escape Typhon's wrath, a situation that Graves [72] sanguinely suggests represents the flight of frightened priests and priestesses from the path of the eruption and ash fall. This general time frame is unquestionably tied to the *Hyksos* period of Egypt during the Second Intermediate Period (Thirteenth Dynasty through the Seventeenth Dynasty), with a beginning date of around 1786-1720 BC to an ending date of about 1550 BC [96, 103]. The Hyksos period is presumed to represent the intrusion into Egypt of possibly Semitic peoples from the Levant and elsewhere [103: 173-182], who actually acceded to the kingship of Egypt (the "Shepherd Kings" so-called by Manetho).

Of particular interest in this regard is a passage from the historian Manetho (as cited by Josephus), dating from the reign of an obscure Egyptian king "Tutimaeus" during the Second Intermediate Period. This passage appears to allude to the Santorin eruption and to the movement of peoples in its wake [193: 79]:

> "Tutimaeus. In his reign, for what cause I know not, a blast of God smote us; and unexpectedly, from the regions of the East, invaders of obscure race marched in confidence of victory against our land."

I am especially struck by the name Tutimaeus, which other translations have recorded as "Timaus" [94, IV: 425). The similarity between this name and Plato's *Timaeus* of the infamous Atlantis dialogue is most intriguing. This possible etiological relationship is even made more intriguing by the fact that Plato's *Timaeus* is an unknown personage who does not appear in any other of his dialogues [108: 450-451]. I expect that future more detailed cosmographic analyses will eventually convincingly couple the Atlantis myth with the destruction of Thera island by Santorin volcano.

Further discussion of these issues and this general time period lie beyond the scope of this paper. My preliminary cosmographic researches indicate that there is great potential for the reconciliation of regional chronologies. For example, the seven year-long famine detailed in the Biblical story of Joseph (Genesis 41-47, R.S.V.) is seemingly related to a similar seven-year drought recorded in the Bamboo Annals during the reign of Shang Dynasty emperor Tang [106: 129]. The date for the drought in China is around 1746-1739 BC according to standard chronology. Likewise, my data generally agree with a late 17th century BC date for the cataclysmic destruction of Thera island. However, there are other events, perhaps including cosmic impacts and/or other eruptions, likely taking place during the period of about 1800 to 1500 BC, thus confounding our understanding. We have much still to learn about this most interesting time period.

12. Giant comets, the Taurids, and coherent catastrophism

Clube and Napier [42, 43] have long pointed out the potential danger to Earth posed by large comets and their debris streams, singling out especially the Taurid meteor streams as a potential hazard. In collaboration with their colleagues [5, 11] they have created the concept of *coherent catastrophism* in which the periodic but regularized disintegration of giant comets (comets whose initial diameters exceed about 100 km) poses a particular threat to planets in the inner Solar System, including Earth. This disintegration process is modeled as taking place over the course of about 20,000 years on the average for each giant comet.

In order to provide support for their model, they have focused on the history of the Taurid meteor streams and likely related objects (e.g., Comet Encke and a number of other Earth-crossing objects). It is their belief that these objects and debris streams represent a late stage in the disintegration of a Taurid giant progenitor comet, and that some known historic impacts are related to the Taurid meteor streams, including the 1908 Tunguska impact.

While my cosmographic research has not yet exhaustively focused on the topic of coherent catastrophism, the general data collected so far do, in fact, provide some support for the reality of giant comets (as defined above) and the historic importance of the Taurid meteor streams. It is noted that the Taurids are composed of a double radiant, the northern of which drifts westward adjacent to the Pleiades and the southern which drifts westward through the Hyades [21]. The Earth passes through these debris streams twice a year, with activity peaks currently in early November and at the end of June.

Various cultures worldwide, but especially in Australia and in North and South America, connect the Pleaides with the origin of fire [66], including the fact that the Pleiades is considered an actual representation of a fire drill. The imagery surrounding the Pleiades and the origin of fire is that of a great meteor storm. I rather expect that the previously described stories from the Popul Vuh and from the Gilgamesh epic (and possibly the Bamboo Annals) encode the first-ever Taurid event, or at least a very early episode of the Taurids. Curiously, a Nahuatl myth states that this event happened in the second year after the Flood [16: 142-145]. In addition, Jewish legends suggest a clear relationship between the Flood, the Ursa Major nova-like event of ca. 2700 BC, and events in the Pleiades, although the Pleiades events may represent novae rather than Taurid meteor showers [69: 162]:

> "The flood was produced by a union of the male waters, which are above the firmament, and the female waters, issuing from the earth. The upper waters rushed through the space when God removed two stars out the constellation Pleiades. Afterward, to put a stop to the flood, God had to transfer two stars from the constellation of the Bear to the constellation of the Pleiades. That is why the Bear runs after the Pleiades. She wants her two children back, but they will be restored to her only in the future world."

The range of dates indicated by these various stories - no earlier than 2805 BC (for the Nahuatl story suggesting the creation of fire two years after the Flood) and not much later than 2650 BC - are remarkably close to the original estimate made by Fred Whipple and Salah Hamid [201] for the breakup of the progenitor to Encke's Comet and the creation of the Taurid streams. Whipple and Hamid came up with a date of about 2700 BC for this event.

The Pleiades asterism (and its parent constellation, Taurus) is a prominent fixture throughout worldwide mythology and early astronomies, which is curious for a asterism with no star brighter than the third magnitude. The significance of the Pleiades goes beyond its utility as a seasonal marker. I suggest that its importance for virtually all cultures is due in large part to the appearance and character of the Taurid streams during the past 4700 years, although the Pleiades may also be unusual in terms of the production of bright nova-like events such as occurred around winter solstice in AD 684 [123, 131].

Further cosmographic analyses should help us to understand just how big of a threat to the Earth that the Taurid streams have posed. The theories of Asher *et al.* [5] and Bailey *et al.* [11] are likely testable through more detailed cosmographic analysis. However, by factoring in precession (in which meteor streams advance calendrically approximately 22 days per millennium), it can be determined that several of the larger impacts of the past 5,000 years do not appear likely Taurid candidates. The Flood Comet impact does not appear to be associated with any modern meteor stream (the equivalent today of ca. 26 January); the Broken Bow impact (which occurred in late fall) would possibly be associated with meteor streams occurring in the period of late July through early September.

The season and dates of the Rio Cuarto impact and those impact(s) which occurred during the period of 2350-2200 BC are presently unknown, although additional texts may be found which encode such information. And it is possible that some of the potential smaller impacts are associated with the Taurid complex. It would prove valuable for Chinese historians to more closely study the original texts so as to attempt to derive firmer lunar and monthly dates to compare against the annual cycle of cometary debris streams.

13. The rates, effects, and risks of cosmic impacts

"Risk" is the degree of probability to which a dangerous element or factor (a "hazard") can create a specified loss. Anyone who has seriously studied the subject of risk in the arenas of health, communication, and public policy realize that meaningful models of risk assessment require the careful gathering, analysis, and interpretation of pertinent data, and likewise require an objectively dispassionate presentation of the results to the public. The consequences for improperly assessing or for miscommunicating risk are profound and potentially harmful in and of themselves.

Currently accepted models of the risk of cosmic impact [34, 136, 137] are carefully and soundly based on cratering rates defined for the Moon, Earth and other planetary bodies, on assumptions regarding the total numbers of Earth-crossing asteroids and long-period comets, and on theoretical models of effects on Earth by different types of impactors. These models suggest that a locally catastrophic 15 megaton (MT) Tunguska-like impact can occur somewhere on Earth on the average roughly of once every 200 years. A 100 MT impact, which is still considered a modest local impact, is expected on the average of about one per 1,400 years. The threshold for a regionally catastrophic impact, as previously defined, lies at about 750 to 1000 MT (1 gigaton), and would be expected to occur by present models on the average of one per 10,000 years. A "globally" catastrophic impact is defined as an impact directly and indirectly capable of causing the death of at least one quarter of the Earth's human population (Chapman and Morrison 1994); the energy threshold for such an impact has been put at around 300 gigatons (300,000 MT), or roughly the equivalent of 1,500,000 Hiroshima atomic bombs. Present risk models suggest that globally catastrophic impacts occur roughly on the average of once every 300,000 years. However, because of the uncertainties in the data base, the risk models allow that the actual risk could be somewhat higher (up to twice as high), or could also be somewhat lower.

I sincerely hope that my archaeological and historian colleagues can get past the issue of chronology in this paper and the question of whether "standard" chronologies are operative in the Old World or should be revised upward or downward in various regions. Absolute chronology is indeed important, but it is not essential for assessing the general risks of cosmic impact.

Leaving aside for the moment the present imprecision for the dating of some of the impacts described in this paper, the past 5,000 years of human history has been shaped, at least in part, by at least two and likely three or more cosmic impacts of unusual magnitude. These include the globally catastrophic impact Flood Comet impact in 2807 BC, an event which is supposed to take place on the average only once every 300,000+ years; the regionally catastrophic Rio Cuarto impact, possibly of 2000 BC, an event which is supposed to take place in the neighborhood of about once every few tens of thousands of years. In addition; a large local/regional catastrophic impact or swarm of smaller local impacts likely occurred in the Near East and possibly western Asia around 2345 BC, an event which should take place about once every 10,000 years or so. And based on my analysis of the Flood Comet impact [125, 126], it is likely that the Younger Dryas geological event [111: 342-356] was precipitated around 9300 BC by a globally catastrophic oceanic impact of considerably greater magnitude than the Flood Comet impact. The Younger Dryas represents a thousand-year blip of cooler temperatures in the warming trend that began at the end of the last ice age. Notable is the fact that worldwide temperatures plunged drastically up to 7° at the very beginning of the Younger Dryas during a period of less than 20 years. Also notable is the fact that several species of large mammals were extirpated or became extinct at about this time, especially in the New World.

Setting aside the unresolved issue of the Taurid meteor streams and coherent catastrophism, and for the moment assuming a model of simple stochastic flux for cosmic impacts, the risk to human life posed by cosmic impacts currently is underestimated by a factor of between about 15 and 30. Rather than globally catastrophic impacts occurring on the average of about once every 300,000 years, I instead conservatively suggest that such impacts occur on the average of about once every 10,000 to 20,000 years. If we were to simply and to uncritically use the impact data from the past 12,000 years, presuming the reality of the hypothesized Younger Dryas oceanic impact, the rate of globally catastrophic cosmic impact actually increases to that of once every 5,000 to 6,000 years.

Soberingly, the range of 6,000 to 20,000 years suggested here for the average rate of globally catastrophic cosmic impact matches the general incidence of periods of rapid substantive change in climate observed in the palaeoclimatic record of the Pleistocene and Holocene periods [111: 324-373]. This is not to imply that every Quaternary period climate change signals a cosmic impact. However, in those several cases in which change is perceived to be extremely rapid, occurring in the space of a few decades as opposed to centuries, cosmic impact should be considered a likely contributing cause.

Equally sobering is the prevalence of eschatological traditions which tell of multiple catastrophic destructions (followed by new creations) of the world. For most cultures the last such creation matches the last globally catastrophic cosmic impact, that of the Flood Comet impact of 2807 BC. In humankind's role as cosmographic observers of the universe around us, this situation is best explained by assuming that the last three or four large regional or globally catastrophic cosmic impacts, ending with the Flood Comet impact, have become enshrined in traditional cosmogonic histories. In addition, local cosmic cataclysms are common enough and of such striking character so as to reinforce humankinds eschatological traditions about multiple destructions and creations.

In terms of actuarial tables [52: 105], instead of the presently modeled rate of death by cosmic impact at 1 in 20,000 (the same as the chances of dying in a passenger aircraft crash), the adjusted rate would be somewhere between 1 in 1,333 (nearly twice as great as the chances of dying in a firearms accident) and that of 1 in 400 (slightly larger than the chances of death by murder), or about the same as that for a lumberjack dying on the job, which is one of the most dangerous professions. Even the lesser of these various rates has rather profound implications for both cultural and biological evolution.

It is particularly noteworthy that I have not yet been able to positively identify a historic oceanic or large lake cosmic impact during the past 5000 years, other than that of the Flood Comet itself. There are some indiciations of oceanic or large lake impacts at around 2650 BC and around 2280 BC, and possibly at around 1800 BC, but these are equivocal at best. This suggests that we still have much to learn and to model in terms of being able to identify the signatures of such impacts in the archaeological and documentary record.

Oceanic impacts, such as that in 2807 BC, also contribute to the difficulty of identifying the remnants of local and regional terrestrial impacts by resculpting landforms and thus obscuring the very presence of earlier impact craters. Likewise, even if oceanic impacts produce a sea-floor crater and send sea-floor debris into and through the upper atmosphere, the presence of large quantities of water vapour in association with the debris particles may operate like a filter to quickly remove the particulates (as rainfall) and thus to reduce the potential for impact dust to accumulate in recognizable concentrations in polar ice. This may explain why multiple cosmic impacts have not been previously identified in ice core studies.

14. Conclusions

> "Those who are good at discussing antiquity must demonstrate the validity of what they say in terms of modern times; those who are good at discussing Heaven must show proofs from the human world. In discussions of all kinds, men value what is in accord with the facts and what can be proved to be valid. Hence if a man sits on his mat propounding some theory, he should be able to stand right up and put it into practice, and show that it can be extended over a wide area with equal validity", Hsün Tzu, 4th century BC, [196: 163].

The art and science of eliciting underlying meaning from potentially cosmographic oral and written traditions is fraught with difficulty, and one should not be overzealous in such interpretations. Nevertheless, the examples cited here indicate that cosmographic contextual reconstruction and dating is not only possible but can be accomplished with a fair degree of certainty. Cosmography can help to isolate contemporary traditions from widespread regions that together provide enough details to reasonably define previously unknown temporary celestial events, including cosmic impacts.

The *Huainanzi*, a philosophical treatise written in the Han Dynasty during the second century BC, provides us with an important perspective on traditional Chinese cosmology and cosmography, which as with Plato's *Timaeus* quoted at the beginning of this paper, contains much more observational knowledge and wisdom than modern Western scholars have previously recognized [114: 65]:

> "The sun is the ruler of yang. Therefore, in spring and summer animals shed their fur; at the summer solstice stags' antlers drop off. The moon is the fundament of yin. Therefore, when the moon wanes, the brains of fish shrink; when the moon dies, wasps and crabs shrivel up. Fire flies upward, water flows downward; thus the flight of birds is aloft, the movement of fishes is downward. Things within the same class mutually move each other; root and twig mutually respond to each other. Therefore, when the burning-mirror sees the sun, it ignites tinder and produces fire. When the square receptacle sees the moon, it moistens and produces water. When the tiger roars the valley winds rush; when the dragon arises the bright clouds accumulate. When *qilins* wrangle, the sun or moon is eclipsed; when the leviathan dies, comets appear. When the silkworm spins cocoons (shaped like) ear ornaments, the *shang* string (of a stringed instruments) snaps; when meteors fall, the sea suddenly swells up."

Let us assume for the moment that a 'rising dragon' is a logical descriptor for a Tunguska-like impactor with its subsequent bright fireball and upper atmospheric displays ("bright clouds"), and that the 'swelling of the sea' is in fact a tsunami. One can only marvel and wonder how many different cosmic impacts over countless hundreds of years went into the simple observed correlation between the fall of 'meteorites' on land and the billowing of impact fireballs, or the fall of 'meteorites' into the ocean and the subsequent devastating effects of tsunami. Clearly scholars have been aware of cosmic impacts and their destructive tendencies for much more than 2,000 years.

That the effects of cosmic impact have been considerably underestimated by modern science is arguable by our seeming inability to recognize the full extent of the devastation caused by the Rio Cuarto impact, and especially by the complete failure of modern science to even begin to recognize the nature and date of the Flood Comet impact. When we add to this situation the possibility that the rate of cosmic impact has likewise been underestimated, given uncertainties in the numbers of Earth-crossing asteroids and in the numbers of long period comets which may periodically be flung into the inner Solar System, the implications are disquieting.

But by far the greatest risk we face is that of complacency. The cosmic impact risks outlined above are high enough to warrant an immediate vigorous and thorough evaluation of possible regional and globally catastrophic impacts during the Holocene period. Currently, few palaeo-environmental researchers are aware of the Holocene period record of cosmic impacts, and most substantive current texts which attempt to explain climate change during the Quaternary period fail to consider cosmic impacts whatsoever [111]. We need to greatly expand our efforts to identify cosmic impacts in the archaeological and palaeo-environmental record, especially oceanic impacts such as that hypothesized for 2807 BC and for the beginning of the Younger Dryas event at around 9300 BC. We likewise need to increase the effort being expended in the search for potentially hazardous asteroids and comets. And we need to expand our efforts to couple computer simulations of oceanic and terrestrial cosmic impacts with our modeling of global climate change.

Cosmic impacts are inevitable. While we cannot presently prevent the occurrence of cosmic impact, we can at least begin the process of serious contingency planning to reduce the overall consequences of such impacts, in particular oceanic impacts. And by embarking on such a process, we will finally do justice to our many forgotten ancestors who bravely strove to warn us of our impending fate of cosmic impact, a fate which is as natural and as normal as the daily workings of the Solar System itself. In a holistic world, where science, nature, religion, and politics were part of an observational continuum, the myth-makers and story-tellers were the scientists, poets, and the true heroes of our mythological past.

Acknowledgments

I thank Benny Peiser and the organizing committee for the opportunity to participate in the 2nd SIS Cambridge Conference. Benny and Vine Deloria, Jr. provided editorial comments on an earlier draft of this paper. Ronald J. Beckwith prepared the illustrations.

In addition to Benny and Vine, several individuals during the past decade have encouraged aspects of my endeavors in cosmography. They do not necessarily agree with my interpretations or with my vision of the importance of cosmography, but their encouragement has strengthened my resolve to continue. I thank J. Carucci, L.A. Carter-Schuster, A.L. Christenson, B.D. Eilerts, J. Fountain, J.G. Hills, R.J.Hommon, L. Lopez, T.E. Luebben, C. W. Meighan, D.M. Raup, T.J. Riley, R.M. Sinclair, G.F. Somers, and R.E. White. I especially thank D.E. Doyel, L.M. Gregonis, A.E. Griffin, P.B. Griffin, G.J. Gumerman, R.K. Johnson, J.A. Masse, M.J. Masse, A.G. Rankin, and H.D. Tuggle for their support through often difficult and discouraging times.

I dedicate this work to my son, Jeffrey. I challenge his and future generations to have the wisdom and grace to squarely face both the challenges and the joys afforded to the human species on this tiny but most interesting piece of the universe called Earth. As Confucius is alleged to have said, and as I have long attempted to practice: "Our greatest glory is not in never falling, but in rising every time we fall." This sage advice is appropriate not only for individuals, such as myself, but perhaps also for our species as a whole.

References

1. Acker, A., et al., 1992. *The Strasbourg-ESO Catalogue of Galactic Planetary Nebulae: Part II, the Catalogue.* Munich: Karl-Schwarzschild.
2. Ahlström, G.W., 1993. *The History of Ancient Palestine.* Minneapolis: Fortress Press.
3. Allen, R.H., 1963. *Star Names: Their Lore and Meaning.* New York: Dover Publications. {Orig. 1899].
4. Anderson, A.J.O., and C.E. Dibble, trans., 1953. *Florentine Codex: General History of the Things of New Spain---Book 7: The Sun, Moon, and Stars, and the Binding of the Years,* by Fray Bernardino de Sahagún. Santa Fe: School of American Research and University of Utah.
5. Asher, D.J., S.V.M. Clube, W.M. Napier, and D.I. Steel, 1994. Coherent catastrophism. *Vistas in Astronomy* 38:1-27.
6. Aveni, A., ed., 1975. *Archaeology in pre-Columbian America.* Austin: University of Texas Press.
7. Aveni, A., 1980. *Skywatchers of Ancient Mexico.* Austin: University of Texas Press.
8. Aveni, A., 1992. *Conversing with the Planets: How Science and Myth Invented the Cosmos.* New York: Times Books.
9. Aveni, A., 1997. *Stairway to the Stars: Skywatching in Three Great Ancient Cultures.* New York: John Wiley & Sons.
10. Baxter, J., and T. Atkins, 1976. *The Fire Came By: The Riddle of the Great Siberian Explosion.* New York: Doubleday.
11. Bailey, M.E., S.V.M. Clube, G. Hahn, W. M. Napier, and G. B. Valsecchi, 1994. Hazards due to giant comets: Climate and short-term catastrophism. In T. Gehrels, ed., *Hazards due to Comets & Asteroids*, pp. 479-533. Tucson: University of Arizona Press.
12. Bailey, M. E., D. F. Markham, S. Massai, and J. E. Seriven 1995. The 1930 August 13 Brazilian Tunguska Event. *The Observatory* 115: 250-253.
13. Bauval, R., and A. Gilbert 1994. *The Orion Mystery.* New York: Crown.
14. Bellwood, P. 1978. *Mans Conquest of the Pacific.* New York: Oxford University Press.
15. Bierhorst, J., 1988. *The Mythology of South America.* New York: William Morrow.
16. Bierhorst, J., 1992. *History and Mythology of the Aztecs: The Codex Chimalpopoca.* Tucson: University of Arizona Press.
17. Bird, C. F. M, and D. Frankel 1991. Problems in constructing a prehistoric regional sequence: Holocene southeast Australia. *World Archaeology* 23(2): 179, Fig. 2.
18. Birrell, A. 1993. *Chinese Mythology: An Introduction.* Baltimore: Johns Hopkins University Press.
19. Black, J., and A. Green 1992. *Gods, Demons and Symbols of Ancient Mesopotamia.* Austin: University of Texas Press.
20. Blong, R. J. 1982. *The Time of Darkness: Local Legends and Volcanic Reality in Papaua New Guinea.* Seattle: University of Washington Press.
21. Bone, N. 1993. *Meteors.* Sky & Telescope Observers Guide. Cambridge, MA: Sky Publishing.
22. Brown, A. G. 1997. *Alluvial Geoarchaeology.* Cambridge: Cambridge University Press.
23. Brown, N. O. , trans. 1953. *Theogeny.* The Library of Liberal Arts. New York: Bobbs-Merrill.
24. Budge, E. A. W., 1967. *The Egyptian Book of the Dead.* New York: Dover Publications. [Orig. 1895].
25. Budge, E. A. W.,1973. *Osiris & the Egyptian Resurrection.* New York: Dover Publications. [Orig. 1911].
26. Budge, E. A. W., 1994. *Legends of the Egyptian Gods.* New York: Dover Publications. [Orig. 1912]
27. Butzer, K. W., and C. L. Hansen 1968. *Desert and River in Nubia: Geomorphology and Prehistoric Environments at the Aswan Reservoir.* Madison: University of Wisconsin Press.
28. Campbell, J. 1988. *The Inner Reaches of Outer Space: Metaphor as Myth and as Religion.* New York: Harper & Row.
29. Carter, L. A., and G. F. Somers 1990. *Here Today Lava Tomorrow: Archaeological Work in Hawaii Volcanoes National Park, 1987 to 1989.* Honolulu: Pacific Area Office, National Park Service.
30. Carucci, J. 1992. *Cultural and Natural Patterning in Prehistoric Marine Foodshell from Palu, Micronesia.* Unpublished Ph.D. dissertation (Ann Arbor, MI: University Microfilms).
31. Ceram, C. W. 1951. *Gods, Graves, & Scholars: The Story of Archaeology.* New York: Alfred A. Knopf.
32. Chang, K. C. 1986. *The Archaeology of Ancient China* (4th Edition). New Haven: Yale University Press.
33. Chapman. C. R. 1996. Review of D. Steel Rogue Asteroids and Doomsday Comets (New York: John Wiley 1995). *Meteoritics & Planetary Science* 31: 313-314.
34. Chapman, C. R., and D Morrison 1994. Impacts on the Earth by asteroids and comets: Assessing the hazard. *Nature* 361: 40-44.
35. Charlesworth, J. H., ed. 1983. *The Old Testament Pseudepigrapha: Vol. I, Apocalyptic Literature and Testaments.* New York: Doubleday.
36. Christie, A. 1968. *Chinese Mythology.* Feltham, Middlesex: Paul Hamlyn.
37. Chu, You-Hua 1987. Multiple Shell Planetary Nebulae. In S. Torres-Peimbert ed, *Planetary Nebulae.* Dordrecht, Netherlands: Kluwer Academic Publishers.
38. Chyba, C. F., P. Thomas, and K. Zahnle. 1993. The 1908 Tunguska explosion: Atmospheric disruption of a stony asteroid. *Nature* 361: 40-44.
39. Clark, D. H., and F. R. Stephenson 1977. *The Historical Supernovae.* Oxford: Pergamon Press.
40. Clark, E. E. 1966. *Indian Legends from the Northern Rockies.* Norman and London: University of Oklahoma Press.
41. Clayton, P. A. 1994. *Chronicle of the Pharaohs.* London: Thames and Hudson.

42. Clube, V., and B. Napier, 1982. *The Cosmic Serpent*. New York: Universe Books.
43. Clube, V., and B. Napier , 1990. *The Cosmic Winter*. Oxford: Basil Blackwell.
44. Collon, D. 1987. *First Impressions: Cylinder Seals in the Ancient Near East*. London: Bristish Museum Press.
45. Cotterell, A. 1981. *The First Emperor of China*. New York: Holt, Rinehart and Winston.
46. Coyne, G. V., S. J., and R. M. Sinclair, eds. 1995. *The Inspiration for Astronomical Phenomena (INSAP). Vistas in Astronomy* 39(4).
47. Crown, P. L. 1994. *Ceramics and Ideology: Salado Polychrome Pottery*. Albuquerque: University of New Mexico Press.
48. Dalley, S. 1989. *Myths from Mesopotamia*. Oxford: Oxford University Press.
49. Daniélou, A. 1991. *The Myths and Gods of India*. Rochester, VT: Inner Traditions International. [Orig. 1964].
50. Decker, R., and B. Decker 1989. *Volcanoes*, revised and updated. New York: W. H. Freeman.
51. Deloria, V., Jr. 1995. *Red Earth, White Lies: Native cans and the Myth of Scientific Fact*. New York: Scribner.
52. Desonie, D. 1996. *Cosmic Collisions*. New York: Henry Holt.
53. Dickinson, O. 1994. *The Aegean Bronze Age*. Cambridge: Cambridge University Press.
54. Dimmitt C., and J. A. Van Buitenen 1978. *Classical Hindu Mythology: A Reader in the Sanskrit Pur_nas*. Philadelphia: Temple University Press.
55. Drews, R. 1993. *The End of the Bronze Age: Changes in Warfare and the Catastrophe ca. 1200 B.C.*. Princeton: Princeton University Press..
56. Dundes, A., ed. 1988. *The Flood Myth*. Berkeley: University of California Press.
57. Dunstan, W. E. 1998. *The Ancient Near East*. New York: Harcourt Brace.
58. Edwards, I. E. S. 1978. Pyramids: Buildings for Eternity. In J. B. Billard, ed., *Ancient Egypt: Discovering its Splendors*. Washington, D.C.: National Geographic Society.
59. Eisenman, R. H., and M. Wise 1992. *The Dead Sea Scrolls Uncovered*. Shaftesbury, Dorset: Element.
60. Erdoes, R., and A. Ortiz 1984. *American Indian Myths and Legends*. New York: Pantheon.
61. Ellis, W. 1979. *Journal of William Ellis*. Rutland, Vt: Charles E. Tuttle. [Orig. 1825].
62. Faulkner, R. O. 1990. *The Ancient Egyptian Book of the Dead*. Austin: University of Texas Press. [Orig. 1972].
63. Feder, K.L., 1996. *Frauds, Myths, and Mysteries: Science and Pseudoscience in Archaeology* [2nd Edition]. Mountain View, CA: Mayfield Publishing Co.
64. Frankfort, H. 1978. *Kingship and the Gods*. Chicago: University of Chicago Press. [Orig. 1948]
65. Frazer, J.G., 1919. *Folk-Lore in the Old Testament*. London: Macmillan.
66. Frazer, J. G., 1930. *Myths of the Origin of Fire*. London: Macmillan.
67. Freidel, D, L. Schele, and J. Parker 1993. *Maya Cosmos*. New York: William Morrow.
68. Gehrels, T., ed. 1994. *Hazards due to Comets & Asteroids*. Tucson: University of Arizona Press.
69. Ginzberg, L. 1937. *The Legends of the Jews*. Vols. I-VII. Philadelphia: The Jewish Publication Society of America. [Orig. 1909].
70. Gleadow, R. 1968. *The Origin of the Zodiac*. New York Athenaeum.
71. Goetze, A. 1969. Hittite myths, epics, and legends. In J. B Pritchard, ed., *Ancient Near Eastern Texts Relating to the Old Testament*, pp. 120-128. Princeton, N. J.: Princeton University Press.
72. Graves, R. 1992: *The Greek Myths*. London: Penguin Books. [Orig. 1960].
73. Gray, J. 1969. *Near Eastern Mythology*. London: Hamlyn.
74. Grayson, A. K. 1969. In J. B. Pritchard, ed., *Ancient Near Eastern Texts Relating to the Old Testament*, pp. 501-503. Princeton, N. J.: Princeton University Press.
75. Grayson, D. K. 1993. *The Deserts Past: A Natural Prehistory of the Great Basin*, Fig. 9-4. Washington, D.C.: Smithsonian Institution Press.
76. Grazia, A. de, R. E. Juergens, and L. V. Stecchini, eds. 1966. *The Velikovsky Affair*. New Hyde Park, N.Y.: University Books.
77. Grieve, R. A. F., and E. M. Shoemaker 1994. The record of past impacts on Earth. In T. Gehrels, ed., *Hazards due to Comets & Asteroids*, pp. 417-462. Tucson: University of Arizona Press.
78. Hadingham, E. 1975. *Circles and Standing Stones*. London: Heinemann.
79. Hadingham, E., 1984. *Early Man and the Cosmos*. Norman: University of Oklahoma Press.
80. Hart, G., 1990. *Egyptian Myths*. British Museum Publications, Austin: University of Texas Press.
81. Hawkins, G. S. 1965. *Stonehenge Decoded*. New York: Doubleday.
82. Hills J. G., and M. P. Goda 1993. The fragmentation of small asteroids in the atmosphere. *Astronomical Journal* 105: 1114-1 144.
83. Hills J. G., I. V. Nemchinov, S. P. Popov, and A. V. Teterev 1994. Tsunami generated by small asteroid impacts. In T. Gehrels, ed., *Hazards due to Comets & Asteroids*, pp. 779-789. Tucson: University of Arizona Press.
84. Hobson, C. 1987. *Exploring the World of the Pharaohs*. London: Thames and Hudson.
85. Hodge, P. 1994. *Meteorite Craters and Impact Structures of the Earth*. Cambridge: Cambridge University Press.
86. Hoffman, M. A. 1991. *Egypt Before the Pharaohs*. Austin: University of Texas Press.
87. Holcomb, R. T. 1987. Eruptive history and long-term behavior of Kiluaea Volcano. In R. W. Decker, T. L. Wright, and P. H. Stauffer, eds., *Volcanism in Hawaii*, Vol. 1, pp. 261-350. U. S. Geological Survey Professional Paper 1350
88. Holt, J. D. 1985. *The Art of Featherwork in Old Hawaii*. Honolulu: Topgallant.
89. Ho Peng Yoke 1962. Ancient and Mediaeval observations of comets and novae in Chinese sources. *Vistas in Astronomy* 5: 127-225.
90. Ho Peng-Yoke and Ang Tien-se 1970. Chinese astronomical records on comets and guest stars. *Oriens Extremus* 1/2: 63-99.
91. Ions, V. 1968. *Egyptian Mythology*. New York: Paul Hamlyn.
92. James, P. 1993. *Centuries of Darkness*. New Brunswick, N. J.: Rutgers University Press.
93. Johannes, R. E. 1981. *Words of the Lagoon: Fishing and Marine Lore in the Palau District of Micronesia*. Berkeley: University of California Press.
94. Josephus, F. n.d. *The Complete Works of Josephus*, Vols. I-IV. National Library Edition. New York: Bigelow, Brown & Co.
95. Kamakau, S. M. 1991. *Tales and Traditions of the People of Old: Na Moolelo a ka Poe Kahiko*. Honolulu: Bishop Museum Press. [Orig. 1865].
96. Kitchen, K. A. 1991. The chronology of ancient Egypt. *World Archaeology* 23: 201-208.
97. Kovacs, M. G. 1989. *The Epic of Gilgamesh*. Stanford, CA: Stanford University Press.
98. Kramer, S. N. 1969a. Lamentation over the destruction of Ur. In J. B. Pritchard, ed., *Ancient Near Eastern Texts Relating to the Old Testament*, pp. 455-463. Princeton, N. J.: Princeton University Press.
99. Kramer, S. N., 1969b. Lamentation over the destruction of Sumer and Ur. In J. B. Pritchard, ed., *Ancient Near Eastern Texts Relating to the Old Testament*, pp. 611-619. Princeton, N. J.: Princeton University Press.
100. Kramer, S. N., 1969c. The curse of Agade: The Ekur avenged. In J. B. Pritchard, ed., *Ancient Near Eastern Texts Relating to the Old Testament*, pp. 646-651. Princeton, N. J.: Princeton University Press.
101. Krupp, E. C. 1991. *Beyond the Blue Horizon: Myths & Legends of the Sun, Moon, Stars, & Planets*. Oxford: Oxford University Press.
102. Krupp, E. C., 1997. *Skywatchers, Shamans & Kings*. New York: John Wiley.
103. Kuhrt, A. 1995. *The Ancient Near East, c. 3000-330 BC,*. London: Routledge.
104. Ladefoged, T., G. F. Somers, and M. M. Lane-Hamasaki 1987. *A Settlement Pattern Analysis of a Portion of Hawaii Volcanoes National Park*. Tucson: Western Archaeological and Conservation Center Publications in Anthropology 44.

105.Lambert, W. G., and A. R. Millard 1969. *Atra-hasis: The Babylonian Story of the Flood.* Oxford: Clarendon Press.

106.Legge, J. 1865a. *The Chinese Classics, Vol III, Part I, Chap IV, The Annals of the Bamboo Books.* London: Henry Frowde (Hong Kong 1939 reprint).

107.Legge, J. 1865b. *The Chinese Classics, Vol III, Part I, Section II, The Body of the Volume* [Shu Ching]. London: Henry Frowde (Hong Kong 1939 reprint).

108.Levison, R. B. 1967. *A Plato Reader.* Boston: Houghton Mifflin.

109.Lévi-Strauss, C. 1969. *The Raw and the Cooked: Introduction to a Science of Mythology*, Vol I. New York: Harper & Row.

110.Lichtheim, M. 1975. *Ancient Egyptian Literature, Vol. 1: The Old and Middle Kingdom.* Berkeley: University of California Press.

111.Lowe, J. J., and M. J. C. Walker 1997. *Reconstructing Quaternary Environments,* Second Edition. Essex: Longman.

112.Luce, J. V. 1969. *Lost Atlantis: New Light on an Old Legend.* New York: McGraw-Hill.

113. Lurker, M. 1982. *The Gods and Symbols of Ancient Egypt.* London: Thames and Hudson.

114.Major, J. S. 1993. *Heaven and Earth in Early Han Thought: Chapters Three, Four, and Five of the* Huainanzi.. Albany: State University of New York Press.

115.Malinowski, B. 1926. *Myth in Primitive Psychology.* London: Routledge and Kegan Paul.

116.Mallory, J. P. 1989. *In Search of the Indo-Europeans.* London: Thames and Hudson.

117.Mallory, J.P., 1997. The homelands of the Indo-Europeans. In R. Blench and M. Spriggs, eds., *Archaeology and Language I: Theoretical and Methodological Orientations*, pp. 93-121. London: Routledge.

118.Martin, P.S., and R. G. Klein, eds. 1984. *Quaternary Extinctions: A Prehistoric Revolution.* Tucson: University of Arizona Press.

119.Masse, W.B., 1981. Prehistoric irrigation systems in the Salt River Valley, Arizona. *Science* 214: 408-415.

120.Masse, W.B., 1985. Review of R. J. Johannes Words of the Lagoon: Fishing and Marine Lore in the Palau District of Micronesia (Berkeley: University of California Press 1981). *Copeia* 1985(3): 797-798.

121.Masse, W.B., 1989. *The Archaeology and Ecology of Fishing in the Belau Islands, Micronesia.* Unpublished Ph.D. dissertation (Ann Arbor, MI: University Microfilms).

122.Masse, W.B., 1991. The quest for subsistence sufficiency and civilization in the Sonoran Desert. In P. L. Crown and W. J. Judge, eds., *Chaco and Hohokam: Prehistoric Regional Systems in the American Southwest*, pp. 195-223. Santa Fe: School of American Research Press.

123.Masse, W.B., 1995. The Celestial basis of civilization. *Vistas in Astronomy* 39: 463-477.

124.Masse, W.B., 1998a. Sky as environment: Solar eclipses and Hohokam sociopolitical cosmography. In D. E. Doyel and J. S. Dean, eds., *Environmental Change and Human Adaptation in the Ancient Southwest.* Salt Lake City: University of Utah Press. [In press]

125.Masse, W.B., 1998b. Noahs Flood and Occams Razor: Part I An environmental analysis of Flood mythology. Ms in the authors possession.

126.Masse, W.B., 1998c. Noahs Flood and Occams Razor: Part II Modeling the Flood Comet impact. Ms. in the authors possession.

127.Masse, W.B., and R. Soklow 1998. Black suns and dark times: The cultural response to total solar eclipses in the ancient Puebloan Southwest. In J. Fountain *Proceedings of the 1996 Oxford V Archaeoastronomy Conference.* Carolina Academic Press. [In press]

128.Masse, W. B., and D. Tuggle 1998. The Date of Hawaiian Colonization. In C. M. Stevenson, G. Lee, and F. J. Moran eds. *Easter Island in Pacific Context.* Easter Island Foundation Occasional Paper 4. Los Osos, CA. [In press].

129.Masse, W. B., D. Snyder, G. J. Gumerman 1984. Prehistoric and historic settlement in the Palau Islands, Micronesia. *New Zealand Journal of Archaeology* 6: 107-127.

130.Masse, W. B., L. A. Carter, and G. S. Somers 1991. Wahaula Heiau: The symbolic and regional context of Hawaii Islands Red Mouth temple. *Asian Perspectives* 30(1): 19-56.

131.Masse, W. B., R. K. Johnson, and H. D. Tuggle n.d. *Dialog of Earth & Sky: The Appropriation of the Heavens by Ancient Hawaiian Royalty.* Honolulu: University of Hawaii Press (in preparation).

132.Mazar, A. 1992. *Archaeology of the Land of the Bible:10,000-586 B.C.E.* New York; Doubleday.

133.Melville, A. D., trans., 1986. *Ovid Metamorphoses.* Oxford: Oxford University Press.

134.Menzel, D. H., and J. M. Pasachoff 1995. *A Field Guide to Stars and Planets.* Peterson Field Guide Series, 2nd edition. Boston: Houghton Mifflin.

135.Metraux, A. 1946. *Myths of the Toba and Pilagá Indians of the Gran Chaco.* Memoirs of the American Folklore Society, vol. 40. Philadelphia.

136.Morrison, D., C. R. Chapman, and P. Slovic 1994. In T.Gehrels, ed., *Hazards due to Comets & Asteroids*, pp. 59-91. Tucson: University of Arizona Press.

137.Morrison, D. 1997. In P. J. Thomas, C. F. Chyba,and C. P.McKay, eds. *Comets and the Origin and Evolution of Life*, pp. 243-258. New York: Spring-Verlag.

138.Motz, L., and C. Nathanson 1988. *The Constellations: An Enthusiasts Guide to the Night Sky.* New York: Doubleday.

139.Mulholland, M. T. 1988. Territoriality and horticulture: A perspective for prehistoric southern New England. In G. P. Nicholas, ed., *Holocene Human Ecology in Northeastern North America*, pp. 137-166, Fig. . New York: Plenum.

140.Naeye, R. 1993. The hole in Nebraska. *Discover* Magazine, April issue, p. 18.

141.Needham, J., and W. Ling 1959. *Science and Civilisation in China*, Volume 3. Cambridge: Cambridge University Press.

142.Nequatewa, E. 1967. *Truth of a Hopi.* Flagstaff, AZ: Northland Press. [Orig. 1936].

143.Numbers, R. L. 1992. *The Creationists.* New York: Alfred A. Knopf.

144.Oates, J. 1986. *Bablyon* (revised edition). London: Thames and Hudson.

145.Olivier, C. P. 1925. *Meteors.* Baltimore: Williams & Wilkins.

146.Oppenheim, A. L. 1969. Babylonian and Assyrian Historical Texts. In J. B. Pritchard, ed., *Ancient Near Eastern Texts Relating to the Old Testament*, pp. 265-317. Princeton: Princeton University Press.

147.Palmer, T. 1994. *Catastrophism, Neocatastrophism and Evolution.* Society for Interdisciplinary Studies. Nottingham: Nottingham Trent University.

148.Pankenier, D.W., 1984. *Mozi* and the dates of Xia, Shang, and Zhou: A research note. *Early China* 9-10: 175-183.

149.Pankenier, D.W., 1995. Astrological origins of Chinese dynastic ideology. *Vistas in Astronomy* 39: 503-516.

150.Parpola, Asko 1994. *Deciphering the Indus Script.* Cambridge: Cambridge University Press.

151.Patten, D. 1966. *The Biblical Flood and the Ice Epoch.* Seattle: Pacific Meridian Publishing Co.

152.Pearson G. W., et al. 1986. High-precision 14C measurement of Irish oaks to show the natural 14C variations from A.D. 1840-5210 B.C. In M. Stuiver,and R.S. Kra eds., Proceedings of the 12th International 14C Conference. *Radiocarbon* 28(2B): 911-934.

153.Pearson, G. W., and M. Stuiver 1986. High-precision calibration of the radiocarbon time scale, 500-2500 B.C. In M. Stuiver,and R.S. Kra eds., Proceedings of the 12th International 14C Conference. *Radiocarbon* 28(2B): 839-862.

154.Pellegrino, C. 1994. *Return to Sodom and Gomorrah.* New York: Avon Books.

155.Phillipson, D. W. 1993. *African Archaeology* (2nd edition). Cambridge: Cambridge University Press

156.Porada, E., D. P. Hansen, S. Dunham, and S. H. Babcock 1992. The chronology of Mesopotamia, ca. 7000-1600 B.C. In R. W. Ehrich, ed., *Chronologies in Old World Archaeology*, pp. 77-121. Chicago: Univesity of Chicago Press.

157.Postgate, J. N. 1992. *Early Mesopotamia.* London: Routledge.

158.Quirke, S. 1992. *Ancient Egyptian Religion.* London: British Museum Press.

159.Redford, D. B. 1992. *Egypt, Canaan, and Israel in Ancient Times.* Princeton: Princeton University Press.

160.Renfrew, C. 1987. *Archaeology & Language: The Puzzle of Indo-European Origins.* Cambridge: Cambridge University Press.

161.Rice, M. 1990. *Egypts Making*. London: Routledge.

162.Rohl, D.M. 1995. *Pharoahs and Kings: A Biblical Quest*. New York: Crown Publishers.

163.Sagan, C. 1979. *Brocas Brain*. London: Hodder and Stoughton.

164.Sagan, C., 1995. *The Demon-Haunted World: Science as a Candle in the Dark*. New York: Random House.

165.Sandweiss, D. H. et al. 1996. Geoarchaeological evidence from Peru for a 5000 Year B.P. onset of El Niño. *Science 273: 1531-1533.*

166.Santillana. G. de, and B. Von Dechend 1969. *Hamlets Mill*. Boston: David R. Godine.

167.Sarna, N.H. 1989. *The JPS Torah Commentary: Genesis*. Philadelphia: The Jewish Publication Society.

168.Schaafsma, P., ed. 1994. *Kachinas in the Pueblo World*. Albuquerque: University of New Mexico Press.

169.Schove, D.J., and A. Fletcher 1984. *Chronology of Eclipses and Comets, A.D. 1-1000*. Woodbridge: Boydell Press.

170.Schultz, P.H., and R.E. Lianza 1992. Recent grazing impacts on the Earth recorded in the Rio Cuarto crater field, Argentina. *Nature* 355: 234-237.

171.Sekanina, Z. 1983. The Tunguska event: No cometary signature in evidence. *Astronomical Jounal*. 88: 1382-1414.

172.Shafer, B.E. ed. 1991. *Religion in Ancient Egypt*. Ithaca, N. Y.: Cornell University Press.

173.Shaw, T., P. Sinclair, B. Andah, and A. Okpoko 1993. *The Archaeology of Africa: Food, Metals and Towns*. London: Routledge.

174.Siliotti, A. 1997. *Guide to the Pyramids of Egypt*. New York: Barnes and Noble.

175.Somers, G. F. 1991. The effects of rapid geological change on archaeology in Hawaii. *Asian Perspectives* 30: 133-145.

176.Speiser, E.A. 1969. Akkadian Myths and Epics. In J. B. Pritchard, ed., *Ancient Near Eastern Texts Relating to the Old Testament*, pp. 60-119. Princeton, N. J.: Princeton University Press.

177.Spencer, A.J. 1993. *Early Egypt*. London: British Museum Press.

178.Stager, L.E. 1992. The periodization of Palestine from Neolithic through Early Bronze times. In R. W. Ehrich, ed., *Chronologies in Old World Archaeology*, pp. 22-41. Chicago: Univesity of Chicago Press.

179.Steel, D. 1995. *Rogue Asteroid and Doomsday Comets*. New York: John Wiley & Sons.

180.Steward, J.H., and L.C. Faron 1959. *Native Peoples of South America*. New York: McGraw Hill.

181.Stonely, J. 1977. *Cauldron of Hell: Tunguska*. New York: Simon and Schuster.

182.Story, R. 1976. *The Space Gods Revealed: A Close Look at the Theories of Erich von Däniken*. New York: New English Library.

183.Stuiver, M., and B. Becker 1986. High-precision decadal calibration of the radicarbon time scale, A.D. 1950-2500 B.C. In M. Stuiver,and R. S. Kra eds. Proceedings of the 12th International 14C Conference. *Radiocarbon* 28(2B): 863-910.

184.Stuiver, M. and G.W. Pearson 1992. Calibration of the radiocarbon time scale, 2500-5000 B.C. In R. E. Taylor, A. Long, and R. S. Kra, eds. *Radiocarbon After Four Decades: An Interdisciplinary Perspective*, pp. 19-33. New York: Springer-Verlag.

185.Sullivan, W. 1996. *The Secret of the Incas: Myth, Astronomy, and the War Against Time*. New York: Crown Publishers.

186.Tainter, J.A. 1997. *The Collapse of Complex Societies*. Cambridge: Cambridge University Press.

187.Tedlock, D. 1985. *Popul Vuh*. New York: Touchstone (Simon & Schuster).

188.Thom, A. 1971. *Megalithic Lunar Observatories*. Oxford: Clarendon Press.

189.Toon, O.B., K. Zahnle, R.P. Turco, and C. Covey 1994. Environmental perturbations caused by asteroid impacts. In T. Gehrels, ed., *Hazards due to Comets & Asteroids*, pp. 791-826. Tucson: University of Arizona Press.

190.Urton, G. 1981. *At the Crossroads of the Earth and the Sky: An Andean Cosmology*. Austin: University of Texas Press.

191.Valeri, V. 1985. *Kingship and Sacrifice*. Chicago: University of Chicago Press

192.Vitaliano, D.B. 1976. *Legends of the Earth: Their Geologic Origins*. Secaucus, New Jersey: The Citadel Press. [Orig. 1973.]

193.Waddell, W.G. 1940. *Manetho*. Cambridge: Harvard University Press.

194.Walters, D. 1992. *Chinese Mythology: An Encyclopedia of Myth and Legend*. London: The Aquarian Press.

195.Waters, F. 1963. *Book of the Hopi*. New York: Penguin Books.

196.Watson, B., trans. 1963. *Hsün Tzu: Basic Writings*. New York: Columbia University Press.

197.Wasson, R. G. 1968. *Soma: Divine Mushroon ofImmortality*. New York: Harcourt Brace Javanovich.

198.Weigall, A. 1925. *A History of the Pharoahs. Vol. I, The First Eleven Dynasties*. London: Thorton Butterworth.

199.Werner, E.T.C., 1922. *Myths and Legends of China*. London: George G. Harrap.

200.Werner, E.T.C., 1969. *A Dictionary of Chinese Mythology*. New York: The Julian Press.

201.Whipple, F.L., and S.E. Hamid 1952. On the Origin of the Taurid Meteor Streams. *Helwan Observatory Bulletin* 41:224.

202.Whitcomb, J.C., and H.M. Morris 1961. *The Genesis Flood: The Biblical Record and its Interpretation*. Grand Rapids, MI: Baker Book House.

203.Whittle, A. 1996. *Europe in the Neolithic*. Cambridge: Cambridge University Press.

204.Wilbert, J., and K. Simoneau 1982. *Folk Literature of the Toba Indians*. Los Angeles: University of California.

205.Wilkinson, R. H. 1994. *Symbol & Magic in Egyptian Art*. London: Thames and Hudson.

206.Williamson, R.A., and C.R. Farrer, eds. 1992. *Earth & Sky: Visions of the Cosmos in Native American Folklore*. Albuquerque: University of New Mexico Press.

207.Wilson, J. A. 1969. The Admonitions of Ipu-wer. In J. B. Pritchard, ed., *Ancient Near Eastern Texts Relating to the Old Testament*, pp. 441-444. Princeton, N. J.: Princeton University Press.

208.Wise, D. U. 1998. Creationism's geologic time scale. *American Scientist* 86: 160-173.

209.Wooley, L., and P.R.S. Moorey 1982. *Ur of the Chaldees*. London: Herbert Press.

210.Yeomans. D.P. 1991. *Comets: A Chronological History of Observations, Science, Myth, and Folklore*. New York: John Wiley.

211.Young, J.Z. 1971. *An Introduction to the Study of Man*. Oxford: Oxford University Press.

212.Yuan Ke. 1993. *Dragons and Dynasties: An Introduction to Chinese Mythology*. London: Penguin Books.

Table 2: Tentative Concordance of Bronze Age Historical Figures, Myths, Archaeology, and Celestial Events

DATE B.C.	SUGGESTED CELESTIAL EVENT	EGYPT	NEAR EAST, MESOPOTAMIA	CHINA, INDIA	OTHER	CONFIDENCE SCALE* 5= certain 3= likely 1= poor or uncertain
May 2807 B.C.	Globally catastrophic oceanic comet impact in Atantic-Indian Basin, with magnitude of between 10^5 to 10^6 gigatons.	Ended Semerkhet reign, First Dynasty 7th King. "Many portents and a great calamity". Entrenchment, lowering of Nile. Myth of sky goddess Hathor turning into the raging lioness, Sekhmet.	Flood stories of Noah, Atrahasis, Utnapishtim. Massive Flood deposits at Ur, Uruk, Shuruppak, and Kish. Largescale movement of people into hill country of Galilee, Samaria, and Judah.	Ended reign of Nu Wa, consort of Fu Hsi and first Empress of China. Cosmic battle between Gong Gong (God of Water) and Zhu Rong (God of Fire), after which Nu Wa mended the damage to the heavens.	Worldwide Flood myths; boundary event between Middle and Late Holocene geologic period; Beginning of El Niño/Southern Oscillation climatic cycle.	<u>comet impact</u> Date = 5 Event Type = 5 Magnitude = 5 Location = 3
ca. 2700 B.C.	Ursa Major supernova or nova-like event; possible eruptive phase of Owl Nebula. May have exceeded naked eye visible size of apparent diameter of Full Moon.	"Moon waxes unseasonably" during reign of 1st king of Third Dynasty. Likely immortalized as the myth of Horus Behdety, winged solar disk. Possible inspiration for ultimate construction of Djoser's Step Pyramid.	Immortalized in myth as the death of Humbaba, the demon guardian of the cedar forest, at the hands of Gilgamesh and Enkidu.	Celestial portent and birth event of Yellow Emperor, Huang Di. [Bamboo Annals].	Immortalized in the Popul Vuh as the death of Seven Macaw (Big Dipper), who tried to be brighter than the Sun and Moon, at the hands of the hero twins.	<u>Nova-like Event</u> Date = 4 Event Type = 4 Magnitude = 3 Location = 4
ca. 2680 B.C.	Multiple simultaneous supernova or nova-like events.	May have served to reemphasize importance of the circumpolar celestial heavens associated with the Ursa Major nova-like event of 2697 B.C.		Seen as auspicious portent for the Yellow Emperor, Huang Di. [Bamboo Annals].		<u>Simultaneous Novae</u> Date = 4 Event Type = 2 Magnitude = 1 Location = 1
ca. 2650 B.C.	Likely first-ever Taurid meteor storm, which appears to have lasted full force for three days.	Immortalized in Egyptian mythology as the killing of Osiris by Set. Beginning of use of *benben* stones (meteorites) as an important architectural element and symbol.	Immortalized in Gilgamesh epic as the attack on Uruk by Ishtar using the "Bull of Heaven".	In 57th year of Yellow Emperor, Huang Di, the heavens were "wrapt in mist" for three days and nights. [Bamboo Annals].	Immortlization in Popul Vuh as the killing of the 400 boys by son of Seven Macaw. Many myths worldwide note the origin of fire from the Pleiades, the "fire-drill" asterism.	<u>Taurid Storm</u> Date = 4 Event Type = 5 Magnitude = 4 Location = 5
ca. 2650 B.C.	Possible oceanic or deep lake cosmic impact, or multiple small impacts associated with Taurid meteor storm.	Mention of the Moon wzxing unseasonably during the reign of the first king of the Third Dynasty. Perhaps terminated the reign of Sekhemkhet., the successor to Djoser.	Exploits of Gilgamesh and Enkidu. Death of the giant guardian of the cedar forest, Humwawa. Massive flood deposits at Kish.	Seven days of torrential rainfall. [Bamboo Annals]. Other myths note this was part of a cosmic battle between Chi You (the "Fiery Emperor", a ferocious god in the heavens) and Huang Di.		<u>cosmic impact</u> Date = 4 Event Type = 2 Magnitude = 2 Location = 3

*Confidence scale based on the author's qualitative perception of the reliability and depth of the overall database for each specific event identified in this study.

DATE B.C.	POSSIBLE COSMIC EVENT	EGYPT	NEAR EAST, MESOPOTAMIA	CHINA, INDIA	OTHER	CONFIDENCE SCALE 5= certai n 3= likely 1= poor or uncertain
ca. 2345 B.C.	Terrestrial cosmic impact by fragmenting comet.	Possible inspiration for the Pyramid Texts of the last king of the Fifth Dynasty, Unas, and the use of such tomb texts throughout the Sixth Dynasty.	Possible basis for the cosmic battle in the epic myth of Anzu, lion-headed doorkeeper for Ellil (Dalley 1989:203-227). Also likely basis for the first lament about the destruction of Ur.	Immortalized as the myth of the Divine Archer, Yi, shooting down nine of ten suns which appeared simultaneously in the sky and which threatend to burn up the Earth.	Possibly the basis of the Greek myth of the tragic sun-chariot ride of Phaëton, which threatened to burn up the Earth.	<u>cosmic impact</u> Date = 4 Event Type = 3 Magnitude = 2 Location = 2
ca. 2315 B.C.	Crater supernova or nova-like event; possibly visible for 14 months. [Duplicate record of 2287 B.C. event?].	Possible inspiration for the common reference to stars in the Pyramid Texts of the Sixth Dynasty.		Celestial portent during 42th year of reign of Emperor Yao. [Bamboo Annals].		<u>Nova-like Event</u> Date = 4 Event Type = 3 Magnitude = 3 Location = 4
ca. 2297-2265 B.C.	Possible oceanic or large lake cosmic impact(s) in or near China.			Possible inspiration for the battle(s) between god of water, Kung Kung, and Kun and his son Yu who attempted to control devastating floods throughout China.		<u>cosmic impact</u> Date = 3 Event Type = 1 Magnitude = 1 Location = 3
ca. 2287 B.C.	Crater supernova or nova-like event. [Duplicate record of 2315 B.C. event?].	Possible inspiration for the common reference to stars in the Pyramid Texts of the Sixth Dynasty.		Celestial portent during 70th year of reign of Emperor Yao. [Bamboo Annals].		<u>Nova-like Event</u> Date = 4 Event Type = 3 Magnitude = 2 Location = 4
ca. 2254 B.C.	Scorpius supernova or nova-like event.	Possible inspiration for the common reference to stars in the Pyramid Texts of the Sixth Dynasty.		Celestial portent during at the beginning of the reign of Emperor Shun. [Bamboo Annals].		<u>Nova-like Event</u> Date = 3 Event Type = 4 Magnitude = 4 Location = 4
ca. 2240 B.C.	Local probable Tunguska-like atmospheric cosmic impact in China.			Occurring during ceremonies being performed by Shun. Caused Shun to name and introduce his ultimate successor, Yu. [Bamboo Annals].		<u>cosmic impact</u> Date = 4 Event Type = 3 Magnitude = 3 Location = 3

Table 2: Tentative Concordance of Bronze Age Historical Figures, Myths, Archaeology, and Celestial Events

DATE B.C.	SUGGESTED CELESTIAL EVENT	EGYPT	NEAR EAST, MESOPOTAMIA	CHINA, INDIA	OTHER	CONFIDENCE SCALE* 5= certain 3= likely 1= poor or uncertain
ca. 2188 B.C.	Local terrestrial cosmic impact in or near southern portion of the Dead Sea.	Beginning of First Intermediate Period.	Destruction of Sodom and Gomorrah.	Approximate beginning of Hsia Dynasty.		cosmic impact Date = 3 Event Type = 4 Magnitude = 4 Location = 4
ca. 2000 B.C.	Regionally catastrophic terrestrial cosmic impact at Rio Cuarto, Argentina.	Approximate beginning of the Middle Kingdom, or at least the boundary betwen the Eleventh and Twelth Dynasty.	Darkened skies noted as an aspect of the second destruction of Ur lament during the reign of Ibbi-sin.		This event is preserved throughout much of South America as the myth of the "Great Fire".	cosmic impact Date = 3 Event Type = 5 Magnitude = 4 Location = 5
ca. 1800 B.C.	Locally catastrophic oceanic or large lake cosmic impact.		Possibly preserved in Hittite myth about the fall of the Moon.	Possibly related to the destruction, by flooding, of Mohenjo-Daro and Harrapan culture in India.		cosmic impact Date = 2 Event Type = 3 Magnitude = 2 Location = 2
722 B.C.	Large probable bolide heard in China.			Thunder without clouds (Bamboo Annals).		cosmic impact Date = 5 Event Type = 2 Magnitude = 2 Location = 3
686 B.C.	Meteor storm with possible associated Tunguska-like atmospheric cosmic impact.			Summertime meteor storm and regular stars invisible at night (Bamboo Annals).	possible association with the Piscids or Orionids, and perhaps with Taurids.	cosmic impact Date = 5 Event Type = 2 Magnitude = 1 Location = 1
505 B.C.	Possible Tunguska-like atmospheric cosmic impact in Europe or Asia			Milky Way not visible in the [night] sky (Bamboo Annals).		cosmic impact Date = 5 Event Type = 1 Magnitude = 1 Location = 1

*Confidence scale based on the author's qualitative perception of the reliability and depth of the overall database for each specific event identified in this study.

The Soil Record of an Exceptional Event at 4000 B.P. in the Middle East

Marie-Agnès Courty

C.N.R.S.-C.R.A., DEMOS, AGER, Institut National Agronomique P-G, F-78850 Grignon, France

Summary

Further investigations have lead us to re-examine the nature, age, causes and effects of the third millennium catastrophe that we first identified as an abrupt climate change dated at 2200 BC, synchronous to a volcanic event, and inferred to have initiated the collapse of the Akkad Empire. Comparison of natural soil sequences and third millennium BC archaeological deposits across northern Syria have allowed us to identify the regional trace of this exceptional event that is represented by a thin dust layer showing great spatial variability due to differences of preservation conditions and non- uniform manifestations. Deposition of the dust layer occurred just after a disruption of surface soils, possibly caused by a shock wave well-documented in archaeological sites, with rapid propagation of a wild-fire, and was synchronous with the fall-out of black carbon originating from major forest fires in other regions. The event was immediately followed by exceptional heavy rain storms. Radiometric dating provides an age at ca. 4000 B.P. (i.e. 4500-4100 cal B.P. at 2 σ), consistent with the chronological estimation based on archaeological data from reliable contexts: the transition between late Early Dynastic III and Early Akkad (i.e. 2350-2300 cal BC). Its petrographic assemblage indicates ejection of rock fragments from various exogenous geological formations (marine limestone, basalts, gabbros, sandstones) and a great diversity of rounded, well-sorted spherules and vesicular glassy grains deriving from vaporised rocks. Re-interpretation of the previously-found thick tephra-rich deposit allows us to demonstrate that it represents a lateral variation of the dust layer and originated from long distance transport of volcano-sedimentary formations by unusual air turbulence. This exceptional accumulation over 1 km^2 area of exogenous deposits is no longer taken to trace a tephra fall-out from a volcanic explosion but confirms that the event is marked by massive ejecta originating from volcanic terrains.

The exceptional manifestations recorded during this event explain our previous confusion with a volcanic explosion and an abrupt climate change. The origin of this mysterious phenomenon still remains unsolved. A better comprehension of the micro-debris fall-out and factors responsible for spatial variability of the signal should help to test further the hypothesis of an explosion caused by disintegration of an extra-terrestrial projectile.

1. Introduction

The fallout of a massive tephra layer and the synchronous large increase of wind turbulence observed from various soil contexts of north-eastern Syria were previously interpreted by us [29] as exceptional manifestations linked to an abrupt climate change at around 3.9 kyr B.P., based on stratigraphic correlation with Tell Leilan archaeological sequence (29). The strict contemporaneity between the abandonment of Tell Leilan and the uncommon aridification lead us to propose a causal linkage between a unique deterioration of environmental conditions in northern Syria with a profound destabilisation of socio-economic systems that could have played a crucial role in the collapse of the Akkad Empire domination over northern Mesopotamia.

Synchronous social collapse, as previously documented in Egypt, Indus and the Agean region, and evidence for an aridity peak at 3.9 kyr B.P. from west Africa to west Asia, served to support the global dimension of the climate disturbance and the related socio-economic crisis [11, 16, 28]. Both aspects have, however, remained severely disputed because of the difficulty in obtaining high chronological resolution for relating different strands of evidence in real time and the common hostility towards viewing the trajectory of ancient societies as punctuated by a series of natural catastrophes [4, 12, 13, 30].

Evidence from the Gulf of Oman has recently reinforced the hypothesis of a unique and global scale event marked by the twinned manifestation of "tephra fallout/dust pulse" [28]. It was suggested, however, that long distance correlation should be considered cautiously given the uncertainties of carbon dating [12].

Surprisingly, the dating evidence for less than about 140 years between fallout of the volcanic ash and the dust pulse obtained in the Gulf of Oman [12] has not been used to confute the climate/historical linkage suggested by us in 1993 [29] on the basis of our previous interpretation of late third millennium soils. The search for large-scale connections and a precise quantitative estimate of the aridity increase seems to have concealed important questions regarding the essence of the ca. 3900 B.P. unique event: (1) the significance of a volcanic deposit showing an uncommon spatial variability and its linkage with the dust spike; (2) and the succession in real time of the related manifestations, reasons for their spatial variability and refined correlation with historical events?

Here, we report the progress of our investigation in northern Syrian extended to additional archaeological sites and natural soil sequences that now allow us to establish the existence of the dust layer on a regional scale. The research strategy was adapted to provide the quality of data necessary for debating the exact nature of linkages between natural phenomena and human behaviour on a time-scale compatible with the one for societal phenomena. This paper presents arguments for the detrital origin of the previously-found tephra and for their deposition by exceptional long-distance air turbulence. We speculate on the special circumstances that produced at the same time a massive dust increase generated by pulverisation of the local soils together with an exogenous component, wild-fires and subsequent heavy rains, and envisage a cosmic hypothesis.

2. Research strategy

The original results reported here have greatly benefited from detailed soil investigations carried out successively in various parts of the Upper Jezireh [5, 8, 10]. Homogeneity of geological materials, landform monotony due to extensive gentle flooding and dust accretion in addition to a profusion

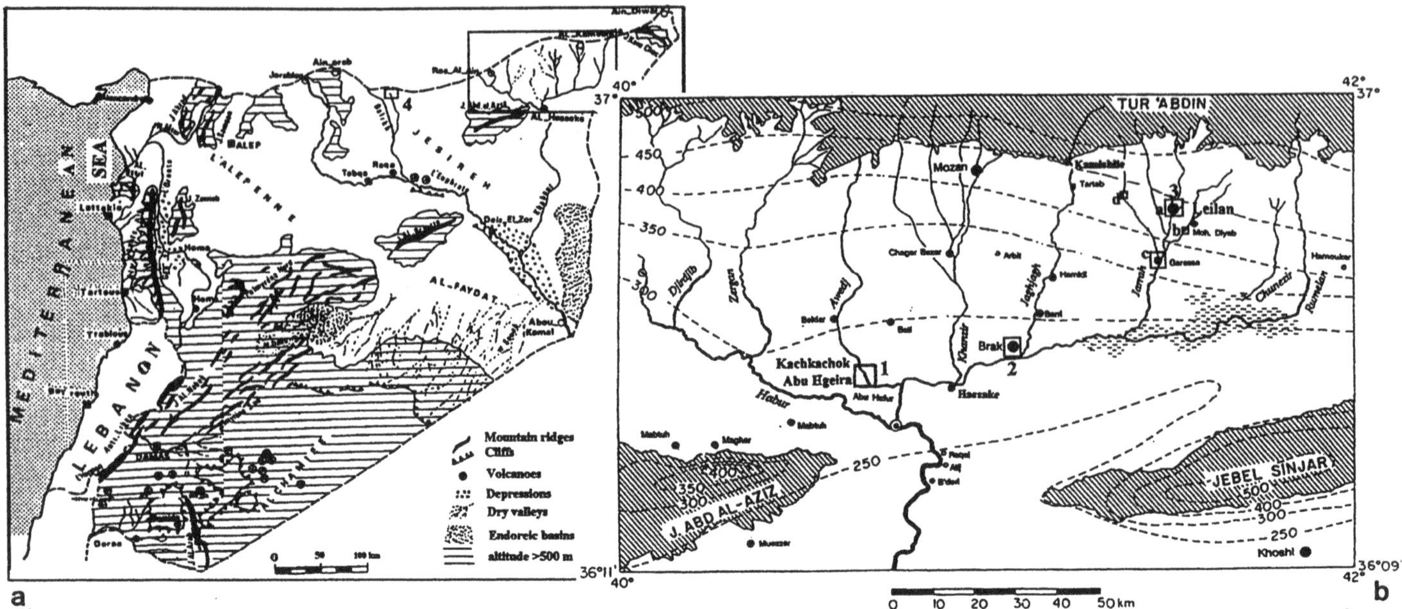

Fig. 1a. Location map of the study areas in the Middle East: Northern Jezireh (large rectangle) and Tell Chuera, upper Balikh valley (4). **Fig. 1b.** Location of the studied areas in the Northern Jezireh region: (1) Abu Hgeira 2 and Abu Fur; (2) Tell Brak; (3a) Tell Leilan; (3b) Abu Fara; (3c) Chura; (3d) Tell Aïd.

of well-preserved archaeological sites have provided optimal conditions to trace spatially from the profile level and the micro-region up to the regional level the unique environmental change that occurred at the end of the third millennium BC. An important research effort was devoted to identify ideal situations offering excellent preservation for the soil signal of this unique event and high resolution dating. The best record was obtained from natural soils subsequently buried below archaeological deposits. In addition comparison with soil sequences from active depositional environments has allowed us to obtain a better understanding of the nature and intensity of geomorphic factors that obscure the signature of short-term events. Aware of the inherent difficulties in precisely bracketing abrupt environmental changes by radiometric dating, microstratigraphic studies have been simultaneously performed on archaeological sites to determine the exact chronological position of this exceptional event in occupation sequence, establish its strict contemporaneity between distant sites and, in an improved fashion test the existence of direct linkages between natural phenomena and cultural changes.

We report, here, results from four micro-regions located in north-eastern Syria (Fig. 1) that have been studied at different levels depending upon the practical facilities operating in the field (Table 1).

3. Analytical procedure

The progress of investigations has lead us to elaborate a flexible analytical procedure designed to detect, quantify and explain all kinds of diagnostic signatures specific to this unique event. The inventory was achieved through a systematic comparison with soils and sediments formed during the Holocene, their modification under the influence of successive climate changes and their transformation by human activities. Considering the great homogeneity of parent materials and a weak differentiation of soils, the study of soil and sediment properties was performed under the petrographic microscope by comparing 900 large thin sections (13 x 7 cm) prepared from undisturbed samples.

N°	Region	Benchmark site	Nature of investigations
1	Lower Awedj valley	Abu Hgeira 2, 2nd millennium	micro-regional study: systematic deep trenches; comparison of buried and surface soils
2	Lower Jagh Jagh valley	Tell Brak, 5th-2nd millennium	study of occupation deposits from the second half of the 3rd millennium
3a	Central Jarrah valley	Tell Leilan, 5th-2nd millennium	(1) extensive study of 3rd millennium occupation deposits; (2) micro-regional study: systematic deep trenches; comparison of buried and surface soils
3b	Central Jarrah valley	Abu Fara, 2nd millennium	micro-regional study: systematic deep trenches; comparison of buried and surface soils
3c	South-central Jarrah valley	Qarassa, 3rd-2nd millennium	micro-regional study: systematic deep trenches (surface soils)
3d	Western Jarrah valley	Tell Aïd, 4th-2nd millennium	micro-regional study: systematic deep trenches; comparison of buried and surface soils
4	Upper Balikh valley	Tell Chuera, 4th-3rd millennium	study of occupation deposits from the second half of the 3rd millennium

Table 1. Summary of the investigations carried out in the different regions studied (see locations in Fig. 1).

The difficulty in recognising with the naked eye clear manifestations of the event has caused us to adopt a back-and-forth strategy between the field and the laboratory in order to determine its exact stratigraphic position and refine later sampling. Rapid examination in thin sections of the deposit micro-facies has provided the most secure diagnostic by allowing simultaneous study of mode of deposition, petrography of each deposit and its soil microfabrics. Under the petrographic microscope, various millimetre-sized rock fragments and uncommon particles have appeared to be specific to this event, thus offering the possibility of performing a preliminary control in the field by using a portable binocular. Large amounts of specimens obtained by random sampling have, therefore, been collected, immediately processed through water-sieving and observed under the

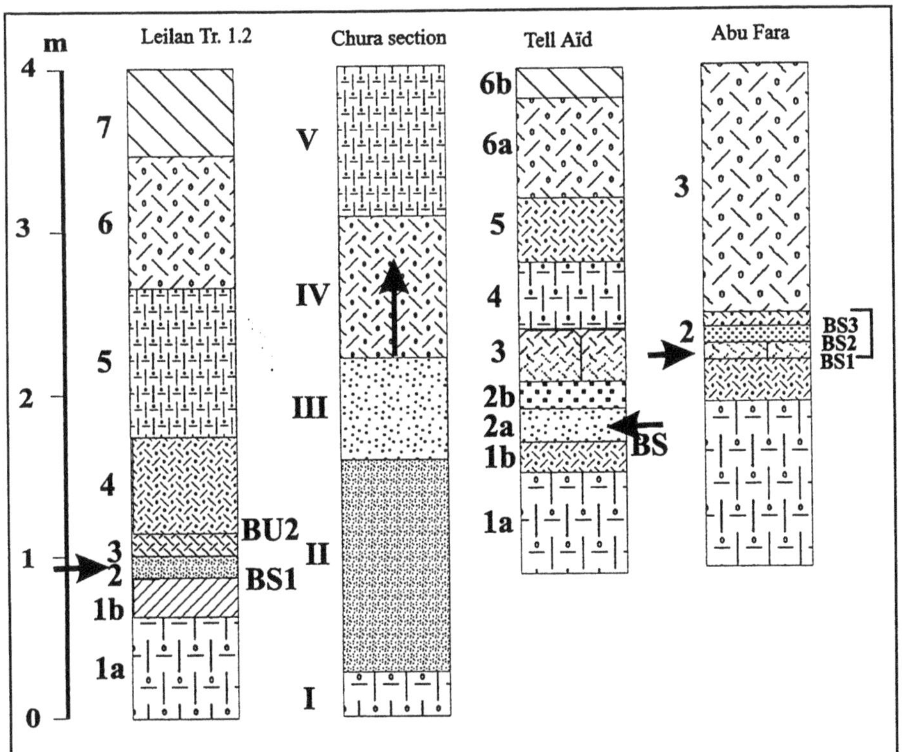

Fig. 2. Selection of natural sequences illustrating variability of the stratigraphic position of the thin dust layer in the region of Tell Leilan. Leilan Tr. 1.2 upslope position (3a): (1a) natural soil; (1b) buried burnt upper horizon; (BS1) lower burnt soil surface; (2) the thin dust layer with high spherule content; (3) upper burnt unit, colluvium from 1b, BS1 and 2; (4) single phase mud flow; (5) 2nd millennium B.C. cummulic soil; (6) late Holocene cummulic soil; (7) surface soil. Chura section, active depression (3c): I to IV are stratigraphic units resulting from intermittent flooding and weak pedogenesis; relicts of unusual mineral grains appear at the contact III/IV and throughout IV, with evidence of burning and uncommon calcitic features; (V) last flooded unit, strongly pedogenised. Tell Aïd (3d): (1a) natural soil; (1b) buried burnt upper horizon; (BS) burnt soil surface (2a) micro-layered dust unit with high spherule content; (2b) micro-aggregated coarse unit, colluvium from 1b, BS and 2a; (3) single phase mud flow; (4) cummulic soil; (5) weakly pedogenised slope colluvium; (6a) active soil; (6b) surface horizon. Abu Fara (3b): (1a) natural soil; (1b) buried burnt upper horizon; (2) micro-stratified sequence showing alternation of thin dust layers and burnt surfaces; (3) Second millennium BC. archaeological layers. The arrow indicates the stratigraphic position of the thin dust layer.

binocular. Abundance of the diagnostic particles from the different fractions -sized estimated in the field has allowed us to select the best preserved situations for performing detailed stratigraphic investigations and obtaining high quality samples for various analyses.

Radiometric dating through AMS and conventional methods was restricted to distinct *in situ* burnt ancient surfaces for natural soils and to concentrations of charred remains from thin ashy layers in occupation deposits in order to obtain reliable age estimations. X-ray analysis of the clay fraction (2) and measurement of the black carbon content (soot) was performed from bulk samples collected from the distinct burnt ancient surfaces [6, 23].

100 g to 1 kg of bulk sample was water-sieved using distilled water, applying minimum hand pressure in order to minimise risk of pollution and preserve as far as possible the most fragile particles. Eight fractions were collected: (1) > 2mm, (2) 2-1 mm, (3) 1 mm-500 µm, (4) 500-200 µm; (5) 200 - 100 µm; (6) 100 - 50 µm; (7) 50-25 µm; (8) < 25 µm. An inventory of petrographic assemblages was performed under the binocular for fractions 1 to 6 and under the petrographic microscope for fraction 7. $\delta^{13}C$ and $\delta^{18}O$ isotope measurement was performed on carbonate spherules extracted from

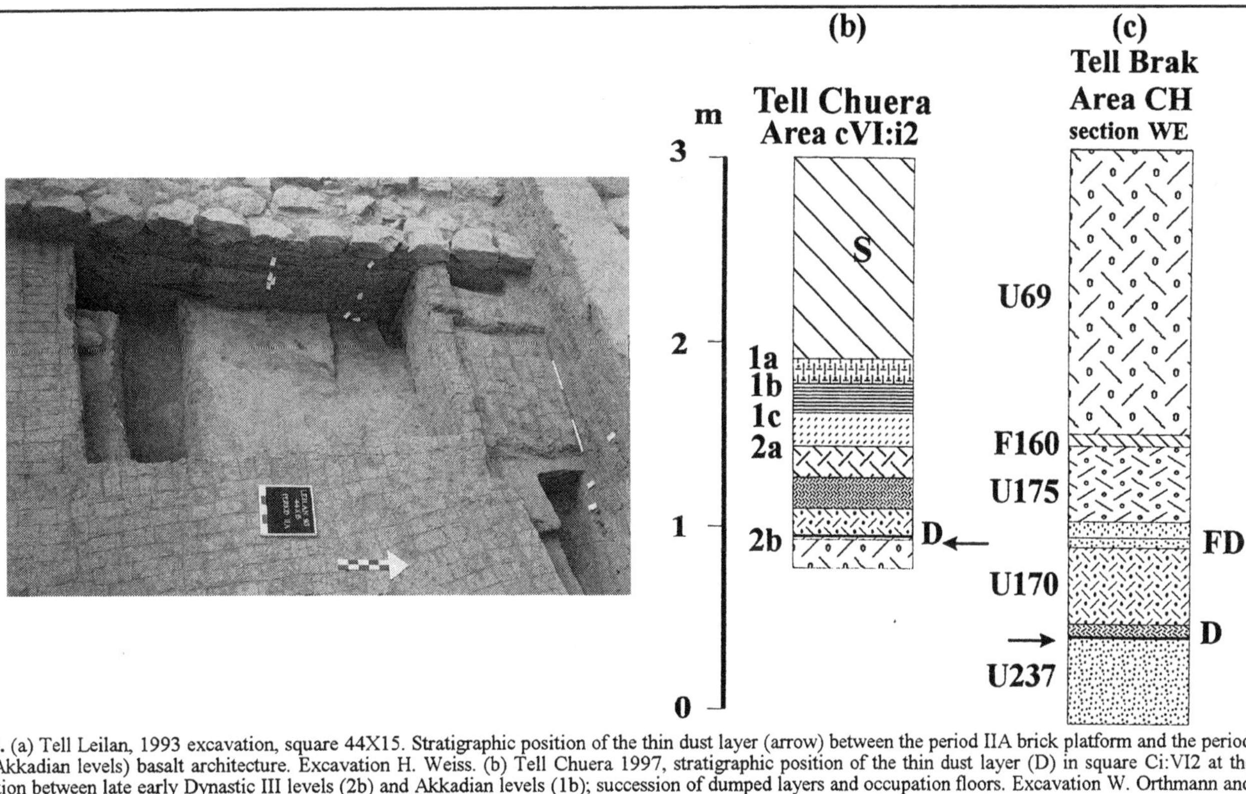

Fig. 3. (a) Tell Leilan, 1993 excavation, square 44X15. Stratigraphic position of the thin dust layer (arrow) between the period IIA brick platform and the period IIB (Akkadian levels) basalt architecture. Excavation H. Weiss. (b) Tell Chuera 1997, stratigraphic position of the thin dust layer (D) in square Ci:VI2 at the transition between late early Dynastic III levels (2b) and Akkadian levels (1b); succession of dumped layers and occupation floors. Excavation W. Orthmann and A. Pruss. (c) Tell Brak 1997, stratigraphic sequence from area CH, West-East section, showing an expanded signal of the event in a narrow passage area between mud brick walls; D: first occurrence of the dust with evidence of explosion; FD: fine dust layer; U237: late early Dynastic III levels; U170, U175, F160: transition EDIII/early Akkad; U69: pre-Akkadian floor preparation. Excavation D. and J Oates.

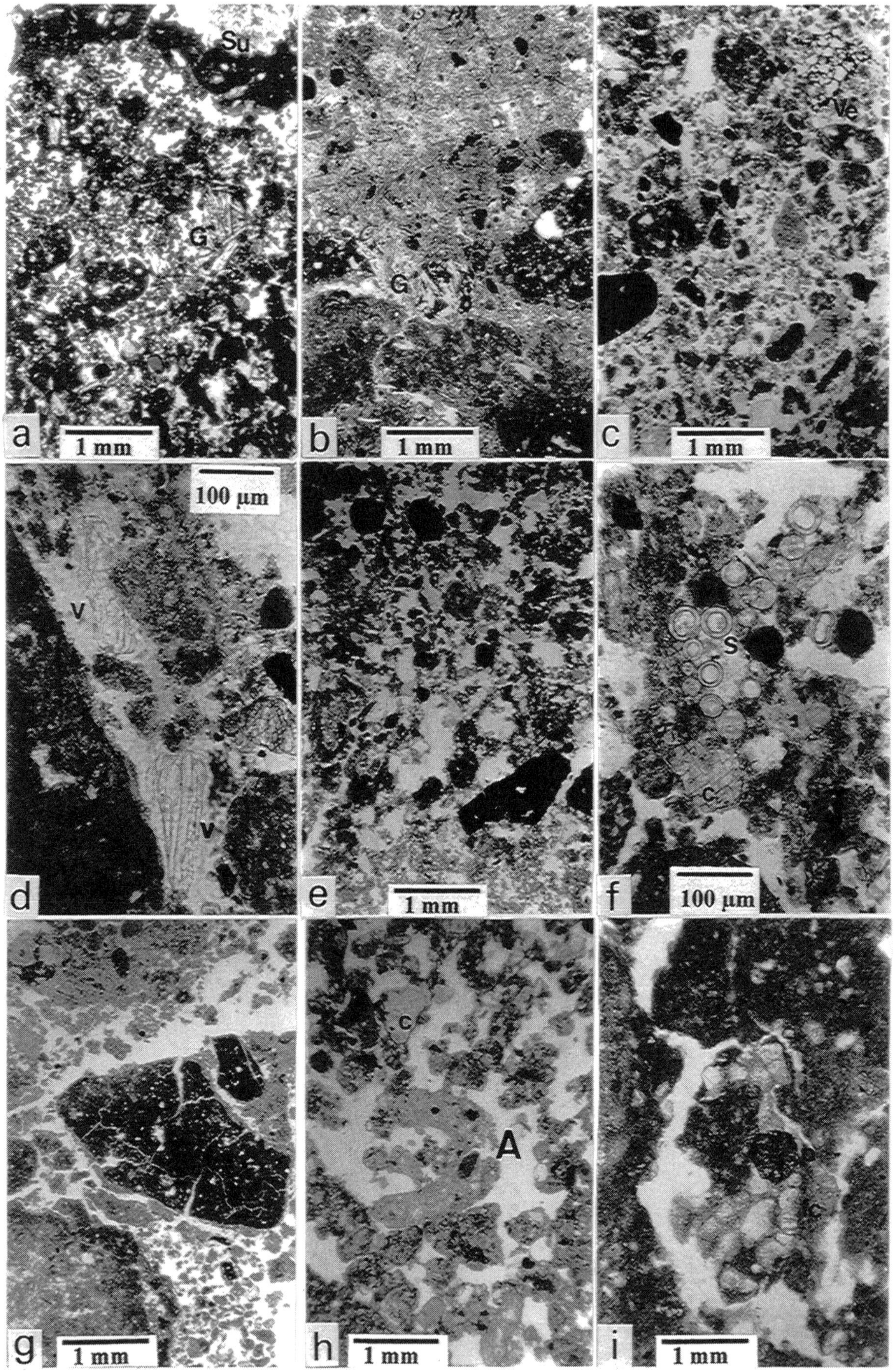

Fig. 4. Micro-photographs showing various aspects of the thin dust layer in different contexts. (a) Abu Fur, burnt soil sealed below second millennium occupation deposits, exploded fabric and exogenous particles. (b) Tell Chuera Ci:VI2, level 2b, exploded fabric and exogenous particles. (c) Tell Brak, CH, lower dust layer, exploded fabric and exogenous particles. (d) Enlargement of (c) showing tephra particles. (e) Tell Brak, CH, upper dust layer. (f) Enlargement of (e) showing high concentration of small sized calcite spherules. (g) Leilan Tr. 1.2, burnt soil surface showing an exploded fabric. (h) Section north of Tell Leilan, exploded structure with elongated glass-rich aggregate (A). (i) Chura section, bottom level IV, relict features of the thin dust layer: concentration of exogenous particles and fine crack filling by fibro-radiated calcite.

fractions 5 and 6 of the layer showing the highest concentration of these specific particles.

X-ray analysis was performed on the total residue for fractions 7 and 8 by using a conventional detector and from individual grains for fraction 4 to 6 by using a linear detector. Uncommon grains sorted under the binocular were studied under SEM equipped with EDS microprobe on clods (secondary electrons) and from polished thin sections (back-scattering).

4. Stratigraphic variability

Further field investigations performed simultaneously on natural soils and archaeological sites have allowed us to extensively trace over the studied region the thin dust layer (Fig. 2 & 3) first identified as a 0.5 cm-thick unit in the late Akkadian period IIb stratum at Tell Leilan (cf. Phase 1 in Weiss et al., 1993). In spite of a major research effort we have, however, been unable to detect more traces of the 60 cm thick tephra deposit originally found at Abu Hgeira 2, and correlated with a major volcanic eruption [29]. On the contrary, re-examination of the stratigraphic sequence from Abu Hafur 2, located 2 km farther south from Abu Hgeira 2, has lead us to conclude that the aeolian unit formerly correlated to the thick tephra deposit [29] presents, in fact, the facies and petrographic characteristics of the thin dust layer (Fig. 4a). The latter contains high amounts of unusual mineral grains with a petrographic assemblage broadly similar to the ones of the tephra-rich layer, although the different mineral species occur in different proportions with a great variability not only between sites but also over short distances. The most striking difference lies in the abundance of well-sorted spherules, predominantly made of fibro-radiated calcite spherules (Fig. 5a), whereas sand-sized tephra grains are exceptional (Fig. 4d). Occurrence of these particles is associated with unique manifestations that are not always clear in the field, however unquestionable they may be in thin sections.

The thin dust layer displays in the field rapid spatial variations of the micro-stratigraphy that appears to be best preserved in upslope micro-undulations, but is often gradually obscured along gentle slopes and can even totally vanish in large, low lying depressions. The most complete micro-stratigraphy, as observed at Abu Habur 2 (Fig.1b) and at several study sites of Tell Leilan region (Fig. 1b & 2), shows a one centimetre thick, greyish yellow, loose dust layer directly over a distinct burnt soil surface marked by a millimetric ashy grey layer rich in partially-melted *graminae* phytoliths and carbonised lignified residues. Below the calcinated vegetation layer, a dense network of charred threads forms the relics of the original *graminae* root mat, fossilised by rapid carbonisation and preserved from subsequent disturbances. The original micro-aggregated structure of the surface soil horizon, produced by intense bioturbation, displays an uncommon juxtaposition of angular large aggregates with sharp fissures and loosely packed fine aggregates derived from fragmentation of the former (Fig. 4g).

The burnt soil surface and the associated dust layer are overlaid by a massive and homogeneous soil unit (see Fig. 2) that contains fragments of slaking crusts, and abundant fine charred residues. The uncommon sharpness of the contact with this upper unit as compared to the usual regional soil stratigraphy shows rapid deposition by a mud flow that allowed the burnt soil and the associated dust layer to be instantaneously sealed.

An attempt to trace the diagnostic stratigraphic layer along trenches has shown a rapid and important spatial variability due to partial reworking or complete erosion during the subsequent episode of mud flow (Fig. 2). Therefore, the exact stratigraphic position of the event in sequences from depositional environments that have remained continuously active during the Holocene period, such as active flood plain, cannot be identified in the field but only by detecting petrographic markers of the dust layer under the petrographic microscope.

The large extension of the burnt soil that can be traced at local to micro-regional scales across all types of depositional environments, and recognised from different regions on the basis of stratigraphic correlation, allows us to avoid possible confusion with human related fires. Its close association with the unusual dust layer and the subsequent mud -flow furthermore corroborates the exceptional nature of this event.

Complementary information is provided by archaeological sites where lack of a vegetation cover and juxtaposition of areas showing contrasted configuration have offered original conditions to record the event as compared to natural soils. The event is traced by presence over a wide region (Fig. 1b and Fig. 3) of unusual mineral grains that are strictly identical to the ones of the dust layer associated with the burnt soil surface in natural soils. In open-air and unroofed areas (Fig. 3a), the mineral grains are concentrated in a centimetre-thick layer and embedded in a loose grey ashy fine mass. The absence of similar particles in the underlying formations helps to identify the exact stratigraphic position of the dust layer. The upper limit is, often, not so clear because the mineral grains have been finely incorporated in the local materials that have subsequently been re-used for constructions (Fig. 3b). Evidence of trampling and re-flooring after the dust accumulation indicates that human activities were not interrupted by the event. Comparison of the associated ashy layer from the different sites did not allow us to identify specific micro-fabrics that would have facilitated a discrimination between micro-stratified ashy layers common to archaeological sites [7].

Narrow open areas that were initially used as passages

Site	Lab. n°	Date (years B.P. ± 1 σ)	cal B.P (range at ± 2 σ)
Tell Brak CH97.1	Beta-110930, AMS	4040 ± 50 BP	4555-4350 BP
Tell Chuera CVI:2a-110	Lyon-473 (OxA), AMS	3980 ± 55 BP	4560-4267 BP
Tell Leilan Tr 1.2 lower burnt soil	GifA 96233, AMS	3790 ± 90 BP	4419-3874 BP
Tell Leilan Tr 1.2 upper burnt soil	GifA 96236, AMS	3980 ± 70 BP	4782-4180 BP

Table 2. Radiocarbon dates. Beta, Beta Analytic, Miami, Florida; Lyon, Centre de datation par le radiocarbone, Lyon, France and Oxford Laboratory, Oxford, England; Gif, Laboratoire des Sciences du Climat et de l'Environnement CNRS-CEA, Gif-sur-Yvette, France. Calibration was with RADIOCARBON CALIBRATION PROGRAM REV. 3.0.3 by the University of Washington.

Site	Tell Brak	Tell Leilan	Tell Chuera
Archaeological period	late EDIII-transition EDIII/early Akkad	late Leilan IIa-transition Leilan IIa/Leilan IIb	late EDIII-transition EDIII/early Akkad
Chronological range (years ± 1 σ)	4240 ± 100/ 3680 ± 100 BP	4020 ± 65 / 3905 ± 65 BP. (from Weiss et al., 1993)	2300 BC. ca. 4250 cal. BP.

Table 3. Age estimation of the dust layer based on archaeological data.

between mud-brick constructions, however, display a distinct succession of micro-facies that cannot be confused with anthropogenic ones (cf. Fig. 3c and 4b & 4c). The first occurrence of unusual mineral grains is associated with a structureless micro-facies made of loosely-packed sub-rounded to subangular aggregates, mixed with fine ashes and carbonised plant residues. Lack of compaction indicates instantaneous desegregation that strongly contrasts with the progressive accumulation generally found in passage areas [3]. Another uncommon feature is the joint occurrence of aggregates arising from fragmentation of the adjacent constructions and biogenic aggregates that are only present in the surface horizons of the surrounding natural soils. Exceptional elongated aggregates made of densely packed flaky, acicular calcite crystals embedded in a yellow isotropic glassy matrix are also observed (Fig. 4h). The sequence, then, shows a decrease of the ash component and ends with a ca. 2 cm thick dense, structureless, yellowish grey layer made of densely packed silt-sized aggregates with a few coarser particles similar to the unusual mineral grains of the lower dust layer (Fig. 4e). This layer contains high amounts of well-sorted spherules, predominantly made of the fibro-radiated calcite type, that are petrographically identical to the ones often found in the dust layer and in the tephra-rich accumulation, although of smaller size (Fig. 4f). The fine textured dust layer is sealed by various occupation deposits, such as well-prepared floors and dumped materials intentionally laid down for preparing a new phase of construction. Therefore, it represents the last unusual manifestation of the event that clearly occurred during a phase of continuous occupation.

5. Chronology of the event

The synchronous occurrence of the various manifestations recorded both in natural landscapes and archaeological sites is fully confirmed by the convergent series of C^{14} radiometric dating (Table 2). Besides aspects related to variability of C^{14} activity during this time period, the reverse order of radiometric ages as compared to the stratigraphic succession of the two burnt layers in the natural sequence close to Tell Leilan (Table 1) warns of the difficulty of using measurement of C^{14} activity for defining the exact age of the event.

Comparison between the three cultural sequences reported also shows a consistent chronostratigraphic position of the event that correlates with the EDIII/Akkad transition both at Tell Brak [18] and Tell Chuera (19), equivalent to the Leilan IIa/Leilan IIb transition at Tell Leilan [27], (Table 3). For the three sites studied, and more particularly for Tell Leilan in which we have performed since 1993 systematic investigations on the third millennium occupation sequences, the series of floors and/or construction deposits that separate the first occurrence of the dust layer *in situ* and the mature Akkad phase provides a secure record to position the event a few decades before the Akkad period, i.e. at ca. 2300-2350 BC on the basis of historical sources [19]. The dust layer has, thus, been easily traced over a 600 m^2 area opened during the 1993 season: exactly located just above the Leilan IIa brick platform and well-sealed below huge basalt blocks brought to the site during Leilan IIb phase, i.e. Akkad period (Fig. 3a). This highly-secured stratigraphic context allows us to re-adjust the exact age of the event to be 150 years older than the first age estimation (2200 BC) as given in Weiss et al., 1993.

6. Evidence for regional wild fires

A significant increase of the black carbon content that exactly coincides with the distinct burnt soil surface associated with the dust layer was measured at three localities in the region of Tell Leilan (Fig. 5). Intensity of the black carbon increase is typical of a high temperature fire that produced high amount of smoke, such as the ones related to burning of a dense savannah or tropical forest [6]. This increase cannot be explained by combustion of the local biomass which at that time consisted of a graminae cover in cultivated fields, an open shrubby steppe in non-cultivated areas and gallery forest along river channels. Rapid decrease of the black carbon content in later soil horizons, however rich in charred plant remains (Fig. 5a), or in occupation layers with high amounts of fine charred residues (Fig. 5c), confirms that endemic local fires, either natural or human-related, do not produce significant black carbon increases. Moreover, absence of detectable clay alteration in the topmost part of the burnt soil surface proves that combustion of the local vegetation occurred instantaneously and that the soil heating has remained surficial [2]. These contradictory facts can be reconciled by relating the black carbon increase to an exogenous input that would have arrived, together with the mineral grains, via air transport from major biomass-

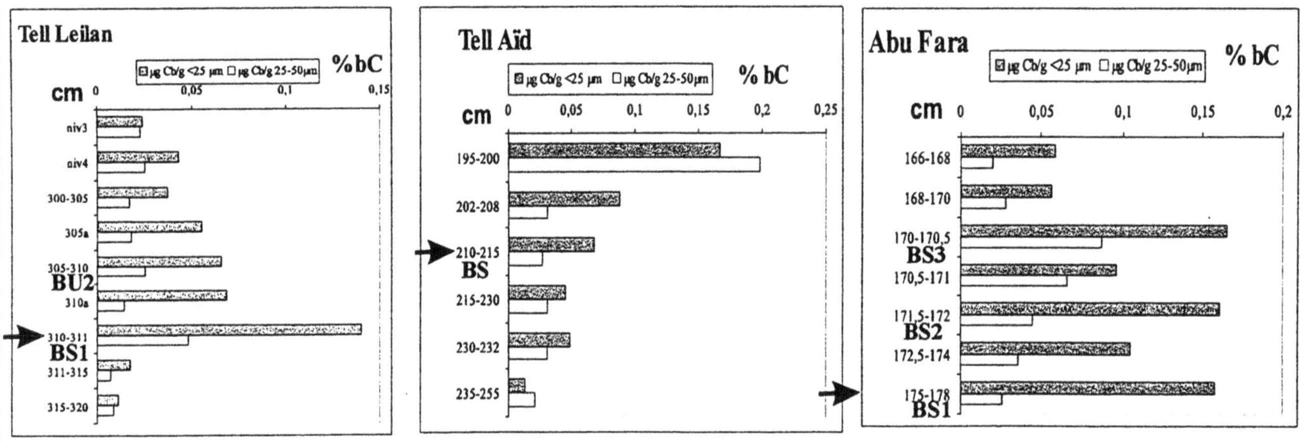

Fig. 5. Measurement of black carbon content (a) Tell Leilan tr. 1.2: marked increase at the exact stratigraphic position of the dust layer; (b) Tell Aïd: marked increase just above the burnt soil surface; (c) Abu Fara: high content in the three burnt surfaces associated with the thin dust layer, decrease in the upper ashy archaeological deposits (cf. section in Fig. 2).

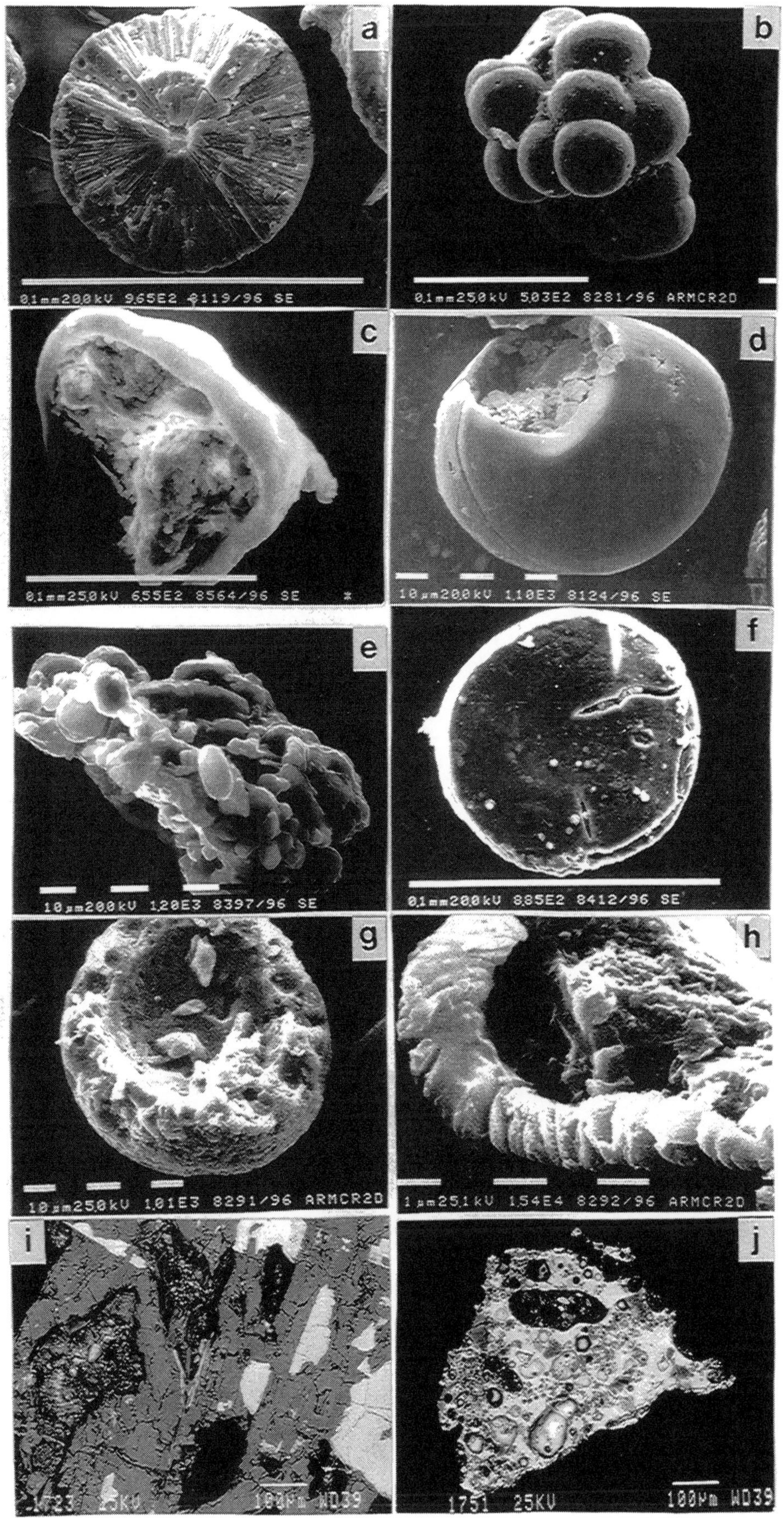

Fig. 6. SEM views of the thin dust layer at Tell Leilan Tr. 1.2. (a) single fibro-radiated calcite spherule, sectioned when air transported and (b) multispherules strongly bound together by silicate cements; (c) translucent, dull orange, tailed-shape, ankerite spherule; (d) silica spherule with Mg carbonate fill; (e) KCl white aggregate; (f) phosphate spherule with KCl small spherules; (g) alumino-silicate spherule; (h) enlargement of (g) showing the ribbon-shape tail made of nickel-rich tiles; (i) angular gabbro fragment with potassium rich glass as infilling of the feldspar fine cracks; (j) complex vesicular glass particle that contains metallic nickel inclusions.

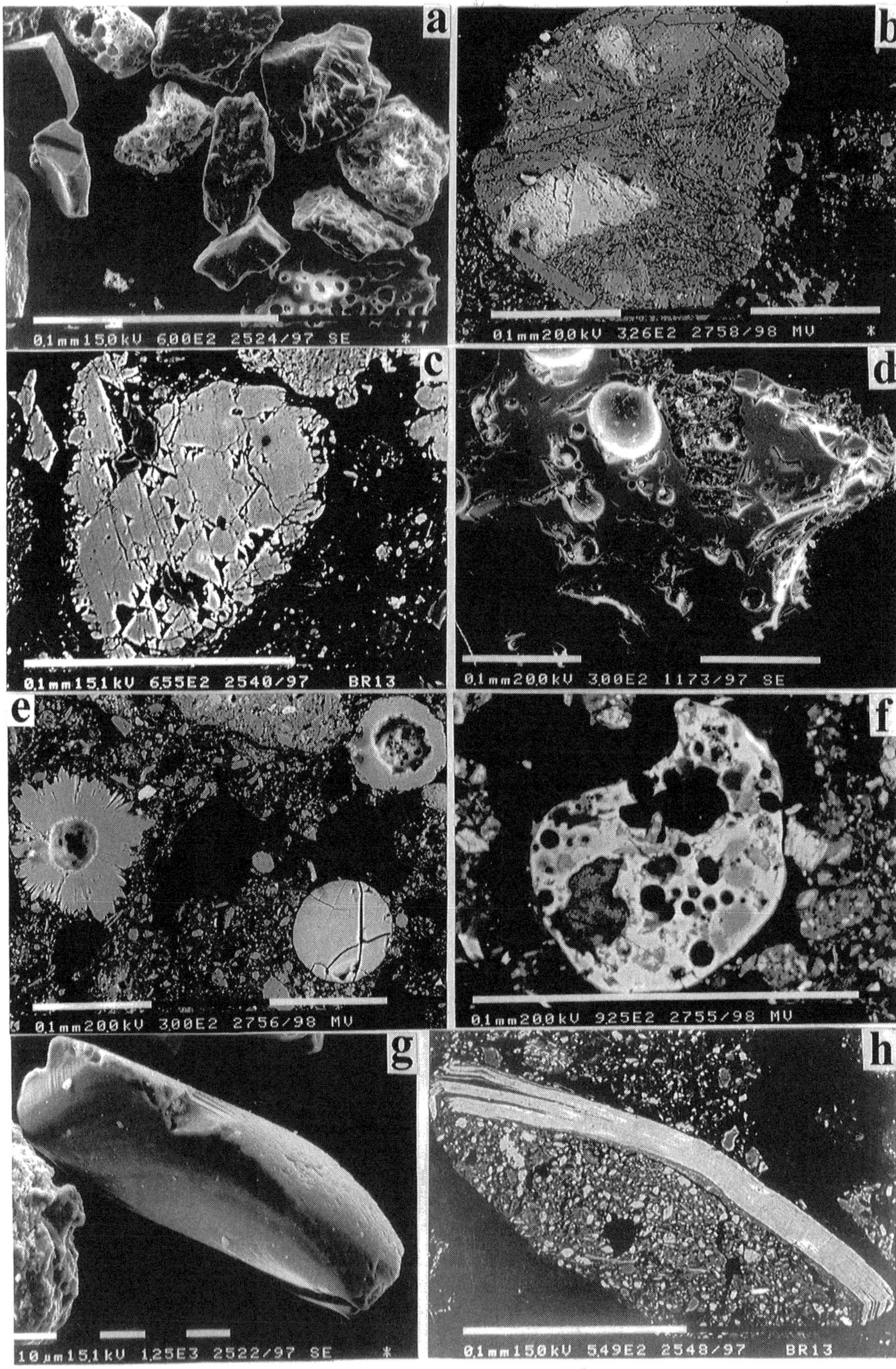

Fig. 7. Petrographic composition of the thin dust layer at Tell Brak.(a) Angular coarse silt (50-25 µm): quartz (1), feldspar (2), vesicular silica grain (3); (b) various highly fissured angular quatzitic sandstone; (c) carbonate fragment showing an uncommon dissolution produced by CO_2 degassing; dark zones correspond to local replacement of carbonates by silica ; (d) carbon-rich (~50%) angular, black, vesicular grain with vesicle infilling made of Mg-Ca rich alumino-silicate glass and pure silica glass; (e) to (f) Specific aspects of the fine-textured dust layer: (e) 50-100 µm spherules, fibro-radiated calcite (1), phosphate (2), silica/carbonate (3); (g) diatom with a pitted weathered surface; (h) elongated fibrous amphibole embedded in fine dust made of acicular crystals.

burning in forested regions [14].

A significant black carbon increase of similar intensity as the one found in natural soils was also measured from occupation deposits of Tell Leilan at a stratigraphic position that exactly coincides with the thin dust layer, well-defined by the sharp increase of exogenous grains [1].

7. Petrography of the thin dust layer

The great variability of the micro-debris assemblages observed between distant sites and at short distance concerns the relative amount, particle-size distribution and properties of the different types of coarse particles identified: spherules, glass particles, fragments of igneous rocks, exogenous carbonates, salts and micro-fossils.

The most comprehensive characterisation of the spherule component was achieved on the dust layer traced over Tell Leilan region (Fig. 1) in which the spherule proportion locally reached 50% of the total dust mass. They are well sorted with an average size between 100-50 µm and a minor proportion in the fraction 200-100 µm. Five mineralogical types have been recognised: carbonate spherules are largely dominant, mainly formed of calcite (Fig. 6a, 6b, 6c & 7e) and in a minor proportion of Mg-rich carbonates (Fig. 7e), amorphous phosphate (Fig. 6f & 7e), silicates (Fig. 6d), alumino-silicates (Fig. 6g), spherules of iron oxides and chromites.

Pure calcite spherules present a granular or a smooth surface and are either white, grey , black, or bi-coloured (internal white part with an external black part, or the reverse); single spherules are predominant (Fig. 6a) and occur together with a minor proportion of multispherules strongly bound together by silicate cements (Fig. 6b). The proportion of major elements is identical for spherules of different colour. KCl is often present as micron-sized crystals or small spherules included in the silicate cement or bound to the surface of the calcite spherules (Fig. 6f). The calcite spherules always display a radiating pattern of densely crystallised needles that show epitaxic growth features and a crystalline continuity from the centre to the outer part (Fig. 6a). Common presence on the sectioned face of the broken spherules of small craters juxtaposed to small calcite spherules of exactly the same size that resisted water-sieving (Fig. 6a) shows that fragmentation, surface transformation caused by ejection and trapping of carbonate droplets occurred during the process of spherule crystallisation and were not caused by laboratory treatment.

Calcite spherules rich in Mg are yellow, generally present an irregular mammilated surface and can display either a continuous fibrous-radiated crystalline fabric or a concentric one. They occur together with spherules of dolomite and ankerite rich in Si that are translucent, dull orange, tailed-shape, showing a smoothened outer part and a finely crystallised inner part (Fig. 6c). The major element composition of the inner and outer part are broadly similar except for the presence of chlorine (~ 1%) in the outer part and in the twisted tail. These orange spherules are extremely fragile and often break during laboratory treatment, losing the twisted tail which can be originally more than 200 µm long. A specific type of orange spherule with the twisted tail, but empty, was observed under the binocular but was severely damaged during careful sampling. The carbonate spherules

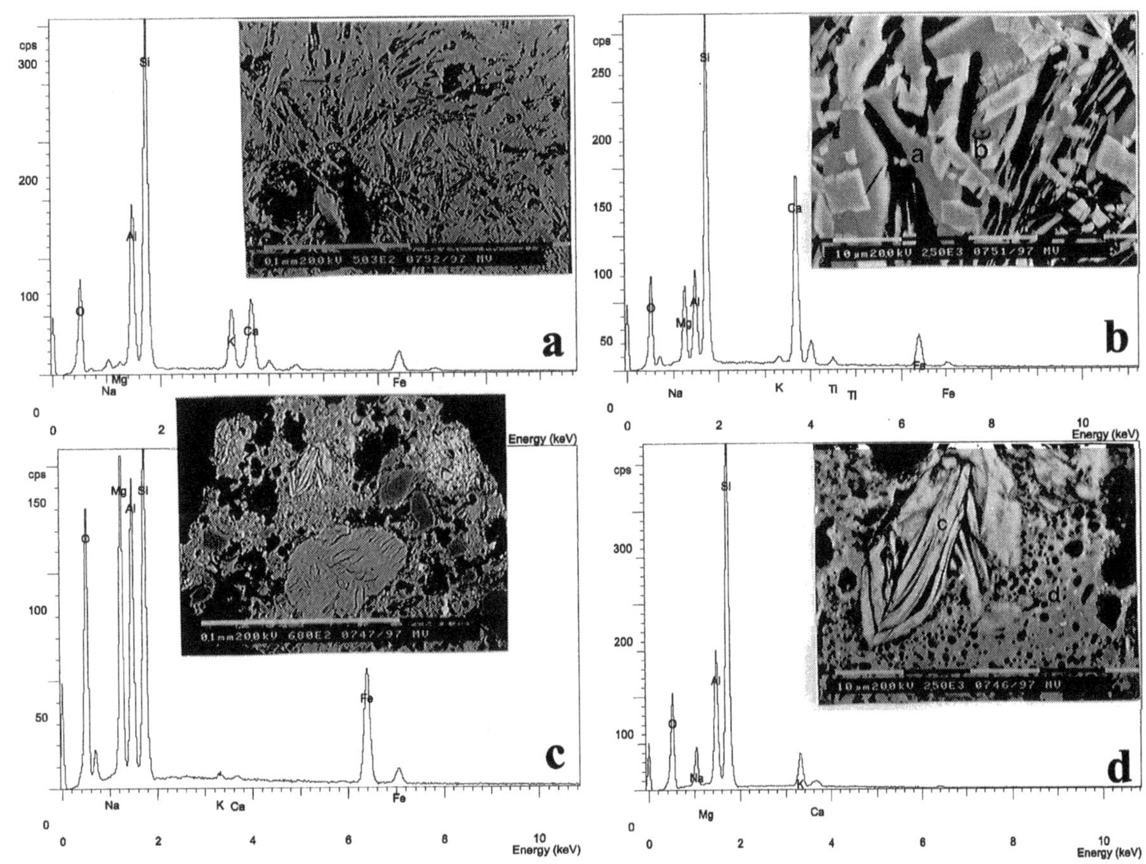

Fig. 8. SEM views and EDS analysis of unusual grains of the thin dust layer at Tell Chuera. (a) Greyish yellow vesicular grain made of needle-shape pyroxene crystals; (b) detailed view of (a) showing the micron-sized acicular pyroxene crystals floating in glass; (c) greyish yellow vesicular grain showing relicts of sands (quartz, amphibole, K-feldspar, argilite) embedded in a K-rich glass; EDS analysis: composition of the fibrous amphibole; (d) detailed view of the grain in (c) and the glassy matrix; EDS analysis: composition of the potassic glassy matrix.

present the following isotopic composition:

$\delta^{13}C$ = -14.7 +/- 0.2 ‰ (PDB)

$\delta^{18}O$ = 23.3 +/- 0.2 ‰ (SMOW).

Pure silica spherules are generally translucent to white, well-rounded, with a smooth surface. In polished section they display a concentric fabric or are undifferentiated, with empty vesicles of various sizes. Some silica translucent spherules are empty with an external silica collar, very fragile and easily broken during sampling. Similar white spherules with a concentric fabric display a compound composition of silicates and carbonates (Fig. 6d & 7e). Spherules of alumino-silicates are generally smaller (80-30 µm), yellowish brown to dark brown, formed of densely-packed needle-shaped crystals of pyroxene, with a great variability of their major element composition, that are embedded in a porous calco-potassic glass (Fig. 6g). Exceptionally, these alumino-silicate spherules present on their external part a ribbon-shape tail made of densely packed nickel-rich tiles (Fig. 6h) that are coated and cemented by alumino-silicates of similar composition to the host spherules, except for the presence of chlorine in traces (~ 1%).

Phosphate spherules are generally pale yellow to yellowish brown, smooth, undifferentiated, finely fissured and often present as micron-sized spherules of KCl strongly attached to the surface (Fig. 6f).

The presence of KCl has also been identified in the form of white grains made of densely packed and strongly cemented ovoid particles made only of this highly soluble salt (Fig. 6e).

The spherules and associated particles are embedded in a pale yellowish brown fine mass formed of very fine silt-sized carbonate particles mixed with clay and with abundant quartzitic, silt-sized, grains. Although apparently similar to the ones of the surrounding soils, various subtle differences suggest that a large part of this fine mass is not of local origin: the abundance of dolomite grains that are never present in the local soils; the angular shape of the quartz and feldspar grains (Fig. 7a) generally coated by reddish brown clay; the abundance of vesicular silica grains; and the common presence of fresh muscovite, biotite and fibrous amphibole and bassanite in the form of silt-sized crystals.

An important component of the micro-debris layer is represented by fragments derived from igneous rocks that range from a few millimetre-sized polycrystalline grains to smaller individual phenocrysts of pyroxene, olivine or Ca-feldspar. The great proportion of coarse crystallised igneous rocks of gabbro type made of large, unweathered olivine, plagioclase and pyroxene (Fig. 6i) contrasts with the local volcanic rocks (20). The latter consists of vesicular basalts with strongly ferruginised olivine embedded in a groundmass of finely-crystallised feldspars and a low amount of finely crystallised pyroxene. In addition to this petrographic difference, the igneous rock fragments of the dust layer differ from the basalt fragments of the local soils by their angular shape and fresh faces, and also by presence of a potassium-rich glass present as infilling of fine cracks and vesicules of the feldspars. Moreover, they occur together with unusual, millimetre-sized to a few hundred microns, pale yellow to yellowish grey vesicular grains that are either made entirely of undifferentiated glass (Fig. 6j) or are formed of a gabbro core embedded in glass or consisting of micron-sized acicular pyroxene crystals floating in glass (Fig. 8 a & b).

The geochemical composition of the glass is generally not uniform within a grain and greatly varies between vesicular grains of the same dust layer and also between the different sites studied. Two types of glass are commonly juxtaposed: (1) a Ca-Mg rich alumino-silicate glass with variable proportions of Fe and the presence of uncommon inclusions such as pure nickel grains (Fig. 6j), zinc or copper, that can be geochemically defined as a pyroxene glass; and (2) a potassium-rich alumino-silicate glass, generally less dense and devoid of metal-rich inclusions, that is geochemically similar to the one present within the vesicles of the gabbro grains (Fig. 6i) and in the vesicular grains with pyroxene needles (Fig. 8a). The latter K-rich glass also forms the fine matrix of other vesicular grains that contain various sand-sized inclusions (quartz, amphibole, K-feldspar, argilite) and appears to be highly transformed sandstones (Fig. 8c & 8d).

Unusual fragments of various sandstones are also present, generally characterised by highly fissured angular grains (Fig. 7b).

Carbonate grains, either limestone fragments or bioclasts, often display an uncommon weathering pattern resulting from dissolution along cleavage planes that might be produced by CO_2 degassing, also associated with the local replacement of carbonates by silica (Fig. 7c).

A common component of the dust layer for all the sites reported here consists of unusual angular, black, vesicular grains that are made predominantly of carbon (~50%) although they do not morphologically resemble either carbonised plant residues or fossil combustibles (Fig. 7d). They are finely fissured, dense and often contain phosphate inclusions. Their large vesicles are often filled by a Mg-Ca rich alumino-silicate glass similar to the type 1 glass described above and a fine vesicular pure silica glass.

The fine-textured dust layer recognised as the last episode of the event at specific localities showing an expanded signal presents a similar petrographic assemblage to the one of the initial dust layer, differing only by the smaller size of the particles and their higher abundance (Fig. 10e & 10f). This greater proportion has allowed a better characterisation of diatoms that were identified in thin sections, occurring either as individual grains embedded in the dust or as aggregates cemented by calcite. In addition to their uncommon presence, SEM observations have revealed a pitted weathered surface and the impossibility of identifying surface features diagnostic of the species (Fig. 10g). Moreover, the very fine dust appears to have been transported as sand-sized complex grains that have suffered minimal fragmentation. This is indicated by the good preservation of fragile fibrous grains (Fig. 10h).

The comprehensive inventory of the dust layer petrography as provided by contexts offering a well-preserved signal has allowed us to identify relict traces of the event in depositional environments that were severely reworked by the subsequent mud-flow as well as from an active flood plain. They comprise rare millimetric fragments of igneous rocks, a marked increase in the coarse silt-sized fraction of individual mineral species derived from the former (pyroxene, olivine, feldspars), occasionally slightly altered calcitic spherules and the presence of uncommon calcitic features along cracks that might derive from pedogenic transformations of the spherule-rich dust layer (Fig. 5i).

8. Petrographic re-examination of the thick tephra facies

The 50 cm thick tephra-rich layer found in the buried soil of Abu Hgeira 2 (Fig.1) was originally interpreted as an accumulation derived from a volcanic plume after long-distance transport that would possibly originate from volcanic centres of the Anatolian-Caucasian area [29]. A series of anomalies not compatible with the volcanic hypothesis has led us to re-examine this interpretation: 1) the restricted distribution of this deposit in spite of its notable thickness; 2) lateral transition between the tephra-rich deposit and the thin dust layer; 3) the failure to confirm the presence of similar tephra grains in the thin dust layer; and 4) the facies and petrography of the tephra-rich layer, which are uncommon for a supposed passive fall-out.

A refined examination of the trenches dug across the Awedj flood plain [5] has failed to detect the presence of tephra. Evidence for their progressive weathering along the restoration of a stable soil cover as observed at Abu Hgeira 2 sequence from layers 4 to 7 (Fig. 9) shows that pedogenic weathering is unlikely to have been responsible for their total disappearance as we originally supposed (29). Recognition of the thin dust layer at Abu Hafur 2 (Fig. 4a), located 2 km farther south, now allows us to establish that the thin dust layer represents a spatial variation of the tephra-rich deposit. The latter is, thus, concluded to have accumulated over a restricted area, not exceeding a few km^2, possibly favoured by the existence of a small basin at this specific location.

A better comprehension of the tephra-rich layer characteristics reinforces this stratigraphic correlation. Further interpretation concerns, firstly, the pseudo-sand micro-facies at the very bottom of the tephra-rich layer (2a, Fig. 9), just above the third millennium BC. Calcisol, which we originally interpreted as evidence of saltation aeolian transport [29]. The puzzling feature was the abundance of uncommon, millimetre-sized sub-rounded aggregates associated to the 50% sand-sized tephra particles embedded in calcitic silt (Fig. 10b). None of these aggregates resemble the ones normally found in the colluviated soils or the wind-blown deposits common in the studied region [5]. A first type of aggregate (type 1) is characterised by a lenticular microfabric typical of freezing that is totally unusual for the studied region (Fig. 10a). It contains tephra gains in the same proportion and size range as the surrounding matrix. A second type of aggregate (type 2) displays an irregular laminated facies similar to the one of slaking crusts created by heavy rain spells (Fig. 10c) as originally interpreted [29]. However, its petrographic composition, made of fine silt-sized tephra particles, polished diatoms with a pitted surface, and quartz embedded in calcitic silt clearly differs from that of the surrounding soil matrix (Fig. 10d). Less common aggregates show a high concentration of siliceous sponge spicules (type 3, Fig. 10e) whereas others display strong ferruginisation and moderate weathering of the tephra particles (type 4) that contrasts with the fresh aspect of the tephra from the juxtaposed materials (Fig. 10f). The close imbrication of these various aggregates within the loose matrix, their morphology and their good sorting indicate that both components have been deposited together after air transport. Preservation of the frost-induced microfabric for type 1 aggregates suggests that they have been transported as frozen materials. The mineralogical composition of type 2 aggregates rules out a formation *in situ*, following deposition of the tephra-rich layer. They are more likely to be relicts of a sedimentary facies initially present in the formation that supplied the tephra particles and associated components. Composition of type 3 aggregates, totally uncommon for the region, also supports this interpretation. Furthermore, the close juxtaposition of tephra particles at different weathering stages achieved under subaerial conditions (cf. type 4

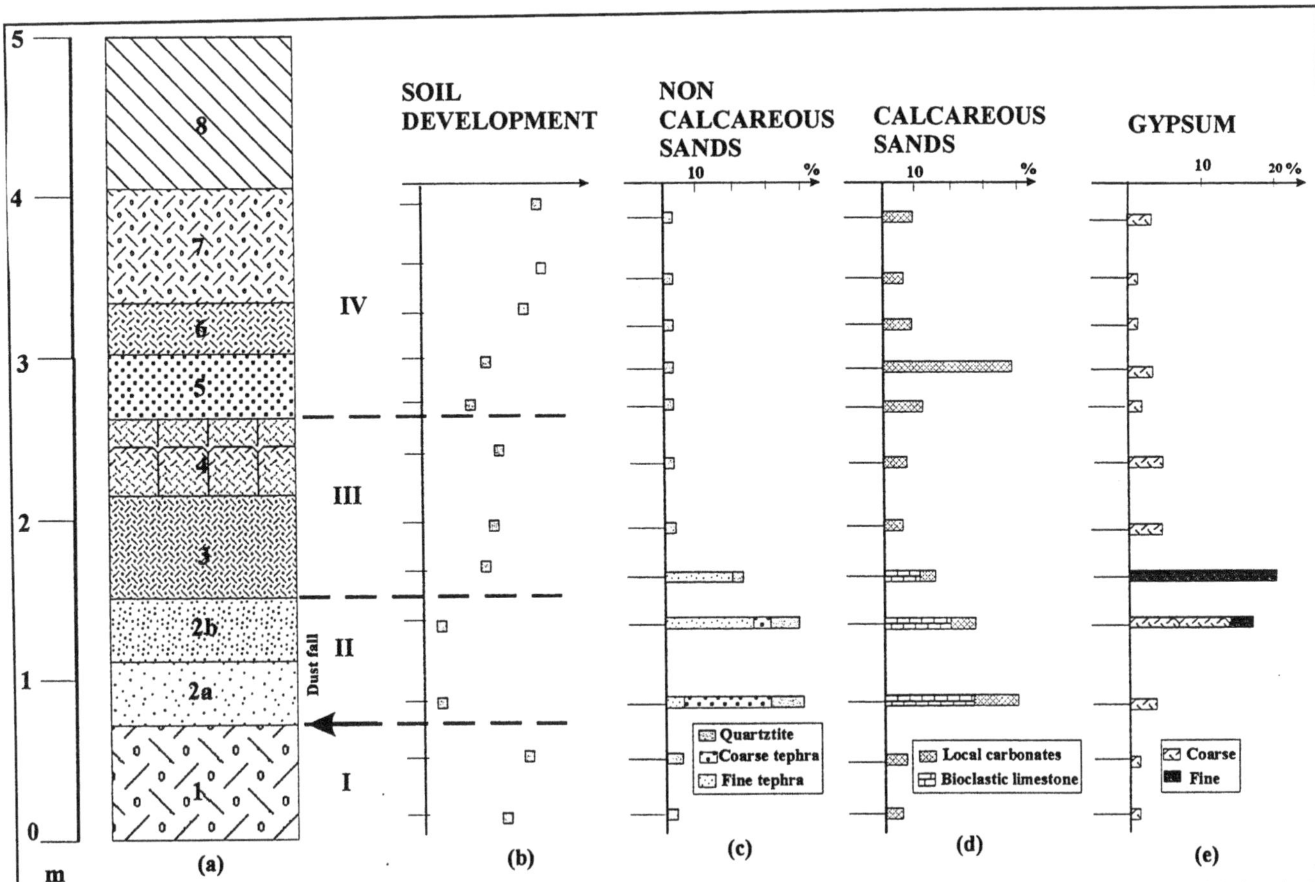

Fig. 9. Abu Hgeira 2- (a) Stratigraphic sequence: (1) 3rd millennium BC. soil; (2a) coarse textured tephra-rich deposit; (2b) fine textured tephra-rich deposit; (3) single phase mud flow; (4) weakly pedogenised wind/water layer; (5) coarse-textured colluvium; (6) weakly pedogenised wind/water layer; (7) well-developed buried soil; (8) 2nd millennium BC. occupation deposits; (b) Profile of the soil development index: (Biological activity + Calcitic features/ Sedimentary features); (c),(d), (e) Profile of non-calcareous sands, calcareous sands and gypsum abundance quantified from thin sections.

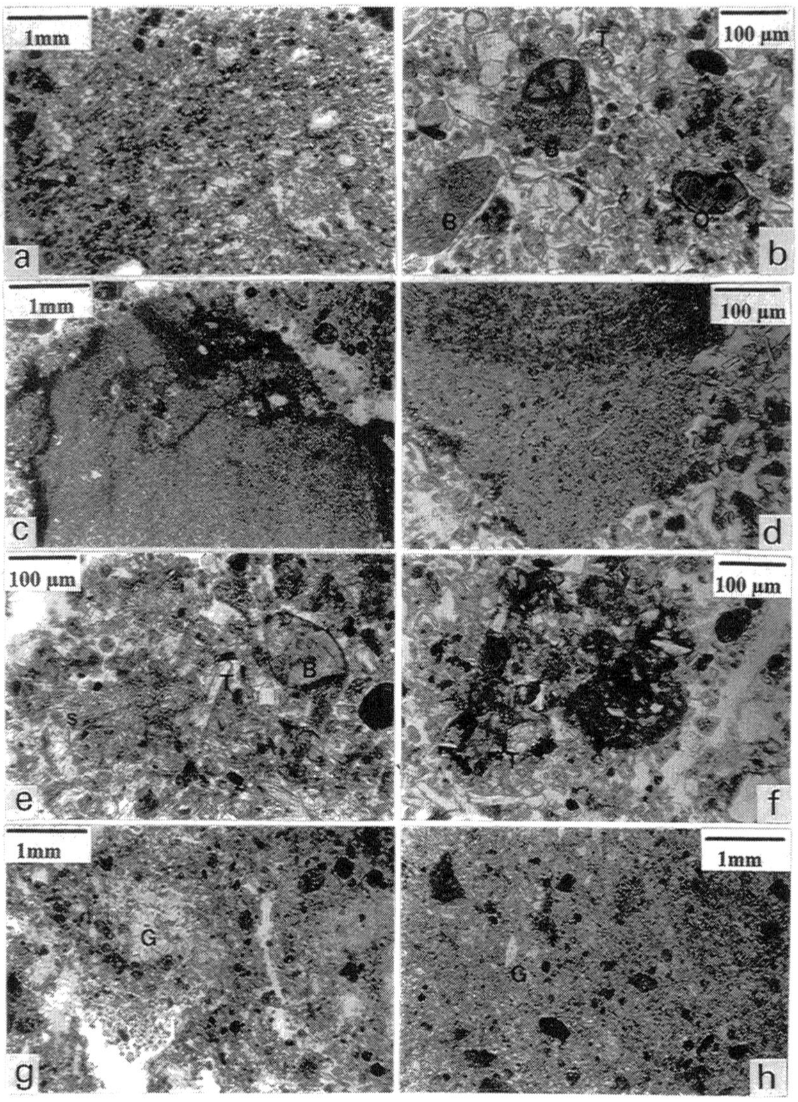

Fig. 10. Micro-photographs showing specific aspects of the tephra-rich deposit, Abu Hgeira 2. (a) type 1: tephra-rich aggregate with lenticular microfabric typical of freezing grains; (b) bioclasts and tephra sand-sized grains embedded in calcitic silt ; (c) type 2: surface-crust like aggregate; (d) higher magnification of (c) showing its composition made of fine silt-sized tephra particles and diatoms in calcitic silt; (e) type 3: aggregate with high concentration of siliceous sponge spicules; (f) type 4: ferruginised aggregate with weathered tephra particles; (g) detrital and partly recrystallised coarse gypsum in layer 2a; (h) detrital fine textured gypsum in layer 2b.

aggregates) proves that the tephra component cannot derive from a freshly exploded volcanic plume as we originally concluded [29].

Our previous study focussed on the geochemistry of the tephra particles [29]. Further investigation of the petrographic composition of the non-tephra component and refined examination of the glass geochemistry brings complementary information on the original deposit and transformation that occurred just before its transportation. The petrographic composition of the tephra-rich layer (2a, Fig. 9) shows that a great diversity of exogenous particles has been simultaneously transported and deposited with the sand-sized tephra particles: mainly sand-sized grains derived from bioclastic and dolomitic marine limestone (Fig. 10b and Fig. 11f); in a minor proportion, sand-sized grains originating from a large variety of volcanic, igneous and metamorphic rocks (granite, green schists, amphibolite, metabasalts, gabbro, peridotite), salts, particularly calcium sulphate (gypsum), barium and strontium sulfate (Fig. 9 & 11h); and fibro-radiated calcite spherules generally present as fragments (Fig. 11c). Outcrops of bioclastic marine limestone are not present locally and the amount of this type of rock fragment never exceeds 1% in parent materials of the local soils derived from the alluvial fans formed by repeated erosion of the Taurus foothills [5].

Similarly, local soils and sediments do not contain fragments of dolomitic limestone, characterised by a high magnesium content, or fibro-radiated calcite spherules and both are, thus, concluded to be of exogenous origin. The petrographic diversity of the aluminosilicate sand-sized fraction strongly contrasts with the normal petrographic assemblage of the local sediments that is dominantly formed of quartzitic grains and finely crystallised basalts identical to the local lava flows. Gypsum occurs at the bottom of the tephra-rich layer (2a) as individual euhedral crystals or aggregates of euhedral crystals (Fig. 10g), clearly different from the pedogenic gypsum that is often present in the local soils as channel infilling produced by the capillary rise of sulphate-rich perched ground water. The sedimentary origin of the gypsum is confirmed by the rounded contours of the crystals as seen under SEM which is typical of wind abrasion (Fig. 11g). Evolution towards angular fine sand-sized crystals in layer 2b (Fig. 9 & & 10h) indicates fragmentation of gypsum crystals along cleavage planes as the result of intense air turbulence. This is also confirmed by the simultaneous abrasion and fragmentation of tephra particles from layer 2a to 2b (Fig.9). The occurrence of strontium sulphate also as individual rounded particles attests for its sedimentary origin (Fig. 11h). Such an important aeolian input of sulphate salts is unique to the 4 kyr. B.P. event as compared to Holocene soils and earlier deposits of the studied region [5].

Observations under BSE (backscattered electron microscopy) and X-ray analysis, furthermore, reveals the occurrence of two types of glass. The first one (Fig. 11a), that is restricted to the tephra particles themselves and homogenous throughout the layer, corresponds to the volcanic glass *strict sense* as previously measured [29]. The second glass occurs as thin coatings around all sand-sized grains of layer 2a (Fig. 11a, 11c, 11e), as thin threads and domains across porous grains and aggregates (Fig. 11f, 11g, 11h) and as scattered domains in the packing porosity. Analyses with an energy dispersive X-ray spectrometer indicates major geochemical differences as compared to the typical volcanic glass (large dominance of calcium and reduced amount of potassium, abundance of Mg, lack of Na, greater abundance of iron) and significant variability of the major element composition across and between individual coatings or domains. The puzzling pattern of the Mg-Ca glass suggests that it was injected as vapour throughout the volcano-sedimentary deposit and rapidly solidified before air transportation. Theresence of Mg-Ca glass coatings around highly fissured grains (see for example Fig. 11c, 11e) suggests that the glass injection and its solidification

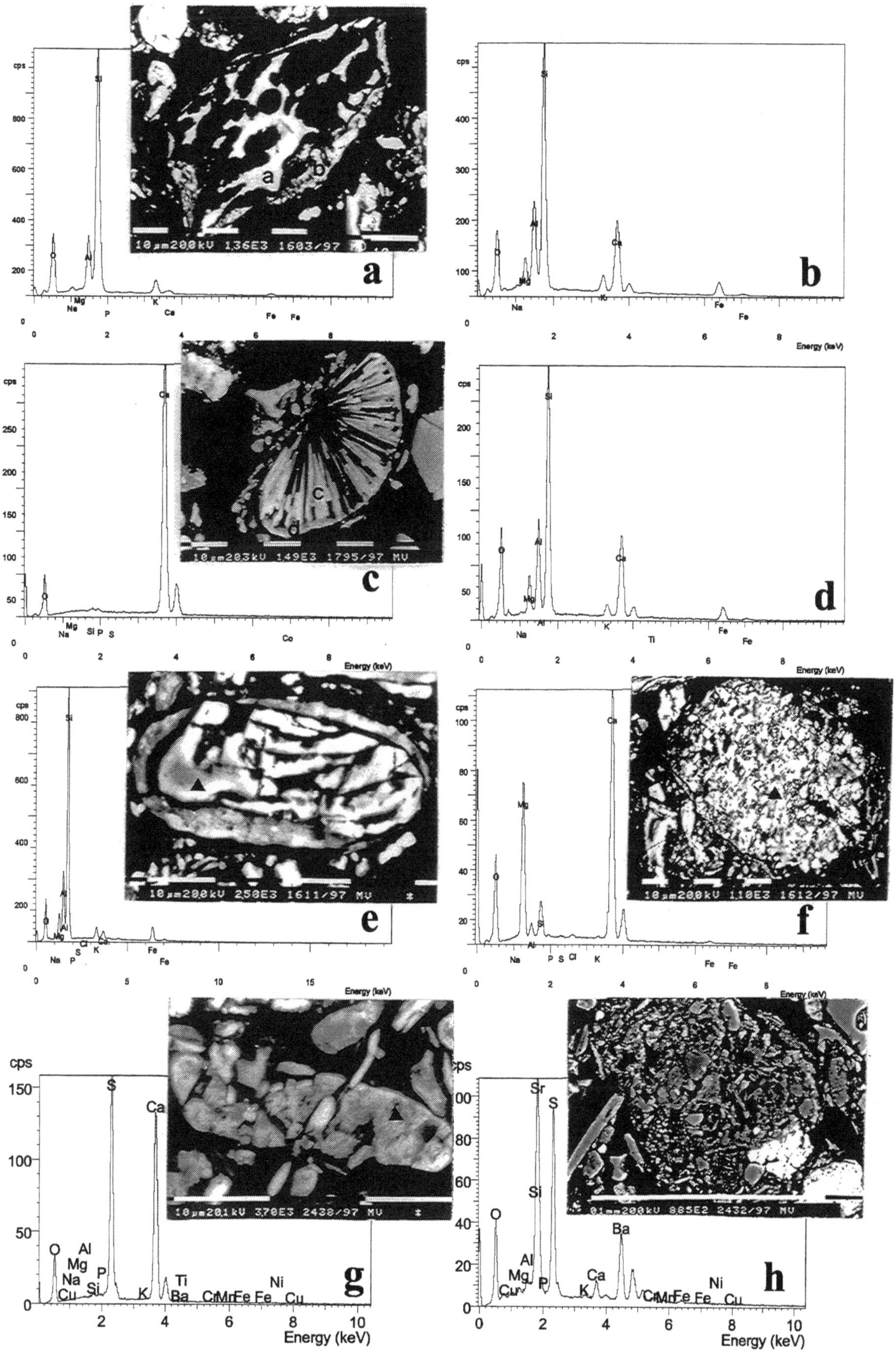

Fig. 11. SEM views and EDS analyses of selected particles from the tephra-rich deposit, Abu Hgeira 2. (a) EDS spectra and view of a typical tephra grain coated by Ca-Mg glass; (b) EDS spectra of Ca-Mg glass coating around the tephra grain viewed in (a); (c) sectioned fibro-radiated calcite spherule with a glass coating; (d) EDS spectra of the Ca-Mg glass around the fibro-radiated calcite spherule viewed in (c); (e) Ca-rich glass coatings around highly fragmented grains; (f) view and EDS spectra of a highly fragmented dolomite grain; (g) photo: threads of magnesio-calcic silicate glass accross a highly porous sulfate grain (EDS spectra); (h) salt inclusion (barium and strontium sulfate) within a compound, exogenous grain.

occurred instantaneously after their disintegration and therefore contributed to preserve the original shape of the grains by holding the fragments together.

9. Discussion

The interpretation of the facies and features listed above lead us to conclude that the tephra-rich layer is, from a sedimentological view-point, a secondary volcano-sedimentary deposit that is, itself, derived from volcano-sedimentary formations. Its deposition is, therefore, not linked to a volcanic explosion but to an unusual air turbulence that induced violent reworking of the initial volcano-sedimentary formations and their transport over long distance. Lack of mixing with soil and sedimentary materials from northern Syria suggests transport as a dust plume of initial high energy that strongly resembles a volcanic one and explains our previous confusion. However, the unique petrographic composition rules out a misinterpretation with an aeolian accumulation that would derive from violent dust bowls generated from the Syrian desert by horizontal wind turbulence during exceptional droughts [5]. Consequently, the seminal air turbulence is unlikely to be linked to a climate disturbance and remains, so far, unexplained.

Evolution from layer 2a to 2b, which was originally interpreted as an abrupt increase in aeolian deflation and aridity following the tephra fall [29], is now concluded to express the inherent properties of the dust plume: firstly accumulation of the coarser debris (layer 2a) followed instantaneously by deposition of the fine dust load that would derive from pulverisation of the initial one (layer 2b). The first evidence of a regional climate disturbance is expressed by deposition of the overlying layer 3. Its sedimentary facies, typical of torrential mass flow indicates occurrence of exceptional heavy rains that contributed to seal instantaneously the tephra-rich layer, and preserve it from subsequent pedogenesis. The rapidity of the fossilisation suggests a causal linkage between production of the dust plume and the subsequent heavy rains, possibly related to a cloud condensation nuclei effect [8].

The refined investigations reported here reinforce the correlation we previously established between the thick tephra-rich deposit identified at Abu Hgeira 2 and the presence of a dust layer with an unusual petrographic assemblage [29]. The spatial stratigraphic linkage between the two facies is now reinforced by the occurrence of similar micro-debris that have never been observed in the local soils and geological formations, particularly calcite spherules, acid rocks, gabbros, dolomite and marine bioclasts. Moreover, the two stratigraphic units indicate an uncommon airborne origin as compared to the regional aeolian component and, therefore, can be considered as two distinct facies of the same event. The normal aeolian component comprises either sand-sized grains locally reworked by saltation from bare surface soils during prolonged droughts marked by he influence of unstable air masses [5] or fine silt-sized particles that are issued from atmospheric dust loading during arid phases marked by increased southerly penetration of summer northwesterly winds [31]. The latter are generally mixed with the local aeolian dust, homogenised during soil development, and therefore leaves just a small increase of the illite-kalonite clay component that can be quantified only by using subtle X-ray treatments [10].

The dust layer reported here not only forms a distinct stratigraphic unit that attests for rapid deposition, but also presents a petrographic assemblage with a clear and important exogenous component, appearing as an uncommon airborne sediment that originates from long distance transport of sand-sized particles. In the absence of an analogue for this mode of airborne transport, not only from geological records but from experiments [21], the large size of exogenous micro-debris, particularly for the tephra-rich deposit, indicates that the source materials were originally propelled at a great velocity. Furthermore, absence of a cementing agent around and within multi-minerals sand-sized particles suggests that a thin hydrated film prevented them from disintegration during transport. The reduced mixing with local soils for the tephra-rich deposit, the abundance of large sized exogenous aggregates, the well-preserved pseudo-sand facies and its significant thickness as compared to opposed characteristics for the dust layer might result from a higher water content for the former, even possibly in a frozen state as indicated by the diagnostic fabric observed.

The reverse proportion of exogenous limestones, that include both bioclasts and dolomite fragments, versus fibro-radiated calcite spherules, of the tephra-rich layer as compared to the dust layer, and their presence in the former mainly as fragments, is one of the puzzling differences between the two facies. The absence of similar spherules in the local geological deposits, soils or archaeological sediments proves that they also have an exogenous origin. However, they cannot be mistaken for marine micro-fossils as shown by their specific morphology, crystallography, size and colour, and confirmed by their too-light isotopic values for marine carbonates. They share some similar properties, except for their smaller size, with calcite spherules recognised in the Cretaceous/Tertiary Chicxulub ejecta deposits and assigned as condensation products formed by the large asteroid impact that vaporised high amount of carbonates [22]. However, the occurrence of multi-spherules accompanied by silicate cements between individual elements in the dust layer indicates that the spherules formed from an heterogeneous medium that is more likely to be a fluid than a vapour. The geochemical and morphological diversity of carbonate-rich spherules, their close association with the silica and alumino-silicated spherules, together with presence of various exogenous rock fragments, suggest that all these fragments are genetically linked. They represent a broad spectrum of ejecta reflecting both the geological complexity of the source areas, formed of sedimentary, volcanic, metamorphic and igneous rocks, and differential thermic alteration of the original rocks. The predominance of weakly altered rock fragments in the tephra-rich deposit compared to the abundance of highly altered rocks and spherules in the dust layer strengthens the hypothesis of a higher water content, possibly in the form of ice crystals, that would have partly preserved the original properties of the ejected rocks. The latter would have been transported as condensed materials, therefore explaining its deposition on a restricted area. In contrast, the extension of the dust layer and its great spatial variability would trace the propagation of a high energetic ejecta plume where both liquid and vapour phases would have coexisted, formed of rocks, sediments and soils ejected from a wider area. The resence of a hydrated form of calcium sulphate in the tephra-rich layer as opposed to the dehydrated calcium sulphate (bassanite) in the dust layer is also consistent with the hypothesis of different conditions for the two parts of the ejecta plume.

The two distinct depositional episodes identified in sites offering an expanded signal suggest a first stage consisting of deposition of the coarser fragments from the high energetic ejecta plume, later followed by settling of the finer debris and neoformed products from condensed droplets, i.e. the calcite spherules. The synchronous occurrence of wildfires and physical disruption of surface soils and constructions during the first phase provide evidence for instantaneous ignition by fallout-induced lightning over a wide area and a low intensity small-scale explosion.

The diversity of these manifestations are inconsistent with the ones which would be induced by volcanic explosion.

These might, eventually, produce physical disturbances and wild-fires but only in the close vicinity of the explosion centre and are always characterised by passive deposition of the plume ejecta after its long distance transport [24]. Interpretation of the sedimentary facies of the tephra-rich deposit has allowed us to establish the detrital origin that the potassic-rich component presents as typical tephra particles. The calco-magnesian glass component showing a wide diversity of forms, from vesicular grains to spherules, has, however, never been identified as typical tephra shards. Therefore, its presence cannot be linked with a volcanic explosion that would have produced a tephra plume.

The loosely-packed structure of the lowermost part of the tephra-rich facies (layer 2a) is interpreted as reflecting the same manifestation as the one responsible for the uncommon mechanical breakdown of the soil aggregates below the thin dust layer. The thick exploded facies observed locally in archaeological sites reflects the more important effect of the physical disruption on the high mud brick walls as compared to the flat topography of the surrounding landscapes. The synchronous incorporation of aggregates issued from natural soils suggests that the physical fragmentation was associated with violent air turbulence similar to the ones contemporaneous with deposition of the tephra-rich layer at Abu Hgeira 2. The spherule-rich upper unit in archaeological sites would represent the very last phase of the event marked by deposition of fine-textured dust just following the surficial shock, before the return to normal conditions. Specific areas of archaeological sites, thus, appear to offer an expanded signal of the event, as compared to the one recorded by natural landscapes. This has made it possible to capture a series of short phases within a time interval that might not have exceeded a few hours.

10. Concluding remarks

Evidence collected over a wide area of northern Syria indicates the occurrence of an explosive phenomenon at ca. 4000 B.P. (i.e. 2350 cal B.C.) that differs in many aspects from a volcanic explosion and dispersion of its volcanic ejecta plume. The explosion is hypothesised to have affected complex geological terrains consisting of sedimentary, igneous, metamorphic and volcanic rocks, closely associated with water bodies rich in various types of salts. The strong contrast observed between the two distinct facies identified for this event is concluded to reflect heterogeneity of the ejecta produced from the source areas, both in term of water content and energy of the initial pulverisation.

Although the source of the ejected materials remain to be identified, petrography of the micro-debris ejected refutes an origin from northern Syria. The evidence presented here for a wide distribution of the benchmark dust layer should motivate extension of research to the entire Middle-East, the Mediterranean basin and even to regions further afield to define in a complete fashion the conditions of the ejecta dispersion, the synchronous phenomena, particularly the wild-fires and the surface shock, and the location of the explosion. This process should also greatly benefit from further detailed geochemical investigations of the unique dust spike identified at the same period in other regions, particularly the Gulf of Oman and Lake Van and suspected to relate to the same event [16, 28].

Moreover, our data suggest the need to revise certain concepts of palaeoclimatology that widely accept a straightforward causative linkage between a dust spike and an abrupt aridification. Our own previous misinterpretation of an unusual air turbulence as a climate induced wind intensification illustrates the difficulty in avoiding such a confusion. The related mechanisms appear to be rather similar but involve different time intervals: a few seconds for the first; and a few decades for the aridity signal. Such a high time resolution is certainly difficult to obtain in most sedimentary records, particularly in calm water bodies that might blur a very short signal into an annual dust input. Furthermore, this type of depositional environment is expected to attenuate synchronous physical disturbances that are, however, of major importance in defining exactly the conditions of dust deposition. The great diversity of the records reported here illustrate completeness of information provided by various terrestrial contexts that specifically reacted to an exceptional situation. The record from archaeological sites, when disentangled from possible confusion with human actions, offers the advantage of presenting locally an expanded signal, ideal in tackling the succession of very short episodes.

The refined investigations reported here fully-confirm the abruptness of the event, the exact date of which is, unfortunately, obscured by constraints inherent in the calibration programmes of C^{14} radiometric ages. The blind test successfully performed on the three archaeological sites that gave exactly the same chronological position, i.e. at the transition between the Early Dynastic period and the early Akkadian one, shows the great potential of this event to be used as a reliable stratigraphic marker competitive with historical documents. Refined examination of archaeological data, particularly at regional and inter-regional levels, should also help to determine the consequential effects of this unique event on social behaviour. The revised dating and interpretation reported here cancel our previous evidence that supported the causative linkage between this event, an abrupt climate change and collapse of the Akkadian Empire. Documenting the consequences of such a massive and instantaneous atmospheric dust load on the regional and global climate is beyond the scope of this paper. The evidence we provided for exceptional heavy rains suggests significant short-term effects consistent with climate perturbation known to be induced by massive smoke and dust injection [25].

Although our data refute the synchronous occurrence of a volcanic explosion and provide an alternative explanation for the coexistence of a thick tephra accumulation and micro-debris from volcanic rocks, importance of the spherule component in the ejecta produced remains a puzzling question. Abundantly reported from past geological records, the presence of spherule-rich layers as possible diagnostic markers of extra-terrestrial impact has been fiercely debated, but it appears they are often associated with volcanic events [15, 17, 26]. Therefore, unambiguous evidence, such as shocked quartz, the presence of tektites, an iridium anomaly, or the occurrence of silicon carbide and diamond are generally preferred to recognise impact-induced ejecta layers [14]. None of these fingerprints has so far been retrieved from the 4 kyr. B.P. dust layer. However, the occurrence of multi-site ignition together with black carbon production resulting from extensive biomass burning, also reported to be caused by asteroid impacts [32], provides indirect evidence to support an extra-terrestrial hypothesis. Various puzzling particles of the dust assemblage still require to be further elucidated, i.e. the carbon-rich black vesicular grains and the metal grains possibly linked to impact-induced lightning [9], and also the nature of the phenomenon responsible for the propagation of a surface shock wave and local explosion.

Acknowledgements

Our sincere gratitude for continuing support goes especially to the Syrian Directorate-General of Antiquities and Museums, Damascus, Professor Dr Sultan Muhesen. We are very grateful to A. Suleiman for providing field assistance at Abu Hgeira. We thank J. Oates, A. Pruss and H. Weiss for their authorisation respectively at

Tell Brak, Tell Chuera, Tell Leilan and surrounding sites during excavations under their supervision. We thank H. Cachier and S. Ruellan at the LSCE, Gif-sur-Yvette, France for black carbon analyses, M. Chaussidon and C. France-Lanord at CRPG, Nancy, France for stable isotope analyses, M. Hardy at INRA, Orléans, France, for X-ray clay analysis, N. Fedoroff at INA P-G, Grignon, France, for assistance with Soil Micromorphology, A.-M. Jaunet and M. Pernes at CNRA, Versailles, France for SEM and X-ray technical assistance, R. Rocchia at the LSCE, Gif-sur-Yvette, France, for Neutron Activation analyses and N. Tisnerat at the LSCE, Gif-sur-Yvette, France, for AMS C^{14} measurement.

References

1. Aid-Boudjera, M. (1995) *Archivage des combustions du passé dans les sédiments marins et continentaux.* D.E.A. Université Pierre et Marie Curie, Paris.
2. Cachier, H. and Ducret, J. (1991) Influence of biomass burning on equatorial African rains. *Nature* **352**, 228-230.
3. Cammas, C., Wattez, J. et Courty, M.A. (1996) L'enregistrement sédimentaire des modes d'occupation de l'espace, *The colloquium of the XIII International Congress of Prehistoric and Protohistoric Sciences. 3- Paleoecology*, (L. Castelleti and M. Cremaschi, eds.). A.B.A.C.O. edizioni, Forli, Italie, 81-86.
4. Cleuziou, S. (1994) La chute de l'empire d'Akkadé: hommes et milieux au Moyen-Orient. *Les Nouvelles de l'Archéologie* **56**, 45-58.
5. Courty M.A. (1994) Le cadre paléogéographique des occupations humaines dans le bassin du Haut Khabour (Syrie du nord-est). Premiers résultats. *Paléorient* **20(1)**, 21-59, CNRS (ed.), Paris.
6. Courty, M.A., Cachier, H., Hardy, M. and Ruellan, S. (1998) Soil record of exceptional wild-fires linked to climatic anomalies (inter-tropical and Mediterranean regions). *Actes du XVIeme Congrès Mondial de Science du Sol, Montpellier 1998.*
7. Courty M.A., Goldberg P. and Macphail R.I. (1989) *Soils and micromorphology in archaeology*. Cambridge Manuals in Archaeology. Cambridge University Press.
8. Courty, M.-A. and Weiss, H. (1997) The scenario of environmental degradation in the Tell Leilan region (N.E. Syria) during the late third millennium abrupt climate change, in NATO ASI Series, Vol. I. 49, *Third Millennium BC abrupt Climate Change and Old World Collapse*, N. Dalfes, G. Kukla and H. Weiss (eds). Springer-Verlag. Berlin. Heidelberg, 107-148.
9. Essene, E.J. and Fisher, D.C. (1986) Lightning strike fusion: extreme reduction and metal-silicate liquid immiscibility, *Science* **234**, 189-193.
10. Gaffie S., Bruand A. and Courty, M.A. (1997) A multilevel approach of structuration in time of soil systems in north-eastern Syria. *International Symposium on Soil System Behaviour in Time and Space, Vienna 1997.* Austrian Soil Science Society, Vienna. 145-148.
11. Gasse F. and Van Campo E. (1994) Abrupt post-glacial events in West Asia and North Africa monsoon domains, *Earth Planet. Sci. Lett.* **126**, 435-456.
12. Glassner J.-J. (1994) La chute de l'empire d'Akkadé, les volcans d'Anatolie et la désertification de la vallée du Habur, *Les Nouvelles de l'Archéologie* **56**, 49-51.
13. Kerr, R.A. (1998) Sea-floor dust shows drought felled Akkadian Empire, *Science* **279**, 325-326.
14. Koeberl, C. (1994) Tektite origin by hypervelocity asteroidal or cometary impact: target rocks, source craters, and mechanisms, *Geological Society of America, Special Paper* **293**, 133-151.
15. Koeberl, C. and Reimold, W.U. (1995) Early Archaean spherule beds in the Barbeton Mountain Land, South Africa: no evidence for impact origin, *Precambrian Research 7, 1,* 1-33
16. Lemcke, G. and Sturm, M. (1997) $\delta^{18}O$ and trace element measurements as proxy for the reconstruction of climate changes at Lake Van (Turkey): preliminary results, in: NATO ASI Series, Vol. I. 49, *Third Millennium BC abrupt Climate Change and Old World Collapse*, N. Dalfes, G. Kukla and H. Weiss (eds). Springer-Verlag. Berlin, Heidelberg, 653-678.
17. Melosh and Vickery, M. (1991) Melt droplet formation in energetic impact events, *Nature* **350**, 494-497.
18. Oates, J. (1994) Excavation at Tell Brak, *Iraq* **56**, 174-175.
19. Orthmann, W., Pruss, A. and Der Palast, F. (1995) Ausgrabungen in Tell Chuera, *Nordost-Syrian I*, 127-172, Saarbruecken
20. Panomarev B.Y. and Bishev A.S (1966) *The geological map of Syria*, V.P. Ponikarov (ed.). Sheets J-37-V, XI. Syrian Arab Republic. Ministry of Industry. Department of Geological and Mineralogical Research.
21. Pettijohn, F.J., Potter, P.E. and Siever, R. (1987) *Sand and Sandstone,* Springer-Verlag New York.
22. Pope, K., Ocampo A.C., Fisher, A.G., Morrison, J. and Sharp, Z. (1996) Carbonate condensates in the Chicxulub ejecta deposits from Belize. *Proceedings of the XXVII Lunar and Planetary Science Meeting*, 1045-1046.
23. Ruellan, S. (1996) *Archivage sédimentaire et glaciaire du carbone particulaire*. D.E.A. Université Pierre et Marie Curie, Paris.
24. Sigurdsson, H. (1990) Evidence of volcanic loading of the atmosphere and climatic response, *Palaeogeography, Palaeoclimatology Palaeoecology* **89**, 277-289.
25. Toon, O.B., Pollack, J.B., Ackerman, T.P., Turco, R.P., McKay, C.P. and Liu, M.S. (1982) Evolution of an impact-generated dust cloud and its effects on the atmosphere, *Geological Society of America Special paper* **190**, 187-200.
26. Wei, W. (1995) How many impact-generated micro spherule layers in the upper Eocene? *Palaeogeography, Palaeoclimatology, Palaeoecology* **114**, 101-110.
27. Weiss, H. (1990) Tell Leilan 1989: New data for Mid-Third Millennium Urbanization and State formation, *Mitteilungen der Deutschen Orient-Gesellschaft*, **122**, 193-218. Berlin, Sonderdruck Aus.
28. Weiss, H. (1997) Late third millennium abrupt climate change and social collapse in West Asia and Egypt, in: NATO ASI Series, Vol. I. *49 Third Millennium BC abrupt climate change and old world collapse*, N. Dalfes, G. Kukla and H. Weiss (eds). Springer-Verlag. Berlin. Heidelberg, 711-723.
29. Weiss, H., Courty, M.A., Wetterstrom, W., Meadow, R., Guichard, F., Senior, L. and Curnow, A. (1993) The origin and collapse of Third Millennium North Mesopotamian Civilisation, *Science* **261**, 995-1004.
30. Wilkinson T.J. (1994) The structure and dynamics of dry-farming states in Upper Mesopotamia. *Current Anthropology* **35(5)**, 483-505.
31. Wilkinson, T. (1998) Akkadian Empire: where to look? *Science* **279**, 1283-1284.
32. Wolbach, W.S., Gilmour, I., Anders, E., Orth, C.J. and Brooks, R. (1988) Global fire at the Cretaceous-Tertiary boundary, *Nature* **334**, 665-669.

Hints that Cometary Debris Played some Role in Several Tree-Ring Dated Environmental Downturns in the Bronze Age

M. G. L. Baillie

Science Palaeoecology Centre, School of Geosciences, Queen's University, Belfast, Northern Ireland

Summary

In 1988 a series of environmental downturns, indicated by 'narrowest ring' events in Irish oak tree-rings, were found to occur in the proximity of large acid anomalies in the Camp Century and Dye 3 ice cores from Summit, Greenland. The original suggestion that these events were caused directly by large explosive volcanic eruptions has continued to cause controversy as workers in various disciplines find it difficult to believe that volcanoes are capable of causing downturns of sufficient duration to cause socio-political changes - this could be called the 'volcanoes cannot collapse civilisations' school of thinking. While I believe this school may be too narrow in its thinking, and that multiple volcanoes could indeed have the power to disrupt agricultural societies, the inability of volcanologists to date volcanoes with high temporal precision limits all such discussion. In order to generalise the debate, the problem seems to be that few workers can adequately conceive of, or prefer not to contemplate, multi-causation; what would happen if several significant volcanoes erupted within the span of a few years? What would happen if an eruption was merely part of a larger tectonic event involving massive earthquakes and undersea outgassing from the estimated 10,000 gigatons of methane held in the ocean beds? What would happen if a series of Tunguska, or super-Tunguska, class impacts by cometary fragments triggered either tectonic effects and/or volcanism and/or outgassing? The simple answer is that no-one knows, while few, if any, have even considered the possibilities. This article will take a brief look at some of the possibilities within the time window of the last 5000 years.

1. General statement of the problem of abrupt environmental change

We can look back from the perspective of the late twentieth century and ask ourselves a simple question: when was the last abrupt global change. On reflection, the answer would probably be that the last major change, known about in any detail, was the warm-up at the end of the Younger Dryas more than 10,000 years ago. Ice core evidence suggests that the change from effectively glacial conditions to effectively modern, Holocene, conditions took just 'a few years' [1]. If pressed we might concede that the ice cores also suggest that there was an abrupt return to cold conditions for a short period around 8200 BP; the event looks severe enough to probably be global in extent [29]. If one was then to re-define the question in terms of the last 5,000 years, the likely answer would be in the negative, namely that there has not been a significant, abrupt, global change during that interval. The best known change in the period relates to the Little Ice Age, within the last millennium, however, that change is ill-defined chronologically and is not regarded as abrupt. It would appear that significant, abrupt, global change, with the possible exception of the AD 540 tree-ring event [4, 5], is missing from recent millennia.

The problem lies in the resolution and geographical spread of available records. Let us actually try to answer the question. Looking back from a modern perspective, the first significant event we come to is 'the year without a summer' in AD 1816 [27]. This was the year when there were poor weather conditions around the north Atlantic following the eruption of Tambora in 1815. The problem is whether this was actually a global event. We lack a sufficiently widespread grid of records to decide if the effects really were global in extent. Moreover, when tree-ring records are interrogated, there is no universal growth downturn associated with 1816. On closer examination, the ice records suggest that there was an earlier eruption of an unknown volcano in 1809-10 [16, 25]. Some trees appear to respond to this earlier eruption, others to Tambora [5]; the event is complex and not necessarily global.

If we move back to the traumatic events of AD 1740-41 we find that some widespread records show the event, though not all. The extreme cold in northern Europe was sufficiently severe to give rise to a demographic crisis, whose effects seem to have been particularly bad in Ireland [19, 38]. However, as far as this author is aware, no-one has searched to see if there were severe effects on human populations in the rest of the world. Moreover, there is no known physical cause for the event. It is not known whether it was due to simple climate perturbation, volcanic activity, solar variation or extraterrestrial dust. In fact we have to go back to 1600-01 to see what appears to be an abrupt event - at least one that shows up very widely in chronologies of tree-ring maximum density (the event shows up as the most dramatic low density event with a negative value of maximum density (-2.72) which is more than double the next worst value, that for the year 1641).

This is the widespread cooling event associated with the eruption of Huaynaputina, Peru [28] which shows up in most of the Greenland ice records as a significant acidity peak [16, 25]. Whether this abrupt environmental event had any widespread effect on human populations is currently unknown. One could follow this pathway back through the AD 1340s, where global tree-rings suggest a widespread downturn at the time of the Black Death [8], to the episode at AD 540 which really does look like a profound global downturn with trees, and human populations, affected around the northern, and into the southern Hemisphere [4, 5, 9].

This quick review of the last two millennia raises some interesting points. What is an abrupt global downturn; how should it be defined? Is it good enough to recognise an effect in a wide grid of tree-ring chronologies and/or ice-core records or is it important to factor in a human dimension? Of the events noted, we have the advantage of knowing that in the AD 1340s and the AD 540s human populations were widely affected, as well as tree-ring chronologies. As we have seen in AD 1601 there were widespread tree-ring effects but no adequate gathering of information on humans exists, largely because of the lack of a major plague, the key factor in the two earlier events.

Problems really start when we attempt to extend our record of abrupt events. Back to around the time of Christ dendrochronology, human history, and the ice-core records are in sufficiently good control that it is possible to accumulate the close clustering of information which can allow an abrupt global change to be defined. Looking further back we can pretty certainly see an abrupt global event around 44-42 BC. Indeed, this may be, in reality, the oldest global event on which we have a reasonably secure handle. The event shows up in all the ice-core records from Greenland at circa 50 BC [19, 25, 50]. It also shows up in historical records from the Near East to China and involves crop failures and famines [5, 45]; it is also there in selected tree-ring chronologies [42]. There may possibly be another such event around 208-204 BC at the start of the Han dynasty in China [5], and just possibly another in the window 430-410 BC [6].

For any date significantly before 430 BC we start to run out of dating control on most of the records other than tree-rings. Earlier than the later first millennium BC, the ice-core records are not absolutely dated, nor are they all that well replicated [2, 16, 50]. The number of tree-ring chronologies begins to become restricted. Human history rapidly loses absolute chronological control. By the time we get back before 800 BC, there are no absolutely-dated historical records. Indeed before that, most dating is provided by archaeological typology and radiocarbon analysis.

Radiocarbon under most circumstances lacks the ability to provide absolute dates; yielding instead calibrated age ranges which are normally many decades, up to centuries, in extent. Even high-resolution radiocarbon 'wiggle matching' exercises seldom define calendar ranges better than a half century [24, 30]. Most pollen records are dated by radiocarbon. Volcanoes are dated by radiocarbon. Thus, by the time we come to study the second and third millennia BC it is essentially impossible to see abrupt events - they may have happened but our ability to see them is diminished by the tendency for dates to be smeared through time. The sheer scale of the problem is not always appreciated by workers in the fields of archaeology or palaeoecology. Because there are no *known* abrupt environmental events between 500 BC and 5000 BC, it is assumed that there *were* no abrupt environmental events in that period. The real answer is that we don't know whether there were or not; they may simply be invisible due to poor chronological control. There is an unfortunate tendency to think that purchasing a few more radiocarbon dates will solve such chronological problems, the simple answer is that this is not the case. In order to adequately see abrupt environmental events, and assess their consequences for human populations, we would require the ancient historical and/or archaeological records to be precisely dated; in order that they could be interrogated at the same level of resolution as the tree-ring records. A few extra radiocarbon dates will not bring about such a chronological revolution, indeed, it is not immediately clear what will.

To summarise: because in the second and third millennia BC all the evidence, with the exception of evidence from the tree-ring chronologies, is smeared through time, it is currently impossible to prove the existence of any abrupt global events. In the text that follows, I will look at several issues relating to this problem. It will become apparent that while there are hints of abrupt global changes, currently none are proven.

2. The tree-ring events

The essential details of the Irish narrowest-ring findings have been published elsewhere [3, 5, 10]. Suffice to say that a series of extreme growth downturns in Irish sub-fossil oak series, at widely-spaced dates including 2345, 1628, 1159, 207 BC and AD 540, appeared to be associated with the effects of large acid-producing volcanoes as witnessed in the Camp Century [25] and Dye 3 [26] ice cores.

It was noted that several of these dates fell close to, if not actually at (given the limitations of archaeological and ancient historical dating), known periods of human trauma involving, for example, the start of the Bronze Age in the British Isles (around 2300 BC), dynastic changes in China (around 1600 BC, around 1100 BC and at 206 BC) and Egypt (around 1600 BC and around 1100 BC), as well as the Justinian Plague of AD 542 [5]. Such assertions do not sit comfortably with many archaeologists and there has been some criticism of these suggestions [14]. Much of the criticism arises due to the poor chronological control on almost all archaeological and volcanological evidence, rendering precise synchronisation of events impossible. Simple logic would suggest that any environmental factor which causes three, or possibly even as few as two, failed harvests will almost certainly cause the failure of an agriculture-dependent civilisation. A global event caused by a widespread dust-veil could easily bring down such societies simultaneously. It has not escaped notice that the robust, circa five-century long, Shang dynasty in China and the robust, circa five-century long, New Kingdom 'dynasty' in Egypt both occur over essentially the same interval in the mid-to-later second millennium BC [5].

The information given so far would appear to constrain the discussion to factors such as the limiting effects of known volcanoes, compared with earlier examples [39], limitations on chronology [5, 14], other likely factors such as acidification [23], etc. However, for the purposes of this paper I wish to turn to fresh environmental, and other, information which hints at a previously unsuspected cause.

3. New directions in environmental understanding

While the search was continuing for evidence to support or falsify the notion that the events, such as those at 1628 BC, 1159 BC and AD 540, represented abrupt environmental downturns with profound effects on human populations, several new items of information opened up a fresh line of thinking. It became known that, in fact, no good evidence existed for a volcanic event in the window AD 536-545 [5, 50]. To this could be added a strong circumstantial case for the happenings around AD 540 having a possible extraterrestrial component [4, 5, 9]. Another environmental event, involving descriptions consistent with atmospheric loading, at 44 BC, coincident with Caesar's death, was known to have been associated with the observation of a red, daylight comet. It was also known that a broad series of references to 'stones falling from the sky' by Livy, bracketed the dust-veil, and narrowest-ring event, of 208/207 BC (22; and see below). Subsequently, it was found that references to comets were to be found associated with the reigns of the Chinese emperors Chieh (the last of the Xia) and Chòu (the last of the Shang), i.e. at the start and end of the circa five-century-long Shang dynasty, traditionally believed to span the period between circa 1600 BC and circa 1100 BC [41]. Such a set of pieces of slender evidence still allows one to pose the question, whether the already specified environmental downturns have anything to do with comets or extraterrestrial cometary debris? Is it possible that the severity of the downturns - necessary if they are to have the unusual effects of collapsing dynasties, etc., - might be due to 'unusual' i.e. extra-terrestrial contributions or causes? Such an idea would initially seem like special pleading were it not for the fact, long known, that the Chinese emperors, indeed dynasties, ruled with the 'Mandate of Heaven' - a concept which has been much argued over but which could as easily encompass comets and bombardment by cometary debris as it could encompass volcanic dust veils [35] or even planetary conjunctions [36; and see this volume]. A much more extensive look at this circumstantial case has been published elsewhere [9], for our purpose here it is sufficient that the possibility of bombardment(s) during the Bronze Age be raised so that additional evidence may be assessed.

Tree-ring event	**1628 BC**	**1159 BC**	**540AD**
suggested	*volcano*	*volcano*	*dust veil*
possible link	*comet*	*comets* China Greece Babylon	*bombardment event?*
possible link	*Exodus*	*David*	*Arthur*
associated with	*plagues of* Egypt	*plagues at* time of David *plague at* Troy	*plague of* Justian

Table 1: The circumstatial case which allows the linking of three of the narrowest-ring dates with pre-existing traditions

To paraphrase the debate so far, Irish trees and Greenland ice-cores indicate significant events at 1628-23 BC and 1159-41 BC; starting some 469 years apart. The Shang dynasty in China and the New Kingdom in Egypt, both around five centuries long, are essentially bracketed - to within decades (less than a century by anyone's standards) - by the dates of these tree-ring events. Kuniholm *et al.* [30] have shown that they have tree-ring anomalies starting exactly 470 years apart in Anatolian trees; the events falling in the seventeenth and twelfth centuries BC, and close to the Irish dates on substantial radiocarbon evidence. In addition, a number of authors have suggested that the 1628 BC date, which may relate to the massive eruption of Santorini in the Aegean, may also relate to the descriptions given in the story of the biblical Exodus [13, 40, 48]. Interestingly, a second catastrophic series of events, involving famine, plague and tectonic effects, is described in the Old Testament at the time of King David, (2 Samuel 21-24) the start of which is placed by Archbishop Ussher exactly 470 years after the date of the Exodus. Given that the early section of the Old Testament is not dated by any independent means, it has to be possible that the early Hebrew stories are preserving a memory of the same spaced events as those which show up in the Irish and Anatolian trees. But we have already seen that there are references to comets at the start and end of Shang dynasty [41] and we have 'angels in the sky' at both the Exodus and David episodes. In addition, according to Sagan and Duyran, we have a comet associated with the fall of Troy and another (or possibly the same one) in twelfth century Babylonian records. The essential elements of this story are encapsulated in Table 1.

4. Environmental context for the 2nd millennium BC events

In 1991 the narrowest-ring dates from 3195 BC to AD 540 were designated as 'marker dates' [3] on the basis of the hints of their global character - if they really were in any way involved with the Chinese and Egyptian dynastic changes then they had to be potentially significant and could be expected to show up in other records. In 1994 Briffa published a preliminary temperature reconstruction, deduced from Fenno-scandian pines, for several millennia BC [12]. Unfortunately that tree-ring chronology is not absolutely dated and cannot yet be compared directly with other tree-ring chronologies, though it is placed in time to within about ±50 years on the basis of multiple radiocarbon analyses. From the point of view of this discussion, the interesting thing about Briffa's tentative reconstruction was that there were two major temperature troughs around 1600 BC and around 1100 BC, see Fig. 1.

These temperature troughs are sufficiently wide that, notwithstanding the flexibility in the dating of the chronology, the 1628 and 1159 BC events pretty well have to be associated with them in some way. Put most simply, either

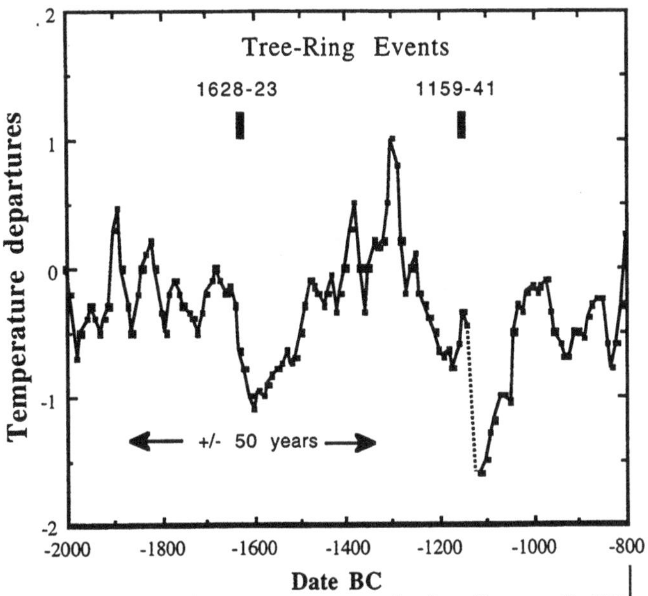

Fig. 1: The provisional temperature reconstruction from Fennoscandia [12] with the dates of the narrowest ring events from Baillie and Munro [10]. Note that the tree-ring based temperature reconstruction is floating by circa ±50 years.

the two environmental events sit in the troughs or they caused the troughs. At the present time, with the Fennoscandian chronology still floating, it seems most reasonable to look around for independent evidence. Fortunately such evidence, involving other temperature sensitive trees, exists at least in outline.

5. American pines

Long prehistoric tree-ring series from temperature sensitive sites in the western United States are in existence but are not yet available at annual resolution. The first of these relates to upper tree-line bristlecone pines originally published by LaMarche and Harlan [32]. LaMarche drew attention to the likely temperature sensitivity of these particular high altitude pines and published a smoothed 20-year record in 1974 [31]. This data set is only available as a five-year-smoothed curve. In addition, there are two curves available for high-altitude foxtail pine from the Sierra Nevada, where the data can be retrieved at circa 5-year resolution [15]. It can be envisaged that the combination of these records forms a reasonable working temperature record for high-altitude western America. If we proceed to 'fit' the general shape of the Fennoscandian pine temperature reconstruction to this American curve, at around the dating proposed by Briffa [12], the result is the curve shown in Fig. 2. Also in Fig. 2 I have indicated the two events deduced from the oak records. It is apparent that this first look suggests that indeed the two

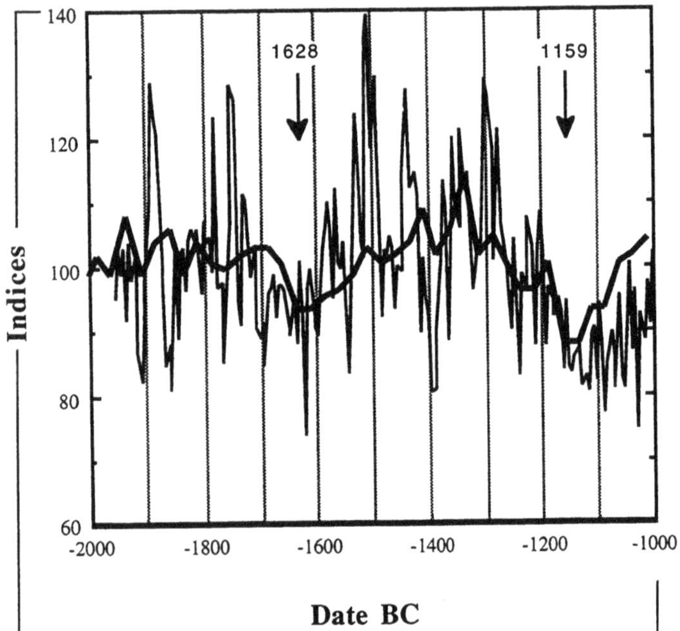

Fig. 2: The Fennoscandian temperature curve from Figure 1, moved back by 25 years, (heavy line) together with a composite American bristlecone [31] and foxtail [15] growth record. This diagram suggests that the Irish narrowest-ring events may sit within pre-existing temperature troughs.

environmental events tend to sit in, rather than cause, the downward trends in both curves.

If this situation were borne out when the Fennoscandian chronologies are absolutely dated, we would then know that the abrupt downturns recorded in the oaks fall inside longer term cooling trends. From the point of view of this paper the significance of such a finding would be the likelihood that the cooling had an external cause, relating either to solar activity or to dust loading from cometary debris. If this latter option were the cause then logic would suggest that the abrupt environmental events themselves could have been triggered by bombardments. This is exactly the sort of scenario envisaged by Clube and Napier [17]. The earth's orbit intersects the annulus of dust and debris left by a comet which disintegrated some 20,000 years ago. Over time it picks up dust loading causing cooling. As its orbit intersects the core of the meteor stream the likelihood of impacts with more solid fragments of cometary debris increases. Such impacts may well cause tectonic effects giving rise to the volcanic acid signals which provide the 'signposts' in the temperate oak tree-ring records. This scenario can be tested by detailed analysis of the Greenland ice-cores, once realistic attempts have been made to define the real ages of the acid layers in the ice using tephra markers [7]. With the cores better dated, they can then be searched for relevant extra-terrestrial debris/tracers.

6. A case study: The Irish tree-ring event at 2354-45 BC, the "Hekla 4" event

In this section I want to look at the information available in the vicinity of the Hekla 4 eruption in Iceland, currently best-dated to 2310+/-20 Cal. BC. To start we can lay out several pieces of information as bald statements.

i) Irish oak chronologies show a notable 'narrowest ring' event at 2345 BC, which starts at 2354 BC; effects are also seen in English chronologies [10].

ii) Physiological information from two of the available Irish trees indicate some sort of inundation of the shores of Lough Neagh, the major body of water in Northern Ireland [5, 9].

iii) The Icelandic volcanic event, Hekla 4, has been high-precision wiggle matched (multiple stratified radiocarbon analyses) at 2310+/-20 CalBC (calibrated radiocarbon years) [24].

iv) Conventional biblical chronology places The Flood at 2349 'BC', deduced by internal study of Old Testament texts by James Ussher, Archbishop of Armagh (King James Bible).

v) The World Chronicle section of the Irish annals describes variously human trauma and 'lakes breaking out' in the 2350/2341 'BC' period [44].

vi) In the seventeenth century William Whiston, successor to Newton at Cambridge, convinced himself that the Flood of 2349 'BC' was brought about by water from a comet, plus water from within the Earth [43, 47].

vii) Marie-Agnès Courty now suggests that an anomalous petrographic assemblage in Syria, which sounds as though associated with extraterrestrial impact, dates to around 2350 BC [18; and see this volume].

viii) This latter event is placed at the start of the Akkad dynasty. Later texts refer to the concept of the 'Curse of Akkad' wherein 'flaming potsherds fall from the sky' [46].

ix) James Legge in the nineteenth century was moved to write that the first emperor of China, Yaou, came to the throne in 2357 BC and was noted for the major floods in his reign [33].

x) A circumstantial case suggests that other narrowest-ring events in the Irish oaks seem to mark episodes which could have a cometary relationship [9; and below].

The problem for scholars is how to handle such a range of hard and soft information. On a scale of 1 to 10 where 10 means 'highly reliable' and 1 means 'hardly worthy of notice' these ten particular items might be notionally scored as

i) 10 tree-ring events

ii) 5 damage effects in tree-rings

iii) 8 radiocarbon wiggle-match

iv) 1 Ussher's Flood date

v) 1 Irish World Chronicle date

vi) 2 Whiston's cometary theory

vii) 5 petrographical assemblage

viii) 2 ancient Sumerian records

ix) 1 very early Chinese records

x) 3 cometary relationships

In other words, most of the information has low reliability. This still leaves the question why the cluster appears to hang together so closely in time. Perhaps we have to elaborate a little. The difficulty appears to be mostly chronological. If all the sources had equally reliable chronologies we might now know that in 2354 BC the Earth was impacted by cometary debris (Courty's glassy balls and anomalous petrology). That this resulted in earthquakes (the Lough Neagh level change being due to a tectonic seiche and/or an abrupt change in water table), volcanism (Hekla 4 and Syrian tephra) and widespread flooding (Ireland, Near East and China).

Unfortunately, in the real world, this coherent packaging of information is not allowable and an initial assumption would be that the package is due to this author 'sucking in' information to create an artificial horizon. The problem with *that* assumption is that this author is not responsible for any of the dates from, or relating to, Whiston/Ussher, Legge, Courty or Hekla 4. In addition, there is more than one way of looking at the package. Most people would find it 'interesting'. People who are comfortable with catastrophism or environmental determinism (CATs) will tend to accept it, in some cases uncritically. People who are uncomfortable with catastrophes (UNCATs) will reject it out of hand and point to the weakness of the various strands of 'evidence'.

How do we handle this type of situation in order to break out of the impasse? The question, after all, is very straightforward: did cometary debris have detrimental effects on planet Earth in the immediate vicinity of 2350 BC? Interestingly this date is in the vicinity of one of Clube and Napier's periods of risk where the earth may have intersected an annulus of cometary material. All we need, to move any debate on that question forward, is a positive identification of extraterrestrial material in the Syrian deposits with backup extraterrestrial material from appropriate levels in the Greenland ice cores (a debate on the detailed chronology of the ice-cores continues [7, 49]) or from terrestrial peat or lake deposits containing the Hekla 4 tephra as a dating control. Let us imagine that several of these sources were actually to supply information on extraterrestrial material. Then we would have the situation where as each piece of relevant information comes in, CATs will tend to be increasingly convinced. Whereas UNCATs will continually point out the less than perfect dating associated with the ice cores, varved deposits, radiocarbon wiggle-matching and archaeological deposits. Because none of these are likely ever to be tied down to the exact calendrical precision of the tree-ring sequences, arguments on cause and effect will endure.

7. An exemplar argument on cause and effect

To show the type of argument on such issues, we can use the straightforward case of the 2354-2345 BC tree-ring event, the postulated inundation of Lough Neagh, and the Hekla 4 eruption. There is no doubt about the tree-ring event or its dating, or to put it more clearly, there is no doubt that there was a detrimental tree-ring event and that it occurred between 2354-2345 BC. However, while the radiocarbon wiggle-match on stratified peat across the Hekla 4 horizon is good, at 2310+/-20 CalBC, it is not well enough dated to say if that eruption caused the effects in the Irish trees i.e. with its existing dating it could have erupted in 2354, but equally it could have erupted around 2345 (being secondary to some unknown prime mover) or, indeed, it could have erupted anytime between about 2350 and 2270 CalBC.

This takes us to the postulated inundation. The postulation is based on intrinsically dated, tree-ring anomalies/damage and appears to be reasonable at this stage in the debate. Indeed it is backed up by recent work on diatoms in peat to the south of Lough Neagh which shows evidence of significant flooding at around the time of Hekla 4 (R. Lamb, *pers. comm*; undergrad thesis Dept Archaeology and Palaeoecology, Queen's University, Belfast). Thus the question reduces to the relationship between the actual eruption of Hekla 4 and the inundation. There are two basic possibilities.

a) Hekla 4 erupted in 2354 BC and triggered the inundation.

b) The inundation was caused by something else (e.g. another distant volcano, some tectonic movement, or an impact event) and Hekla 4 erupted merely as part of a tectonic 'package', possibly considerably after the event.

So, even with the good dating evidence available in the 2354/Hekla 4 debate we cannot currently prove a connection. There is however a way forward. It is highly likely that the Hekla 4 eruption deposited identifiable tephra fragments on the Greenland ice-cap. If we search the GISP2 record [50] we see that there are acid spikes at 2398 'BC' and 2310 'BC'. We then have to consider the dating accuracy of the GISP2 core. The workers themselves give a maximum error of ±1% i.e. ±43 years in the 23rd century BC, though Zielinski and Germani argue that the error is much less than this, probably only a 'few' years [49]. I have suggested [7] that in the 2nd/3rd millennia BC, the GISP2 record is mis-dated by around 40 years with the dates given by the American ice-core workers being too old by this amount. Thus a scenario exists whereby I would suggest that the Hekla 4 tephra might reasonably be found in the 2398 'BC' layer in GISP2, whereas the ice-core workers would probably be more comfortable with Hekla 4 tephra in their 2310 'BC' layer, closer to the central point of the radiocarbon-based date for Hekla 4 itself. This situation could be moved forward by the ice-core workers were they to analyse all the relevant acid layers looking for identifiable Hekla 4 shards.

One could reasonably ask whether any of the strange tephra in Syria has the same chemistry as any rogue shards which might occur in the Irish peat deposits in the same layer as the Hekla 4 tephra. This would help link other strands of evidence and might give clues to likely sources of some of the other tephra. It might provide support for the idea of widespread tectonic activity causing multiple eruptions, irrespective of cause. But if Hekla 4 tephra was found in the ice at an ice-core date 2390 'BC', what then? Ironically, we are nearly as far on in the debate now as we would be if we had the additional information! It is basically a value judgement with CATs already on the verge of acceptance of an impact-induced catastrophe and UNCATs, presumably, totally unconvinced and probably unconvinceable. For example. Imagine that Courty does prove to have extraterrestrial material in Syria.

Peiser [37] tells us that there may well have been a simultaneous destruction of sites at various times in the Bronze Age.

> "Schaeffer analysed and compared the destruction layers of more than 40 archaeological sites in the near and Middle East, from Troy to Tepe Hissar on the Caspian Sea and from the Levant to Mesopotamia. he was the first scholar to detect that all had been totally destroyed

several times in the early, Middle and late Bronze Age, apparently simultaneously."

But, it is impossible to tell if all the sites mentioned by Peiser or earlier by Mandelkehr [34] actually were destroyed simultaneously, thus the 'space debris' located by Courty, if it turns out to be such, will be classed as 'local' to one site. It may have destroyed one region but it cannot be proven to have destroyed others at the same time. The change seen in the archaeological record in Britain and Ireland around this time, that is the end of the later Neolithic and the arrival of Beakers/Beaker folk, is (and will probably remain) too poorly dated to be able to see if population movement/ cultural change in that area might be temporally related to the postulated catastrophe in Syria, or to the event in the Irish tree-rings. Thus critics of the idea of a catastrophic impact having real widespread effects on human populations will remain unconvinced.

We come dangerously close to having to conclude that it may be impossible to prove that an impact had effects on human populations. We may prove an actual impact, but linking its effects to human cultural or archaeological records may be impossible other than on a highly limited scale. Archaeology almost certainly cannot help in such debates because of its poor time control, perhaps even clouding the issues after the manner of Buckland *et al.* and their treatment of the 'Santorini = 1628 BC' debate [14]. This excursion tells us that it is going to be difficult to move this debate forward; endless circumstantial evidence can be accumulated but it will not convince sceptics. However, should we allow such issues to be an actual hindrance? Is it possible that catastrophists can co-exist with non-catastrophists in a discussive mode? Currently there appears to be an impasse wherein non-catastrophist archaeologists/ ancient historians appear to demand absolute proof of any catastrophic vector/effect before they would be willing to consider the possible implications. Thus a situation is envisaged where catastrophists continue to look for evidence to prove their case while others ignore the issue and its implications until the case is proven. As an example, relevant to this issue, I indicate below how ancient history and mythology can be used to provide circumstantial support for a postulated bombardment event which would be acceptable to many CATs. UNCATs must decide for themselves whether there is anything in it.

8. The 207 BC event

As noted above, the possibility exists that some of the original narrowest-ring events in the Irish oaks may have a 'bombardment' dimension. With that in mind we can go back and look at the records from around one of the events, that at 208/207 BC, in a little more detail. We already know that it represents, to within one or two years, the start of the Han dynasty in China. The date 208 BC represents a year when the Chinese failed to see the stars clearly [45]. In northern Europe, we have the end of the Irish prehistoric bog oak chronology, consistent with some environmental change around 200 BC, while in Germany we have the start of a 'wash-out' phase of deposition of river gravel oaks [5, 11]. LaMarche's bristlecone pine record shows a frost ring in 206 BC [5]. However, it was Forsyth's comments on Livy and the references to repeated assertions of 'stones falling from the sky' which would seem to hint at a possible extraterrestrial forcing factor for the environmental event. At face value it looks impressive:

> "[...] Livy tells us that in the year 206 B.C. two suns were seen at Alba, and adds that light appeared in the night at Fregellae. [...] Livy also tells us that the decision to consult the (Sibylline) books was precipitated by frequent showers of stones seen in 205 BC." [22].

Unfortunately it is known that Livy regularly mentions such factors for years between 344 BC and 91 BC and this of course reduces the significance of his comments at this point in time - maybe he was just of a catastrophic turn of mind? However, there is no doubt that his tightest cluster of references to 'stones falling from the sky' relates to the period 218 to 202 BC (In fact if the dates of the references to such apparitions are looked at in detail it seems that Livy might just be recording real changes in frequency of such sightings.

There are two very early references at 344 and 258 BC and then a long gap before a cluster of ten 'regular' references between 218 and 202 BC. Another ten references are scattered across 194 to 125 BC, with a final cluster of four between 102 and 91 BC. I am grateful for this information which was pointed out by John D. Morgan III). It is interesting that several of the comments relating to rings round the sun, etc., would be consistent with atmospheric loading, while 'light appearing during the night' would sit comfortably with the reported happenings on the night of the Tunguska impact event in June 1908. Forsyth also provides some confirmatory evidence from another author, Appian, who refers to 'direful prodigies sent by Jupiter (which) had appeared in Rome in 204 BC.' [22].

Why the mention of Jupiter? The traditional explanation is that the reference to Jupiter is simply because Jupiter is the sky god. However, once one begins toying with the issue of bombardment, it seems at least possible that the reference is because the stones really were falling from the sky after the manner of Jupiter's thunderbolts. Still, if that was all there was to this issue, it would have no serious substance, but, fortunately, there is more. The point of Forsyth's article was that it had never been fully understood just why the oracle was consulted and the Magna Mater, the goddess Cybele, was brought back to Rome in 204 BC. My belief is that there is more to this issue than even Forsyth realised, and this relates to Cybele, herself. The situation appears simple; stress in Rome - apparitions in the sky - bring back Cybele - everything improves. The detail is more interesting. As Frazer [21] puts it:

> "The worship of the Phrygian Mother of the Gods was adopted by the Romans in 204 BC towards the close of their struggle with Hannibal. For their drooping spirits had been cheered by a prophecy, alleged to be drawn from that convenient farrago of nonsense, the Sibylline Books, that the foreign invader would be driven from Italy if the great Oriental goddess were brought to Rome."

The important point to realise is that the Magna Mater, Cybele, was a 'small black stone which embodied the mighty divinity'; Cybele was herself a meteorite, but, not just any meteorite - she was brought from Phrygia. Phrygia has an interesting attribute, in mythology, part of it is called 'Burnt Phrygia'. Phrygia in earlier times had been visited by Aigis:

> "Aigis, a frightful earth-born she-monster who breathed forth flames from her mouth. She appeared first in Phrygia and burned up that part of the country that was called Burnt Phrygia. From there she went to the Tauros Mountains of Cilicia and to the Lebanon Range of Syria, burning up all the forests as she went." [20]

So, Cybele, apparently an extraterrestrial object, was brought from Phrygia which has associations which would again do justice to a Tunguska, or super-Tunguska, type event. In fact, other references in mythology tell us about the area:

"[...] that part of Lydia and Phrygia called Katakekaumenê (Burnt Lydia and Burnt Phrygia) was said to have been scorched either by Zeus's thunderbolts or by Typhon's fiery breath [...]" [20]

Thus the Sybilline Books' 'goddess of choice' to save Rome at a time of extraterrestrial bombardment may well have been exactly the right goddess to call on in the circumstances. What better goddess to bring relief from possible extraterrestrial bombardment that a goddess who is herself an extraterrestrial object with catastrophic associations. Perhaps there is even more to the myths than one could reasonably expect. Note how Aigis burns up all the forests as far as 'the Lebanon range of Syria'. Maybe Courty's 'anomalous petrographic assemblage' (this volume) in Syria is not so unexpected after all?

What is particularly interesting about these excursions into mythology is that they provide pre-existing links between several of the events which seem to relate broadly to those very 'marker dates' suggested by the narrowest-ring events in the Irish oaks. We have just seen a reference to Typhon's fiery breath in relation to Burnt Phrygia. The battle between Zeus and the 'dragon/monster' Typhon is paralleled by another god who battles a dragon/monster, i.e. Apollo kills Python (20). But, Apollo, who killed Python, was present at the fall of, and indeed sent plague against, Troy, which is traditionally linked to the twelfth century BC. So, starting at 207 BC we find a link to Burnt Phrygia which hints at a bombardment event which might encompass Courty's anomalous event circa 2350 BC. But that 'link' refers us back to happenings in the twelfth century BC in the vicinity of the 1159 BC event. But, and this is why this story seems so strange, introducing Apollo also introduces Python. Incredibly, Python was an enormous serpent which crawled forth from the slime left by the waters of *the Flood* which we have already seen is dated by Ussher to 2349 'BC'. Thus, an excursion triggered by mention of the goddess Cybele's arrival in Rome in 204 BC provides tenuous mythological links between events around three of the original Irish narrowest-ring dates. As I have argued in detail elsewhere [9], it really begins to look as though 2345 BC, 1628 BC, 1159 BC, 207 BC and AD 540 may be 'marker dates' after all - marked by extraterrestrial visitations, after the style of Clube and Napier.

Acknowledgements

Much useful information used in this article has been gleaned at conferences on dendrochronology, volcanology and catastrophism attended as part of more mainstream environmental research funded by EC Environment Research Programme, Contracts EV-5V-CT94-0500, and ENV4-CT95-0127 Climatology and Natural Hazards, 2K and 10K.

References

1. Alley, R. B., Shuman, C. A., Meese, D. A., Gow, A. J., Taylor, K. C., Cuffey, K. M., Fitzpatrik, J. J., Grootes, P. M., Zielinski, G. A., Ram, M., Spinelli, G. and Elder, B. 1997, Visual stratigraphic dating of the GISP2 ice core: basis, reproducibility, and application. *Journal of Geophysical Research*. **102**, No. C12, 26,411-26,423.
2. Alley, R.E., Meese, D.A., Shuman, C.A., Gow, A.J., Taylor, K.C., Grootes, P.M., White, J.W.C., Ram, M., Waddington, E.D., Mayewski, P.A., and Zielinski, G.A. 1993, Abrupt Increase in Greenland Snow Accumulation at the End of the Younger Dryas Event, *Nature* **362**, 5279.
3. Baillie, M.G.L. 1991, Marking in Marker Dates; towards an archaeology with historical precision, *World Archaeology* **23**, 233-43.
4. Baillie, M.G.L. 1994, Dendrochronology raises questions about the nature of the AD 536 dust-veil event, *The Holocene* **4**, 212-217.
5. Baillie, M.G.L. 1995, *A Slice Through Time: dendrochronology and precision dating*. Batsford, London.
6. Baillie, M.G.L. 1996, Chronology of the Bronze Age 2354 BC to 431 BC. *Acta Archaeologica* **67**, 291-298.
7. Baillie, M.G.L. 1996, Extreme environmental events and the linking of the tree-ring and ice-core records. In J.S. Dean, D.M. Meko and T.W. Swetnam, eds., Tree Rings, *Environment and Humanity*, Proceedings of the International Conference, Tucson, Arizona, 17-2 May 1994. Tucson, *Radiocarbon*, 703-711.
8. Baillie, M.G.L. 1998, *Putting abrupt environmental change back into human history*. Linacre Lecture Series 1998 (in press).
9. Baillie, M.G.L. 1998, *Exodus to Arthur: close encounters with comets and the fiery dragons of myth*. Batsford, London.
10. Baillie, M.G.L. and Munro, M.A.R. 1988, Irish tree-rings, Santorini and volcanic dust veils, *Nature* **332**, 344-6.
11. Becker, B. and Schirmer, W. 1977, Palaeoecological Study on the Holocene Valley Development of the River Main, Southern Germany, *Boreas* **6**, 30321.
12. Briffa, K.R. 1994, Mid and late Holocene climate change: evidence from tree-growth in northern Fennoscandia in Funnell, B.M and Kay, R.L.F. Eds *Palaeoclimate of the Last Glacial/Interglacial Cycle.* Special Publication No **94/2**, NERC, London, 61-65.
13. Bruins, H.J. and van der Plicht, J. 1996, The Exodus Enigma. *Nature* **382**, 213-214.
14. Buckland, P.C., Dugmore, A.J. and Edwards, K.J. 1997, Bronze Age Myths? Volcanic activity and human response in the Mediterranean and North Atlantic regions, *Antiquity* **71**, 581-593.
15. Caprio, A.C. and Baisan, C.H. 1992, Multi-millennial tree-ring chronologies from foxtail pine in the southern Sierras of California. Abstract in *Bulletin of the Ecological Society of America* **73**, 133.
16. Clausen, H. B., Hammer, C. U., Hvidberg, C. S., Dahl-Jensen, D. and Steffensen, J. P. 1997, A comparison of the volcanic records over the past 4000 years from the Greenland Ice Core Project and Dye 3 Greenland ice cores, *Journal of Geophysical Research* **102**, No. C12, 26,707-26,723.
17. Clube, S.V.M. and Napier, B. 1990, *The Cosmic Winter*, Blackwell, Oxford.
18. Courty, M-A. 1998, The soil record of an exceptional event at 4000 B.P. in the Middle East, in: Bailey, M.E., Palmer, T. and Peiser, B.J., eds., *Natural Catastrophes During Bronze Age Civilizations*. Oxford.
19. Drake, M. 1968, The Irish Demographic Crisis of 1740-41, in T. Moody (Ed) *Historical Studies VI* (Dublin June 1965), Routledge and Kegan Paul, London, 101-124
20. Fontenrose, J. 1959, *Python: a Study of Delphic Myth and its Origins*. University of California Press, Berkeley.
21. Frazer, J.G. 1930, *The Golden Bough,* McMillan and Co. London.
22. Forsyth, P.Y. 1990, Call for Cybele, *The Ancient History Bulletin* **4(4)**, 75-78.
23. Grattan, J.P. & Gilbertson, D.D. 1995, Acid-loading from Icelandic tephra falling on acidified ecosystems as a key to understanding archaeological environmental stress in northern and western-Britain, *Journal of Archaeological Science* **21**, 851-9.
24. Hall, V.A., Pilcher, J.R. and McCormac, F.G. 1994, Icelandic volcanic ash and the mid-Holocene Scots pine (Pinus sylvestris) decline in the North of Ireland; no correlation, *The Holocene* **4**, 79-83.
25. Hammer, C.U., Clausen, H.B. and Dansgaard,W. 1980, Greenland Ice Sheet Evidence of Post-Glacial Volcanism and its Climatic Impact, *Nature* **288**, 230-5.
26. Hammer, C.U., Clausen, H.B., Friedrich, W.L. and Tauber, H. 1987, The Minoan Eruption of Santorini in Greece Dated to 1645 BC?, *Nature* **328**, 517-9.
27. Harington, C.R., ed., 1992, *The Year Without a Summer? World Climate in 1816*, Canadian Museum of Nature,Ottawa.
28. Jones, P.D., Briffa, K.R. and Schweingruber, F.H. 1995, Tree-ring evidence of the widespread effects of explosive volcanic eruptions, *Geophysical Research Letters* **22 (11)**, 1333-1336.
29. Klitgaard-Kristensen, D., Sejrup, H. P., Haflidason, H, Johnsen, S. and Spurk, M. 1998, A regional 8200 cal. yr BP

cooling event in northwest Europe, induced by final stages of the laurentide ice-sheet deglaciation? *Journal of Quaternary Science* **13 (2)**, 165-169.

30. Kuniholm, P.I., Kromer, B., Manning, S.W., Newton, M., Latini, C.E. and Bruce, M.J. 1996, Anatolian tree rings and the absolute chronology of the eastern mediterranean, 2220-718 BC, *Nature* **381**, 780-783.
31. LaMarche, V.C. Jr., 1974, Palaeoclimatic inferences from long tree-ring records. *Science* **183**, 1043-1048.
32. LaMarche, V.C. Jr. and Harlan, T.P. 1973, Accuracy of Tree-Ring Dating of Bristlecone Pine for Calibration of the Radiocarbon Time Scale, *Journal of Geophysical Research* **78**, 8849-58.
33. Legge, J. 1879, *The Sacred Books of China.* Clarendon Press, Oxford.
34. Mandelkehr, M.M. 1983, An Integrated Model for an Earth-wide Event at 2300 BC. Part I: The Archaeological Evidence, *S.I.S. Review* **5**, 77-95.
35. Pang, K.D. & Chou, H.-H. 1985, Three Very Large Volcanic Eruptions in Antiquity and Their Effects on the Climate of the Ancient World, EOS, *Transactions, American Geophysical Union* **66**, 816.
36. Pankenier, D. 1995, Astrological Origins of Chinese Dynastic Ideology, *Vistas in Astronomy* **39(4)**, 503-516.
37. Peiser, B., 1997, Comets and disaster in the Bronze Age, *British Archaeology* **30**, Dec 1997, 6-7.
38. Post, J., 1977, *The Last Great Subsistence Crisis in the Western World,* John Hopkins University Press.
39. Pyle, D.M., 1990, The Application of Tree-Ring and Ice-Core Studies to the Dating of the Minoan Eruption. in Hardy, D.A. (ed.) *Thera and the Aegean World III* Vol.3, The Thera Foundation, London, 167-73.
40. Rampino, M.R., Self, S. and Stothers, R.B., 1988, Volcanic Winters, *Annual Review of Earth Planet Science* **16**, 73-99.
41. Sagan, C. and Druyan, A., 1985, *Comet.* Michael Joseph, London.
42. Scuderi, L.A., 1990, Tree-Ring Evidence for Climatically Effective Volcanic Eruptions, *Quaternary Research* **34**, 6785.
43. Steel, D., 1995, *Rogue Asteroids and Doomsday Comets*, John Wiley and Sons, New York.
44. Warner, R.B., 1990, The 'Prehistoric' Irish Annals: Fable or History, *Archaeology Ireland* **4(1)**, 30-3.
45. Weisburd, S., 1985, Excavating Words: A Geological Tool, *Science News* **127**, 91-6.
46. Weiss, H., 1996, in Dalfes, H.N., Kukla, G. and Weiss, H. (eds), *Third Millennium BC Climate Change and Old World Coll*
47. Whiston, W., 1696, *A New Theory of the Earth from its Original to the Consummation of all Things, Wherein the Creation of the World in Six days, The Universal Deluge, and the General Conflagration, as Laid Down in the Holy Scriptures, are Shown to be Perfectly Agreeable to Reason and Philosophy,* London.
48. Wilson, I., 1985, *The Exodus Enigma*, Guild Publishing, London.
49. Zielinski, G. A., and Germani, M.S., 1998, New Ice-Core Evidence Opposing a 1620s BC Age for the Santorini Minoan Eruption, *Journal of Archaeological Science* **25**, 279-289.
50. Zielinski, G. A., Mayewski, P. A., Meeker, L. D., Whitlow, S., Twickler, M. S., Morrison, M., Meese, D. A., Gow, A. J. and Alley, R. B., 1994, Record of volcanism since 7000 B.C. from the GISP2 Greenland ice core and implications for the volcano-climate system. *Science* **264**, 948-952.

Comparative Analysis of Late Holocene Environmental and Social Upheaval:

Evidence for a Global Disaster around 4000 BP

Benny J Peiser

Liverpool John Moores University, School of Human Sciences, Trueman Street, Liverpool L3 3AF, UK

"C.F.A. Schaeffer presumes that the catastrophes which caused the end of civilisations in Eurasia originated in devastating earthquakes which shook the world. He mentions that many sites show that the destructions have been contemporary with 'climatic changes, which seem to have brought about transformations in the occupation and the economy of the country'. Schaeffer does not seem to have been struck by the connection between those two important contemporaneous events: earthquakes and climatic changes. As we have previously seen, those two events are closely connected with cosmic catastrophism: both are inevitable results of huge meteoric impacts" [René Gallant, *Bombarded Earth*, 1964, pp. 214/15].

1. Introduction

At some time around 2300 ± 200 BC, a large number of the major civilisations of the world collapsed. The Old Kingdom in Egypt [26, 39], the Akkadian Empire in Mesopotamia [292, 294], the Early Bronze Age societies in Anatolia [169], Greece [158] and Israel [226, 227, 228], as well as the Indus Valley civilisation in India [203, 204, 205], the Hilmand civilisation in Afghanistan [155], and the Hongshan culture in China [177] the first urban civilisations in the world all fell into ruin at more or less the same time.

The reasons for these widespread and apparently simultaneous disasters which coincided with major social and climatic changes elsewhere have long been a fascinating mystery. Traditional explanations included warfare and socio-political disintegration [265], 'system collapse' [215], widespread seismic activity [232], and, more recently, climate change [294, 52]. But the apparent absence of direct archaeological or written evidence for causes, as opposed to effects, has left many researchers resigned to the opinion that no definite explanation can be found.

Some decades ago, the hunt for clues was largely assumed by natural scientists. Concentrating on the late 3rd millennium BC collapse, researchers began to find evidence that natural causes, rather than human actions, may initially have been responsible. There began to be talk of volcanic activity, large earthquakes [232, 157, 260], and climate change [21]. Some of this material has now found its way into standard historical accounts of the period.

The most recent publication, *Third Millennium BC Climate Change and Old World Collapse* [52] incorporates 33 papers by archaeologists and climatologists. In a concentrated effort, the contributors assess the nature and magnitude of the environmental and cultural changes which befell many parts of the world during the late 3rd millennium BC.

The majority of these papers venture to interconnect the archaeological, geological and climatological records in order to prove, or, in the case of Karl Butzer [39] to disprove the occurrence of an abrupt climatic downturn at c. 2200 BC. The papers range from studies on Nile flood fluctuation and their impact on ancient Egyptian society, the environmental and social changes in the Near and Middle East, and abrupt climate change in north and tropical Africa during the late Holocene, to the collapse of the Indus Valley civilisations and environmental changes in prehistoric Europe at roughly the same period of time.

The volume thus presents a comprehensive review of the palaeo-environmental data for major climatic and cultural shifts during the late Holocene. While the authors document convincing evidence for major climatic punctuations, they fail to assess the most reliable signals of abrupt climate change, those in the tree-ring and ice-core records. What is more, the publication simply ignores Schaeffer's main findings (and those of other archaeologists) of significant tectonic activity during the same period of time. The extensive evidence for seismic site destruction at the end of the Early Bronze Age in Greece, Anatolia, Israel and Mesopotamia appears to be incompatible with a climate model of civilisation collapse and points to more complex causes for upheaval.

Over the past 15 years, an alternative agent for natural disaster has been much discussed and a growing number of scientists are beginning to regard it as the most probable single explanation for widespread and simultaneous cultural collapse, not only during the Bronze Age but at other times as well. The new theory has been advanced largely by British scientists such as Victor Clube and Bill Napier [47, 48], Sir Fred Hoyle and Chandra Wickramasinghe [119], Mark Bailey [13, 14], Duncan Steel [257, 258], and Gerrit Verschuur [283]. The theory maintains that massive cultural and historical disasters were caused by the impact of meteor showers and other types of cosmic debris on the Earth.

The hunt for natural causes for these catastrophes and civilisation collapses began when the French archaeologist Claude Schaeffer published his book *Stratigraphie Comparée et Chronologie L'Asie Occidentale* in 1948. Schaeffer [232] analysed and compared the destruction layers of more than 40 archaeological sites in the Near and Middle East, from Troy to Tepe Hissar on the Caspian Sea and from the Levant to Mesopotamia. He was the first scholar to detect that all of these settlements had been entirely destroyed and or abandoned several times in the early, middle and late Bronze Age, apparently simultaneously. Since the damage did not show signs of military or other human involvement, and in any case was too excessive and widespread, he argued that repeated earthquakes might have been responsible.

"Our enquiry has shown that these successive crises which opened and closed the principal periods of the 3rd and 2nd millennia were not caused by the action of man. On the contrary, compared with the magnitude of these general crises and their profound effects, the exploits of conquerors and the schemes of political leaders seem very insignificant" [232, p. 565].

When *Oxford University Press* published his work, Schaeffer was not taken seriously. During the last two decades, however, natural scientists have found widespread and unambiguous evidence for abrupt climate change, sudden sea level changes, catastrophic inundations, widespread seismic and tectonic activity and evidence for a dramatic drop in lake levels at around 2300 ± 200 BC. Areas such as the Sahara and the Dead Sea region, which were once settled or farmed, became deserts. Tree rings show disastrous growth conditions at c. 2350 BC, while sediment cores from lakes and rivers in Europe, America, Asia and Africa show a catastrophic drop in water levels at the same time. In Mesopotamia, vast areas of land appear to have been devastated, inundated, or totally burned.

Scholars who, following Schaeffer, favour earthquakes as the principal cause of civilisation collapse, argue that the world can expect vast earthquakes every 1,000-2,000 years, leading to widespread abandonment of sites and settlements (Nur, this volume); while scholars who prefer to attribute climate change as the principal cause of the late 3rd millennium civilisation collapses argue that severe droughts caused agriculture to fail and that Early Bronze Age societies inexorably fell apart as a result.

Yet, what was the cause of the earthquakes, eruptions, tidal waves, fire-blasts and climate changes which have been detected in the geological and climatological records?

Since the early 1980s, when Luis Alvarez and his colleagues linked the extinction of the dinosaurs to a cosmic impact catastrophe, research in the field of astronomical neo-catastrophism has quickened its pace. During the course of the last decade, most scientists have gradually accepted the idea that global disasters can be caused by extra-terrestrial bodies. This acceptance went with the assumption that cosmic-induced cataclysms were restricted to primordial times, millions of years before the origins of the human race. Yet an increasing number of astronomers claim that a series of cosmic disasters on a much smaller yet global scale punctuated the Earth in more recent, i.e. historic and late prehistoric times.

These authors envisage trains of cometary debris which repeatedly encounter the Earth, the largest and most dangerous representatives of which may be between one and several hundred metres in diameter. Such bodies, depending on their cometary constitution and hence their cohesive strength, can have catastrophic effects on the ecological system in a variety of ways. According to recent research findings, the encounters between cometary debris and the Earth are capable of producing (although they do not always do so) three types of natural disaster: i) low or high level multi-megaton explosions of fireballs which destroy natural and cultural features on the surface of the Earth by means of floods, blasts and seismic damage and which can destroy regions up to the size of a large country; ii) massive high level influx of cosmic dust high above the stratosphere which causes a dramatic drop of global temperature, (this cosmically induced climatical disaster can last for decades and consequently lead to the suspension of agriculture); iii) and massive high-level influx of cosmic chemicals (associated with dust) with as yet incalculable biochemical potentials which may be harmful to DNA and can trigger evolutionary mutation.

During the last decade, scientific findings have identified plausible causes of cosmic catastrophes. More importantly, new evidence has led to a radical reduction of time-scales for cosmic disasters. As a result, research has just started to detect cosmic catastrophes in historical times.

Neo-catastrophism became accepted as a new scientific paradigm during the early 1980s, when Luis Alvarez and his colleagues discovered abnormally high concentrations of iridium in the clay layer which marks the Cretaceous-Tertiary (K/T) boundary. Since iridium in these extreme concentrations is lacking in the Earth crust, it was linked to iridium rich cosmic agents which, subsequently, were made responsible for the extinction of the dinosaurs. Most scientists have gradually accepted the possibility of global disasters caused by extra-terrestrial bodies. The most important factor for this paradigm shift, however, was the testability of this catastrophe hypothesis. If, as was claimed, a global disaster of cosmic origin had ever struck the Earth at the end of the Cretaceous period, the smoking gun [i.e. massive impact crater] and the body [i.e. an iridium layer around the globe] had to be detected. Since both pieces of evidence have been discovered over the last ten years, the cosmic origin of at least some aspects of the K/T boundary catastrophe is now generally accepted by the scientific community [95]. Their willingness, however, depended on the assumption that cosmic-induced cataclysms were restricted to distant times, many millions of years before the origins of the human race.

This picture has changed rapidly. Since 1991, more than 10 international conferences, dealing exclusively with cosmic disasters and impact hazards, have been organised. Dozens of astronomers around the globe are now involved in research projects, focusing on past, present and future impact catastrophes [11, 15, 92].

Research in many other scientific fields has already been affected by the emergence of neo-catastrophism. Over the last few years, scholars have started to investigate the implications of catastrophic events on societal evolution [31], cultural anthropology [196, 197], human social behaviour and the development of religion (Heinsohn this volume). Above all, neo-catastrophist astronomers have argued that both the emergence and the collapse of human cultures, e.g. the Neolithic Revolution, the origin of metallurgy [118], the onset and collapse of the Bronze Age civilisations [112], the construction of the Egyptian pyramids [112, 257], the multiple destructions during the Bronze Age civilisations in the Near and Middle East [47], the rise of Monotheism (Heinsohn, this volume), and even the collapse of the Roman Empire [48, 19], etc., might be associated with episodes of increased meteoric activity and multiple impacts. Such episodes punctuating the evolution of human cultures are looked upon as primary agencies determining the rise and fall of ancient civilisations.

The new findings of cometary astronomy and archaeological research, therefore, point to a new scientific hypothesis which deserves careful scrutiny - specifically that some of the multiple destructions of the Bronze Age cultures might be related to cosmically-induced catastrophes.

During the last two decades a major paradigm shift towards neo-catastrophism has also occurred in the field of archaeology. A growing number of archaeologists and climatologists have documented strong evidence for seismic, climatical and ecological disasters at the end of the Early, Middle and Late Bronze Age in many parts of Greece [310], Anatolia [169], the Near East [294], Egypt [26], China [190, 191, 192] and India [203].

Extra-terrestrial catastrophes, on the other hand, can lead to a variety of ecological effects and physical damage (Napier 1998, this volume). One should, however, bear in mind that only eighty years after the cosmic catastrophe over the Tunguska river, it is quite difficult to detect any hard evidence for large-scale destruction even at the epicentre of the explosion. In fact, more than 90% of all meteorites are not composed of iron, but huge stone projectiles, which normally break up before impacting (i.e. atmospheric impacts), thus destroying by force of explosion, without causing any impact crater on the ground [47].

Yet the likelihood of large-scale atmospheric impact events makes it much more difficult to detect unambiguous evidence for the occurrence of celestial disasters during the Holocene. Not only is the smoking gun (i.e. the cosmic agent or a *major* impact crater) lacking; but the body itself (corresponding destruction layers of continental or global extent) is still missing. Consequently, scholars doubt whether a continental or even global assault during the last 10,000 years ever took place.

Even so, a number of scholars hypothesise that environmental disasters caused by cometary punctuations with effects similar to the Tunguska event in 1908 or, on a larger scale, the impact of many pieces of comet Shoemaker-Levy 9 on Jupiter, did occur during the Bronze Age producing both isolated and multiple multimegaton catastrophes. Such impacts may have affected the climate, civilisations and belief systems on a global scale. The particular characteristics and the degree of damage afflicted upon the eco-system, however, would differ considerably from continent to continent and from region to region. Rather than looking for iridium anomalies, we should expect to find regional (or even global) patterns of synchronous destructions or abandonment of sites as well as other archaeological and historical evidence for natural disasters.

It is beyond doubt that Bronze Age cultures around the world repeatedly experienced ecological disasters [21]. Moreover, Schaeffer stressed that the major destruction layers of the Near and Middle East, which were detected throughout a territory of 5000 kilometres in diameter, were not caused by the actions of man.

But if cometary or meteorite debris really punctuated the earth during historical times, can we link any of the abrupt environmental changes and destruction layers with such punctuations?

Moe Mandelkehr focused on the destructions and cultural breaks of the early Bronze Age cultures and produced strong evidence for global destruction levels and climatological changes at the end of the Early Bronze Age [155, 156, 157]. British geologist Richard Huggett even linked this evidence to the Biblical flood story [120].

Whilst the relatively new search for archaeological evidence of impact catastrophes is producing growing information about such events in historical times, there are still notable limitations with regard to the identification, let alone the accurate dating of hypervelocity impact structures. This great uncertainty regarding reliable dating became apparent when a team of French geologists analysed the geological character of the recently-discovered chain of impact craters in Aorounga (Chad, Africa). According to the French team, the geological evidence in situ strongly indicates that the impact configuration was formed during the Holocene period, i.e. ca. 12,000 - 3500 BP. In comparison, research by planetary geologist Adriana Ocampo, who analysed the craters by means of radar images, appears to suggest a date during the Late Devonian, i.e. 360 million years B. P.

2. The historical record

During the last 150 years, however, it has become increasingly unfashionable to use the historical records of early mankind as evidence for the reconstruction of societal evolution. Yet these records provide a rich source of transmitted information about natural phenomena and catastrophic events. The main reason for this ignorance stems from the failure of modern scholarship to decipher the historical background of the origins and social function of ancient myths and legends.

The strongest evidence for historical impact events derives from ancient written documents and traditions [281, 47]. Ancient and medieval traditions frequently report extraordinary fireball events and major disasters, and often both phenomena are linked together. But can we really link historical impact catastrophes with mythological catastrophe legends? According to the generally-accepted view among scholars, myths and legends about astral objects, sky gods and their battles represent astronomical movements which occur in a periodic manner and which are observable today (e.g. the movement of the sun, the moon and the planets, eclipses, constellations etc.). If we accept this actualistic paradigm, myths and legends and their playful re-enactment in ancient rituals incorporated the function to enhance the vitality of society, the fertility of the crops or the stability of the sun by way of 'sympathetic magic'. According to this view, ancient people modelled their myths according to the same sky that we can see today. Yet the classical scholar Lewis Farnell stressed as early as 1919 that normal events and regular occurrences in nature would hardy have been incorporated in myths:

> 'What is normal in nature and society rarely excites the myth-making imagination, which is more likely to be kindled by the abnormal, some startling catastrophe, some terrible violation of the social code' [75].

3. Methodological problems and limitations

The research on natural disasters which might have been caused by cosmic impacts during the Holocene is not only of historical interest. The results of these investigations have important ramifications for a better understanding of our cosmic environment and the hazards our own civilisations face due to comets, asteroids and extraordinary meteor showers. Whilst geologists and palaeontologists have attempted to discover a pattern of impacts and mass extinctions in the geological record, archaeologists, climatologists and cultural anthropologists face a similar challenge with regard to the historical and prehistorical time frame (i.e. after 10,000 yrs BP).

There has long been a debate, which is still ongoing, about the effects of abrupt environmental downturns on human and societal evolution. Until recently, it has proved difficult to establish the actual occurrence of a natural catastrophe of hemispheric or global extent during the Holocene. But even more complicated to assess are the environmental and social knock-on effects of such punctuations. This comparative study seeks to draw upon the widest-possible range of research findings which have been published in the course of the last two or three decades. I will review this evidence selected from thousands of investigations from around the world and will evaluate the impact of the 3rd millennium BC environmental changes on cultural and societal evolution. The mass of data derived from these studies will then be analysed in order to produce a pattern of evidence.

The data for this study were obtained in a wide variety of publications and scientific papers. There are literally thousands archaeological and paleo-climatological studies which focus on the environmental changes at the mid/late Holocene boundary. Of this vast number of research papers some 500 were selected and reviewed. The selection of these studies is biased in that only those publications were included which provided evidence for distinct signals in the sedimentary, climatological, archaeological or geological records. Such distinct *unconformities* suggest *major changes* at the mid/ late Holocene and document *abrupt* rather than gradual events. These data are thus evidence for abrupt transitions triggered by punctuations which lead to sudden alterations of the environmental equilibrium.

The question, however, raised by many of these studies is whether or not the discovered events are linked since many of the data are scattered in time. It is hypothesised that the clustering of these unconformities over several hundred years is due to dating inaccuracies rather than the result of a random distribution of significant environmental punctuations.

We are now in a much better position to compare and analyse the growing collection of reasonably-detailed excavation reports and palaeo-environmental studies on civilisation collapse and climate change. During the last fifty years or so, archaeological and palaeo-climatological research has accumulated a vast amount of proxy data which provide indirect evidence and measurements of past environmental and social turbulence. Such proxy data embrace documentary evidence of tree-rings, ice-cores, varves, glacier movements, pollen remains, and many other sources.

In reviewing these data and possible effects of environmental punctuations on societal evolution we need to consider the current methodological shortcomings which make the detection of any large scale (let alone global) ecological and social upheaval extremely difficult. The main problems in interpreting proxy data are those of i) dating, ii) magnitude, iii) causation, and iv) societal effects.

Due to generally inaccurate dating methods, most of the dates found in the literature on abrupt environmental events cannot be correlated to *absolute* years in which abrupt climatic downturns occurred and which have been detected in tree ring records. The dissimilar temporal resolutions (and thus quality) of the environmental data on the one hand and the much more precise methods of tree-ring and ice-core dating on the other hand currently prevents any accurate interpretation of most of the paleo-environmental proxy data.

Because of the well-known problems associated with conventional radiocarbon dating, a comparative analysis limited to studies which are based on ^{14}C datings do not permit a precise identification of the timing of an abrupt paleo-environmental event.

It follows that any claim of a global climate change must be clearly detectable in tree-ring records (as well as in ice-core records if significant volcanic activity was involved). In short, a global climatic downturn must be discernible in tree rings from the period in question.

There is general consensus that the most reliable proxy data for the reconstruction of abrupt environmental change during the last 5000 years are those from tree-rings, ice-cores and pollen deposits from varved lakes. This evidence may be dated accurately to the year. However, the vast majority of proxy data and archaeological strata have been dated by means of the radiocarbon method. But the ^{14}C-method is commonly recognised for its lack of dating precision. This uncertainty cannot be underestimated, particularly in view of the large fluctuations which have occurred in atmospheric carbon-14 [121].

Isotopic, palynological and other microfossil ^{14}C data can thus only help confirm sudden changes of climate and help assess whether there is general agreement with the estimates derived from tree-rings or Greenland ice-cores.

The second problem of any comparative analysis of proxy data is that of magnitude. Only detailed local or regional climate information has been obtained, and only this is considered adequate for a causal link with societal change. “The problems involved in studying a large area such as an entire country or even a continent are obvious. Weather and climate patterns may vary markedly within national boundaries, the influence of these patterns will vary according to the varieties of regional landform and even local topography, and it will be necessary to take account of regional variations in economic and social structures” [121]. Due to the inherent problems of dating, interpretation and stringent methodology, attempts to detect corroborative evidence for hemispheric or even global environmental disasters during the last 5000 years have so far been unsuccessful.

The third problem is related to the difficulties in detecting the actual causes of past catastrophes. Most environmental disasters which have been witnessed and studied by modern humans occur on a strictly local or regional level. Such calamities are triggered by natural agencies such as volcanic eruptions, seismic activity, catastrophic storms or giant floods which affect only relatively small geographical areas. None of these currently observable activities, however, are thought capable of having significant hemispheric or even global effects.

Even more complicated are the attempts to correlate the effects of abrupt climatic change with societal change. The links between climate and human societies seem far too complex as to allow a simple correlation between both variables. “Statistically tested correlations between gross climatic variables and data on economic activities more closely linked with atmospheric changes may be more useful. [...] In order to be useful these tests must be based on a detailed model of causality which specifies the nature of the hypothesised links between climate and human activity. [...] in reality this is not feasible because data on several of the key variables are lacking. However, in many cases the problem is not only that of inadequate data but also an imperfect understanding of the economic and social processes involved” [121].

Given the methodological limitations, it would be unreasonable to underestimate the current uncertainties. Instead it is the aim of this paper to present the proxy data and environmental variables for an environmental disaster around 2350 BC which, at face value, appear to be interrelated in time and location. It is this rather elementary accumulation of proxy data from around the world which eventually will need to be critically assessed once a viable method to do so is available.

One of the most complicated problems of palaeo-catastrophic research is the difficulty of drawing a distinction between the evidence for local, regional, hemispheric or global disasters. The most sensitive indicators of significant and abrupt shifts in the geological and climatological records are always site-specific. In order to distinguish global climate disasters from local or regional events, it is necessary to scrutinise and compare the results of research conducted around the world. It follows that any study investigating natural catastrophes of global extent will try to find a widespread pattern of stratigraphic and cultural breaks, climate anomalies and vegetational change in various types of environmental proxy data and a clustering of events in a short period of time.

What is therefore required is the compilation of a comprehensive data collection which does not exclude any source of data. Only such an extensive accumulation of data will enable the researcher to gain a complete picture of the evidence. This, then, might enabled us to start contemplating which natural triggers are capable of causing the unconformities which have been detected.

4. Holocene climate history

During the last couple of decades, detailed information about the main patterns of Holocene climate, particularly at the mid/late Holocene boundary (c. 2300 ± 200 BC), has emerged as more and more sites suitable for palaeo-ecological reconstruction have been excavated and analysed by researchers. Despite this growing data base, the global distribution of such sites and the climatological information which they reveal remain uneven due to the limitations and constraints of scientific research in many underdeveloped countries. Nonetheless, research on mid/late Holocene (MLH) environments and their climates has now been carried out in almost all corners of the globe and a comprehensive semblance of palaeo-climate and palaeo-environments has emerged. From this vast amount of proxy data it has become manifest that Holocene climate and the development of vegetation and human cultures were repeatedly punctuated by episodes of abrupt climatic downturns.

Abrupt climatic change has been defined as a sudden shift in temperature or precipitation which is so severe that it causes a significant disturbance or aberration of the ecological balance [255]. Such punctuations are detectable in sedimentary and climatological records. There are, however, certain difficulties with palaeo-ecological reconstruction since any abrupt climate change has 'an impact on nearly every aspect of our environment', which means that 'the lines of evidences of past events are innumerable' [139].

It is hypothesised in this paper i) that the abrupt climate change at the Mid/Late Holocene transition is contemporary with the same climatic downturn detected at the Atlantic/Sub-Boreal boundary, ii) that both markers evidence the same environmental event, iii) that this climatic downturn must be evident in dendrochronological records, and that iv) this event is synchronous with the narrow tree-ring signal around 2350 BC. The theory of a notable climatic downturn at the Atlantic/Sub-Boreal (ASB) boundary was first developed by Sten Florin in the late 1930s. By 1945, Knut Faegri believed that this cool spell was momentous enough to have triggered the formation of some Norwegian moraines at that time [170].

Frenzel [82] reviewed the accumulating evidence of climate change at the Atlantic-Sub-Boreal transition and concluded that a distinct and abrupt cold spell led to an interruption of the Hypsithermal. According to Frenzel [82], this abrupt climatic transformation coincided with the end of the Atlantic and the onset of the Sub-Boreal period. But when exactly did the Sub-Boreal period supervene the Atlantic period and did it really occur simultaneously with the mid/late Holocene (MLH) transition? Due to the inherent constrains and limitations of current dating methods, researchers have failed to arrive at a consensus as to the absolute dates of both the ABS and MLH transitions [105]. The current estimates range from 5000 4000 BP, and most of the hundreds of papers published each year on mid and late Holocene climate date both the ASB and MLH transitions somewhere within this time frame.

In order to base our dating of abrupt environmental change on a firm foundation, we need to use the most reliable signals for such events. The tree-ring and ice-core records will therefore be of crucial importance for the assessment of abrupt climate change. The analysis of annual growth rings in trees in particular will be used to assign precise calendar dates to the abrupt environmental and climatic events around 2350 BC. This method can help establish the accurate dating of this climatic downturn in remains of dead trees which must have been affected by such a disastrous drop of temperature. Ring-width patterns when compared with control chronologies can reveal the growth anomalies in witness trees which are most likely related to climate change and will thus be applied in our attempt to refine the dating of these events.

This paper will focus particularly on the ecological crisis with its pernicious effects on human cultures and civilisations around the world which occurred at c. 2350 BC since it is clearly detectable in climatological, geological and archaeological records.

5. Lake-level changes

The Late Holocene climatic history of the world has been studied for more than half a century, yet much of the distinct pattern of climate change remain unaccounted for, particularly with regard to the factors and mechanisms which triggered these abrupt climatic up- and downturns during the last 10,000 years.

Some of the most direct evidence for natural disasters and climatic shifts is derived from lake profiles and the reconstruction of lake-level changes [108]. The exploration of past lake-level changes can be based on the analysis of a variety of data, including sediment cores from lakes, the characteristics and distribution of aquatic organisms in lake sediments, raised beaches and other evidence of high or low lake stands.

Since lakes alter their size in response to variations in the water budget, their sediments are perfect indicators of past climate change. If many lakes in a certain region or continent display similar stratigraphical signals in their sedimentary records, these patterns are clear indications for past climate changes. Because notable lake-level fluctuations correlate with distinct climatic fluctuations, and since late-Holocene hydrologic dynamics are well-correlated with abrupt climate changes, global lake dynamics offer an important insight into the sensitivity of the world's water resources to sudden climate change.

Most of these lake-level and palaeo-climatic records are based on ^{14}C ages. Yet the problems with these dating methods are well-known: The radiocarbon method precludes a very precise dating of abrupt changes because (i) the standard error is high (10^2 yr) compared with the duration of some climatic shifts, (ii) several plateaus in ^{14}C age (expressed as yr BP) vs. calendar age (expressed as cal BP) plots occurred during the period considered, and (iii) causes of age bias in lake sediments are numerous [91]. In short, radiocarbon dates alone, even when they cluster around a particular period of time, cannot provide us with the dating precision necessary to corroborate the actuality of an abrupt and catastrophic episode in the geological record.

Despite these general problems, however, the detection and timing of abrupt punctuations of lake levels can be determined by studying the profiles of lake sediments, since significant anomalies and disruptions in the constant mean sedimentation rate are transparent indicators for abrupt ecological shifts. In addition, variations in lacustrine organic substances can be used as indicators of distinct episodes of past climate change since organic matter in lakes is influenced directly not only by local and regional but even more so by global climate change.

It has been known for some time that the water-level of many African lakes dropped abruptly during the 3rd millennium BC. The water-level of Lake Kivu and Lake Tanganyika in East Africa began to drop around 4000 BP [111] whilst the water level of Lake Bosumtwi in Ghana saw the onset of a critical decline c. 3800 BP [103, 267].

The cause for this abrupt drop in lake levels at the end of the mid Holocene which has been detected in the vast majority of African lakes has been associated with the sudden onset of an arid period around 4000 - 3800 BP. Recent research by Casanova & Hillairemarcel [40] of sediment cores of Lake Tanganyika, situated in the Western Branch of the East African Rift System has confirmed this punctuation of the water budget of the lake at c. 3800 BP. The arid conditions not only closed the Tanganyika basin from any inlet from rivers, but also lowered the lake-level by 75-150 m.

Limnological research into the history of the lakes of the Ethiopian Rift and the Afar region has revealed that a big and rapid fall in their water-level occurred at c. 4000 BP [90]. Researchers detected an identical signal while analysing Holocene fluctuations of the water level of Kenya's Lake Turkana which also began to drop abruptly at the same time [219, 188]. The lowest water level of Lake Victoria is dated to ca. 3700 BP at which time Lake Nabugabo (in the northern shore of Lake Victoria) was cut off from the main lake [268, 269, 270].

Investigations of lake-level fluctuations of West African lakes have exposed a similar pattern of decline during the late 3rd millennium BC. Sediment cores show that Lake Chad reached a minor high-stand between c. 6000 - 4000 BP which was followed by a recession at the end of this period [224]. At around 4000 BP, Lake Chad and the Bar-el-Ghazal River linked to the lake encountered a dramatic dry spell [91].

Lake Bilma, Lake Fachi and Lake Dibella are three palaeolakes along the southern and eastern margins of the Ténéré desert in the western Sahara which dried out around 4000 BP. At the same time, two palaeolakes in the Sahel part of the Chad basin, Lake Termit and Lake Bougdouma, experienced extreme episodes of dryness [91].

After comparing the results of various African lake level studies during the 1960s and 70s, Hamilton concluded in 1982: 'The most important general change was probably a switch to lower levels at around 4000 BP, a time which also witnessed expansion of the Sahara and the demise of the Saharan Neolithic' [107].

Twenty years on, this picture has become even more visible and reveals a striking pattern of abrupt lake level change at this crucial period of climate and cultural history. Scientists are now convinced that they are dealing with a dramatic climate shift, the 4000 yr BP event [which is] recorded throughout much of tropical Africa by dropping lake level [219].

Lake Sumxi and Lake Bangong lie in western Tibet, the highest, coldest and driest part of the Tibet Plateau. Lake-level research conducted at both lakes reveal that episodes of maximum dry spells started at c. 3800 BP. At this time, Lake Bangong experienced an extremely low-water stand [74]. Corroborative evidence for the global nature of this environmental downturn comes from China. The analysis of sediments and pollen of three 5-m long cores taken from Lake Manas in western China has revealed that the onset of a significant dry spell started here also around 3800 BP and lasted for some 300 years. According to Rhodes *et al.* [216] this environmental shift is roughly in phase with similar downturns which have been observed in Tibet.

At exactly the same time, c. 3800 BP, the water level of three salt lakes in the Thar desert of north-west of India, Lake Lunkaransar, Lake Didwana and Lake Sambhar dropped abruptly [91]. In addition, all three lakes witnessed a noticeable change of their salinity at the same time some evidence even suggests that the lakes first became saline then [248, 250]. Many researchers have attributed this salinity-event to a sharp decline of precipitation. 'This is, without doubt, one phenomenon that might lead to increased salinity. But, it is also true that at ca. 2000 B.C. there is geological evidence for major changes in surface, and probably subsurface, drainage in the northern Rajasthan, the eastern Punjab, Haryana and the western Uttar Pradesh. This event led to the desiccation of the Sarasvati Riverine system, the rejuvenation of the Sutlej River and the creation of the Yamuna River' [203].

Recent palaeoenvironmental research on these saline lakes has further corroborated this picture. Work carried out at altogether five lake sites has revealed major shifts in the climate and the vegetational belts during the early Holocene. 'The taxa indicative of both summer and winter precipitation in the mid-Holocene declined during the late Holocene, at the same time as falling lake levels around 4000 yr BP, a time when other lakes, at Sambhar, Lunkaransar and Pachpadra, also became ephemeral' [128].

Osborne *et al.* [186] have analysed the environmental history of Waigani Lake, a small basin adjacent to the Laloki River floodplain in lowland Papua New Guinea. These proxy data from Waigani Lake are the first palaeoenvironmental records from the lowlands of New Guinea and thus provide important information about the nature and geographical extend of the crisis at the mid/late Holocene boundary. It is worth noting that the investigation of plant macrofossils, geochemistry and mineralogy has revealed that peat started to accumulate at the start of the late Holocene, which is tentatively dated to c. 4400 BP.

Harrison, Prentice and Guiot [108] analysed the sedimentary stratigraphy of 37 European lakes (11 from S Sweden, 10 from Estonia, 11 from S France and 5 from Greece). One of the main findings of their comparative study shows a spectacular low stand of water-levels of almost all of the 37 North and South European lakes in question. At this time, the European lakes had dropped to their lowest level during the entire Holocene period, indicating that a dramatic dry spell at around 4000 BP. had affected most of the water levels of these lakes. A recent comparative study of Holocene water-level changes of altogether 83 northern European lakes has confirmed this pattern with its onset of a dry spell starting around 4000 BP [305, 306].

A collection of lake-level records from the Alps foreland and the Jura mountains show a distinct regression of water-levels at c. 4000 BP [151, 152, 153]. Ariztegui *et al.* [10] have found a well-preserved pattern of variations in the type and quality of the organic matter at discrete intervals in sediment cores of Lake St Moritz, Switzerland. Organic matter production and preservation appear to dominate from the early Holocene until an abrupt cessation approximately 3800 BP, evidenced by a strong signal in the lake's sediments. A similarly abrupt transgression of the water-level of Lake Gosciaz in Poland has been dated to c. 3900 BP [194].

Lake-level fluctuations at the same period of time have also been detected in Swedish lakes. The most pronounced episode of natural acidification of the lakes in southern Finland began c. 4000 BP [274]. The southern margin of Fjallsjokull in southeastern Iceland provides evidence for the sudden creation of new ponds which started to become filled c. 4200 BP [225]. A detailed reconstruction of the water-levels of Lake Ljustjarnen in central Sweden shows a distinct drop in the water stand at roughly the same time [4].

The marker date for this abrupt drop in lake-levels which crops up most often in the literature is 4000 BP [108]. Since this marker date is based on relatively vague and often uncalibrated radiocarbon dating, it is of little use for any attempt to pinpoint an abrupt event at an absolute year date. It is therefore fortunate that we possess much more reliable proxy data for lake level change which come from the analysis of *yearly* mud layers of Lake Saki in the Crimea. A careful analysis of this precise record of yearly lake sedimentation shows that the marked regression of lake levels during the late Holocene occurred at 2300 BC, indicating a sharp climatic downturn at this time [139].

The Holocene history of the Lake Michigan basin is characterised by substantial fluctuations in lake-levels. Much of the geomorphological research on the Upper Peninsula of Michigan has focused on the late Quaternary evolution of the Great Lakes and their fluctuating lake-levels which were determined by a complex interplay of climate change as well as episodes of glaciation and deglaciation of the Laurentide Ice Sheet. With regard to the mid-late Holocene boundary, there is hard evidence that a marked episode of lake-level fluctuations, vegetational change and beach-ridge development along the northern shore of Lake Michigan occurred around 3900 BP [200]. The timing of this event correlates with other recorded punctuations of lake levels in the USA described on Maniyoulin Island [145] and south-eastern Wisconsin [141] as the Nipissing II highstand and have been tied to the onset of cool weather conditions at the end of the Hypsithermal around 4000 BP [200, 24].

Interestingly, the abrupt climate change from the arid prairie period to the dry hardwood period (leading to a drop in lake levels) in Minnesota has also been dated to c. 4000 BP [235]. Minnesota has thousands of lakes which hold the key to the reconstruction of precipitous climatic shifts in their sedimentary records. Two of these waters, Williams and Shingobee Lakes, have revealed sedimentary and pollen evidence for five such events during the entire Holocene period. Our interest is obviously focused on the shift from the prairie to the hardwood period around 4000 BP. What Schwalb *et al.* [235] discovered in the cores of these two lakes resembles in many ways the findings in other North-American lake sediments. A marked change in pollen features indicates not only vegetational transformation from mainly prairie grasses to forests but also the onset of lower temperatures at c. 4000 BP.

The study of sediment cores of the Medicine Lake basin (South Dakota, USA) reveals a number of sharp indicators for lake-level changes during the Holocene. Lake levels were relatively high (at least 10 m higher than today) but saline conditions predominated through the early to mid Holocene. At the end of the mid Holocene, an abrupt drop in lake-level occurred. According to Garces *et al.* [87] the Medicine Lake record is consistent with the palaeoclimatic interpretation for the entire northern Great Plains, an area in which a more arid early and middle Holocene was followed by an abrupt transition to a cooler and wetter late Holocene. 'The sedimentary succession in Medicine Lake documents a much more fluctuating and abrupt climate behavior than suggested by the vegetation palaeorecord and suggest a late Holocene climate punctuated by arid episodes' [87].

Sediment lithology and mineralogy, as well as plant macrofossil and stable isotope stratigraphies of lake sediment cores, were used to reconstruct the late Holocene hydrologic changes at Kenosee Lake, a relatively large, hyposaline lake in south-eastern Saskatchewan, Canada. All indicators outline that the onset of an early, low-water, saline phase of lake history began around 4100 BP [278].

A series of palaeoenvironmental investigations indicate the occurrence of similar fluctuations of lake-levels in South and Central America. Interestingly, wet conditions in Mexico also lasted until the mid-Holocene after which the climate became drier. A number of lakes in the Basin of Mexico, the Upper Lerma and the Zacapu basin were all particularly low at the end of the mid-Holocene [184]. A first period of a marked shift in lake-levels and lake sediments is dated to c. 4000 - 3600 BP. According to O'Hara *et al.* [184] these environmental changes were more dramatic than those associated with either interstadial/glacial or glacial/interglacial shifts in the climate.

The records of Lake Santa Teresa, located in the south-west of Venezuela, show that the lake became significantly smaller after 3900 BP and forests surrounding the lake retreated [25].

To summarise, we can conclude that detailed lake-level data from around the world show a marked transgression of water-stands at c. 2300±200 BC. Since most of the lakes and lake systems mentioned above are not interconnected geographically, it would appear likely that the abrupt fluctuation of their water regime in different parts of the globe was more or less synchronously. This suggestion would agree well with the evidence of a major climatic shift at the mid-late Holocene transition.

6. Changes of rivers discharge

TThe reconstruction of ancient river beds has shed important ight on the fluctuations of floodwaters in European, Asian and African rivers during the Holocene. Abrupt shifts in the water-level of rivers obviously reflect meteorological and climatic changes. Since substantial spells of global cooling or warming often bring about abrupt changes in precipitation, river beds and water-levels will suddenly change their features. In the fluvial environment, the level of river discharge and sedimentary loading as well as the rate of soil formation will be immediately affected by drastic climate shifts [255].

Of all the great rivers of the world none has attracted more interest and paleo-environmental research than the river Nile. Its sedimentary history during the Holocene has been studied for many years and has revealed substantial information about past climate changes. A reconstruction of the northwestern edge of the Nile Delta provides consequential data for an abrupt environmental transformation at the mid/late Holocene boundary. Analysis from a core taken at the edge of Lake Maryut near Alexandria shows that Holocene sedimentation here started c. 2400 BC, which coincided with a westward shift of the course of the Canopic Nile distributary [97].

The mid and late Holocene development of the rivers Tigris and Euphrates has been reconstructed on the basis of proxy data derived from annual layers of Lake Van. This lake functions as a main drainage and its history thus reflects the catchment of both rivers. According to Butzer [38] there is evidence for a distinct episode of low rainfall in c. 2350 BC. At the same time, the south of Mesopotamia was badly hit by another calamity: the sudden onset of salinisation of vast areas of the agricultural landscape at 2350 BC led to a dramatic drop of crop yields and a shift from wheat to the more salt-tolerant barley.

Geological research in the South Loup River Valley in central Nebraska has revealed three distinct episodes of floodplain changes. The first episode occurred some time before 3500 BP., following deep incision of the valley [164]. Research findings from the North and Middle Loup River Basins show that this punctuation is synchronous throughout the Loup River Basin. While the causes for these abrupt changes may be multiple, the effects of climate change on flood regimes and water tables explain most of this sudden change in fluvial behaviour.

Martin [159] found evidence for major changes in the alluvial development of the Republican River in south-central Nebraska and that of the central Great Plains at the start of the late Holocene. What the alluvial and climatic records revealed was that synchronous incision of the river around 4200 BP coincided with a major climate shift.

Other sedimentary indicators of alluvial activity, such as extreme palaeo-floods, can also reveal major punctuations of river beds. Ely [66] has recently produced a regional synthesis of palaeo-flood chronologies on rivers in Arizona and southern Utah. His research suggests that the largest floods over the last 5000 years cluster into distinct time periods that are related to regional and global climatic fluctuations. Interestingly, high-magnitude floods in this area of the world were frequent during the entire mid Holocene climate optimum until they came to an end c. 2200 BC. The start of the late Holocene c. 2200 BC is marked by sharp decreases in the occurrence of large floods on these rivers [66].

Vanderhammen *et al.* [277] studied the Holocene dynamics of alluvial sedimentation of the Caqueta River Basin of Colombian Amazonia and found that a strong signal in the sedimentation cycle started c. 3800 BP. With the onset of this period, more organic sediments were formed due to reduced river discharge. According to the authors, this particular decrease in alluvial activity is related to the onset of drier climates around 3800 BP in the Andes and Amazonia, a climate shift which corresponds with the start of a general dry phase in other parts of north-western South America.

7. Sea level changes

When the ice-sheets of the last glaciation at the end of the Pleistocene began to melt, due to an abrupt warming period starting at around 10,000 BP, it triggered an immense rise of the world sea-level [201]. This upsurge continued for more than 5000 years and came to a hold at the end of its mid-Holocene highstand some time between c. 4500 BP [140] and 4000 BP [238]. At around this period of time, with the onset of the late Holocene, the mean sea-level (MSL) of the world oceans suddenly began to drop a phenomenon which has been associated by many researchers with an abrupt downturn of temperatures which triggered an episode of glacial resurgence around 2300 ± 200 BC.

During the last couple of decades concerns about alleged greenhouse effects and possible catastrophic consequences by rising sea-levels have increased [23]. As a result of these fears, many studies have attempted to reconstruct past climate fluctuations and their correlation with sea-level change. While there is general agreement about significant sea-level oscillations during the early and mid Holocene, there is no consensus about the absolute dates for and universality of these fluctuations. Mandelkehr [156] has emphasised the fact that investigators of Holocene sea-level change are divided in a number of competing schools. Yet an increasing number of palaeo-environmental investigations have revealed evidence for rapid transitions of MSL at the mid/late Holocene chronozone. As with most other fields of mid/late Holocene research, the absolute date for this boundary has not been established beyond reasonable doubt and ranges between c. 2500 to 2300 BC. This is to a large extent due to the inherent inaccuracies of the currently available dating methods [21]. According to Jerardion [125] 'the current reconstructions of sea-level history provide information on gross changes, but lack detail as to the exact course and timing of sea-level change.'

However, there is general agreement among climatologists that the mid/late Holocene transition was characterised by a sudden climatic downturn which should be detectable in the most reliably dated proxy data, the tree ring record. Since tree ring records are based on a yearly count and thus the most trustworthy dating tool we have currently at hand, I have used dendrochronological findings as my main basis for establishing the absolute date for the onset of the mid/late Holocene transition.

Rhodes Fairbridge was one of the first scholars to realise that changes in levels of Holocene shorelines around the world indicated that significant changes of mean sea- level had taken place during the last 10,000 years, and that these discernible fluctuations correlated with significant climatic shifts [69, 70, 71, 72, 73]. Such changes, Fairbridge suggested, may be caused by both episodes of notable glaciation and deglaciation, thereby affecting ocean-water volume.

Indeed, relative sea-levels (RSL) in many parts of the globe were significantly higher at the end of the mid-Holocene and began to decrease at c. 2300 ± 200 BC [238]. The end of this mid-Holocene highstand and a subsequent drop of sea-levels at the late Holocene boundary have been detected on coastlines around the world [57, 56, 212]. Mandelkehr [157] has listed some 20 scientific reports about sudden decrease of sea levels in many different continents which investigators have dated between c. 2500 and 2300 BC. This pattern of sea-level discontinuities at c. 2500/2300 BC emerged from research during the 1960's and 1970's and accumulated from a wide range of geographical locations, such as Florida, Brazil, the Netherlands, the Pacific coast of the United States, Panama, France, England, the west coast of Africa, Japan, Hawaii, Australia and New Zealand [157]. Palaeo-environmental research conducted during the last 15 years has provided further evidence for the general accuracy of this pattern.

Geological research of sea level fluctuations in western Mauritania (West Africa) exposed evidence for a drop of the sea by c. 3.5 m at about 4300 BP [64]. At the same time, a marked uplift occurred at the coast of Morocco whilst evidence for subsidence was discovered on the coast of Senegal [179].

At a number of localities on the west coast of South Africa, geological and archaeological investigations by Miller *et al.* [171] show evidence for a mid-Holocene high stand of the sea of about 2 to 3 m above modern mean sea-level. With the onset of the late Holocene, a significant drop in RSL took place. More recent investigations along the western coastal region of the Cape of South Africa have confirmed these findings and exposed a major shift in sea levels at c. 4300 BP. A sequence of sediments taken from Lake Verlorenvlei, today a freshwater lake 15 km inland, testify to the fact that a sea-level highstand led to marine conditions during the mid Holocene. After c. 4300 BP, with falling sea-levels and the retreat of the sea, marine conditions disappeared in this area and were replaced by fresh water [168].

Goy *et al.* [100] found a marked gap in the sedimentary record of the coastlines of southern Iberia, Spain at c. 4000 BP and has linked the geological unconformity to unusual sea-level fluctuations. A study on the coastal evolution of the south-Atlantic coast of SW Spain and the outlet of the Guadalquivir River revealed similar results. A distinct episode of sedimentation and marsh deposition began at c. 4200 BP [223]. Regnauld *et al.* [212] came to a very similar conclusion after examining the coastlines of northern Brittany, France. Whilst Holocene sea-levels along the French Channel coast are distinguished by an almost continuous rise, three marked periods of decreases (c. 10,000 BP, 4000 BP, and 3000 BP) were detected. The Gironde estuary in France was also punctuated at c. 4000 BP according to research conducted by Allen and Posamentier [3]. Healy *et al.* [110] discovered a conspicuous peak in the rate of sedimentation in the Kieler Bucht of the Western Baltic at c.

4000 ± 500 BP.

Shennan *et al.* [243] analysed palaeo-climatic proxy data from isolated sea-water basins in northwest Scotland in order to quantify Holocene sea-level fluctuations in this part of northern Europe. Corroborating the main findings of many previous sea-level studies, the British researchers confirmed that the maximum Holocene sea-level occurred during the mid Holocene climatic optimum (within the range +6.6 m OD to +9.3 m OD) between c. 6600 BP and 4000 BP before falling to its present level.

Recent studies on the Holocene evolution of the Rhine-Meuse delta in central Netherlands has enabled researchers to evaluate the avulsion parameters for the mid and late Holocene. According to Tornquist [273], avulsion frequency was especially large prior to 2300 BC which indicate a strong correlation with the mid Holocene sea-level high stand.
Geological and archaeological investigations in central West Greenland have revealed a distinct RSL fall at around 4000 BP, which coincided with the first foundation of Eskimo settlements throughout Greenland [208].

Sedimentological, archaeological and seismic evidence from the lower Apalachicola River region of northwest Florida indicate a major environmental event starting c. 4000 BP. Enclosing the modern delta and estuary is a barrier island chain which began to develop 4000 years ago. Archaeological evidence from the barriers indicates a minor higher-than-present sea-level during the late Holocene. The evidence thus suggests that these deltas were punctuated by major retreats, possibly in response to sea-level fluctuations at around this time [58]. The timing of the highstand is consistent with evidence from other locations in the south-eastern United States.

A marked increase of the sea-level and change in sedimentation of the estuarine system of the coastal plain of South Carolina is dated to c. 4200 BP when the earliest known marine shell middens were deposited along the Broad River and Port Royal Sound [49].

The dune systems on the piedmont of eastern Colorado and adjacent High Plains have been repeatedly re-activated during the past 20,000 years. Forman *et al.* [79] have found evidence for a significant episode of aeolian activity at the end of the mid Holocene which the authors have dated to c. 4500 BP. The occurrence of major vegetation change and the aeolian depositional sequence at this time has been linked to a reduction in plant coverage associated with a major drought.

A recent study on sea level fluctuations along the Atlantic coast of Nova Scotia comes to very similar results, suggesting a rapid rise and fall of sea-level at the end of the mid-Holocene, starting c. 4200 BP [236]. It should be emphasised that the onset of this oscillation at the mid/late Holocene transition has been recognised along a number of US coasts [237] and many other parts of the world, such as West Africa [94], eastern South America [57], Georgia [56], and Vietnam [133]. Consequently, Scott *et al.* [237] have suggested that this sudden acceleration was triggered by and coincided with the end of the mid-Holocene warming period which led to similarly rapid environmental shifts throughout the world.

At roughly the same time, evidence for a rapid, short-term sea-level change has been detected along the south-east coast of Delaware Bay [77]. A distinct episode of sea-level fluctuation also occurred c. 2300 BC at Murrells Inlet, South Carolina, Chezzetcook and Bale Verte, Nova Scotia and Montmagny, Quebec. Scott and Collins [238] hypothesise that a eustatic sea-level oscillation of about 2-10 m occurred at the end of the mid-Holocene on the south-east coast of North America. Interestingly, the authors link this sea-level oscillation and climatic event with the collapse of the Egyptian Old Kingdom.

Similar evidence for the end of the mid-Holocene sea level highstand at c. 4000 BP comes from the Lago Roca-Lapataia valley, a palaeo-fjord in Tierra del Fuego, southern Argentina [99] and from the Persian Gulf. Geological mapping of the Persian Gulf coast north of Al Jubail, in Saudia Arabia, has revealed the onset of coastal uplift at around 3812 145 BP [220].

Recently, researchers discovered five distinct chronostratigraphic units in coastal dune fields near Cape Arnhem, northern Australia. The unit which is of greatest interest to this study dates from c. 4000 BP and has been associated by Lees *et al.* [143] with the disturbance of this material by rising sea-level and climatic variations in the late Holocene.

There is, moreover, additional geological evidence for notable punctuations of coastal equilibrium at the mid/late-Holocene chronozone. In order to reconstruct the development of coastal dunes in northern Australia, researchers analysed quaternary litho-stratigraphic units from the Cape Arnhem Land dune field on the western side of the Gulf of Carpentaria. The episodic nature of dune evolution which has been detected in other coastal regions of northern Australia [142] is also evident in the sediment cores of the Cape Arnhem samples. Of the five separate episodes of dune emplacement, dated between 19,000 and 2100 BP, the marked late Holocene unit C was dated to. c. 3800 BP and has been associated with the period of late Holocene aridity [143] which was identified in numerous other sites in Australia [142]. Once again, the onset of this exceptional dry phase is tentatively dated c. 4000 BP, coinciding with the mid/late Holocene transition [245, 246].

The study of coastal relief and deposits of the Vietnam shelf islands (Tkhanlam, Kaoptyaotyai, Koto, Kaotkhaotyai, Dongkho, Fongwong, Timatao Re, Che, Mung, Tyam, Kondao, Baikan, Fukuok, Tkhotyu, Tkhom) has provided further evidence for a major transgressive episode at the mid/late Holocene boundary. Korotky *et al.* [133] have found evidence of a distinct episode of sea-level change at the end of the mid Holocene which they date to c. 4100 BP. Regressions were accompanied by intensive aeolian activities, downslope processes and erosional downcut of river valleys in many parts of Vietnam. The mid Holocene warming is also well represented by sea-level records from many parts of the Pacific which show a contemporary maximum ca. 5000-4000 BP [183].

As if this enormous quantity of proxy data was not enough, palaeo-environmental studies along Japanese and Korean coastlines strongly underpin this picture. Here too, the so-called Post-glacial Climatic Optimum of the mid-Holocene with its MSL highstand comes to an abrupt end at c. 4000 BP and is followed by a cool spell [195].

Admittedly, in a small number of cases it seems difficult to select the right event-horizon for our late 3rd millennium BC event when the signal is bracketed between a number of other signals close by. For example, Jerardino [125] has found altogether three minor but important sea-level fluctuations since the mid-Holocene sea-level highstand at the south-western Cape. These oscillations consisted of two regressive events at about 4200 BP and 3500 BP, and one transgressive episode at 4000 BP. While Jerardino claims that the negative oscillations were closely-timed with the global climatic changes experienced in the Southern Hemisphere during neoglacial episodes, it remains difficult to decide which of the two regressive episodes is the one associated with the 2300 BC climatic downturn.

Since this problem only occurs in a relatively small number of cases, it does not pose a real obstacle to testing our main hypothesis. In fact, if we take the above reviewed findings into account, a striking pattern of sea-level fluctuations at 2300 ± 200 BC in most areas of the world emerges. In my view, there can be little doubt that these discernible shifts and transformations in sea-level, which occurred at roughly the same time around the world, were caused by a major climatic downturn which led to a new round of glacier build-up. Indeed, many sea-level experts (such as Rhodes Fairbridge) agree that the discrepancies regarding the absolute dates for these events are presumably due to the variation of data quality and dating methods, or may be related to differences in responses in different temperature zones .

> 'This catalog [of sea level changes] could be continued with evidence of this turning point at the end of the mid-Holocene from around the world. It is emphatically not a local or regional phenomenon' [73, p. 138]

This is why some of the world's leading sea-level experts [238], have recently suggested that the evidence for sea level oscillation at the mid/late Holocene boundary coincided with and were triggered by an episode of abrupt cooling which also led to the collapse of Early Bronze Age civilisations.

8. Changes of forests and vegetation

The reconstruction of the Holocene evolution of forests is mainly based on palynological data, i.e. the analysis of tree pollen samples. Fossil pollen data provide us with information about the quantity and quality of pollen productivity and the feature of vegetation communities. Since plant communities are very sensitive to abrupt climatic change, evidence of significant pollen fluctuations provides a good signal of abrupt vegetation-pollen-climate change [91].

Over the last couple of decades, the data have accumulated to such an extent that they have enabled researchers to reconstruct tree-limit fluctuations and changes of tree-types with respect to exact geographical and altitudinal positions and their connection with climate change.

For the reconstruction of palaeo-environments and the climatological history of the Holocene, the fluctuation of forest expansion and recession, as well as the emergence and extinction of forests, therefore provides valuable proxy data for climate shifts and ecological downturns [138].

In this section I will undertake a comparative examination of the synchronicity of major palynological horizons in the Holocene records of forests and vegetations from around the world. This comparison will help us establish whether there is sufficient evidence to document a marked transition in the nature and extent of forests at the mid/late Holocene and whether this change emanated from abrupt climate change. This set of data will also allow an appraisal of the overall hypothesis that a climate disaster during the late 3rd millennium BC was of global extent with catastrophic ecological consequences.

Whilst the study of Holocene developments in the Sahara desert has yielded strong evidence for a climatic and cultural downturn at the boundary of the mid/late Holocene chronozone, similar results have recently been yielded by researchers investigating the climatic and vegetational history of the equatorial regions of Africa. It should no longer come as a surprise to notice that pollen studies concerning the evolution of African forest zones found that the mid-Holocene forest expansion came to an end around 4000 BP, 'marked by a decline in rainfall and the reduction of the rainforest' [154].

According to Vincens *et al.* [285], pollen data from a core of Lake Sinnda located in the Niari valley in Southern Congo indicate that this region, which is currently covered by savannas, was occupied by semi-deciduous forests during the early and mid Holocene until they disappeared in c. 4000 BP. The researchers have linked the sudden disappearance of these forests and the transformation of the region into grassland savannas to the onset of the late Holocene arid phase which later also led to the drying up of the lake. A more recent reconstruction of climate history around Lake Sinnda in the Guineo-Congolian region of Congo has confirmed that the driest phase of the late Holocene started here c. 4000 BP, commencing with the opening of dense forest and its replacement by a short grass savanna [2].

Jolly *et al.* [127], who studied the Holocene history of the tropical montane forest in Burundi, Africa, detected four distinct stages of forest development. The strongest signal for a major shift in climate change discovered in the pollen samples is dated to c. 3800 BP., 'when a sharp decrease of arboreal pollen percentages and a noticeable increase of Celtis, a semi-deciduous tree indicate the opening of montane forest due to a drier climatic event.'

Recent pollen studies connected with the forest history of the Swedish Scandes Mountains show a distinct change of vegetation starting at around 4000 BP. Prior to the onset of this cold spell, pine trees reached an altitudinal position of almost 200 metres higher than their current level and began to decline as a result of a climatic downturn at the end of the mid-Holocene [136]. Research on peat sediments in Jotunheimen, southern Norway, presents evidence for infrequent but large-magnitude debris-flow events during the mid and late Holocene. Particularly low colluvial activity during the mid Holocene was terminated by a succession of major debris-flow events starting around 2300 BC. This pattern of colluvial deposition demonstrates significant landscape instability at c. 2300 BC which coincided with a general climatic deterioration at the start of the late Holocene [163]. Similar evidence for major vegetational changes was recently discovered at Lake Lilompolo, north-eastern Finland. Mäkelä [149] found that at c. 3700 BP climatic conditions deteriorated so that pine began to retreat whilst mountain birch started to appear.

The polar desert of North Greenland is an extreme environment on the shores of the Arctic Ocean, where low temperatures in the growing season constrain the growth of vascular plants. Pollen samples taken from a polar desert lake in eastern North Greenland document abrupt fluctuations in the vegetation history of this region during the last 7000 years. A recent study of lake sediments by Founder and Abrahamsen [80] found a particularly significant transition to colder summers anda longer duration of snow cover at c. 2350 BC which caused certain types of vegetation to disappear from the region. The authors emphasis that this is not a local phenomenon since the sediments of Lake Klaresø in the interior part of this region as well as sediments from Kap Inglefield in northwestern Greenland also show a marked decline of Salix arctica at roughly the same time [80].

Studies of peat and organic-rich sediments along the coast of Co. Donegal, north-west Ireland, found a sharp decline in the level of *Pinus* pollen, a process which began c. 4000 BP and ended with the complete disappearance of *Pinus* around 3600 BP [242]. Coastal peat analysis elsewhere provides further evidence that these coastal areas were once well-wooded and that *Pinus* was an important woodland component prior to the end of the mid Holocene c. 4000 BP. At the same time, c. 4000 BP, new genera of trees (*Abies, Carpinus betulus, Taxus*) suddenly appeared in southern Italy [288].

The fluctuations of precipitation can also be reconstructed by analysing the development of peat-bog evolution. At the Lagoni peat-bog in Mercurago, Northern Italy, a series of radiocarbon-dated peat-cores showed the onset of a climatic downturn which started around 4000 BP. Evans and Trevisan [68] found that at this crucial time peat started to recede which indicates reduced precipitation. Interestingly, the period which followed this natural crisis is associated with major changes in human settlement patterns, both at a local and regional scale. Archaeologists have wondered for a long time why Early Bronze Age lake-dwellings were suddenly abandoned at this time and why settlement patterns shifted from the low plains of the Po valley to the upper plain and hills of the pre-Alpine system. The synchronicity of these events with the environmental changes suggest that natural disasters could have been responsible for the mass migrations which took place in many parts of the world at the same time.

Palaeo-environmental studies on Gramousti lake and Rezina marsh in the north-west of Greece show a distinct vegetational change starting around 4000 BP which resulted in the total depletion of the montane forest and the dominance of herbaceous vegetation at both sites [301].

Major changes in the nature and character of forests at c. 2200 ± 200 BC have also been detected in Asia. Pollen diagrams from the Sandynallah swamps of the Nilgiri hills, southwest India, show a sharp peak of Michelia, a semi-deciduous tree at c. 3800 BP, indicating rapid climate change at this time [263].

Palynological research conducted in the Wharau Road Swamp in Northland, New Zealand, has revealed evidence for significant climate changes around 4000 BP [65].

Abrupt climatic shifts can also have significant implications for the planktonic vegetation in oceans. Recent research on planctic foraminiferal assemblages from the palaeo-oceanographic records of the north-eastern South China Sea has revealed that 'during the late Holocene a short-lived cooling event occurred at about 4000 ^{14}C yr ago' [289].

Abundant evidence exists in North America for a significant change in vegetation and forests at the mid/late Holocene (MLH) boundary. In most parts of North America, this transition is characterised by the end of the warm mid-Holocene warm period and the onset of the cold late-Holocene. This climate shift has been particularly well detected in pollen diagrams and the analysis of forest change.

A marked period of water-level rise and peatland expansion at Lake Kerr, Michigan started around 3900 BP. At the same time, the forests of the Sylvania Wilderness Area of the Upper Peninsula of Michigan began to change from dry *Pinus*-dominated woodland to moister *Tsuga*-dominated hardwoods. The most rapid growth of the peatland started c. 3900. Brugam and Johnson [43] therefore concluded that a manifest climate shift triggered the emergence of these moisture-demanding forests at c. 3900 BP.

A pollen study conducted at two sites of the Cascade Range in central Oregon provides further data which corroborate the global nature of the late Holocene cooling periode. According to Sea and Whitlock [239], conditions before the start of the late Holocene were much warmer and drier and had led to the establishment of a closed forest composed of *Pseudotsuga* and *Abies*. During this period, Gold Lake Bog in the High Cascades was surrounded by closed forest of *Pinus* and *Abies*. In the late Holocene, however, the onset of cooler and more humid conditions, which Sea and Whitlock tentatively date to c. 4000/4500 BP., began to prevail and late Holocene vegetation was established. The authors have also compared their findings with those published from others studies in the Pacific Northwest and found that, at this turning point, 'major patterns of vegetational change at individual sites were a response to large-scale changes in the climate system that affected the entire region' [239].

In the lower Hudson Valley, a distinct change of forest decline set-in around 4100 BP. A second episode of neoglacial cooling and vegetational change in the Hudson Highlands has been dated to c. 3600 BP [150].

Fredlung [81] has analysed the geomorphic thresholds for the grassland system of Black Hills, South Dakota and found that these were correlated with periods of abrupt climatic change. The findings reveal that a prolonged, mid-Holocene period of landscape stability was abruptly terminated at the start of the late Holocene, which he believes dates to c. 3600 BP.

A pollen analysis from a small lake near Chibougamau in central Quebec provides further evidence for major vegetation and climate change at the end of the mid Holocene. According to research by Garralla and Gajewski [88], the features of boreal forests in central Quebec began to change significantly around 3900 BP. Tree families such as *Picea, Ericaceae*, and *Sphagnum* suddenly increased in abundance at the expense of *Betula*, suggesting the onset of cooler and wetter climatic conditions.

The same picture of abrupt vegetational changes at this conspicuous time comes from Gajewski *et al.* [86] who analysed the pollen contents of sediment cores from three sites in north-western Quebec. The pollen diagrams from these sites in the boreal forest and the forest-tundra show that a mid Holocene period dominated by *Alnus crispa* 'ended synchronously at all sites around 4000 BP. From then on *Picea* pollen dominated the pollen spectra' [86].

The study by Hebda [109] of the evolution of Holocene vegetation in British Columbia confirms that modern vegetation only began around the same time with *Abies, Pinus* and *Picea mariana* predominating at various sites after 4000 BP. At around the same time, a marked decline of hemlock in Quebec has been detected by Bhiry and Filion [30].

Pollen analyses from swamp forest sites on the coast of central Chile reveal the disappearance of halophytic *Chenopodiaceae* and the increase of *Cyperaceous* and Rubiaceous taxa, indicating increasingly wet conditions starting 4000 BP [284]. Lacustrine sediments from the saline lake Negro Francisco, in Southern Chilean Altiplano, confirm that at this time a distinct spell of increased moist climate also affected large areas of South America [102].

A very similar pattern was detected during palaeoclimatological research on the Holocene development of Columbian savannas. Here, pollen records show a marked change from herb savanna to woodland savanna around 3700 BP. This palaeo-ecological transition is also evidenced in other savanna regions of northern South America and has been linked to a marked change in climatic conditions at this chronozone [25].

Pollen records of Holocene sediment-cores from the Costa Rican Cordillera de Talamanca show significant changes in forest vegetation starting with the onset of the late Holocene, which Islebe *et al.* [122] have tentatively dated to c. 4500 BP. The pollen evidence of their research in Costa Rica indicates that this period of the late Holocene experienced drier environmental conditions than today.

In conclusion, I would like to summarise the global evidence for an abrupt shift in forest evolution and vegetational environments at the end of the mid-Holocene by referring to one of the most recent publications on this subject. Palaeo-hydrological and palaeo-environmental studies from five different locations in northern Scotland have confirmed this distinct and large-scale climate shift at the mid-late Holocene transition [8]. Radiocarbon dates place the onset of the abrupt climate change to c. 3900 BP. 'This event appears to be synchronous with changes inferred for other regions, suggesting that it reflects a continental-scale, or possibly even global-scale, change in climate' [8].

9. Glacier resurgence

Evidence from geological and palaeo-ecological analysis of lake-sediments and terrestrial deposits shows that a number of glaciers in the Southern and Northern hemispheres expanded significantly at the mid/late Holocene (MLH) transition. Some of these reached their greatest post-glacial extent during this marked period of cold [170].

The knowledge about exceptional glacier resurgence at the MLH boundary emerged more than thirty years ago. During the last three decades, further investigations have substantiated this understanding and many studies of glacier oscillation during the Holocene have accumulated enough data to identify a distinct climatic downturn and glacier resurgence at c. 2300 ± 200 BC.

Marked advances of some of the glaciers in the Alps, in particular, are associated with the abrupt climate change at the MLH boundary. The sudden expansion of the Oberaar Glacier is dated to c. 4500 BP [93], while the Cadlimo Glacier, near the St. Gotthard Pass, advanced at about the same time.

In western Norway, researchers of glacier fluctuations during the Holocene registered the disappearance of the Jostedalsbreen ice cap during the warm interval between 8000 and 5000 BP. The glaciers began to reappear around c. 5000 BP, with a marked peak for glacier extension which began at c. 4000 BP [229]. Marked Holocene glacial variations have also been detected in sediment cores from the proglacial Lake Linnevatnet in western Spitsbergen. Increasing coal content, incipient laminations and rising sedimentation indicate that the resurgence of the Linnébreen glacier began at c. 2400 BC, probably as a response to the late-Holocene cooling [264]. The same picture can be found in Iceland. Several episodes of neoglaciation have been recognised during the mid and late Holocene, one distinct episode dating to c. 4200 BP [104]. According to Weidick *et al.* [290], the Greenland Ice Sheet re-advanced during the onset of the Late Holocene, dated by Rasch and Jensen [208] to c. 4000 BP.

In Alaska, the ice from the Mount Fairweather Range spread out 50 km past the Reid Glacier at around 4200 B.P. [96] and an exceptional glaciation occurred at c. 3600 B.P. in the Kenai Mountains of Alaska [299]. Microfaunal and microfloral data from two gravity cores, and stable-isotope results from one of these cores taken on Fram Bank near Prydz Bay, Antarctica indicate changes in sea-ice patterns and oceanographic conditions starting around 3800 BP [211]. Glacial resurgence at around this time has also been detected during glacial-geological research on the Windmill Islands, East Antarctica. The glaciological and geological evidence of this study suggests that the re-advancement of the Law Dome ice-margin occurred after c. 4000 BP [98].

A study of glaciers from the Harding and Grewingk-Yalik Icefields in Alaska shows that these glaciers advanced several times into coastal forests during the late Holocene. The records of fluctuations are derived from 16 land-terminating and seven tidewater glaciers in three fjord systems. The research shows that one of the major episodes of Holocene glacier expansions occurred c. 3600 BP.

With reference to the San Rafael Glacier on the western side of the South American Andes, Heussner [113] detected evidence for maximum post-glacial advances at some time before 3740 ± 400 BP. According to British geologists who studied the Holocene evolution of the Fjallsjokull basins in southeastern Iceland, there is evidence for a marked episode of sedimentation and glaciation in this region of Iceland spanning c. 4200 BP to c. 3800 BP [225].

Glacial geological studies on the Windmill Islands and along the Budd Coast at the margin of the Law Dome, East Antarctica, show that Loken Moraines comprise reworked glacial and coastal marine sediments and ice. The evidence suggests that a glacial re-advance occurred around 4000 BP [98].

The occurrence of a major climatic downturn at the mid/late Holocene transition can also be established by other proxy data. Sediment cores raised from the Okinawa Trough and South China Sea were analysed by Jian *et al.* [126] for planktonic foraminifers. These species are very sensitive to changes in sea-surface temperature and therefore function as a palaeo-oceanographic and climatic monitor. The most conspicuous change during the entire Holocene is the P. obliquiloculata minimum zone around 4200 BP which coincides with a widespread cooling-event throughout the western Pacific.

10. Deserts

For a long time, scientists have reflected on and researched the evolution of arid land vegetation and the origin of the world's deserts. The plant and animal fossil records in these areas show conclusively that many of these dry zones once enjoyed wetter climates and different vegetation [249].

Of all the deserts in the world, the Sahara is by far the largest, incorporating almost 10 million square kilometres. Since the 1960s, geological, climatological and environmental research has accumulated a vast amount of data (some 2000 radiocarbon dates of palaeo-environmental proxy samples) which indicate that an abrupt weather-shift to a drier climate affected large parts of northern Africa at around 2300±200 BC [282, 199].

The moist period of the Mid-Holocene (the so-called Neolithic Wet Phase), which had a beneficial effect on the natural and cultural environment of the Sahara for almost 2000 years, came to a sudden end with the onset of 'the severe Post-Neolithic Arid Phase which struck the central Sahara between c. 2500 and 1000 BC' [174]. This ecological punctuation, however, was not limited to the north of the African continent. At exactly the same time, the mid-Holocene Kabangian wet phase of central Africa was replaced by a sudden dry spell lasting from about 4000 to 2000 BP [63].

With the onset of a sudden climatic regression at c. 4000 BP, large parts of habitable areas of northern Africa turned into desert and became part of a greatly-enlarged Sahara. This downturn triggered the collapse of a number of North African cultures, such as the Ténéré culture, and led to momentous mass migrations and the subsequent foundation of new settlements and cultures in regions more favourable for human habitation [282].

The analysis of sand dune movements and dynamics in the Sahara desert during the Holocene shows a marked phase of dune-building around 4000 BP [266]. During the Neolithic period, in particular between 6500 and 4000 BP, the Sahara was populated by a variety of herding groups and tribes who lived amidst agricultural landscapes and savanna-like ecosystems. After more than 5000 years of existence and development, these Saharan cultures vanished at around 4000 BP, when an episode of increased desiccation turned most of the Sahara into a region unsuitable for human habitation [253, 254]. During their research on the environmental history of predynastic settlements of Hierakonpolis, Egypt, Hoffman and Mills [115] found additional evidence for the termination of a moist Neolithic interval at c. 2500 BC.

> 'Recently, Ritchie [221] has provided evidence from pollen studies from a site on the High Plains of Algeria showing that a relatively more humid type of vegetation, consisting dominantly of pine and oak woodlands, prevailed before 4000 years ago and that the present-day steppe vegetation followed some time after that date. Similarly, pollen studies from buried lake muds in north-west Sudan, in the hyperarid part of the Eastern Sahara, show that a relatively deep stratified lake occurred surrounded by tropical savanna woodland vegetation during the early to mid-Holocene. [...] The lake dried out completely by 4500 BP and was covered subsequently by aeolian sediments (Ritchie et al. 1985). The desert scrub-grassland also disappeared at the same time from all habitats except at oases and wadis' [249].

To conclude, we can observe that the climatic shift to drier conditions in particular zones of the world hastened the process of aridification at the end of the postglacial climatic optimum.

11. Extinction of species

Ever since Luis Alvarez and his colleagues hypothesised that the demise of the dinosaurs some 65 million years ago was triggered by the impact of a giant asteroid, research regarding the link between cosmic catastrophes and mass extinction has mushroomed.

It is widely assumed that the last in a series of mass extinction episodes occurred at the Pleistocene/Holocene boundary c. 8000 BC. Only few scholars are aware of the conspicuous fact that 4000 years after the end of the Ice Age, yet another extinction episode, albeit of limited extent, led to the demise of a variety of species.

During the transition from the mid to late Holocene, abrupt climatic changes resulted in the modification of key environments and have even led to the extinction of a number of species. While the exact mechanisms of this rapid transformation and the role of natural factors remain uncertain, it has become evident that several species which were present up until c. 2300 BC, became extinct at the mid/late Holocene transition.

One of the most extraordinary discoveries of a case of mid/late Holocene extinction was detected when, during the early 1990s, Russian scientists published the results of their research on mammoth finds on the Wrangel Island, north of the East-Siberian coast [279, 280]. It came as a total surprise when the Russian scientists produced twenty radiocarbon dates of mammoth teeth and tusk from the Wrangel Island which ranged from 6200 to 2200 BC [160]. The results of this research show that mammoths inhabited this part of the world more than 6000 years after the estimated extinction of the species on the European and American continent. Yet even more startling is the fact that the woolly mammoth died out at exactly the same time as the great ecological and cultural disaster which is the main subject of this paper. It is, therefore, not far-fetched to suggest that the reason for the demise of the mammoth is correlated in time and cause with the catastrophe which led to the collapse of mankinds first civilisations.

Rock art from the western and central Sahara which depicts animals and dates from the Neolithic period [36, 254], together with the discovery of animal skeletons all over the Sahara [134], documents in great detail that previous to 4000 BP there was enough surface water for animals to pass the terrain that is now desert and for at least some human occupation [138]. Analysis of relict aquatic fauna of the surface waters of Meidob, Sudan and Wadi Howar, Eastern Sahara, revealed that fish existed in these waters before 4000 BP but became extinct at around this time [137, 60].

Fossils of many extinct giant lemurs have recently been discovered in a system of caves in the Ankarana Massif of northern Madagascar [247]. Radiocarbon dates of bones of the giant lemur *Megaladapis* confirm that this species, in contrast to previous estimates, did not become extinct at the end of the Pleistocene but was present in northern Madagascar until the mid Holocene. What is particularly interesting about these research findings is the fact that the lemur *Hapalemur simus* was present during the mid-Holocene and that the extinction of this and other species, such as *Babakotia radofilai,* appears to have occurred after c. 2400 BC. Simons *et al.* [247] therefore conclude: 'New data from the Ankarana and other sites on the island add to the consensus that major biotic changes occurred on Madagascar in the late Holocene.'

12. Volcanic activity

In 1993, *Science* published the results of archaeological research on the Habur Plains of North-Eastern Syria which was conducted by Harvey Weiss, Marie-Agnès Courty et al [294]. Their findings suggested that a severe drought disaster befell ancient Mesopotamia around 2200 BC and which continued for almost 300 years. According to the authors, it was this natural catastrophe rather than political reasons which resulted in the large-scale abandonment of Early Bronze Age settlements and the subsequent collapse of the Akkadian Empire. The conspicuous soil and ash evidence from Tell Leilan and other sites appeared to suggest a volcanic trigger for this abrupt climatic downturn. However, the lack of a marked tephra layer, the absence of a acidity signal in Greenland ice-cores at c. 2200 BC and doubts that a volcano could perturb the climate over such a large area for 300 years led the team of researchers to question the volcanic origin of this disaster [294].

In 1994, at an international meeting under the auspices of the NATO Scientific and Environmental Affairs Division, more than forty researchers compared the Syrian findings with evidence of synchronous environmental punctuations from many other parts of the world [52]. By now it had dawned on some researchers that the abrupt climate change which had hit Early Bronze Age Mesopotamia most probably coincided with a global environmental disaster [291, 293].

The hypothesis of a global climatic downturn at c. 2200 BC raised two issues: Given the nature and extent of such an event, it should be clearly detectable in tree-ring records dating from this period; and if the event was volcanic in origin, it should have left a signal in the Greenland ice-cores.

With regard to both the dendrochronological and the ice-core evidence, which are the most reliable indicators of abrupt climate change, the answer is rather unambiguous:

There is neither tree-ring nor ice-core evidence for a *significant* climatic deterioration at c. 2200 BC. Yet a distinct narrow tree-ring signal has been detected in Irish and English bog oaks at 2345 BC [20, 21]. This event has been linked to the volcanic eruption of the Icelandic volcano Hekla-4.

Recent findings have substantiated this idea. Investigations in the Over Wyre region of Lancashire, England yielded a discrete layer of microscopic tephra, geo-chemically assigned to the Icelandic Hekla-4 eruption of 2350 BC. The tephra was recorded in peat which formed during a period of freshwater flooding [295]. Two layers of volcanic tephra deposited around 4000 BP were also detected in western Ireland [62]. At exactly the same time, a complex layer of acidish ash, most probably from the Hekla 4 ash-fall, was deposited in Shetland where Bennett *et al.* [28] also found evidence for major synchronous landscape and climate changes.

Despite all the Hekla-4 evidence discovered in the British Isles, it would appear more than questionable whether this regional event could have had global impact, particularly given that it did not leave an acidity signal in the Greenland ice-cores. Nevertheless, there is evidence for volcanic activity in other parts of the globe.

The volcanic sediments of the Toro-Ankole volcanic field represent geological evidence for volcanism in the Fort Portal area of Eastern Africa, dated to c. 4100 BP by Osmaston [187].

In the Wharau Road Swamp in Northland, New Zealand, organic sediment began to accumulate around 2300 BC as a result of damming by a basaltic lava flow from the nearby Mount Te Puke which erupted at around that time [65].

13. Inundations and flood disasters

We now turn our attention to geological evidence detected at around 2350 BC which seems in many ways unrelated to and indeed incompatible with climatic change. In other words, here we are dealing with proxy data which are unlikely to be the result of climate change and which needs to be explained by alternative triggers.

There are hundreds of written records from around the world concerning the submergence of land by floods, tsunamis and inundations during the last 5000 years. Of all the myths and legends of mankind, none has generated more interest and stimulated more research than the story of the great flood .

Whilst many geological and climatic events are known to trigger widespread inundation, it is interesting to note that one of the oldest traditions links the Biblical flood to the impact of cosmic debris. According to the Babylonian Talmud (ca. 2nd century CE), the biblical flood was caused by two celestial objects which punctuated the earth: 'When God decided to bring about the Flood, he took two stars from Khima and threw them onto the earth and thus brought about the Flood' (Tract. Brakhot 59).

Recent research has confirmed that the impact of an asteroid in one of the oceans can indeed produce considerable flood damage by means of destructive tidal waves. Such impacts can cause widespread destruction over many thousands of kilometres of coast. An impact by an asteroid 300 400 metres in diameter anywhere in the Atlantic would cause rampant flood disasters and would devastate the coasts on both sides of the ocean by tsunamis over 100 metres high [114].

Here, I will focus on recent excavations which have revealed evidence for significant floods at around 2300 BC.

In many parts of the world, records of such flood disasters are not restricted to ancient traditions and documents but have been detected by archaeological and geological research. Particularly well attested is the flood catastrophe which destroyed many Neolithic villages along the Yellow Sea coast in China [43]. Until c. 2300 BC, most of these settlements (which include Guojiacun, Baigangzi, Dapanjia, Guandimiaopu, Tasitun, Qinghuagong, Wujiacun, Liqiangtun, Dantuozi, Dgang, Houwa and others [307] of the Liaodong peninsula in north-eastern China (close to the Korean border) were located near to the sea coast. Many other sites, which were part of the splendid Longshan culture, were situated on similarly low terraces of comparatively flat fields further inland. At around 2300 BC, most of these coast-line settlements were submerged by the sea. This inundation, which coincided with the collapse of the Longshan culture, invaded 15km into the land and led to the destruction and abandonment of many other sites along the coast [43]. New villages, established shortly after this disaster, were consequently founded on much higher terraces. Later sites of four to three thousand years ago, that is, late Neolithic to early Bronze Age, are mostly located on comparatively higher terraces or even several hundred meters higher on the top of hills [307]. At roughly the same time (c. 2300 2000 BC), extraordinary flood disasters are reported from many other parts of China [190].

Extraordinary floods also devastated a number of settlements and cities of the Indus Valley civilisation during the late 3rd millennium. In one of the major centres of this high culture, Lothal, five major flood layers sealed the cultural remains of consecutive settlements, the first of which is dated by Seth [240] to c. 2350 BC.

In 1867 the largest recorded flood disaster occurred on the Nepean River, Australia. When Saynor and Erskine [231] studied sediment cores along the river, they were able to identify high-level flood deposits dating to the 19th century inundation. While analysing the sediment cores, the researchers also found evidence for another, even more devastating palaeoflood. 'A radiocarbon date of 3756±72 years BP was obtained by tandem accelerator mass spectrometry on small fragments of charcoal contained in the high-level Slackwater deposits (SWDs). Therefore, at least one palaeo-flood larger than any historic flood occurred during the late Holocene. SWDs indicated that the largest palaeoflood had a peak height at least 8.0 m higher than the 1867 flood' [231].

Archaeological and stratigraphical research along Mobile Bay, Alabama, the Mississippi Delta, south-west Florida, the St. Vincent Island and in the lower Apalachicola River region of north-west Florida exhibits almost identical results. The excavations of sites near Mobile Bay show that these early settlements were inundated by the sea at c. 4000 BP [116]. This episode of submergence of Mobile Bay coincided with a marked migration of and major inundation along the Mississippi Delta [131] and along the Apalachicola River Delta in north-west Florida.

The evolution of the Connecticut River estuary during the Holocene was punctuated several times by episodes of submergence of the coastline and the sediment-trapping efficiency of the estuary and tidal river system. One of the most notable episodes of submergence, detected by Patton and Horne [193], began about 4000 BP in the southern estuary. Recent research has detected geological evidence for a major palaeo-flood event which affected the Black Bear Creek basin in north-central Oklahoma and which has been dated to c. 2000 BC [167].

Major environmental and climatic changes very often lead to

significant transformations in the nature and sedimentation of soils and the consequent use of land. Deciphering the effects of such environmental shifts on land-use requires detailed information regarding the processes underlying soil and landscape evolution. A study of the relation between landscape evolution, shallow groundwater hydrology and water table fluctuations of a glaciated landscape in Iowa, USA analysed data from 128 soil cores collected from a 32-ha site. Stratigraphical evidence found by Steinwald and Fenton [259] suggests that the upper stratum was produced by slope alluvium which was deposited after c. 4300 BP.

Evidence for a significant alluviation event has been uncovered during excavations in the Argive plain and dated to the end of the Early Helladic II period (ca. 2200/2300 BC) [276]. Floodplain deposits of up to 3 metres thick and stretching up to 15 kilometres inland have been detected between Tiryns and Mycenae. It is evident that these deposits represent a unique disaster because since that time no other alluviation event has affected the Argive plain. While the eastern coastlands of Attica near Athens show particular stability for most parts of the Holocene, typified by well-developed soil formation, a distinct alluvial event was detected which dated to the Early Bronze Age [29]. Similar regional geomorphology was discovered for the Southern Argolid peninsula. Here a major alluvial event, (the so-called Pikrodafni alluvium), dated by the content of Late Helladic II pottery sherds to ca. 2200/2300 B.C. was recognised in late Quaternary deposits [276].

14. Widespread seismic activity

No discussion of late 3rd millennium BC environmental change and civilisation collapse is complete without a survey of those Early Bronze Age settlements which fell into ruin due to seismic activity. Earthquake damage has been a frequent explanation for settlement destruction, destruction layers or abandonment of sites. Despite the often fragmentary and sometimes inconclusive character of archaeological findings, it is feasible to recognise specific features of earthquake effects in archaeological works and thereby to distinguish these peculiarities and those from other natural or anthropogenic effects of site damage and destruction [260].

Interestingly, of all the various factors and data scrutinised for clues related to the environmental and social upheavals at the end of the Early Bronze Age, seismic and tectonic activity is the subject matter most neglected by scholars. Yet most archaeologists are only too aware that Claude Schaeffer's voluminous *Stratigraphie Comparée et Chronologie de l'Asie Occidentale* is teeming with archaeological evidence for extensive earthquake damage detected in Bronze Age settlements throughout the Near and Middle East [232].

Claude Schaeffer, the 20th century's most eminent French archaeologist, was the first researcher to present evidence for widespread seismic catastrophes in large parts of Asia minor and the Levant at around 2300 BC. Based on a comparative study of destruction layers in more than 40 sites, he ordered and classified earthquake horizons as synchronous and interrelated benchmarks in archaeological stratigraphy and chronology. Evidence for major earthquake damage in Early Bronze Age strata had been detected in many Anatolian and Near Eastern settlements, such as Troy, Alaca Hüyük, Boghazköy, Alishar, Tarsos, Ugarit, Byblos, Qalaat, Hama, Megiddo, Tell Hesi, Beit Mirsim, Beth Shan, Tell Brak and Chagar Bazar [85].

Most scholars, however, have refrained from taking Schaeffer's main research-findings into consideration. The recent and most comprehensive textbook on 3rd millennium BC civilisation collapse fails to mention his research altogether [52]. One looks in vain for *any* reference to his theory of Early Bronze Age collapse. This reticence is even more remarkable in view of the fact that Schaeffer was also, to my knowledge, the first archaeologist to claim that a distinct shift in climate was synchronous with civilisation collapse.

> 'Au Caucase et dans certains régions de l'Europe protohistorique, des changements de climat semblent, à cette période, avoir amené des transformations dans loccupation et léconomie du pays' [232, pp. 555/556].

There is, of course, a plausible reason for the selective perception of the available evidence by most scholars. Of all the proxy data available in the vast literature on late 3rd millennium environmental change, the evidence for seismic and tectonic activity appears to be the most inconclusive. Moreover, it seems impossible to integrate widespread seismic activity into any scenario of abrupt climate change. Indeed, most researchers would agree that there seems to be no natural phenomenon capable of triggering abrupt climate change *and* extensive seismic activity at the same time. Yet almost 35 years ago, René Gallant [84] focused on this apparent anomaly in Schaeffer's work and suggested that cosmic impacts would easily account for the synchronicity of climate change and seismic activity.

> 'C.F.A. Schaeffer, Gallant wrote, presumes that the catastrophes which caused the end of civilisations in Eurasia originated in devastating earthquakes which shook the world. He mentions that many sites show that the destructions have been contemporary with climatic changes, which seem to have brought about transformations in the occupation and the economy of the country. Schaeffer does not seem to have been struck by the connection between those two important contemporaneous events: earthquakes and climatic changes. As we have previously seen, those two events are closely connected with cosmic catastrophism: both are inevitable results of huge meteoric impacts' [84, pp., 214/15].

Some twenty years later, Mandelkehr (1988) published further evidence which confirmed a pattern of geological perturbations at c. 2300 BC. According to his findings, the most significant aspect of the geological evidence is the crustal movements that apparently began at about the same time around 2300 BC at many regions of the Earth (Mandelkehr 1988, 11). In this section, I will present the most important palaeo-seismic data collated by Mandelkehr and will add to that more recent research findings.

During the 1950s and 60s, a number of geological investigations along the east coast of North America found that the Atlantic coast experienced differential warping at c. 2300 BC (Mandelkehr 1988). While the regional uplifts in question appear to be a gradual geological upward movement, the onset of these discontinuities at this particular time is certainly conspicuous.

Strong earthquakes leave geological evidence in form of surface faulting, folding or (in the case of earthquakes in urban areas) in form of site destruction. These features are often preserved in the archaeological record and can therefore be detected by archaeo-seismological research (Stiros and Jones 1996). The presence of historical seismicity and tectonic features are recognised as the most reliable criteria for identifying earthquake damage. However, lack of such unambiguous characteristics cannot be considered conclusive evidence for a lack of earthquake destruction (Vittori et al. 1991).

Geological evidence for ancient earthquakes may be preserved in the archaeological record and, therefore, palaeoseismological studies may detect and date them. Historical seismicity and the presence of well-developed tectonic geomorphic features are often recognised as reliable criteria for identifying active faulting. However, extra-terrestrial impacts are capable of triggering seismic activity in geological areas which are normally not prone to tectonic activity.

Crustal deformation at around the same time has also been detected along the Gulf of Mexico. At Vera Cruz, Mexico, the so-called Palo Hueco Culture which had existed here since c. 4000 BC 'was found to be sealed throughout its extent by a culturally sterile sand cap after 2400 BC, apparently corresponding to a major inundation shortly after that time [...]. The conclusion reached by the investigator was that the Mexican coast experienced a long term subsidence at the same time that the Florida region apparently underwent uplift' [157].

Nunn's research [182] has established that the islands of the south-central Lau Ridge of the South Pacific experienced a number of seismic events, one of which appears to coincide with the mid/late Holocene transition. On the island of Lekeba, four coseismic-uplift events during the Holocene were identified, the latest of which occurred c. 3800 BP.

Research by Worsley *et al.* [302] on the Holocene vegetational evolution around Lake Hendry in northern Quebec, Canada, shows that the development of a thin peat cover coincided with the emergence of land around 4300 BP, following an episode of isostatic uplift. Analysis of stratigraphic sections of the Montague Harbour in British Columbia has documented a distinct episode of sea-level transgression caused by tectonic subsidence of the area c 3600 BP [213]. At about the same time, tectonic sea-level changes have also occurred in other parts of western British Columbia. Evidence for tectonically-induced events at c. 3600 BP in southern areas of Vancouver Island, Canada, have recently been reported by Matthews and Clague [162]. These punctuations are thought to have caused coastal subsidence and are consistent with large-scale (magnitude M>8) earthquakes [213].

A particularly manifest episode of seismic activity at the mid/late Holocene boundary has been detected by Forman *et al.* [79] in sediment cores of the Wasatch fault zone, in north central Utah. The stratigraphy in two trenches excavated across fault scarps is characterised by a distinct earthquake stratum which buried a soil developed on a middle Holocene layer. The researchers have dated this palaeo-earthquake at the mid/late Holocene transition tentatively to c. 4300 BP.

Other evidence for extensive tectonic activity during the late 3rd millennium BC comes from the Oquirrh fault zone, a normal fault that bounds the east side of Tooele Valley in central Utha. A recent study by Olig *et al.* [185] on scarp morphology suggests that the most recent surface-faulting earthquake in this region occurred during the 3rd millennium BC, an event tentatively dated to c. 4400 BP. The researchers also found that, at two sites at the Big Canyon and Pole Canyon, trenches exposed faulted Lake Bonneville sediments and thick wedges of fault-scarp derived colluvium associated with this event. Since a bulk sediment sample from fluvial deposits which buried the fault scarp of the last major earthquake yielded a radiocarbon age estimate of c. 4340 BP, it is highly probable that this massive earthquake coincided with similarly high levels of seismic activity in other parts of the world.

15. Conclusions

At some time around 2300 BC, a large number of major civilisations collapsed. At the same time, there is widespread evidence for abrupt and widespread environmental catastrophes. Sudden sea-level changes, catastrophic inundations, widespread seismic activity and earthquake damage, changes in glacial features and a signal for an abrupt climatic downturn have been detected at c. 2350. A survey of some 500 excavation reports, research papers and scientific abstract on late 3rd millennium BC civilisation collapse and environmental change show a distinct pattern of environmental and social upheaval at this time.

A large number of sites and cities of the first uraban civilisations in Asia, Africa and Europe appear to have collapsed at around the same time. The proxy data detected in the marine, terrestrial, biological climatological and archaeological records point to sudden environmental, climatic and social upheavals which appear to coincide with simultaneous sea- and lake-level changes, increased levels of seismic activity and widespread flood disasters. The main problem in interconnecting this vast amount of data is the application of incoherent and imprecise dating methods in different areas of geological and climatological research. It is hypothesised that the globally detected evidence for the sudden environmental and social upheavals at the start of the late Holocene are interconnected and that chronological deviations are primarily due to imprecise dating methods. Neither a seismic nor a climatic explanation for the natural and social punctuations appear capable to account for these events since they are evidenced by a great diversity of ecological alterations and an enormous variety of damage features. The punctuation of extra-terrestrial debris, on the other hand, can have catastrophic effects on the ecological system in a variety of patterns which match the main features documented in this comparative analysis.

References

1. Adams, R.M. (1995) Contexts of civilizational collapse: A Mesopotamian View, in: Yoffee, N. and Cowgill, G.L., eds., *The Collapse of Ancient States and Civilizations*, 3rd ed., Tucson and London: University of Arizona Press, pp. 20-43.
2. Alexandre, A., Meunier, J.D., Lezine, A.M., Vincens, A., Schwartz, D. (1997) Phytoliths: indicators of grassland dynamics during the late Holocene in inter-tropical Africa, *Palaeogeography Palaeoclimatology Palaeoecology* 136(1-4), pp.213 - 229.
3. Allen, G.P., Posamentier, H.W. (1993) Sequence stratigraphy and facies model of an incised valley fill the Gironde Estuary, France, *Journal of Sedimentary Petrology* 63(3), 378-391.
4. Almquist Jacobson, H. (1995) Lake level fluctuations at Ljustjarnen, central Sweden and their implications for the Holocene climate of Scandinavia, *Palaeogeography Palaeoclimatology Palaeoecology* 118(3-4), 269-290.
5. Ambraseys, N.N. (1971) Value of historical records of earthquakes, *Nature* 232, 375.
6. Ambraseys, N.N. (1978) Studies in historical seismicity and tectonics, in: Brice, W.C., ed., *The Environmental History of the Near and Middle East since the Last Ice Age*, New York: Academic Press, pp. 185-210.
7. Amiran, D.H.K. (1991) The climate of the ancient Near East: The early third millennium BC in the northern Negev of Israel, *Erdkunde* 45, 153-161.
8. Anderson, D.E., Binney, H.A. and Smith, M.A. (1998) Evidence for abrupt climate change in northern Scotland between 3900 and 3500 calendar years BP, *Holocene* 8(1), 97-103.
9. Anderson, T.W. and Lewis, C.F.M. (1985) Postglacial water-level history of the Lake Ontario Basin, in: Karrow, P.F. and Calkin, P.E., eds., *Quaternary Evolution of the Great Lakes*. Geological Association of Canada Special Paper 30, 231-251.
10. Ariztegui, D., Farrimond, P., Mckenzie, J.A. (1996) Compositional variations in sedimentary lacustrine organic-matter and their implications for high alpine Holocene environmental changes - Lake St Moritz, Switzerland, *Organic Geochemistry* 24(4), 453-461.
11. Asher, D.J. and Clube, S.V.M. (1993) An extraterrestrial influence during the current glacial-interglacial, *Quarterly Journal of the Royal Astronomical Society* 34, 481-511.

12. Asher, D.J., Clube, S.V.M., Napier, W.M. and Steel, D.I. (1994) Coherent Catastrophism, *Vistas in Astronomy* 38, 1-27.
13. Bailey, M.E., Clube, S. V. M. and Napier, W.M. (1990) *The Origin of Comets*, Oxford: Pergamon Press.
14. Bailey, M.E. (1995) Recent Results in Cometary Astronomy: Implications for the Ancient Sky, *Vistas in Astronomy* 39, 647-671.
15. Bailey, M.E., Clube, S.V.M., Hahn, G., Napier, W.M. and Valsecci, G.B. (1994) Hazards due to giant comets: climate and short-term catastrophism, in: Gehrels, T., ed., *Hazards due to Comets and Asteroids*, Tucson: University of Arizona Press, pp. 479-536.
16. Baillie, M.G.L. (1982) *Tree-Ring Dating and Archaeology*, London: Croom-Helm.
17. Baillie, M.G.L. (1991a) Marking in Marker Dates: towards an archaeology with historical precision, *World Archaeology* 23, 233-243.
18. Baillie, M.G.L. (1991b) Suck-in and smear: two related chronological problems for the 1990's, *Journal of Theoretical Archaeology* 2.
19. Baillie, M.G.L. (1994) Dendrochronology raises questions about the nature of the AD 536 dust-veil event, *The Holocene* 4(2), 212-217.
20. Baillie, M.G.L. (1995a) Dendrochronology and the Chronology of the Irish Bronze Age, in: Waddell, J. and Shee-Twohig, E., eds., *Ireland in the Bronze Age*, Stationery Office, Dublin, 30-37.
21. Baillie, M.G.L. (1995b) *A Slice through Time: Dendrochronology and Precision Dating*, London: Batsford.
22. Baillie, M.G.L. (1998) Book Review: Third millennium BC Climate Change and Old World Collapse, *Journal of Archaeological Science* 25(2), 185-186.
23. Barth, M.C. and Titus, J.G., eds., *Greenhouse Effect and Sea Level Rise*, New York: Van Nostrand Reinhold.
24. Bartlein, P.J., Webb, T., and Fleri, E.C. (1984) Holocene climate ge in the northern Midwest: pollen-derived estimates, *Quaternary Research* 22, 361-374.
25. Behling, H. and Hooghiemstr, H. (1998) Late Quaternary palaeoecology and palaeoclimatology from pollen records of the savannas of the llanos Orientales in Columbia, *Palaeogeography, Palaeoclimatology, Palaeoecology* 139, 251-267.
26. Bell, B. (1971) The Dark Ages in Ancient History, *American Journal of Archaeology* 75, 1-26.
27. Ben-Menahem, A. (1981) Variations of slip and creep along the Levant rift over the past 4500 years, *Tectonophysics* 80, 183-197.
28. Bennett, K.D., Boreham, S., Sharp, M.J., Switsur, V.R. (1992) Holocene history of environment, vegetation and human settlement on Catta Ness, Lunnasting, Shetland, *Journal of Ecology* 80, 241-273.
29. Bintliff, J. (1992) Erosion in the Mediterranean Lands: A Reconsideration of Pattern, Process and Methodology, in: Bell, M. and Boardman, J., eds., *Past and Present Soil Erosion: Archaeological and Geographical Perspectives.* Oxbox Monographs 22 (Oxford), pp. 125-143.
30. Bhiry, N. and Filion, L. (1996) Characterization of the soil hydromorphic conditions in a paludified dunefiled during the mid-Holocene hemlock decline near Quebec City, Quebec, *Quaternary Research* 46(3), 281-297.
31. Boulding, K.E. (1992) Punctuationism in Societal Evolution, in: Somit, A. and Peterson, S.A., eds, *The Dynamics of Evolution: The Punctuated Equilibrium Debate in the Natural and Social Sciences*, Ithaca & London: Cornell University Press.
32. Brice, W.C., ed., *The Environmental History of the Near and Middle East since the Last Ice Age*, London/New York/San Francisco: Academic Press.
33. Broshi, M. and Gophna, R. (1986) The settlements and population of Palestine during the Early Bronze Age II-III, *Bulletin of the American School of Oriental Research*, 41-54
34. Brugam, R.B. and Johnson, S.M. (1997) Holocene lake-level rise in the Upper Peninsula of Michigan, USA, as indicated by peatland growth, *Holocene* 7(3), 355-359.
35. Bryson, R. and Swain, A.M. (1981) Holocene variations of monsoon rainfall in Rajasthan,*Quarterly Research* 16, 135-145
36. Butzer, K.W. (1958) Studien zum vor- und frühgeschichtlichen Landschaftswandel in der Sahara, *Akademie der Wissenschaften. Math-naturw. Klasse* 1, pp. 67-88.
37. Butzer, K.W. (1984) Long-term Nile-Flood Variations and Political Discontinuities in Pharaonic Egypt, in: Clark, J.D. and Brandt, S.A., eds., *From Hunters to Farmers: The Causes of Food Production in Africa,* Berkeley: University of California Press.
38. Butzer, K.W. (1995) Environmental Change in the Near East and Human Impact on the Land, in: Sasson, J. M. (1995) *Civilizations of the Ancient Near East,* vol. 1, New York: Simon & Schuster, pp. 123-151.
39. Butzer, K.W. (1997) Sociopolitical Discontinuity in the Near East c. 2200 BCE: Scenarios from Palestine and Egypt, in: Dalfes, H.N., Kukla, G., and Weiss, H., eds., *Third Millennium BC Climate Change and Old World Collapse* [NATO ASI series: Ser. I, Global environmental Change; Vol 49], Berlin, Heidelberg, New York: Springer, pp. 245-298.
40. Casanova, J. and Hillairemarcel, C. (1992) Late Holocene hydrological history of Lake Tanganyika, East Africa, from isotopic data on fossil stromatolites, *Palaeogeography Palaeoclimatology Palaeoecology* 91(1-2), pp. 35-48.
41. Caseldine, C., Hatton, J., Huber, U., Chiverrell, R., Woolley, N. (1998) Asessing the impact of volcanic activity on mid-Holocene climate in Ireland: the need for replicate data, *Holocene* 8(1), 105-111.
42. Caskey, John L. (1960) The early Helladic Period in the Argolid, *Hesperia* 29, 285-303.
43. Cheng-hui, C., Yen-chou, L., and Cheng-the, S. (1978) Development of natural environment in the southern part of Liaoning province during the last 10,000 years, *Scientia Sinica* 21(4), 516-532.
44. Clube, S.V.M. (1994a), Hazards from Space: comets in history and science, in: Glen, W., ed., *The Mass-Extinction Debates: How Science Works in a Crisis*, Standford: University of Stanford Press, pp. 159-169.
45. Clube, S.V.M. (1994b) Revelation and catastrophe during the Christian Era: a basis for historical interpolation and future extrapolation, in: *Chronology and Catastrophism Review Special Issue: Proc. 1993 Cambridge Conference,* Manchester: SIS, pp. 66-73.
46. Clube, S.V.M. (1996) Evolution, Punctuational Crises and the Threat to Civilization, in: *Earth, Moon and Planets* 72(1-3), 433-440.
47. Clube, S.V.M. and Napier, W.M. (1982) *The Cosmic Serpent*, London: Faber and Faber.
48. Clube, S.V.M. and Napier, W.M. (1990) *The Cosmic Winter*, Oxford: Blackwell.
49. Colquhoun, D. J. and Brooks, M. J. (1987) New Evidence for Eustatic Components in Late Holocene Sea Level, in: Rampino, M. R., Sanders, J. E., Newman, W. S., Königsson, L. K., eds., (1987) *Climate: History, Periodicity, and Predictability* (New York: Van Nostrand Reinhold), pp. 143-156.
50. Cowgill, G.L. (1995) Onward and upward with collapse, in: Yoffee, N. and Cowgill, G.L., eds., *The Collapse of Ancient States and Civilizations*, 3rd ed., Tucson and London: University of Arizona Press, pp. 244-276.
51. Crossland, R.A. and Birchall, A., eds. (1978) *Bronze Age Migrations in the Aegean* (Noyes).
52. Dalfes, H.N., Kukla, G., and Weiss, H., eds., 1997, *Third Millennium BC Climate Change and Old World Collapse* [NATO ASI series: Ser. I, Global environmental Change; Vol 49], Berlin, Heidelberg, New York: Springer.
53. Dever, W.G. (1980) New Vistas on the EB IV (MB I) Horizon in Syria-Palestine, *Bulletin of the American School of Oriental Research* 237, 35-64.
54. Dever, W.G. (1986) Pastoralism and the End of the Urban Early Bronze Age in Palestine, in: Bar-Yosef, O. and Khazanov, A., eds., *Pastoralism in the Levant: Archaeological Materials in Anthropological Perspectives*, Madison, Wisconsin: Prehistory Press, pp. 83-92.
55. Dever, W.G. (1989) The Collapse of the Urban Early Bronze Age in Palestine: Toward a Systematic Analysis, in: Miroschedji, P., *L'urbanisation de la Palestine à lâge du Bronze ancien*, BAR International Series 527, pp. 247-255.
56. De Pratter, C. and Howard, J.D. (1981) Evidence for a sea-level lowstand between 4500 and 2400 years b.p. on the southeast coast of the united States, *Journal of Sedimentary Petrology* 51, 1287-1296.
57. Dominquez, J.M.L., Martin, C. and Bittencourt, A.C.S.P. (1987) Sea-level history and Quarternary evolution of river mouth-asociated beach-ridge plains along the east-southeast

Brazilian coast a summary, in: Nummendahl, D., Pilkey, O.H. and Howard, J.D., eds., *Sea-Level Fluctuations and Coastal Evolution. SEPM Special Publication* 41, 115-128.

58. Donoghue, J.F. and White, N.M. (1995) Late Holocene sea-level change and delta migration, Apalachicola River region, Northwest Florida, USA, *Journal of Coastal Research* 11(3), 651-63.
59. Dubertret, L. and Dunand, M. (1954-55) Les gisements ossiferes de Khibet El-Umbachi et de Hebariye (Safa), *Les Annales Archaeologiques de Syrie* 4/5, 8-76.
60. Dumont, H.I. and el Moghraby, A. I. (1993) Holocene evolution of climate and environment, and stone city ruins in Northn Darfur, Sudan: is there a relationship?, in: Krzyzaniak, L., Kobusiewicz, M. and Alexander, J. (1993) *Environmental Change and Human Culture in the Nile Basin and Northern Africa until the Second Millennium B.C.* Poznan: Poznan Archaeological Museum, pp. 381-397.
61. Duthie, H.C., Yang, J.R., Edwards, T.W.D., Wolfe, B.B. and Warner, B.G. (1996) Hamilton Harbour, Ontario: 8300 years of limnological and environmental change inferred from microfossil and isotopic analyses, *Journal of Paleolimnology* 15, 79-97.
62. Dwyer, R.B. and Mitchell, F.J.G. (1997) Investigation of the environmental impact of remote volcanic activity of North Mayo, Ireland, during the mid-Holocene, *Holocene* 7(1), 113-118.
63. Eggert, M.K.H. (1993) Central Africa and the archaeology of the equatorial rainforest: reflections on some major topics. In: Shaw, T., Sinclair, P., Andah, B. and Okpoko, A., eds., *The Archaeology of Africa: Food, Metals and Towns.* London/New York: Routledge, pp. 289-29.
64. Einsele, G., Herm, D. and Schwarz, H.U. (1974) Sea Level Fluctuation During the past 6000 yr at the Coast of Mauritania, *Quaternary Research* 4, 282-289.
65. Elliot, M.B., Striewski, B., Flenly, J.R., Kirkman, J.H., Sutton, D.G. (1997) A 4300 year palynological and sedimentological record of environmental change and human impact from Wharau Road Swamp, Northland, New Zealand, *Journal of the Royal Society of New Zealand* 27(4), 401-418.
66. Ely, L.L. (1997) Response of extreme floods in the southwestern United States to climatic variations in the late Holocene, *Geomorphology* 19(3-4), pp.175-201.
67. Esse, D.L. (1989) Secondary state formation and collapse in Early Bronze Age Palestine, in: Miroschedji, P., *L'urbanisation de la Palestine à lâge du Bronze ancien*, BAR International Series 527, pp. 81-116.
68. Evans, S.P. and Trevisan, M. (1995) A soil water-balance bucket model for paleoclimatic purposes. Model application to late Holocene climate, *Ecological Modelling* 82(2), 131-138.
69. Fairbridge, R.W. (1961a) Eustatic changes in sea-level, in: Ahrens, L. H., Press, F., Rankama, and Runcorn, S. K. (eds.) *Physics and Chemistry of the Earth.* Vol. 4, London: Pergamon Press, pp. 99-185.
70. Fairbridge, R.W. (1961b) Convergence of evidence on climatic change and ice ages, *New York Academy of Science Annals* 95(1), 542-579.
71. Fairbridge, R.W. (1962) World sea-level and climatic changes, *Quaternaria* 6, 111-134.
72. Fairbridge, R.W. (1984) The Nile floods as a global climatic/solar proxy, in: Mörner, N.A. and Karlen, W. (eds.) *Climatic Changes on a Yearly to Millennial Basis.* Dordrecht: Reidel, pp. 181-190.
73. Fairbridge, R.W. (1987) The Spectra of Sea Level in a Holocene Time Frame, in: Rampino, M.R., Sanders, J.E., Newman, W.S., Königsson, L.K., eds., (1987) *Climate: History, Periodicity, and Predictability* (New York: Van Nostrand Reinhold), pp. 127-142.
74. Fan, H., Gasse, F., Huc, A., Li, Y.F., Sifedine, A., Soulimarsche, I. (1996) Holocene environonmental changes in Bangong Co Basin (Western Tibet), *Palaeogeography Palaeoclimatology Palaeoecology* 120(1-2), 65-78.
75. Farnell, L.R. 1919, The value and methods of mythological study, in: *Proceedings of the British Academy* XX, 30-55.
76. Finkelstein, I. and Gophna, R. (1993) Settlement, demographic, and economic patterns in the Highlands of Palestine in the Chalcolithic and Early Bronze periods and the beginning of Urbanisation, *Bulletin of the American School of Oriental Research* 289, 1-22.
77. Fletcher, C.H., Vanpelt, J.E., Brush, G.S., Sherman, J. (1993) Tidal Wetland record of Holocene movements and climate history, *Palaeogeography Palaeoclimatology Palaeoecology* 102(3-4), 177-213.
78. Forman, S.L., Oglesby, R., Markgraf, V., Stafford, T. (1995) Paleoclimatic significance of late Quaternary eolian deposition on the piedmont and High Plains, central United States, *Global and Planetary Change* 11(1-2), 35-55.
79. Forman, S.L., Nelson, A.R., McCalpin, J.P. (1991) Thermoluminescence dating of fault-scarp-derived colluvium - deciphering the timing of paleoearthquakes on the Weber segment of the Wasatch fault zone, north central Utah, *Journal of Geophysical Research-Solid Earth And Planets* 96(1), pp.595-605.
80. Founder, S. and Abrahamsen, N. (1988) Palynology in a polar desert, eastern North Greenland, *Boreas* 17, 195-207.
81. Fredlung, G.G. (1996) Late Quaternary Geomorphic history of Lower highland Creek, Wind Cave National Park, South-Dakota, *Physical Geography* 17(5), 446-464.
82. Frenzel, B. (1966) Climatic Change in the Atlantic/sub-Boreal transition on the Northern Hemisphere: botanical evidence, in: *World Climate from 8000 to 0 B.C.* Royal Meteorological Society, pp. 99-123.
83. Fritts, H.C., Lofgren, G.R. and Gordon, G.A. (1981) Reconstructing seasonal to century time scale variations in climate from tree-ring evidence, in: Wigley, T.M.L., Ingram, M.J. and Framer, G., eds., *Climate and History: Studies in past climates and their impact on Man.* Cambridge: Cambridge University Press, pp. 139-161.
84. Gallant, R. (1964) *Bombarded Earth*, London: John Baker
85. Gammon, G. (1980)., Bronze Age Destructions in the Near East, *SIS Review* IV(4), 104-108.
86. Gajewski, K, Payette, S., Ritchie, J.C. (1993) Holocene vegetation history at the boreal-forest - shrub-tundra transition in north-western Quebec, *Journal of Ecology* 81(3), 433-443.
87. Garces, B.V.L., Kelts, K., Ito, E (1995) Oxygen and carbon-isotope trends and sedimentological evolution of a meromictic and saline lacustrine system - the Holocene medicine lake basin, North-American great-plains, USA, *Palaeogeography Palaeoclimatology Palaeoecology* 117 (3-4), 253-278.
88. Garralla, S. and Gajewski, K. (1992) Holocene vegetation history of the Boreal forest near Chibougamau, central Quebec, *Canadian Journal of Botany* 70(7), pp.1364-1368.
89. Gasse, F. and Street, F.A. (1978) Late Quaternary lake-level fluctuations and environments of the Northern Rift Valley and Afar Region (Ethiopia and Djibouti, *Palaeogeography Palaeoclimatology Palaeoecology* 24, 279-325.
90. Gasse, F., Rognon, R. and Street, F.A. (1980) Quaternary history and the Afar and Ethiopian Rift lakes. In: Williams, M. A. J. and Faure, H., eds., *The Sahara and the Nile*, Rotterdam: Balkema, pp. 361-400.
91. Gasse, F. and Van Campo, E. (1994) Abrupt post-glacial climate events in West Asia and North Africa monsoon domains, *Earth and Planetary Science Letters* 126, 435-456
92. Gehrels, T, ed. (1994) *Hazards due to Comets and Asteroids*, Tucson: University of Arizona Press.
93. Gfeller, C., Oeschger, H, and Schwarz, U. (1961) Bern radiocarbon dates II, *Radiocarbon* 3, 15-25.
94. Giresse, P. (1989) Geodynamique des lignes de rivage Quaternaires du continent Africain et applications, in: Scott, D.B., Pirazzoli, P.A. and Honing, C.A., eds., *Late Quaternary sea-level correlation and application.* NATO ASI Series, series C, 256, pp. 131-152.
95. Glen, W., ed. (1994) *The Mass-Extinction Debates: How Science Works in a Crisis*, Standford: University of Stanford Press.
96. Goldthwait, R.P. (1966) Evidence from Alaskan glaciers of major climatic changes, in: *World Climate from 8000 to 0 B.C.*, Royal Meteorological Society, pp. 40-53.
97. Goodfriend, G.A., Stanley, D.J. (1996) Reworking and discontinuities in Holocene sedimentation in the Nile Delta - documentation from amino-acid racemization and stable isotopes in mollusk shells, *Marine Geology* 129 (3-4), 271-283.
98. Goodwin, I.D. (1996) A mid to late Holocene re-advance of the Law Dome ice margin, *Antarctic Science* 8(4), 395-406.
99. Gordillo, S., Coronato, A.M.J., Rabassa, J.O. (1993) Late Quaternary evolution of a sub-antartic paleofjord, Tierra del Fuego, *Quaternary Science Review* 12(10), 889-897.

100.Goy, J.L., Zazo, C., Dabrio, C.J., Lario, J., Borja, F., Sierro, F.J., and Flores, J.A. (1996) Global and Regional Factors controlling Changes of Coastlines in Southern Iberia (Spain) during the Holocene,*Quaternary Sciences Reviews* 15,773-780.

101.Grieve, R.A.F. and E.M. Shoemaker (1994) The record of past impacts on Earth, in: Gehrels, T., ed., *Hazards due to Comets and Asteroids*, Tucson: University of Arizona Press, pp. 417-462.

102.Grosjean, M., ValeroGarces, B.L., Geyh, M.A., Messerli, B., Schotterer, U., Schreier, H., Kelts, K. (1997) Mid- and late-Holocene limnogeology of Laguna del Negro Francisco, northern Chile, and its palaeoclimatic implications, *Holocene* 7(2), 151-159.

103.Grove, A.T. (1993) Africas Climate in the Holocene. In: Shaw, T., Sinclair, P., Andah, B. and Okpoko, A., eds. *The Archaeology of Africa: Food, Metals and Towns,* London/New York: Routledge, pp. 32-42.

104.Gudmundsson, H.J. (1997) A review of the Holocene environmental history of Iceland, *Quaternary Science Reviews* 16(1), 81-92.

105.Gunn, J.D. (1997) A Framework for the Middle-Late Holocene Transition: Astronomical and Geophysical Conditions, *Southeastern Archaeologists* 16, 134-151.

106.Gupta, S.K. (1975) Silting of the Rann of Kutch During Holocene, *Indian Journal of Earth Sciences* 2(2), 163-175.

107.Hamilton, A.C. (1982) *Environmental History of East Africa*, London/New York: Academic Press.

108.Harrison, S.P., Prentice, I.C., Guiot, J. (1993) Climatic controls on Holocene lake-level changes in Europe, *Climate Dynamics* 8, 189-200.

109.Hebda, R.J. (1995) British Columbia vegetation and climate history with focus on 6 ka BP, *Geographie Physique et Quaternaire* 49(1), 55-79.

110.Healy, T., Erlenkeuser, H., Winn, K, and Werner, F. (1987) Theoretical post-glacial sedimentation rates for a semi-enclosed sea; example for the Kieler Bucht, Western Baltic, *Boreas* 16, 207-213.

111.Hecky, R.E. and Degens, E.T. (1973) *Late Pleistocene-Holocene chemical stratigraphy and paleolimnology of the Rift Valley lakes of Central Africa*, Tech. Rept. Woods Hole Oceanogr. Inst. Massachusetts.

112.Heinsohn, G. (1990) Destruction layers in archaeological sites: the stratigraphy of Armageddon, in: Zysman, M. and Whelton, C., eds., *Catastrophism 2000*, Toronto: Heretic Press, pp. 213-247.

113.Heusser, C.J. (1961) Some comparisons between climatic changes in north-western North America and Patagonia, *New York Academy of Science Annals* 95, 642-657.

114.Hills, J.G. & Mader, C.L. (1997) Tsunami produced by the impacts of small asteroids. Annals of the New York Academy of Sciences 822, 381-394.

115.Hoffman, M.A. and Mills, J.O. (1993) Problems of assessing environmental impact on the Predynastic settlements of Hierakonpolis, in: Krzyzaniak, L., Kobusiewicz, M. and Alexander, J. (1993) *Environmental Change and Human Culture in the Nile Basin and Northern Africa until the Second Millennium B.C.*, Poznan: Poznan Archaeological Museum, pp. 359-370.

116.Holmes, N.H. and Trickerey, B.E. (1974) Late Holocene sea level oscillations in Mobile Bay, *American Antiquity* 39, 122-124.

117.Hood, M.S.F. (1974) Northern Penetration of Greece at the End of the Early Helladic Period and Contemporary Baltic Chronology, in: R. A. Crossland and A. Birchall (eds.) *Bronze Age Migrations in the Aegean* (Noyes).

118.Hoyle, F. (1993) *The Origin of the Universe and the Origin of Religion*, Rhode Island & London: Wakefield.

119.Hoyle, F. and Wickramasinghe (1996) *Our Place in The Cosmos*, London: Phoenix.

120.Hugget, R. (1998) *Catastrophism: Asteroid, Comets, and other dynamic Events in Earth History*, London: Verso.

121.Ingram, M.J., Farmer,G. and Wigley, T.M.L. (1981) Past climate and their impact on Man: a review, in: Wigley, T.M.L., Ingram, M.J. and Framer, G., eds., *Climate and History: Studies in past climates and their impact on Man*, Cambridge: Cambridge University Press.

122.Islebe, G.A., Hooghiemstra, H. Vantveer, R. (1996) Holocene vegetation and water-level history in 2 bogs of the Cordillera-de-Talamanca, Costa-Rica, *Vegetatio* 124(2), 155-171.

123.Issar, A.S., Govrin, Y., Geyh, M.A., Wakshal, E., and Wolf, M. (1992) Climate changes during the Upper Holocene in Israel, *Israel Journal of Earth Science* 40, 219-223.

124.Jacobsen, T. and Adams, R.M. (1958) Salt and Silt in Ancient Mesopotamian Agriculture, *Science* 128, 1251-1253.

125.Jerardino, A. (1993) Mid Holocene to late Holocene sea-level fluctuations the archaeological evidence at Tortoise Cave, south-western Cape, South Africa, *South African Journal of Science* 89(10), 481-488.

126.Jian, Z.M., Li, B.H., Pflaumann, U., Wang, P.X. (1996) Late Holocene cooling event in the western Pacific, *Science in China Series D-Earth Science* 39(5), 543-550.

127.Jolly, D. Bonnefille, R., Roux, M. (1994) Numerical interpretation of a high-resolution Holocene pollen record from Burundi, *Palaeogeography Palaeoclimatology Palaeoecology* 109(2-4), 357-370.

128.Kajale M.D., and Deotare B.C. (1997): Late Quaternary environmental studies on salt lakes in western Rajasthan, India: a summarised view, *Journal of Quaternary Science* 12(5), 405-412.

129.Karlstrom, T. (1964) *Quaternary Geology of the Kenai Lowland and Glacial History of the Cook Inlet Region, Alaska*. U.S. Geological Survey, Special Paper 443.

130.Kaufman, H. (1995) The Collapse of ancient states and civilizations as an organizational problem, in: Yoffee, N. and Cowgill, G.L., eds., *The Collapse of Ancient States and Civilizations*, 3rd ed., Tucson and London: University of Arizona Press, pp. 219-235.

131.Kolb, C.R. and Van Lopik, J.R. (1966) depositional environments of the Mississippi River Deltaic Plain, South-eastern Louisiana, in: Shirley, M.L. and Ragsdale, J.A., eds., *Deltas in their Geological Framework*, Houston.

132.Korhola, A.A. and Tikkanen, M.J. (1997) Evidence for a more recent occurrence of water chestnut (*Trapa natans* L.) in Finland and its palaeoenvironmental implications, *The Holocene* 7(1), 39-44.

133.Korotky, A.M., Razjigaeva, N.G., Ganzey, L.A., Volkov, V.G., Grebennikova, T.A., Bazarova, V.B., and Kovalukh, N.N. (1995) Late Pleistocene-Holocene coastal develoment of islands off Vietnam, *Journal of Southeast Asian Earth Sciences* 11(4), 301-308.

134.Kröpelin, S. (1993) The Gilf Kebir anad Lower Wadi Howar: contrasting Early and Mid-Holocene environments in the Eastern Sahara, in: Krzyzaniak, L., Kobusiewicz, M. and Alexander, J. (1993) *Environmental Change and Human Culture in the Nile Basin and Northern Africa until the Second Millennium B.C.*, Poznan: Poznan Archaeological Museum, pp. 249-258.

135.Krzyzaniak, L., Kobusiewicz, M. and Alexander, J. (1993) *Environmental Change and Human Culture in the Nile Basin and Northern Africa until the Second Millennium B.C.*, Poznan: Poznan Archaeological Museum.

136.Kullmann, L. (1987) Sequences of Holocene forest history in the Scandes, inferred from megafossil Pinus sylvestris, *Boreas* 16, 21-26.

137.Kuper, R. (1988) Neuere Forschungen zur Besiedlungsgeschichte der Ostsahara, *Archäologisches Korrespondenzblatt* 18(2), 127-142.

138.Lamb, H.H. (1974) Climate, vegetation and forest limits in early civilized times. *Phil. Trans. R. Soc. London* A., 276, 195-230.

139.Lamb, H. H. (1982) *Climate History and the Modern World*, London & New York: Methuden.

140.Lambeck, K (1993) Glacial rebound and sea-level change: an example of a relationship between mantle and surface processes, *Tectonophysics* 223, 15-37.

141.Larsen, C.E. (1985) Lake level, uplift, and outlet incision, the Nipissing and Algoma Great Lakes. In: Karrow, P.F. and Calkin, P.E., eds., *Quaternary Evolution of the Great Lakes*, Geological Association of Canada, Special Paper 30, 63-78.

142.Lees, B.G. (1992) Geomorphological evidence for late Holocene climatic change in northern Australia, *Australian Geography* 23, 1-10.

143.Lees, B.G., Stanner, J., Price, D.M., Yanchou, L. (1995) Thermoluminescence dating of dune podzols at Cape Arnhem, northern Australia, *Marine Geology* 129 (1-2), 63-75.

144.Lees, B.G., Lu Yanchou, and Head, J (1990) Climatic implications of chenier dates in northern Australia, *Radiocarbon* 29(3), 311- 317.

145.Lewis, C.F.M. (1969) Late Quaternary history of the lake levels in the Huron and Erie basins, *12th Conference on Great Lakes Research, Ann Arbor, MI. Proceedings of the International Association for Great Lakes Research*, pp.147-148.

146.Lillios, K.T. (1993) Regional settlement abandonment at the end of the Copper Age in the lowlands of west-central Portugal, in: Cameron, C.M. and Tomka, S.A., eds., *Abandonment of settlements and Regions: Ethnoarchaeological and archaeological approaches*, Cambridge: Cambridge University Press.

147.Lillios, K.T. (1997) The Third Millennium BC in Iberia: Chronometric Evidence for Settlement Histories and Sociocultural Change, in: Dalfes, H.N., Kukla, G., and Weiss, H., eds. (1997) *Third Millennium BC Climate Change and Old World Collapse* [NATA ASI Series I: Global Environmental Change, vol. 49], Heidelberg: Springer, pp. 173-192.

148.Macumber, P.C. and Head, M.J. (1991) Implications of the Wadi al-Hammeh sequences for the terminal drying of Lisan Lake, Jordan. Palaeogeography, Palaeoclimatology, Palaeoecology 84, 163-173.

149.Mäkelä, E. (1998) The Holocene history of *Betula* at Lake Lilompolo, Inari Lapland, northeastern Finland, *Holocene* 8(1), 55-67.

150.Maenza-Gmelch, T.E. (1997) Holocene vegetation, climate, and fire history of the Hudson Highlands, southeastern New York, USA, *The Holocene* 7(1), 25-37.

151.Magny, M. (1992) Holocene lake-level fluctuations in Jura and the northern subalpine ranges, France; regional pattern and climatic implications, *Boreas* 21, 319-334.

152.Magny, M. (1993) Solar influences on Holocene climatic changes illustrated by correlations between past lake-level fluctuations and the atmospheric ^{14}C record, *Quarternary Research* 40, 1-9.

153.Magny, M. (1995) Successive ocean and solar forcing indicated by Younger Dryas and early Holocene climatic oscillations in the Jura, *Quarternary Research* 43, 279-285.

154.Maley, J. (1993) The climatic and vegetational history of the equatorial regions of Africa during the upper Quaternary, in: Shaw, T., Sinclair, P., Andah, B. and Okpoko, A., eds. *The Archaeology of Africa: Food, Metals and Towns*. London/New York: Routledge, 43-52.

155.Mandelkehr, M.M. (1980/81) An Integrated Model for an Earthwide Event at 2300 BC. The Archaeological Evidence, *SIS Review* V(3), 77-95.

156.Mandelkehr, M.M. (1987) An Integrated Model for an Earthwide Event at 2300 BC. Climatology, *Chronology and Catastrophism Review* IX, 34-44.

157.Mandelkehr, M.M. (1988) An Integrated Model for an Earthwide Event at 2300 BC. Geological Evidence, *SIS Review* X, 11-22.

158.Manning, S.W. (1997) Cultural Change in the Aegean c. 2200 BC, in: Dalfes, H.N., Kukla, G., and Weiss, H., eds. (1997) *Third Millennium BC Climate Change and Old World Collapse* [NATA ASI Series I: Global Environmental Change, vol. 49], Heidelberg: Springer, pp. 149-172.

159.Martin, C.W. (1992) The response of fluvial systems to climate change - an example from the central Great Plains, *Physical Geography* 13(2), pp.101-114.

160.Martin, P.S. and Stuart, A.J. (1995) Mammoth extinction: Two continents and Wrangel Island, *Radiocarbon* 37(1), 7-10.

161.Masse, W.B. (1995) The Celestial Basis of Civilization, *Vistas in Astronomy* 39(4), 463-478.

162.Matthewes, R.W. and Clague, J.J. (1994) Detection of large prehistoric Earthquakes in the Pacific Northwest by microfossil analysis, *Science* 264, 688-691.

163.Matthews, J.A., Dahl, S.O., Berrisford, M.S., Nesje, A., Dresser, P.Q., Dumayne Peaty_L , (1997) A preliminary history of Holocene colluvial (debris-flow) activity, Leirdalen, Jotunheimen, Norway, *Journal of Quaternary Science* 12 (2), 117-129.

164.May, D.W. (1992) Late Holocene valley bottom aggradation and erosion in the south Loup River valley, Nebraska, *Physical Geography* 13(2), pp.115-132.

165.McCarthy, F.M.G., Collins, E.S., McAndrews, J.H., Kerr, H.A., Scott, D.B., and Medioli, F.S. (1995) A Comparison of postglacial Arcellacean (Thecamobeian) and pollen succession in Atlantic Canada, *Journal of Paleontology* 69(5), 980-993.

166.McDonald, M.M.A. (1993) Cultural adaptations in Dakhleh Oasis, Egypt, in the Early to Mid-Holocene, in: Krzyzaniak, L., Kobusiewicz, M. and Alexander, J. (1993) *Environmental Change and Human Culture in the Nile Basin and Northern Africa until the Second Millennium B.C.* Poznan: Poznan Archaeological Museum, 199-209.

167.McQueen, K.C., Vitek, J.D., and Carter, B.J. (1993) Paleoflood analysis of an alluvial channel in the south-central Great Plains: Black Bear Creek, Oklahoma, *Geomorphology* 8, 131-146.

168.Meadows, M.E., Baxter, A.J., and Parkington, J. (1996) Late Holocene environments at Verlorenvlei, Western Cape Province, South Africa, *Quaternary International* 33, 81-95.

169.Mellaart, J. (1968) The End of the Early Bronze Age in Anatolia and the Aegean, *American Journal of Archaeology* 62, 9-33.

170.Mercer, J.H. (1966) Glacier resurgence at the Atlantic/sub-Boreal transition, *Quarterly Journal of the Royal Meteorological Society*, (?) 528-534.

171.Miller, D.E., Yates, R.J., Parkington, J.E, Vogel, J.C. (1993) Radiocarbon-dated evidence relating to a mid Holocene relative high sea-level on the south-western Cape coast, South Africa, *South African Journal of Science* 89 (1), 35-44.

172.Mörner, N.A. (1976) Eustatic changes during the last 8000 years in view of radiocarbon calibration and new information from the Kattegatt region and other northwestern European areas, *Palaeogeography, Palaeoclimatology, Palaeoecology* 19,63-85.

173.Mörner, N.A. and Karlen, W., eds., (1984) *Climatic Changes on a Yearly to Millennial Basis*. Dordrecht: Reide.

174.Muzzolini, A. (1993) The Emergence of a food-producing economy in the Sahara, in: Shaw, T., Sinclair, P., Andah, B. and Okpoko, A., eds. *The Archaeology of Africa: Food, Metals and Towns,* London/New York: Routledge, pp. 227-239.

175.Neev, D. , Bakler, N., and Emery, K.O. (1987) *Mediterranean Coast of Israel and Sinai*. New York: Tayler and Francis.

176.Neev, D. and Emery, K.O. (1995) *The Destruction of Sodom, Gomorrah and Jericho*, New York, Oxford: Oxford University Press.

177.Nelson, S.M., ed. (1995), *The Archaeology of Northeast China*, London/New York: Routledge.

178.Nesje, A., Asbjørn R.A.A., Kvame, M., Sønstegaard, E. (1994) *Norsk Geologisk Tidskrift*, 74(2), 71-76.

179.Newman, W.S. (1968) Coastal stability, in: R. W. Fairbridge, (ed.), *Encyclopedia of Geomorphology*, (New York), pp. 150-156.

180.Nesje, A., Aa, A.R., Kvamme, M., Sonstegaard, E. (1994) A record of late Holocene avalanche activity in Frudalen, Sogndalsdalen, western Norway, *Norsk Geologisk Tidsskrift* 74(2), 71-76.

181.Nissenbaum, A. (1994) Sodom, Gomorrah and the other Cities of the Plain a Climatic Perspective, *Climatic Change* 26(1), 435-446.

182.Nunn, P.D. (1995) Holocene tectonic histories for 5 islands in the south-central Lau Group, South Pacific, *Holocene* 5(2), 160-171.

183.Nunn, P.D. (1997) Environmental changes on Pacific islands: controversy, certainty and conjecture, *Journal of Quaternary Science* 12(5), 443-450.

184.O'Hara, S.L., Metcalfe, S.E., Streetperrott, F.A. (1994) On the arid margin - the relationship between climate, humans and the environment - a review of evidence from the highlands of central Mexico, *Chemosphere* 29(5), 965-981.

185.Olig, S.S., Lund, W.R., Black, B.D. (1994) Large mid Holocene and late Pleistocene earthquakes on the Oquirrh fault zone, Utah, *Geomorphology* 10(1-4), 285-315.

186.Osborne, P.L., Humphreys, G.S., Polunin, N.V.C (1993) Sediment deposition and late Holocene environmental change in a tropical lowland basin - Waigani Lake, Papua New Guinea, *Journal of Biogeography* 20(6), 599-613.

187.Ostmaston, H.A. (1967) Plant fossils in volcanic tuffs near Ruwenzori, *Palaeoecology of Africa* 2, 25-26.

188.Owen, R.B., Barthelme, J.W., Renaut, R.W. and Vincens, A. (1982) Paleoclimnology and archaeology of holocene deposits north-east of Lake Turkana, Kenya, *Nature* 298, 523-529.

189.Pachur, H.J. and Braun, G. (1980) The Paleoclimate of the central Sahara, Libya and the Libyan Desert, *Paleoecology of Africa* 12, 351-363.

190.Pang, K.D. (1987) Extraordinary Floods in Early Chinese History and Their Absolute Dates, *Journal of Hydrology* 96, 139-155.

191.Pang, K.D. and Srivastava, S.K. (1988) Climatic Impacts and Volcanic Eruptions: Interferences from Ice Core, Tree Ring and Historical Data, *Eos* 69(44), 1062.

192.Pang, K.D., Keston, R. and Srivastava, S.K. (1989) Climatic and Hydrologic Extremes in Early Chinese History: Possible Causes and Dates, *Eos* 24, 1095.

193.Patton, P.S. and Horne, G.S. (1992) Response of the Connecticut River estuary to late Holocene sea-level rise, *Geomorphology* 5(3-5), pp.391-417.

194.Pazdur, A., Fontugne, M.R., Goslar, T., and Pazdur, M. (1995) Late-glacial and Holocene water-level changes of the Gosciaz lake, central Poland, derived from carbon isotope studies or laminated sediment, *Quarternary Science Review* 14, 125-135.

195.Pearson, R. (1977) Peleoenvironment and human settlement in Japan and Korea, Science 197, 1239-1243.

196.Peiser, B.J. (1994) Catastrophism and Anthropology, in: *Chronology and Catastrophism Review Special Issue: Proc. 1993 Cambridge Conference*, Manchester: SIS, pp. 130-134.

197.Peiser, B.J. (1996) Catastrophe Games. Playful re-enactment of traumatic events, in: Pfister, E., Niewerth, T. and Steins, G., eds., *Games of the World between Tradition and Modernity*, Sankt Augustin: Akademia, pp. 112-124.

198.Peiser, B.J. (1997) Comets and Disaster in the Bronze Age, *British Archaeology* 30, 6-7.

199.Petite-Maire, N., Beufort, L. and Page, N. (1997) Holocene Climate Change and Man in the Present Day Sahara Desert, in: Dalfes, H.N., Kukla, G., and Weiss, H., eds. (1997) *Third Millennium BC Climate Change and Old World Collapse* [NATA ASI Series I: Global Environmental Change, vol. 49], Heidelberg: Springer, pp. 297-308.

200.Petty, W.H., Delcourt, P.A., and Delcourt, H.R. (1996) Holocene lake-level fluctuations and beach-ridge development along the northern shore of Lake Michigan, USA, *Journal of Paleolimnology* 15, 147-169.

201.Pirazzoli, P.A. (1996) *Sea-Level Changes: The Last 20.000 Years* (Chichester, New York, Brisbane, Toronto, Singapore).

202.Porter, S.C. (1981) Glaciological evidence of Holocene climatic change, in: Wigley, T.M.L., Ingram, M.J. and Framer, G., eds. (1981) *Climate and History: Studies in past climates and their impact on Man*. Cambridge: Cambridge University Press, pp. 82-110.

203.Possehl, G.L. (1997) Climate and the Eclipse of the Ancient Cities of the Indus, in: Dalfes, H.N., Kukla, G., and Weiss, H., eds. (1997) *Third Millennium BC Climate Change and Old World Collapse* [NATA ASI Series I: Global Environmental Change, vol. 49], Heidelberg: Springer, pp. 193-244.

204.Raikes, R.L. (1964) The End of the Ancient Cities of the Indus, *American Anthropologist* 66, 284-299.

205.Raikes, R.L. (1979) The Mohenjo-doro floods: the dabete continues, in: Taddei, M., ed., *South Asian Archaeology 1977*, Naples: Instituto Universario Orientale, Seminario di Studi Asiatici, Series Minor VI, pp. 561-566.

206.Raikes, R.L. and Dales, G.F. (1977) The Mohenjo-doro floods reconsidered, *Journal of the Palaeontological Society of India* 20, 251-260.

207.Rampino, M.R., Sanders, J.E., Newman, W.S., Königsson, L.K., eds., (1987) *Climate: History, Periodicity, and Predictability* (New York).

208.Rasch, M. and Jensen, J.F. (1997) Ancient Eskimo dwelling sites and Holocene relative sea-level changes in southern Disko Bugt, central West Greenland, *Polar Research* 16(2), 101-115.

209.Rast, W.E. (1987) Bad edh-Dhra and the origin of the Sodom saga, in: Pardee, D.L., Toombs, L., and Johnson, G. (eds.), *Archaeology and Biblical Interpretation*, Atlanta: John Knox Press, pp. 185-201.

210.Rast, W.E. and Schaub, R.T. (1982) The Dead Sea expedition, Bad Edh-Dhra and Numeira, *Newsletter of the American Schools of Oriental Research* 4, 4-12.

211.Rathburn, A.E., Pichon, J.J., Ayress, M.A., DeDeckker, P. (1997) Microfossil and stable-isotope evidence for changes in Late Holocene palaeoproductivity and palaeoceanographic conditions in the Prydz Bay region of Antarctica, *Paleogeography Palaeoclimatology Palaeoecology* 131(3-4), 485-510.

212.Regnauld, H., Jennings, S., Delaney, C. and Lemasson, L. (1996) Holocene sea-level variations and geomporphological response: an example from northern Brittany (France), *Quaternary Science Reviews* 15, 781-787.

213.Reinhardt, E.G., Easton, N.A. and Patterson, R.T. (1996) Foraminiferal evidence of late Holocene sea-level change and Amerindian site distribution at Montague Harbour, British Columbia, *Géographie physique et Quaternaire* 50(1), 35-47

214.Renfrew, C. (1972) *The Emergence of Civilisation. The Cyclades and the Aegean in the Third Millennium B. C.* London.

215.Renfrew, C. and Whitehouse, R. (1979) *Problems in European Prehistory*. Cambridge: Cambridge University Press.

216.Rhodes, T.E., Gasse, F., Lin, R.F., Fontes, J.C., Wei, K.G. Bertrand, P., Gibert, E., Melieres, F., Tucholka, P., Wang, Z.X., Cheng, Z.Y. (1996) A late Pleistocene-Holocene lacustrine record from Lake Manas, Zunggar (Northern Xinjiang, Western China), *Palaeogeography Palaeoclimatology Palaeoecology* 120(1-2), 105-121.

217.Richard, S. (1980) Toward a Consensus of Opinion on the End of the Early Bronze Age in Palestine-Transjordan, *Bulletin of the American School of Oriental Studies* 237, 5-34.

218.Richard, S. (1987) The Early Bronze Age in Palestine: The Rise and Collapse of Urbanism, *Biblical Archaeology* 50, 22-43.

219.Ricketts, R.D. and Johnson, T.C. (1996) Climate change in the Turkana basin as deduced from a 4000 year long delta O^{18} record. *Earth and Planetary Science Letters* 142, 7-17.

220.Ridley, A.P. and Seeley, M.W. (1979) Evidence for recent coastal uplift near Al Jubail, Saudi Arabia, *Tectonophysics* 52, 319-327.

221.Ritchie, J.C. (1984) Analyse Pollinique de sédiments Holocène supérieurs des hauts plateux du Maghreb oriental, *Pollen et Spores* 26, 489-496

222.Ritchie, J.C., Eyles, C.H., Haynes, C.V. (1985) Sediment and pollen evidence for an early to mid-Holocene humid period in the eastern Sahara, *Nature* 314, 352-355.

223.Rodriguez-Ramirez, A., Rodriguez-Vidal, J., Caceres, L., Clemente, L., Belluomini, G., Manfra, L., Improta, S. and De Andres, J.R. (1996) Recent Coastal Evolution of the Donana National Park, *Quaternary Sciences Reviews* 15, 803-809.

224.Rognon, P. and Williams, M.A.J. (1977) Late Quaternary climatic changes in Australia and North Africa: a preliminary interpretation, *Palaeogeography Palaeoclimatology Palaeoecology* 21, 285-327.

225.Rose, J., Whiteman, C.A., Lee, J., Branch, N.P., Harness, D.D., and Walden, J. (1997) Mid- and late-Holocene vegetation, surface weathering and glaciation, Fjallsjokull, southeast Iceland, *Holocene* 7(4), 457-471.

226.Rosen, A.M. (1989) Environmental change at the end of the Early Bronze Age in Palestine, in: Miroschedji, P. de, ed., *L'urbanisation de la Palestine à lâge du Bronze ancien*, pp. 247-255.

227.Rosen, A.M. (1997) Environmental change and human adaptation failure at the end of the Early Bronze Age in the southern Levant, in: Dalfes, H.N., Kukla, G., and Weiss, H., eds., *Third Millennium BC Climate Change and Old World Collapse* [NATO ASI series: Ser. I, Global environmental Change; Vol 49], Berlin, Heidelberg, New York: Springer, pp. 25-38.

228.Rosen, A.M. (1995) The Social Response to Environmental Change in Early Bronze Age Canaan, *Journal of Anthropological Archaeology* 14, 26-44.

229.Rousseau, D.D., Limondin, N., Magnin, F. and Puissegur, J.J. (1994) Temperature oscillations over the last 10,000 years in western Europe from terrestrial mollusc assemblages, *Boreas* 23, 66-73.

230.Sampson, A. (1996) Cases of Earthquakes at Mycenaean and Pre-Mycenaean Thebes, in: Stiros, S. and Jones, R.E., eds., *Archaeoseismology*, Athens: Institute of Geology and Mineral Exploration & The British School of Athens, pp. 113-117.

231.Saynor, M.J. and Erskine, W.D. (1993) Characteristics and implications of high-level slackwater deposits in the Fairlight Gorge, Nepean River, Australia, *Australian Journal of Marine and Freshwater Research* 44(5), 735-747.

232.Schaeffer, C. F. A. (1948) *Stratigraphie Comparée et Chronologie de L'Asie Occidentale: IIIe et IIe millénaires* (London: Oxford University Press).

233.Schultz, P.H. and Beatty, J.K. (1992) Teardrops on the Pampas, *Sky & Telescope* 83(4), 387-392.

234.Schultz, P.H. and Lianza, R.E. (1992) Recent grazing impacts on the Earth recorded in the Rio Cuarto crater field, Argentina, *Nature* 355, 234-237.

235.Schwalb, A., Locke, S.M., and Dean, W.E. (1995) Ostracode delta[18] and delta ^{13}C evidence of Holocene environmental changes in the sediments of two Minnesota lakes, *Journal of Paleolimnology* 14, 281-296.

236.Scott, D.B., Brown, K., Collins, E.S., and Medioli, F.S. (1995a) A new sea-level curve from Nova Scotia: evidence for a rapid acceleration of sea-level rise in the late mid-Holocene, *Canadian Journal of Earth Science* 32, 2071-2080.

237.Scott, D.B., Gayes, P.T. and Collins, E.S. (1995b) Mid-Holocene precedent for a future rise in sea level along the Atlantic coast of North America, *Journal of Coastal Research* 11(3), 615-622.

238.Scott, D.B. and Collins, E.S. (1996) Late mid-Holocene sea-level oscillation: A possible cause, *Quaternary Science Reviews* 15 (8-9), 851-856.

239.Sea, D.S., Whitlock, C. (1995) Postglacial vegetation and climate of the Cascade Range, central Oregon, *Quaternary Research* 43(3), 370-381.

240.Seth, S.K. (1978) Desiccation of the Thar Desert, in: Brice, W.C., ed., *The Environmental History of the Near and Middle East since the Last Ice Age*, London/New York/San Francisco: Academic Press, 280-305.

241.Shaw, T., Sinclair, P., Andah, B. and Okpoko, A., eds. (1995) *The Archaeology of Africa: Food, Metals and Town*, London/ New York: Routledge.

242.Shaw, J. and Carter, R.W.G. (1994) Coastal peats from northwest Ireland: implications for lat-Holocene relative sea-level change and shoreline evolution, *Boreas* 23, 74-91.

243.Shennan, I., Innes, J.B., Long, A.J., Zong, Y.Q (1994) Late Devensian and Holocene relative sea-level changes at Loch Nan Eala, near Arisaig, northwest Scotland, *Journal Of Quaternary Science* 9(3), 261-283.

244.Short, S.K., Andrews, J.T., Williams, K.M., Weiner, N.J., Elias, S.A. (1994) Late Quaternary marine and terrestrial environments, north-western Baffin Island, north-west Territories, *Geographie Physique et Quaternaire* 48(1), 85-95.

245.Shulmeister, J. (1992) A Holocene record from lowland tropical Australia, *Holocene* 2, 107-116.

246.Shulmeister, J. and Lees, B.G. (1995) Pollen evidence from tropical Australia for the onset of an ENSO dominated climate at circa 4000 yrs B.P., *Holocene* 5, 10-18.

247.Simons, E.L., Burney, D.A., Chatrath, P.S., Godfrey, L.R., Jungers, W.L., Rakotosamimanana, B. (1995) Ams c-14 dates for extinct lemurs from caves in the Ankarana Massif, northern Madagascar, *Quaternary Research* 43(2), 249-254.

248.Singh, G. (1971) The Indus Valley Culture seen in the context of postglacial climatic and ecological studies in northwest India, *Archaeology and Physical Anthropology in Oceania* 6, 177-189.

249.Singh, G. (1988) History of aridland vegetation and climate: a global perspective, *Biological Reviews* 63, 159-195.

250.Singh, G., Joshi, R.D., and Singh, A.B. (1972) Stratigraphic and radiocarbon evidence for the age and development of three salt lake deposits in Rajasthan, India, *Quaternary Research* 2, 496-505.

251.Singh, G., Joshi, R.D., Chopra, S.K. and Singh, A.B. (1974) Late Quaternary history of vegetation and climate of the Rajasthan Desert, India, *Philosophical Transactions of the Royal Society of London, Biological Sciences* 267, 467-501.

252.Singh, I.B., Rajagopalan, G., Agarwal, K.K., Srivastava, P., Sharma, M., Sharma, S. (1997) Evidence of middle to late Holocene neotectonic activity in the Ganga Plain, *Current Science* 73(12), 1114-1117.

253.Smith, A. B. (1980) The Neolithic tradition in the Sahara, in: Williams, M. A. J. and Faure, H., eds., *The Sahara and the Nile*. Rotterdam: Balkema, pp. 451-465.

254.Smith, A. B. (1993) New Approaches to Saharan rock art of the Bovidian Period, in: Krzyzaniak, L., Kobusiewicz, M. and Alexander, J. (1993) *Environmental Change and Human Culture in the Nile Basin and Northern Africa until the Second Millennium B.C.* Poznan: Poznan Archaeological Museum, pp. 77-89.

255.Starkel, L. (1984) The reflection of abrupt climatic changes in the relief and sequence of continental deposits, in: Mörner, N.A. and Karlen, W., eds., (1984) *Climatic Changes on a Yearly to Millennial Basis*. Dordrecht: Reidel, pp. 135-146.

256.Steel, D.I., Asher, D.J., Napier, W.M., Clube, S.V.M. (1994) Are impacts correlated in time?, in: Gehrels, T., ed., *Hazards due to Comets and Asteroids*, Tucson: University of Arizona Press.

257.Steel, D.I. (1995) *Rogue Asteroids and Doomsday Comets: The Search for the Million Megaton Menace that threatens Life on Earth*, New York/ Chichester/ Brisbane/ Toronto/ Singapore: John Wiley & Sons.

258.Steel, D.I. (1996) The Limitations of NEO-Uniformitarianism, *Earth, Moon & Planets* 72, 279-292.

259.Steinwand, A.L. and Fenton, T.E. (1995) Landscape evolution and shallow groundwater hydrology of a till landscape in central Iowa, *Soil Science Society Of America Journal* 59(5), 1370-1377.

260.Stiros, S. and Jones, R.E., eds., (1996) *Archaeoseismology*, Athens: Institute of Geology and Mineral Exploration & The British School of Athens.

261.Stiros, S. (1996) Identification of earthquakes from archaeological data: methodology, Criteria and Limitations, in: Stiros, S. and Jones, R.E., eds., *Archaeoseismology*, Athens: Institute of Geology and Mineral Exploration & The British School of Athens, pp. 129-151.

262.Stuiver, M. and Becker, B. (1986) High-precision Decadal Calibration of the Radiocarbon Scale, AD 1950-2500 BC, *Radiocarbon* 28, 863-910.

263.Sutra, J.P., Bonnefille, R. and Fontugne, M. (1997) A new pollen diagram from the Sandynallah swamps (Nilgiri hills, South West India), *Geographie Physique et Quaternaire* 51(3), 415-426.

264.Svendsen, J.I. and Mangerud, J. (1997) Holocene glacial and climatic variations on Spitzbergen, Svalbard, *The Holocene* 7(1), 45-57.

265.Tainter, J.A. (1988) *The Collapse of Complex Societies*, Cambridge: Cambridge University Press.

266.Talbot, M. R. (1980) Environmental responses to climatic change in the West African Sahel over the past 20 000 years, in: Williams, M. A. J. and Faure, H., eds., *The Sahara and the Nile*. Rotterdam: Balkema, pp. 37-62.

267.Talbot, M. R., Livingstone, D. A., Palmer, P. G., Maley, J., Melack, J. M., Delibrias, G. and Gulliksen, S. (1984) Preliminary results from sediment cores from Lake Bosumtwi, Ghana. *Palaeoecology of Africa* 16, 173-192.

268.Temple, P. H. (1964) Evidence of lake-level changes from the northern shoreline of Lake Victoria, Uganda. In: Steel, R. W. and Prothero, R. M., eds., *Geographers and the Topics: Liverpool Essays*, London: Longman, pp. 31-56.

269.Temple, P. H. (1966) Lake Victoria levels, *Proceedings of the East African Academy* 2, 50-58.

270.Temple, P. H. (1967) Causes of intermittent decline of the level of Lake Victoria during the late Pleistocene and Holocene. In: *Liverpool Essays in Geography*, London: Longman, pp. 43-63.

271.Ters, M. (1987) Variations in Holocene Sea Level on the French Atlantic Coast and Their Climatic Significance, in: Rampino, M. R., Sanders, J. E., Newman, W. S., Königsson, L. K., eds. (1987) *Climate: History, Periodicity, and Predictability* (New York), pp. 204-237.

272.Toon, O.B., Zahnle, K., Turco, R.P. and Covey, C. (1994) Environmental perturbations caused by impacts, in: Gehrels, T., ed., *Hazards due to Comets and Asteroids*, Tucson: University of Arizona Press, pp. 791-826.

273.Tornqvist, T.E. (1994) Middle and late Holocene avulsion history of the river Rhine (Rhine-Meuse delta, Netherlands), *Geology* 22(8), 711-714.

274.Tolonen, K., Liukkonen, M., Harjula, R., and Pätilä, A. (1986) Acidification of small lakes in Finland documented by sedimentary diatom and chrysophycean remains, in: Smol, J.P., Battarbee, R.W., David, R.B., and Marilainen, J., eds., *Diatoms and Lake Acidity*, Dordrecht: Dr Junk, pp. 169-199.

275.Trump, D. H. (1980) *The Prehistory of the Mediterranean*, Yale University Press.

276.Van Andel, T. H., Runnels, C. N. and Pope, K. O. (1986) 5000 Years of land Use and Abuse in the Southern Argolid, Greece, *Hesperia* 55, 103-128.

277.Vanderhammen, T., Urrego, L.E., Espejo, N., Duivenvoorden, J.F., Lips, J.M. (1992) Late glacial and Holocene sedimentation and fluctuations of river water level in the Caqueta river area (Colombian Amazonia), *Journal of Quaternary Science* 7(1), 57-67.

278. Vance, R.E., Last, W.M. Smith, A.J. (1997): Hydrologic and climatic implications of a multidisciplinary study of late Holocene sediment from Kenosee Lake, southeastern Saskatchewan, Canada, *Journal of Paleolimnology* 18(4),365-393. 279. Vartanyan, S.L., Garutt. V.E. and Sher, A.V. (1993) Holocene dwarf mammoth from Wrangel Island in the Siberian Arctic, *Nature* 362, 337-340.

280. Vartanyan, S.L., Arslanov, K.A., Tertychnaya, T.V. and Chernov, S.B. (1995) Radiocarbon dating evidence for Mammoth on Wrangel Island, Artic Ocean, until 2000 BC, *Radiocarbon* 37(1), 1-6.

281. Velikovsky, I. (1973) *Worlds in Collision*, London: Victor Gollanz.

282. Vernet, R. (1993) Le Néolithique récent dans le sud-ouest du Sahara. In: Krzyzaniak, L., Kobusiewicz, M. and Alexander, J. (1993) *Environmental Change and Human Culture in the Nile Basin and Northern Africa until the Second Millennium B.C.* Poznan: Poznan Archaeological Museum, pp. 91-102.

283. Verschuur, G.L. (1996) *Impact! The Threat of Comets and Asteroids*, Oxford: Oxford University Press.

284. VillaMartinez, R. and Villagran, C. (1997) Vegetational history of the swamp forests of the central Chilean coast during the mid- and late Holocene, *Revista Chilena de Historia Natural* 70(3), 391-401.

285. Vincens, A., Buchet, G., Elenga, H., Fournier, M., Martin, L., Denamur, C., Schwartz, D., Servant, M, Wirrmann, D. (1994) Pollen evidence for major vegetation change around Lake Sinnda (Niari Valley, southern Congo) in relation to the late Holocene dry phase, *Comptes Rendus de l'Academie des Sciences*, Serie II 318(11/2), 1521-1526.

286. Vittori, E., Labini, S.S., Serva, L. (1991) Palaeoseismology - Review of the state-of-the-art, *Tectonophysics* 193(1-3), 9-32

287. Warren, P. (1980) Problems of Chronology in Crete and the Aegean in the Third and Earlier Second millennium B.C., *American Journal of Archaeology* 84, 498ff.

288. Watts, W.A., Allen, J.R.M., Huntley, B., and Fritz, S.C. 91996) Vegetation History and Climate of the last 15,000 Years at Laghi di Monticchio, Southern Italy, *Quaternary Science Reviews* 15, 113-132.

289. Wei, K.Y., Lee, M.Y., Duan, W., Chen, C., Wang, C.H. (1998) Palaeoceanographic change in the northeatern South China Sea during the last 15,000 years, *Journal of Quarternary Science*, 13(1), 55-64.

290. Weidik, A., Oerter, H., Reeh, N., Thomas, H.H., & Thorning, L. (1990) The recession of the Inland Ice margin during the Holocene climatic optimum in the Jakobshavn Osfjord area of West Greenland, *Palaeogeography, Palaeoclimatology, Palaeoecology* 82, 389-399.

291. Weiss, H. (1996a) Desert Storm, *The Sciences* 36(3), 30-36.

292. Weiss, H. (1996b) Causality and Chance: Late Third Millennium Collapse in Southwest Asia, in: Roualt, O., ed., *La Djezireh et lEuphrate Syriens*, Paris: ERC.

293. Weiss, H. (1997) Late third millennium abrupt climate change and social collapse in West Asia and Egypt, in: Dalfes, H.N., Kukla, G., and Weiss, H., eds., *Third Millennium BC Climate Change and Old World Collapse*, Berlin, Heidelberg, New York: Springer, pp. 711-723.

294. Weiss, H., Courty, M.-A., Wetterstrom, W., Meadow, R. Guichard, F., Senior, L., Curnow, A. (1993) The origin and collapse of Third Millennium North Mesopotamian Civilization, *Science* 261, 995-1003.

295. Wells, C., Huckerby, E., Hall, V. (1997) Mid- and late-Holocene vegetation history and tephra studies at Fenton Cottage, Lancashire, UK, *Vegetation History and Archaeobotany* 6(3), 153-166.

296. Whitaker, J.H.M. (1997) The geology of Sodom and Gomorrah, Jericho and Lots wife, *Geology Today* 9/10, 193-197.

297. Wiencke, Martha H (1989) Change in Early Helladic II, *American Journal of Archaeology* 93, 495-509.

298. Wigley, T.M.L., Ingram, M.J. and Framer, G., eds. (1981) *Climate and History: Studies in past climates and their impact on Man*. Cambridge: Cambridge University Press.

299. Wiles, G.C. and Calkin, P.E. (1994) Late Holocene, high resolution glacial chronologies and climate, Kenai Mountains, Alaska, *Geological Society of America Bulletin* 106, 281-303.

300. Williams, M. A. J. and Faure, H., eds., (1980) *The Sahara and the Nile*. Rotterdam: Balkema.

301. Willis, K.J. (1992) The late Quaternary vegetational history of northwest Greece. A comparative study of 2 contrasting sites, *New Phytologist* 121(1), pp.139-155.

302. Worsley, P., Gurney, S.D., Collins, P.F. (1995) Late Holocene mineral palsas and associated vegetation patterns - a case-study from Lac Hendry, northern Quebec, Canada and significance for European Pleistocene thermokarst, *Quaternary Science Reviews* 14(2), 179-192.

303. Yoffee, N. and Cowgill, G.L., eds. (1995), *The Collapse of Ancient States and Civilizations*, 3rd ed., Tucson and London: University of Arizona Press.

304. Yoffee, N. (1995) The Collapse of ancient Mesopotamien states and civilization, in: Yoffee, N. and Cowgill, G.L., eds., *The Collapse of Ancient States and Civilizations*, 3rd ed., Tucson and London: University of Arizona Press, pp. 44-68.

305. Yu, G. and Harrison, S.P. (1995a) Holocene changes in atmospheric circulation patterns as shown by lake status changes in northern Europe, *Boreas* 24, 260-268.

306. Yu, G. and Harrison, S.P. (1995b) *Lake Status Records from Europe: Data Base Documentation*. Boulder: World Data Center-NOAA.

307. Yu-lin, X. (1995) The Houwa site and related issues, in: Nelson, S. M., ed., *The Archaeology of Northeast China*. London/New York: Routledge, pp. 65-88.

308. Zachos, K. (1996) Tracing a destructive earthquake in the southwestern Peloponnese (Greece) during the Early Bronze Age, in: Stiros, S. and Jones, R.E., eds., *Archaeoseismology*, Athens: Institute of Geology and Mineral Exploration & The British School of Athens, pp. 169-185.

309. Zangger, E. (1994) Landscape Changes around Tiryns during the Bronze Age, *American Journal of Archaeology* 98, 189-21 2

310. Zangger, E. (1996) Prehistoric earthquakes and their consequences, as preserved in Holocene sediments from Volos and Argos, Greece, in: Stiros, S. and Jones, R.E., eds., *Archaeoseismology*, Athens: Institute of Geology and Mineral Exploration & The British School of Athens, pp. 245-248.

311. Zielinski, G.A., Mayewski, P.A., Meeker, L.D., Withlow, S., Twickler, M.S., Morrison, M., Meese, D.A., Gow, A.J., and Alley, R.B. (1994) Record of volcanism since 7000 B.C. from the GISP2 Greenland ice core and implications for the volcano-climate system, *Science* 264, 948-52.

312. Zoller, H. (1960) Pollenanalytische Untersuchung zur Vegetationsgeschichte der insubrischen Schweiz, *Schweizer Naturfreundegesellschaft. Denkschrift* 83(2), 45-156.

The End of the Bronze Age by Large Earthquakes?

Amos Nur

Department of Geophysics, Stanford University, 397 Panama Mall, Mitchell, Stanford, California, USA

Summary

Twentieth century geophysical data about the geography of active tectonic faults especially at geological plate boundaries, the location of earthquakes, the geography of ground motion intensity, and earthquake frequency-magnitude statistics in the eastern Mediterranean, show that most of the sites that collapsed at the end of the Bronze Age must have experienced destructive earthquakes repeatedly in their past. Recent and historical evidence shows also that these massive earthquakes reoccur every few hundred years in bursts, or "storms" of large events that sweep across broad portions (500 to 2000 km in length) of the eastern Mediterranean, and over short periods of time (ca. 50 years).

This suggests that large earthquakes could have (and probably did) contribute to both the physical and political collapse of the great centres of civilization at the end of the Bronze Age. This probably began by an earthquake storm that unzipped the plate boundaries in the eastern Mediterranean between 1225 BC to 1175 BC. The earthquakes in this ca. 50 year long storm could have rendered many of the urban centers militarily vulnerable, thus inviting attacks not by powerful, distant Sea People but by opportunistic indigenous or neighbouring populations. These attacks may have led in turn to the political and social collapse of the centres, followed by a dark age of recovery and rebuilding often lasting a few hundred years.

1. Introduction

This paper addresses two basic questions about the catastrophic physical destruction that marked the end of the Bronze Age in the eastern Mediterranean. First, could earthquakes have been responsible for the sudden demise of entire cities, regions, and perhaps even empires? And second, if earthquakes were involved in that demise, how were they involved?

Figure 1 is the map by Robert Drews showing the destruction in the eastern Mediterranean at the end of the Bronze Age [11]. Drews names forty-seven sites with archaeological evidence for some level of catastrophic collapse. Presumably, the destruction occurred between 1225 BC and 1175 BC, meaning that all of these places collapsed in a period of about fifty years. But what caused this collapse?

It was Claude Schaeffer who first made the sweeping statement that earthquakes were responsible for this destruction [28]. Although Schaeffer was a great archaeologist, this idea got him into trouble with his peers, and he probably suffered because of the ridicule heaped on him [14]. However by the end of this article it will become clear that careful analysis exonerates Schaeffer for what he proposed at a time when the information he had at his disposal was much less than that available today.

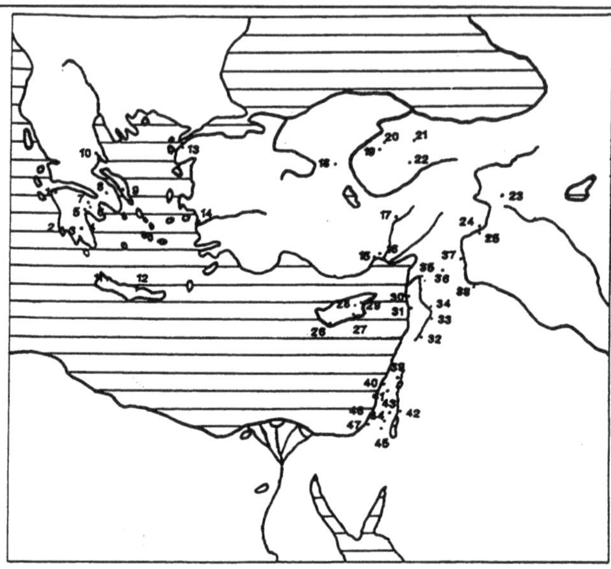

FIGURE 1. The Eastern Mediterranean: Major sites destroyed in the Catastrophe

GREECE
1. Teichos Dymaion
2. Pylos
3. Nichoria
4. The Menelaion
5. Tiryns
6. Midea
7. Mycenae
8. Thebes
9. Lefkandi
10. Iolkos

CRETE
11. Kydonia
12. *Knossos*

ANATOLIA
13. Troy
14. *Miletus*
15. Mersin
16. Tarsus
17. Fraktin
18. Karaoglan
19. Hattusas
20. Alaca Höyük
21. Maşat
22. Alishar Höyük
23. Norşuntepe
24. Tille Höyük
25. Lidar Höyük

CYPRUS
26. Palaeokastro
27. Kition
28. Sinda
29. Enkomi

SYRIA
30. Ugarit
31. Tell Sukas
32. Kadesh
33. Qatna
34. Hamath
35. Alalakh
36. Aleppo
37. *Carchemish*
38. Emar

SOUTHERN LEVANT
39. Hazor
40. Akko
41. Megiddo
42. Deir 'Alla
43. Bethel
44. Beth Shemesh
45. Lachish
46. Ashdod
47. Ashkelon

* At sites in italics destruction in the Catastrophe is probable but not certain.

Figure 1. The Eastern Mediterranean: The major sites that were destroyed in the Catastrophe that marks the end of the Bronze Age (from Drews [11]).

2. Twentieth century earthquakes in the eastern Mediterranean

Figure 2 shows all the earthquakes in the eastern Mediterranean of magnitude 7 and larger which occurred between ca. 1900 and 1980 [3]. The map is thus a snapshot of seismic activity over a period of approximately eighty years in Greece, Turkey, Syria, Israel and Lebanon. The data is not based on inferences from writings but was recorded by modern seismic instruments, which have existed since shortly after the turn of the century. Most of this data was not well known at the time Schaeffer wrote his 1948 book.

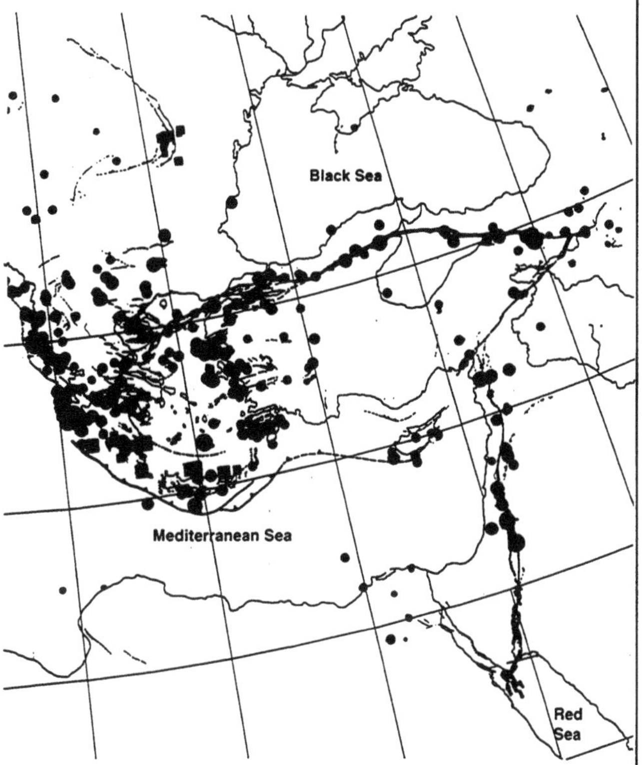

Figure 2. The Eastern Mediterranean: Instrumentally recorded earthquakes of M6.5 in the period ca. 1900-1980 (after Armijo et. al. [3]).

Another thing we now - know and which was not clear in Schaeffers time - is that when an earthquake strikes, there is sliding along a well defined major fault. We do not know this from stories or excavations, but from seismic measurements of events in this century. The slippage by the earthquakes in Fig. 2 on faults in this region provided the basis for a model that has changed earth sciences completely in the second half of this century: Plate Tectonics. The main faults along which most of the slip occurs are today recognized as boundaries of tectonic plates. This means that these plates move more or less as rigid bodies whereas most of the deformation in the crust is localized at the boundaries between these plates.

Figure 3 [18] shows the main plates, their boundaries, and the directions that the plates are moving in the area of interest here. For example, Arabia is moving to the north relative to Africa (which is also moving to the north, but more slowly), the Turkey-Aegean plate is moving west, and the Europe plate is moving southeast. This model is largely based on geological and geophysical knowledge derived from measurements, so it is "hard" science.

Obviously plate motions did not just start in this century. The configuration of these faults and their seismic activity goes back hundreds of thousands, or perhaps even a few million years, depending on how we extrapolate the recent seismic data backward. The motion of the plates is on the order of centimetres per year, so the landscape and the plate bounding fault pattern do not change very fast. Thus, as far as human history and prehistory is concerned, this century's earthquake pattern is pretty much what humans have experienced in this area ever since man first came into the world.

The epicentres of the ca. 1900-1980 earthquakes in Fig. 2 are easily located using seismograms recorded by a network of seismographs. It is also possible to tell from these seismograms the strength - or *magnitude* - of these earthquakes. Knowing the magnitude of an earthquake, it is possible to calculate the maximum ground motion - or *intensity* - that can be expected from it as a function of its magnitude and the location of and distance to its epicentre. Figure 4 reproduces a map published by Karnik [20]. The grey regions in Karnik's map give the best estimate of the Intensity VII or greater as a function of location due to the earthquakes over the last eighty years.

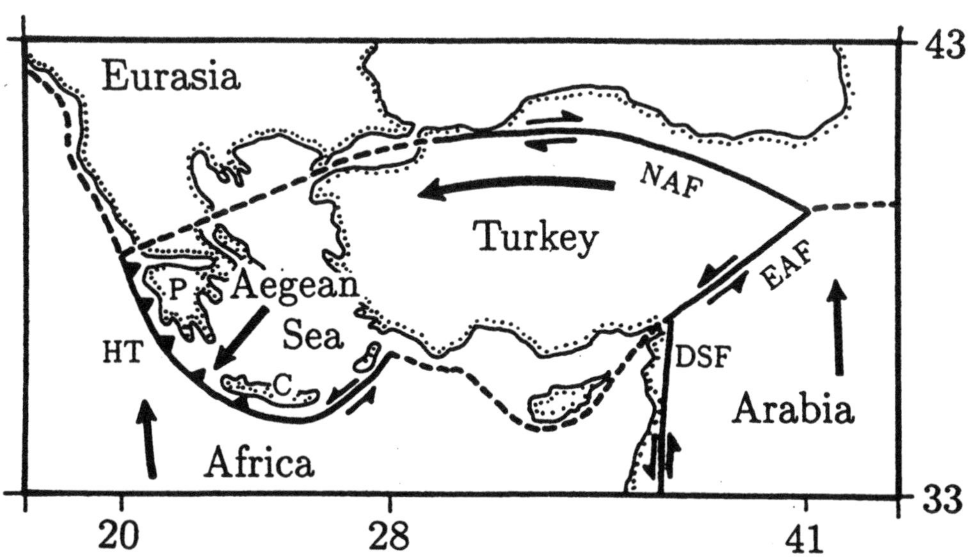

Figure 3. Sketch of the plates in Turkey and the Aegean, showing the relative motions between the Africa, Arabia, Turkey, and Eurasia, accommodated by strike-slip faulting on the North (NAF) and east (EAF) Anatolian fault zones and by shortening in the HellenicTrench (HT). Black arrows are the approximate directions of motion of Arabia, Turkey, the southern Aegean, and Africa relative to Eurasia. DSF is the Dead Sea Fault zone, C is Crete, and P is Peloponnese (from Jackson [18]).

Another thing that is known about the earthquakes recorded during this century is that the distribution of their magnitude obeys a simple statistical rule; in any given region, there are many small earthquakes and a few large ones. For example, figure 5 shows a cumulative frequency function of earthquakes in Greece over a 60-year period which had a magnitude of 4.5 or greater [24]. This function is logN = A+BM, where N is the cumulative number of events, B is a constant (typically approximately equal to 1) for the region, and A is a constant that is related to the size of the area sampled and the length of time over which the data has been collected.

If calculations were made for the entire Mediterranean, the figure would have a similar slope, but the number of quakes would rise, since the area covered would be larger. If calculations were made for a smaller area - say, only Crete - the figure would show roughly the same slope, but the total number of events would be smaller. Similarly, if we collected data for eight hundred years instead of eighty, we would again get the same curve, except that it would be shifted up, because the time window would be longer, so we

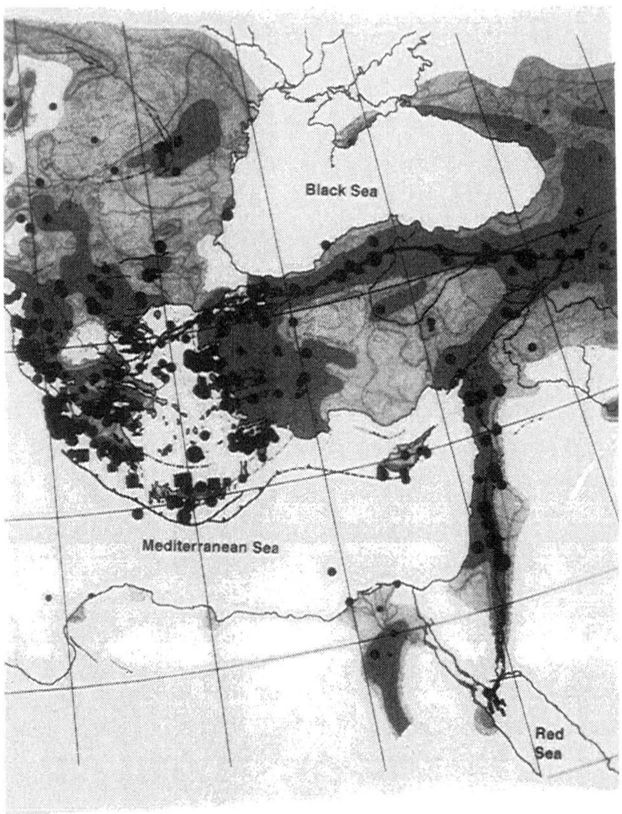

Figure 4. Map showing the earthquakes of Fig. 2 superimposed on the maximum intensity of seismic ground motion in the eastern Mediterranean during ca. 1900-1980, based on this seismicity (after Karnik [20]).

would have ten times more earthquakes. Furthermore, the probability is that the longer the period one examines, the greater the chance of finding a *bigger* earthquake. This is because very large earthquakes are much less frequent than small ones, so if the sampling time window or the area investigated increase, so would our chances of capturing a large one. However, it is important to note that there is an upper limit to earthquake magnitude.

We know from modern earthquake studies that there is a relationship between the magnitude of an earthquake and the length of its associated fault rupture. A magnitude 10 or 10.5 earthquake would run all the way around the Earth. Probably that has never happened, and never will. Many earth scientists feel that a magnitude of 9.0 or 9.5 is the upper limit, and in many regions perhaps even lower. We know for example that in Turkey and Greece there were earthquakes in the past that were greater in magnitude than 8.0, so perhaps the cutoff there is between 8.2 and 8.5.

3. Past earthquakes

The point is that even a short catalogue of eighty years tells us a lot about the seismicity in an area that can be extrapolated over much longer periods of time. The only thing it cannot tell is *when* the big earthquakes struck, since the data is time-wise all statistical. The longer we wait inside the dark shaded region, the more likely it is that we are going to be hit by a strong earthquake. The longer a civilization exists the greater the probability that it will experience a devastating earthquake. Therefore, given enough time in a seismically active region, it is virtually certain that the buildings and facilities of a city or nation which lie in the shaded region in Fig. 4 have experienced damaging and often destructive earthquakes. We don't need archaeological records or historical accounts to tell us that this is practically a certainty. The only thing we don't know is when these earthquakes happened, because there is no regular pattern, so far as we know, to the occurrence of large events.

To bring the point home, consider the monastery in Fig. 6 just outside ancient Jericho. A number of historians and archaeologists have questioned whether it is scientifically acceptable to mention earthquakes as a cause of Jerichos destruction. In contrast, since the city sits right on top of a plate boundary, earth scientists don't debate this question. For us it is a fact that Jericho *was* destroyed by earthquakes, and repeatedly. We just don't know when - whether in Joshua's time or some other time. But there is no escaping the fact that the kind of construction seen in Fig. 6 (which is about as good as they have in that part of the world) is very poor as far as earthquakes are concerned, since it is heavy and does not have resistance to shearing in ground motion.

The picture in Fig. 6 was taken right after the M6.25 Jericho earthquake in 1927, but if there had been photography when earthquakes hit there in 1548 AD or 363 AD, the scene would be virtually identical: the monastery is almost two

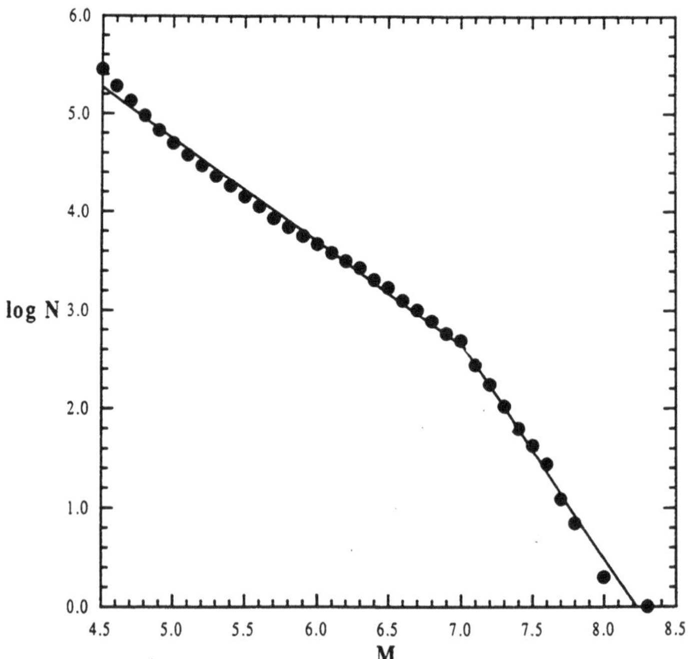

Figure 5. Reconstructed earthquake magnitudes M vs. their number N (on a log scale) in Greece for the past 2,500 years (from Papazachos & Papazachou [24]).

thousand years old and surely must have been destroyed repeatedly before. The city is only three miles from the active fault that marks the boundary between the Arabian and African plates, so in the last two thousand years it must

Figure 6. A photo of an ancient monastery outside Jericho shortly after the 1927 M6.25 Jericho earthquakes. This structure, which has been repaired by now, must have suffered similar punishment 4-5 times since it was first built in ca. 200 AD.

have collapsed at least four or five times - and it will again in the future. We cannot predict exactly when, but that it *will* collapse is certain. The 1927 earthquake, which caused all this damage, was only 6.25 in magnitude, a fairly moderate quake. And looking at the frequency-magnitude relationship (Fig. 5), there is no reason not to believe that even larger earthquakes hit Jericho's area in the past.

4. Intensity and destruction

Consider Karnik's map again, but this time without the earthquakes and focus just on the intensity (Figure 7).

The dark shaded regions show earthquake shaking in this century of intensity greater than VII. This intensity is considered the threshold for significant damage. In ancient times the destruction in these regions was probably greater because construction methods then were much worse. In antiquity, there was no awareness of the repeatability of earthquakes so far as we can tell, and construction methods were earthquake-wise primitive - often just heavy stones piled on top of one another - so intensity VII then was probably more like VIII today.

Figure 8 shows the forty-seven sites that were destroyed in the Bronze Age catastrophe according to Drews [11] superimposed over the intensity VII ground motion map of Figure 7. Most of these sites fall within the high intensity region or close to it, and must have been badly damaged by an earthquake, many repeatedly, in their past.

Consider two examples that illustrate this point. Figs. 9 and 10 from Ambraseys (who pioneered many of the concepts considered here) and Finkel [1, 2] show data for two earthquakes that occurred on the North Anatolian Fault. The first (Fig. 9), on August 17, 1668 probably had a magnitude between 7.5 and 8.0. Added to the original map is the location of the capital of the Hittite empire, Hatussas. For a geophysicist, this map can only mean one thing: if Hatussas ever experienced an earthquake like the one in 1668, it could not have escaped significant damage, considering the building methods in those days. The only thing we don't know is *when* such an earthquake hit Hatussas. We need archaeology

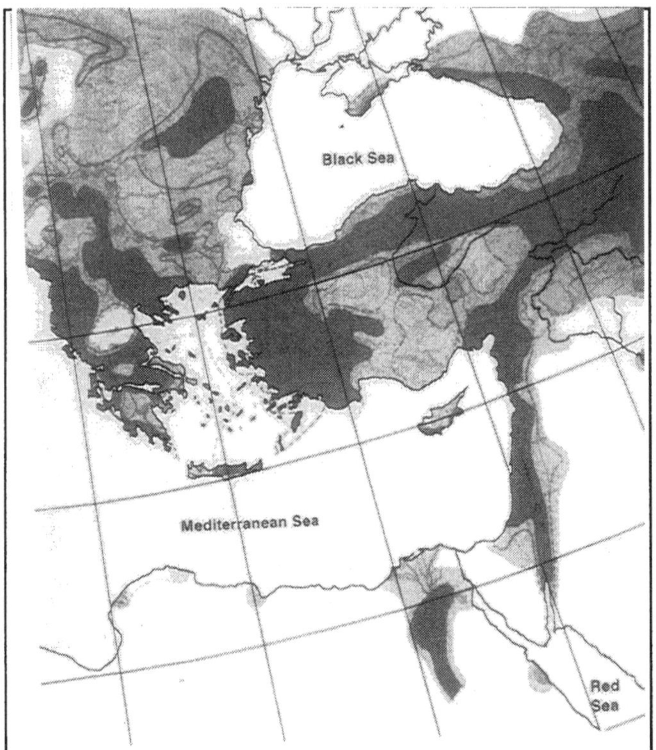

Figure 7. Modified map of Fig.4 showing only the maximum intensity of seismic ground motion in the eastern Mediterranean during 1910-1980 (after Karnik [20]). The maximum local intensity within the dark shaded regions during this period is greater than VII.

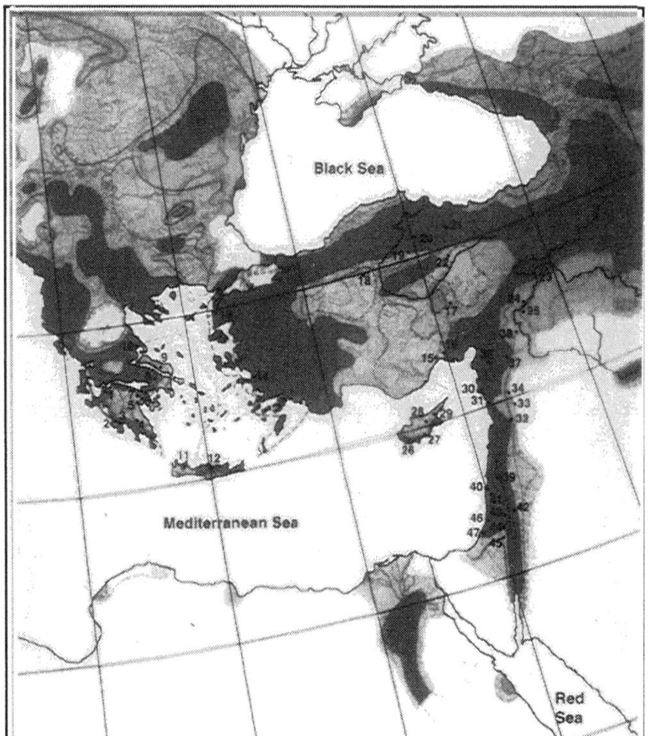

Figure 8. Superposition of the sites destroyed at the end of the Bronze age in 1200 BC and earthquake intensity I>VII in our century. The coincidence suggests that the end of the Bronze age could be related to earthquakes.

and history to pinpoint the time.

The second example (Fig. 10) is the earthquake that struck in western Turkey on August 9, 1912. The extent of the intensity VII region was estimated from reports at the time of damage in villages and towns. Marked is also the location of the ancient city of Troy. Note that it is not only inside the intensity VII outline but actually inside the intensity VIII contour, which indicates heavy damage. Of course, Troy was only an archaeological ruin in 1912, but we can be

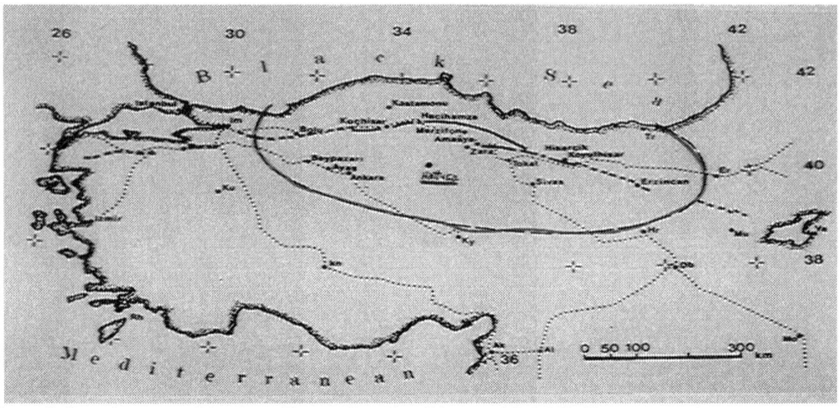

Figure 9. Location map of the Anatolian earthquake of August 17, 1668. Place names given in full are sites affected by this earthquake. The dashed ellipse shows the approximate intensity I>VII (from Ambraseys & Finkel, 1988). Added also is the location of the ancient Hittite capital Hatussas, which falls well within the damage zone of the 1668 earthquake.

completely certain that in antiquity this region was visited by similar earthquakes - probably every few hundred years. Therefore, if this kind of earthquake struck while Troy was a thriving city, there is no way it could have escaped severe, and possibly total, destruction. Finally a quake of this type of intensity VIII today may well have been a IX at Troy then, since ancient construction methods were poorer than today's.

We see the same principle operating in modern times. For example, in 1993 in Armenia, where construction is relatively poor, ten thousand people were killed by a M6.8 earthquake. By contrast, in the 1989 M7.0 Loma Prieta earthquake in California, where anti-seismic construction methods are good, only fifty people died. Most of those people were killed in the collapse of a single poorly built freeway section.

5. Earthquake storms

So, the connection between Troy and earthquakes is not at all far fetched: Blegen [5] and Schaeffer [28] who independently suggested it years ago had the right idea. There is no way that Troy, like Jericho and Hatussas, could have escaped the wrath of earthquakes. One question is, why was Schaeffer [14] subjected to so much ridicule and Blegen [25] to resistance? One reason is that Schaeffer thought that to explain the almost simultaneous destruction requires a single huge earthquake. Many at the time believed that such huge (say magnitude 9) earthquakes never happen. But they do happen. For example, the length of rupture of the huge M9 or so earthquake in Chile in 1971 was approximately 2,000 kilometres. Of course, Chile is in a different geological setting, and one can't tell if this huge event could happen in the eastern Mediterranean. The point is, however, that huge earthquakes occasionally do happen. For example, if we compare the size of the region of destruction at the end of the Bronze Age according to Schaeffer with the size of the Chilean earthquake, the two are comparable. So it is possible at least in principle that one single earthquake could have devastated all of Schaeffer's Bronze Age civilizations. Schaeffer was not completely off base.

Nevertheless, it is unlikely that only one earthquake was involved, since from the existing evidence it appears that the destruction at the end of the Bronze Age was spread over a period of approximately fifty years from about 1225 BC to about 1175 BC. So the big question is, how could earthquakes have been responsible for that distributed destruction? It turns out, as we know today, that large earthquakes in the eastern Mediterranean are clustered in time in the form of "earthquake storms". An example of this is the North Anatolian Fault (Figure 11) [29] in this century. A 30 year storm consists of a series of earthquakes in 1939, 1942, 1943, 1944, 1951, 1957, and 1967 along the fault. Each of these earthquakes ruptured a segment of the North Anatolian fault until the entire length of this plate boundary was unzipped.

Figure 11 shows the amount of displacement or slippage, in metres, that occurred on the fault during each of those earthquakes. During this thirty year North Anatolian earthquake "storm", slippage on the order of two to four metres occurred along the

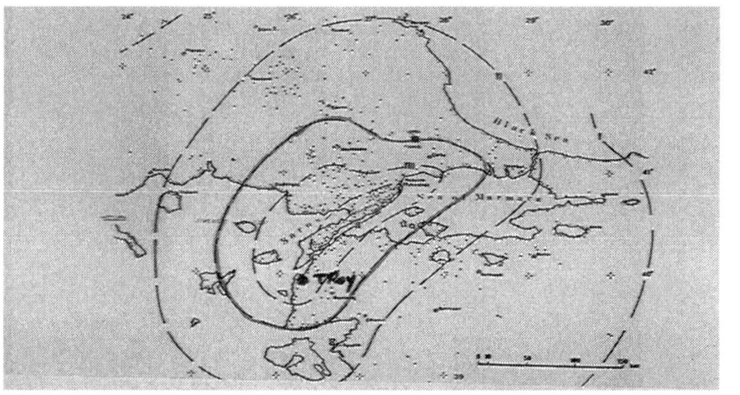

Figure 10. Isoseismal map of the Saros-Marmara earthquakes of August 9, 1912, M=7.4 (from Ambraseys & Finkel, 1987). The site of ancient Troy falls within the intensity I>VIII. Had this kind of earthquake occurred when Troy was inhabited, the damage would have been very significant.

entire plate boundary. It is known from geophysical plate motion reconstructions that the relative motion between the plates that must be accommodated on the North Anatolian Fault is about one to two centimetres per year, or equivalently one to two metres per hundred years. Thus the slip of 2-4 metres during the storm released strain that has accumulated over 200 to 400 years or so. We imagine therefore that strain gradually accumulates over prolonged periods, lasting a few hundred years during which this fault (and similar cases elsewhere) is relatively quiet, and is then released by earthquakes over short periods of time. This is what happened along the North Anatolian Fault in the twentieth century, and must have happened there in the past. One can easily imagine how ancient Hatussas could have been caught in one of these storms.

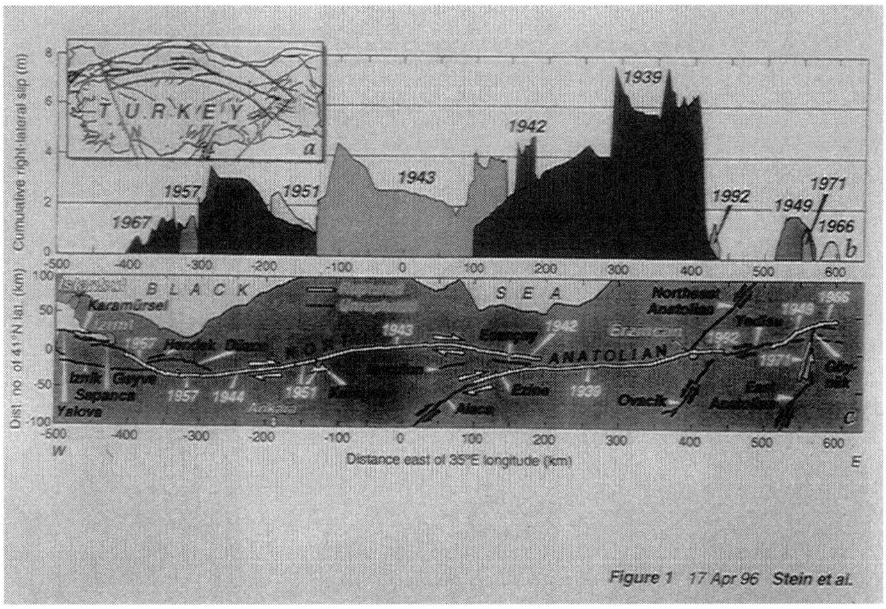

Figure 11. The mid 20th century earthquake storm that unzipped the north Anatolian fault during a ca. 50 year period.

An even greater earthquake storm and a seismic crisis occurred in the middle fourth century AD during a thirty-year period between 350 and 380 AD [16]. A series of significant earthquakes that caused a lot of damage (uncovered in many archaeological excavations) hit the Holy Land, Cyprus, NW Turkey, Crete, Corinth, Reggio Calabria, Sicily, and northern Libya. This active period was preceded and followed by seismically relatively quiet periods, probably for about three hundred years.

What happens geologically is that earthquake rupture of the plate boundary in one place loads other portions of the plate boundaries. This stress redistribution brings them closer to failure and in turn causes their rupture by earthquakes in a relatively short period of time. This is followed by longer periods when the region gradually strains again but doesn't quite give. Eventually, when the whole system is strained enough, the storm begins again.

There is other evidence that supports the notion that earthquakes were involved in the destruction of the Bronze Age. First of all, we can look at skeletons. There are a dozen or so examples of humans from that period who were crushed by collapsed buildings (Table 1). This strongly suggests the action of earthquakes, since people do not usually bury their dead in rubble.

Table 1. End of Bronze Age Earthquake Destruction and Crushed Skeletons (ca. 1200 BC)

site	excavator	crushed skeletons
Knossos	Evans	Yes
Troy	Blegen	Yes
Mycenae	Mylonas	Yes
Tiryns	Kilian	
Thebes	Sampson	Yes
Pylos	Blegen	
Gla	Iakovidis	
Midea	Åström	Yes
Kynos	Dakoronia	
Jericho?	Garstang	Yes
Ugarit	Schaeffer	?
Megiddo	Kempinski	Yes

Local geology is another factor that must be taken into account. Fig. 12 shows the approach to the "Lion Gate" entrance of the ancient city of Mycenae. Everything in the picture probably looks impressively solid to an historian or an archaeologist. But not to an earth scientist: the walls of the city are built on top of a geological fault scarp (Fig. 12). The fault scarp seen here has been polished by sliding during past earthquakes. Because it is a steep rock surface this scarp has been used by the builders of the city as part of the very wall that was supposed to protect Mycenae! One can easily imagine what will happen (and probably did happen) to this wall during even a modest earthquake: it will topple down, and the city will become (as it probably has in its past) defenseless.

In spite of all of this evidence, there is still a great deal of resistance among historians and archaeologists to the idea that earthquakes have played a key role in the destruction of the Bronze Age civilizations. Robert Drews,

Figure 12. The city wall of Mycenae includes in its lower part and as its foundation a natural fault scarp dipping here at 60 degrees or so. Even a very modest earthquake on this fault would cause the wall to tumble, and render it ineffective for defense against attack.

for example, dismisses the idea totally. He states that the only people who have suggested earthquakes as a cause for this destruction have been the archaeologists who excavated the particular sites where they were working, e.g. Blegen et al. [5] at Troy, Mylonas [23] at Mycenae, Schaeffer [28] at Ugarit, Sampson [27] at Thebes, Evans [13] at Knossos, Kilian [22] at Tiryns, and Davies [8] and Kempinski [21] at Meggido. One of Drew's criticisms is that these scientists were limited in their vision because they only looked at their own sites. But as I have shown here to the contrary, their visions actually fit precisely with what we know today about earthquakes.

6. Earthquakes, war, and Dark Ages

In 1996 Elizabeth French [14] wrote:

> "Archaeologists of my generation, who attended university in the immediate aftermath of Schaeffer's great work in 1948, were brought up to view earthquakes, like religion, as an explanation of archaeological phenomena to be avoided if at all possible. Thus, it is only recently that an earthquake at Mycenae has begun to be a serious hypothesis."

So, the answer to the first question raised in the beginning of this paper is that earthquakes could have been, and probably were, responsible for some of the sudden physical collapse at the end of the Bronze Age. The second question remains, however: even if earthquakes were responsible for the physical damage and occasional total collapse of structures, could they by themselves have destroyed entire societies? Probably not. What earthquakes did however was render the cities and centres vulnerable and at times defenseless against human attack. These ancient societies were always in conflict - the inhabitants fighting among themselves or with neighbours or with raiders from far away - and the occurrence of large earthquakes when they occurred made them vulnerable and exposed to attack. The aftermath of earthquakes at times probably even *invited* attack. In other words, earthquakes triggered or enabled war, which then finished these places off.

Below are three examples of this thesis (Table 2). First, there is the indigenous uprising in Sparta in the fifth century BC, described so well by both Thucydides [30] and Diodorus [9]. When an earthquake destroyed the defenses of the city, there was an immediate uprising of the oppressed and enslaved poor in the countryside. The attackers probably realized the opportunity they had to take back their grain, gold, daughters, and whatever else had been taken from them, because the city suddenly became vulnerable, and the small group of ruling families that owned everything couldn't defend what they had amassed.

The second example is the story of the siege of Jericho by the Israelites under Joshua and the collapse of the city's walls. There is an ongoing debate over whether the siege is historical or mythic, but let that be. The point is that the biblical description of the physical collapse of the city's defenses was probably rooted in an oral transmission of an earthquake and its destruction that provided an opportunity for attack. Joshua was probably not a great geophysicist, but as he realized that Jericho's walls collapsed by an earthquake, he recognized the opportunity that the earthquake knocked down the walls for him. It is highly unlikely that trumpets brought down the walls, since air waves don't cause much collapse. Ground waves, on the other hand, do. The third example, well documented by Josephus [19], is the 31 BC earthquake in Judea that breached the walls of Jerusalem. When the Arabs in Transjordan heard about it, they immediately attacked the city. According to Josephus the outcome was a close call - Herod the Great barely managed to lead his army to victory.

Table 2. Examples for links between earthquakes and wars

1. Indigenous uprising - Sparta, 464 BC (Ref: Thucydides).
2. Attack by invaders during siege - Jericho, 1300 BC? (Ref: The Old Testament).
3. Initiate attack by external enemy - Jerusalem, 31 BC (Ref: Josephus).

The Bronze Age post destruction evidence is also consistent with earthquakes. After the destruction that marks the end of the Bronze Age, the so-called Dark Ages set in, a period during which no large buildings were erected, and the written language of linear B disappeared. Imagine the destruction by earthquake of a facility that was spectacular but not essential to the community. The repair of palaces and temples right after an earthquake must have had a low priority, especially if the economy was weak, and could easily have taken a hundred or two hundred years - if repaired at all. Consequently one would expect after an earthquake, especially in ancient times, a long period when there is no building or restoring of most monumental structures.

In addition to the absence of major repair one would also expect to find squatters everywhere, because when earthquakes strike, most people actually survive, even if the buildings and the infrastructure do not. Following the destruction of the heart of the ancient city by an earthquake, survivors would probably collect stones from the rubble to built rudimentary shelters here and there right in the abandoned spaces - e.g., avenues, plazas, and public buildings in which they lived as squatters. It is much later that archaeologists have to try to figure out what happened, or what these rudimentary shelters were. These squatters could have lived in their temporary shelters for generations before the economy rebounded to the point where the rebuilding of massive structures could take place (if and when it did).

One can imagine also that following earthquake destruction of the monumental centres and subsequent looting by the enslaved indigenous population, neighbours, or attacking armies that the educated class would have left. And they would have taken literacy with them. The scribes, who typically in antiquity made up less than one percent of the population, would be out of work. Some would have died, others would have left, and still others must have switched to other jobs that would bring home some bread. Quite possibly, they didn't bother to train their children to read and write, since those skills were now useless.

7. Conclusion

In conclusion, large earthquakes could have and probably did contribute to the physical and political collapse of the great population centres at the end of the Bronze Age. This probably happened by a storm of earthquakes that swept the eastern Mediterranean between 1225 BC to 1175 BC. If true, these earthquakes physically damaged many of the urban centres involved. This damage rendered these centres militarily vulnerable or defenseless, thus inviting attacks not so much by powerful, distant, and scheming Sea People, but by indigenous or neighbouring populations. These attacks led in turn to political and social collapse of the centres, followed by a dark age of recovery and rebuilding lasting a few hundred years (and just in time for another earthquake storm.)

Acknowledgements

I thank Dawn Burgess, Eric Cline, and Julie Baker for their many editorial and other corrections, Paul Weisser for editing this paper, and Hagai Ron for many fruitful discussions.

References

1. Ambraseys, N. N. and C. Finkel, 1987. Seismicity of Turkey and neighboring regions, 1899-1915, *Annales Geophysicae*, **5**, 701-725.
2. Ambraseys, M. N. and C. F. Finkel, 1988. The Anatolian Earthquake of 17 August 1668, in: *Historical Seismograms and Earthquakes of the World*, 173-180. W. Lee, H. Meyers, and K. Shimazaki (eds.), San Diego: Harcourt Brace.
3. Armijo, R., A. Deschamps, and J. P. Poirier, 1986. *Carte Sismotectonique -Europe et Bassin Mediterraneen*, Paris: Institut de Physique du Globe de Paris.
4. Åström, P., 1968. The destruction of Midea. *Atti e memorie del 1o Congresso Internationale di Micenologia. Incunabula Graeca,* **25**, 54-57.
5. Blegen, C. et al., 1950-1958. *Troy: Excavations Conducted by the University of Cincinnati, Vols.1-4*, Princeton: Princeton University Press.
6. Blegen, C. and M. Rawson, 1966. *The Palace of Nestor at Pylos in Western Messenia Vol. I: The Buildings and Their Contents*, University of Cincinnati: Princeton University Press.
7. Dakoronia, P., 1996. Earthquakes of the Late Helladic III Period (12th century BC) at Kynos (Livanates, Central Greece)". In *Archaeoseismology*, 41-44. S. B. Stiros (ed.), Exeter, Great Britain: The Short Run Press.
8. Davies, G., 1986. *Megiddo*, Cambridge: Lutterworth Press.
9. Diodorus Siculus. In *Catalogue of Ancient Earthquakes in the Mediterranean Area up to the 10th Century*, 113. E. Guidoboni. Bologna: Istituto Nazionale di Geofisica.
10. Di Vita, A., 1990. Sismi, Urbanistica e Cronologia Assoluta: Terremoti e Urbanistica Nelle Città di Tripolitania fra il I Secolo A. C. e il IV D.C., *L'Afrique dans l'Occident Romain. Collect. Ecole Franç. Rome*. **134**, 423-494.
11. Drews, R., 1993. *The End of the Bronze Age*, Princeton: Princeton University Press.
12. Evans, A., 1928. *Palace of Minos II*. London: Macmillan.
13. Evans, A., 1964. *The Palace of Minos at Knossos*. New York: Biblo and Tannen.
14. French, E. B., 1996. Evidence for an Earthquake at Mycenae, in: *Archaeoseismology*, 51-54. S. B. Stiros (ed.), Exeter, Great Britain: The Short Run Press.
15. Garstang, J. and J.B.E. Garstang, 1940. *The Story of Jericho*. London: Hodder and Stoughton.
16. Guidoboni, E., 1994. *Catalogue of Ancient Earthquakes in the Mediterranean Area up to the 10th Century*, Bologna: Istituto Nazionale di Geofisica.
17. Iakovidis, S. E., 1986. Destruction horizons at Late Bronze Age Mycenae, *Filia Epi eis G. E. Mylonas A.*, 233-260.
18. Jackson, J., 1993. Rates of active deformation in the eastern Mediterranean. In *Recent Evolution and Seismicity of the Mediterranean Region*, 53-64. E. Boschi, E. Mantovani, and A Morelli (eds.), Boston: Klutwer Academic.
19. Josephus, Flavius, 1982. *The Jewish War*. G. Cornfield (ed.), Grand Rapids, Michigan: Zondervan Publishing House.
20. Karnik, V., 1986. *Seismicity of the European Area*. Praha: Academia.
21. Kempinski, A., 1993. *Megiddo: A City-State and Royal Centre in North Israel*, Tel-Aviv: Hakibutz Hameuchad Publishing House.
22. Kilian, K., 1993. Earthquakes and Archaeological Context at 13th Century BC Tiryns, in *Archaeoseismology*, 63-68. S. B. Stiros (ed.), Exeter, Great Britain: The Short Run Press.
23. Mylonas, G., 1975. Excavations at Mycenae, *Praktika Archaeologikis Etaireias*, 158-161.
24. Papazachos, B. and C. Papazachou, 1997. *The Earthquakes of Greece*, Thessaloniki, Greece: P. Ziti and Co.
25. Rapp, G. Jr., 1987. Assessing archaeological evidence for seismic catastrophes, *Geoarchaeology*, **1**, 365-379.
26. Rapp, G., 1982. Earthquakes in the Troad, *Troy: The Archaeological Geology*, Princeton: Princeton University Press, 43-58.
27. Sampson, A., 1996. Cases of Earthquakes at Mycenaean and Pre-Mycenaean Thebes, in: *Archaeoseismology*, 113-118. S. B. Stiros (ed.), Exeter, Great Britain: The Short Run Press. 113-118.
28. Schaeffer, C. F. A., 1948. *Stratigraphie Comparée et Chronologie de l'Asie Occidentale*, London: Oxford University Press.
29. Stein, R., A. Barka, and J.H. Dietrich, 1997. Progressive failure on the North Anatolian fault since 1939 by earthquake stress triggering, *Geophys. J. Int.* **128**(3), 594-604.
30. Thucydides, 1914. *History of the Peloponnesian War*. R. Crawley (trans.), W. R. Connor (ed.). London: J. M. Dent.

Landscape Analysis and Stratigraphical and Geochemical Investigations of Playa and Alluvial Fan Sediments in Tunisia and Raised Bog Deposits in Sweden:

A possible correlation between extreme climate events and cosmic activity during the late Holocene.

Lars Franzén

University of Göteborg, Physical Geography, Department of Earth Sciences, PO Box 460, 405 30 Göteborg, Sweden

Thomas B. Larsson

University of Umeå, Department of Archaeology, 901 87 Umeå, Sweden

Summary

Since the last glacial retreat climate has changed more or less regularly. While most of these shifts could be regarded as relatively normal, there are multiple indicators showing that rapid changes took place several times. The causes of such strong and sudden climate fluctuations have been the subject of much debate, some of the most popular explanations presented include volcanism, sudden pole shifts and extraterrestrial factors such as cosmic dust veils and direct asteroid or comet collisions with the earth.

This work presents some results from two separate investigations, one in Tunisia, the other in Sweden, in which the aim was to detect such events and find possible causes. Both studies showed that climatic downturns happened around 1500 BC and 1000 BC, involving general cooling, extreme amounts of precipitation, followed by flooding. From many other sources of information it is obvious that these events were sudden and occurred world-wide. A compilation of available data for the Holocene period of the major event around 1000 BC was made, including data on climate in general, glaciers, river activity, lake level changes, geomorphology, sea level changes, vegetation and human activity, all of which would show up if an episode of this magnitude had really happened. The material, although scattered in time e.g. due to uncertainties in the dating methods used, was surprisingly unequivocally pointing to excess water, cooling and indicating that two natural catastrophes with global dimensions took place somewhere within the period of 1600-1400 BC and in another bracketed within the dates of 1200-800 BC. In the latter period, dendrochronological data, which could be regarded as a reliable dating tool, and with a good resolution, points to one low peak of tree growth at 1159 BC and another one at a few years after 950 BC.

From an archaeological point of view the period bracketed within about these years is generally known for a sharp decline in human culture and cultural activities in many parts of Europe and the Near and Middle East. The onset of a so-called Dark Age followed this natural disaster. Many attempts have been made throughout the years to explain this phenomenon, which is best described as a hiatus, the list including droughts, plagues, warfare, political and economical recession followed by famine etc.

One of the most interesting features found in our stratigraphical records described below, are small glassy spherules (5-100 μm diameter) with a very varying composition, from pure Fe-types, to types with the same basic composition as those found in the K/T-transition [55]. These spherules, along with other indicators such as signs of severe ground frost in Tunisia, make us propose that cosmic activity could offer an explanation for the observed changes. We even suggest that relatively large asteroids or comets (c. 0.5 km diameter) hit somewhere in the eastern Atlantic, possibly at the shelf of the Atlantic west coast of Africa/Europe around 1500 BC and 1000-950 BC mainly affecting the British Isles, Central and SE Europe, the North African countries and the Near and Middle East.

Regarding other episodes found in our material there are indications of strongly increased atmospheric circulation in rythmically appearing periods with high peaks centred around c. 2100 BC, 1500 BC, 950 BC, 400 BC, 400 AD, 1000 AD and 1600 AD, the 400 BC and 400 AD ones being weaker than the others. Harmonic analysis points to a periodicity of c.550 years or 1100 years.

1. Introduction

The present paper reports some of the findings achieved during a project in which the aim has been to investigate the late Holocene climate and landscape development in Central Tunisia, with the aid of landscape analysis and analyses of playa deposits. While previous studies present a rather smooth impression of the climate development, the results from our investigation show that climate has often changed sharply since 5000 BP. Some of these changes seem to have been more or less abrupt. Two of the most significant episodes which are clearly shown in the Tunisian stratigraphy seem to have occurred around 1500 BC and 1000 BC calender years (^{14}C). Other less significant events recorded in Tunisian stratigraphies show a remarkable correspondence to European events as recorded from other sources of climate data. Parallel to the Tunisian study, raised bog deposits in Sweden have been examined. These studies unequivocally show rapidly changing atmospheric conditions including events of dust deposition, increased and chemically altered precipitation and periods of severe cold spells during the last 6000 years.

Comparisons with the climate record from other sources indicate that the 1000 BC event, which is the most marked one, can be traced in many other parts of the world including Europe, the Americas, the Near East and the Antarctic. A representative sample of these global observations are given below.

Regarding the climate development in general, Lamb [58], found a particularly sharp change occurring at c.1200 BC recorded from a large number of different environments. From an archaeological point of view, this period overlaps the fall of various civilizations and migrations of the so called 'barbarian' cultures of northern and central Europe and Asia.

According to Neumann and Parpola [72] a cool and rainy period began about 900 BC lasting through the millennium. Textual evidence from this period (c. 750-650 BC) e.g. letters by Assyrian administrators, contain many climatologically interesting details such as descriptions of extreme temperatures, exceptional rainfalls, severe floodings, persistent and thick snow-cover, and ice.

Frenzel [41] stated that there was a marked change in Alaska around 1000 BC as well as in western and central USA, towards cooler and moister conditions. A compilation from various sources made by Clapperton (1993), of the South American late Holocene climate development, suggests that at least two significant periods of climate deterioration occurred around 5000 BP and 3000 BP, corresponding closely to the neo-glacial interval known world-wide [48]. Van Geel [93] presented curves of local moist conditions from the late Atlantic period through the sub-Atlantic showing a fluctuating unstable climate with one distinct deterioration occurring at 2850 BP.

In Lapland, northernmost Sweden, Karlén [52, 53] distinguished four pronounced cold periods, marked by glacier expansion and low tree limits, one of the most distinct ones at 2800-2200 BP. Glacier advancement during this period has also been reported from the Himalayas [84], North America [78, 22, 85, 86] and South America [83, 69, 70, 71, 84].

In their compilation of Holocene ^{14}C-dates Wendland and Bryson [95], a frequency analysis indicated a major global climatic discontinuity around 2700 BP i.e. a botanic discontinuity averaging at 2760 BP and a cultural date at 2510 BP. Correlation between the data obtained by Wendland and Bryson indicate that culture tended to lag behind botanic discontinuities by about 50 years. Assuming that vegetation lags behind climatic changes by 100 years as proposed by Cole [21] the mean cultural lag after a climatic event would be about 150 years.

Regarding fluvial and lacustrine events, which would show up if the assumed period was characterised by increased humidity, according to Starkel [89] the oldest part of the sub-Atlantic period in Poland was characterised by a violent increase in the amount of water in rivers and lakes. Similar observations have also been made in Central Europe where more than 1350 trunks of sub-fossil oak have been dredged out from fluvial deposits along the valleys of the larger rivers within South-central Europe. The trunks have been used to calibrate ^{14}C-datings [8]. Like the Polish material the trunks were believed to associate with major flooding events. Looking at the Rhine material alone, it could be concluded that the outstanding majority of oak trunks found there have ages centred around 1000 BC.

In the Ripple Brook catchment, central England, radiocarbon dating of floodplain deposits showed a dramatic increase in sediment deposition during the late Bronze Age and early Iron Age (2900-2300 BP). Deforestation and cultivation of the catchment slopes and resultant soil erosion was suggested as one possible explanation [13]. It was estimated that during this period the total suspended sediment output must have increased by around 500%, '[...] which is probably an underestimation because of the natural tendency of floodplain accumulation rates to decrease because of the increased flood heights required to reach bankfull stage if sufficient bed deposition does not occur'. Calculations of the sediment budget showed that the annual erosion values of 20-30 tons/km^2 prior to 2950 BP suddenly increased to values of around 140 tons/km^2.

Regarding climate's role Brown and Barber concluded that the effects of any climatic change on soil erosion would largely be dependent on the soil conditions and vegetation cover, and hence the initiation of this event was blamed on human activity.

Regarding North-Africa and the Middle East, peak streamflows were recorded for Tigris-Euphrates [54], in Iraqi Kurdistan [100] and the Nile [16]. In South America the Amazon Basin was subjected to what was believed to be lacustrine phases, forming hypothetical *Lago Amazonicas* [26, 81]. The topmost sediments, however, belonging to a phase centred around 2800 BP show cross-bedding structures and ripple marks being typical indicators of running water.

As for lake level changes, several indicators of extreme levels have been identified from Europe and other parts of the world. Lake levels in southern Sweden according to Digerfeldt [25] seemed to have been higher after c.1000-900 BC than before. In central Europe a dramatic climate deterioration was observed soon after 1000 BC. According to Gams and Nordhagen [43] in a number of lakes the water levels were raised far over normal i.e. the lakes Saalachsee, Ammersee, Federsee and Bodensee in addition to a number of Swiss lakes. The raised lake levels resulted in the build up of new high beach terraces and the widespread abandonment of lake pile dwellings. New lakes were formed near München, Tölz, Memmingen and Ravensburg.

In Northern Ireland there is evidence of an increase in lake levels at c. 1000-600 BC. In the small lake of Loughnashade there was a significant high stand around 950 BC [94], corresponding with a major gap in the Irish bog oak chronology mentioned below [3]. According to Weir [94] lake levels on the average appear to have been higher after c. 1000-900 BC than before, judging from the sediment, after a period of fairly dramatic changes.

In easternmost Turkey, Lake Van, the largest soda lake on Earth, to judge from the carbonate isotope composition [92], rose almost 80 metres in a period bracketed within the years 971 and 902 BC (based on varves). A simple calculation shows that this would require an excess precipitation (evapo-transpiration not accounted for) of between something like 19,000 and 38,000 mm. In central Africa high lake levels were reported from several sites [16]. As for North and South America, several authors report high lake levels during this period e.g. Great Salt Lake [68], Waldsea Basin, Canada [64], Lake Titicaca [96, 18] and Lago Cardiel in western Argentina [46]. In the Antarctic, a distinct peak in sedimentation was reported from Midge Lake, Beyers Peninsula, Livingstone Island between 3200 and 2700 BP and culminating around 2900 BP [11].

Regarding the response of the natural environment to rapid climatic changes one would expect vegetation to be a good indicator, hence pollen diagrams and tree ring chronologies would serve this purpose. The European series of tree-ring data are based on bog oak data from Germany and Ireland. The bog oak trees grew in forested swamps, often on the base of the peat, under conditions near the limits of the ecological amplitude for oaks, hence being sensitive to changes regarding hydrology and nutrient supply. During dry periods growth increased while during times of moister conditions the trees reduced their growth rate or even died.

In the Irish bog oak series there are significant gaps, i.e. in the first centuries BC and around 1000 BC, the earlier event showing up as a pronounced break of several decades of oak growth on blanket bogs. According to Baillie [1] 'no matter how many samples were processed, none could be found which related to these important gaps [...] it was impossible to go out and find naturally preserved timbers of these specific ages'. Regarding the '1000 BC gap', according to Baillie [3] and Weir [94] it is possible that a single episode of particularly wet conditions occurred, centred around 950 BC, or to use Mike Baillie's own words regarding bog oak growth, '[...] .if oak trees have been growing continuously on the surface of a bog for 1500 years (e.g. Garry Bog, Co. Antrim) and suddenly that bog no longer supports trees then something must have changed' [3]. Even if the extinction of oaks on a single bog could be explained by e.g. local hydrological conditions, the synchronous death of bog oaks on a much larger regional scale could not happen unless we are looking for the effect of a widespread environmental change [65, 66]. Similar observations to those of Baillie's regarding break of bog oak growth are presented from Errand Island, France where the trees survived until 950 BC [97].

According to the tree-ring chronology of the Campito Mountain in North America, the period 2900-2300 BP was cold, marked by narrow ring widths [57]. Davis [23] compiled data from USA (CA, OR, ID, UT, AZ and TX) of the *Homeric-Greek Minima* cold wet period. The radiocarbon dates given were calibrated here [90] and the mean of the calibrated ages was calculated at 1 1027 (980,960,940) 902 calender years BC.

Peat growth in wetlands is a significant climatic indicator. Since the last glacial retreat large amounts of peat have been stored in the peatlands of the earth, hence being one of the largest terrestrial carbon sinks [31, 32, 39]. Fig.1 shows the mean rate of peat growth of some 100 Central Swedish mires from 7000 BP [27]. According to several European and other sources, peat growth seems to have been especially significant in the cooler episode that began in the late second and early first millennium BC. There are pointers to this from all over the world. In southern Sweden where bog stratigraphies were thoroughly examined by Granlund [47], the study of so called *Recurrence surfaces* (RY), the most significant one, RY III, found in most bogs occurred sometime between 1000 and 500 BC. In northern and central Sweden the mean rate of peat growth, as recorded from some 100 mires, rose suddenly c. 3000 BP from approx. 0.4 mm/year to 0.8 mm/year with peak values of 0.9 mm/year around 2900 BP.

The stratigraphical level described above, also known as the *Grenzhorizont*, is clearly visible not only in Sweden but in many other ombrotrophic sites in Europe, the former Soviet Union, Canada and elsewhere, as distinct transition from a lower, strongly decomposed peat-layer known as *Schwartztorff* to a weakly decomposed upper one, *Weisstorff*. The strong humification of the older, black Sphagnum peat in generally accepted to originate from a long period of weathering without further peat growth, under relatively warmer and drier climate, while the younger white peat, would indicate much more humid conditions.

An extreme case of increased peat growth during this period was recorded at Steng Moss, Northumberland, where the accumulation rose from a value of 0.65 mm/year in the period of 1320-820 BC, to 4.47 mm/year, in the period of 832-765 BC, more than a six-fold increase. Barber [5] concluded that this must have been a catastrophic decline to cooler and/or wetter conditions 2850-2550 BP.

Looking at the dating of the major transition zone described above we find that it took place around and after 1000 BC, and there are also indications that the increased rate of peat growth started after a period of deterioration, i.e. there is a hiatus of a few hundred years between the underlying dark peat and the overlying light peat. In the immediate transition zone, the rich abundance of remains from cotton-grass (*Eriophorum vaginatum*) in many stratigraphies is another indicator of changed geo-chemical conditions of the mire surfaces. The hiatus is partly confirmed by the observations made by Godwin [44] from bogs in England and Wales where the peat below trackways [80] had radiocarbon ages averaging at 2959 BP, the trackways gained average ages of 2604 BP and the overlying wet re-growth of peat at 2575 BP. According to Barber [5] this '[...] shows conclusively that we are dealing with a phenomenon which, whatever the details of its effect on the bog plants, certainly elicited a response from man.'

From the Antarctic, there are similar observations as those from Europe regarding peat growth. Björck et al. [10], show that the growth rate of a moss peat bank on Elephant Island, South Shetlands suddenly increased, with a distinct maximum at 3200-3100 BP.

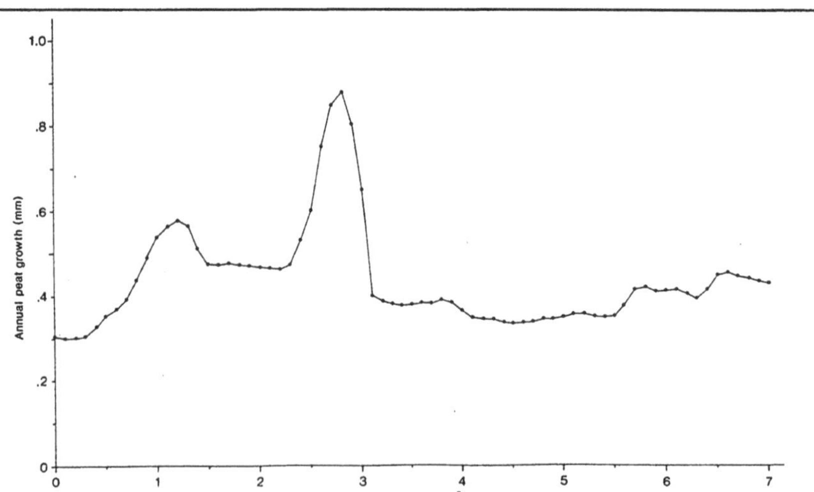

Figure 1: Rate of peat growth of some 100 mires in central Sweden during the last 7000 years

The last, but not the least important evidence of the c. 1000 BC climate event is found from cultural and textual sources. It seems natural to suggest that a climatic event or shift suggested here would not have passed without any trace in the history of human societies, climate being a decisive factor for most human activities. Although textual sources from this period are sparse for obvious reasons, there are other pieces of evidence of a Dark Age, that if filtered and puzzled together gives a rather good picture of what might have happened. A few examples are given below.

Regarding Dark Ages in Scandinavia a decline in the number of metal finds has been recorded, from around 1000 BC although it has been associated with a 'temporary crisis' ascribed to decreasing metal supplies. Other indicators are the break in long-practised burial traditions and the retreat from some marginal lands [56]. The display of personal wealth in burials was heavily reduced, and tumuli building came to an end. According to Kristiansen, increasing warfare and continuous ecological degradation led to settlement clustering, imposing new constraints on the economy.

On the Hungarian Plain, a vast area of low-lying land where a number of large rivers converge, disastrous floods occurred sometime after 1300 BC causing the eruption of the Bronze age people (Hungarian Bronze Age - Bronze IV) and migration to a drier climate [12]. According to Carpenter [17], the area must have become almost totally submerged by an overwhelming flood early in the first millennium BC during which period the population suddenly disappeared after a rich and active phase, or to use the words of Åberg (c.f. Carpenter [17], 'an unexpected and sudden end [...] after which the country is without any discoverable sign of occupation and seems deserted.'

According to Burgess [15] the mild climatic episode came to an end in the second millennium, the Plenard period, c. 1250-1000 BC. Conditions deteriorated sharply, initiating a decline which continued over the next few centuries. A colder, wetter, more unsettled stormy period began in which temperatures fell nearly 2C, reducing the growing season by more than five weeks [91]. According to Lamb et al. [59] the period around 900-450 BC appeared to have been characterised by abnormal wetness, especially in the western parts of the British Isles. An overall decline of crop ripening levels of about 150 m between the 12th and 7th centuries BC seems entirely feasible on the Medieval evidence according to Parry [75]. There was also a general contraction of settlement, abandonment of marginal lands, gained in high altitudes during the warmer period, a general retreat from the uplands, and a far-reaching dislocation of settlement and agricultural systems [15]. According to Burgess there are also indications that new crops, more intensive farming methods and food collection and storage measures were introduced. The population during these years as indicated by finds of axes and other metal work was reduced to less than half [15].

In the Mediterranean, the decline of the Mycenaean culture, also known as the Greek Dark Age, is one of the unsolved mysteries of the ancient world. In the efforts to find a possible explanation to the general decline which, according to common knowledge, took place sometimes after 1200 BC, several causes have been suggested, none of which has proved satisfactory [51]. To quote Desborough [24], he described the decline:

> 'During these generations the changes that came about are little short of fantastic. The craftsmen and artists seem to have vanished almost without trace: there is little new stone construction of any sort, far less any massive edifices; the metal workers techniques revert to primitive, and the potter, except in the early stages, loses his purpose and inspiration; and the art of writing is forgotten. But the outstanding feature is that by the end of the twelfth century the population appears to have dwindled to about one-tenth of what it had been little over a century before. This is no normal decline, and the circumstances and events obviously have a considerable bearing on the nature of the subsequent Dark Ages, and must be in part at least a cause of its existence.'

One of the first attempts to give a climatological explanation to the decline as presented by Rhys Carpenter [17] who tried to explain it by supposed temporary northward displacement of the arid zone and hence a severe widespread and long-lasting drought. Carpenter's proposals that the Mycenaean decline was due to a prolonged drought was rejected by most meteorologists, with a few exceptions (e.g. Bryson et al. [14]). The strongest evidence against the drought theory is based on paleo-ecological studies. Pollen diagrams from Jordan, Syria, Turkey, Iran and Greece, dated independently of the archaeological chronology, show no signs of any long-term dry spell between 1200 and 900 BC.

Although Carpenter's main point was the drought, he argued that this was succeeded by a cool and wet period i.e. 'at the end of droughts of this kind it frequently occurs that it is succeeded by a compensatory onset of unusually heavy rains.' According to Carpenter the onset of cold and a much wetter climate could not have been entirely abrupt and the timing of this event is unknown 'unless we may argue from the incidence of Protogeometric vases that it should be dated somewhat later than the year 1000 B.C.' Regarding house construction Carpenter stated for the east Mediterranean 'that there was a shift in architecture from flat to pitched gable roofs, indicative of heavy rainfall.' An interesting hypothesis regarding the decline of Mycenae was recently presented by Zangger [101] in his monograph *The Flood from heaven - deciphering the Atlantis legend.* Zangger argued that Troy, situated close to the mouth of the Dardanelles, was destroyed sometime during the Dark Ages, by extensive floods coming down from the surrounding mountains.

The dating of the Greek Dark Age is a matter of debate. In recent works (e.g. James [51]) the timing for the collapse of Mycenae has been re-evaluated. Based on paleo-environmental data from sediment cores of the Osmanaga Lagoon near Pylos in Messina and Lake Voulkaria in western Greece, James [51] argues:

> '[...] the calibrated radiocarbon dates from these cores show that olive cultivation became important around 1500 B.C., and continued to be so for several hundred years. [...] The Osmanga core also shows that olive cultivation continued well after 1200 B.C., indeed peaking around 950 B.C. - in the middle of the supposed Dark Age! [...]. Instead, if the collapse of Mycenae occurred in the late 10th century instead of the 12th, then the swift decline of olive cultivation would provide a perfect match with the post-palatial pastoral society as postulated by Snodgrass.'

Dark ages could also be traced from other parts of the world e.g. SE Asia. Spriggs [88] made a thorough compilation of calibrated ^{14}C-dates for Island Southeast Asia. From this compilation the dated material could be separated (charcoal, shells etc.). If we accept the occurrence of charcoal produced by human settlements as one evidence for activity, from this compilation we find some very interesting features around 1000 BC, i.e. there is a marked hiatus in the number of charcoal datings between c .950 BC and c. 800 BC. The absence of charcoal during this period could hence indicate a general cultural decline analogous to that of Mycenae. A similar compilation of calibrated ^{14}C-datings made by Gosden et al. [45], for the Bismarck Arcipelago, NE New Guinea, shows a similar distribution, i.e. no charcoal between c. 1000-950 BC and c.750 BC.

In conclusion, the above indicators equivocally point at a climate event beyond the normal, followed by a cultural crisis which could be traced world-wide.

2. Peat deposits

Peat deposits in raised bogs, dated by ^{14}C, show different atmospheric conditions during their formation e.g. fallout of condensation nuclei such as mineral particles and/or chemical substances, all of which could be a result of global, regional and local events. Marine aerosols, dry or wet deposited [29, 30, 49], could potentially reflect the conditions of atmospheric circulation during the time of deposition i.e. a strong circulation would increase the share of marine components in the peat [98, 99]. Mineral particles in the peat stratigraphies could potentially show events such as dust storms [34, 28, 35, 36, 37, 38] and volcanic eruptions [76, 77] but also substances from extraterrestrial activity e.g. input of micro-meteorites and/or tektites.

Several ombrotrophic peat stratigraphies in southern Sweden have been examined for their mineralogical and chemical composition. Macro-fossils in the peat, mainly remains from cotton-grass (*Eriophorum vaginatum*) have been dated with ^{14}C.

The stratigraphical investigations of the peat, which have mainly been done in 5-cm intervals giving a high time resolution (40-100 years), include rate of decomposition, mineral content (ash content), microscope analysis of mineral particles and the concentration of major cations (Na, Ca). The rate of decomposition, which on a regional scale is believed to reflect the climate development [47, 42] has been studied by a new method, using a Minolta Chroma Meter. The reflectance and relative colour of dried peat samples have been used. The mineral content depends both on influx of mineral particles and chemical substances trapped in the peat. Mineral content has been examined by standard methods i.e. combustion of the peat. Microscope analyses of mineral particles, which generally make up for less than 10 per cent of the ash weight, have been made with a high magnifying stereo microscope.

The occurrence of volcanic ash and cosmic particles have been of special interest. A polarising microscope has been used to distinguish glass from ordinary mineral particles such as quartz and mica. Chemical analyses of the peat ash have mainly been done for calcium. Calcium, being the main substance in most peats makes up for average value of 25 per cent of the total ash weight [40]. Sodium has also been studied, but being a more mobile ion it is a coarser instrument.

The studies, so far unpublished, have shown some interesting results:

- The rate of decomposition co-varies more or less with mineral content which is natural. If the influx rate of minerals is kept at a constant rate, this means that strongly decomposed peat would give a higher ash content per weight unit due to a slower rate of growth, and vice versa.

- Microscope analyses of minerals show a background influx of small mineral grains (1-100 μm), mainly quartz and mica, and in specific layers, volcanic ash, according to Persson [76, 77], but also a rare abundance of glass spherules (Figs. 2, 3) at certain levels i.e. mainly in the immediate transition from strongly to weakly decomposed peat recurrence surfaces [47]. These kind of stratigraphical changes, which are found in different levels of the bog sequences are believed to reflect the transition from warm and dry to cold and humid conditions. The glass spherules have been examined with SEM/EDS and most of them have the same basic composition of those found in the Tunisian material as described below, but also of tektites associated with cosmic activity (Tab.1a: Beloc). There are also black magnetic spherules dominantly made up by iron (Fig. 4). The major influx of spherules was found in two

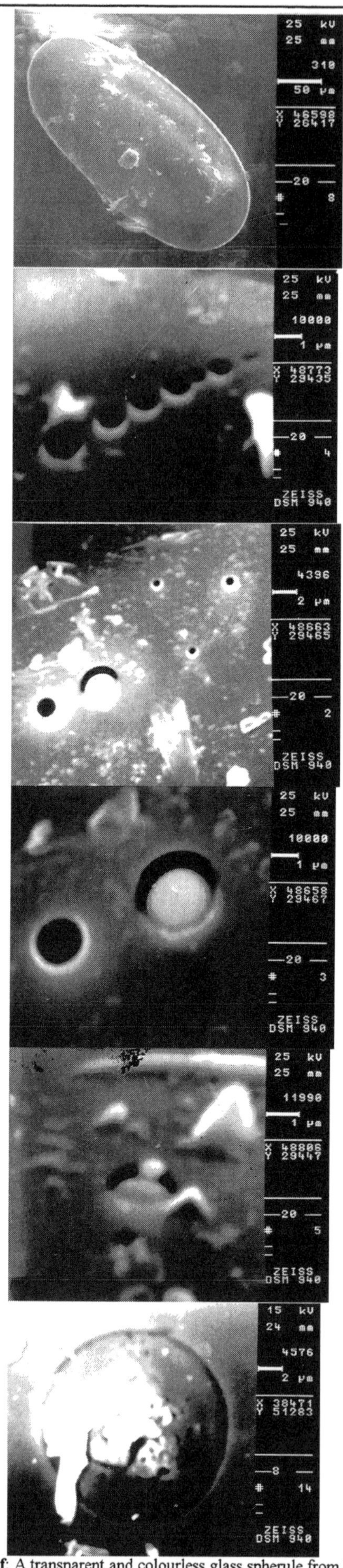

Figure 2 a-f: A transparent and colourless glass spherule from bog peat (Tab. 2, Oval) The surface is patterned with micro impact structures (b). The bright spots (c, d) and the object (e) are microspherules entirely made up of iron. The ring structure (f) is a shock wave ring from a micro impactor.

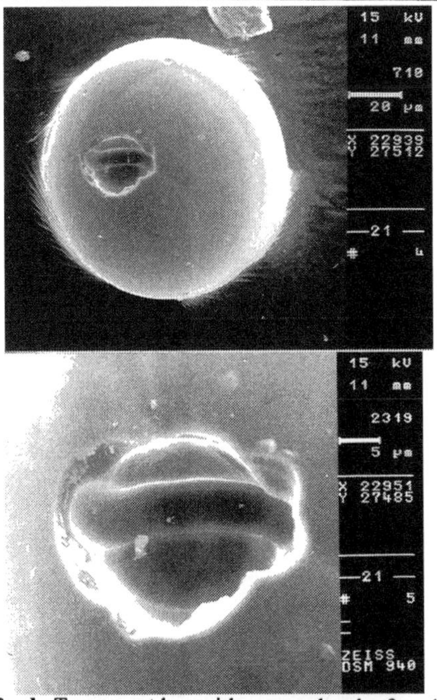

Figure 3 a-b: Transparent brownish green spherules from bog peat

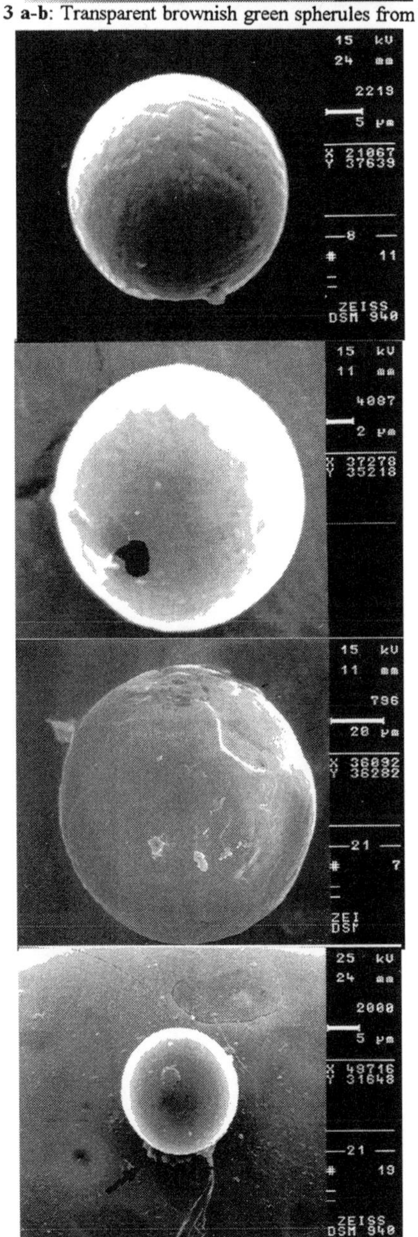

Figure 4 a-d: Iron typr spherules from bog peat and the Tunisian playa sediments. Spherule (a) has very marked solidification structures, (b) is hollow, (c) is made up of titanium and iron and below (d) there is a cluster of very small spherules of glass types.

transition layers dated to c. 1400 BC and 900 BC. (Table 2: 52-53, 57).

- Chemical analyses of the peat ash show a high concentration of calcium, as expected, but also show that the general calcium concentration increases with depth. Theoretically, calcium could be trapped as a result of the oxic/anoxic transition between the acrotelm and the catotelm situated ca. 20 cm below the surface, i.e. Ca is kept dissolved in the acrotelm but precipitates in the catotelm, due to chemical reactions not well known. The precipitation of Ca this way is e.g. indicated by the abundance of small calcite crystals (<20 microns) in the catotelm which do not occur in the acrotelm. Calcium also shows a rather varying concentration throughout the stratigraphies (Fig.7) with an apparent cyclicity. Harmonic analysis of the time series, based on ^{14}C-datings shows their strongest peaks at c.550 and 1100 years. The co-variation between Ca and rate of decomposition is also strong, with a 20 cm lag, i.e. the Ca peaks are found c. 20 cm below the recurrence surfaces. One of the highest Ca-peaks is associated with the major

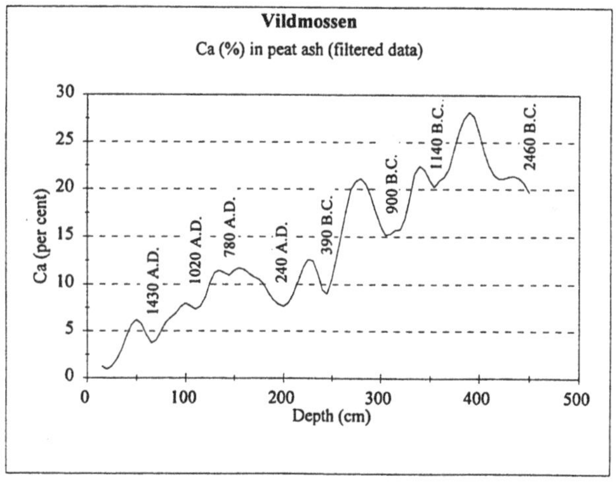

Figure 5: Typical development of calcium concentration in a south Swedish bog stratigraphy.

spherule event described above. The origin of the calcium is not yet known, the most probable explanation being increased influx of marine aerosols. An alternative and/or complementary explanation is deposition of marine carbonate-rich sediments from shallow areas of the ocean, which could imply impact events.

3. The Tunisian study

The Tunisian study was started in 1988 in connection with the first in a row of annual Swedish Physical geography field expeditions in this area. These studies have been concentrated in the Sidi Bouzid region situated in the central part of the country (Fig.6).

During interpretation of SPOT satellite images and field work several interesting phenomena have been detected, e.g. signs of extreme fluvial activity. In a research project, emanating from these observations, special interest has been drawn to a playa (salt lake) and its surroundings, a large wadi, Oed Seguia, which was created during the great floodings of 1969 [19] and 1990, and a small ravine, Oed Dahkla, situated in the mountainous area of Djebel El Kebar.

It should be pointed out that the ravines in the investigated area are modern phenomena, cut out due to a combination of extreme rains and increased erosion caused by deforestation

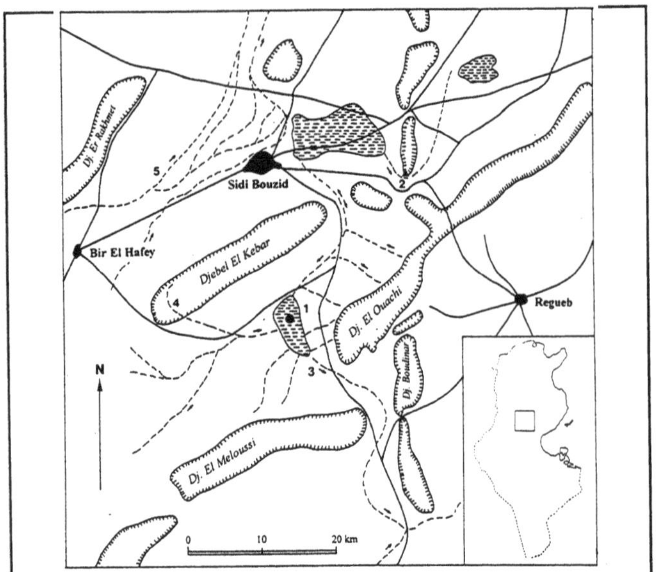

Figure 6: The Sidi Bouzid region of central Tunesia. 1/Chott Nejla (Ez Zaf Zaf), black dot showing site of excavation, 2/ Oed Seguia, triangle showing site of well, 3/ Outlet of Chott Nejla with sand ridges, 4/ Oed Dahkla, 5/ Oed Fekka. Wadis (tempirary water courses) are shown as dashed lines.

and over-grazing.

Since many signs of cultural acivity have been found in the investigated deposits, during the years of field work in Tunisia, co-operation between the Department of Physical Geography at Göteborg University and the Department of Archaeology, Umeå University was established in 1996. The 1997 field expedition confirmed these archaeological findings as belonging to periods as old as 9000 BC [62]. Among these findings, flint tools, ostrich egg shells, 'escargotières' and fire places, were found, exposed at deflation surfaces at the sand ridges of the southern outlet of Chott Nejla. Radiocarbon dating of these findings (Tab.2:24,25) (ostrich egg shells, mollusc shells and charcoal from fire places) showed a more or less continuous Stone Age settlement in this area until 1000 BC. Unfortunately the material on the deflation surfaces are only an 'integration' of the activity on the site, i.e. deep wind erosion has left a stone pavement made up by fire cracked stones, flint tools, pieces of pottery etc. which have been distributed at different levels of the stratigraphies hence the remaining sand hills (yardangs) offer potential excavation sites.

Alluvial fan, valley fillings and playa deposits in desert margin areas offer an alternative archive in the investigation of landscape and climate development [50] in areas where e.g. lakes are rare and peat-land more or less non-existent. Events as those found in bog stratigraphies are harder to locate owing to the much larger input of material from local sources. In spite of this, thorough laboratory preparations and microscope analyses have shown that influx of external material could be detected, such as volcanic glass and micro-spherules.

Chott Nejla

The playa investigated, Chott Nejla, is a ca. 20 km long and 10 km wide depression situated between the three mountain ridges Djebel El Kebar, Djebel El Meloussi and Djebel El Boudinar, 20 km S of the regional capital, Sidi Bouzid (Fig.6).

With present climatic conditions the playa is normally dried out, with a halophile vegetation. During periods of heavy precipitation such as 1969 [19, 33] and 1990, due to swelling clays at the surface, the playa is transformed into a fresh water lake with a maximum depth of less than 10 m. The southern end of the playa, which has the lowest threshhold (320 m a.s.l.), acts as the lake outlet. At the outlet, the water is forced into a narrow 'straight' between two parallel ridges of dune sand hardened by calcrete. The

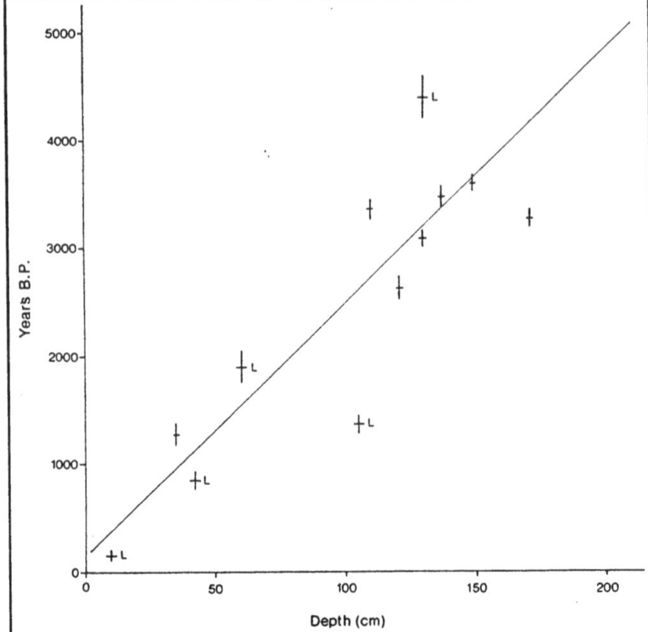

Figure 7. Datings from the Chott Nejla excavation site (L = luminescence datings, others are ^{14}C on carbonate nodules and shells.

total sediment sequence of Chott Nejla is more than 500 m, deposited since late Tertiary.
Three excavation pits, each two metres deep, were dug down

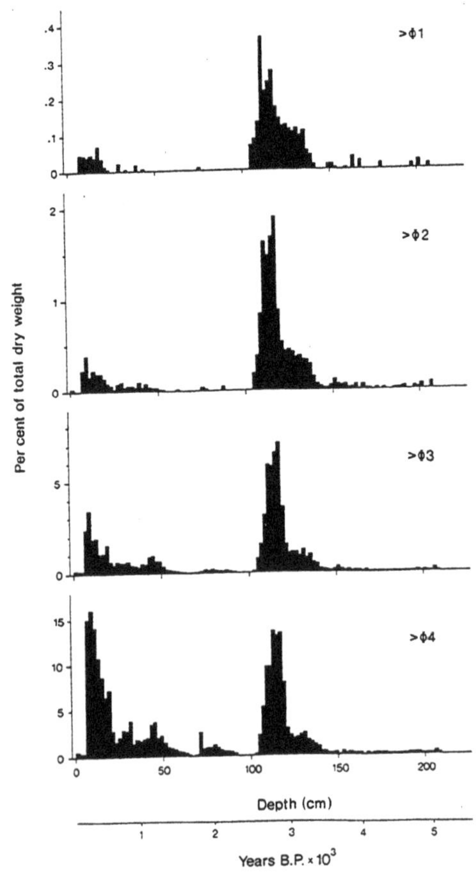

Figure 8. Relative share, in different sizes, of material insoluble in hydrochloric acid (HCI) from the Chott Nejla stratigraphy.

to the ground water level. Samples were collected in 2 cm intervals, analysed and dated with different methods (14C, TL and OSL) (Table 2:26-38) (Fig.7).

In the playa sediments, exceptional pluvial events are indicated by the occurrence of layers with anomalous coarse particles (Fig.8).

The location of the sampling sites, in the central part of this closed basin, would normally only permit silt and clay to deposit, unless the playa was overfilled and water running through it. The initiation of pluvial phases is often indicated by lower coarse strata underlying clays indicating a more calm sedimentation environment such as in a lake. The major transition in the playa sequence is found at a depth of 1.3 m where a coarse gypsum sediment is abruptly replaced by a carbonate rich sand followed by a 10 cm clay sequence. Another transition layer is found at a depth of 0.9 m. In the lower transition layer the mass occurrence of shells from large land snails (*Eobania sp., Sphincterochila sp.*) indicate the sudden death of a vast part of the population. ^{14}C-dates on these shells (Table 2:32-33) gained ages of around 1300-1550 BC (cal). In this same layer a twenty-fold increase of magnetic susceptibility suddenly occurs (Fig.9). Microscope studies of the insoluble part (HCl) of the sediments showed the occurrence of small glassy spherules above the transition zone (Fig.10), the highest concentration immidiately above the gypsum.

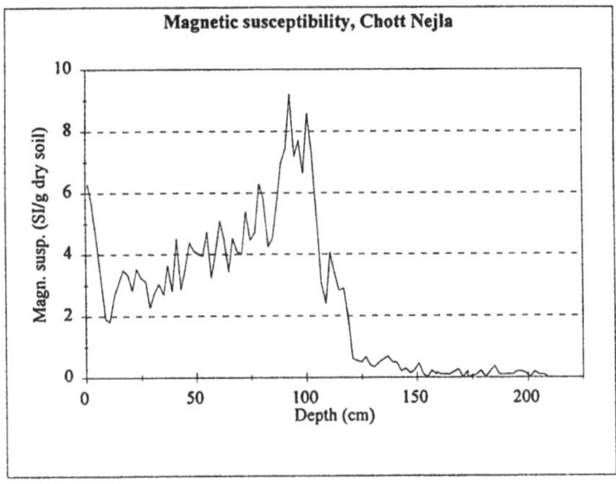

Figure 9. Magnetic susceptibility of the sediments in Chott Nejla. The highest values are found in the lowest lake stage layers.

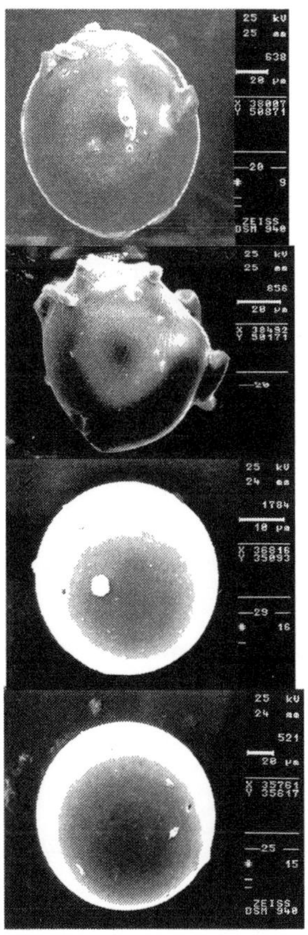

Figure 10 a-d. Various types of glassy spherules from the Tunesian playa sediments.

The Tunisian spherule composition differs much within the population but most of them have the same basic composition as those found in the Swedish peat samples i.e. high concentrations of Si, Al, Ca, and Fe (Tab.1c). Like the Swedish findings many of the spherules are magnetic, being almost entirely made up of Fe (Tab.1d).

Unfortunately, pollen analyses of the playa sediments, giving valuable information on the vegetation composition, have not been done yet, due to lack of a suitable reference material. Since 1990, pollen traps have therefore been used in the investigated area and in a nature reserve 100 km south of Sidi Bouzid, to establish a recent pollen flora.

The ridges at the outlet of Chott Nejla (Fig. 6:3) are of special interest showing many signs of cultural activity which could be interpreted as the remains of an unusually long settlement history reaching over the period of 9000-100 0 BC.

This is uncommon for the Capsian culture, where settlement sites seem to change within the whole of the Maghreb region. A flint technological series of development, with tools of different fabric and purposes, has been found. Most of these flints are associated with a hunting economy.

Oed Seguia

One of the most exciting sites in the area is a wide and deep ravine, Oed Seguia, situated some 15 km SE Sidi Bouzid (Fig. 6:2).

The ravine, with its 5-7 m high walls, was cut out during the great flood of Tunisia in 1969 [19] and was further extended upstream during another flooding occasion 1990. The erosion during these events exposed several phenomena of geological and archaeological interest [63].

The archaeological features present in Oed Seguia are of three categories: (i) finds of a few flint artefacts in the soil at the bottom of the wadi, (ii) pits filled with charcoal and fire-cracked stones and culture layers in general buried beneath coarse debris flow sediments.

The few flint tools are manufactured according to the Capsien culture indicating a wide chronological frame from c. 9000-1000 BC [4, 20, 67]. One of the finds - a bi-facially worked arrowhead - is probably of a late date, belonging to the final phase of the Capsien tradition (ca. 3000-1000 BC). The functional or temporal relation between the flint implements and the other structures are at this stage of research very uncertain. The tools are found as stray-finds on the wadi floor and the pits with charcoal are preserved in situ, as sections hanging in the wadi-wall.

The pits with charcoal and fire-cracked stones are concentrated in a certain area in the wadi system, and not spread randomly. More than 100 pits have been identified only in the exposed wadi-walls of this area, on both sides of the 300 m wide wadi, and they seem to be clustered around and downstream of a natural well coming out of the mountainside on the northern wadi wall. According to local shepherds the water is bitter, not used for drinking and is considered poisonous for humans and animals (although occasionally we have seen sheep and goats drinking from it). Samples of water were collected and analysed for its chemical composition. Compared with the recommended highest concentrations acceptable in Swedish drinking water, ionwise, (according to the National Health Authorities), the content of sodium is four times, magnesium 10 times, sodium 11 times, aluminium three times and sulphate 14 times higher, the former most probably leading to at least severe diarrhoea if consumed. Immediately beside the well, erosion has today exposed a stratigraphy containing culture layers buried below very coarse debris flow sediments indicative of heavy rainfall. The culture layers contain charcoal, pieces of wood and bones from domestic animals. Radiocarbon dating of these findings unequivocally point at a calibrated age of c. 1000 BC (Tab. 2:18-22)

Regarding the pits, they range in size between 0.8 and 1.5 metres in diameter and the depth is around 0.3-0.5 metres. They occur on the wadi-walls at different levels, from the bottom of the wadi, situated directly on top of a calcrete crust, up to 4 metres above this crust, in the unstructured silty sediments, indicating a continuity in heating and cooking at certain spots despite the immense rate of sedimentation.

Stratigraphically the pits change gradually in appearance, from the bottom-most, which are the largest, containing large stones and much charcoal, towards the topmost which are much smaller, with few and small stones, if any, and containing mostly ash and very little charcoal. Charcoal from several pits at different levels and localities was investigated for origin and age. Most charcoal samples originated from Mediterranean species of pine and juniper (Thomas Bartholin, Copenhagen, personal communication). The ^{14}C-datings showed up with some very peculiar results. Regardless of stratigraphical location they had the same age (calibrated centre ages gave a span of 1680-1280 BC), and it even appeared that stratigraphically higher pits gained higher ages than lower. This could be explained by the fact that the last thing to burn on a piece of wood would be the core. Pine and juniper, in this area, could become as old as 1000 years meaning that most ages received by charcoal dating, for the timing of the pits, would be too high. During sieving and microscope preparation of charcoal from the bottom-most pit a small piece of charred herb was found which gained a ^{14}C-date of 2495 80 BP. Situated in the ^{14}C-plateau, calibrated ages gave this sample the possible span of 1Σcal BC 791-410, 2Σ cal BC 808-394. This finding put us into a dating dilemma, are the pits from c. 1500 BC or younger. The small piece gaining a younger age could be a root fragment and hence the pits could be older. Regarding the charcoal found near the well this gained the same age as the bone fragment indicating that the older age of the pits could be correct. The wood buried one metre below the culture layers under coarse sediments (Tab.2:23) with an age of c.1500 BC is another indicator that there may have been two separate episodes with flooding. Further studies of charcoal from the hearths will be made with the hope of finding charred seeds etc. which could reveal the true age. Under all circumstances the ^{14}C-datings, with the same apparent age for all dated structures, irrespective of stratigraphical location, suggests a very short time-span of perhaps a few hundred years or less, meaning that the rate of sedimentation must have been enormous during this period. Most likely, these pits should be interpreted as hearths and/or cooking-pits. The presence of great amounts of fire-cracked stones in the structures signals that their use was related to heating; either warming the people living there or heating their food. In many cases the prehistoric ground levels corresponding to certain hearths/cooking-pits are visible in the wadi wall as a strip of soot and charcoal, starting in the hearth and slowly fading out to a very thin, gray line some 4-6 metres away. This shows both the natural spread of particles from the fire-place and also, perhaps, that the human activities were concentrated in the areas close to the hearths.

The examples of the use of heated stones for cooking and warming purposes are many, and from Europe we can take the burnt mounds from the British Isles or from south Scandinavia as representatives of this technology [6, 7, 60, 61, 74]. An interesting feature in these cases is the presence of water (bog, lake, stream or well) close to the cooking-pits and accumulations of fire-cracked stones [79]. In many examples from the British Isles stone-lined tanks were used for heating the water with hot stones from an adjacent hearth [6].

The case from South Central Tunisia, as discussed here, seems to have one vital thing in common with the European *burnt mounds*, except for their equal spacing in time, and that is the close proximity to water. The well is quite close to the agglomeration of hearths/cooking-pits and it is likely that the well was the primary factor for the location of the hearths in this particular area of the landscape.

As mentioned above, the water is of bad quality and it is not recommendable for drinking neither for humans nor animals. If this was the case already 3500-3000 years ago we don't know for certain, but if so, we must ask the question why the well and the area close to it were of any interest for the people living there.

The possible reasons could have been many, even if the water was undrinkable. One explanation could be that the well served a ritual purpose during the late second millennium BC for the bands of hunters (and perhaps also cattle-breeders) who lived in this neighbourhood. The water from the well could have been used for healing, libation, or for ritual consumption in small quantities by shamans or other types of ritual leaders at particular ceremonial occasions. In that case the hearths and cooking activities around the well are not functionally associated with the water or the well. Instead, they can be interpreted as the remains of large gatherings of people for ritual ceremonies, perhaps a few times a year when hunting bands from a wide region met and strengthened the social bonds. This seasonal gatherings of bands is well known among hunter-gatherer societies [9, 82, 87].

The possible non-utilitarian (ritual) purposes of this well are probably as many as the utilitarian. The connection between wells and ritual activities can be seen in the myths of many people of the earth, and, for example, in Scandinavia *Mimers well* is an important feature in Viking Age mythology (sagas).

Geologically, the sediments of Oed Seguia are mainly unconsolidated medium sized sands and silts showing very little sedimentation structures, indicating a rapid settlement history. This is also confirmed by the charcoal dating, mentioned above. Lumps and pieces of charcoal from the bottom of the sediment sequence date to c. 1700-1200 BC (cal) while an extensive layer of charcoal and ash in the topmost layers, c. 1 m below the present ground level gained an age of only 1000 years (Tab. 2:15-16) coinciding with the oldest event found in Oed Dakhla. The base of the unconsolidated sediment sequence is a calcrete crust c.10 cm thick, most probably a former soil surface. On top of this, the first layer to appear is a structured, deformed pseudo-breccia i.e. folded sands containing lumps and balls of former soil sequences. Most of these lumps are angular while others are well rounded indicating bottom stream transportation. Rounded balls of clay are common in present wadi bottoms, being transported and eroded in occasional floods. However, several of the lumps and balls exposed here near the well, in sediment layers isochronical to the charcoal of the hearths, were made up of loose dune sand, and they easily fell apart when removed from their burial site.

We can find no other explanation for the transportation of loose dune sand in the form of aggregates, at the bottom of a wild stream, than frost i.e. at the time of transportation they were water saturated and deep frozen, or in other words they had been eroded from a dune sand area subjected to deep ground-frost. This indicator of severely cold conditions would certainly explain both the increased need for heating sources and the choice of location near a bedrock well. The present temperature of the water is c. 18°C with very little annual amplitude, and it could be expected that even after a long period of severe cold, running water would be found here.

Without more archaeological and geological fieldwork in this area, the interesting question concerning the relationship between the hearths and the well (functional or ritual?) is impossible to answer in a satisfactory way.

Oed Dahkla

Oed Dakhla (Fig.6:4) is a medium size ravine situated in the mountainous area of Djebel El Kebar 20 km SW Sidi Bouzid. The ravine, which was cut out in a valley filling during the same flooding as Oed Seguia in 1969, is appr. 1 km long, 50 m wide and has 4-6 m steep walls. The sediments are situated on top of an old organic soil horizon gaining a ^{14}C-date (cal.) of 10,550 BC (Tab. 2:39). The valley fillings are thick layers of homogeneous silty sediments lacking structures, and they frequently contain long trunks of wood. The dating performed on buried wood and fragments of charcoal from herbs (Tab. 2:42-50) show that the filling occurred during two separate episodes, the first ca. 1000 AD reaching down to the old soil horizon, and the second c. 1600 AD. No sediments in the time span of 10,550 BC and 1000 AD are found in this place, except for a thin humus/ash horizon (Tab. 2:40-41) situated immediately on top of the old soil horizon, indicating that the valley was emptied sometime during this period. Some logs, from both episodes, were obviously burning during burial, which indicate that the sedimentation was associated with forest fires in the area. Both layers contain logs sedimented in a vertical position, and large boulders indicative of debris flows. No signs of cultural activity were found in Oed Dakhla although some roman inscriptions in a rock near the present valley floor indicate that the valley was empty during Roman times.

4. Conclusions

In conclusion, the climate history, as indicated in playa and alluvial fan deposits in Tunisia, and peat in Sweden, points to a very turbulent late Holocene with several sudden and dramatic events. The most prominent of these events are centred around 2100 BC, 1600-1400 BC, 1000 BC, 1000 AD and 1600 AD, however, their origin and extent are still unknown. The cultural and geological traces of the major two events, 1600-1400 BC and 1000 BC, indicate a sudden transition to extremely cold and humid conditions. The spherules found, which are the only hard evidence from these events could be of volcanic origin, but according to the literature available they could not be associated with any known major volcanism. Their heterogeneous composition points to another formation mechanism, maybe comet or asteroid impacts in ocean shelf sediments. Hence we take the liberty to suggest that relatively large extra-terrestrial bodies hit somewhere in the eastern North Atlantic, probably on the shelf of the Atlantic coast of North-Africa or southern Europe around 1600-1400 BC and 1000-950 BC mainly affecting the Mediterranean parts of Africa and Europe, but also globally.

Acknowledgements

The authors would like to thank the authorities of Sidi Bouzid, Tunisia for their permission to undertake the Tunisian study, and CRDA, Sidi Bouzid for valuable help with the excavations. The project is mainly financed by the Swedish Natural Science Research Council.

References

1. Baillie, M.G.L. 1985. Irish dendrochronology and radiocarbon calibration. *Ulster Journal of Archaeology*, 48, 11-23
2. Baillie, M.G.L. 1992. Dendrochronology and past environmental change. *Proceedings of the British Academy* 77, 5-23
3. Baillie, M.G.L. 1993. Dark ages and dendro-chronology, *Emania* 11, 5-12
4. Balout, L. 1955. *La Prehistoire du lAfrique du Nord.* Paris
5. Barber, K.E. 1982. Peat-bog stratigraphy as a proxy climate record, in: Harding, A. (Ed.) *Climate Change in Later Prehistory,* Edinburgh University Press, pp. 103-113
6. Barber, J. 1990. Burnt mound material on settlement sites in Scotland, in: *Burnt Offerings* (Ed. V. Buckley), Dublin, pp.92-97
7. Barfield, L. and Hodder, M. 1987. Burnt mounds as savannas, and the prehistory of bathing, *Antiquity* 61, 370-379
8. Becker, B. 1982. Dendrochronologie und Paläo-ökologie subfossiler Baumstämme aus Fluss-ablagerungen. Ein Beitrag zur nacheiszeitlichen Auen-entwicklung im südlichen Mitteleuropa, *Mitt. der Kommission für Quartärforschung der Österr. Akad. der Wissenschaften* 5
9. Bettinger, R.L. 1991. *Hunter-Gatherers. Archaeology and Evolutionary Theory.* New York
10. Björck, S., Malmer, N., Hjort, C., Sandgren, P. Ingólfsson, O., Wallén, B., Lewis-Smith, R.I. and Liedberg-Jönsson, B. 1991, Stratigraphic and palaeo-climatic studies of a 5500-year-old moss bank on Elephant Island, Antarctica, *Arctic and Alpine Research* 23(4), 361-374
11. Björck, S., Håkansson, H., Zale, R., Karlén, W. and Liedberg-Jönsson B. 1991, A late Holocene lake sediment sequence from Livingston Island, South Shetland Islands, with palaeoclimatic implications. *Antarctic Science* 3(1), 61-72
12. Brooks, C.E.P. 1949. *Climate through the Ages. A Study of Climatic Factors and their Variations.* Ernest Benn Ltd., London
13. Brown, A.G. and Barber, K.E. 1985. Late Holocene paleoecology and sediment history of a small lowland catchment in central England. *Quaternary Research*, 24, 87-102
14. Bryson, R.A., Lamb, H.H. and Donley, D.L. 1974. Drought and the decline of Mycenae. *Antiquity* 48, 46-50
15. Burgess, C. 1985. Population, climate and upland settlement. In Spratt, D. and Burgess, C. (Eds.) Upland settlement in Britain. The second Millennium B.C. and after. *BAR British Series* 143, 195-230
16. Butzer, K.W., Isaac, G.L., Richardson, J.L. and Washbourn-Kamau, C. 1972. Radioactive dating of East African lake levels. *Science* 175, 1069-1076
17. Carpenter, R. 1966. *Discontinuity in Greek Civilization.* Cambridge: University Press
18. Clapperton, C.M. 1993. *Quaternary Geology and Geomorphology of South America.* Elsevier, Amsterdam
19. Clarke, F., 1973. The great Tunisian flood, *Journal of Research of the U.S. Geological Survey* 1(1), 121-124
20. Clarke, G. 1977. *World Prehistory in New Perspective.* Cambridge
21. Cole, H.S. 1969. *Objective reconstruction of the palaeoclimatic record through the application of eigenvectors of present-day pollen spectra and climate to the late-Quaternary pollen stratigraphy.* Ph.D. thesis, Dept. of Meteorology, University of Wisconsin, Madison

22. Curry, R.R. 1971. Holocene climatic and glacial history of the central Sierra Nevada, California, in: Schuman, S.A. and Bradley, W.C. (Eds.) United States Contributions to Quaternary Research, *Geological Society of America Special Paper* 123, 1-47
23. Davis, O.K. 1992. Rapid climatic change in coastal southern California inferred from pollen analysis of San Joaquin Marsh. *Quaternary Research* 37, 89-100
24. Desborough, V.R.dA. 1972. *The Greek Dark Ages*. Ernest Benn Ltd. London
25. Digerfeldt, G. 1988. Reconstruction and regional correlation of Holocene lake-level fluctuations in Lake Bysjön, South Sweden. *Boreas* 17, 165-182
26. Frailey, C.D., Lavina, E.L., Rancy, A. and de Souza Filho, J. P. 1988. A proposed Pleistocene/Holocene lake in the Amazon basin and its significance to Amazonian geology and biogeography. *Acta Amazonica* 18(3-4), 119-143
27. Franzén, L. 1987. Peat in Sweden - a method to calculate the resources. *GUNI Report* 21
28. Franzén, L. 1989. A dustfall episode on the Swedish west coast, October 1987. *Geografiska Annaler* 71(3-4), 263-267
29. Franzén, L.G. 1990. Transport, deposition and distribution of marine aerosols over southern Sweden during dry westerly storms. *Ambio* 19(4), 180-188
30. Franzén, L.G. 1991. The changing frequency of gales on the Swedish west coast and its possible relation to the increased damage to coniferous forests of southern Sweden. *International Journal of Climatology* 11, 769-793
31. Franzén, L.G. 1992. Can Earth afford to lose the wetlands in the battle against the increasing greenhouse effect? *International Peat Journal Special Edition* 1(3), 1-18
32. Franzén, L.G. 1994. Are wetlands the key to the ice age cycle enigma? *Ambio* 23(4-5), 300-308
33. Franzén, L. and Lindqvist, S. 1988. Precipitation - a case history from the Sidi Bouzid area. In Franzén, L. (ed.) Field course in the Sidi Bouzid area, Central Tunisia, April 1988 (Lindqvist et al.) *GUNI Report* 26
34. Franzén, L. and Hjelmroos, M. 1988. A coloured snow episode on the Swedish west coast, January 1987 - a quantitative and qualitative study of air-borne particles. *Geografiska Annaler* 70A, 235-243
35. Franzén, L., Hjelmroos, M. and Kållberg, P. 1992. Long range transport of airborne solid particles into northern Fennoscandia - the yellow snow episode of March 1991, in the light of the Chernobyl accident of 1986. *Arctic Centre Publications* 4, 69-72
36. Franzén, L.G., Hjelmroos, M., Kållberg, P., Brorström-Lundén, E., Juntto, S. and Savolainen, A.-L. 1994a. The yellow snow episode of northern Fennoscandia, March 1991 - a case study of long-distance transport of soil, pollen and stable organic compounds. *Atmospheric Environment* 28(22), 3587-3 604
37. Franzén, L.G., Mattsson, J.O., Mårtensson, U., Nihlén, T. and Rapp, A. 1994. Yellow snow over the Alps and Subarctic from a dust storm in Africa, March 1991. *Ambio* 23(3), 233-235
38. Franzén, L., Hjelmroos, M., Kållberg, P., Rapp, A., Mattsson, J.O. and Brorström-Lundén, E. 1995. The Saharan dust episode of south and central Europe, and northern Scandinavia, March 1991, *Weather* 50(9), 313-318
39. Franzén, L.G., Chen, D. and Klinger, L.F. 1996: Principles for a climate regulation mechanism during the late Phanerozoic era, based on carbon fixation in peatforming wetlands. *Ambio*.25(7), 435-442
40. Fredriksson, D. 1996: Peat resources in Sweden. In: Lappalainen, E. (Ed.) *Global Peat Resources*. IPS, pp.137-148
41. Frenzel, B. 1966. Climatic change in the Atlantic/sub-Boreal transition on the Northern Hemisphere: botanical evidence. In: Sawyer, J.S. (Ed.) *Proceedings of the International Symposium on World Climate 8000 to 0 B.P.*, pp. 99-123
42. Frenzel, B. 1983. Mires - repositories of climatic information or self-perpetuating ecosystems. In Gore, A.J.P. (Ed.) *Ecosystems of the World. Mires: Swamp, Bog, Fen and Moor*, 4A General studies. Elsevier, Amsterdam, pp. 35-65
43. Gams, H. and Nordhagen, R. 1923. Postglaciale Klimaänderungen und Erdkrustenbewegungen in Mitteleuropa. *Landeskundliche Forschungen* 25
44. Godwin, H. 1966. Introductory address, In: Sawyer, T.S. (Ed.), *Proceedings of the International Symposium on World Climate 8000 to 0 B.P.*, pp. 3-14
45. Gosden, C., Allen, J, Ambrose, W., Anson, D., Golson, J., Green, R., Kirch, P., Lilley, I., Specht, J. and Spriggs, M. 1989. Lapita sites of the Bismarck Archipelago. *Antiquity* 63, 561-586
46. Gradin, C. 1980. Secuencias radiocarbónicas del sur de la Patagonia, Argentina. *Relaciones Sociedad Argentina de Antropologia* 14, 177-194
47. Granlund, E. 1932. De svenska högmossarnas geologi. *Sveriges Geologiska Undersökning* (The Swedish Geological Survey) C 373, 193 p. (in Swedish)
48. Grove, J. 1988. *The Little Ice Age*. Methuen, London
49. Gustafsson, M.E.R. and Franzén, L.G. 1996: Dry deposition and concentration of marine aerosols in a coastal area, SW Sweden. *Atmospheric Environment* 30(6), 977-989
50. Hjelmroos, M. and Franzén, L.G. 1989. Desertation and soil damage in Central Tunisia during the Holocene. *Pact* 33, 33-38
51. James, P. 1991. *Centuries of Darkness. A challange to the conventional chronology of Old World archaeology*. Jonathan Cape, London
52. Karlén, W. 1976. Lacustrine sediments and tree-limit variations as indicators of Holocene climatic fluctuations in Lapland, northern Sweden. *Geografiska Annaler* 58(1-2), 1-34
53. Karlén, W. 1988. Scandinavian glacial and climatic fluctuations during the Holocene. *Quaternary Science Reviews* 7, 199-209
54. Kay, P.A. and Johnson, D.L. 1981. Estimation of Tigris-Euphrates streamflow from regional Palaeo-environmental data. *Climatic Change* 3, 251-263
55. Koeberl, C. and Sigurdsson, H. 1992. Geochemistry of impact glasses from the K/T boundary in Haiti: relation to smectites and a new type of glass. *Geochim. Cosmochim. Acta* 56(5), 2113-2129
56. Kristiansen, K. 1982. The formation of tribal systems in later European prehistory: Northern Europe, 4000-500 B.C. In Renfrew, C., Rowlands, M.J. and Segraves, B.A. (Eds.), *Theory and Explanation in Archaeology*. New York and London, pp.241-280
57. LaMarche, V.C.Jr. 1978. Tree-ring evidence of past climatic variability. *Nature* 276, 334-338
58. Lamb, H.H. 1995. *Climate, History and the Modern World*. Routledge
59. Lamb, H.H., Lewis, R.P.W. and Woodroffe, A. 1966. Atmospheric circulation and the main climatic variables between 8000 and 0 B.C.: Meteorological evidence. In: Sawyer, J.S. (Ed.), *Proceedings of the International Symposium on World Climate 8000 to 0 B.P.*, pp.174-217
60. Larsson, T. B. 1989. A spatial approach to socioeconomic change in Scandinavia: Central Sweden in the first millennium B.C. In: Gibson, B. And Geselowitz, M. (Eds.), *Tribe and Polity in late Prehistoric Europe. Demography, Production and Exchange in the Evolution of Complex Social Systems* , pp. 97-115. Plenum Press. New York
61. Larsson, T. B. 1990. Skärvstenshögar - the burnt mounds of Sweden. In: Buckley, V. (Ed.), *Burnt Offerings*, pp. 142-153). Dublin
62. Larsson, T.B. and Franzén, L.G. (in press). *The Capsien Culture at Ez Zaf Zaf, Chott Nejla, Central Tunisia*. Preliminary Report, Univ. of Umeå, Univ. of Göteborg, Sweden
63. Larsson, T.B. and Franzén, L.G. (in press). *The anomalous dwelling sites of Oed Sequia, Central Tunisia. A comparison between fire cracked stones in Tunisia and Northern Europe*. Preliminary Report, Univ. of Umeå, Univ. of Göteborg, Sweden
64. Last, W.M. and Schweyen, T.H. 1985. Late Holocene history of Walsea Lake, Saskatchewan, Canada. *Quaternary Research* 24, 219-234
65. Leuschner, H.-H., Delorme, A., Tüxen, J and Höfle, H.-Chr. 1986. Über Eichenwaldhorizonte in küsten-nahen Mooren Ostfrieslands, *Telma* 16, 61-82
66. Leuschner, H.-H. and Delorme, A. 1988. Tree-ring work in Göttingen - absolute oak chronologies back to 6255 B.C., *Pact II.5 Wood and Archaeology*, 123-132
67. MacBurney, C. B. M. 1961. The Stone Age of Northern Africa. London.McKenzie, J.A. and Eberli, G.P. 1987. Indications for abrupt Holocene climatic change: Late Holocene oxygen isotope stratigraphy of the Great Salt Lake, Utah. In Berger,

W.H. and Labeyrie, L.D. (Eds.) *Abrupt Climatic Change*, D. Reidel Publ. Comp., pp.127-136
68. McKenzie, J. and Eberli, G.P. 1987. Indications for abrupt Holocene climatic change: late Holocene oxygen isotope stratigraphy of the Great Salt Lake, Utah. In: Reidel, N. (Ed.), *Abrupt Climatic Change* SI Ser. C 216, 127-136
69. Mercer, J.H. 1970. Variation of some Patagonian glaciers since the late-Glacial: *II Americian Journal of Science* 269, 1-25
70. Mercer, J.H. 1982. Holocene glacier variations in southern Patagonia. *Striae*, 18, 35-45
71. Mercer, J.H. and Palacios, O. 1977. radiocarbon dating of the last glaciation in Peru. *Geology* 5, 600-604
72. Mercer, R.J. 1991. The Highland zone: reaction and reality 5000 B.C.-2000 A.D. *Proceedings of the British Academy* 76, 129-150
73. Neumann, J. and Parpola, S. 1987. Climatic change and the eleventh-tenth-century eclipse of Assyria and Babylonia. *Journal of Near Eastern Studies* 46(3), 161-182
74. O'Kelly, M. 1954. Excavations and experiments in ancient Irish cooking-places. *Journal of the Royal Society of Antiquaries of Scotland* 84, 105-155
75. Parry, M. 1978. *Climatic Change, Agriculture and Settlement.* Folkstone
76. Persson, C. 1966. Försök till tefrokronologisk datering av några svenska torvmossar. *GFF* 88(3), 361-394 (in Swedish)
77. Persson, C. 1967. Nya C14-dateringar av horisonter med vulkanisk aska i svenska torvmossar. *GFF* 88(4), 49 (in Swedish)
78. Porter, S.C. and Denton, G.H. 1967. Chronology of neoglaciation in the North American Cordillera. *American Journal of Science* 265(3), 177-210
79. Power, D. 1990. *Fulachta fiadh in County Cork.* (Ed. V. Buckley), pp. 13-17, Dublin
80. Raftery, B. 1990. *Trackways through time. Archaeological investigations on Irish Bog Roads, 1985-1989.* Headline Publishing, Dublin
81. Räsänen, M., Neller, R., Salo, J. and Jungner, H. 1991. Recent and ancient fluvial depositional systems in the Andean forelands of the Peruvian Amazon. In Räsänen, M. 1991. History of the fluvial and alluvial landscapes of the western Amazon Andean forelands. *Annals of the University of Turkuensis* Ser.A 75(3), 1-22
82. Renfrew and Bahn 1991. *Archaeology. Theories, Methods and Practice.* Thames and Hudson
83. Rodbell, D.T. 1992. Lichenometric and radiocarbon dating of Holocene glaciation, Cordillera Blanca, Peru. *The Holocene* 2(1),19-29
84. Röthlisberger, F., 1986, *10,000 Jahre Gletscher-geschichte der Erde.* Sauerländer, Aarau
85. Scuderi, L.A. 1984. *A dendroclimatic and geomorphic investigation of Late-Holocene glaciation, southern Sierra Nevada, California,* Ph.D. dissertion, Univ. of California, Los Angeles
86. Scuderi, L.A. 1987. Glacier variations in the Sierra Nevada, California, as related to a 1200-year tree-ring chronology. *Quaternary Research* 27, 220-231
87. Service, E. R. 1966. The Hunters. *Foundations of Modern Anthropology Series.* New Jersey
88. Spriggs, M. 1989. The dating of the Islands Southeast Asian Neolithic: an attempt at chronometric hydiene and linguistic correlation. *Antiquity* 63, 587-613
89. Starkel, L. 1966. Post-glacial climate and the moulding of European relief. In: Sawyer, J.S. (Ed.), *Proceedings of the International Symposium on World Climate 8000 to 0 B.C.* pp.15-33
90. Stuiver, M. and Reimer, P.J. 1986. A computer program for radiocarbon age calibration. *Radiocarbon* 28, 1022-1030
91. Taylor, J.A. 1980. Culture and environment in prehistoric Wales. *Brittish Arcaeol. Rep.* 76
92. Valeton, I. 1978. A morphological and petrological study of the terraces around Lake Van, Turkey. In: Degens, E.T. and Kurtman, F. (Eds.) *The Geology of Lake Van*, Ankara 1978, pp. 64-80
93. Van Geel, B. 1978. A palaeoecological study of Holocene peat bog sections in Germany and the Netherlands. *Review of Palaeobotany and Palynology*, 25, 1-120
94. Weir, D.A. 1993. Dark ages and the pollen record. *Emania* 11, 21-30
95. Wendland, W.M. and Bryson, R.A. 1974. Dating climatic episodes of the Holocene. *Quaternary Research* 4, 9-24
96. Wirrman, D., Mourguiart, P. and de Oliveira Almeida, L.F. 1988. Holocene sedimentology and ostracodes distribution in Lake Titicaca- Paleohydrological interpretations. *Quat. S. America and Antarctic Peninsula*, 6, 89-128
97. Visset, L., Girardclos, O. and Lambert, G.N. 1994. La foret de chenes sur tourbe, a lile dErrand, dans les marais de Briere (Massif amorican-France). Palynologie et premiers resultats dendrochronologiques. *Quaternaire* 5(2), 69-77
98. Witting, M. 1947. Katjonbestämningar i myrvatten. *Botaniska Notiser* 4, 287-304, (in Swedish)
99. Witting, M. 1948. Preliminärt meddelande om fortsatta katjonbestämningar i myrvatten sommaren 1947. *Svensk Botanisk Tidskrift* 42, 116-134, (in Swedish)
100. Wright, H.E.Jr. 1960. Climate and prehistoric man in the eastern Mediterranean. In: Braidwood, R.J. and Howe, B. (Eds.) *Prehistoric investigations in Iraqi Kurdistan - Studies in Ancient Oriental Civilization* 31, 71-97
101. Zangger, E. 1992. *The Flood from Heaven. Decipering the Atlantis Legend.* Morrow, New York

Table 1: Spherules in bog and playa deposits and for a comparison spherules associated with the K/T impact event [55]. Note that, for technical reasons, the results have been normalised.

a/ Tektites from Beloc K/T and bog spherules of various composition

	Beloc	Beloc	**Beloc mean**	1 Oval	2	3	4	5	**Sw. mean**
SiO_2	44.4	67.5	**63.1**	47.3	31.7	60.2	63.0	40.6	**48.6**
TiO_2	0.7	0.6	**0.7**	0.6	0.8	0.2	0.2	1.6	**0.7**
Al_2O_3	11.7	15.1	**15.2**	14.3	14.7	12.0	8.5	11.2	**12.2**
FeO	4.8	4.3	**5.4**	<0.1	2.9	14.5	13.8	12.0	**8.7**
MnO	0.1	0.2	**0.1**	0.3	0.2	0.2	0.2	3.5	**0.9**
MgO	4.4	2.3	**2.7**	4.0	10.1	0.3	1.2	6.5	**4.4**
CaO	30.7	5.0	**7.3**	30.9	36.3	9.4	11.4	12.4	**22.1**
NaO	2.0	3.5	**3.6**	1.1	1.3	0.8	1.1	1.8	**1.2**
K_2O	0.5	1.8	**1.6**	1.6	1.9	2.1	0.5	0.4	**1.3**

b/ Other Swedish bog spherules

	1	2	3	4	5	6	7
SiO_2	66.5	57.0	59.8	73.9	20.9	18.0	57.5
TiO_2	0.1	0.2	1.0	0.2	0.2	0.2	0.2
Al_2O_3	10.6	11.3	6.2	0.4	<0.1	<0.1	3.7
FeO	11.0	16.5	18.3	0.2	37.3	31.0	10.4
MnO	0.2	0.2	0.2	<0.1	<0.1	0.1	0.5
MgO	2.0	1.3	1.3	4.5	0.6	0.6	23.4
CaO	7.9	13.0	12.4	6.2	10.7	13.5	2.4
Na2O	0.5	0.0	0.1	14.0	0.5	0.5	1.8
K_2O	1.1	0.5	0.3	0.4	0.1	0.1	0.1
Cr_2O_3	-	-	-	-	30.1	36.3	-

c/ Tunisia (perfectly glossy glass spherules)

	1	2	3	4	5	6	7	8	9	**Mean**
SiO_2	48.3	39.9	77.8	53.0	57.8	50.3	37.1	40.4	52.0	**50.6**
TiO_2	0.5	0.5	0.2	0.6	0.1	0.4	0.5	0.4	0.4	**0.4**
Al_2O_3	20.2	16.3	<0.1	18.8	5.5	19.8	13.1	19.2	10.4	**14.1**
FeO	5.2	4.7	0.1	6.1	18.3	5.2	4.7	3.8	19.8	**6.0**
MnO	<0.1	<0.1	0.1	0.1	0.2	0.1	0.0	<0.1	0.1	**0.1**
MgO	11.5	13.8	3.2	11.0	1.5	6.5	13.6	18.5	3.7	**10.0**
CaO	9.8	19.6	9.3	7.2	15.3	12.8	15.3	10.6	12.2	**12.5**
Na_2O	2.3	3.6	9.3	0.9	0.5	2.7	13.0	4.2	0.0	**4.6**
K_2O	2.3	1.6	0.1	2.5	0.5	2.1	3.1	3.1	0.5	**1.9**

d/ Tunisia (glossy iron-type spherules, mainly black)

	1	2	3	4	5	6	7	8	9	**Mean of 2-9**
SiO_2	1.4	5.9	11.9	11.9	9.3	14.0	0.6	7.6	18.1	9.8
TiO_2	42.9	0.5	0.1	0.4	1.2	0.1	0.1	0.0	0.4	0.4
Al_2O_3	1.5	3.8	5.0	5.4	4.4	5.6	0.0	2.0	8.3	4.3
FeO	50.0	84.9	77.3	77.3	82.0	75.6	98.8	85.4	67.6	81.0
MnO	0.6	0.3	0.2	1.2	0.2	1.0	0.5	0.7	0.3	0.6
MgO	3.6	2.1	4.4	2.4	2.6	2.7	0.1	3.3	3.5	2.6
CaO	0.1	1.5	0.6	0.5	0.1	0.3	0.0	0.3	0.5	0.5
Na_2O	0.0	0.9	0.3	0.5	0.2	0.4	0.1	0.6	0.7	0.5
K_2O	0.1	0.2	0.2	0.4	0.1	0.5	0.0	0.2	0.7	0.3

Table 2: Radiocarbon and luminesence datings

Nr.	Lab.nr.	Material	^{14}C-date B.P.	Calibrated date	Note
Oed Seguia, ravine with fire places 35°00.19' N, 9°39.73' E					
1/	Ua-4558	Charcoal	3235±70	1Σ B.C. 1594 (1510) 1421 2Σ B.C. 1676 (1510) 1326	Pit no 1 from bottom
2/	Ua-11033	Charcoal	2495±80	1Σ B.C. 791 (760, 680, 660, 640, 550) 410 2Σ B.C. 808 (760, 680, 660, 640, 550) 394	Small piece of charred herb from pit 1
3/	Ua-4560	Charcoal	3320±55	1Σ B.C. 1675 (1610, 1560, 1540) 1519 2Σ B.C.1737 (1610, 1560, 1540) 1449	Pit no 2 from bottom (W)
4/	Ua-11029	Charcoal	3375±60	1Σ B.C.1738 (1680) 1531 2Σ B.C.1865 (1680) 1516	"
5/	Ua-11031	Charcoal	3390±65	1Σ B.C.1744 (1680)1546 2Σ B.C.1876 (1680) 1517	"
6/	Ua-4559	Charcoal	3170±70	1Σ B.C. 1513 (1420) 1328 2Σ B.C.1598 (1420) 1264	Pit no 3 from bottom
7/	Ua-4561	Charcoal	3040±55	1Σ B.C.1390 (1290, 1280, 1270) 1208 2Σ B.C. 1415 (1290, 1280, 1270) 1119	Pit no 4 from bottom
8/	Ua-11033	Charcoal	3300±100	1Σ B.C. 1683 (1530) 1442 2Σ B.C. 1869 (1530) 1328	"
9/	Ua-4562	Charcoal	3230±60	1Σ B.C. 1525 (1510) 1424 2Σ B.C.1627 (1510) 1395	Pit no 5 from bottom
10/	Ua-11026	Charcoal	2490±55	1Σ B.C. 775 (760, 680, 650, 550) 423 2Σ B.C. 797 (760, 680, 650, 550) 403	Charred herbs 1 m below surface above pits
11/	Ua-4557	Charcoal	3075±75	1Σ B.C.1414 (1380, 1350, 1320) 1219 2Σ B.C.1510 (1380, 1350, 1320) 1117	Charcoal 6 m below surface
12/	Ua-11942	Charcoal	2190±65	1Σ B.C. 366 (200) 160 2Σ B.C. 391 (200) 45	Charcoal found beside flint arrow head 0.5 m below surface
13/	Ua-11943	Charcoal	2435±60	1Σ B.C. 759 (490, 440, 420) 402 2Σ B.C. 783 (490, 440, 420) 390	Charcoal piece 1.5 m below surface 4 m above bottom
Oed Seguia, south side, 34°59.73' N, 9°40.04' E					
14/	Ua-11027	Charcoal	3310±55	1Σ B.C. 1672(1600, 1570, 1530) 1516 2Σ B.C. 1734 (1600, 1570, 1530) 1441	Pit, 4 m below surface
15/	Ua-4554	Charcoal	1055±60	1Σ A.D. 899 (1000) 1024 2Σ A.D. 883 (1000) 1155	Charcoal mixed with ash and peat 1 m below surface
16/	Ua-4555	Charcoal	1495±60	1Σ A.D. 539 (600) 638 2Σ A.D. 428 (600) 662	
17/	Ua-4556	Charcoal	1070±65	1Σ A.D. 894 (990) 1021 2Σ A.D. 869 (990) 1153	
Well in Oed Seguia, 35°00.16'N, 9°39.49'E					
18/	Ua-3529	Bone	2840±50	1Σ B.C. 1035 (994) 915 2Σ B.C. 1126 (994) 847	Hip bone from cow
19/	Ua-3530	Bone	2845±55	1Σ B.C. 1045 (1000) 915 2Σ B.C. 1157 (1000) 844	Vertebra from ass
20/	Ua-3013	Wood	2855±70	1Σ B.C. 1118 (1000) 913 2Σ B.C 1255 (1000) 833	Small piece of wood from some bush
21/	Ua-3021	Charcoal	2820±70	1Σ B.C. 1035 (980, 970, 930) 867 2Σ B.C. 1160 (980, 970, 930) 817	Charred herb found in culture layer below bones
22/	Deb-2029	Charcoal	2850±40	1Σ B.C. 1035 (999) 926 2Σ B.C. 1122 (999) 904	Charcoal horizon in sand
23/	Deb-2007	Wood	3240±50	1Σ B.C. 1522 (1513) 1442 2Σ B.C. 1608 (1513) 1417	Wood buried in silt partly charred. 1 m below culture layer.
Chott Nejla, outlet, 34°49.16' N, 9°31.11' E					
24/	Ua-11945	Shell	9685±95	1Σ B.C. 9034 (8990) 8638 2Σ B.C. 9053 (8990) 8524	Shell from 'escargotiére' *Sphincterochila candidis*
25/	Ua-11760	Egg shell	8350±65	1Σ B.C. 7484 (7430) 7302 2Σ B.C. 7505 (7430) 7103	Ostrich egg shell
Chott Nejla, excavation site for stratigraphical studies, 34°52.49'N, 9°29.30' E					
26/	Ua-1515	Shell	1280±100	1Σ A.D. 659 (720, 740, 770) 883 2Σ A.D. 603 (720, 740, 770) 984	34-36 cm below surface *Sphincterochila sp.*
27/	Ua-1519	Shell	8815±330	1Σ B.C. 8092 (7920) 7503 2Σ B.C. 8583 (7920) 7047	84-86 cm below surface small fragment of shell
28/	Ua-1516	Shell	2615±100	1Σ B.C. 834 (800) 606 2Σ B.C. 971 (800) 410	120-122 cm below surface *Sphincterochila sp.*
29/	Ua-1517	Carbonate	3465±100	1Σ B.C. 1889 (1750) 1645 2Σ B.C. 2030 (1750) 1519	Carbonate nodule 136-138 cm below surface
30/	Ua-1518	Carbonate	3585±70	1Σ B.C. 2022 (1920) 1783 2Σ B.C. 2133 (1920) 1740	Carbonate nodule 148-150 cm below surface
31/	Ua-1520	Carbonate	3270±75	1Σ B.C. 1622 (1520) 1439 2Σ B.C. 1734 (1520) 1400	Carbonate nodule 170-172 cm below surface
32/	Ua-13268	Shell	3315±65	1Σ B.C. 1677 (1600, 1560, 1530) 1515 2Σ B.C. 1741 (1600, 1560, 1530) 1431	128-130? cm below surface *Sphincterochila sp.*
33/	Ua-13269	Shell	3060±70	1Σ B.C. 1405 (1310) 1212 2Σ B.C. 1443 (1310) 1115	128-130? cm below surface *Sphincterochila sp.*
Luminescens) (OSL on feldspar			B.P.	cm below surface	
34/	R-912801		200±50	8-12	
35/	R-912802		900±100	32-36	
36/	R-892801		1800±200	58-62	
37/	R-912803		1400±100	104-108	
38/	R-892803		4700±400	128-132	
Oed Dahkla, ravine in Djebel el Kbar, 34°54.17' N, 9°20.55' E					
39/	Ua-3021	Soil	10575±95	1Σ B.C. 10668 (10550) 10425 2Σ B.C. 10775 (10550) 10273	Old soil horizon forming base of younger sed.
40/	Deb-2023	Humus	4730±50	11Σ B.C. 3625 (3610, 3512, 3392, 3389) 3378 21Σ B.C. 3639 (3610, 3512, 3392, 3389) 3365	Humus and ash above Ua-3021
41/	Ua-3016	Humus	4475±75	1Σ B.C. 3339 (3260, 3250, 3100) 2930 2Σ B.C. 3366 (3260, 3250, 3100) 2914	Same material as Deb-2023
42/	Ua-4552	Charcoal	970±50	1Σ A.D. 1016 (1032) 1159 2Σ A.D. 988 (1032) 1206	Charcoal in silt 6 m below surface
43/	Deb-2029	Wood	860±40	1Σ A.D. 1164 (1214) 1230 2Σ A.D. 1045 (1214) 1278	Wood buried in silt 5 m below surface
44/	Deb-2030	Wood	870±40	1Σ A.D. 1161 (1195) 1226 2Σ A.D. 1040 (1195) 1276	Wood buried in silt 7 m below surface
45/	Ua-3014	Charcoal	400±60	1Σ A.D. 1440 (1470) 1627 2Σ A.D. 1419 (1470) 1648	Charred herbs in 1 layer 1 m below surface
46/	Ua-3015	Charcoal	845±70	1Σ A.D. 1129 (1220) 1277 2Σ A.D. 1026 (1220) 1292	Charred herbs in layer 2 m below surface
47/	Deb-2003	Wood	270±40	1Σ A.D. 1637 (1651) 1663 2Σ A.D. 1515 (1651) 1954	Wood buried in silt 2 m below surface
48/	Deb-2006	Wood	400±40	1Σ A.D. 1445 (1473) 1616 2Σ A.D. 1434 (1473) 1635	Wood near surface
49/	Deb-2014	Wood	360±40	1Σ A.D. 1471 (1511, 1600, 1616) 1635 2Σ A.D. 1445 (1511, 1600, 1616) 1648	Wood buried in silt 1 m below surface
50/	Ua-11030	Wood	525±65	1Σ A.D. 1397 (1420) 1441 2Σ A.D. 1303 (1420) 1476	Surface wood from lump 2 m below surface
Raised bog Lyngmossen (Fig.2)					
51/	Ua-13265	Peat	1720±60	1Σ A.D. 248 (340) 412 2Σ A.D. 149 (340) 441	*Eriophorum vaginatum* fibre 205 cm below surface. Layer with spherules
52/	Ua-13266	Peat	3150±65	1Σ B.C. 1500 (1410) 1322 2Σ B.C. 1523 (1410) 1261	*Eriophorum vaginatum* fibre 365 cm below surface. Layer with spherules.
53/	Ua-13267	Peat	3180±70	1Σ B.C. 1515 (1430) 1396 2Σ B.C. 1606 (1430) 1267	*Eriophorum vaginatum* fibre 375 cm below surface.
Raised bog Vildmossen (Fig.2). All these samples were taken in the theoretical level of a calcium event i.e. 20 cm above Ca-peaks.					
54/	Ua-11604	Peat	1255±55	1Σ A.D. 684 (780) 873 2Σ A.D. 662 (780) 892	*Eriophorum vaginatum* fibre 150 cm below surface.
55/	Ua-11605	Peat	1805±55	1Σ A.D. 139 (240) 325 2Σ A.D. 84 (240) 387	*Eriophorum vaginatum* fibre 200 cm below surface.
56/	Ua-11606	Peat	2335±60	1Σ B.C. 406 (390) 374 2Σ B.C. 523 (390) 206	*Eriophorum vaginatum* fibre 250 cm below surface.
57/	Ua-11607	Peat	2760±70	1Σ B.C. 987 (900) 822 2Σ B.C. 1046 (900) 801	*Eriophorum vaginatum* fibre 310 cm below surface. Layer with spherules.
58/	Ua-11608	Peat	2950±55	1Σ B.C. 1258 (1150, 1130) 1038 2Σ B.C. 1371 (1150, 1130) 993	*Eriophorum vaginatum* fibre 390 cm below surface.
59/	Ua-11609	Peat	3930±55	1Σ B.C. 2496 (2460) 2332 2Σ B.C. 2568 (2460) 2206	*Eriophorum vaginatum* fibre 375 cm below surface

Solar Forcing of Abrupt Climate Change around 850 Calendar Years BC

Bas van Geel

The Netherlands Centre for Geo-ecological Research, University of Amsterdam, Kruislaan 318, 1098 SM Amsterdam, The Netherlands

Oleg M. Raspopov

St.Petersburg Branch of the Institute of Terrestrial Magnetism, Ionosphere and Radiowave Propagation, Russian Academy of Sciences, PB 188, St.Petersburg 191023, Russia

Johannes van der Plicht

Centre for Isotope Research, University of Groningen, Nijenborgh 4, 9747 AG Groningen, The Netherlands

Hans Renssen

The Netherlands Centre for Geo-ecological Research, Free University, De Boelelaan 1085, 1081 HV Amsterdam The Netherlands

Summary

A sharp rise in the ^{14}C-content of the atmosphere, which occurred at the Bronze Age-Iron Age transition in NW-Europe (around 850 calendar years BC), was contemporaneous with indications for an abrupt climate change from relatively warm and continental to cooler and wetter conditions (the transition of the Sub-boreal to the Sub-atlantic). Archaeological and palaeoecological evidence for the abandonment of low-lying areas in The Netherlands can be explained due to a rise of the water table and the extension of fens and bogs. There is palaeoecological, geological/ geomorphological and archaeological evidence for synchronous climatic change elsewhere in Europe and on other continents. The discussed ^{14}C-oscillation was caused by changes in solar activity. An important effect of a reduced solar activity is an increase in the cosmic-ray flux leading to a higher production of ^{14}C in the upper atmosphere. This increased cosmic ray flux can cause a reduction in the amount of solar energy reaching the earth surface through mechanisms such as a reduction in the ozone layer density, the development of an aerosol layer, an atmospheric veil and increased cloudiness and precipitation. All these factors amplified the effects of a reduced solar activity, which appears to be the forcing mechanism behind the change to cooler and wetter conditions at middle and high latitudes in both hemispheres. We explain the recorded synchronous change to dryness in tropical areas as the effect of a decrease of the latitudinal extension of the Hadley Cell circulation and a possible associated weakening of monsoonal systems. Furthermore an expansion of the Polar Cells and a repositioning of the main depression tracks at mid-latitudes towards the equator may have occurred. Our evidence supports the idea that solar/cosmic ray forcing is an important factor in the present climate and may also dominate climatic changes in the near future.

1. Introduction

The NW-European Holocene (Flandrian) peat deposits form a valuable archive for the reconstruction of climate change. Changes in the species composition of peat forming vegetation, especially those of mosses (*Sphagnum* species) in raised bogs, are indicative for climate change [3, 25, 51].

Kilian *et al.* found evidence for climatic cooling around 850 calendar years (cal) BC, which is around 2750 BP on the radiocarbon time scale (see fig. 1). Based on archaeological, palaeoecological and geological/ geomorphological evidence, van Geel *et al.* [54] concluded that climate change around 850 cal BC occurred in both hemispheres. Van Geel and Renssen [55] explained the evidence for this climatic change with a climate model. Here we present evidence for abrupt climate change around 850 cal BC, and we tentatively explain this event as the result of a temporarily reduced solar activity and related increase of cosmic ray flux, causing both increased radiocarbon production and climate change to cooler and wetter conditions in the temperate zones. In various regions on earth the effect of the 850 cal BC change of climate was a catastrophe from the human point of view, whereas in other areas the change was an advantage for prehistoric man.

Determinations of radiocarbon ages in BP (Before Present = 1950) are based on the half-life of ^{14}C of 5568 years and on the assumption of a constant atmospheric ^{14}C level. Changes in the atmospheric ^{14}C level in the past appear as fluctuations in the calibration curve. Pronounced fluctuations are called wiggles and these often hamper the precise dating procedure in calendar years by calibration. Van Geel and Mook [53] proposed 'wiggle matching' of a series of dates from organic deposits with the calibration curve as a strategy to overcome the problems of inaccurate dating of single radiocarbon dates. Kilian *et al.* [25] applied this high resolution wiggle match dating (WMD) technique to raised bog peat at the Sub-boreal - Sub-atlantic transition. The climate change during that transition has already been well recognised by the pioneers in bog studies and Holocene climate change (L. von Post, R. Sernander, C.A.Weber, H. Godwin, F. Overbeck; see reference 2). However, they did not know the mechanism to explain climate change, and the study of fossils and climate change still had a rather descriptive character. Sernander ([41]; criticised by Sörbom [45]) combined the palaeoclimatological evidence based on raised bog deposits with a story given in the Edda. He suggested that the so-called Fimbul-winter (according to the saga, the sun, moon and stars were swallowed by a wolf and a period of three continuous, extremely severe winters followed) might go back to experiences during the Sub-boreal/Sub-atlantic transition.

WMD allowed Kilian *et al.* [25] to find a relationship between the recorded climate induced changes of the spectrum of peat forming plants and a sharp rise of the atmospheric radiocarbon content ($\Delta^{14}C$)*. Van Geel *et al.* [54] collected archaeological and palaeoecological evidence for climate change around 850 cal BC but no clear mechanism explaining the relation between $\Delta^{14}C$ and climate change was understood. The possible relationship between fluctuations in $\Delta^{14}C$ and climate change was suggested by one of the pioneers of ^{14}C studies, Hessel de Vries [10], who noticed a major increase of $\Delta^{14}C$ during the Little Ice Age, a well known period of climatic deterioration and reduced solar activity (the Maunder minimum). Indeed for the last millennium, the known solar activity minima correspond to increased cosmogenic isotope (^{14}C and ^{10}Be) production [4], and possibly climate change as recorded in ice cores [49]. A climate model for the 850 cal BC climate change was suggested by van Geel and Renssen [55]. In the present paper we give an overview of evidence for climate change around 850 cal BC and we propose a possible explanation of the observed phenomena.

2. Evidence for climate change around 850 BC raised bogs in Europe

The history of vegetation and climate can be reconstructed from peat samples over periods of thousands of years. An important example is the Sub-boreal/Sub-atlantic transition in NW-Europe, which is often characterised by a sudden change in peat forming mosses. At many sites we find slightly decomposed *Sphagnum papillosum* and *S. imbricatum* (indicating oceanic conditions) on top of more decomposed *Sphagna* of the section *Acutifolia*. Kilian *et al.*, in 1995 [25] and van Geel *et al.*, in 1996 [54], studied this transition in continental European raised bogs and found that this most important change in peat forming vegetation, indicating a transition to a more oceanic climate, was synchronous with a sudden and sharp rise in the atmospheric ^{14}C content. More recently, Van Geel *et al.* [56] wiggle-matched radiocarbon dates in a raised bog core from Carbury Bog in Ireland. This core also showed a sharp transition from more decomposed peat, rich in *Sphagnum* section *Acutifolia* and ericaceous rootlets, to *Sphagnum imbricatum* peat (more oceanic conditions) at the start of the sharply rising atmospheric radiocarbon content around 850 cal BC.

3. The end of settlement sites in the northern Netherlands in hydrologically marginal areas and migration to the salt marshes

For the northern Netherlands van Geel *et al.* [54 ,56] found archaeological evidence that climate change around 850 cal BC must have had dramatic effects for prehistoric farmers. Based on excavations in the West-Friesland area (province of Noord-Holland) it was concluded that the settlement phase suddenly stopped during the Late Bronze Age and no settlements were present until Medieval time. The period of the last circa 140 "radiocarbon years" of the Bronze Age settlements (between 2760 ± 35 BP and 2620 ± 20 BP, and corresponding to only circa 50 calendar years) was characterised by houses that were built on artificial mounds. The houses must be there due to the ground water level being too high. All the radiocarbon dates of that last wet phase of the settlements correspond with the first circa 50 years of the period of sharply rising ^{14}C content of the atmosphere (see fig. 1). The very last date of the settlement phase in West-Friesland corresponds to the steepest rise of $\Delta^{14}C$, when *Sphagnum imbricatum* starts to be an important peat former in the raised bogs elsewhere in the landscape. In West-Friesland the highest, driest, parts (the farming areas) disappeared and the human population could not stay there because the area became too wet. Apart from the archaeological information also the study of preserved organic deposits under a West-Frisian Medieval dike near Enkhuizen experienced a rise in the water table [52, 56]. At that site Enkhuizen-dijk a sandy soil that formed during the Bronze Age habitation phase was covered by an organic deposit representing a shallow fresh water pool, formed after a rising water table. AMS-radiocarbon dating [56] of the start of the formation of the organic deposit showed that the sudden rise of the water table occurred somewhere between 906 and 820 calendar years BC, which means - in combination with the archaeological radiocarbon dates from the area - that the area became inundated (with fresh water) during the period of that sudden increase of ^{14}C.

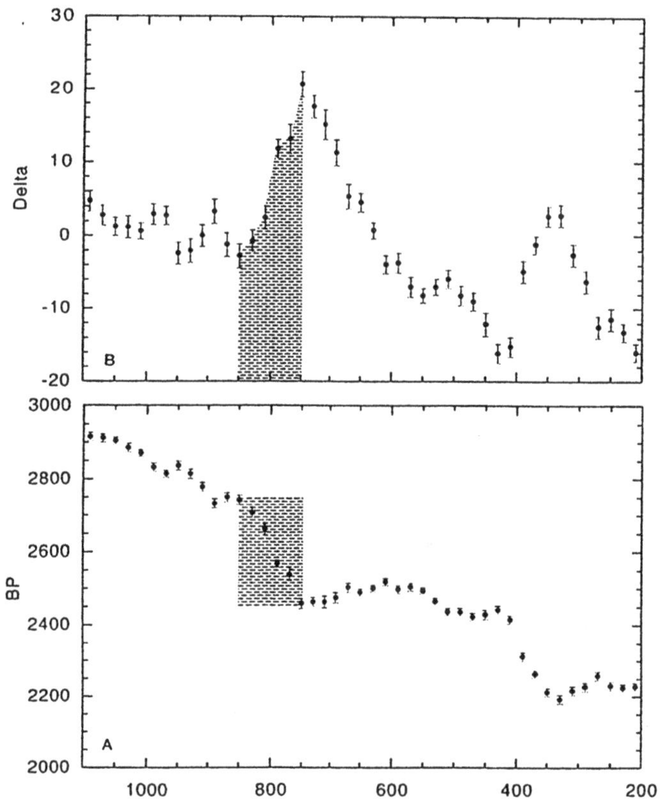

Figure 1A: The 14C calibration curve between 1100 and 200 calendar years BC, and **Fig. 1B**: the corresponding former changes of the atmospheric radiocarbon content (detrended Δ 14C). The horizontal calendar time axis is based on dendrochronologically dated wood samples. The phase of palaeoecological, geological and archaeological evidence for climate change is shown by vertical hatching. It is characterised by a considerable rise in the ^{14}C-content of the atmosphere. The actual sharp increase of cosmic rays and related climate change may have started 10-20 years earlier than the start of the rise in Δ C14 (cf. Bard *et al.* [4]). For the period 8100 to 5200 calendar years BP, Finkel and Nishiizumi [13] showed that peaks of the cosmogenic isotope 10B precede Δ ^{14}C peaks even by circa 100 years.

Another clear example is derived from the remaining part of the former raised bog Fochtelooërveen, which is situated in the northern Netherlands, about 25 km distance SSW from the city of Groningen. The present peat deposits are a remnant of a former extensive bog, of which most of the western part has been removed during the last few centuries

* $\Delta^{14}C$ is the relative deviation of the measured ^{14}C activity from the international standard, after correction for the isotope fractionation and radioactive decay [47]. Note that in the present paper the $\Delta^{14}C$ is 'detrended' as well: the long term geomagnetic field effect on ^{14}C production is subtracted.

for the production of fuel. The sandy soils below the peat were exposed again after peat digging, and in this area there is archaeological evidence that settlements died out at the transition Bronze Age/Iron Age, apparently when a rise in the water table stimulated peat growth. AMS radiocarbon dates from the remaining nearby bog deposits showed that peat growth started around 800 cal BC, when $\Delta^{14}C$ rose steeply [56].

Waterbolk [57, 58] used the distribution of the Ruinen-Wommels pottery type in the Northern Netherlands to reconstruct migration pathways. He concluded that the population of Late Bronze Age settlements that were present in relatively low areas (which were already marginal from a hydrological point of view) moved to the salt marshes along the coast of the provinces of Friesland and Groningen. They had to leave their original settlement sites because pools, mires and raised bogs suddenly appeared and increased in surface area, causing loss of cultivated land. Van Geel *et al.* [54] postulated that the appearance - for the first time at around 2650 BP - of those salt marshes was also one of the effects of the sudden climate change. As a consequence of colder conditions a temporary halt in the sea level rise may have occurred, after thermic contraction of ocean water and/or, reduced velocity and pressure on the coast by the Gulf Stream. Moreover, after this climate change in the temperate zone, more water accumulated in glaciers, as ground water in soils and in fens and bogs.

4. Lake-side villages in the alpine region

Becker *et al.* in 1985 [5] published dendrochronological data from lake-side villages in the area north of the Alps. They inferred that no new villages were built for about half a millennium after 850 cal BC. Magny [31] and also Gross-Klee and Maise [18] showed that preserved settlements along lakes in the alpine area were present mainly during phases of relatively low values of $\Delta^{14}C$, when water levels in the lakes were relatively low. During phases of rising $\Delta^{14}C$ lake-side settlements stopped, apparently because the inhabitants had to move as a consequence of the rising water tables. The end of habitation as a consequence of the quickly rising water table in West-Friesland chronologically, and also from a causal point of view, parallels the end of the last Bronze Age lake-side village in the alpine region. In South Sweden the start of lake Bjäresjö, ^{14}C-dated around 2690 BP, on top of a fen deposit [16] reflects a similar phenomenon of a sudden and sharp rise of the water table.

5. The earliest records of Scythians in Europe

Only recently have ^{14}c dates become available to establish Scythian chronologies. At present this is a subject of collaboration between Western and Russian scientists, likely to yield more evidence for ^{14}C climate/archaeological tele-econnections. New radiocarbon dates of the earliest Scythian grave monuments in Europe and the amount of evidence for abrupt climate change around 850 calendar years BC may indicate that extreme climatological conditions in Central Asia triggered the migration of Scythians to S.E. Europe and W.Asia. The oldest dates of the Kelermess monuments (NW Caucasus) are 2690 ± 150 and 2610 ± 60 BP [59], and thus may also correspond to the period of the sharply rising radiocarbon content of the atmosphere.

6. Evidence from North Africa

Franzén and Larsson [14] described geological and archaeological evidence for an abrupt transition from aridity to very humid conditions, leading to a severe flooding catastrophe in Central Tunisia, with indications for severe ground frost in the initial phase. A radiocarbon date of herbaceous material at one of the sites (2500 ± 80 BP) may indicate that the climate change occurred during the period of rapid increase of the radiocarbon content of the atmosphere, around 800 cal BC.

7. Evidence for climate change in the Central African rain forest belt

Reynaud-Farrera *et al.*, in 1996 [36] (for more details and extra ^{14}C dates see van Geel *et al.* [56]) studied the Holocene deposits of Lac Ossa in Cameroon. They recorded a drastic change in the vegetation cover (from rain forest to a more open savannah landscape) as a consequence of an arid crisis after ca 2730 BP. This change to dry conditions has also been observed elsewhere in the central African rain forest belt [12, 17]. The climate shift and related change in vegetation from dense forest to savannahs was also important for prehistoric man. A population of farmers, coming from the south, migrated into the area. The contrast between this change to dryness in Central Africa and the contemporary increase of precipitation in the temperate zone is discussed below.

8. Sun, climate and ^{14}C: a discussion

The isotope ^{14}C is produced in the upper atmosphere under the influence of cosmic rays, which are ionised nuclei and electrons, of both galactic and solar origin. The most energetic cosmic rays are of galactic origin. The decay of the radioactive isotope ^{14}C is a constant factor, but the production of this cosmogenic isotope is subject to change due to fluctuations in the cosmic ray flux. As a consequence, when interpreting radiocarbon dates one has to realise that the radiocarbon scale is not a linear calendar time scale. The fluctuations in the cosmic ray flux on Earth are mainly caused by changes in the solar wind, which is a low density ionised gas ejected from the sun, which strongly influences the magnetic field strength around the Earth. The ^{14}C calibration curve [48] is based on ^{14}C measurements of dendrochronologically dated tree rings and it reflects the fluctuations of the atmospheric ^{14}C content in the past. Fig. 1 shows the calibration curve between 1100 and 200 calendar years BC, together with the relative ^{14}C changes in the atmosphere. Note that we subtracted the long term geomagnetic factor in this plot (detrended $\Delta^{14}C$). When solar activity is high, the extended solar magnetic field sweeps through interplanetary space, thereby more effectively shielding the Earth from cosmic rays and reducing the production of ^{14}C. Low solar activity allows more cosmic rays to enter the Earth's atmosphere, producing more ^{14}C. Therefore the ^{14}C record is a good proxy for the solar radiant output [4]. Thus increases of the atmospheric ^{14}C content will have been caused by increases of the cosmic ray flux. Reductions of the ^{14}C content in the atmosphere, however, may be related to declines of cosmic ray flux but also 'old' carbon, escaping as CO_2 from reservoirs (mainly oceans, and in theory also peat deposits), will have contributed to declines of the former atmospheric $^{14}C/^{12}C$ ratio.

The combination of a continuous increase of atmospheric carbon dioxide in modern time, due to the use of fossil fuel, and an over-all temperature increase on Earth for the last century, is considered as evidence for an enhanced greenhouse effect, which in the near future is expected to play a dominant role [22]. However, there was no constant increase of temperature during the period of rising CO_2. In the Northern Hemisphere the years between approximately 1940 and 1965 were characterised by a temporary decline of the mean land air temperature [15]. From the archives and instrumental measurements in historical time it is evident that a less active sun results in an increased cosmic ray flux reaching the earth. Climatic cooling and also temperature rises seem to be related to changing solar activity. In 1997, Hoyt and Schatten [23] reviewed the possible sensitivity of the Earth's climate to solar changes and showed that there are many arguments for the supposition that the sun is a driving force behind climate change. Phenomena related to changes in the sun's surface, like sunspots, aurorae, faculae, plages, and geomagnetic disturbances show a cyclic behaviour of about 11-years. When sunspots are most numerous on the solar disc, the sun shines with a greater intensity. Extra energy leaves the sun's surface during a sunspot maximum from faculae, bright areas surrounding active sunspots. The length of the sunspot cycle is not constant but varies from ca 9 to 12 years, or even longer. During periods of relatively long solar cycles the sunspot maxima are less pronounced, whereas periods of short solar cycles are characterised by a more intense solar radiation during the sunspot maxima phases. Changes in the solar-cycle lengths seem to be a good proxy for solar-irradiance variations, and Friis-Christensen and Lassen [15] showed that the variation of the solar cycle length closely matches the long-term variations of the Northern Hemisphere land air temperature during the past 130 years. Therefore the present global temperature increase which started after 1965 seems to parallel solar activity and is not necessarily the effect of an enhanced green house effect.

The sun's radiant output varies over decades and longer time scales and these variations may also have played a significant role in climate change in the past. For modelling past solar irradiance, the physical causes are uncertain; nevertheless, records of sunspots started in the 17th century, and these can be used as a substitute for the facular influence. Both Lockwood and Skiff, in 1990 [28], and Baliunas and Vaughan, in 1985 [1], found that the variations in the total radiative output from solar-type stars exceeded the observed solar-constant variations during late historical time by a factor of nearly 4. This suggests that the sun may have undergone irradiance variations several times larger than any we have seen during the past decades [23]. During the 'Little Ice Age', the climate was notably cooler throughout the world. One of the coldest phases corresponded with the Maunder Minimum (1645-1715), an interval characterised by few sunspots and aurorae (geomagnetic storms).

Using $\Delta^{14}C$ as a proxy, one can study the mechanism of solar forcing of climate (through the action of cosmic rays) back many thousands of years. Considering the abrupt and intense increase of the ^{14}C content of the atmosphere between 850 and 760 calendar years BC (fig.1), the sun's activity must have been much lower than observed in recent decades. Also another independent cosmogenic isotope (^{10}Be) shows an increase around this time, as observed in ice cores [6]. The reduction of solar energy output may have been comparable to the situation during the Maunder Minimum, or it was even more pronounced. According to Lean *et al.* [27] the sun during the Maunder Minimum was 0.25% less bright than it was during the solar minimum of 1985-1986. According to Hoyt and Schatten [23] such a variation apparently was large enough to have an impact on climate in historical time. From 1880 to 1940 the amplitude of the solar irradiance variations was only about 0.14%. For that reason physicists and climatologists often denied the possibility of solar forcing of climate change. It was almost generally accepted that variations in solar irradiance would need to be nearly 10 times larger to be effective. Present climate models may underestimate the sensivity of climate to solar variations by failing to properly model the solar spectral variations, the vertical spatial energy input distribution, the geographical climate responses and especially the related effects of changes in cosmic ray flux. The last mentioned factor seems to be of crucial importance and therefore we give an overview of the physical reason, which associates the solar variability and climate change. Attempting to explain a physical link on the basis of the relationship "solar wind - magnetosphere - ionosphere atmosphere" is difficult because of a very large difference in the solar wind energy and the energy of the atmospheric processes (4 orders of magnitude). Thus it is necessary to develop another approach to the problem solution: the solar irradiance remains the main source of the energy affecting the atmosphere, but some agents controlled by solar activity must act directly on the atmosphere and change the amount of solar energy reaching the Earth surface. What is needed is a positive feedback mechanism, explaining how relatively small variations in solar activity can cause significant climate changes.

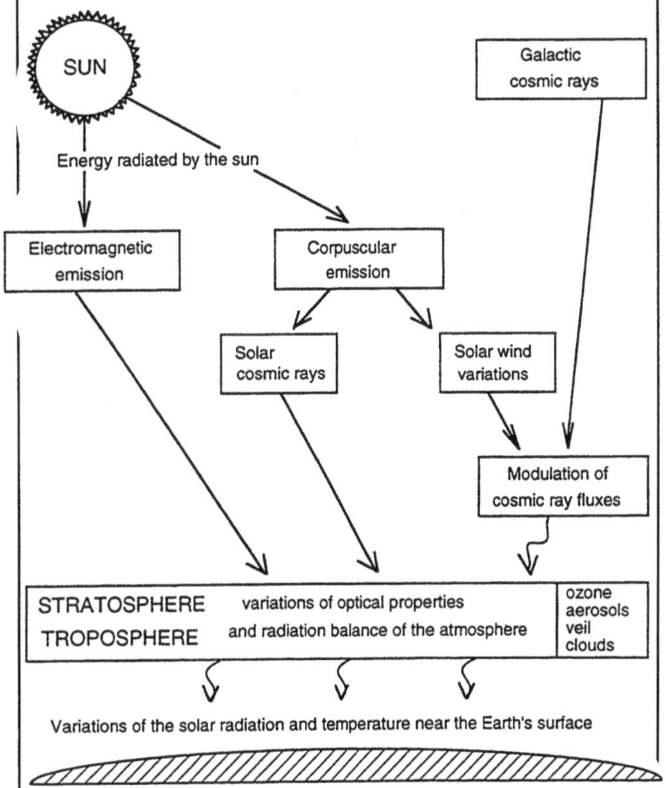

Figure 2: Scheme showing the influence of solar activity and cosmic ray fluxes on atmospheric processes and climate.

Pudovkin and Raspopov, in 1992 [33], and Raspopov *et al.*, in 1997 [37], proposed and developed the physical mechanism of the influence of the solar activity on climate and climatic variables (such as temperature and precipitation), the main idea of which is the change of optical parameters and radiation balance of the atmosphere by action of variation of cosmic ray fluxes (figure 2).

Galactic and solar cosmic ray fluxes penetrate into the stratosphere and troposphere and cause physical-chemical reactions, but these fluxes are modulated by the solar activity. Experimental data indicate that the intensity of galactic cosmic ray fluxes is changed by several percent during geomagnetic storms (Forbush decrease). During the 11-year solar cycle the change of galactic cosmic ray fluxes could be about 10% on the Earth surface and about 50% in the stratosphere. Even larger changes in cosmic ray intensity could be expected during secular cycles of solar activity, like the Maunder minimum and the period between 850 and 760 cal BC as discussed in the present paper. Recent experimental data support the physical mechanism proposed. The data demonstrate the change of the ozone layer density, the development of cloudiness, the formation of an aerosol layer in the stratosphere and an atmospheric veil during periods of increased cosmic ray fluxes in the stratosphere. Shumilov *et al.* [42, 43], Stephenson and Scourfield [46] and Kodama *et al.* [26] showed that solar proton events (known as SPE), which generate the solar cosmic rays, may produce ozone mini-holes at the high latitudes (the decrease of the total ozone content is about 10 - 15%) and these are accompanied by a decrease of the temperature in the stratosphere (down to -2.4° C). Modeling experiments performed by Haigh [19, 20] suggest that such an ozone depletion and stratospheric cooling could significantly affect the atmospheric circulation. Considering the period between 850 and 760 cal BC, van Geel and Renssen [55] inferred from the results of Haigh that a stratospheric cooling could have led to a weakening of the stratospheric winds and an equator ward displacement of the tropospheric jet streams. Moreover, they postulated that it could ultimately have resulted in a decrease of the latitudinal extent of the Hadley Cells and a weakening of the monsoons. Furthermore, it is feasible that these changes occurred in conjunction with an expansion of the Polar Cells and a repositioning of the main depression tracks at mid-latitudes [55].

In 1977, Dmitriev and Lomakina [11] demonstrated on the basis of the American satellite data that after the solar flares an increase of cloudiness is observed during two days at the middle latitudes, and after this a decrease of cloudiness below the initial level occurs. This result was confirmed by Pudovkin and Veretenenko [34, 35] using the data of cloud formation during the temporal interval of magnetic storms, and by Svensmark and Friis-Christensen [50] who used the cloudiness data on the temporal scale of the solar cycle. Shumilov *et al.* [44] showed that after SPE the development of an aerosol layer is observed in the atmosphere at an altitude of 12 to 18 km. In 1991, Schuurmans [40] reported that after SPE a decrease of the atmospheric temperature (about 1.4° C) was observed at altitudes between 5.5 and 11.7 km during 10 days. This effect is associated with the development of clouds and aerosols. The data mentioned above concerning the changes of optical and thermodynamical properties of the atmosphere related to cosmic ray flux variations in the atmosphere permitted Raspopov *et al.* [37-39] to estimate the possible decrease of the Earth's surface temperature during the Maunder minimum of solar activity. The estimation was 0.6 - 0.7° C which corresponds with the temperature change observed in reality. With the physical mechanism proposed, it is possible to explain the physical link between the solar cycle length and the global Earth surface temperature [38]. At the same time the proposed mechanism permits the explanation of the decrease in the Earth's surface temperature during excursions, inversions and low value periods of the main geomagnetic field [39].

The sharp rise of $\Delta^{14}C$ between 850 and 760 calendar years BC, and the palaeoecological/geological and archaeological evidence for a contemporaneous climate change to cooler and wetter conditions fit remarkably well with the mechanism as proposed here: a sharp increase of galactic cosmic rays (as a consequence of a suddenly reduced solar wind) must have had a strong impact on aerosol formation, cloudiness and precipitation, and it resulted in a temperature decline, a dramatic rise of the ground water table and a fast extension of marshes and mires in the temperate zones. Moreover, the inferred decrease in the latitudinal extent of the Hadley Cells and the associated weakening of the monsoons in the tropics is in good agreement with the paleoaecological evidence of dry conditions in tropical Africa.

In the northern Netherlands hydrological changes caused problems for farmers. For those living in areas like West-Friesland - already marginal from a hydrological point of view - the event around 850 cal BC must have been a climatic crisis, having the characteristics of a disaster. Depopulation of these marginal areas and migration to salt marshes took place. The present opinion among archaeologists (L.P. Louwe Kooijmans, personal communication, 1995) is that during the Holocene external climate forcing did not play a significant role for prehistoric man (the ecological determinism axioma). However, combining evidence from palaeoecology, dendrochronology, archaeology, (palaeo)climatology and isotope geophysics, we feel that the evidence for solar forcing of climate change around 850 cal BC is strong, and the impact of that climate change on human habitats situated in marginal areas is evident.

Major wiggles in the ^{14}C calibration curve, like the one starting around 850 cal BC, occur in the ^{14}C calibration curve every circa 2400 years. In addition, evidence for a similar oscillation, linked to re-occurrences of cold climatic phases, has also been found in Greenland ice cores [8, 32] and in terrestrial records from Europe, North America and the Southern Hemisphere [21]. One can speculate whether this periodicity only has its origin in the internal dynamics of the sun (like the 11 year sunspot cycle), or if external (non-solar) factors played a role. Here we come to some questions. Is it possible that periodically occurring cometary dusting in the space between the sun and the earth surface would influence the amount of solar wind in a negative way and would lead to temporary increases of cosmic rays? Or would the earth's magnetic field, and thus the amount of cosmic rays penetrating the atmosphere, be influenced by such dustings or during periods of increased meteoric activity? Or would solar wind be influenced by recurrent encounters of the sun's surface with streams of cometary debris?

In 1995, Bond and Lotti [7] showed that during the Last Glaciation (Weichselian, Devensian) nearly synchronous increases in rates of iceberg discharges occurred at the edges of Icelandic and NE-American icecaps. This phenomenon occurred every 2000 to 3000 years, and they suggest that those icebergs were discharged when air temperatures dropped below a critical threshold. A possible cause was unknown, but Bond and Lotti [7] (see also Kerr [24]) made a comparison between these cyclic iceberg discharges and the evidence for regular climatic fluctuations during the Holocene as given by Denton and Karlen [9] and Magny [30]. In addition, Lowell *et al.* [29] showed a symmetry between upper Weichselian Southern Hemisphere palaeoclimatic data (glacial pulses) and those of the Northern Hemisphere (called Dansgaard-Oeschger cold events), the fluctuations being too frequent to be explained by Milankovitch orbital seasonal forcing (would occur in an opposite sense at mid-latitudes in the two hemispheres). Lowell *et al.* mention the need for a more complete understanding of the production and redistribution of atmospheric water vapor and the accompanying effects on global climate and they suppose that the interhemispheric synchrony may implicate varying concentrations of atmospheric water vapor as the immediate source of the late Weichselian climatic changes.

Knowing the effect of decreased solar wind and related increases of cosmic ray flux and cloud formation, and observing the circa 2400 years periodicity in the $\Delta^{14}C$ values during the Holocene, we postulate that - periodically - sudden and strong increases of cloudiness, precipitation (snow) and declining temperatures as a consequence of solar/cosmic ray forcing played a crucial role in the regularly occurring iceberg discharges as recorded in North Atlantic deep sea cores and the synchronous events in the Southern Hemisphere as mentioned by Lowell *et al.* [29]. Finkel and Nishiizumi [13] present a detailed record of the cosmogenic isotope ^{10}Be from the GISP2 ice core for the period 40,000 to 3300 years BP. Fluctuations in the ^{10}Be values during the Weichselian (reflecting changes in cosmic ray flux, mainly caused by heliomagnetic variations) evidently parallel the Dansgaard-Oeschger warm/cold cycles in the $\delta^{18}O$ record from the same ice core: relatively warm phases show low ^{10}Be values and peaks in the ^{10}Be record occur during the the cold phases. In 1995, Bond and Lotti [1] suggested the existence of a mechanism operating within the atmosphere that caused rapid oscillations in air temperature above Greenland and in calving from different North Atlantic ice sheets. The new ^{10}Be records are supporting evidence for our idea (as explained above) that solar/cosmic ray forcing caused the Dansgaard-Oeschger temperature cycles and related iceberg discharges.

Finally, we wonder how Man may have experienced the abrupt climate change around 850 cal BC. For those who were living during the period of the shift, the appearance of an atmospheric veil and the sudden increase in cloudiness and related increased precipitation and temperature decline were probably radical changes with an agro-economical, and in some marginal areas even an emotional impact. Rutger Sernander's idea about a real climatological event as the background for the Fimbul-winter of the Edda saga indeed might be true but, unfortunately, can never be tested.

9. Conclusions

Periods of relativily low solar irradiance are also characterised by a high intensity of the cosmic ray flux on Earth. Both interrelated factors enforce each other in their effects on climate. From the combination of archaeology, raised bog studies, isotope geophysics and physical aspects of meteorology we found evidence for solar/cosmic ray forcing of climate change at the transition of Bronze Age to Iron Age in NW-Europe. Our evidence supports the idea that solar/ cosmic ray forcing is an important factor in the present climate and may also dominate climatic changes in the near future.

Acknowledgements

The authors thank Peter Langdon and Sarah Morriss, University of Southampton, for correction of the English text and Dr R.C. Finkel, Lawrence Livermore National Laboratory, California, for sending a preprint of the publication about the ^{10}Be concentrations in the GISP2 ice core.

References

1. Baliunas, S.L. and Vaughan, A.H., 1985, Stellar activity cycles, *Annu. Rev. Astron. Astrophys.*, **23**, 379-412.
2. Barber, K.E., 1981, *Peat stratigraphy and climatic change. A palaeoecological test of the theory of cyclic peat bog regeneration*, Balkema, Rotterdam.
3. Barber, K.E., Chambers, F.M., Maddy, D., Stoneman, R. and Brew, J.S., 1994, A sensitive high-resolution record of late Holocene climatic change from a raised bog in northern England, *The Holocene*, **4**, 198-205.
4. Bard, E., Raisbeck, G.M., Yiou, F. and Jouzel, J., 1997, Solar modulation of cosmogenic nuclide production over the last millennium: comparison between ^{14}C and ^{10}Be records, *Earth Planetary Sci. Lett.*, **50**, 453-462.
5. Becker, B., Billamboz, A., Egger, H., Gassmann, P., Orcel, A. and Ruoff, U., 1985, Dendrochronologie in der Ur- und Frühgeschichte - Die absolute Datierung von Pfalbausiedlungen nördlich der Alpen im Jahrringkalender Mitteleuropas, *Antiqua*, **11**, 1-68.
6. Beer, J., Siegenthaler, U., Bonani, G., Finkel, R.C., Oeschger, H., Suter, M. and Wölfli, W., 1988, Information on past solar activity and geomagnetism from ^{10}Be in the Camp Century ice core, *Nature*, **331**, 675-679.
7. Bond, G.C. and Lotti, R., 1995, Iceberg discharges into the North Atlantic on millennial time scales during the Last Glaciation, *Science*, **267**, 1005-1010.
8. Dansgaard, W., Johnsen, S.J., Clausen, H.B., Dahl-Jensen, D., Gundestrup, N. and Hammer, C.U., 1984, North Atlantic climatic oscillations revealed by deep Greenland ice cores, *Geophys. Monogr.*, **29**, 288-298.
9. Denton, G.H. and Karlén, W., 1973, Holocene climatic variations: their pattern and possible cause, *Quat. Res.*, **3**, 155-205.
10. de Vries, H., 1958, Variation in concentration of radiocarbon with time and location on Earth, *Proc. K. Ned. Akad. Wet.*, **B61**, 94-102.
11. Dimitriev, A.A. and Lomakina, E.Y., 1977, Cloudiness and solar X-ray emission. in *Solar activity effects in the lower atmosphere*, 70-75, L.R. Rakipova (ed.), Hydrometeoizdat, Leningrad (in Russian).
12. Elenga, H., Schwartz, D. and Vincens, A., 1994, Pollen evidence of late Quaternary vegetation and inferred climate changes in Congo, *Palaeogeogr,, Palaeoclimatol., Palaeoecol.*, **109**, 345-356.
13. Finkel, R.C. and Nishiizumi, K., 1997, Beryllium 10 concentrations in the Greenland Ice Sheet Project 2 ice core from 3-40 ka, *J. Geophys. Res.*, **102; C12**, 26699-26706.
14. Franzén, L.G. and Larsson, T.B., 1997, Landscape analysis, stratigraphical and geochemical investigations of playa and alluvial fan sediments in Tunesia and raised bog deposits in Sweden: a possible correlation between extreme climate events and cosmic activity during the late Holocene, *Abstr. 2nd SIS Cambridge Conf. Natural catastrophes during Bronze Age civilisations.* p. 19; also personal communication.
15. Friis-Christensen, E. and Lassen, K., 1991, Length of the solar cycle: an indicator of solar activety closely associated with climate, *Science*, **254**, 698-700.
16. Gaillard, M.-J. and Berglund, B.E., 1988, Land-use history during the last 2700 years in the area of Bjäresjö, Southern Sweden, in *The Cultural Landscape: past, present and future*, 409-428, H.H. Birks, H.J.B. Birks, P.E. Kaland and D. Moe (eds.) Cambridge University Press, Cambridge.
17. Giresse, P., Maley, J. and Brenac, P., 1994, Late Quaternary palaeoenvironments in the Lake Barombi Mbo (West Cameroon) deduced from pollen and carbon isotopes of organic matter, *Palaeogeogr., Palaeoclimatol., Palaeoecol.*, **107**, 65-7 8.
18. Gross-Klee, E. and Maise, C., 1997, Sonne, Vulkane und Seeufersiedlungen, *Jahrb. Schweiz. Ges. Ur- u. Frühgesch.*, **80**, 85-94.
19. Haigh, J.D., 1994, The role of stratospheric ozone in modulating the solar radiative forcing of climate, *Nature*, **370**, 544-546.
20. Haigh, J.D., 1996, The impact of solar variability on climate, *Science*, **272**, 981-984.
21. Harvey, L.D.D., 1980, Solar variability as a contributing factor to Holocene climatic change, *Progress in Physic. Geogr.*, **4**, 487-530.
22. Houghton, J.T., Meira Filho, L.G., Callander, B.A., Harris, N., Kattenberg, A. and Maskell, K. (eds.), 1995, *Climate change 1995; The science of climate change.* Contribution of working group 1 to the second assessment report of the Intergovernmental Panel on Climate Change, Cambridge University Press, Cambridge.
23. Hoyt, D.V. and Schatten, K.H., 1997, *The role of the sun in climate change*, Oxford University Press, Oxford.
24. Kerr, R.A., 1996, Millennial climate oscillation spied, *Science*, **271**, 146-147.
25. Kilian, M.R., van der Plicht, J. and van Geel, B., 1995, Dating raised bogs: new aspects of AMS ^{14}C wiggle matching, a reservoir effect and climatic change, *Quat. Sci. Rev.*, **14**, 959-966.
26. Kodama, M., Kohmo, T. and Kanzawa, H., 1992, Stratospheric sudden cooling after solar proton event over Syowa Station, *J. Geomagn. Geoelectr.*, **44**, 361-366.
27. Lean, J., Skumanich, A. and White, O., 1992, Estimating the sun's radiative output during the Maunder Minimum, *Geophys. Res. Lett.*, **19**, 1591-1594.
28. Lockwood, G.W. and Skiff, B.A., 1990, Some insights on solar variability from precision stellar astronomical photometry, in *Conference on the climate impact of solar variability*, 8-15, K.H. Schatten and A. Arking (eds.), NASA conference publication CP-3086.
29. Lowell, T.V., Heusser, C.J., Andersen, B.G., Moreno, P.I., Hauser, A., Heusser, L.E., Schlüchter, C., Marchant, D.R. and

Denton, G.H., 1995, Interhemispheric correlation of Late Pleistocene Glacial events, *Science*, **269**, 1541-1549.
30. Magny, M., 1993, Solar influences on Holocene climatic changes illustrated by correlatons between past lake-level fluctuations and the atmospheric ^{14}C record, *Quat. Res.*, **40**, 1-9.
31. Magny, M., 1993, Un cadre climatique pour les habitats lacustres préhistoriques?, *Compte Rendu de l'Acad. Sci. Paris*, **316**, II, 1619-1625.
32. O'Brien, S.R., Mayewski, P.A., Meeker, L.D., Meese, D.A., Twickler, M.S. and Whitlow, S.I., 1995, Complexity of Holocene climate as reconstructed from a Greenland ice core, *Science*, **270**, 1962-1964.
33. Pudovkin, M.I. and Raspopov, O.M., 1992, The mechanism of action of solar activity on the state of lower atmosphere and meteorological parameters, *Geomagn. Aeronomy* (English ed.), **32**, no 5, 593-608.
34. Pudovkin, M.I. and Veretenenko, S.V., 1992, The influence of geomagnetic disturbances on the intensity of the flux of direct solar radiation, *Geomagn. Aeronomy*, **32**, no 1, 148-152 (in Russian).
35. Pudovkin, M.I. and Veretenenko, S.V., 1995, Cloudiness decreases associated with Forbush decreases of galactic cosmic rays, *J. Atmosph. Terrestr. Physics*, **57**, 1349-1355.
36. Reynaud-Farrera, I., Maley, J. and Wirrmann, D., 1996, Végétation et climat dans les forêts du Sud-Ouest Cameroun depuis 4770 ans BP: analyse pollinique des sédiments du Lac Ossa. *Comptes Rendus de l'Acad. Sci. Paris*, **322**, série IIa, 749-755.
37. Raspopov, O.M., Shumilov, O.I. and Kasatkina, E.A., 1997, Influence of galactic and solar cosmic rays on the atmosphere: the main factor of solar variability influence on the meteoparameters. *Biophysics* (in press).
38. Raspopov, O.M., Shumilov, O.I., Kasatkina, E.A., Jacoby, G. and Dergachev, V.A., 1997, *The cosmic ray influence on cloudy and aerosol layers of the Earth and connection of solar cycle lengths to global surface temperature*, Reprint 1694, Phys.-Techn. Inst., St.-Petersburg, pp. 20.
39. Raspopov, O.M., Shumilov, O.I., Kasatkina, E.A., Dergachev, V.A. and Creer, K.M., 1997, *Impact of cosmic ray flux variations caused by changes in geomagnetic dipole moment on climate variability*, Reprint 1693, Phys.-Techn. Inst., St.-Petersburg, pp. 41.
40. Schuurmans, C.J.E., 1991, Changes of the coupled troposphere and lower stratosphere after solar activity events, *J. Geomagn. Geoelectr.*, Suppl., **43**, 767-773.
41. Sernander, R., 1910, Die schwedischen Torfmoore als Zeugen postglazialer Klimaschwankungen, *Die Veränderungen des Klimas seit dem Maximum der Letzten Eiszeit,* Stockholm.
42. Shumilov, O.I., Henriksen, K., Raspopov, O.M. and Kasatkina, E.A., 1992, Arctic ozone abundance and solar proton events, *Geophysic. Res. Lett.*, **19**, 1647-1650.
43. Shumilov, O.I., Kasatkina, E.A., Henriksen, K. and Raspopov, O.M., 1995, Ozone "mini-holes" initiated by energetic solar protons, *J. Atmospheric Terrestr. Physics*, **57**, 665-671.
44. Shumilov, O.I., Kasatkina, E.A., Henriksen, K. and Vashenyuk, E.V., 1996, Enhancement of stratospheric aerosols after solar proton event, *Ann. Geophysicae*, **14**, 1119-1123.
45. Sörbom, G., 1966, Rutger Sernanders klimatförsämringsteori, *Tor*, **11**, 83-115.
46. Stephenson, J.A.E. and Scourfield, M.W.J., 1992, Ozone depletion over the polar cap caused by solar protons, *Geophys. Res. Lett.*, **12**, 2425-2428.
47. Stuiver, M. and Pollach H.A., 1977, Discussion: reporting ^{14}C data, *Radiocarbon*, **19**, 355-363.
48. Stuiver, M., Long, A. and Kra, R. S., 1993, Calibration 1993, *Radiocarbon*, **35**, 215-230.
49. Stuiver, M., Braziunas, T.F., Grootes, P.M. and Zielinski, G.A., 1997, Is there evidence for solar forcing of climate in the GISP2 oxygen isotope record?, *Quat. Res.*, **48**, 259-266.
50. Svensmark, H. and Friis-Christensen, E., 1997, Variation of cosmic ray flux and global cloud coverage - a missing link in solar-climate relationships, *J. Atmosph. and Solar-Terrestr. Physics*, **59**, 1225-1232.
51. van Geel, B., 1978, A palaeoecological study of Holocene peat bog sections in Germany and the Netherlands, *Rev. Palaeobot. Palynol.*, **25**, 1-120.
52. van Geel, B., Hallewas, D. P. and Pals, J .P., 1983, A Late Holocene deposit under the Westfriese Zeedijk near Enkhuizen (Prov. of N-Holland, The Netherlands): palaeoecological and archaeological aspects, *Rev. Palaeobot. Palynol.*, **38**, 269-335.
53. van Geel, B. and Mook, W.G., 1989, High-resolution ^{14}C dating of organic deposits using natural atmospheric ^{14}C variations, *Radiocarbon*, **31**, 151-156.
54. van Geel, B., Buurman, J. and Waterbolk, H.T., 1996, Archaeological and palaeoecological indications of an abrupt climate change in The Netherlands, and evidence for climatological teleconnections around 2650 BP, *J. Quat. Sci.*, **11**, 451-460.
55. van Geel, B. and Renssen, H., in press, Abrupt climate change around 2650 BP in North-West Europe: evidence for climatic teleconnections and a tentative explanation., in *Water, Environment and Society in Times of Climatic Change*, A. Issar and N. Brown (eds.), Proc. Sede Boker Workshop, Kluwer, Dordrecht.
56. van Geel, B., van der Plicht, J., Kilian, M.R., Klaver, E.R., Kouwenberg, J.H.M., Renssen, H., Reynaud-Farrera, I. and Waterbolk, H.T., in press, The sharp rise of ^{14}C ca. 800 cal BC: possible causes, related climatic teleconnections and the impact on human environments, *Radiocarbon*.
57. Waterbolk, H. T., 1959, Nieuwe gegevens over de herkomst van de oudste bewoners der kleistreken, *K. Ned. Akad. Wet., Akademiedagen*, **11**, 16-37.
58. Waterbolk, H. T., 1966, The occupation of Friesland in the prehistoric period, *Ber. Rijksd. Oudheidk. Bodemonderz.*, **15/16**, 13-35.
59. Zaitseva, G. I., Possnert, G., Alekseev, A. Y., Sementsov, A. A. and Dergachev, V. A., in press, The first radiocarbon dates of monuments in European Scythia, *Radiocarbon*.

Can European Prehistory Detect Large-Scale Natural Disasters?

Euan W. MacKie

Hunterian Museum, University of Glasgow, University Avenue, Glasgow G12 8QQ, UK

1. Introduction

Most European prehistorians would probably assert that there is no evidence for sudden, continent-wide disruptions of the archaeological record. Unlike the situation in the Mediterranean and the Near East the rural, stone-using, and later bronze-using, agriculturists from the fourth, third and second millennia BC north of the Alps did not build great cities the overthrow of which would have left mounds containing layers of ash among tumbled masonry. In fact there are numerous elaborate individual drystone buildings - the Neolithic chambered burial cairns, some of which are architecturally very sophisticated - which have survived intact from the third and even the fourth millennium and which clearly argue against anything really cataclysmic happening since at least as early as 3000 BC.

Most archaeologists like myself would probably say that the astronomers will have to come up with some very clear extra-terrestrial evidence of the impact of swarms of large meteorites at specific dates in the past before they would be prepared seriously to check the archaeological record at those dates for signs of major disruptions. Moreover we have to study sites individually and cannot be sure, even if a major disruption is detectable at one, that it was not just an isolated incident. The dating techniques we make use of are, with the one exception of dendrochronology, quite incapable of telling us that events at a number of sites over a wide area were truly contemporaneous and were not scattered across several decades at least.

What follows therefore is a preliminary and fairly light-hearted attempt to play the devil's advocate as it were (in terms of the current beliefs of my own profession) and to try to foresee what might happen if, thanks to the astronomers, those circumstances came to pass. The three phenomena now described briefly are effectively chosen at random, being known to me at first hand.

2. The end of the Upper Palaeolithic Magdalenian culture

The late Upper Palaeolithic in Europe defines the last of the hunter-gatherer cultures who lived during the last major glacial period [6]. Those in the Dordogne in central France have long been famous for their spectacular cave paintings, and also because - near the village of Les Eyzies - not only can a number of these painted caves be seen but also several of the contemporary habitations. These last are the rock shelters, shallow overhangs under cliff faces which were inhabited in some cases for tens of thousands of years. Many of the famous excavated ones - like Laugerie Haute, Laugerie Basse, La Madeleine and Le Moustier - can be visited and the sections cut through the occupation layers are carefully preserved and labelled. The rock shelters which were inhabited up to the end of this period show a long series of closely spaced fine dark floor layers between thin paler strata of small rock chippings; many centuries of almost continuous occupation seem indicated, with no obvious rock falls or other disruptions. The final Magdalenian layer - dated by radiocarbon to about 10,000 or 11,000 years ago, is similarly level. Shortly after this culture ended the last glacial period itself evidently came to an end and the climate began to warm up. The following period - of forest-dwelling hunter-gatherers - is termed the Mesolithic. Detailed study of these rock shelters has, naturally, concentrated on dating the sequences discovered and building up cultural correlations between the sites [4].

On top of the Magdalenian floor in at least two of the rock shelters is a massive and thick layer of huge blocks of rock fallen from the cave roof and embedded in volcanic ash; the ash appears to rest directly on the last Palaeolithic occupation with no obvious signs of a sterile layer between. Here then is an example of several well documented living sites under cliff overhangs which appear to have been destroyed and filled in at the end of the Palaeolithic age by some violent natural cataclysm nearby. Should the circumstances outlined in the second paragraph of this article come to pass this would be one obvious archaeological phenomenon ripe for reappraisal.

Figure 1: Stratigraphy at Laugerie haute rock shelter, Dordogne.

3. Neolithic and Bronze Age rock carvings in north Britain.

The nature and purpose of the 'cup-and-ring rock' carvings of north-western Iberia and upland Britain has long been a matter of controversy, simply because it is so hard to say anything definite about them except that they were carved on to natural rock surfaces by the pecking technique. The absence of hard evidence naturally encourages untrammelled speculation. Various indirect dating methods imply that they belong to the latter part of the Neolithic period and to the beginning of the Bronze Age - say approximately 2500 1500 BC. Many hypotheses have been advanced to explain them - for example that they mark territorial boundaries or the nearness of copper deposits - and most archaeologists would probably put the idea that they represent comets visible in daylight - first advanced by Victor Clube and Bill Napier [2] - well down the list in terms of plausibility. However if we suppose for a moment that a plausible astronomical hypothesis exists which infers from the study of mathematically retrojected orbits that there probably were large swarms of unusually bright comets near the Earth at specific dates in the third millennium BC then this explanation for the cup-and-ring carvings would become plausible.

In this case one particular site ten miles west of Glasgow (now unfortunately removed to allow quarrying) might provide an interesting if provisional test of the comet explanation. Carefully examined and described in 1984 [5], it turned out to have two very clear periods of carving - almost the only such site where this has been inferred with a high degree of confidence. The earlier phase, with more eroded carvings, seems to belong to the main phase of cup-and-ring markings, probably in the latter part of the third millennium BC. However at some stage the 'raft' of metamorphosed sandstone (smoothed by glaciation much earlier) on which the carvings were done was quarried and many carvings destroyed. On the freshly exposed flat surfaces more carvings, much fresher in appearance, were done including the largest and finest double disc symbol on the site.

Figure 2: Period 2 rock carving (7th century BC?) at Greenland, Dumbartonshire, Scotland.

Of course dating the later carvings is extremely difficult (and the author cannot claim to have achieved a consensus on this) but circumstantial evidence is available which suggests that this happened quite late - probably at the beginning of the Iron Age in about the 6th or 7th centuries BC. The construction of a small timber-framed hillfort (see next section) nearby could explain the damage to the outcrop which preceded the fresh carvings. If this is so, and if the comet explanation for the designs is plausible, then evidence should one day appear for a period of high activity in the sky about 26 centuries ago. However it must be said that the acceptance of such a late date for the second phase of carvings requires more evidence, which is still probably available at the site.

4. The vitrified forts of central and north-eastern Scotland

A vitrified fort is a ruined hilltop stronghold which has been affected by fire to such an extent that parts of the stone rubble ramparts have become fused together; temperatures up to 1000° centigrade are indicated. Ever since the phenomenon was first studied in an antiquarian manner (in the 1790s) the belief that vitrification was a peculiar, late prehistoric method of constructing an enormously strong rampart by fusing it with fire has remained popular [1]; it has proved very difficult to dislodge despite the mounting archaeological evidence which contradicts it [3].

Briefly it is now clear that vitrified forts are the destroyed remains of strongholds defended by a timber-framed stone walls; in these various kinds of frameworks of logs and beams are incorporated within the rubble core of the wall, presumably to strengthen it. The phenomenon is known in the continental and south English Iron Age but only in Scotland it seems was this timber framing set inside walls containing drystone rubble without earth. The likeliest explanation for the vitrification (supported by practical experiments) is that the wooden huts and wall top palisades of the stronghold were set on fire (by natural means or by an attack) and that the heat gradually carbonised many of the beams inside the wall core. Eventually parts of the wall collapsed and the wind then fanned the exposed charcoal to a red heat, creating by accident a sort of charcoal blast furnace which fused and even melted large parts of the wall core. Excavation of these sites always shows that the burning and vitrification occurs at the *end* of their occupation, rendering them uninhabitable.

Enough vitrified timber-framed forts have been dated by radiocarbon now to make it clear that they belong broadly to the first millennium BC, approximately between about 800 and 200 BC [3, p. 445]; dating by thermoluminescence does however suggest a wider spread). They cannot all be contemporary and evidence of a huge and widespread conflagration. Nearly every discussion about the phenomenon in the archaeological literature so far has concentrated on the phenomenon of vitrification itself - whether it was deliberate construction (now disproved) or caused by a destructive fire.

Because of this interminable argument (some scientists unfamiliar with the archaeological evidence still seem to believe in the 'creative vitrification' theory) there has been hardly any discussion of the phenomenon of the timber-framing of stone walls itself. Why was this done over a few centuries in the late Bronze and early Iron Ages when drystone walls were perfectly good defences (as is shown by numerous later examples)? Moreover after a few had burned it must have been obvious that the risk from fire to this type of wall was severe, yet it continued to be used at least until the third century BC. This is shown by the well-dated timber-framed *dun* (small round stone stronghold) in Strath Oykell in Sutherland.

If a theory of violent disruption of the landscape in the early and middle of the first millennium BC becomes plausible because of good astronomical evidence we may have an explanation. It is well understood that in the geologically more unstable Near East many stone walls in Bronze Age cities were rendered more earthquake-proof by incorporating wooden beams in their faces and cores; presumably this made them more flexible during a tremor. Could the Scottish (and other British and European) timber-framed hillforts have been built for the same reason - fear of earthquakes? Perhaps, but no archaeologist will at present give this idea the slightest credence until independent evidence for a period of quakes at that time is produced.

References

1. Brothwell, D, Bishop, A C & Woolley, A.R., 1974, Vitrified forts in Scotland: a problem in interpretation and primitive technology, *Journal Archaeol. Sc.* **1**, 101-07.
2. Clube, S.V.M & Napier, W.M., 1982. *The Cosmic Serpent*, Faber, London.
3. Harding, D., ed., 1976, *Hillforts: Later prehistoric Earthworks in Britain and Ireland*. London.
4. Laville, H., Rigaud, J-P. & Sackett, J., 1980, *Rock shelters of the Perigord: geological stratigraphy and archaeological succession*. London.
5. MacKie, E. W. & Davis, A., 1991, New light on Neolithic rock carving; the petroglyphs at Greenland (Auchentorlie), Dunbartonshire, *Glasgow Archaeol. Journ.*, **15** (1988-89), 125-56.
6. Sonneville Bordes, Denise de, 1974, The Upper Palaeolithic c. 33,000 - 10,000 BC, in: *France before the Romans*, eds. S. Piggott, G.E. Daniel & C. McBurney. London.

The Catastrophic Emergence of Civilization: The Coming of Blood Sacrifice in the Bronze Age

Gunnar Heinsohn
Universität Bremen, FB 11, P.O. Box 330 440, 28334 Bremen, Germany

Fig. 1: Sacrifice of a person representing a celestial body among the Andean Antis Indians. The half-decayed previous sacrifice is suspended, looking like a statue, in a sort of wooden shrine or golgatha standing on a hill. This is where the gestures of ritual propitiation are performed in front of the corpse turning into a celestial deity in human form. (Copperplate engraving from 1723 adapted from J. Campbell, *Historical Atlas of World Mythology. Volume II. Part 1: The Sacrifice*, New York et al.: Harper & Row, 1988, p. 42). Compare also figure 9 below.

Summary

Modern scholarship has come to no consensus about blood sacrifice cults and deities in human or animal form. Some scholars even insist that it will be impossible ever to comprehend these core elements of religion and that it is a violation of a sacred secret to go on asking questions: "The unprovability of the sacred, which is expressed, however inadequately, in phrases such as the 'original experience' of the sacred, together with the reality of the shedding of blood belongs to the basics that science needs to accept and cannot prove completely" [102].

The peoples of antiquity did not have any such qualms. Their texts and pictures show us lavishly decorated actors who clearly represent destructive celestial bodies and who are participating in blood sacrifice (figure 4). Later turned into idols, the actors portrayed the causes of deluges and world-wide conflagrations. The sacrificial victims were pierced through, decapitated, chopped into small pieces or castrated (figure 2), representing the breaking-up of a comet. The rain-like fall of ash and of meteorite material was imitated by ritual smearing with soot and by throwing stones and grains. Even the tremendous fright suffered by people who escaped the cataclysms is said to have entered the rituals. Their participants had to utter panic-stricken screams and grown men - after taking laxatives - had to dirty themselves.

As the belief in a world of tiny and harmless changes, imposed in particular by Lyell and Darwin, dominated science until the seventh decade of our century, the statements of the ancients, which sounded so totally different, had to be dismissed as fantasies, and the origin of the major religions had to be declared unsolvable mysteries. Only outsiders insisted on taking the evidence seriously and were frequently laughed out of court or ignored. Only since the neo-catastrophist theses of the origin of geological ages through cosmic catastrophes and the destruction of the dinosaurs by meteorites has conventional thinking slowly begun to turn away from suppression of the subject. Meanwhile, renowned scientific journals are making great efforts to reconstruct the history of our "asteroid stricken planet."

However, many of the theories prevalent today concerning sacrifice and gods were formed before this turn towards fact-oriented science, yet only now are things becoming

Fig. 2: Left: Sacrificial tree of Diana on Tauris [sketch of part of a relief] (C. Bötticher, *Der Baumkult der Hellenen*, Berlin: Weidmannsche Buchhandlung, 1856, table 31): Heads of sacrificed humans and animals who met their deaths enacting forces of nature, have been raised on a tree to enable the propitiatory gestures of the sanctifying and healing community to be carried out below them. In the background we see a statue of Diana, or rather Artemis, which points to a later origin than the Mycenaean seal ring (figure 9, below) because freshly slaughtered images of the deity and those made in stone now exist side-by-side. The sword on the right side of the statue here perhaps points to a female executor of the sacrificial act. (Cf. also O. Dickinson, *The Aegean Bronze Age*, Cambridge: Cambridge University Press, 1994, p. 265, as well as N. Marinatos, *Role and Sex Division in Ritual Scenes of Aegean Art*, in Journal of Prehistoric Religion, 1, 1987, pp. 23 ff.) Cf. in the middle the altar scene combined from two similar Minoan seals. (Cf. B. Bergquist, „Bronze Age Sacrificial Koine in the Eastern Mediterranean, in: J. Quaegebeur (ed.), *Ritual and Sacrifice in the Ancient Near East: Proceedings of the International Conference Organized by the Katholieke Universiteit Leuven from the 17th to 20th of April 1991, Leuven*: Uitgeverij Peeters & Department Orientalistiek Leuven, 1993, p. 24.) Right: Oscillum (plaquette depicting humans) from the Jupiter sacrifice on the top of Mons Albanus (circa 20 km south-west of Rome): Many writers of antiquity considered the figurines to be later substitutes of the human sacrifice practiced in more primitive times. (After H.H. Scullard, *Römische Feste: Kalender und Kult* (1981), Mainz am Rhein: Philipp von Zabern, 1985, p. 176 [illustration on page 177].)

transparent. Neither attack nor flight nor negotiation were suitable reactions to the events of the Bronze Age. Their excessive panic was thrown back at them, expressing itself as paralysing rigidity, or uncontrolled eruptions of aggression. In this situation, the boldest found the way to creating rituals and thus became the first priests of the Bronze Age. Like small children, entire communities were engaged in ritual play-acting as a process of healing from the overwhelming experiences. They regained their mental balance in a discharge of the accumulated fury by killing one or several impersonators of celestial bodies, but this had to be paid for with guilt feelings. The victims' corpses were elevated, while wearing full astral costume, so that gestures of lamentation, bowing down and reconciliation could be performed. With this "adoration," people created celestial deities, who now had taken the shape of animals, humans, or - depending on their beastly costumes - mixed creatures. When, towards the end of the Bronze Age catastrophes, these freshly slaughtered celestial fighters were represented in wood or stone, the statues of gods had arrived.

1. Scholarly perplexities concerning the rise of blood sacrifice and high civilization in the Bronze Age

"Although scholars have pondered about the beginnings of civilization [sacrificial priest-kingship in the Bronze Age] as long as there has been a concern for history, in terms of empirical information and results, this study is still in its infancy" [100, p. 278].

"Human sacrifice is the most difficult to understand./ It shows clearly how difficult the problem of sacrifice is; here - if anywhere - beautiful theories fall short" (Jan de Vries, *Perspectives in the History of Religion*, Berkeley et al.: University of California Press, 1977, pp. 198/200)

"Sacrifice provides us with a particularly sublime example for the conservatism of the sacred act. If it were possible to elucidate its prehistory and to penetrate as far as its early significance, this would be a great gain both for *Greek and for general comparative history of religion*" [86, p. 907].

"What legal or pious fiction or what economic or social pressure conditioned this attitude [the acceptance of priest-kingship and blood-sacrifice in Bronze Age *Mesopotamia* - G.H.] we shall probably never know" [89, p. 97].

"The difficult question of the character of sacrifice by slaughter and burnt offering *in the Egyptian cult* still awaits a solution" [63, p. 69].

"Since the rise of the *comparative or historical study of religions* in the latter part of the 19th century, attempts have been made to discover the origins of sacrifice. These attempts, though helpful for a greater understanding of sacrifice, have not been conclusive" [27, p. 791].

"The [Bronze Age, G.H.] appearance of writing, the elaborate buildings, the use of imported materials, sophisticated art-works and the population increase reflected in the size of the settlements, all signal the emergence of substantial urban communities with developed socio-economic structures. Nothing else can explain the materials, techniques and scale of building. *But their precise origins as well as specific character, remain a matter of speculation. The* evidence points to a highly evolved political system in place: the figure dominating so many of the pictorial scenes probably stands at the head of this society, and is presumably the *ruler,* the fact that he appears repeatedly as *the most prominent actor in ceremonial contexts* suggests that the important ideological and religious activities of the state are in his care and under his control" (A. Kuhrt, *The Ancient Near East c. 3000-330 BC, 2* volumes, London & New York: Routledge, 1995, vol. 1, p. 25; - emphasis in the last five quotations G.H.).

The debate on the emergence of human and animal sacrifice in Mesopotamia's Early Bronze Age has been going on for a hundred years. Only recently, a consensus has been slowly forming that these sacred executions were, indeed, common in the Ancient Near East [39]. This position, however, is not all that new. It had already been put forward at the beginning of the controversy when the practice of blood ritual was clearly recognised and openly discussed, for example by Ménant [84].

Two years after Ménant's article of 1887, the very same glyptic evidence (figure 3) was interpreted quite differently. The slaughtering men depicted on the seals were no longer identified as priests, i.e., as sacred executioners (the most concise definition of priests), but as slaughtering deities [113]. This interpretation raised the issue: how could man get the idea of a killing deity in human (or animal) form without having seen real humans active in such sacred slaughtering? This question could not be answered satisfactorily. In 1927/1928, Mesopotamian human sacrifice was established beyond doubt [118] when Woolley discovered the

Royal Cemetery at Ur: "The ritual of burial included human sacrifice; the number of victims might vary from a mere half-dozen to seventy or eighty, but a certain number had to accompany the owner of the tomb" [120]. These discoveries at last archaeologically confirmed the very existence of human sacrifice. However, the killed vassals did not allow the establishment of meaningful relationships with the scenes shown on the seals because they could not be identified as deities. The slaughtered vassals must have belonged to a different category of blood ritual.

The subject moved from the hands of narrow specialists to students of religion. They discovered a world-wide "combat myth/of gods and heroes who encounter and defeat dragons, monsters, demons, and giants" [28, pp.1,9]. During such duels, deities were trying to exterminate each other. Thus, many of the ritualistic scenes could be categorised under the "combat type of sacrifice," as was done by Fontenrose [28]. Illustrations and sagas were found all over the world, depicting or describing these deities in anthropomorphic, bestiomorphic or mixed forms. Psychoanalytical explanations were offered. "The primeval precedent" of the bloody cults was located in "a common pattern" of the human soul: "So we may look upon the whole combat in all its forms as the conflict between Eros and Thanatos. It is that opposition between life instincts and death instincts that Freud was the first to formulate, albeit tentatively, as the central principle of all living organisms from the beginning; though it was seen dimly and expressed in dramatic or metaphysical terms by poets and philosophers before him. But in life the two kinds of instincts, though opposed, are always mingled. Thus do the fantasies of myth disguise the fundamental truths of the human spirit" [28, pp. 3,474].

Such an overly simplistic adoption of Freudian thought very much reflected the *Zeitgeist*. It was soon elaborated into a whole school of universal and eternal sacrifice driven by the psyche of man [10, 14, 37, 38]. But why, then, do we not have evidence of human sacrifice being practised everywhere throughout history? Obviously, this variety of a psychoanalytic approach overargued its case:

Fig. 3*: Mesopotamian Bronze Age seals depicting scenes of human sacrifice.* (All illustrations except bottom right from P. Amiet, *La glyptique Mésopotamienne archaique*, Paris, Editions du Centre National de la Recherche Scientifique, 1980, passim; bottom right from D. Collon, *First Impressions. Cylinder Seals in the Ancient Near East*, London, British Museum Publications, 1987, p. 176.)

> "Like Girard, Burkert postulates that the sacred is violence transcended and that sacrifice is a violent act making this transcendence possible. However, while Girard explains violence with a murky metaphysics of desire, Burkert founds it in genetic platitudes. According to him, the biological program acquired by the human species at the hunting stage involved the development of a strong intra-specific violence that originally found its outlet in the violence exerted on game./ Sacrifice is thus a 'ritualized' equivalent of hunting" [111].

Other voices were also heard. These dissenting thinkers said, the rituals not only "reflect" [77] common psychic drives but ecological and social circumstances that pre-exist the sacred acts and exist quite independently from them. Of course, one would like to know more about such circumstances. In Assyriology, the cause of the ritual is vaguely perceived as a "power not of this world" [58]. Students of mythology even concede "supernatural objects" [65] or a general threat of chaos [34]. Are these supernatural causes of fear then supposed to be more powerful than ordinary earthquakes, storms, volcanoes, eclipses, and tidal waves?

To answer the question as to the character of a "power not of this world," archaeology proves helpful by coming up with more specific evidence: A bronze relief mounted on the gate of the New Year's Chapel in Assur carried the inscription: "The figure of Assur going to battle against Tiamat is that of Sennacherib" [89]. Assyria's Great King Sennacherib playing the celestial God Assur and battling the celestial serpent Tiamat seems to allow for a solution of the above mentioned controversy. The ritualistic scenes, indeed, depicted activities of gods and, yet, these deities were played by humans. There even exist illustrations which might perfectly fit the inscription from Assur (see figure 4 below, upper part).

"Assur against Tiamat" most probably presents other versions of "Enki against Kur" [69: pp.169 f], "Ninurta against Asag" [69: pp.170 f], "Marduk against Tiamat" [98, pp.64 ff] or "Gilgamesh against Huwawa" [98, pp.78 ff], etc. These celestial combat myths may well portray the duel between a cosmic destroyer and a cosmic saviour of the world. The winning hero keeps floods away and puts the sky into order. This is best documented in the Marduk-Tiamat combat of "The Creation Epic" (tablet IV):

> "When Tiamat opened her mouth to consume him,
> He drove in the Evil Wind that she close not her lips.
> As the fierce winds charged her belly,
> Her body was distended and her mouth was wide open.
> He released the arrow, it tore her belly,
> It cut through her insides, splitting the heart.
> Having thus subdued her, he extinguished her life. [...]
> He split her like a shellfish into two parts:
> Half of her he set up and ceiled it as sky,
> Pulled down the bar and posted guards.
> *He bade them to allow not her waters to escape. [...]*
> He constructed stations for the great gods,
> Fixing their astral likeness as the Images.
> He determined the year by designating the zones:
> He set up three constellations for each of the twelve months.
> After defining the days of the year by means of (heavenly) figures,
> He founded the station of Nebiru i.e. Jupiter to determine their (heavenly) bands,
> *That none might transgress or fall short*" [98, p. 67; emphasis added].

Since these celestial deities also pose prominently in sacrificial drama one has to ask why human (or animal) actors were needed to play non-human powers. Was there an event - or a repetition of events - that caused grown-up people to perform collective, ritualistic procedures very much in the same way children feel compelled to play? Also: Why are gods killed in ritualistic combats? What is a god? Why is the king also a priest? And why do his fellow men suddenly accept priests as their superiors?

In the Neolithic (Late Stone Age) "cults and shrines were generally of a domestic kind./Public temples manned by whole-time priests were still something for the future" [18]. Rituals basically focused on reconciliation with the deceased. This practice extended to animals killed for food or in defence. The attempt to shed guilt feelings, which arise after the death of beings close to oneself, indeed represents an eternal and universal propensity of man. These attempts can be traced back to *Homo sapiens neanderthalensis* known not only for burying his dead, but also for ritually decorating them [87]. Below, it will be shown that this universal propensity is also incorporated in the grand blood rituals of the Bronze Age. But, it cannot explain their sudden post-Neolithic appearance.

Fig. 4 a-b: Above: Terra-cotta plaquette from Old Babylonian Khafajah. A human, dressed up as a celestial god with a horned crown kills another person, garbed in star-shaped mask and scale-covered skirt who plays the celestial serpent.(H. Müller-Karpe, *Handbuch der Vorgeschichte. Dritter Band: Kupferzeit. Dritter Teilband: Tafeln*, München: C.H. Beck, 1974, table 233, illustration 2.) Below: Toltec duel of ballgame players with star- or comet-shaped feathered

2. Bronze Age catastrophes and the rise of blood sacrifice

"[Priest-]Kingship was lowered from Heaven [on earth]/ after the flood had swept thereover" [57: pp. 71/77].

"Any explanation of sacrifice is, in fact, a theory of religion in miniature" (J. Z. Smith et al. (eds.), *The Harper Collins Dictionary of Religion*, San Francisco: Harper Collins, 1995, entry Sacrifice, p. 948).

"The various cosmogonies that I have consulted agree unanimously that the earth was destroyed by fire and by water. It appears that differences in the traditions are only caused by differences in the countries which these people inhabit./We have further noted that these old celebrations rituals are linked with a theology in an allegorical language which we call mythology. It generally presents the misfortunes that befell the earth in symbols and images which only show the happy and unhappy deeds of the gods, their wars, their encounters, their victories, and the good and the evil that they did mankind." [7: p.533].

"Although scholars have pondered about the beginnings of civilization [in the Bronze Age] as long as there has been a concern for history, in terms of empirical information and results, this study is still in its infancy" [100].

Though it is generally accepted that priest-kings and "temples had the role of catalysts in the formation of cities," [80: p.155] the causes behind the emergence of these Bronze Age cult places are described as "various and highly contingent" [80: p.392] but not yet sufficiently intelligible. So much for modern researchers! What about the people who lived when civilization, i.e., the Bronze Age with blood sacrifice and its staff of priestly royalty began? They do not leave us uninformed. The earliest extant text on the subject known from Mesopotamia, "The Deluge," first published in 1914, relates the emergence of cult centres to some disaster that had struck the earth:

"My mankind, in its *destruction* I will ...

I will return the people to their settlements

After the ... of kingship had been lowered from heaven,

After the exalted tiara and the throne of kingship had

been lowered from heaven,

He perfected the rites and the exalted divine laws...,

Founded the five cities in ... pure places,

Called their names, apportioned them as *cult centres*

[69, p. 149; 98, p. 43, emphasis added]

The "destruction," after which "cult centres" with their sacrificial compounds seem to have emerged and priest-"kingship" was lowered from heaven, was by no means the last cataclysm to terrify the Ancient Near East. Soon after the cult centres were established "the Flood swept thereover. After the flood had swept thereover [...] the kingship was in Kish" [57]. When Kish was systematically excavated between 1923 and 1933, the Anglo-American team discovered a flood stratum upon which the remains of the Early Dynastic I period, i.e., the first period dominated by priest-kings, was found [36]. Meanwhile, excavations have "shown that the Archaic Sumerian or Early Dynastic civilization of the early third millennium follows notable flood levels at several important sites: Shuruppak, Kish, and Uruk among them. [...] These levels close the last prehistoric period" [103, p.114].

The first analysis of material taken from Woolley's flood deposits at Ur in Southern Mesopotamia was initiated by Watelin, the excavator of Kish. It resulted in "fresh-water silt containing just those elements which are to be expected from the waters of the Euphrates" [119, p.15]. However, the silt caused surprise because of "the absence of fresh-water molluscs as well as by the absence of contemporary marine organisms, and by the presence of terrestrial molluscs in one specimen only" [83]. This curious sterility was not to be expected for the tremendous flood of riverine origin required to accumulate up to 3.70 meters of silt. When, two decades later, geologists found that the water of the Shatt el Arab, around which the flood-yielding sites are located, is practically silt-free [76], the riverine theory could no longer be sustained. Even if there had been silt the velocity of a riverine flood "would incidentally inhibit the deposition of silt" [99, p. 60]. "The great recorded depth of the deposits at Ur, over 3 meters, and at Shurrupak, probably about 60 cm, are significant as they would require lagoon-like conditions for a fairly long time" [99, p. 53]. Therefore, an aeolian (wind blown) origin of the deposit was considered. The absence of lamellation seemed to point in that direction. Yet, there was "a complete absence of any larger rounded particles which usually occur in aeolian deposits" [82, p. 73]. In addition, it was found that the flood deposits differed "from the present desert sand" [119, pp. 160 ff].

Flood research ended in perplexity. Scholars had to go back to the drawing board where they now focus on the "impounding of the water through the operation of some hitherto unidentified phenomenon" [99, p. 62]. A subsidence of Mesopotamia by "many tens of meters in vertical extent" plus a "subsequent" uplift [99, p. 62] is a scenario more recently suggested by geology. Such a huge geological catastrophe might account for the depth of the deposits but still falls short of explaining the absence of marine molluscs, etc. Therefore, it is cautiously labelled "one of possibly many explanations" [99, p. 62]. With the riverine flood theory more or less debunked, the pendulum may swing back to the 20's and 30's when a cometary origin of many clay deposits on the Earth was discussed in geology [64] and astronomy [92, 93, 121]. More than three quarters of a century later, mainstream space research is whole-heartedly subscribing to the supposedly "new paradigm" (Dale P. Cruikshank) of silicon, water, salt, formaldehyde, methyl alcohol, hydrogen sulphide, methylamine, ethylene, ketene, methyl diacetylene, methyl isocyanide, butatrienylidene and carbon monoxide falling from the sky. It is a new science, "*bio-astronomy* which looks to the heavens for the chemical forerunners of life. And the tidings of cosmic fertility are strikingly at odds with the old view of comets as portents of doom" [8].

A celestial body called Inanna/Ishtar/Nintu, etc. - not always distinguishable from the Great Serpent - was singled out - e.g., in the "Atra-Hasis" epic - as the main culprit behind flood disasters: "How did I, with them, command total destruction, she wept after the flood's might came upon the peoples [like a battle array]" [73]. Ishtar was the most feared and most venerated deity of the Ancient Near Eastern pantheon - as depicted in the *Gilgamesh Epic* (Tablet XI; lines 103 - 130):

"The Annunaki lifted up the torches;
setting the land ablaze with their flare;
Stunned shock over Adad's deeds overtook the heavens
and turned to blackness all that had been light.
The [...] land shattered like a [...] pot.
All day long the South Wind blew [...],
blowing fast, submerging the mountain in water,
overwhelming the people like an attack.
No one could see his fellow,
they could not recognise each other in the torrent.
The gods were frightened by the Flood,
retreated, ascending to the heaven of Anu.
The gods were cowering like dogs, crouching by the outer wall.
Ishtar shrieked like a woman in childbirth,
the sweet-voiced Mistress of the Gods wailed:
'The olden days have alas turned to *clay,*
because I said evil things in the Assembly of the Gods!
How could I say evil things in the Assembly of the Gods,
ordering a catastrophe to destroy my people?
No sooner have I given birth to my dear people
than they fill the sea like so many fish!'
The gods - those of the Annunaki- were weeping with her,
the gods humbly sat weeping, sobbing with grief (?),
their lips burning, parched with thirst.
Six days and seven nights
came the wind and flood, the storm flattening the land.
When the seventh day arrived, the storm was pounding,
the flood was a war- struggling with itself like a woman
writhing (in labour)" [68].

Another hymn may suffice to illustrate the catastrophic prominence of Ishtar in ancient Mesopotamia:

"Proud Queen of the Earth Gods, Supreme Among the
Heaven Gods
Loud Thundering Storm, you pour your rain over all the
lands and all the people.
You make the heavens tremble and the earth quake./
You flash like lightning over the highlands; you throw
your firebrands across the earth.
Your deafening command, whistling like the South Wind,
splits apart great mountains.
You trample the disobedient like a wild bull; heaven and
earth tremble./
Your frightful cry descending from the heavens devours
its victims. /
Holy Inanna, the riverbanks overflow with the
flood-waves of your heart" [117].

In Hebrew sources we read about clashes of a deity (Sisera) which most probably was another *alter ego* of Ishtar: "They fought from heaven; the stars in their courses fought against Sisera" [63]. The Greek philosopher and historian, Aristotle, also conveys that in those days "the planets were gods" [5]. The period termed Bronze Age by modern scholarship, according to Aristotle, was divided by periodic catastrophes that destroyed civilizations [4]. His predecessor, Plato, reported three "disastrous floods which preceded the destructive deluge in the time of Deucalion" [95]. In another dialogue, Plato assigned these conflagrations to celestial powers:

"There have been and there will be many and diverse destructions of mankind, of which the greatest are by fire *and* water, and lesser ones by countless other means. For in truth

Table 1: Archaeologically attested evidence of Catastrophes in Bronze Age Mesopotamia [19, 45, 82, 104, 114, 115].

(The period presented here is conventionally dated in the late 4th to the early 3rd millennium; in the chronology based on stratigraphical evidence, however, it is placed between the 10th and 6th centuries [44, 47].)

The Myth of the dying Redeemer God, the Virgin Birth and the Madonna with Child [61] starts being popular in the so-called Late Sumerian culture [48].
Last destruction layer with **collapse of the Babylonian Ziggurat tower** of Kish (red stratum) and downfall of the Old Akkadian Empire.
Bronze Age/Early Dynastic IIIb and Old Akkadians with Iron knives [81] (Chagar Bazar, Tell Asmar)
Flood or destruction layer found in Ur and Kish; sterile deposit below lime stone temple in Uruk
Bronze Age/Early Dynastic II/IIIa with start of archaic cuneiform writing
Flood or destruction layer found in Ur and Kish
Bronze Age/Early Dynastic I with continuation of priest-kingship and pictographic writing
Flood or destruction layer found in Ur, Shurrupak (Fara), and Kish
Bronze Age/Uruk Period with start of priest-kingship, pictographic writing and Ishtar symbols
Flood or destruction layer found in Ur, Kharabeh Shattani and (probably) Kish
Chalcolithikum/Ubaid (last Stone Age stratum, where seals are used but writing and priest-kingship are still missing)

the story that is told in your country Greece as well as ours Egypt, how once upon a time Phaeton, son of Helios, yoked his father's chariot, and, because he was unable to drive it along the course taken by his father, burnt up all that was upon the earth and himself perished by a thunderbolt - that story, as it is told, has the fashion of a legend, but the truth of it lies in the occurrence of a shifting of the bodies in the heavens which move around the earth by fierce fire, which recurs at long intervals" [96]

In the 5th Book of the *Sibylline Oracles* [91] the prominence of Venus-Morning Star is emphasized at the finale of the fight between the gods (verses 512-516):

"The threat of a burning 'sun' amongst the stars I saw
And the terrible wrath of a 'moon' wrapped in lightning;
The stars had battle in their faces; God let them fight;
In place of the 'sun' long flames interwove.
The Morning Star directed the battle, mounting the
back of Leo" [71: pp. 11, 13].

Kugler, in his time, and probably still today, the first name in cuneiform Babylonian mathematics and astronomy, gave the following opinion, in 1927, on the *Book of the Sibyl*:

"The 'insane finale' revealed itself as a poetic treatment of actual natural events, according to a completely unitary plan. ... Two enormous meteors of the apparent size and form of the sun and the moon appear threateningly in the sky with their characteristic accompanying features. This causes uproar among the stars, and the actual fight between the stars begins. The Morning Star (Venus), standing on the back of Leo, introduces it. ... The constellations, which dominated the dawning morning sky at the start of the fight, finally depart to Okeanos and thereby set fire to the Earth" [71, pp. 9 ff].

More than half a century was to pass until an approximation of this view was reached by a mainstream classical scholar [25].

At Uruk, for example, the earliest "Inanna symbols in the form of animal figurines" [97] were dug up in the Eanna complex directly above a destruction layer and indicate how intensely the survivors concerned themselves with her. I propose now to give a survey of archaeologically tangible Bronze Age catastrophes (see Table 1 below) and then to address the question how mankind was able to cope with the cataclysms, or why they produced sacrificial cults, temples, and priests.

After the Bronze Age, catastrophes of global extent cannot be shown convincingly to have occurred. The end of the Bronze Age, with its dramatic climate change, caused by a "sudden and sharp increase of Carbon in the atmosphere" [33], is dated in one of the most thoroughly studied territories - the Netherlands - to 2650 years before the present, i.e. the 7th century BCE. According to preserved contemporary reports, the next time a major cataclysmic event was observed from the Earth was only in 1178 AD when an asteroid struck the Moon and possibly created the Giordano Bruno Crater [41]. In the early 14th century, which after the so-called Little Ice Age (from 1301/03) was struck by the Black Death (1348-52), contemporary witnesses reported strikingly large numbers of terrifying celestial portents. It is, however, unlikely, that more happened than the passing of the Earth through the dust tail of a comet. Nevertheless, renowned astronomers have tried to explain the predisposition for a plague epidemic with dangerous chemical substances in this dust [51].

The most recent serious cosmic blow to hit the Earth happened in 1908 above the Tunguska River in Siberia. A mysterious "blast wave" with a strength equivalent to twenty hydrogen bombs, caused by a bolide that entered the atmosphere above western China and descended 2000 km further to the north-east, flattened many hundreds of square kilometres of forest [32]. A "fellow sitting on his porch in the factory town of Vanavara sixty kilometers south of the point of touchdown was 'thrown several meters, felt a sensation of heat, and lost consciousness'" [70, 105].

Since the beginning of the nineties, the Solar System is being systematically searched for asteroids, or for "doomsday comets" [108], which might hit the Earth. Such bodies with a diameter of about one to two kilometres are considered to be extremely rare, a hit being expected only every 100,000 years. In the years 1991 and 1992, asteroids have passed the Earth at a distance of little more than 170,000 km, which is considered to be "quite a close shave" [66]. Even without catastrophes, it is estimated that every year up to 200.000 tons of cosmic material, including water, drops on the Earth [108: p. 191]. The identity of some 12 small meteorites found on earth with rock material analysed on Mars in July 8, 1997 further strengthens the view of the basic veracity of ancient mythology dealing with planets mostly of Venus, Mars and Mercury struck by comets or other celestial bodies on a scale that caused pillars of fire and tails of smoke visible on earth by night and day respectively and actively showering Earth itself.

Since Comet Shoemaker-Levy hit Jupiter in July 1994, even more attention is being paid. As this comet had a diameter of only 500-700 m, yet "trailed behind it gigantic fireballs and enormous quantities of dust, some of which is still visible in Jupiters atmosphere" [42], the planning against danger from space has been extended in the meantime.

The increased attention is due to the fact that bodies of up to 500m diameter occur ten times as frequently as lumps with a diameter of a kilometre. Somewhat smaller bodies can hit the Earth every 50-100 years [54]. This is the reason why four powerful observatories, distributed over the globe, are to search for such impact candidates [42]. At the same time it is planned to develop rockets which are able to deflect such bodies from Earth.

3. Sacrifice as collective play-therapy for the catastrophically beclouded Bronze Age communities

"But what is this 'additional spiritual element' [Johan Huizinga], which comes into play in *ritual* acts and which is 'very difficult to describe' [52]? According to Huizinga, it is a mystical realisation, but he does not give any further details, and we remain without a satisfactory answer to this question, though it is crucial to our culture./ Ritual play receives its sacred character from being a remembrance of an elementary event in primeval times and a re-awakening of that profound emotion. In those creative primeval times (as well as in every other creative period of history), the real world itself, or part of it, must have revealed itself to mankind who represented its essence in sacred acts and not in an order already shaped by man. /We would have to grasp the aspect of reality that made such a profound impression on those people who lived close to a natural state, that caused them to make killing the object of their most important cultural activities" [60, p. 85].

"We can only have a vague guess as to the backgrounds of human sacrifice" (Michael M. Rind, *Menschenopfer: Vom Kult der Grausamkeit*, Regensburg: Universitätsverlag Regensburg, 1996, p.18).

Boulanger, in 1766, was one of the earliest to ask the question: what may have gone on in the minds of people who escaped the cosmic catastrophes? As modern religious scholars simply ignored these cataclysms, the bizarre forms of antique religion had to remain inexplicable to them:

"Senseless, due to their misfortune, they only waited for their destruction, and they lacked the strength to feel their complete suffering. These are the traits with which OVIDUS and SENECA draw the physical and moral image of the Deluge; they show us men gathered on the mountain peaks, timid, astonished, out of their minds, and senseless with all their misfortune" [7, pp. 529 ff].

In 1931, more than 160 years after the publication of Boulanger's book, the cuneiform epic regarding the king and shepherd Etana was published. He was one of the earliest rulers of Kish, which is one of the first five cult centres built in Mesopotamia after the Flood. The introduction is a - certainly priestly - explanation for the origin of priest-kingship. It does not mention "intimidated", but "beclouded," i.e. confused people, who are given "counselling," even healing, through the new institution of priest-kingship and its performances.

On the basis of the references from *The Deluge* epic to an evolution from "destructions" to cult-centres it should be possible to reconstruct the original connection between the catastrophes suffered by Bronze Age man and the deadly sacrificial rituals he performed under the guidance of priests during this first period of high culture [46]. "Destruction," according to *The Deluge* epic, *preceded* the lowering of kingship from heaven. *After* this disaster, according to a late (Old-Babylonian) version of the *Epic of Etana*, 'gods' considered an innovative remedy for man:

The great Annunaki, who decree the fate,
Sat down, taking counsel about the land.
They who created the regions, who set up the establishment,
The Igigi were too lofty for mankind,
A stated time for mankind they decreed.
The *beclouded people*,* in all, had not set up a king.
At that time, no tiara had been tied on, nor crown,
And no scepter had been inlaid with lapis;
The shrines had not been built altogether.
The Seven Igigi had barred the gates against the settler's settlements.
Scepter, crown tiara, and (shepherd's) crook
Lay deposited before Anu in heaven,
There being *no counselling* for its people.
(Then) kingship descended from heaven
[98, p. 114, emphasis added].

* Instead of "beclouded," another English translation chooses "teeming." This also fits into the context, if "teeming" is read in the sense of overflowing, exaggerated, or in the sense of a proliferous or flourishing imagination and spiritual state of mind; cf. Dalley, S., Myths from Mesopotamia: Creation: The Flood, Gilgamesh, and Others (1989), Oxford University Press, p. 190

The best minds devoting their analytical skill to this myth somehow sensed that the somewhat enigmatic statement on "the beclouded people" in want of "counselling" must provide a clue to the rationale behind Bronze Age man's readiness to accept priest-kings: "The first line of the quotation intimates that the people were lost, lacking all direction, moving, as it were, in a fog, because there was no king" [29] This attempt to comprehend the origin of kingship is insufficient. We are not told that people were beclouded "because there was no king." A "destruction" had "beclouded" the people. They were not confused because they lacked a king. Instead, some "power not of this world" (Jacobsen), so to speak, had caused their confusion and, therefore, put them in need of a kind of counselling for which they saw no use before.

Students of Mesopotamian religion were hampered in their analysis of priest-kingship, i.e., the emergence of urban civilization, because they could not bring themselves to connect the "destruction" (from *The Deluge* epic) with the destruction layers found right underneath the earliest Bronze Age settlements. They acknowledged "the thunderstorm and the yearly flood" [56] as possible irritations of nature. But they could not recognise the catastrophic "black cloud" of Gilgamesh's Ishtar flood [68] as a real and most terrifying phenomenon and, therefore, could not make any sense of the "decree" for the "beclouded people." Seasonal disturbances of nature could not figure as Bronze Age innovations which alone could have assigned them explanatory power for the emergence of post-Neolithic cult centres.

Myths about great men inventing sacrifice, prayers, and gods after they had survived destructive floods, etc. are by no means restricted to Mesopotamia. This area, however, is rich in written material. In Chaldaea, Ziusudra is such a hero sacrificing after a flood [98, p. 44]. Assyrians have Utnapishtim [98, p. 95], and Hebrews Noah [35 (a)] in a similar role. In India, Manu invents sacrifice after the flood [106]. In Greek traditions Perseus, Deucalion, Megaros, Aiakos, etc., started to sacrifice after a flood [110]. Also the transformation of the poor hut of Philemon and Baucis into a temple after the flood [90] represents a way in which Greek memory kept the connection between priesthood and sacrifice on the one hand, and an overwhelming cataclysm on the other. Diodorus reports of Samothrace that survivors of a natural catastrophe "after being saved, marked the water level that saved them all around the island and erected altars on which they sacrifice to this day" [21]. An old-Egyptian text, dealing with catastrophe-bringing Sekhmet, ends: "And the majesty of this god said: 'Your sins shall be forgiven! For the sacrificial victims have removed the need for sacrificial killing.' - This is the origin of animal sacrifice" [9]. Last, but not least, the Algonquins of North America have their hero Nanaboush start the practice of praying after a flood [3]. In China, fighting dragons in the sky are seen as responsible for a flood after which the first dynasty of the Shang priesthood emerged [28, pp. 419 ff].

What is the meaning of "becloudedness" and what are the methods of counselling or healing of those who set themselves up as the first priest-kings by virtue of their ritual ideas? "Senseless with their misfortune," Boulanger describes the post-catastrophic mood and continues: "It seems that there was a time when the unhappy inhabitants of the Earth must have felt disgust and weariness with their homes having been destroyed under the most dreadful circumstances and their life being full of horror and dread" [7, p. 180]. Up to this point he is clear, but then his understanding of rituals fails him, he does not comprehend their horror and sadness. For he believes that the "thousands of customs" of religious man tend to "depress his mind, poison his days with constant tears, and make his life miserable" [7, p. 180].

We shall see in a moment that child's play is another human activity that frequently appears shocking and yet is performed with pleasure and repeated with perseverance, because it relieves distress. However, as a full understanding of this mental activity of playing was only achieved by Freud in 1920 [31], we cannot very well take Boulanger to task. "Senseless with misfortune" - in modern idiom - means to be so traumatised or mentally damaged that the self-healing powers alone are not sufficient to become again capable of facing reality.

Such an overwhelming state occurs especially when the threat cannot be influenced by either attack, flight or negotiation. This is the type of situation we must assume to have existed. A graphic example is given by the Prophet *Amos*: "Though they dig into hell, thence shall mine hand take them; though they limb up to heaven, thence will I bring them down: And though they hide themselves in the top of Carmel, I will search and take them out thence; and though they be hid from my sight in the bottom of the sea, thence will I command the serpent to bite them" [2].

Since there is evidence of floods, destruction, and climatic change one may well ask what impact they had on man? How did he react to them? Can we imagine his "becloudedness?" Though answers have to remain somewhat speculative, we have a body of psychiatric and psychological scholarship to direct us. Those who survived the mechanical impact of disaster could still die from fear. Those who managed to transform fear into fury were still helpless because of the absence of meaningful adversaries to turn to. Those who were not struck dumb by events and did not try to commit suicide resorted to flight that achieved nothing. Men cried out against the sky like wild animals; they threatened the forces of nature with outstretched fists. During storms with thunder and lightning, even male chimpanzees rush up hillsides with sticks in their hands and rage against the celestial attack [74]. Survivors might even try to release tension by indecent exposure to the heavens. It seems quite probable that the defensive erection of certain primates, which may be repeated in statues of Hermes with erect phalluses, indicates that it must have been part of the rituals [15].

When a heavy truck thunders past us so that we jump aside, startled, we may feel the urge to imitate the loud noise of the engine with our own voice. Children do this with less inhibition, and feel better for it. In the act of this play they anthropomorphise with their human voice the inanimate noise of combustion. It was this playful self-healing mechanism that the heroes had to use comprehensively and systematically to bring their fellow men back to their senses. We immediately understand that an imitated engine howling has an unpleasant effect on the outsider. What he hears is not nice. The real thing was bad enough. Is it not enough that this is behind us? Why is the unpleasant repeated and the pleasure principle violated?

Like children who try to master an overwhelming trauma by re-enacting it in their play, the terror-stricken communities of the Bronze Age - this author claims - produced heroes who introduced the sacrificial dramas for their beclouded kinsmen in a very similar manner. In 1920, Freud described the enigmatic activity of a boy of one and a half years:

> "He was greatly attached to his mother, who had not only fed him herself but had also looked after him without any outside help. This good little boy, however, had an occasional disturbing habit of taking any small objects he could get hold of and throwing them away from him into a corner, under the bed, and so on, so that hunting for his toys and picking them up was often quite a business. As he did this he gave vent to a loud, long-drawn-out 'o-o-o-o', accompanied by an expression of interest and satisfaction. His mother and the writer of the present account were agreed in thinking that this was not a mere injection but represented the German word *'fort'* 'gone'. I eventually realised that it was a game and that the only use he made of any of his toys was to play 'gone' with them. One day I made an observation which confirmed my view. The child had a wooden reel with a piece of string tied round it. It never occurred to him to pull it along the floor behind him, for instance, and play at its being a carriage. What he did was to hold the reel by the string and very skilfully throw it over the edge of his curtained cot, so that it disappeared into it, at the same time uttering his expressive 'o-o-o-o'. He then pulled the reel out of the cot again by the string and hailed its reappearance with a joyful *'da'* 'there'. This, then, was the complete game - disappearance and return. [...] The child cannot possibly have felt his mother's departure as something agreeable or even indifferent. How then does his repetition of this distressing experience as a game fit in with the pleasure principle? [...] On an unprejudiced view one gets an impression that the child turned his experience into a game from another motive. At the outset he was in a *passive* situation he was overpowered by the experience; but, by repeating it, unpleasuarable though it was, as a game, he took an *active* part. [...] It is clear that in their play children repeat everything that has made a great impression note the element pressure in this term - G.H. on them in real life, and that in doing so they abreact the strength of the impression and, as one might put it, make themselves master of the situation. [...] The unpleasurable nature of an experience does not make it unsuitable for play. If the doctor looks down a child's throat or carries out some small operation on him, we may be quite sure that these frightening experiences will be the subject of the next game; but we must not in that connection overlook the fact that there is a yield of pleasure from another source. As the child passes over from the passivity of the experience to the activity of the game, he hands on the disagreeable experience to one of his

playmates and in this way revenges himself on a substitute" [30].

A powerful *impression* can only be withstood when its pressure can be compensated. Children are, therefore, under an emotional compulsion to play all the time. If their games are destroyed or they are forbidden to play, they become mentally ill. Modern child therapy is, therefore, set up as play therapy whose success is measured by whether children can again play games actively and thereby become the stage directors of their own arrangements. Any new information is attached to a pressure that needs to be relieved so that space can be made to take in what has been unknown until then.

"Deities and other representations used in sacrifice have the role of concepts that allow participants to classify and to interpret impressions" [111, p. 347] is the supposition of Valeri, following Loisy [78]. But that is not what the origin of a game is all about. The classification and interpretation with the aid of concepts is beyond this primarily compulsive mental process. It liberates for analysis but is not analysis itself. The repetition, organised by the priest, of the sacrificial game, once it has been created, does not turn it into a system of conceptual analysis.

Having understood the child's compulsion to play games we now understand why early Bronze Age cultures played "deluge" and "world conflagration." These particular events provided overwhelming impressions, which must have led to mentally-caused chaotic disorder within the communities if they could not be healed. The first priests became the heroes of mankind because they knew how to stage the sacred-healing *Agon*. We are not surprised, then, to learn that besides the roles for the people concerned there was also a need for actors to appear as destructive or saving celestial bodies: "The figure of Assur, who wages battle against Tiamat, is that of King Sennacherib" [89]. Boulanger knew all this, but he was unable to comprehend why there was a need for it:

> "This shows why people imagined that the perpetuation of the *mysteries* was of the greatest importance for the human race. It shows why all mysteries, moreover, were based on an astronomical ceremonial; for there really was one of the priests, during the ceremonies, who was called *the Creator of the World*, another called himself *Sun*, a third played the part of the *Moon*. In the mysteries of *Mythras* of the *Persians* [85: pp.51 ff and 86-133], the different categories of initiates had the names of the Signs of the *Zodiac*, and of the different constellations in the sky" [7, p. 232].

Lucian reports similarly about the training given to the artists used in late classical catastrophic plays: "A dancer needs to be able to represent the following: Deukalion the hero of the Flood, the great shipwreck of life during his time, a *larnax* which preserves the remainder of the human race, and the men who have been resurrected from stones" [79].

The major sacrificial cults turn out to be collective healing rituals for communities that have become half crazy, or "senseless with misfortune" (Boulanger), through the action of global catastrophes. The rage directed against the victoriously aggressive nature, which mankind cannot constructively discharge by attack, flight or negotiation, is, so to speak, being pushed back at them, down their throats. Within the ritual, it is used in an organised fashion. Now the collective, in need of healing, does to human or animal actors representing the force of nature "the unpleasantness which itself has experienced, and in this way takes revenge upon the person of this representative" (Freud). By slaughtering the representatives of falling celestial bodies, the sickening rage is burned up: "In mythology, the first murderer is identical with the saviour" [109]. This sacred-healing side of the ritual turns it into the blood sacrifice, the

Fig. 5: Above: The celestial goddess, Maia, in front of her celestial child (here Hermes), who lies between cattle (bestiomorphized celestial bodies) in a so-called swing cradle which becomes the prototype for the birth of the Jesus Child in a manger. Below: The celestial child deity Hermes with his virginal mother (here called Iris) in the firmament. This motive is repeated in the representations of Mary as Queen of Heaven with the Christ Child (both figures from the 5th century BCE. Adapted from K. Schefold, together with F. Jung, 1981, *Die Göttersage in der klassischen und hellenistischen Kunst*, Hirmer, München, p. 47.)

mysterious origin of which we have set out to discover: "For the sacrificial victims have removed the need for sacrificial killing" [9]. The healing effect of the slaughter is also attested in an inscription in a Mithraeum (Santa Prisca, Rome) regarding Mithras' sacrifice of a bull which is interpreted as a duel between Sun and Moon: "You have saved us, too, by shedding the eternal blood" [85: p.199]. Cosmic collisions of inorganic lumps of matter have been viewed frequently, from an anthropomorphizing or bestiomorphising angle, as duels between warriors, beasts, or fabulous creatures. The end of the catastrophe is seen as the victory of a celestial fighter or a deity, when another "loses" or "dies." This is why the cosmic or "divine" aspect of the overpowering impression needs to be included as a duel in the healing play enacted by men: "In this context, the killed being - man or animal - represents the deity in the same sense in which on our theatre stage an actor represents King Lear, for example, and is identical with him for the duration of the play" [60].

To play these roles, living humans and animals were adorned or masked and dressed up to play these two celestial bodies. In their duel there is again seen under an anthropomorphizing aspect a victor and a loser. A central concept is frequently that of a loser who brings redemption for heaven and earth by his "death," i.e. one who ends the catastrophic period. These dying redeemer gods [107] populate, in legends and iconography, the pre-Christian era of Eurasia. They are frequently remembered as heavenly children [88] or youths who were born from a virgin [75]. This notion appears to be based on the observation that two celestial bodies of different size, or two differently-sized parts of the same celestial body, inspired the image of the single mother. Christian iconography has borrowed many of its motives from this material (see figures 5 and 6).

It is difficult to decide whether it was determined since the inception of the major sacrifice cults who would die in the role of the falling celestial body and who would be victorious as the one that stayed in its orbit. At a later stage, however, the victims were selected, prepared and trained. Even when

Fig. 6: Athens mythical primaeval king, Pandion, who saved the city with his heroic death, is carried away by the goddess Athena (image taken from black-figured vase of around 500 BCE. Adapted from K. Schefold, together with F. Jung, 1988, *Die Urkönige Perseus, Bellerophon, Herakles und Theseus in der klassischen und hellenistischen Kunst*, Hirmer, München, p. 72.*)*. As Pandion is equated with Heracles and had a temple and a festival in Athens, it appears that an initially humanized celestial deity was later reinterpreted as a human king. His celestial transport over the sea represents a scene that later becomes the prototype for the Pietà of Christian iconography.

a bull was used as a representative of a celestial body, it was hoped that "the animal would proceed to the sacrifice willingly, even 'voluntarily'" [12, p. 101]. In both cases, the community that participates in the great play is able to release in the act of killing the aggression born out of helplessness with regard to the catastrophe. The slaughter of the sacrificial victim is therefore both the conclusion and the cathartic climax of the ceremony. Men liberate themselves in this bloody act from the fury that until then had been turned inward and that had caused them helpless numbness, psychosomatic suffering or aggressiveness which endangered their fellow men.

Everything that has made an overpowering impression needs to be played out in the healing ceremonies. This includes the destructive natural events as much as the disturbing extreme behaviour of fellow humans observed during the cataclysm. One of the most embarrassing enigmas of the research into the subject of sacrifice is the administering of "powerful laxatives" [38, p. 20] to some participants in the ritual. It is no longer difficult to understand this procedure: to their own horror and that of their fellow men, some adults, under the impression of the catastrophic events, lost control over their bowels. "Dirtying ones pants with fear" was, therefore, part of the sacrificial play.

Scholars are also astonished when they learn of a staged ecstasy (*ekstasis*), in which someone "leaves his usual path, his level-headedness" [12, p. 178], as well as studied frenzy (*mania*). With these ritual techniques, the panic and neurotic reactions of people under the impression of catastrophic events are rendered. As it is not as easy during religious processions to produce erections that remind of aggression as it is to provoke diarrhea, artificial *phalloi* were carried (figure 7): "There are many variants of processions with outsize artificial phalluses; those who carry them have to hide their civil identity by smearing themselves with soot or bran, or by wearing masks" [12, p. 171].

The *phalloi*, which are well attested as far as China [17], are - like the smearing with soot or ashes - considered to be as mysterious as the laxatives. Assuming that the horrifying impression of the catastrophes led to spontaneous erections, then the survivors would have been so deeply impressed that these extreme gestures were incorporated into the ceremonies as was the smearing with soot or ashes as a re-enactment of the falling ashes.

As early as 1932, the English eccentric, Beaumont, pointed to the symbolic association between the tail of the comet and the *phalloi*: "It is definitely instructive, and it is curious that the Great War of 1914 was heralded in by an enormous comet, Delevans, in appearance like a huge illuminated phallus" [6, p. 85]. In his study he sees the phallus cults as having been determined more by the celestial loss of a tail rather than the impressions left by panic-induced and publicly exhibited erections. However, this possible background shall not be left unexplored. What speaks against it

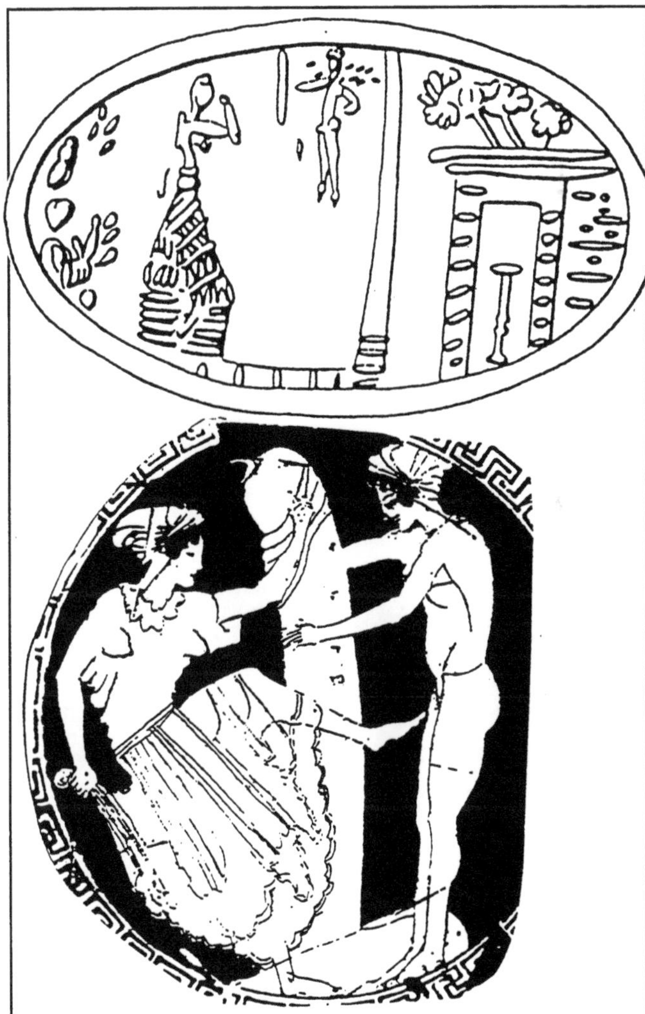

Fig. 7: Above: Woman or goddess below youthful, staff-bearing god or Hermes, whose position is interpreted as either flying or hanging, and whose end, or sacrifice, is completed by castration (adapted scene from a golden, late Mycenaean or early Archaic seal ring from Knossos/Crete. H. Müller-Karpe, 1980, *Handbuch der Vorgeschichte. Vierter Band: Bronzezeit. Dritter Teilband. Tafeln*, Beck, München, plate 204, ill. 18.*)*. Below: Carnivalesque phallus worship, practiced by women in classical times, possibly originated in an earlier scene of mourning, in which the drama of the cosmic loss of a tail was felt by the inhabitants of the Earth to be an act of redemption and was re-enacted in a bloody ritual (red figured beaker from the 5th century BCE; adapted from C. Johns, 1989, *Sex or Symbol: Erotic Images of Greece and Rome* (1982), British Museum Press, London, p. 43*)*.

above all is the voluminous size of the cultic *phalloi* used in ceremonial processions and the castration ritual.

Besides the dirtying of the clothes and the smearing with ashes, research into sacrifice has found mysterious the cutting off and burning of hair [23, pp. 344ff.] of the sacrificial victims. Could this, once more, point to a flaming comet, whose tail lit up on approaching the Sun? "[Nestor]...offered up his earnest prayers to Athene as he began the sacrifice by throwing a lock from the victims head on the fire" [50: III, 446] tells us. "With amazing obstinacy, scholars have hunted for demons who demand hair" [13] mocks Burkert. He himself recognizes in this, chosing an etymological approach, a minimizing "act of the beginning" [13] of the sacrifice. Cutting off the forelock, he says, damages the heifer, but does not really hurt it. There is not much to be said against a soothing "start." But then there is the gilding of the horns. By having its horns rendered shiny, the heifer plays a part that goes beyond it. The hair, too, is not just cut off, it is burned as well. This turns the whole procedure into a real scene with a cosmic precedent that we need to inquire into. After all, the celestial deity *Pallas Athene* [Homer, III, 435] participates in the ritual. From Hittite sources we know, moreover, that the gilding of the horns can be removable [72], so that it can be re-used for the next bull and in this way forms a precursor of the bull statue. In ancient Egypt, the cosmic destroyer, Seth, was seen "transformed into a red bull, fighting Horus" [63, p. 72]. In the ritual, this "sacrificial animal enacted the enemy of the god and the king" [63, p. 72].

Was the ritual burning of the hair preceded by a cosmic loss of "hair" which was interpreted as the beginning of the end of a falling celestial body? Homer's description of Athena does not contradict an astral scene: "like a meteor ... as a warning to ... some great army ... comes blazing through the sky and tossing out innumerable sparks. Thus Pallas Athene flashed down to earth and lept into their midst. ... Trojans and Achaean men-at-arms were awe-struck at the sight" [49:,IV, 75-79]. The history of astronomy yields this report:

> "Bright jets or streams of light are projected from the nucleus, and their action betokens the character of the gases. Some jets resemble rams horns and are due to iron vapour, others like goats horns streaming away in a curve, indicate hydro-carbon. ... [Seen were] figures as disks, upright cones, horses manes, torches, swords, spears, bearded faces and one called horror producing. The bearded faces recall the red-bearded and mischievous Satyrs always intent upon worrying mankind, the followers of Dionysus, the Serpent God, and comets are seen not unlike the monstrous human visage" [6: p. 84].

Even the burnt offering (*holocaust*) is not well understood. Could it, too, be interpreted as a healing re-enactment of a celestial burning? Again, this is suggested by texts and rituals concerning Seth (figure 8). Pieces of geese - they symbolise "the soul of Seth" [63: p.73] - or gazelles - they represent Seth himself - "were placed on the fire and burned; that it was not mere roasting is evident from the texts. The deity Sekhmet, Horus, etc. shall take pleasure in the total annihilation of his or her enemy, which is why his limbs are turned to ashes [...]. To fully appreciate the idea it is necessary to go back to the myths which tell of the annihilation of Seth; there, we hear again and again how he was thrown into the fire and turned to ashes" [63, pp.74 ff.].

In ancient Egypt, Seth/Apophis animals - gazelles, red goats, hippopotamus, turtles and bulls - could be replaced or complemented in rituals by humans who had been sentenced to death. Above all offences such as the desecration of tombs, temples, holy precincts or other sacred institutions could lead to the obligation of having to play the "evil one" in a ritual and of "getting burned" [116]. Anyone who treated blood sacrifice in any way with disrespect had to pay for it by assuming the role of the sacrificial victim:

> "By his deeds the violator himself proved that he was a manifestation of Seth or of Apophis, the embodiment of

Fig. 8: Ancient Egyptian sacrifice of a gazelle which represents Seth/Apophis, who was destroyed by Horus, and whose parts were then turned to ashes in a burnt offering (holocaust); P. Derchain, 1962, *Rites Egyptiens. I: Le sacrifice de loryx*, Fondation Egyptologique Reine Elisabeth, Bruxelles, p. 9.)

> disorder. The felon was put on the same level as the animals which were slaughtered in the temple or in the mortuary rites as a symbol of the defeat of evil. Seen from this aspect, the punishment was not simply an act of revenge, though this would have played a part frequently. It was, rather, a method of encountering cosmic disorder" [116].

The ritual throwing of barleycorn [23, pp. 261ff.] or - in the Tereus sacrifice at Megara - "of pebbles instead of sacrificial barley" [23, p. 281] is an enigma to researchers into sacrifice, just as much as laxatives, erections, ashes and burning of hair: "When we observe what the rites consist of grains of sand thrown into the wind and what they are aimed at, we ask ourselves in amazement where these ideas came from and, above all, why men remained so faithful to them" [22].

"Simultaneous throwing towards a common goal is an age-old gesture of aggression" [13] and prepares for killing within the sacrificial procedure, Burkert believes, and therefore thinks the throwing of stones is explained. The same author knows, however, that certain "stone heaps" [12, pp. 76 ff] had the character of holy places as early as Minoan times (Bronze Age). In any event, "the veneration of stones" is found "in all the religions of the Ancient World and is attested in the Near East to the present day" [67]. Such holy stones - *massebes* - "in which the deity manifested himself" [27] are also known in ancient Israel: "And he took of the stones of that place, and put them for his pillows, and lay down in that place to sleep. / And he dreamed, and behold a ladder set up on the earth, and the top of it reached to heaven .../And Jacob rose up early in the morning, and took the stone that he had put for his pillows, and set it up for a pillar" [35(b)].

Levantine-Syrian shrines of the Middle and Late Bronze

Age contained "*in addition to* the altar for the sacrificial cult, the *massebes* and the *ashera* to mark the holy place" [101]. The Phoenician-Carthaginian goddess Tanit, too, was associated with a holy "stone" [53]. Equally, pre-monotheistic Yahwe of the Ancient Israelites is addressed as stone [35(c)]. Mesopotamian Ninurta - yet another cosmic manifestation of Mercury - "fights with the support of 'good' stones against 'evil' stones" [67]. In the largest temple of the Hellenic world, there is also a meteorite that is being worshipped: "What man is there that knoweth not how that the city of the Ephesians is a worshipper of the great goddess Diana, and of the image which fell down from Jupiter?" [1]. Even now, one billion Muslims turn five times daily towards a holy stone that fell from the sky. The name Kaaba for the cubic structure, the south-eastern corner (*al-Rukn*) of which contains the black lump of rock that, according to tradition, comes from the Archangel Gabriel, might be a corruption of the old-Semitic word for star - *kakkabu*. The Arabic name for the Polar Star - *al-Rukaba* seems to point in the same direction [16, 59]. Likewise, hail accompanied by fire (*barad*) as mentioned among the Egyptian plagues of Exodus does not stand for ice grains, but for a hail of hot stones:

"And the Lord sent thunder and hail, and the fire ran along upon the ground. / So there was hail, and fire mingled with the hail, very grievous, such as there was none like it in all the land of Egypt since it became a nation. / And the hail smote throughout the land of Egypt all that was in the field, both man and beast; and the hail smote every herb of the field, and brake every tree of the field" [26].

Can the ritual throwing of barleycorn [49, I, 458; 49, II, 410; 50: III, 447] or stones be made to yield its mystery by being explained as a re-enactment of the dread-inducing meteorite showers that fell on Earth? Religious scholars will not want to agree. Nearly 200 years after the recognition of the existence of these cosmic phenomena by scientists, they are still caught up in that reluctance which, even at the time they happened, made the recognition of the facts so difficult:

"From the time of Aristotle, who asserted that a meteorite, which fell on Aegospotami when a comet was glowing in the sky, had been lifted from the ground by the wind and carried in the air and dropped over that place, until the year 1803 when, on April 26, a shower of meteorites fell on lAigle in France and was investigated by Biot for the French Academy of Sciences, the scholarly world - and in the meantime there lived Copernicus, Galileo Galilei, Kepler, Newton, and Huygens - did not believe that such a thing as a stone falling from the sky was possible at all. ... Only shortly before 1803, the Academy of Sciences of Paris refused to believe that, on another occasion, stones had fallen from the sky. The fall of meteorites on July 24, 1790 in southwest France was pronounced 'un phénomène physiquement impossible' [a physically impossible phenomenon]. Since the year 1803, however, scholars have believed that stones fall from the sky" [112].

Ordinary people knew, but scientists needed to be informed first. Once more they paid for their straitlaced academic evolutionism with perplexity and humiliation. They have in front of them all the elements required for understanding sacrifice, yet they are not allowed to draw a sensible conclusion. Some see "men dance or hop or snarl through the shrub in some sort of rituals" [43]. But even someone who - like Burkert - can feel the cataclysmic element, does not succeed in fitting it correctly into the ritual: "What in reality is set in motion in the 'unspeakable sacrifice' is not the order of nature, but that of community life and of the corresponding inner life. The emotional shock that is unleashed is so tremendous that the cosmos appears to swing along in the same rhythm" [14, pp. 131ff.].

This assessment is not far removed from the enthusiastic academics and creative artists who revealed to the present author over decades and with grandiose Marxist gestures of wisdom that the global catastrophes of the past were allegories for the class struggle which might even reach as far as cosmic space. In the meantime, there is some movement on the left. The willingness to recognise cosmic facts has increased noticeably: "From the geological point of view, the impact craters are functional equivalents of wars and revolutions in the history of mankind" [20].

Quite similar to Burkert - the first name in religious studies concerning Ancient Greece - Jacobsen - the first name in religious studies concerning Ancient Mesopotamia - decided in favour of a shaking of heaven by the ritual on Earth, in order to avoid having to investigate the reverse idea. Based on the cataclysmic language in the so-called "Cursing of Agade" - the first capital of a major empire - he attributes to the destruction of the stepped tower "cosmic effects" and influences "on nature" [53] instead of asking for cosmic or other natural causes for the collapse of this enormous building.

The original texts themselves do not in any way indicate a reversal from cosmic cause and terrestrial effect to terrestrial cause and cosmic effect. First something happens in the sky, and then rituals are set up. In a Late Bronze Age sacrificial text of the east Anatolian Hurrians (the ancestors of the Kurds), a star falls in the sky, whereupon its representative - in this case a bird - is killed:

"But when on that day, at night time,
a (shooting) star falls, the ritual lord goes
to the old temple; he bows to the deity
and goes behind the blood ritual;
and others perform the blood ritual with a bird" [40].

Burkert, too, knows the myths of the raging skies and the restless stars, but he has accepted the teaching that this is astrophysically unthinkable. He also understands the mental excitement of those who perform the sacrifice. The ritual - this he sees clearly - "creates anxiety situations in order to overcome them, and thereby provides a primeval model for overcoming anxiety" [12: p.396]. It serves "to overcome anxiety situations imposed from outside" [11]. Burkert denies neither human worries nor images of a disorderly cosmos, but tries to do justice to both. So while, in his view, the terrified souls of mankind are capable of shaking the cosmos, they could never have been terrified by horrifying cosmic events. Why they enact deluges, world conflagrations and celestial bodies that lose their bleeding tails after collisions, cannot even be guessed at from this point of view. It must remain unexplored that people became mentally ill through superhuman - but in no way supernatural - events and therefore looked for unconventional ways to heal themselves. Yet the Greek word for medication - *pharmakeia* (cf. pharmacy) - refers to the *pharmakos* who attests "in the middle of Greek civilization that human sacrifice existed as a hint, as a possibility, maybe even as a fixed institution" [12: p.139]. Only later the human *pharmakos* was replaced by a sacrificial ram.

If the major blood sacrifice rituals and the accompanying ceremonies came into the world as collective healing plays, they may be able to shed light on the "ceremonial healing" [24] which presents ethnologists with a particular enigma. In these rites, people with mental illness are not brought to the usual local healers, but the great religious ceremony of the tribe is performed for them out of turn. The Navaho (Arizona and New Mexico), for example, perform the *Mountain Chant*. The sand pictures that are prepared within this ceremony, which are representations of "deities representing the forces of nature, perhaps celestial bodies" [94] are on this occasion not only prepared for the community of up to two thousand people but are at the end given the sick person as medication: the coloured sands are swept up and thrown over them. It is thought inexplicable why this process is expected to provide mental healing. In the light of the view developed here, the Navaho must have preserved a vague memory that religious ceremonies, long ago, were invented for healing purposes. Why an age-old religious 'collective therapy' should be used to heal an individual of today appears, then, much less unintelligible than what modern scholars more or less perplexed call "ceremonial healing." As long as the original ceremony remains misunderstood, of course its modern therapeutic use can only be thought absurd.

After the termination of disastrous and widespread catastrophes at the beginning of the Iron Age, movements arose which propagated the abolition of human sacrifice. Wherever they succeeded, to the chagrin of the faithful, freshly slaughtered idols were no longer available. Conservative forces replaced them with more durable images like carved tree-trunks or stone sculptures still retaining the shackles which had been necessary to hold up dead victims. In front of such idols rituals were performed for the release of any unbearable tension, whether its source was cosmic or not.

Fig. 9: Lion flanked column of gibbet type with two animal heads - or masks - tied to its top. From Late Bronze Age Mycenaean gold ring (Ashmolean Museum, Oxford; *from* H. Müller-Karpe, 1980, *Handbuch der Vorgeschichte. Vierter Band: Bronzezeit. Dritter Teilband. Tafeln*, Beck, München, table 205, no. 17.*)*. The - as yet - enigmatic sacredness of columns, pillars, posts etc. may have been due to reconciliatory rites performed before victims fastened to them after - or whilst - being killed in sacrificial rituals, in which the victims may have played celestial bodies defeated in a cosmic combat (theomachy). Asking forgiveness from these freshly slaughtered idols the members of the sacrificially healed communities turned themselves into worshippers of idols who could not help but also sanctify the pillars, crosses etc. to which the victims were fettered, chained, pulled up or attached in some other manner. The decline of blood sacrifice after the termination of the Bronze Age with its series of cataclysms eventually brought about the idols that are so numerous in the Iron Age. They were made of wood, bone, clay, stone or metal and originally exhibited shackles the enigmatic fettered gods (Karl Meuli).

Occasionally, ritual sacrifice was even carried out in front of such statues. These regressions - falling back on fresh victims in front of 'petrified' ones - add to the difficulty of explaining the origin of idols. Because of their shackles carved into wood or stone they were identified as a peculiar brand of "fettered gods" [86]. In actual fact, the earliest sculptors of dead idols only seem to have tried to reproduce the slaughtered ones as unadulterated as possible. It is not by chance therefore, that post-Babylonian Judaism with its radical criticism of blood sacrifice also turned against its petrified version, i.e., it prohibited idols [46].

References

1. *Acts of the Apostles* 19: 35.
2. *Amos* 9: 2f.
3. Andree, R., 1891, *Die Flutsagen*, Friedrich Vieweg, Brauschweig, pp. 71 ff./153.
4. Aristotle, *Fragments*, 18.
5. Aristotle, *Metaphysics*, 1074 bl.
6. Beaumont, W.C., 1932, *The Mysterious Comet. Or: The Origin, Building up, and Destruction of Worlds by Means of Cometary Contacts*, Rider, London.
7. Boulanger, N.-A., 1767, *Das durch seine Gebräuche aufgedeckte Altertum: Oder Kritische Untersuchung der vornehmsten Meynungen, Ceremonien und Ein-richtungen der verschiedenen Völker des Erdbodens in Religions- und bürgerlichen Sachen* (1766), Anton Ferdinand Rösens Buchhandlung, Greifswald, pp. 529 f.
8. Broad, W.J., 1997, Gifts of the Comet [summary of articles from *Science*, March 28, 1997], *The New York Times - Science Times*, April 1.
9. Brugsch, H., 1881, *Die neue Weltordnung nach Vernichtung des sündigen Menschengeschlechts nach einer altägyptischen Untersuchung*, Calvary u. Co., Berlin, p. 187.
10. Burkert, W., 1996, *Creation of the Sacred: Tracks of Biology in Early Religions*, Harvard University Press, Cambridge/ Mass. & London
11. Burkert, W., 1981, Glaube und Verhalten: Zeichengehalt und Wirkungsmacht von Opferritualen, *Le Sacrifice dans L'Antiquité*, J. Rudhardt, O. Reverdin (eds.), Vandoeuvres, Foundation Hardt Pour l'Étude de L'Antiquité Classiques, Entretiens, Tome XXVII, Genève.
12. Burkert, W., 1977, *Griechische Religion der archaischen und klassischen Epoche*, Kohlhammer, Stuttgart et al.
13. Burkert, W., 1990, Griechische Tragödie und Opferritual, in Idem, *Wilder Ursprung: Opferritual und Mythos bei den Griechen*, Wagenbach, Berlin.
14. Burkert, W., 1972, *Homo Necans: Interpretationen altgriechischer Opferriten und Mythen*, Walter de Gruyter, Berlin und New York.
15. Burkert, W., 1979, *Structure and History in Greek Mythology and Ritual*, University of California Press, Berkeley et al., pp. 40 ff.
16. Cardona, D., 1988, The Kaaba, *Kronos*, XII, p. 25.
17. Chang, K.C., 1983, *Art, Myth, and Ritual: The Path to Political Authority in Ancient China*, Harvard University Press, Cambridge/Mass. and London. pp. 116, 118.
18. Clark, G., 1977, *World Prehistory in New Perspective*, Cambridge University Press, Cambridge et al.
19. Dalfes, N., Kukla, G. and Weiss, H. (eds.), 1997, *Third Millennium BC Climate Change and Old World Social Collapse*, Springer, Berlin.
20. Davis, M., 1996, Cosmic Dancers on Historys Stage. The Permanent Revolution of the Earth Sciences, *New Left Review*, no. 217, p. 84.
21. Diodoros *FGrHist* 548.
22. Durkheim, E., 1981, *Die elementaren Formen des religiösen Lebens* (1915), Suhrkamp, Frankfurt am Main, p. 469.
23. Eitrem, S., 1977, *Opferritus und Voropfer der Griechen und Römer*, Kristiania: J. Dybwad, 1915; Reprint Georg Olms, Hildesheim & New York.
24. Ellenberger, H.F., 1973, *Die Entdeckung des Unbewussten*, Hans Huber, Bern, Stuttgart & Wien, p. 58.
25. Engelhardt, W.v., 1979, Phaetons Sturz - ein Naturereignis?, *Sitzungsberichte der Heidelberger Akademie der Wissenschaften. Mathematisch-Naturwissenschaftliche Klasse*, p. 161-199.
26. *Exodus*, 9: 23/24/25
27. Faherty, R.H., 1991, Sacrifice, *The New Encyclopedia Britannica: Macropedia*, vol. 26, Encyclopedia Britannica Inc., Chicago et al., p. 795.
28. Fontenrose, J., 1980, *Python. A Study of Delphic Myth and its Origins* (1959), University of California Press, Berkeley et al.
29. Frankfort, H. 1978, *Kingship and the Gods. A Study of Ancient Near Eastern Religion as the Integration of Society and Nature* (1948), University of Chicago Press, Chicago and London, p. 237.

30. Freud, S., 1984, Beyond the Pleasure Principle (1920), *The Pelican Freud Library. Volume 11. On Metapsychology*, Penguin Books, Harmondsworth, pp. 284ff., emphasis added.
31. Freud, S., 1947, Jenseits des Lustprinzips (1920), Idem, *Gesammelte Werke*, vol. XIII, S. Fischer, Frankfurt am Main, pp. 11-15.
32. Ganapathy, J., 1983, The Tunguska Explosion of 1908: Discovery of Meteoric Debris Near the Explosion Site and at the South Pole, *Science*, vol. 220, pp. 1158 ff.
33. Geel, B. van, Buurman, J. and Waterbolk, H.T., 1996, Archaeological and Palaeological Indications of an Abrupt Climate Change in the Netherlands, and Evidence for Climatological Telecommunications around 2650 BP, *Journal of Quaternary Science*, vol. 11, p. 275.
34. Geertz, C., 1966. Religion as a Culture System, *Anthropological Approaches to the Study of Religion*, M. Banton (ed.), Tavistock, London.
35 (a). *Genesis*, 8: 20.
35 (b). *Genesis*, 28: 11/12/18.
35 (c). *Genesis*, 49: 24.
36. Gibson, M., 1976-80, Kis. B. Archäologisch, *Reallexikon der Assyriologie*, Vol. 5, Walter de Gruyter, Berlin and New York, p. 618.
37. Girard, R., 1972, *La violence et le sacré*, Paris, Bernard Grasset, Paris.
38. Gould, J., 1985, On Making Sense of Greek Religion, *Greek Religion and Society*, P.E. Easterling, J.V. Muir (eds.), Cambridge University Press, Cambridge et al.
39. Green, A.R.W., 1975, *The Role of Human Sacrifice in the Ancient Near East*, Scholars Press, Missoula/Montana
40. Haas, V., 1993, Ein hurritischer Blutritus, *Religionsgeschichtliche Beziehungen zwischen Kleinasien, Nordsyrien und dem Alten Testament: Internationales Symposium Hamburg 17.-21. März 1990*, B. Janowski, K. Koch and G. Wilhelm (eds.), Universitätsverlag, Freiburg/Schweiz and Vandenhoeck & Ruprecht, Göttingen, p. 75.
41. Hartung, J., 1976, Was the Formation of a 20-km-Diameter Impact Crater on the Moon Observed on June 18, 1178?, *Meteoritics*, vol. 11, p. 187.
42. Hawkes, N., 1995, How Great Is the Threat from Comets? Danger from Deep Space, *The Times*, October 2, p. 18.
43. Heinrich, K., 1986, *Dahlemer Vorlesungen 2: Anthropomorphe*, Stroemfeld/Roter Stern, Basel und Frankfurt, p. 39.
44. Heinsohn, G., 1996, *Assyrerkönige gleich Perserherrscher! **Die Assyrienfunde** **bestätigen** das Achämenidenreich*, Mantis, Gräfelfing.
45. Heinsohn, G., 1990, Destruction Layers in Archaeological Sites: The Stratigraphy of Armageddon, *Catastrophism 2000*, M. Zysman and C. Whelton (eds.), Heretic Press, Toronto, pp. 213-247.
46. Heinsohn, G., 1997, *Die Erschaffung der Götter: Das Opfer als Ursprung der Religion*, Rowohlt, Reinbek.
47. Heinsohn, G. and Illig, H., 1997, *Wann lebten die Pharaonen? Archäologische und technologische Grundlagen für die Neuschreibung der Geschichte Ägyptens und der übrigen Welt*, Mantis, Gräfelfing.
48. Helck, W., 1971, *Betrachtungen zur großen Göttin und den ihr verbundenen Gottheiten*, R. Oldenbourg, München and Wien, pp. 28 ff.
49. Homer, *Iliad.*
50. Homer, *Odyssee.*
51. Hoyle, F. and Wickramasinghe, N.C., 1979, *Diseases from Space*, J.M. Dent & Sons, London et al.
52. Huizinga, J., 1944, *Homo Ludens: Versuch einer Bestimmung des Spielelementes der Kultur* (1938), Akademische Verlangsanstalt Pantheon, Basel.
53. Huss, W., 1994, *Die Karthager* (1990), Beck, München, p. 367.
54. Irwin, A., 1995, Scientist Calls for Meteor Monitor, *The Times Higher Education Supplement*, October 6.
55. Jacobsen, T., 1987, *The Harps that once...*, Yale University Press, New Haven & London:, chapter: "The Cursing of Akkade", pp. 366 f.
56. Jacobsen, T., 1972, Mesopotamia, *The Intellectual Adventure of Ancient Man. An Essay on Speculative Thought in the Ancient Near East*, H. Frankfort et al. (eds.), University of Chicago Press, Chicago and London, p. 127.
57. Jacobsen, T., 1939, *The Sumerian King List*, University of Chicago Press, Chicago, p. 77.
58. Jacobsen, T., 1976, *Treasures of Darkness. A History of Mesopotamian Religion*, Yale University Press, New Haven and London, p. 3.
59. Jastrow, M., 1912, *Die Religion Babyloniens und Assyriens*, vol. II/2, Ricker, Gießen, p. 153 f.
60. Jensen, A.E., 1991, *Mythos und Kult bei den Naturvölkern:Religionswissenschaftliche Betrachtungen* (1951), with an introduction by E. Haberland, dtv, München, p. 232.
61. Jeremias, A., 1930, *Das Alte Testament im Lichte des Alten Orients*, 4th edition, J.C. Hinrichs'sche Buchhandlung, Leipzig, p. 790.
62. *Judges* 5:20.
63. Junker, H., 1910, Die Schlacht- und Brandopfer im Tempelkult der Spätzeit, *Zeitschrift für ägyptische Sprache und Altertumskunde*, 47.
64. Keilhack, K., 1920, Das Rätsel der Lößbildung, *Zeitschrift der deutschen geologischen Gesellschaft*, 72, nos. 6 and 7.
65. Kirk, G.S., 1970, *Myth. Its Meaning and Function in Ancient and Other Cultures*, University of California Press, Berkeley and Chicago, p. 283.
66. Knight, S., 1991, Watching Out for Doomsday. An Asteroid Heads Towards the Earth: Sarah Knight Meets the Scientists Who Are Watching the Sky, *Élan*, July 26-28, p. 16.
67. Korpel, M.C.A., 1994, Stone, *Dictionary of Deities and Demons in the Bible (DDD)*, K. van der Toorn, B. Becking and P.W. van der Horst (eds.), E.J. Brill, Leiden et al., col. 1547.
68. Kovacs, M.G., 1989, *The Epic of Gilgamesh*, Stanford University Press, Stanford/CT, pp. 100 ff.
69. Kramer, S.N., 1981, *History Begins at Sumer. Thirty-Nine Firsts in Man's Recorded History*, University of Pennsylvania Press, Philadelphia.
70. Krinov, E.L., 1963, The Tunguska and Sikhote-Alin Meteorites, *The Moon, Meteorites, and Comets*, B.M Middlehurst and G.P. Kuiper, G.P. (eds.), The University of Chicago Press, Chicago, p. 210 (retranslated from the German translation).
71. Kugler, F.X., 1927, *Sibyllinischer Sternkampf und Phaethon in naturgeschichtlicher Beleuchtung*, Aschendorffsche Verlagsbuchhandlung, Münster i.W.
72. Kühne, C., 1993, Voropfer im alten Anatolien, *Religionsgeschichtliche Beziehungen zwischen Kleinasien, Nordsyrien und dem Alten Testament: Internationales Symposium Hamburg 17.-21. März 1990*, B. Janowski, K. Koch and G. Wilhelm (eds.), Universitätsverlag, Freiburg/Schweiz and Vandenhoeck & Ruprecht, Göttingen, pp. 276 f.
73. Lambert, W.G. and Millard, A.R., 1969, *Atra-Hasis. The Babylonian Story of the Flood*, Clarendon Press, Oxford, p. 95.
74. Lawick-Goodall, .J. v., 1971, *Wilde Schimpansen*, Rowohlt, Reinbek, pp. 48f.
75. Leeming, D.A., 1987, Virgin Birth, *The Encyclopedia of Religion*, M. Eliade (ed.), Macmillan/Collier Macmillan, New York and London, vol. 15, p. 272 ff.
76. Lees, G.M. and Falcon, M.L., 1952, The Geographical History of the Mesopotamian Plains, *Geographical Journal*, CXVIII, Part I.
77. Lienhardt, G., 1961, *Divinity and Experience. The Religion of the Dinka*, Clarendon Press, Oxford, p. 170.
78. Loisy, A.F., 1920, *Essai historique sur le sacrifice*, E. Nourry, Paris.
79. Lucian, *Salt.* 39.
80. Maisels, C.K., 1990, *The Emergence of Civilization*, Routledge, London and New York.
81. Mallowan, M.E.L., 1980, The Early Dynastic Period in Mesopotamia, *The Cambridge Ancient History. Third Edition. Volume I. Part 2. Early History of the Middle East* (1971), Cambridge University Press, Cambridge et al., p. 305.
82. Mallowan, M.E.L., 1964, Noahs Flood Reconsidered, *Iraq*, XXVI.
83. Malycheff, V., 1931, Analyse des Limons de Kish et d'Ur, *L'Anthropologie*, XLI, p. 271.
84. Ménant, J., 1887, Le sacrifice humain, *Recherches sur la Glyptique Orientale*, 1, pp. 150 ff.
85. Merkelbach, R., 1984, *Mithras*, Hain, Königstein/TS.
86. Meuli, K., 1975, Die gefesselten Götter, *Gesammelte Schriften*, vol. II, Schwabe & Co., Basel, pp. 1034 ff.

87. Müller-Karpe, H., 1974, *Geschichte der Steinzeit*, Beck, München, pp. 249ff.
88. Norden, E., 1924, *Die Geburt des Kindes: Geschichte einer religiösen Idee*, Teubner, Leipzig.
89. Oppenheim, A.L., 1977, *Ancient Mesopotamia*, revised edition completed by Erica Reiner, University of Chicago Press, Chicago, p. 185.
90. Ovid, *Metamorphoses*, 8:620-674.
91. Parke, H.W., 1988, *Sibyls and Sibylline Prophecy in Classical Antiquity*, ed. by B.C. McGing, Routledge, London and New York.
92. Penniston, J.B., 1929, Note on the Origins of Loess, *Popular Astronomy*, 39, pp. 429 ff.
93. Penniston, J.B., 1943, Additional Note on the Origin of Loess, *Popular Astronomy*, 51, pp. 170 ff.
94. Pfister, O., 1932, Instinktive Psychoanalyse unter den Navaho-Indianern, *Imago*, 18 (1), p. 86.
95. Plato, *Critias*, 112 A.
96. Plato, *Timaeus*, 22 C-D.
97. Porada, E., 1965, The Relative Chronology of Mesopotamia. Part I: Seals and Trade (6000 - 1600 B.C.), *Chronologies in Old World Archaeology*, R.W. Ehrich (ed.), The University of Chicago Press, Chicago and London, p. 156.
98. Pritchard, J.B. (ed.), 1969, *Ancient Near Eastern Texts Relating to the Old Testament*, Princeton University Press, Princeton.
99. Raikes, R.L., 1966, The Physical Evidence of Noah's Flood, *Iraq*, XXVIII.
100. Redman, C.L., 1978, *The Rise of Civilization. From Early Farmers to Urban Societies in the Acient Near East*, W.H. Freeman and Company, San Francisco, p. 278.
101. Reichert, A., 1977, Massebe, *Biblisches Reallexikon*, K. Galling (ed.), J.C.B. Mohr, Tübingen, p. 206.
102. Ross, J., 1996, Griechische Passion: Walter Burkerts Anthropologie des Opfers in der Antike, *Frankfurter Allgemeine Zeitung: Geisteswissenschaften*, March 27, p. N 5.
103. Sandars, N.K., 1972, *The Epic of Gilgamesh*, Revised Edition, Penguin Books, Harmondsworth, p. 14.
104. Schaeffer, C.F., 1948, *Stratigraphie comparée et chronologie de l'Asie Occidentale (IIIe et IIe millénaires)*, Oxford University Press, London.
105. Shaw, H.R., 1994, *Craters, Cosmos and Chronicles. A New Theory of Earth*, Stanford University Press, Stanford/CA, p. 429.
106. Shulman, D., 1988, The Tamil Flood Myths and the Cankam Legend (1978), *The Flood Myth*, A. Dundes (ed.), University of California Press, Berkeley et al., p. 296.
107. Smith, J.Z., 1987, Dying and Rising Gods, *The Encyclopedia of Religion*, M. Eliade (ed.), Macmillan/Collier Macmillan, New York and London, vol. 4, p. 521 ff.
108. Steel, D., 1995, *Rogue Asteroids and Doomsday Comets*, Wiley & Sons, London.
109. Stucken, E., 1907, *Astralmythen: Religions-geschichtliche Untersuchungen* (1896-1907), Eduard Pfeiffer, Leipzig, p. 433.
110. Usener, H., 1899, *Die Sintfluthsagen*, Friedrich Cohen, Bonn, pp. 51 f.
111. Valeri, V. 1985, *Kingship and Sacrifice*, University of Chicago Press, Chicago and London, p. 70.
112. Velikovsky, I., 1977, *Worlds in Collision* (1950), Pocket Book edition, New York, p. 56 f.
113. Ward, W.H., 1889, Human Sacrifice on Babylonian Seals, *American Journal of Archaeology*, 5, pp. 34 f.
114. Watkins, T., 1987, Kharabeh Shattani: An Halaf culture exposure in Northern Iraq, *Préhistoire de la Mésopotamie*, Centre National de la Recherche Scientifique (ed.), Edition du Centre National de la Recherche Scientifique, Paris.
115. Weiss, H. et al., 1993, The Genesis and Collapse of Third Millennium North Mesopotamian Civilization, *Science*, 261, August 20, pp. 995 ff.
116. Willems, H., 1990, Crime, Cult and Capital Punishment (Moalla Inscription 8), *Journal of Egyptian Archaeology*, **76**, p. 52.
117. Wolkstein, D. and Kramer, S.N., 1983, *Inanna. Queen of Heaven and Earth. Her Stories and Hymns from Sumer*, Harper & Row, New York et al., p. 95.
118. Woolley, L., 1934, *Ur Excavations. Vol. II. The Royal Cemetery*, Oxford University Press, Oxford.
119. Woolley, L., 1955, *Ur Excavations. Vol. IV: The Early Periods*, Oxford University Press, Oxford.
120. Woolley, L. 1982, *Ur 'of the Chaldees'. The final account, Excavations at Ur*, revised and updated by P.R.S. Moorey, The Herbert Press, London, p. 60.
121. Zysman, M., 1996, The Greater and Lesser Swarms. Crustal Formation: A Meteoric Hypothesis, paper given at the *Symposium on Cosmic Catastrophism and Ancient History*, Deerfield Beach, Florida, July 12 - 14.

Heaven-Sent: Understanding Cosmic Disaster in Chinese Myth and History

David W. Pankenier
Lehigh University, Dept. MFL, Maginnes Hall, 9 W Packer Avenue, Bethlehem, PA 18015, USA

1. Introduction

From the very beginning of their civilization the ancient Chinese were acute observers of their natural surroundings. Few celestial or terrestrial phenomena, especially irregular occurrences, were deemed inconsequential, since all were thought to convey vital intelligence from the supernatural realm. Out of this intersection between phenology and human society there gradually developed a conviction that the Chinese State and culture enjoyed a genetic relationship with the supreme power residing in the sky. Indeed, the ideology underpinning the archaic kingship in the late Bronze Age, as well as during the later unified empire, held that political legitimacy was conferred directly by a deified Heaven (lit. Tian, 'sky').

Dynastic successions being violent affairs, usurpation had to be rationalized as justifiable regicide, so that the earliest historical texts we possess are often colored with moral outrage at the depredations of the final kings of the Xia (mid-20th to mid-16th c. BCE) and Shang (mid-16th to mid-11th c. BCE) dynasties. In some cases this is the result of the Confucian moralizing that characterized the late 1st millennium BCE when many of the texts assumed their present shape. In some cases, however, accounts of seasonal dislocations, calendrical mismanagement, natural calamities and the like may actually have a historical basis. [1] When it comes to the received texts, it is by no means an easy task to separate records of *real* astronomical and meteorological anomalies from the standard litany of disasters to be found in *stereotyped* portrayals of "evil last rulers" whose depredations, by definition, had to elicit warnings from an anthropomorphic Heaven.

Throughout Chinese history, divination, and especially portentology, provided the interpretive mechanisms that allowed the ruling elite to gauge Heaven's disposition at any given time. As early as the late Neolithic the cosmos was likened to the flat plastron and rounded carapace of a turtle, the latter arching over the former like the firmament overarching the square earth. [2] It is this correspondence that probably explains the Shang preference for searing and cracking the plastron shells of turtles to divine the will of the departed ancestors who had joined the Lord Above (Shangdi) in the other world. From Shang times on, natural and man-made disasters came to play a crucial role in Chinese cosmo-political thinking as indexes of perturbations in what would later be conceived as a dynamic equilibrium among the complementary influences of *yin* and *yang* and the five elemental forces (wood, fire, earth, metal, water). Their complex interactions were thought to give rise to all phenomena. This view of things is, of course, a late theoretical rationale for what those ancient Chinese thought they were about, but there can be no doubt that similar, albeit more rudimentary, conceptions of the mechanisms of divination and propitiation were already at work in the 2nd millennium BCE. [3]

For the systematic interpretation of astrological omens, over time an elaborate scheme of astral-terrestrial correspondences grew up based on a parallelism between the celestial and terrestrial topography, with the Milky Way corresponding to the Yellow River flowing through the North China plain. [4] From the point of view of this mature system, the two realms were merely different manifestations of one and the same material essence or *qi*. Qualities pertaining to a given terrestrial locality were thought to be readily imparted to heavenly bodies as they took up temporary lodging in the corresponding astral space. Disturbances on the ground, whether physical or spiritual, just as readily propagated through *qi* to produce dislocations in the corresponding heavenly field; from this it follows, conversely, that untoward events in the skies could be read like texts for guidance about incipient tendencies in the corresponding terrestrial polities below.

This conception of a harmonic resonance between the realms above and below ultimately became enshrined in the imperial ideology. In the words of one scholar:

> "By the third century BC, virtually all Chinese authors recognized correspondences between the realms of nature and man, and they consequently insisted that the true ruler must model himself on Heaven or the cosmic Way. This proposition meant different things to writers of the various philosophical schools, but they all showed a common and deep-rooted belief in the intimate linkage of cosmic principles, astral events, and earthly government. This belief was reflected in the close ties of observational astronomy and the calendrical sciences to political power that emerged in Warring States China and lasted into the present century." [5]

From at least Shang times until the late 19th century, the imperatives of this concept of rulership, with the emperor as sole intermediary between the temporal and supernatural realms, in times of national calamity might prompt the emperor to issue a *mea culpa* assuming personal moral responsibility for the disaster befalling the nation and promising immediate improvement in his stewardship of the mandate to rule, lest Heaven see fit to rescind its recognition of the dynasty's legitimacy. One can hardly imagine a more colossal act of impiety and unfilial conduct than for the emperor to lose the Mandate of Heaven conferred on his revered ancestor generations before. As if that weren't cause enough for concern, from the very earliest enunciation of the doctrine of Heaven's Mandate Chinese ideologues emphasized the Mandate's inconstancy and the need for constant vigilance.

Given the pervasiveness of this response to calamity throughout Chinese history, especially the heaven-sent variety, it is perhaps timely to ask whether such anxiety about preserving Heaven's Mandate is adequately explained by an attitude of reverence for the awesome majesty of the all-encompassing sky and for the *constancy* of natural cycles, or whether such attentiveness to Heaven's "bright manifestations" might reflect an equally early, profound fear and dread with which the ancients may have viewed *unpredictable* cosmic events in the formative period of Chinese civilization.

2. Objectives

This aim of the present essay is twofold. First, it will report briefly on what can be asserted with confidence about the chronology of China in the period Xia through Zhou (20th to 11th centuries BCE) in order to provide more secure chronological benchmarks than the outdated traditional chronology. Second, it will examine certain archetypal examples of disasters in Chinese myth and tradition, both to provide a Chinese perspective on their cultural and ideological significance and to examine how traditional accounts of catastrophe match the historical record. It is hoped these perspectives will enable non-sinologists better to evaluate the Chinese evidence bearing on the nature, extent, and implications of disaster for Bronze Age civilization in East Asia.

3. Chronology of the Early Bronze Age in China

The Bronze Age in China occurred significantly later than in western Asia, essentially spanning the 2nd and the early part of the 1st millennium BCE. Not coincidentally, this corresponds to the formative period of history known as the 'Three Dynasties', a term which in Chinese tradition denotes the period which saw the emergence of the hereditary dynastic state. The latter two of the three dynasties, Shang and Zhou, are fully attested historically and archaeologically. The historicity of Shang was incontrovertibly established as recently as the first half of the present century with the archaeological discovery of the royal cult center and mortuary at Anyang in Henan, together with vast numbers of divination records inscribed on bovine scapulae and turtle shells. The historicity of the first recorded dynasty, the Xia, has still not been conclusively established, although many scholars now hold that the earliest archaeological levels of the palatial foundations at the site of Erlitou in Henan are probably attributable to the Xia. Whether in fact Xia was a dynastic state like Shang is controversial.

For over two thousand years, the historicity of the Three Dynasties was unquestioned, as was that of the mythical Three Sovereigns and Five Emperors who preceded the Xia. The traditional genealogies and chronologies of the earliest dynasties enjoyed canonical status in the minds of orthodox Chinese scholars, even though troublesome gaps and glaring inconsistencies in the record were noted before the compilation of Sima Qian's comprehensive history *The Grand Scribe's Records* early in the first century BCE. Such chronologies and genealogies, fanciful, inflated, and speculative as they are, served traditional Chinese historians adequately enough, but over the past half-century archaeological discoveries, together with research on the Shang dynasty oracle bones and the social and political history of the pre-imperial period, have yielded impressive results which have provoked a fundamental reinterpretation of Chinese history in the archaic period.

It is now clear, to begin with, that the traditional dates and retrospective impressions of the period before 841 BCE are for the most part utterly unreliable, the earlier the date the greater the unreliability. A generation of increasingly sophisticated studies of both transmitted and excavated texts and traditions from the pre-imperial period (before 221 BCE) has brought a vastly improved understanding of intellectual and literary developments during the latter part of the first millennium BCE. Therefore, it is no longer acceptable to cite timeworn received chronologies in discussing the pre-imperial history of China any more than it is to interpret Chinese myths and quasi-legendary traditions without regard to the diverse periods, localities, and intellectual climates that produced them.

For this reason, scholars are well advised not to attempt to mine translated editions of the Chinese Classics or other received texts for chronological and historical evidence bearing on the earliest period without reference to the pertinent sinological scholarship concerning their relevance and reliability. For example, many of the myths about the Three Sovereigns and Five Emperors who were thought to have ruled before the Three Dynasties are now known to some extent to be culturally conditioned artifacts (if not inventions) of the last half of the 1st millennium BCE, first making their appearance in the literature some 2,000 years after the epoch they purport to describe as a direct consequence of contemporaneous intellectual and social developments. We possess little reliable genealogical information for the period before about 1,300 BCE, and no chronological or genealogical information at all for the period before about 2,000 BCE, except for a wealth of new archaeological data on a variety of distinctive Chinese regional cultures in both north and south in the late Neolithic, data that underscore the fanciful nature of traditional conceptions of a unitary state dating back to the dawn of Chinese civilization.

Table I below, which is annotated in some detail with reference to the relevant scholarship, contrasts the traditional chronology for the Three Dynasties period with what are now more reliably known to be the dates of the Xia, Shang and Zhou Dynasties. In what follows, before discussing one pattern of archetypal disaster, I refer to Table I in briefly characterizing how our current understanding of the chronology was arrived at and where further refinement can be anticipated.

4. The Pivotal Date of the Zhou Conquest of Shang

The most important benchmark in establishing the chronology of the 2nd millennium BCE in China is the date of the Zhou Conquest of Shang. The traditional date of 1122 is an artificial construct arrived at by Liu Xin (d. CE 24) in the first quarter of the first century of the present era based on retrospective computations using his newly constructed *Santongli* (The Calendrical Complex of the Triple Concordance System). It is only secondarily bolstered by a smattering of chronological data of dubious provenance gleaned from the imperial archives of the Han Dynasty (206 BCE - CE 220). For two millennia, 1122 was the accepted date of the Zhou Conquest. During the past six decades, scholars have proposed several dozen more dates ranging from 1122 to 1028 BCE or later, many on the basis of questionable or incomplete data, special pleading, or in ignorance of highly pertinent scholarship on the genealogy and relative chronology of the Shang kings. The latter has been more or less firmly established since the 1950s, based on the archaeologically excavated Shang oracle bone inscriptions.

With the publication in English of David N. Keightley's *Sources of Shang History: The Oracle-Bone Inscriptions of Bronze Age China*, [6] for the first time the contents of the royal Shang oracle bone inscriptions from the late 13th to mid-11th centuries BCE were made accessible to a wider historical readership. In analyzing their significance as historical documents, Keightley offered a judicious evaluation of their contribution to research on chronology and concluded, based on all the available evidence, that the final or Yin-Shang period of the Shang Dynasty began about 1200 BCE and ended in mid-11th century. Thereafter, scholarly attention began to focus on the period 1050-1040 BCE as the probable date of the overthrow of the Shang Dynasty by the Zhou.

Because a chronicle known as the *Bamboo Annals* recovered centuries ago from a plundered third century BCE tomb also implied a date of 1050 for the Conquest, that text became the subject of intensive philological study in an effort to determine how much of its contents are authentically early, since it is well known that emendations and interpolations found their way into the text during the reconstruction of the damaged bamboo slips by court scholars after the chronicle's discovery in 281 CE. Proof of the veracity of unusual planetary observations recorded in the *Bamboo Annals*, and their independent corroboration by other Zhou texts, led to the precise dating of the Zhou Conquest to 1046 BCE, at once reaffirming the mid-11th century date for the end of Shang while at the same time rehabilitating portions of a text about whose reliability scholars had previously expressed doubt. [7]

The consensus with regard to the mid-11th century BCE date of the Zhou Conquest of Shang is underscored by the dating of a total lunar eclipse recorded in precise detail in *Yi Zhou shu* as occurring in the first month of the 35th year of the Zhou dynasty founder, King Wen (r. 1099-1050), on the 13th day (*bingzi*) of the sexagenary day cycle. The eclipse occurred during the night of March 12-13, 1065, and its 35th year relative date in King Wen's reign correlates exactly with that of the spectacular planetary massing six years later in May 1059 which, according to the *Bamboo Annals* and other sources, had to have occurred in King Wen's 41st year. [8] This means that we now have at least two independently corroborated astronomical dates for crucial events in mid-11th century with which to establish a baseline for the reconstruction of the precise chronology of late Shang and early Zhou. [9]

In all, three planetary massings involving the five naked-eye planets are thought to have been witnessed and associated by the Chinese with the first three dynastic transitions in the 2nd millennium BCE - the founding of Xia (mid-20th c. BCE), the Shang overthrow of Xia (mid-16th c. BCE), the Zhou conquest of Shang (mid-11th c. BCE). Two of the three, in 1953 and 1059 BCE, were the densest such massings of planets to occur in the past 5,000 years and bear eloquent testimony to the observational acuity of the ancient Chinese. [10]

As a result of the establishment of astronomical benchmarks, other accounts in the Shang oracle bones and received texts concerning meteorological phenomena, prolonged droughts,

Table 1

Event	Traditional Date	*Bamboo Annals*	Reconstructed [11]
Xia Begins [12]	2204	1990	1953
ten suns appear [13]	**1893**	**1607**	**1604**
King Jie [14]	1818-1767	1589-1558	1585-1555
5 planets 'criss-crossed'; stars fell like rain; the earth shook [15]	**1809**	**1580**	**1576 Nov-Dec**
three suns appeared [16]	**1791**	**1561**	**1557**
Shang Begins	1766	1558	1554
5-7 yr. drought [17]	**1766-1762/60**	**1558-1554/52**	**1554-1550/48**
Di Xin	1154-1123	1102-1050	1086-1047
lunar eclipse (King Wen 35th year)			**1065**
5 planets gather; Comet Halley appears (King Wen 41st year) [18]		**1071**	**1059**
Zhou Conquest	1122	1050	1046

meteor showers, etc. can in some cases now be dated more reliably than in the past. In the absence of contemporaneous written records from the first two-thirds of the 2nd millennium BCE, however, the account of Xia in the *Bamboo Annals* and elsewhere still remain uncorroborated and scholars are well advised to take nothing at face value.

5. Multiple Suns, meteor showers, earthquakes, and catastrophic drought

One of the most famous episodes in China's early dynastic history about which numerous traditions persisted into classical times has to do with the simultaneous appearance in the sky of what are described as multiple suns. [19] As in the case of several other traditions there is a contrast between the matter-of-fact recounting of the anomaly in the *Bamboo Annals* and other written accounts found in late Zhou sources. The latter appear to be transcriptions of regional oral traditions concerning cosmogonic events and demiurgic culture heroes in the pre-dynastic period. One scholar, Sarah Allan, attempts to connect the myth of ten suns (or in some accounts sun-ravens) specifically with a Shang cosmological tradition in which ten suns rise each day in sequence from the branches of a Mulberry Tree growing by a solar spring at the eastern limits of the world. The suns are the sons of a goddess Xihe, who may be identical with a consort of the mythical progenitor of the Shang people, Di Ku. Allan hypothesizes that in historical times the Shang ruling elite was still organized totemically, with each of ten groups being identified with one of the ten suns, and with each being denoted by one of the so-called 'heavenly stems' used to name the days of the ten-day week (these ten then combine with the twelve 'earthly branches' to make up the sixty unique terms used to name the days in the sexegenary cycle). This classification scheme is still discernible in the oracle bone inscriptions where the royal Shang ancestors receive cult exclusively on their particular posthumously designated name days.

Nevertheless, the common form of the ten suns myth is rather more suggestive than such a Shang 'totemic' realization suggests. In the traditional accounts the appearance of multiple suns brought an abrupt end to an age of perfect harmony and abundance, when "in the time of Yao (a legendary ruler who immediately preceded the dynastic period), the ten suns came out together, withering the crops of grain and killing the grasses, so that the people had nothing to eat [...]"[20] Threatening monsters also appeared and Yao, alarmed, deputed the hero Archer Yi to shoot down the nine superabundant suns (or sunbirds), leaving only one. Rather than interpreting this version of the myth (with Sarah Allan) as a later rationalization of the tradition after Shang was supplanted by Zhou, it may be that the Shang version is a typically Shang proto-bureaucratic realization of an earlier pre-dynastic myth. At a minimum, the nine extra suns' destruction is difficult to reconcile with the supposed totemic role of all ten suns in Shang political and religious ideology, as is the presence in the *Bamboo Annals* of a record of the observation of ten suns several decades before the founding of Shang. [21] Comparative study of this motif, which is attested predominantly among minority and aboriginal cultures from mainland China, to northeast India, to Taiwan, the Philippines and Indonesia, suggests that it was originally a common cultural feature of Proto-Malayo-Polynesian peoples who originally dispersed outward from the East Asian mainland during the Neolithic. Most of their languages belong to the Austro-Asiatic language family and they may be related to the 3,000 BCE archaeological culture of central and southern Shandong and Jiangsu known as Dawenkou. [22] It is quite possible, therefore, that the myth is predynastic, originating in mainland China among Neolithic east coast cultures known to have included groups directly ancestral to the Shang people.

The dichotomy displayed by the transmitted versions is in some ways reminiscent of the contrasting accounts of the planetary massing of 1059 BCE, which is represented on the one hand by a straightforward record in the *Bamboo Annals*, 'the five planets gathered in (asterism) Room,' [23] and on the other by a mythicized version with parallels in later Zhou textual accounts which preserve early oral traditions relating to the dynastic founding (e.g., 'a great red bird alighted on the Zhou altar to the soil'). The *Bamboo Annals* makes no mention of multiple suns associated with the legendary Yao, but several decades before the end of the Xia dynasty (in the eighth year of the reign of the Xia king, Yin Jia, in about 1607) ten suns are said to have appeared together in the sky: "In the eighth year there was an evil omen in the sky; ten suns came out together. That year [the king] died." [24] No further details are provided, though the implication is, of course, that this was an omen of the king's impending demise.

Not long after this, in the tenth year of the last Xia king, Jie (1576) is the record "the five planets criss-crossed; in the middle of the night the stars fell like rain. There was an earthquake. The Yi and Luo Rivers dried up." [25] And again in King Jie's 29th year (1557) "three suns came out together." [26] Finally, immediately after Xia Jie's overthrow in 1555 by Cheng Tang, founder of the Shang dynasty, there occurred a drought notorious for its severity that persisted for five to seven years until, in an act much celebrated in later Confucian tradition, virtuous King Cheng Tang volunteered to expose himself as a expiatory offering in the sacred Mulberry Grove. According to tradition this selfless act was promptly answered with copious life-giving rainfall that saved the people, and probably the fledgling dynasty, from utter ruin. Some interpret the episode as an act of expiation for the usurpation and regicide of which the Shang founder was actually guilty. What is of particular interest is not so much the curious record(s) of the anomalous suns by themselves, but that there is a series of apparently unrelated observations and natural calamities within a span of some fifty years. Might these observations be consistent with the apparition of a disintegrating comet or meteor storm? Other traditional accounts of the end of King Jie's reign, while they obviously do conform to the imperative in retrospect to malign failed last kings, could conceivably also harbor a grain of truth. For example, in the earliest strata of the 4th century text *Mozi* the seasonal dislocations accompanying the transition from Xia to Shang are described as follows:

> "There was a harsh command from Heaven; the days and months were untimely, heat and cold arrived confusedly, the five grains were scorched to death, ghosts howled within the state, cranes called for more than ten nights running [...]"[27]

6. Cosmic violence: the Yellow Emperor vs. Chi You "The Wounder"

Another ancient Chinese myth remarkably rich in associations potentially relevant to Victor Clube and Bill Napier's cometary impact hypothesis [28] is the story of the conflict between the legendary Yellow Emperor and his adversaries, particularly the monstrous miscreant known as Chi You (Fig. 1). In this myth, as recounted in a wide variety of texts from the 3rd and 4th centuries BCE through the Han dynasty (206 BCE - CE 220), violence on a cosmic scale intrudes repeatedly into the benign existence of the Chinese people under the benevolent rule of the Yellow Emperor. The Yellow Emperor comes to the forefront as mythic culture hero and progenitor of the Chinese people in the Warring States period (5th to 3rd centuries BCE) in stories in which the Chinese dramatized their ideology and self-understanding in the form of tales about the emergence of civilization.

Fig. 1. Chi You, The Wounder. Tomb wall-relief, Yinan, Shandong, late 2nd-3rd cent. CE. After Bodde, *Festivals in Classical China*, 123.

His civilizing accomplishments and the adversaries he encountered are described in the following account from Sima Qian's *The Grand Scribe's Records*:

> "The Yellow Emperor was the son of Shao Dian. His surname was Gongsun and his personal name was Xuanyuan. At birth his spirit was magically efficacious, in infancy he could speak, when young he was wise and equable, when he grew up he was quick-witted and when he reached maturity his perceptions were supremely acute. At that time the lineage of the Divine Husbandman had been in decline for generations. The nobles attacked one another and violently persecuted the common people, and the lineage of the divine husbandman could not send punitive expeditions against all of them. So Xuanyuan practiced the use of weapons in order to punish all those who did not attend the ruler's court, and the nobles all served him as retainers.
>
> But Chi You was the most savage of them all, and none could attack him. The Fiery Emperor sought to invade and bully the nobles by force, so they all took refuge with Xuanyuan. Xuanyuan cultivated his spiritual potency and arrayed his troops. He channeled the five *qi*, planted the five grains, consoled the myriad people, measured the four quarters, and instructed the bears, leopards, and tigers, and with them he battled the Fiery Emperor on the plain of Ban Springs. He fought three battles and finally achieved his goals.
>
> Chi You rose up in rebellion, and rejected the Yellow Emperor's command. So the Yellow Emperor levied the armies of the nobles, did battle with Chi You at Zhuolu, and killed Chi You. The nobles reverenced Xuanyuan as the Son of Heaven, replaced the lineage of the Divine Husbandman with him, and he became the Yellow Emperor.

[...] He was always moving without any fixed residence, using only his army as an escort. The titles of his officials were all taken from clouds, so they became a cloud army. He appointed Grand Observers of the Left and Right to oversee the myriad states. The myriad states were in harmony, and there were numerous offerings to the ghosts, spirits, mountains, rivers, and the *feng* and *shan*. He obtained the precious tripods [symbols of sovereignty]. He manipulated the spirit yarrow to lay out the calendar. He raised up the Wind Monarch, the Mighty Shepherd, the Constant Vanguard, and the Great Vastness to direct the people. He followed the divisions of Heaven and Earth [yin and yang and the four seasons], the divinations of the hidden and the clear, the sayings about life and death [methods of immortality], and those about preservation and destruction [military theories]. In their proper seasons he planted the hundred grains, grasses, and trees; through his simple purity he civilized the birds, beasts, reptiles and insects. On all sides he spread out the sun, moon, stars, water, soil, rocks, metal, and jade. He toiled with mind, body, eyes, and ears. He was sparing in the use of water, fire, and materials. He received the auspicious signs of the virtue of the phase of Earth, so he was called the Yellow Emperor." [29]

It is impossible to do justice here to Mark Lewis's recent detailed study of this cosmic conflict motif in its socio-political context. In summary, however, Lewis concludes,

"in the Warring States the cultic roots of the tales of the Yellow Emperor were gradually suppressed as part of the general reinterpretation of ancient divinities as historical figures. This 'suppression' of the magical, shamanistic roots of these tales [...] reflected the changing character of official religion and the class basis of cultic practice in Warring States and early imperial China. Among the common people shamanic rituals devoted to a host of anthropomorphic spirits continued to flourish throughout the Zhou and indeed the entire history of imperial China, but elite religion and the state cults focused increasingly on ancestor worship, official sacrifices to abstract natural forces or spirit analogues of the bureaucracy, and the commemorative cults of select moral exemplars. The reinterpretation of ancient cult figures as historical personalities who provided models and sanctions for state power, and the devotion of the state cults entirely to abstract deities such as Heaven, Earth, or the Supreme Unity were part of a general re-creation of the spirit world by the emerging elite of the Warring States. It was the first round in a battle for control of the realm of the spirits that would span the history of imperial China, a battle between the historicizing, philosophical intellect of the literate elite and the protean religious practices of the local, peasant communities."[30]

In this influential view the tales of the Yellow Emperor and his battles were reinterpreted as "accounts of the origins of violence in the world of men and its subsequent use by the sages to create a proper political order [...] claims for the indispensability of moral instruction, textual military skills, [judicious application of force,] and the patterns of Heaven [all precedents established by the Yellow Emperor] were used to rationalize and justify the government's mobilization and control of the peasant populace. In this way the tales of the Yellow Emperor became a charter myth for the absolutist state." [31]

In *Sanctioned Violence in Early China*, Lewis's chapters on 'Cosmic Violence' and the 'Social History of Violence' deal exhaustively with the twin motifs of civilizing order vs. chaos, sanctioned application of force vs. the indiscriminate use of violence. By exploring the mythic careers of the principal demiurges, the Yellow Emperor and his chief adversary Chi You, Lewis tries to show how ancient "tales of cosmic battles between rain and drought were recast as accounts of the origins of warfare and punishments. [...] so men of the Warring States, Qin and Han periods knew the Yellow Emperor and Chi You as creators of warfare and punishments." [32] Ultimately, the Yellow Emperor's nemesis Chi You was transformed and domesticated, once the most terrifying archaic embodiment of unbridled violence and evil he finally becomes a god of warfare, inventor of weapons, bringer of storms and rain, and tutelary spirit of the New Year season who eradicates old influences in anticipation of the new, and tomb guardian and apotropaic deity who is invoked to turn his weapons to exorcising evil spirits.

This evolutionary process seems plausible enough, but when the original cosmic conflict and Chi You's career are examined from a slightly different perspective, other interpretive possibilities present themselves which may be even more illuminating. If one of the chief functions of myth is to mediate for a collective the irreconcilable elements of tension in their lives as religious, political, and social beings, the myth of Chi You's subjugation by a deity of order and regularity and his subsequent domestication as the Yellow Emperor's servant or votary may represent an attempt to resolve even more profound existential anxieties than previously suggested.

Before describing in more detail the characteristics of the two demiurges and the nature of their conflict, it might be helpful to briefly recapitulate the seven principal motifs of a worldwide pattern of mythic conflict consistent with the disintegrating comet hypothesis, to which Victor Clube and Bill Napier drew attention in *The Cosmic Winter.* [33]

- There was once a supreme god; this god often constructed heaven and earth.
- Later the "world" was peopled with giants in the sky.
- There was a combat involving two major participants, god and dragon, two dragons, or whatever.Often the outcome of the battle was fire and flood on the earth below.
- The major participants were frequently accompanied by hordes of lesser creatures.
- The combat took place in the sky.
- It was recurrent.
- The combatants were celestial creatures, giants, noisy, chaotic, winged, serpent-like, and they spanned the skies. They might crash into the ground in flames.

It will be helpful to keep in mind these major motifs and their hypothesized association with meteor storms and multiple cometary apparitions as we review the main features of the cosmic conflict at the dawn of Chinese civilization.

7. Characteristics of the Yellow Emperor and Chi You

There is a remarkable consistency in Han and pre-Han texts in the portrayals of the Yellow Emperor and Chi You as divine beings who wielded extraordinary power. We saw above how by the 1st century BCE the Yellow Emperor was credited with a civilizing, indeed, a cosmogonic role in making the earth habitable for the people. In some respects the Yellow Emperor gives evidence of being a popular personification of the very ancient but otiose supreme lord Shangdi. In his divine aspect the Yellow Emperor is described as being dragon-like: "he had the face of a dragon [...] his imminent attainment of the Mandate of Heaven was marked by the appearance of dragons [...] the constellation that bears his name had the body of a yellow dragon and ruled thunder and rain, and [...] he departed on a dragon at the end of his life." [34]

The pole star is the Yellow Emperor's spirit father and astral double and his identification with the celestial Big Dipper is emblematic of his role as bringer of order and regularity. Throughout early Chinese history, the Dipper was employed as a sort of celestial timepiece whose orientation in the sky accurately told the seasons of the year. More than this, the unmoving pivot of the pole, in fact the abode of Shangdi, the Lord Above, symbolized for the ancient Chinese the awesome and unfathomable power of the cosmos - invisible, unmoving, yet controlling all change and transformation in time and space. The constellations, or Celestial Offices, and the planets, the Lord Above's minions, revolved about the unmoving center going about their official business at the proper times seemingly without his intervention. Small wonder that the patterns of the stars were employed by Han court specialists to bring imperial rituals into accord with the actions of Heaven and Earth. Throughout, a principal thrust of such efforts was to achieve regularity and predictability so as to ensure that terrestrial affairs would always be in harmony with the celestial realm. But chaos and irregularity were an ever-present threat. Natural disaster visited by the Lord Above on terrestrial polities was seen as a form of chastisement, as was warfare waged by dynastic paragons explicitly in Heaven's behalf. And so the Yellow Emperor became, in one view, the domesticator of violence in the service of order: "the savage power of inanimate nature as well as that of beasts was brought into the service of the emperor and offered a demonstration of the celestial origins and character of the violence of the sage ruler." [35]

Enter Chi You. Chi You was thought to have first created disorder, thereby corrupting the moral character of the people. Exactly how is not clear, but the indiscriminate use of violence - punishments and the use of weapons, especially 'natural' weapons fire, powerful winds, rain, fog, etc., are implicated. So, for example, one text says:

> "Long ago the Yellow Emperor fought the battle of Zhuolu [with Chi You] to settle the disaster of fire; Zhuan Xu [a legendary successor to the Yellow Emperor] had the battle with Gong Gong to settle the disaster of water; [36] Cheng Tang [of Shang] had the battle at Nanchao to extirpate the chaos of Xia. They rose and fell in sequence, the victor governing the affairs of state. This was received from Heaven." [37]

Another says:

> "The Yellow Emperor battled Chi You at Zhuolu, and Chi You created a great fog that lasted three days. The soldiers did not know what to do, so the Yellow Emperor ordered the Wind Monarch to imitate the North Star and create the 'south-pointing chariot' [a mechanical proto-compass] in order to distinguish the four directions, and they captured Chi You. [...] In the time of the Yellow Emperor the sky was filled with a great fog for three days. While wandering about above the Luo River, the Emperor saw a great fish and killed five sacrificial beasts to offer up to it. Then Heaven sent down a great rain for seven days and seven nights. He received an illustrated book which was the *He tu* [River Diagram]." [38]

8. Ritual Games

But it is with descriptions of the appearance of Chi You and ritualized commemorations of the Yellow Emperor's primal conflict with him that we strike a mother lode of suggestive imagery. Detailed accounts survive from pre-Han and Han times of two types of ritualized combat, one a game of kickball, the other a wrestling match involving 'horn butting', and both performed explicitly in commemoration of the cosmic combat which led to the introduction of warfare and weapons. Both originally figured as important aspects of military training as strength-building exercises in the state of Qin before unification in 221 BCE, but cosmic kickball later assumes a ritual function as a symbol of imperial sovereignty.

The one form of military training, 'head butting', which eventually turned into a ritualized entertainment in connection with the New Year's festivities, is shown in a Han dynasty tomb relief that depicts a pair of men wearing horns and engaged in a wrestling match. A slightly later literary account has this to say about this form of wrestling:

> "In Qin and Han times they said that Chi You's ears were like swords and the hairs of his temples like lances, and he had horns on his head. When he fought with Xuanyuan [the Yellow Emperor] he butted men with his horns, and none could stand against him. Now in Jizhou there is a performance called the 'Chi You game'. The people divide into groups of two or three, put horns on their heads and butt each other. The Han had created the 'horn butting game', and this is probably its vestige." [39]

This horn-butting exercise already had a long history by the Han dynasty as a form of military training along with staged animal combats that also included bull grappling. [40] But in the Han it too became institutionalized as a ritual performance of the rivalry between Chi You and the Yellow Emperor.

The second game, cosmic kickball, was played on a military training field in the palace complex with nine raised steps on the left and a flat area on the right. The game was said "to resemble the state," the game itself being in some sense "a simulacrum or double (*xiang*) of regulating the state." [41] The game involved "a round ball and square walls; it imitates the [Heavenly] simulacra [images in the heavens] with a yin and a yang. Imitating the moon[s] they rush against one another; the two [teams of] sixes facing one another." [42] According to a recent commentator,

> "it is an obscure explanation, but all the lines refer to basic Chinese cosmological ideas. The 'round ball' and 'square walls' evoke the standard description of the universe as a round Heaven over a square Earth. The 'simulacra' or 'doubles' commonly refer to the various astral bodies and constellations which resonate with and reflect corresponding earthly locales or institutions. These phenomena all have a *yin* and a *yang* in that they pass through an annual cycle based on the successive shift from one to the other. One major exception was the moon - or 'moons,' for each month was a moon to itself - which passed through a series of monthly cycles that could be mapped onto the solar year. [...]" [43]

Mark Lewis sees this game of kickball, which became a "fundamental element in the ritual theater of Han imperium" [44] as a ritualized simulation of the regular cosmic cycles that the emperor sought to imitate and maintain congruence with. In his view, as an "archetypal event marking the introduction of organized warfare into human society, it is not surprising that many of the chief rituals pertaining to combat were interpreted as commemorations of the primal conflict." [45] Lewis suggests that this interpretation was a late rationalization of an earlier mythos. But if we recognize in this 'game' a ritualized commemoration of a primal conflict, one need not necessarily conclude that it only lately assumed this character. Viewed from a slightly different perspective, as a ritualized commemoration of the cosmic conflict between the Yellow Emperor and Chi You, the game may actually have evoked the locale of the primal combat - *it took place in the sky*. Consider this description of the invention of the game by the Yellow Emperor from one of the 2nd century BCE Mawangdui silk manuscripts:

> "Thereupon [the Yellow Emperor] took out his axes of punishment and mobilized his troops. The Yellow Emperor in person encountered Chi You and slew him. He stripped off his skin and made it into an archery target; he had men shoot at it and rewarded those who had the most hits. He cut off his hair,

mounted it in Heaven, and called it 'Chi You's Banner.' He stuffed his stomach and made it into a ball; he had men kick it and rewarded those who scored the most. He fermented his bones and flesh, threw them into a bitter meat stew, and had all the men drink it. Thus the Supreme Emperor [= Lord Above or Shangdi] made his prohibitions." [46]

Now, Chi You's Banner figured prominently in astral divination in this period as "a broom star or comet but with curved back" whose appearance presaged royal military expeditions in all directions, according to Sima Qian's monograph on astrology in his *The Grand Scribe'sRecords.* [47] In another remarkable archaeological confirmation of that historical text, examination of the famous Mawangdui cometary atlas drawn on silk shows that out of 29 comets illustrated in the catalogue all but one have military prognostications inscribed beneath them, and one of the 29 types depicted is in fact labeled 'Chi You's Banner' [48] (Fig. 2). Chi You, therefore, was almost certainly a comet. [49] The myth of Archer Yi downing the superfluous suns could then refer to the ritualized shooting of targets made to symbolize the frightening cometary apparitions in the sky in an attempt magically to dispose of the threat.

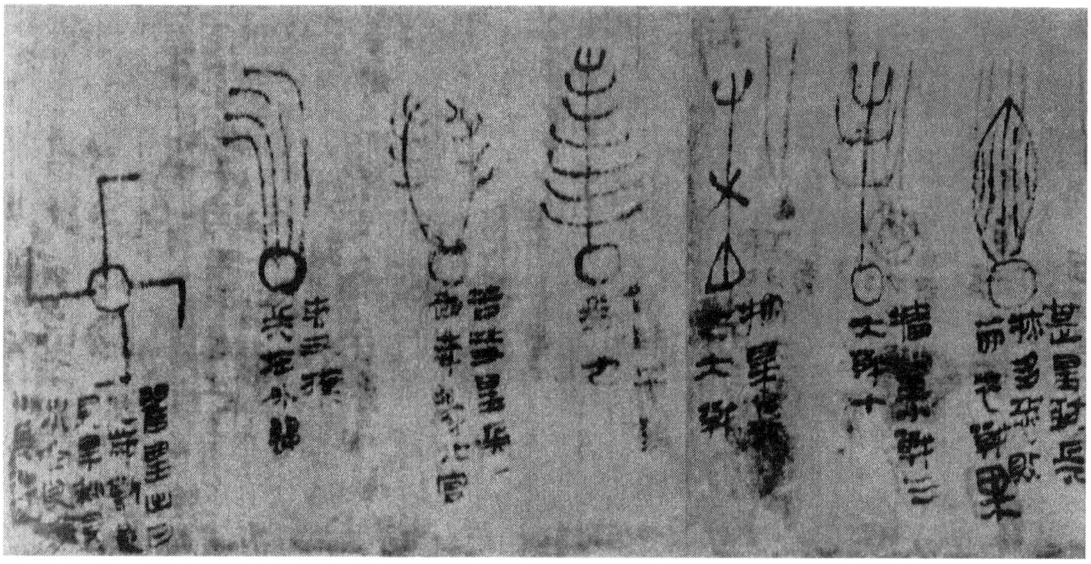

Fig. 2. *Chi You's Banner* (second from left). Mawangdui cometary atlas, early 2nd century BCE. After *Zhongguo gudai tianwen wenwu tuji*, 23.

TABLE II
Elements of primal conflict:
Yellow Emperor vs Chi You

Scenario: Sudden intrusion of chaos on peace and order. A battle to confront fiery disaster: prolonged fog, searing heat, strong winds, drought, followed by heavy rainfall that possibly causes watery catastrophe.

Protagonists: Yellow Emperor: a dragon-like aspect; spirit father and astral double is the Pole Star > emblem of constancy and stability; his own asterism is Xuanyuan (in Leo); plays a cosmogonic role in making the world safe and orderly; benign progenitor of the people; commands celestial spirits, thunder and rain; possessor of Mandate to rule; received revealed treatises from Heaven; introduces measured and judicious use of violence to achieve stability.

Fiery Emperor: causes excess of sun/heat that scorches the earth > bringer of drought; propitiated with sacrifices, including burning of victims; Yellow Emperor battles him repeatedly.

Chi You: most threatening miscreant and initiator of chaos; defeated more than once by the Yellow Emperor; archetypal bestial warrior who kills indiscriminately; hideous aspect > horned, sword-like ears and hair; hair placed in heaven by Yellow Emperor as a comet presaging military conflict; wields weather weapons > winds, fog, clouds, rain; becomes god of weapons and organized warfare associated with correct killing season in autumn consonant with Heaven-ordained pattern; transformed through sacrifice/ritual consumption into Yellow Emperor's votary; becomes exorcist of evil influences and tutelary spirit of cleansing, expiation, rain-seeking rituals; games and tests of strength (bull-grappling, cosmic kickball, horn-butting) ritually reenacting primal combat during year-end ceremonies.

Consistent with Clube and Napier's prediction that "the annual appearance of the [Taurid] meteor stream blazing forth from the sky would make an unmistakable annual event, and one might seek evidence of an annual rite at this date," [50] the ritual commemoration of the cosmic combat between the Yellow Emperor and Chi You figured prominently in exorcistic festivities which preceded the New Year during the Qin and Han Dynasties. [51] According to Mark Lewis, "the figures of the Yellow Emperor and Chi You, like that of the Fiery Emperor, all emerged from rituals used to expel evil influences at the end of the old year and bring on rain at the end of the dry season." [52] There is strong evidence that the Fiery Emperor, who was responsible for repeated calamities of fire that caused severe droughts, is in fact a regional variant of the same demiurge as Chi You. [53] His depredations using weapons of fire could only be countered by equally powerful watery weapons wielded by the Yellow Emperor. Propitiatory sacrifices were already being offered to the Fiery Emperor in the state of Qin in the west as early as the 5th century BCE, so we are indeed dealing with archaic ritual practices. By the Han dynasty the ritual reenactments of the Yellow Emperor's combats with Chi You and the Fiery Emperor became part of the most important elements of communal cleansing and expulsion of evil influences during the festivals concluding the year. [54]

9. Duality of Accounts of Epochal Phenomena in the Sky

As I have shown in the past with respect to the surviving

accounts of unusual planetary observations from the 2nd millennium BCE, two parallel accounts sometimes survive in the literature. On the one hand, there are the straightforward records in the *Bamboo Annals* that mention the behavior of the five planets and locate the events in a coherent chronology. On the other hand, there are numerous accounts appearing in various pre-Qin texts in a kind of imagistic language that bespeaks their origin in and transmission by other means, for example, in popular traditions. In certain key instances it is possible to demonstrate that the two versions of events actually refer to the same phenomena in the sky. [55] Moreover, the account of the appearance of multiple suns in the latter part of the Xia Dynasty occurs in both versions of the *Bamboo Annals*, the so-called current version as well as the genuine version reconstituted from quotations in ancient works, so there can be no question that this record appeared in the original text.

What I would like to propose as a tentative conclusion to this examination of the historical and mythological evidence bearing on the possible observation in ancient China of cometary and impact phenomena is the following. First, there appears to have occurred a series of related astronomical/meteorological phenomena between 1610 and 1550 BCE, contemporaneous with the dynastic transition from Xia to Shang. Assuming they are factual, accounts of these phenomena in the *Bamboo Annals* may point to repeated multiple cometary apparitions, meteor storms, and possible impacts leading to anomalous weather-related consequences on a wide scale in north China, most notably one of the severest droughts in cultural memory. Second, accounts of a cosmic conflict between the Yellow Emperor and Chi You, that is between an established sky-god and order-giver and a fearsome, chaos-bringing intruder bear all the earmarks of a mythicized version of such events; that is, the sudden appearance of a spectacular comet, possibly with mutiple companions, whose passage is attended by the kind of meteorological disruptions to be expected from one or more sizeable impacts. Since the earliest known rituals and sacrifices commemorating this event date at least to the mid-1st millennium BCE, it is possible that the myths originated in the mid-2nd millennium BCE or even earlier. The pre-Shang date of the *Bamboo Annals* record to which the myth may have reference could explain the connection between the ten suns motif and the origin myths of the Shang people, since the cataclysmic astronomical events would have been contemporaneous with the rise of the Shang in the east and their westward expansion as they challenged the constellation of forces that constituted the Xia polity in the central plains area.

On the other hand, the series of unusual observations during the first half of the 16th century BCE could conceivably represent a recurrence of chaotic events still remembered from a time prior to the dispersal of Proto-Malayo-Polynesian peoples outward from the East Asian mainland during the Neolithic, as suggested by the wide geographic dispersal of the multiple sun mythos. Various motifs dating from the Neolithic period may preserve representations of those cometary events (Fig. 3a-c). Therefore, the dating of the Yellow Emperor's battles with Chi You/Fiery Emperor to the time of Yao in the pre-dynastic past could also derive from prehistoric Neolithic traditions.

In either event, the broad spectrum of cultural responses to cataclysmic events - from deep-seated fear and dread, to intense efforts to mediate what they saw through ritual reenactment, mythic recounting, and sacrifice, to the ultimate domestication of the frightening implications of chaotic intrusions into their lives through various forms of deep-play all attest to the profoundly unsettling impact chaotic events in the skies may have had on the minds of those ancient Chinese.

Fig. 3 a-c. Neolithic motifs said to depict lunar phases (a), and solar deities or images (b, c), but suggestive of comets: a) Images appearing on Yangshao painted pottery. b) Petroglyphs from Lianyungang. c) Painted pottery shards from Dahecun. All mid-5th to mid-3rd millennium BCE. After Lu Sixian, *Shenhua kaogu*, 123, 88, 160.

Bibliography

Allan, Sarah. *The Shape of the Turtle: Myth, Art, and Cosmos in Early China* (Albany: SUNY Press, 1991).

Allan, Sarah. "Drought, Human Sacrifice and the Mandate of Heaven in a Lost Text from the *Shang Shu*," *Bulletin of the School of Oriental and African Studies* XLVII.3 (1984): 523-539.

Baillie, M. G. L. *A Slice Through Time: Dendrochronology and Precision Dating* (London: Batsford, 1995).

Bodde, Derk. *Festivals in Classical China: New Year and other Annual Observances During the Han Dynasty (206 B.B. A.D. 220)*, (Princeton: Princeton University Press, 1975).

Boltz, William G. Kung-kung and the Flood: Reverse Euhemerization in the *Yao tien*, *Toung Pao* 67 (1981), 141-153.

Clube, Victor and Napier, Bill. *The Cosmic Winter* (London: Basil Blackwell, 1990).

Keightley, David N. *Sources of Shang History: The Oracle-Bone Inscriptions of Bronze Age China* (Berkeley: University of California Press: 1978).

Kuniholm, Peter, et al. Anatolian tree rings and the absolute chronology of the eastern Mediterranean, 2220-718 BC, *Nature* 381.27 (27 June 1996): 780-782.

Le Blanc, Charles and Mathieu, Rémi. *Mythe et philosophie à l'aube de la Chine Imperiale* (Montréal: Les Presses de lUniversité de Montréal, 1992).

Legge, James. *The Chinese Classics*, vol. 3, *The Shoo King or The Book of Historical Documents* (2nd ed.) (rpt. Taipei: Wenshizhe chubanshe, 1972).

Lewis, Mark Edward. *Sanctioned Violence in Early China* (Albany: SUNY Press, 1990).

Li, Changhao, ed. *Zhongguo tianwenxue shi* (History of Chinese Astronomy) (Beijing: Kexue chubanshe, 1981).

Li, Fuqing. Xiaomie duoyu taiyang de shenhua (The myth of the eradication of superabundant suns), *Lishi yuekan* (Historical Monthly) 3 (1997), 70-76.

Li, Xueqin. A Neolithic Jade Plaque and Ancient Chinese Cosmology, *National Palace Museum Bulletin* XXVII.5-6 (Nov-Dec 1992, Jan-Feb 1993), 1-8.

Loewe, Michael. The *chüeh-ti* games: a re-enactment of the battle between Ch'ih-yu and Hsüan-yüan? in *Divination, mythology and monarchy in Han China* (Cambridge: Cambridge University Press, 1994), 236-248.

Loewe, Michael. The Han view of comets, in *Divination, mythology and monarchy in Han China* (Cambridge: Cambridge University Press, 1994), 61-84.

Lu, Sixian. *Shenhua kaogu* (The Archaeology of Myths) (Beijing: Wenwu chubanshe, 1995).

Major, John S. *Heaven and Earth in Early Han Thought* (Albany: State University of New York Press, 1993).

Mozi jiangu, Xinbian zhuzi jicheng ed. (Taipei: Shijie shuju, 1974), vol. 6.

Pang, K. D. and Srivastava, S. K. Climatic Impacts of past Volcanic Eruptions: Inferences from Ice Core, Tree Ring and Historical Data, *EOS* 69.44 (1988), 1062.

Pang, K. D., Keston, R. and Srivastava, S. K. Climatic and Hydrologic Extremes in Early Chinese History: Possible Causes and Dates, *EOS* 247 (1989), 1095.

Pankenier, David W. Astronomical Dates in Shang and Western Zhou, *Early China* 7 (1981-82): 2-37.

Pankenier, David W. *Mozi* and the Dates of Xia, Shang, and Zhou: A Research Note, *Early China* 9-10(1983-85), 175-181.

Pankenier, David W. The *Bamboo Annals* Revisited: problems of method in using the chronicle as a source for the chronology of early Zhou, Part 1, B*ulletin of the School of Oriental and African Studies* LV, Part 2 (1992), 272-297.

Pankenier, David W. The Cosmo-political Background of Heaven's Mandate, *Early China* 20 (1995), 121-176.

Pankenier, David W. Characteristics of Field Allocation Astrology in Zhou China, in *Cultural Aspects of Astronomy: An Intersection of Disciplines* (Durham: Carolina Academic Press) ms. (in press).

Peiser, Benny J. Catastrophism and Anthropology: The Influence of Neo-Catastrophism on the Interpretation of Flood Rituals and Ceremonies, *Chronology and Catastrophism Review* (Special Issue 1994), 130-134.

Peiser, Benny J. Cosmic Catastrophes and the Ballgame of the Sky Gods in Mesoamerican Mythology, *Chronology and Catastrophism Review* 17 (1995), 29-35.

Weitzel, R. B. Clusters of Five Planets, *Popular Astronomy* 53 (1945), 159-161.

Xi, Zezong. Mawangdui Han mu boshu zhong de huixing tu, The Comet Atlas on Silk from the Han Tomb at Mawangdui), *Zhongguo gudai tianwen wenwu lunji* (Beijing: Kexue chubanshe, 1989), 29-34.

Yang, Kuan. Huang di yu Huang di, in *Zhongguo gu shi yanjiu luncong*, (Taiwan rpt. of *Gu shi bian*), vol. 7, *Gu shi chuanshuo tonglun* (Taipei: n.d.), 199-206.

Zhang, Peiyu. Yin Shang Wu Ding shi de yueshi he lifa (Lunar eclipses and the calendar in the time of Wu Ding of Yin-Shang), *Zhongguo gudai tianwen wenwu lunji* (Beijing: Wenwu, 1989), 17-28.

Zhongguo shehui kexue yuan kaogu yanjiusuo, ed. *Zhongguo gudai tianwen wenwu tuji* (Beijing: Wenwu chubanshe, 1980).

Zhuang, Tianshan. Ancient Chinese Records of Meteor Showers, *Chinese Astronomy* 1 (1977), 197-220.

References

1. David W. Pankenier, "The Cosmo-political Background of Heaven's Mandate," *Early China* 20 (1995), 121-176.
2. See Sarah Allan, *The Shape of the Turtle: Myth, Art, and Cosmos in Early China* (Albany: SUNY Press, 1991); also Li Xueqin, A Neolithic Jade Plaque and Ancient Chinese Cosmology, *National Palace Museum Bulletin* XXVII.5-6 (Nov-Dec 1992, Jan-Feb 1993), 1-8.
3. Pankenier, "The Cosmo-political Background of Heaven's Mandate," 161-170.
4. David W. Pankenier, Characteristics of Field Allocation Astrology in Zhou China, in *Cultural Aspects of Astronomy: An Intersection of Disciplines* (Durham: Carolina Academic Press) ms. (in press).
5. Mark Edward Lewis, *Sanctioned Violence in Early China* (Albany: SUNY Press, 1990), 138.
6. David N. Keightley, *Sources of Shang History: The Oracle-Bone Inscriptions of Bronze Age China* (Berkeley: University of California Press: 1978).
7. According to Keightley, the earliest oracle bone records of lunar eclipses recorded during the reign of Shang King Wu Ding cluster in an 18-year period from 1198-1180 BCE (Keightley, *Sources*, 174). Subsequently, astronomer Zhang Peiyu's analysis concluded that all must fall between 1250 and 1165; see his "Yin Shang Wu Ding shi de yueshi he lifa (Lunar eclipses and the calendar in the time of Wu Ding of Yin-Shang)," *Zhongguo wenwu tianwen wenwu lunji* (Beijing: Wenwu, 1989), 25. Zhang now dates King Wu Ding's reign to 1239-1181 (personal comm. 12/18/97). The 10 (or 9) kings of the historical Anyang phase ruled from about 1239 to 1047, or 192 years. Subtracting this from the 496 years for the Shang dynasty given in the *Bamboo Annals* and Han dynasty texts leaves 304 years for the first part of the Shang dynasty, thus implying a Shang founding date of about 1543. This result is in very close agreement with my date of 1554 for the founding of Shang based on the *Bamboo Annals* records of astronomical observations and internal chronology. For his part, Keightley's analysis of the chronology of late Shang led him to conclude that the dates he assigns to the individual kings are not likely to be in error by more than 25 years (Keightley, *Sources*, 176).
8. For this lunar eclipse, see Pankenier, "The Cosmo-political Background of Heaven's Mandate," 129 and Li Changhao ed., *Zhongguo tianwenxue shi* (History of Chinese Astronomy) (Beijing: Kexue chubanshe, 1981), 21. The Zhou King saw the eclipse as particularly ominous because it occurred in the Vermilion Bird constellation astrologically associated with the Zhou. For the planetary massing of 1059 BCE, see "The Cosmo-political Background of Heaven's Mandate," 124ff.
9. For a discussion of other evidence, astronomical as well as historical, for the chronology of the immediate conquest period, see David W. Pankenier, "The *Bamboo Annals* Revisited: problems of method in using the chronicle as a source for the chronology of early Zhou, Part 1," B*ulletin of the School of Oriental and African Studies* LV, Part 2 (1992): 272-297. This also means, for example, that the attempt to connect a *Bamboo Annals* record of dust-laden rain in the 5th year of the last Shang King Zhou Xin (r. 1086-1047) with an eruption of Hekla 3 in the late 12th century (tree ring date of 1159-1140 BCE) is problematical since it misses the mark by some 50 years. See K. D. Pang and S. K. Srivastava, "Climatic Impacts of past Volcanic Eruptions: Inferences from Ice Core, Tree

Ring and Historical Data," *EOS* 69.44 (1988), 1062.

10. The earliest of these extraordinary events was first computed and described by R. B. Weitzel in 1945 as follows: "The five planets provided a magnificent spectacle. Mercury, Venus, and Mars approximated a triple star; Saturn was somewhat lower and to the left; while Jupiter, more apart, shone above and to the right of them. On the morning of February 26, 1953 B.C. Mercury, Venus, Mars, Jupiter, and Saturn were clustered in a field of three and eight tenths degrees - an exceptionally dense assemblage of planets"; see Weitzel, "Clusters of Five Planets," *Popular Astronomy* 53 (1945): 161; cited in Pankenier, "*Mozi* and the Dates of Xia, Shang, and Zhou: A Research Note," *Early China* 9-10 (1983-85), 183.
11. Dates reconstructed by the author based on astronomical benchmarks and analysis of systematic error in the *Bamboo Annals'* chronology; see Pankenier, "The *Bamboo Annals* Revisited."
12. The densest planetary massing in the past 5,000 years occurred in Aquarius. For analysis of a 4th c. BCE textual account and verification see Pankenier, "*Mozi* and the Dates of Xia, Shang and Zhou," 175-183, and "The Cosmo-Political Background of Heaven's Mandate," 123ff.
13. James Legge, *The Chinese Classics*, vol. 3, *The Shoo King or The Book of Historical Documents* (2nd ed.) (rpt. Taipei: Wenshizhe chubanshe, 1972), Prolegomena, "The Annals of the Bamboo Books," 124.
14. Last king of the Xia Dynasty.
15. Legge, *The Shoo King*, 125. Legge's rendering "the five planets went out of their courses" is a mistranslation. In its entirety the entry for King Jie's 10th year reads, "10th year, the five planets criss-crossed; in the middle of the night the stars fell like rain; the earth shook; the Yi and Luo Rivers ran dry." The curious term *cuoxing* 'criss-crossed; moved crosswise' is used to describe the behavior of the five planets in the late fall of 1576 as the sun overtook the grouping, so that the time and place of observation changed from the NW horizon after sunset to the SE horizon before dawn. For detailed discussion of the term *cuoxing* and the movements of the planets on this occasion, see Pankenier, "Astronomical Dates in Shang and Western Zhou", *Early China* 7 (1981-82), 18ff. Associated with the planetary massing is the only mention in the *Bamboo Annals* during the Xia and Shang dynasties of an impressive meteor shower. The timing suggests the shower in question may have been the Taurids, or perhaps the Geminids. This record of a meteor shower is the earliest of the 147 reports collected by Zhuang Tianshan in his "Ancient Chinese Records of Meteor Showers," *Chinese Astronomy* 1 (1977): 197-220. Unfortunately, Zhuang also misread the *Bamboo Annals* record, placing the observation in King Jie's 15th year.
16. Legge, *The Shoo King*, 126. The record "three suns appeared together" is in the 29th year of King Jie of Xia, and probably relates to the spring or summer of that year. An effort by Pang et al. to correlate Thera/Santorini's eruption (which they date to 1600 ±30 BCE) with a spate of weather anomalies near the end of King Jie of Xia's reign (1585-1555) misses the mark by a considerable margin. Recently, Peter Kuniholm and Stuart Manning have more precisely dated the eruption to 1628 BCE, some 50 years too early to be associated with the Chinese records of unseasonal phenomena; see K.D. Pang, R. Keston, and S. K. Srivastava, "Climatic and Hydrologic Extremes in Early Chinese History: Possible Causes and Dates," *EOS* 247 (1989), 1095; Peter Kuniholm et al., Anatolian tree rings and the absolute chronology of the eastern Mediterranean, 2220-71 8 BC, *Nature* 381.27 (27 June 1996): 780-782. Cf. also M.G.L. Baillie, *A Slice Through Time: Dendrochronology and Precision Dating* (London: Batsford, 1995).
17. Legge, *The Shoo King*, 129. This is probably the most well known drought in early Chinese history. The Shang king Cheng Tang's willingness to offer himself as a propitiatory sacrifice to alleviate the people's suffering is much celebrated in later classical texts. For a discussion of this account and drought sacrifices in general, as well as the theme of atonement for regicide in connection with dynastic overthrow, see Allan, *The Shape of the Turtle*, 19-56, esp. 42ff, and "Drought, Human Sacrifice and the Mandate of Heaven in a Lost Text from the *Shang Shu*," *BSOAS* XLVII.3 (1984): 523-539.
18. The fullest account of this planetary massing and its implications for the chronology of the Zhou Conquest period is in Pankenier, "Astronomical Dates," however, "The Cosmo-Political Background of Heavens Mandate," 121-136 may be more accessible to the non-sinologist.
19. For an in-depth discussion of the original sources, including *Shanhaijing*, *Huainanzi*, and *Chuci*, and their antiquity, see Allan, *The Shape of the Turtle*, 25ff.
20. Allan, *The Shape of the Turtle*, 37.
21. The account of the appearance of multiple suns in the latter part of the Xia Dynasty occurs in both versions of the *Bamboo Annals*, the so-called current version as well as the genuine version reconstituted from quotations in ancient works more or less contemporaneous with the chronicle's discovery. There can be no question, therefore, that this record appeared in the original 3rd century BCE version of the text.
22. Chantal Zhang, "Le Mythe de larcher et des soleils", in Le Blanc and Mathieu, *Mythe et philosophie à l'aube de la Chine Imperiale* (1992), 27-48; Li Fuqing, "Xiaomie duoyu taiyang de shenhua (The myth of the eradication of superabundant suns)," *Lishi yuekan* (Historical Monthly) 3 (1997): 70-76.
23. Legge, *The Shoo King*, 140.
24. Legge, *The Shoo King*, 124.
25. Legge, *The Shoo King*, 125.
26. Legge, *The Shoo King*, 126.
27. *Mozi jiangu*, Xinbian zhuzi jicheng ed., (Taipei: Shijie shuju, 1974), vol. 6, 93.
28. Victor Clube and Bill Napier, *The Cosmic Winter* (London: Basil Blackwell, 1990).
29. Tr. Mark Edward Lewis, *Sanctioned Violence*, 174-75; for a fully annotated translation, see William H. Nienhauser, Jr., ed., *The Grand Scribe's Records, Vol. 1, The Basic Annals of Pre-Han China* (Bloomington, Indiana University Press, 1994), 1-3.
30. Lewis, *Sanctioned Violence*, 195.
31. Lewis, *Sanctioned Violence*, 210-212 (insertions mine).
32. Lewis, *Sanctioned Violence*, 195.
33. Clube and Napier, *The Cosmic Winter*, 185.
34. Lewis, *Sanctioned Violence*, 180.
35. Lewis, *Sanctioned Violence*, 162.
36. The myth of Gong Gong and the cosmic disaster that produced the tilting of the ecliptic and a cataclysmic flood are discussed in detail by John S. Major, *Heaven and Earth in Early Han Thought* (Albany: State University of New York Press, 1993), 26 and *passim*; see also William G. Boltz, "Kung-kung and the Flood: Reverse Euhemerization in the *Yao tien*," *Toung Pao* 67 (1981), 141-153.
37. Lewis, *Sanctioned Violence*, 141.
38. Lewis, *Sanctioned Violence*, 182.
39. Lewis, *Sanctioned Violence*, 158. See also Michael Loewe, "The *chüeh-ti* games: a re-enactment of the battle between Chih-yu and Hsüan-yüan?" in *Divination, mythology and monarchy in Han China* (Cambridge: Cambridge University Press, 1994), 236-248.
40. Lewis, *Sanctioned Violence*, 158; cf. Derk Bodde, *Festivals in Classical China: New Year and other Annual Observances During the Han Dynasty (206 B.B. A.D. 220)*, (Princeton: Princeton University Press, 1975), 201 ff.
41. Lewis, *Sanctioned Violence*, 149.
42. Lewis, *Sanctioned Violence*, 149.
43. Lewis, *Sanctioned Violence*, 149.
44. Lewis, *Sanctioned Violence*, 150.
45. Lewis, *Sanctioned Violence*, 148. For a discussion of Mesoamerican parallels, see Benny J. Peiser, "Cosmic Catastrophes and the Ballgame of the Sky Gods in Mesoamerican Mythology," *Chronology and Catastrophism Review* 17 (1995), 29-35.
46. Lewis, *Sanctioned Violence*, 148 (insertion mine).
47. Lewis, *Sanctioned Violence*, 184; *Shiji*, 27.1335. See Bodde (*Festivals in Classical China*, 121), who also calls attention to the historians explicit identification of the comet of 134 BCE with Chi You's Banner. See also Michael Loewe, "The Han view of comets," in *Divination, mythology and monarchy in Han China* (Cambridge: Cambridge University Press, 1994): 77ff.
48. Xi Zezong, "Mawangdui Han mu boshu zhong de huixing tu, (The Comet Atlas on Silk from the Han Tomb at Mawangdui)," *Zhongguo gudai tianwen wenwu lunji* (Beijing: Kexue chubanshe, 1989): 29-34. For detailed discussion of this and other apparitions of Chi You's Banner, see Michael Loewe,

"The Han view of comets," 61-84.

49. It is noteworthy that the instigator of recurring chaos, Chi You, is associated with cometary phenomena, while the Yellow Emperor is identified with a fixed constellation called Xuanyuan in Leo and his astral double is the Pole Star.
50. Clube and Napier, *The Cosmic Winter*, 197.
51. Bodde (*Festivals in Classical China*, 52) discusses at length the problem of dating rituals such as the *La* in Zhou times and the possibility that there may have been regional variations in their timing. In any case, he concludes they always occurred around the end of the year.
52. Lewis, *Sanctioned Violence*, 194. Lewis (187) also cites direct evidence that Chi You received sacrifice in the state of Qin at the time of the New Year *Zha* rites in the tenth month (October-November).
53. See Sima Qian's account above of the Yellow Emperors career where 'Chi You' and 'Fiery Emperor' are used seemingly interchangeably to refer to a single miscreant. The identity of the two was suggested early on by Yang Kuan in *Gu shi bian*. See his "Huang di yu Huang di," in *Zhongguo gu shi yanjiu luncong*, (Taiwan rpt. of *Gu shi bian*), vol. 7, *Gu shi chuanshuo tonglun* (Taipei: n.d.), 199-206.
54. Lewis, *Sanctioned Violence*, 202; Bodde, *Festivals in Classical China*, 120-127. For a recent re-interpretation of flood rituals and ceremonies elsewhere in the world as a reflexive reaction to cosmic catastrophe, see Benny J. Peiser, "Catastrophism and Anthropology: The Influence of Neo-Catastrophism on the Interpretation of Flood Rituals and Ceremonies," *Chronology and Catastrophism Review* (Special Issue 1994), 130-134.
55. A similar contrast between the tenor of the *Bamboo Annals* and many of the classical texts was noticed and commented on by the great sinologist James Legge over a century ago. According to Legge, this is one of the features of the text that underscore its credibility as having escaped the wholesale rewriting of China's ancient history in heroic terms so evident in many other Warring States texts; see Legge, *The Shoo King*, 178, 182-183.

The Agenda of the Milesian School:

The Post-Catastrophic Paradigm Shift in Ancient Greece

William Mullen

Bard College, Annandal-on-Hudson, New York 12505, USA

1. Introduction

The present monograph consists of two chapters intended to begin a study in the catastrophist reinterpretation of the principal pre-Socratic Greek thinkers - philosophers and proto-scientists - from Thales to Democritus. I conceive of that study as part of a yet larger project for which such a book would be the first of several - an effort to rethink the so-called "Axial Age" connecting cultures from Greece to China in the 6th and 5th century B.C.E. The assumption on which this rethinking is based is that these cultures' simultaneous activities in rewriting the mythical accounts of world-destructions bequeathed to them by immediately preceding generations was essentially conditioned by the fact that human consciousness had only recently emerged from such events into a period of relative celestial and terrestrial stability. My rethinking assumes as its framework the chronological revisions of Heinsohn, Illig and Peiser with respect to Mesopotamian, Egyptian and Greek chronology, according to which the first two of these three are so radically downdated that some of their principal texts assume a place within the same 6th/5th century horizon as the other civilizations usually claimed for the Axial Age - Greece, Israel, Persia, India, China. [1] The rethinking is thus to be conceived as evidenced in the texts of *all* attested civilizations from Greece to China as they emerge from the so-called Bronze Age, during a period of intensive interaction with each other precipitated by catastrophic events they all underwent simultaneously.

In terms of modern scholarship, therefore, the reader of this monograph is asked to entertain theses generated by not one "paradigm shift" but two - that from 19th century uniformitarianism to historical catastrophism, and that from 19th century dating schemes for the ancient world to the much more compressed schemes just cited. The former shift, to historical catastrophism, is well underway; its development is mainly a matter of allowing the larger shift begun in the 1980's by the Alvarez hypothesis of the extinction of the dinosaurs by asteroid impact (a return to the general theory of catastrophism which preceded uniformitarianism in the 19th century) to encroach upon time frames involving human memory and, indeed, the foundation of ancient civilizations as we know them. The latter shift, to radical chronological compression, is less well-known and still hotly disputed by those familiar with it. Readers unfamiliar with it, or unpersuaded of it, will be relieved to know that I invoke it only rarely in this monograph. Most of what I have to say about ancient Greece occurs within an unexceptionable traditional chronology, and Mesopotamia and Egypt enter my argument only when I find scholars too ready to base conjectures about Greek thought on the assumption that, since these high civilization to the east and south of it had existed for two millenia already, any Greek idea resembling one of theirs *must* have been borrowed from them.

As my subtitle indicates, however, is it not only modern "paradigm shifts" I have in my mind. Just as 19th century uniformitarianism succeeded within the scientific community for a century and a half in silencing a whole series of earlier catastrophist questions as unfruitful - questions about fossils, interruptions in geological strata, repeated signs of sudden and widespread upheaval - so, within a comparable period from the mid-6th to the late 5th century B.C.E., Greek intellectuals succeeded in reformulating all accounts of world-destruction in terms that eliminated that terrifying element of random unpredictability which the myths of the poets had expressed in terms of angry gods who acted at their own whims and were thus in need of propitiation through sacrifice. The result was that by the time of Plato and Aristotle the gods of Homer and Hesiod, while still central to popular thinking, had been bracketed by intellectuals and treated as fair game for radical reinterpretation in terms of their own various philosophical systems, to which sacrifice became irrelevant. A negative characterization of these reformulations would point to their cosmological schematizations, far beyond what any modern sense of scientific evidence would warrant, whose purpose was always, in one way or another, to eliminate the random and the saltatory from the cosmos and substitute the orderly and the predictable. A positive characterization would point to the success, albeit limited, of such proto-scientific rejections of the mythical in putting a cap on apocalyptic thinking, with all its dangers of social upheaval - a success, again, not unlike that of the last century and a half of uniformitarianism.

No paradigm is final. Historical catastrophism, if consolidated in the mainstream of science, may eventually be replaced by yet another paradigm able to see unwarranted schematizations on which, for the present, it must rely in order to proceed. Such are the hazards of thinking in the midst of a paradigm shift.

2. Thales

Thales is known to most moderns for having said something like "Everything is water." Whatever doctrine he may have actually propounded about water, he could not have known that someday he would be assigned his place in the history of philosophy by Aristotle as the first of a series of Greek "materialist monists" associated with the elements, to be followed by Anaximenes' doctrine about air and Heraclitus' about fire. [2] Modern scholarship has gone a certain distance in liberating Thales from the burden of Aristotle's presentation, mostly by arguing that Thales was not answering the question "What is everything made of?" so much as the question "What is the origin of all things?" [3] The relative paucity of reliable ancient sources on Thales, however, makes it difficult to locate even the latter answer in a larger context. Did he lay it down as the point of departure for the rest of his investigation of nature, arrive at it at the investigation's conclusion, or venture it as a hypothesis along the way? And more importantly, was he answering the question in terms of the traditional mythology of the cultures around him, or was he transforming those terms into a proposition which we would wish to call philosophical or even scientific?

The range of topics in his doctrines, as received in later antiquity, seems well indicated in Diogenes Laertius' two summary sentences: "He taught that water was the source of everything, and that the cosmos was alive and full of divinities. He is said to have discovered the seasons of the year and to have divided it into 365 days." [4] In the first sentence here we have indications of several rethinkings likely to be interconnected, and each needs to be understood in terms of the traditional thinking it was challenging. Thales was, first of all, making statements, according to Diogenes, about a *kosmos,* a single ordered universe, and elsewhere he is said explicitly to have held that the *kosmos* was a unity [5]; in Homer and Hesiod the term for the universe as a whole does not yet exist. His understanding of the unity of the *kosmos* includes the notion that everything in it had a single source, whereas in Hesiod the things from which everything else came into being were four: Chaos, Earth, Tartaros, and Eros. His sense of unity also includes the notion that the *kosmos* as a whole is "ensouled", i.e. has *psykhe* in it, traditionally understood as the stuff, present in some things and not in others, whose presence made living things alive and whose departure made them dead. Finally, he seems to be affirming that divinity, like life, is more broadly diffused throughout the universe than was traditionally allowed.

These three interconnected propositions about the cosmos are then followed in Diogenes' exposition by a tradition that he was the first to make basic observations about the sun's orderly motion from year to year, one of several such observations - about eclipses, solstices, equinoxes, the zodiac - with which he is recurrently associated. Doctrines we might call cosmological, biological, and theological, are thus set next to astronomical research. I propose that the astronomical research is of a piece with these other doctrines, and in fact provides the necessary context for them. Modern uniformitarian interpretation has found little stimulus to reflection in the evidence that Thales had a research agenda dedicated to spelling out as many instances as possible of the orderly recurrent motions of the heavenly bodies. A catastrophist scenario such as I am assuming here, by contrast, in which he belongs to the first few generations after the cessation of threats from erratic movements in the sky, calls for a rereading of this evidence as the decisive point of departure for the rest of his work.

Scepticism about Thales' achievements as an astronomer has been expressed in recent times by both uniformitarians and catastrophists. [6] For the uniformitarian the only question to be answered is: When do we have the first hard evidence that human beings, after immemorial millennia of watching an orderly sky, first began to formulate with mathematical precision the laws that govern its orderly movements? For the catastrophist, however, the question must be at least twofold: first, when and how did humans watching the skies after the end of the last instabilities began to formulate their realization that all the heavenly bodies were now moving with orderly periodicities; second, by what steps did they then come to develop a fully mathematical account of the laws governing these periodicities? Scepticism as to whether Thales engaged in a fully mathematicized astronomy should be kept separate from skepticism as to whether he was a careful observer of a recently stabilized sky, who proclaimed articulately that none of the bodies moving in it were to be feared because all had unvarying regular motions.

Thales' intuition that the universe was a single ordered unity, having a single origin and with soul and divinity diffused throughout it, was made possible by a prior observation that the heavens were characterized by strict regularity. This observation, self-evident as it has been to us until recently, was in fact a novelty in his time. Inevitably it made his doctrines stand out against the stories of erratic sky-gods at the core of the myth and ritual around which most of his traumatized contemporaries still organized their lives. [7]

3. Problems of dating

It is this novelty in astronomical observation, I suggest, which accounts for the fact that so many of the testimonia are concerned to assert that he was in one way or another "the first". His is the first life in Book I of Diogenes Laertius' *Lives of Eminent Philosophers*, and in the "Prologue" Diogenes explains how he is using his terminology. He begins by dealing at length with the opinions of those who say that "the study of philosophy had its beginnings among the barbarians", and spells out claims to be made for the Persians, Babylonians, Indians, Celts, Phoenicians, Thracians, Libyans and Egyptians (1.1-11). Among the Greeks, he says, the first to call himself a "philosopher", or "lover of wisdom" (*sophia*), was Pythagoras. Before that there were only the poets and the sages, and both groups were called "wise men" (*sophoi* or *sophistai* - 1.12). In the traditional list of Seven Sages he places Thales first. Philosophy itself had a twofold origin, with Anaximander and with Pythagoras; the former was a pupil of Thales, the latter of Pherecydes (1.13). Thales is thus located at the intersection of three antitheses: Greeks and barbarians, sages and poets, wise men and lovers of wisdom.

In his life of Thales proper Diogenes then mentions a number of more concrete "firsts" in addition to the fact that he was first to receive the name of sage (1.22). He defined the Little Bear [8] and was "first to study astronomy and predict eclipses of the sun and fix the solstices... for which Xenophanes and Herodotus admired him, and to which Heraclitus and Democritus bear witness" (1.23). He was first to declare that souls were immortal; first to make a number of specific observations about the sun and the moon; first to discuss nature, or*physis* (1.24). After learning geometry from the Egyptians he was the first to inscribe a right-angled triangle in a circle (1.25). As noted earlier, he is said to have discovered the seasons and to have divided the year into 365 days (1.27). Elsewhere the tradition duplicates many of these firsts and adds others, all in the realm of astronomy, such as that he was the first to say that the moon was lit by the sun. [9] His status is well-summed up in the verse of Timon cited by Diogenes: "Thales, of the Seven Sages the sole sage of astronomy" (1.34).

Modern scholars of the doxographical tradition are sensitive to the needs of any tradition to give itself authority by claiming knowledge of its founders, a practice encountered earlier in the cases of Homer and Hesiod themselves.[10] This sensitivity leads them to be especially sceptical of specific claims about "firstness". Thales' reputation as the first *sophos* to practice astronomy, however, has made it irresistible to many of these same scholars to date him by one of the phenomena he is said repeatedly to have been first to predict, eclipses. [11] The fact that a number of ancient sources, beginning with Herodotus, have claimed that Thales predicted an eclipse which occurred during the battle between the Medes under Alyattes and the Lydians under Cyaxares has led modern scholarship to set 585 BCE as the date of the most distinctive achievement of the first "pre-Socratic philosopher". [12] The various assumptions on which this dating rests have been spelled out by Peiser, who cites also its recent critics and concludes both that Thales could not have possessed the knowledge necessary to predict eclipses in general and that the specific date 585 BCE cannot be used to ground Thales' time of flourishing.[13]

I shall return to the question of Thales' capacity to predict an eclipse. For the present it is sufficient to ask what criteria are left for dating Thales' life if the famous eclipse story is no longer deemed reliable for fixing his adulthood firmly around 585 BCE. Here, as in later cases, our first order of business must be to secure a relative dating of the figures with whom we will be concerned, such that the sequence of contacts and influences among them can be used fruitfully even if an absolute date for each is not to be had. It is sufficient to locate Thales early enough in the Sixth Century to allow for Anaximander, Anaximenes, and Xenophanes to be his immediate Milesian followers and for Thales himself not to have been preceeded by any known Greek philosopher or "physicist" whose influence he could not have escaped.

Peiser's article discusses the various scholarly arguments for dating Thales' birth at 640 (p. 89), 624 (p. 90), and 600 (p. 98), and his association, in Herodotus and later writers, both with the Median Alyattes' battle against the Lydian Cyaxares in 585 and the Lydian Croesus' battle against the Persian Cyrus in 546. [14] An independent date is to be had in the canonization of the Seven Sages in the archonship of Damasias at Athens in 582/1, the year of the establishment of the Pythian Games; though the lists of seven sages vary considerably, all include Thales. [15] Other stories of Thales' political activity also associate him in one way or another with responses to the threat of Persian invasion of the Ionian and Lydian territories. Herodotus tells us that before the Persian conquest of Ionia was accomplished (546-5 BCE) Thales had advised the various Ionian cities to set up a supreme deliberative council in order to respond concertedly to the Persian threat (1.170). He is also said to have frustrated a compact between Miletus and Croesus the Lydian, and thus left his city free to negotiate a later peace with Cyrus the Persian which spared it from destruction (DL 1.25). There is, in short, abundant material to place him in the first half of the Sixth Century BCE, before the Persian conquest of Ionia under Cyrus. This general location suffices to leave standing his position as initiator of the Milesian school of philosophical and physical inquiry in which Anaximander and Anaximenes are the immediate followers, and to which Xenophanes, Pythagoras, and Heraclitus are all soon thereafter significant respondents. He does not, in short, need any particular eclipse to ground him in the first few memorial generations after the last instability as dated by Heinsohn to the late Seventh Century. [16]

Stories of his learning from Egyptians and Babylonians have also been subjected to critical scrutiny. [17] In evaluating such stories the first question must be, What specific kind of knowledge is Thales being alleged to have acquired through this particular contact? In Egypt Thales is said by Diogenes to have acquired geometrical knowledge which he introduced to Greece, and this is the tradition most disputed by Dicks. But Diogenes also has him measuring the height of the pyramids there by the shadows they cast (1.27), and this kind of activity is likely to have been related to the calculation of the solstitial days by the length of the shadows cast by gnomons. [18] While the specific measurement of the solstices by use of the shadows cast by a gnomon is attributed to his first pupil Anaximander, Thales himself is associated with the various kinds of knowledge about the solstices numerous times. [19] It is not necessary, however, to suppose that he could learn to do this only by going to Egypt. With equal plausibility one can imagine that his discovery how to measure solstices by shadows involved a kind of thinking which led him likewise to measure the heights of the pyramids by their shadows (when, that is, they were cast at the time of day when a man's shadow was equal to himself in height), and that on his visit to Egypt he was showing this knowledge off. [20]

Similar distinctions need to be made in assessing modern speculation about Thales' learning from the Babylonians. [21] Since Herodotus' account of the famous eclipse states only that Thales predicted the year of it, not the date, the time of day, or the place of visibility, it has been thought that if Thales were acquainted with Babylonian methods of predicting eclipses they would have sufficed for this achievement. The tablets on which these Babylonian observations and methods are to be found, however, are all subject to Heinhsohnian downdating of Mesopotamian chronology in general, and can no longer be assumed to have predated Thales. [22] In defense of the possibility that the story about Thales' prediction is true, van der Waerden commits a curious *non sequitur* indicative of a general trend among scholars to invoke Mesopotamian help for Greek thinkers as the recourse of choice. "Such a feat", van der Waerden writes [sc. to predict a solar eclipse] "requires the experience of more than forty years, no matter how one proceeds. It is not possible to accomplish it alone. But Thales had no Greek predecessors. The conclusion is inescapable that he must have drawn upon the experience of Oriental astronomers". [23] Thales had no predecessors named or called "first" by Greek tradition; this does not mean that he could not have relied upon the observations amassed by one or two preceeding generations of Greeks, as part of the forty or so years van der Waerden posits. His position as named "first" would have been due to his having been the first Greek to conclude on the basis of the observations amassed that some kind of eclipse cycle did exist.

How probable is such a scenario on the hypothesis of the catastrophist chronology being used here, with the last celestial instabilities occurring in the last quarter of the Seventh Century BCE? Obsessive notation of the movements of the heavenly bodies is unlikely, on this hypothesis, to have been confined to any one culture; Greek and Babylonian astronomers could quite well have come up with eclipse cycles independently as soon as a period of heavenly stability long enough to observe the sun and moon had passed for this purpose. [24] A general theory of eclipse cycles, based either on two or three repetitions of the 18-year cycle frequently invoked, or more loosely on the "forty years or more" posited by van der Waerden, should on this scenario have been possible for the first time only in the first quarter of the Sixth Century BCE, whether in Miletos or in Babylon. On the assumption that a portion of traumatized humanity would have been keeping close watch on the skies every since the last instability, one could argue not merely that the perception of cycles would have been possible at this time but, more strongly, that this is precisely the time frame in which one would expect it to emerge. As scholars have also not failed to point out, such a general theory would be insufficiently precise to determine where on the earth the path of totality would lie, so that it remains unlikely that Thales could by anything more than "a stroke of good luck" [25] have predicted that it would happen in the area where Herodotus' battle took place. But his fame for announcing the existence of an eclipse cycle may then have led him to be associated, rightly or wrongly, with the battle mentioned by Herodotus. [26]

It thus becomes evident, in our first effort to date the Axial Age thinkers with whom we are concerned, that two assumptions must always be questioned, both of them so axiomatic to modern scholarship as normally to be spared questioning of any kind. The first is that the distinctive achievements of civilization, such as astronomical observation, emerged by a process so slow and incremental that ancient claims that a given figure was "first" in any such achievement have always to be set aside. The modern assumption is that since the achievement must have been the result of slow progress any ancient claim that it came about suddenly is in fact likely to be a spurious retrojection motivated only by a desire to secure for a given tradition the authority of a single founder. DeGrazia's notion of "cultural hologenesis" is a useful designation of the new paradigm's approach to these issues: it reminds us of the possibility that a culture could in fact set itself on a new footing in a brief period, and that this process may well have been directed by a few well-remembered figures the historicity of whose achievements need not be automatically doubted. [27] The second assumption is that the Egyptian and Babylonian civilizations preceded all others by a good millenium or two, and thus that any Greek claim to "firstness" can be allowed only if there is no evidence that there were Egyptian and Babylonian predecessors from whom the Greek in question could have learned what he put forward, with whatever adaptations. It can be argued that this second assumption is in fact a subsidiary to the first; the Nineteenth Century assumption of millennia of early slow evolution in self-contained Egyptian and Mesopotamian river civilizations nicely reconfirms the larger Nineteenth Century evolutionary assumption that human civilization in general is the product of long tranquil development. Both assumptions rest on dubious analogies with the Darwinian oversimplification of evolutionary mechanisms which has been exposed and replaced by the model of "punctuated equilibrium".

4. Astronomical observations and innovations

We have heard various indications already that Thales's reputation as an astronomer looms large in surviving traditions about him, and have argued that these traditions need to be reevaluated in the catastrophist paradigm. Before examining specific claims about his astronomical achievements I wish to juxtapose more carefully those texts in which his general reputation in this respect is alluded to. In two of the texts his twin reputations as a sage and as an astronomer are highlighted in complementary ways: Timon says that of all the Seven Sages Thales was the only one to be sage in respect to astronomy, while a statue of him purportedly had inscribed on its base a verse stating that of all the astronomers he was most revered for his wisdom. [28] One of the weightiest of the testimonies, because loaded with names of those who lived shortly after him, is cited by Diogenes Laertius from Eudemus' treatise *On the Astronomers*, who follows a general statement that he was "the first to do astronomy (*astrologesai*)" with his specific achievement in predicting solar eclipses and fixing the solstices; Diogenes Laertius then says that "it was for this that Xenophanes and Herodotus admired him, and Heraclitus and Democritus bear witness to him" (DL 1.23 = D A 1). The vague "this" for which Herodotus admired him may well be the specific acts of prediction and measurement, since what we actually have in Herodotus' text is the statement (1.74) that Thales had predicted the eclipse at the battle of Alyattes and Cyaxares mentioned earlier. Heraclitus, on the other hand, may have something larger in mind; elsewhere a scholiast says that Heraclitus called not only Thales but also Homer an astronomer (*astrologos*, not "astrologer" but "one who discourses about the heavenly bodies") when the poet spoke about the inescapability of fate. [29]

We may understand more what Heraclitus was bearing witness to in Thales if we take seriously Aetius' statement, cited ealier, that "Thales and those who followed him said that the *kosmos* was a unity". [30] The simple language here is hard to evaluate. Aetius may be only contrasting Thales with later *physiologoi* like Anaximander or Democritus who claimed that there were an infinite number of *kosmoi*, and it is not clear on what grounds Aetius rested this contrast. Given, however, that the notion of a single unified *kosmos* is nowhere made verbally explicit in the texts of Hesiod and Homer, it is not impossible that among Thales' firsts was his use of the word *kosmos* to designate an ordered unity, and that his "followers" took this proposition seriously. And since the specific astronomical achievements in terms of which his general reputation as an astronomer is spelled out usually have to do with observations of regular and predictable phenomena (e.g. solstices and eclipses), it may be inferred that Thales' intuition of the ordered unity of the *kosmos* flowed from his sense, supported by observation, that all the motions of the heavenly bodies in it obeyed unvarying laws and were in that sense parts of one whole. The intuition becomes more vivid if we try to imagine anyone having it a few generations earlier, when the celestial and terrestrial environments were still massively unstable; under such conditions "ordered unity" is not a natural or easily available way of describing the world around one. The emergence of the word *kosmos* to designate the world as a whole would most naturally occur in the first generations of stability, neither earlier nor later.

While Diogenes report that Pythagoras was first to use the word *kosmos* to refer to heaven (DL 8.48), the first time we actually encounter it in a pre-Socratic text is in Heraclitus himself, who speaks of a *kosmos* made by neither god nor man but always existent. [31] It is common for scholars to speak of his use of the word here as "transitional" between the earlier Homeric sense of any (local) ordered whole, such as an army well drawn up into battle array, and the later philosophical sense of the universe as a world-system. But the principal point of Heraclitus' asseveration seems to be temporal: the order we see before us did not come into being at some particular time, but has always and will always exist. Since the common rhetorical stance of his statements is to controvert a commonly held opinion, the opinion controverted here would seem to be simply that what order we perceive in the world had not always been there, and this is in fact the opinion one would expect to be commonly held by human beings only recently emerging from a period of global catastrophe. Between the last generation to experience the catastrophe and Heraclitus' position here the necessary intermediary is a first effort to formulate the concept of any order governing heaven and earth at all, even if only temporary or recently emergent. The record is consistent with Thales having been the Greek with whose name that formulation was from early on associated.

Thales' insight into the unity of the order pervading heaven and earth would, moreover, be likelier to be admired by Heraclitus than mere specific predictions. The language of fate and destiny for which Heraclitus called Homer, as well as Thales, an *astrologos* resembles the language in which Heraclitus himself speaks of the unvarying regularity by which the *kosmos* alternates in its Great Year between "solstices" of flood and of conflagration; it is thus a term reserved for speaking of the universe on the greatest poossible scale. In the course of summarizing Heraclitus' general doctrines Diogenes Laertius in fact uses language identical to Aetius' statement about Thales: Heraclitus taught "that the whole is limited, that the *kosmos* is a unity", and that its eternal cycles (*periodoi*) of fire "come about through destiny". [32] Thales' insight into unity would be similarly admired by Xenophanes, about whom Aristotle explicitly says that it was when looking at the sky as a whole (*eis ton holon ouranon apoblepsas*) that Xenophanes affirmed that the One was God (*to hen einai ton theon*). [33] We will come shortly to statements attributed to Thales in which a notion of the divine is brought in. It is sufficient for now to note that the doxographical tradition makes Thales, Xenophanes and Heraclitus alike both in studying the heavens as a whole and in asserting that the *kosmos* is one. The later the thinker the ampler are the testimonia and the more explicit is the connection made between the regularity of the heavenly cycles and the perceived unity of the *kosmos*. But all three thinkers flourished in the Sixth Century BCE and there is no intrinsic implausibility in Thales having been the first to make this connection or in Xenophanes and Heraclitus having admired him for it; in this sense Aetius would be correct in including them among "his followers".

It is essential to keep in mind some such larger sense of Thales' concern with issues of cosmic unity and regularity when one turns to the claims made for his specific astronomical achievements. In evaluating recent critiques of claims involving solstices, equinoxes, eclipses, and the zodiac, whether for Thales or his immediate disciple

Anaximander, one has to keep in mind two distinctions, only one of which was available to scholars working within the uniformitarian paradigm. The first distinction is between drawing attention to the importance of a celestial phenonenon and actually measuring it with any degree of accuracy. If, in the language of many of the sources, Thales was the first *astrologos* - and the term at its most literal means no more than a "man who gave an account of the heavenly bodies" - this does not necessarily mean that each of his accounts of a heavenly body was presented in terms of accurate measurement or mathematically based reasoning; Thales could simply have been defining the set of topics with which any comprehensive astronomical discourse must concern itself. None of Dicks' arguments denying Thales (or his "associate" Anaximander) the capacity to determine the precise moment within the day on which solstice or equinox fell, the eclipse cycles, or the degree of obliquity of the ecliptic, is sufficient to prove that Thales did not begin an inquiry into each of these phenomena. What Dicks is specifically attacking, and on solid grounds, is the imputation to Sixth Century Milesians of the combination of theoretical framework, measuring devices and computational accuracy necessary to predict eclipses (Thales) or equinoxes (Anaximander) as specific moments in time. That the Milesians lacked such abilities in no way proves that they had not begun the process of defining the importance and interrelatedness of solstices, equinoxes, eclipses, and the zodiac. [34]

The second distinction is more radical, since it occurs only within the catastrophist paradigm. [35] Even those who agree that there is overwhelming evidence that the earth experienced global catastrophes within the memory of humanity disagree sharply on whether the agents of these catastrophes were such as to have affected the orbit of the earth, its axis, or the position of its poles, and hence the appearance of the motions of sun and other heavenly bodies. So far those working exclusively within the astronomical profession have tended to confine their scenarios to the impact of comets, asteroids, meteors and meteorites, while those approaching the subject from outside the profession have been readier to hypothesize changes in planetary orbits, including the earth's own. [36] Presumably only the latter, more radical type of scenario would actually change the phenomena with which solstices, equinoxes and zodiac are associated by terrestrial observers. If these more radical scenarios have any truth to them, and if they obtain for the most recent global catastrophes which humanity underwent, then it follows necessarily that during the first period of celestial stability there would have been, from culture to culture, first individuals or groups to draw attention to the new regularities of solar movement throughout the year and to attempt to measure them and theorize about them. Within Greek culture, on this hypothesis, Thales in particular, and the Milesian school in general, would have been remembered for this "first" and honoured accordingly.

If, on the other hand, the more conservative astronomical approach is correct and humanity's most recent global catastrophes were due to impacts of bodies smaller than planets, the nature of Thales' achievement will have to be characterized somewhat differently. Speaking broadly of the Bronze Age as the time of a series of catastrophes repeatedly disrupting early civilizations, we would have to imagine its populations as living in terror of heavenly bodies other than the sun, moon and planets, and as building their systems of astronomical and ritual observations around fear of such bodies, viewed as capricious deities whose irregular actions require constant propitiation. In the course of observing these erratic bodies various cultures may well have evolved specialized castes of sky-watchers capable of noting various periodicities in the orbits of the sun, moon and planets as well, but the social disruption and collapse caused by each recurrent catastrophe may well have recurrently destroyed both the sky-watching castes and the specialized knowledge they accumulated. The end of this Bronze Age series of catastrophes would then see the commencement, or recommencement, of an astronomy whose principal objects were the regular movements of the sun and the other planets themselves, rather than the capricious movements of comets or meteorites. [37] While the vast majority of people would remain fixated, in the first few generations of stability, on myths and rituals generated by the traumatized experience of irregular heavenly deities, a few thoughtful observers would, in Aristotle's phrase about Xenophanes, "look off at the sky as a whole", and in due time would propose a new way of "giving an account of the heavenly bodies" at whose center were the regular movements of the sun, moon and planets. As fear of capricious deities receded over the following generations the reputation of these men for having begun something new and decisively different would be enhanced, even if, occasionally, by means of imputing to them a degree of refinement in fact only reached later.

On either catastrophist model, then, Thales' activity as an astronomer may be summed up as a survey of heavenly phenomena motivated by the perception that they form a unified system characterized by stability and regularity, and are hence in principle susceptible to prediction and measurement. In this light many specific achievements take on a new aspect which renders them more credible.

Consider, for instance, the topic of the solstices. Dicks (1966.31) points out that the term used, *tropai*, (literally "turnings"), is to be found in the texts of both Homer (*Od.* 15.404) and Hesiod (*W & D* 564 and 663); in the former case, according to Dicks, it seems to mean only vaguely "west", in the latter it is a seasonal reference point, like the rising of certain constellations, which the farmer is to use as a guide in timing his activities. [38] Thales, on the other hand, is said not only to have "recognized" the solstices and "predicted" them, but to have been the first to discover the passage (*parodos)* of the sun from solstice to solstice and to discover that its cycle (*periodos*) between the solstices was not always equal, i.e. that there were "slight variations in length of the solar seasons as divided by solstices and equinoxes". [39] He attained, in other words, a degree of precision in measurement unnecessary for farmers and unattempted by them, and his motivation in so doing seems to have been to determine the exact degree of regularity and predictability in the sun's orbit. In this context might be placed other statements about his study of the sun, even if partially anachronistic: that he was the first Greek to have knowledge of the size and nature of the sun, and that in fact he determined that the sun's size was a specific fraction of its orbit. [40] Diogenes Laertius goes into yet further detail, stating that Thales was "according to some the first to declare the size of the sun to be one seven hundred and twentieth part of the solar circle, and the size of the moon to be the same fraction of the lunar circle" (DL 1.24). There are problems of anachronism here in the assumptions that Thales thought in terms of a complete circular orbit and that he used some kind of sexagesimal system, [41] but the impulse behind the "discovery", if Thales really proclaimed it, may be characterized as a quest for signs of orderly symmetry in the heavenly bodies verifiable by measurement: both sun and moon have the same ratio between their diameters and their orbits. It would have been an impulse of the same kind that would have led him to attempt to measure the number of days in the four parts of the year as divided by solstices and equinoxes. If in so doing he discovered a variation in the length of these four periods rather than a symmetry, it is to his credit as a faithful reporter of the phenomena that he did not try to declare an absolute symmetry where he found none. As a searcher for predictability as well as mathematical symmetry, he would have rested content with the observation that the four periods each remained the same from year to year.

Similar concerns for symmetry and unity can be inferred from what is reported about his study of the constellations and the zodiac. On one level his work would seem to have had a practical side: some say he wrote an "Astronomy for Sailors" (*Nautike Astrologia*), and if he "measured out" the stars of the Little Bear it may have been only to point out to sailors that their smaller revolution "provides a more fixed point than the Great Bear". [42] On another level, however, we hear of him, along with Pythagoras, dividing the heavenly sphere into five symmetrical zones: the two circumpolar zones, the two tropics, and the equatorial zone in the middle, with the oblique zodiac confined to the middle three zones. [43] Here too a distinction has to be made between formulating a theoretical account of the ecliptic and measuring its obliquity, on the one hand, and simply drawing attention to the band of the zodiac within which all paths of sun, moon and planets are confined. Thales was certainly capable of the latter, and his motivation in so doing may again have been to point out that the paths of sun, moon and planets are not erratic but operate within certain definable limits.

And it is here, finally, that we should perhaps lay to rest the question of Thales' relation to eclipses. Let us set aside, for argument's sake, the possibility broached earlier that he did indeed, through his own effort or with the help of predecessors, possess enough observational data to understand an eclipse cycle and determine the likelihood of an upcoming eclipse to within the year. His association with eclipses might, then, have begun simply through the fact that his theory of sun and moon had arrived at the point of maintaining that because eclipses of the sun were caused by the passage of the moon in front of it, and because the orbits of sun and moon were regular, eclipses were recurrent phenomena which were in principle predictable.[44] Eclipses were not all that Thales was known for predicting; according to Aristotle he predicted, "through his study of the heavenly bodies", that there would be a large olive crop one year. [45] What is essential to our understanding of his achievement is therefore not whether he could have predicted, or did predict, a given eclipse at a given hour or place, but rather whether, as a result of his theorizing about the regularity of the motions of the heavenly, he became associated in his own lifetime with the notion of predictability in general. Similar speculation has been engaged in by modern scholars to explain Anaxagoras' reputation for predicting the fall of a meteorite at Aegospotami in 467:

> Although it is frequently declared absurd that Anaxagoras could have predicted the fall of a meteorite, he may well have said that the earthen bodies held aloft by the cosmic vortex can sometimes slip and fall to earth. If this assertion (which might loosely be called a general prediction) were widely known when the famous meteorite fell, it would be but a small step to credit Anaxagoras with predicting the event. [46]

In both instances a well-known thinker would have insisted that a certain kind of event held by most people to be erratic is in fact expectable within the general terms of his account of the heavens. [47]

Erratic events in the heavens are terrifying; predictable events need not be so. The former belief is the heritage of the traumatizing catastrophes of the past; the latter is the product of a new determination to survey the heavens as an orderly system. If we need reminding of how many popular Greek superstitions about eclipses there were in the period between Thales and Anaxagoras we will find them amply supplied by a fragment of Pindar:

> Ray of the sun, O you who see many things, O mother of eyes, what are you devising as you steal away the supreme star in full daylight? Why are you rendering helpless the strength of men and the path of wisdom? Rushing along the path of darkness, are you driving forward something unheard of in the past? ... Are you bringing some sign of war, or a withering of crops, or the unspeakable strength of a blizzard, or destructive civil war, or the emptying of the sea onto the plain, or a frost on the earth, or the southwind's heat flowing with angry streams? Or do you intend to flood the earth and create afresh a new race of men? [48]

The connection in the popular mind between eclipses and catastrophic world-destructions could not be made more explicit than Pindar does by capping his list of possible disasters with a universal deluge. It is of precisely this kind of fear of divine wrath that the Milesian school founded by Thales is so remarkably free. For them eclipses, and other phenomena in principle predictable, no longer "rendered helpless the path of wisdom". [49]

5. Divinity, life, water

Thales, then, as the "astronomer most revered for his wisdom" and the "only sage versed in astronomy", seems to have gained a reputation not merely for observing regularities in the motions of the heavenly bodies but also for connecting these observations to larger questions about the nature of the cosmos as a whole. Among these larger questions were the nature of divinity and the nature of soul. On the topic of divinity we have already seen that in saying that the *kosmos* was full of gods Thales seems to be implying that the Homeric Olympians, or even the welter of divinities in the *Theogony*, do not adequately represent the all-pervasive nature of the divine. Diogenes also attributes to him the apothegms that "God is the most ancient of all beings, for he is ungenerated" and that "The divine has neither beginning nor end" (1.35-36).[50] The notion that a god has no ending, i.e., is immortal, is essential to any Hesiodic or Homeric understanding of divinity. The notion that a god has no beginning, on the other hand, marks a new turn of thought, since Hesiod does not even maintain that to be the case with his four gods who "came into being first" (*Theogony* 115-122). "Beginninglessness" (in whatever language Thales himself may actually have expressed it) is destined to become, in the hands of later thinkers such as Heraclitus and Democritus, an ever more potent concept in dividing philosophy from traditional religion. Finally, we should note than according to Seneca Thales held that earth rested on water and that earthquakes were due merely to the water's fluctuations, a proposition which explicitly controverts the traditional belief that, like all other catastrophes, earthquakes were the work of a particular god with will and emotions, in this case Poseidon. [51]

On the relation of the topics of divinity and soul Diogenes says, as we have seen, that Thales taught that "the *kosmos* was animate" (lit. *empsykhon,* 'ensouled') and full of divinities" (DL 1.26). The same connection is made by Aristotle himself: "Some say that soul is intermingled in the universe, whence, perhaps, Thales too thought all thing are full of gods". [52] Diogenes also attributes to the 3rd-2nd c. B.C. poet Choirilos the belief that Thales was the first to say that souls were immortal; while some scholars consider this an anachronistic retrojection of Stoic thinking, [53] we should not be too quick to reject the possibility that Thales was making radically new propositions about the nature of soul. He was well known for doing so in another context, in asserting that even supposedly inanimate substances like amber and magnets had a share of soul, evidently since, though motionless themselves, they had the power to cause motion in other things such as iron. [54] Clearly Thales had set about to redefine traditional notions both of divinity and of soul; and since soul was traditionally the principle that made a living thing alive, he was also implicitly redefining life.

Once one perceives the magnitude of Thales' redefinition of traditional Greek categories of thinking one inevitably asks, as the central focus for speculation, what was his primary motivation in doing so. This is the question to which both ancient and modern scholarship recurs. And the standard beginning for an answer has always been to adduce the other topic for which he is best known, the doctrine which Aristotle gives us, in his own terminology, that the *arche* - "source" or "principle" - of all things is water. The putative connection among the three topics - water, divinity, and life - has been sketched succinctly by McKirahan:

> Water's unceasing mobility... reveals it to be a living substance. If everything is made of water or ultimately arises from water, the life force of water pervades the whole world... Moreover, as a living thing with no beginning in time (everything else owes its beginning to *it*), and apparently no end in time either, water is divine (since for the Greeks the primary characteristic of the divine are immortality and power independent of human will) ... Hence all things, being composed of or arising from water, are divine. [55]

Note that McKirahan elides here the crucial innovation Thales seems to have made in proposing that the divine had no beginning as well as no end. The traditional understanding of the divine by "the Greeks", as shown in the *Theogony*, is that the gods were immortal in the sense that they did not die but not in the sense that they were not born; as we have seen, they all had beginnings, and all but four of them came into being by birth. For the rest, however, the connections inferred here have their own integrity.

Unfortunately, once modern scholars bring water into the discussion they then show a tendency to foreclose further discussion by accepting the view of some ancients that Thales' doctrine of water is one he might have learned from the Egyptians or Babylonians. One of the most important of the many cosmogonic myths in Egypt was that the creation occurred out of Nun, the primeval watery abyss which still surrounds what are now heaven and earth. Likewise, in the *Enuma Elish*, the Babylonian creation epic, Apsu and Tiamat form the primeval waters and are later split to form the waters under the earth and in the sky. [56] Thales is thus discounted as radical innovator; his cosmogonic question "looks backward rather than forward", [57] and once again (as with eclipse cycles) the assumption of several millennia of cosmogonic speculation in Egypt and Mesopotamia is interposed as an obstacle to the possibility that in Thales a genuinely new situation in the environment is begetting genuinely new thoughts.

Thales may have actually traveled to Egypt, and may have had access to both Egyptian and Babylonian learning; he may even have been stimulated in his astronomical or geometrical thinking by these cultures. But to locate his doctrine of water in the mythopoeic cosmogonies of either culture is to denature it entirely, since the testimonia concerning him contain not a phrase to indicate mythical thinking. The Egyptian notion of Nun is always mythical in the sense that it is always associated with the intentionality of specific deities; one god or another arises out of Nun and creates the world in the first of many willed acts. The story of Apsu and Tiamat is likewise part of a creation epic replete, like the *Theogony*, with cosmic battles and victorious deities. Acts of divine intentionality are notable by their absence in Thales' thought. [58] As we have seen, in proposing that earthquakes are caused by fluctuations of the waters on which the earth rests, Thales is unambiguously controverting the traditional belief that earthquakes were caused by an angry Poseidon; the constant nature of water is being adduced to render unnecessary the hypothesis of erratic divine interventions in the steady tenor of things.

A comprehensive answer is thus at hand to the question I began by asking: in what context did Thales asseverate that water was the origin of all things? I agree with McKirahan that water was adduced to give the simplest possible account of the all-pervasive mobility of life in the universe, and of divinity conceived as perpetual life. But it is not enough to assume that Thales' point of departure was simply a vision of life and divinity as all-pervasive. I would argue rather that he was motivated to rethink traditional notions both of life (as limited only to individual living beings as long as they were alive) and of divinity (as limited to individual gods in the Homeric or Hesiodic accounts) by his more fundamental insight into the unity of the *kosmos*, and that this insight flowed from the astronomical observations which led him to conclude that that the motions of the heavenly bodies were in all respects regular and constant. In his theory of earthquakes he even extended this conclusion to the earth, and did so not through direct observation but, in the spirit of science down to the present, through positing the existence of something intelligible but unobserved. [59] His fundamental insight was thus neither mythical nor mystical (as Nietzsche wished to call it [60] but philosophic and, in important senses, scientific. We should not be led astray by his bizarre attribution of life to magnets, or his fuzzy notion that everything is full of gods. These assertions are secondary to the primary work in astronomy, which led him to predicate a universe similar to that of modern physics in that it has no beginning through divine fiat and moves unceasingly according to regularities not subject to divine interruption.

"By their fruits ye shall know them": the essentially scientific nature of Thales' thinking is manifest in the increasingly scientific activity of his "school" of Milesian followers and those who in turn formulated their own systems in response to this school. As we proceed we shall see how many topics in the investigation of nature are held in common by both Milesians and their respondents, and we will be impressed, retrospectively, by how many of those topics the testimonia indicate Thales himself addressed. Not least among the topics on what might be called the Milesian "research agenda" was what seems to be a systematic account of phenomena deemed terrifying by most people. This category includes, in Thales' own work, not only eclipses and earthquakes but also thunder and lightning (*tonitruum sonora miracula* in Apuleius' summary [61], a topic which by the time of Epicurus and his follower Lucretius will have become paradigmatic of all the phenomena of which men need to be given scientific explanations if they are to be freed of religious superstition. In the work of Thales' immediate Milesian and neighboring Ephesian respondents, Xenophanes and Heraclitus, the category of the terrifying also explicitly includes world-destructions, which will be explained as due to unending natural cycles: extremities of wet and dry in Xenophanes, "solstices" of water and fire in Heraclitus. Thales' work is thus a decisive first step in the human effort to overcome obsession with terrifying forces in nature by seeing them as governed by constant laws rather than initiated by capricious gods. His reputation as sage is not undeserved.

6. The Milesians and Xenophanes

The three famous 6th century residents of Miletos - Thales, Anaximander, and Anaximenes - are often said to have constituted a Milesian "school", but this term needs to be defined in order to be useful here. [62] It is impossible to determine whether the association of these men had an institutional setting, or was characterized by either a formal teacher-student relation or the loyalty of adherents of a sect to its founder. The fragmentary nature of the testimonia also make it difficult to say to what extent they shared each other's methods and theories, though some such common ground is evident. Personal predilections as to how one wishes to write the history of Western philosophy may also color the discussion. If, like Nietzsche, one wishes to dramatize the agonistic heroism of each Greek thinker with relation to his predecessors, one will play down schools and play up what seem to be points of difference emphasized by the thinkers themselves. [63] If, like Kahn, one wishes to stress continuity in "the Greek study of nature" from Thales through the Hellenistic schools, one will have no trouble marshalling evidence for "a common set of problems, principles, and solutions".[64] In speaking of a Milesian school here I wish simply to see how many of the problems, or better, projects, held in common over the several centuries for which Kahn makes his case are addressed by more than one of the three Milesians themselves. By a slight broadening of scope one can also usefully inquire how many of these projects are shared by the three great thinkers who grew up in the neighborhood of Miletus, Xenophanes of Colophon, Pythagoras of Samos, and Heraclitus of Ephesus; all would have been exposed to its intellectual currents in their earlier years and well before the destruction of Miletus by the Persians in 494.

Three of the dominant projects for the Milesians are 1) systematically surveying the cosmos, 2) theorizing about its fundamental elements and processes, and 3) redefining traditional understandings of life and divinity. Any account of projects defining a Milesian school is of course bedevilled from the outset by the problems of terminology presented by the doxographers. Again and again one has to ask whether a word used by a doxographer to designate a thinker's concerns was, demonstrably or even possibly, a word that thinker would have used in the doxographer's sense. This is a problem we have already encountered in asking whether Thales could in fact have spoken of a *kosmos*, and we will have to encounter it many times again in the case of each thinker and each term. If, then, I make a preliminary characterization of the Milesian school as undertaking the projects of surveying the cosmos, theorizing about the elements, and redefining divinity, it should be understood that I intend in the case of each thinker to examine what can be known of the Greek words he might have used to designate "cosmos", "elements", "life" and "divinity" respectively. [65]

To these three I would add a fourth project for which no single self-evident Greek term is forthcoming: all three Milesians, not just Thales, gave explanations, in terms strikingly different from traditional religion, of phenomena deemed terrifying by most of their contemporaries. The lack of an obvious term here is best accounted for by the assumption that the common aim of these Milesian explanations was to render the phenomena in question terrifying no longer, by subsuming them into a larger account of the elemental processes by which the cosmos is perpetually governed. It is in relation to what I am calling their "explanations of the terrifying" that I shall ultimately be evaluating their other common projects as well.

A few words at the outset, then, on each of the four projects, to which I shall recur in detail in the cases not only of the Milesians but also of their neighbors. The survey of the cosmos we have already seen extensively in reviewing the traditions about Thales. For all three Milesians it included observation and theorizing about the motions of the heavenly bodies, their origin and composition, their orbits and periodicities, their relation to poles and ecliptic. It also included meteorology, the phenomena of the mid-air between heaven and earth: clouds, mists, wind, rain, thunder, lightning, possibly also rainbows and meteorites Finally the survey takes in the earth itself: the borders of land and sea, the position of the earth, the causes both of its being at rest and of its earthquakes. Whatever terminology they might have used for the object of their survey, it is clear from their consistency in dealing with details that Anaximander and Anaximenes, no less than Thales, set about to theorize about all the phenomena just listed as a unitary set; they wished to see what kinds of explanations might be common to all of them.

The notion of an all-pervasive regularity and predictability is intimately connected, as we have already seen in the case of Thales, with the second project, the redefinition of life and of divinity. Thales had said, according to Diogenes, that the *kosmos* was alive and full of gods, and both Anaximander and Anaximenes also speculated about the nature of the life-stuff and of divinity itself. The topic of life involved the question of the nature of *psykhe* or *pneuma*, the breath without which life was not possible: what was the extent of this breath's location, how was it related to air, how did it cause motion? The topic also led to speculation about the origin of life and its evolution into the kind of human beings we know today. The topic of divinity, likewise, involved the questions of the extent of its location and its relation to one or more elements. But it also involved the potentially more inflammatory question of how true divinity (no longer just *theoi*, gods, but *to theion*, the divine) differed from the accounts given by Homer, Hesiod and the other poets. Did the divine have an origin, and if so was it from one of the principles of the philosophers rather than from the melange of mythic formulations in the *Theogony*? If the capacity to cause motion, in oneself or others, was the essence of life, and divinity was what possessed life most abundantly, would not the highest instance of such capacity, and hence the most divine, be that motion which lacked beginning as well as end -- was in fact birthless as well as deathless? In Xenophanes this inflammatory potential erupts: a poet himself, he attacked the gods of Homer and Hesiod. (Heraclitus did likewise, and included Xenophanes in his attack.[66])

The third project had to do with a kind of explanation made all too familiar through Aristotle's review in the first book of his *Metaphysics*. The "materialist monist" approach there has the earliest thinkers explaining everything in the *kosmos* in terms of what later came to be called "elements": Thales in terms of water, Anaximenes of air, Heraclitus of fire. (In these terms one might add that Xenophanes was a "materialist dualist", since he said all things that come to be are water and earth.[67] Another kind of explanation was in terms of various "opposites" of which material opposites formed only one possible kind: doxographers have the Milesians, like all of their followers, speaking not only in terms of water and fire but also more abstractly in terms of the hot and the cold, the wet and the dry.[68] Finally, in Anaximenes if not earlier, there emerged a kind of explanation in terms not of opposite elements or qualities but of opposite processes: through rarefaction and condensation of air he accounts for the other material elements as well as for the hot and the cold.[69] Associated with all these kinds of explanation through what might be called "elemental processes" was the assumption that the motion underlying the changes they caused was eternal, beginningless as well as endless. Such an assumption of constancy within change at the fundamental level was inseparable from the project of rethinking divinity.

The gods of the poets were not only, as the Homeric formula puts it, "givers of good things" (*doteres eaon*). They also engaged in struggles which devastated the earth as well as the sky, sometimes in a way that was deemed to begin a new world age, as in the battle of Zeus and the Titans in the *Theogony*, sometimes in the course of joining a human fray, as in the final stretch of the fighting in the *Iliad*.[70] Their display of force on such occasions was the supreme instance of the terrifying: Zeus deployed his thunder and lightning to defeat the Titans and Typhon, on which occasion the sea boiled and the whole earth burned; he displayed the same lightning when the Olympians entered the battle at Troy, and the Trojan plain was overcome first by flood and then by fire; in the course of the battle Zeus darkened the sky;[71] Troy's end was caused by Poseidon's earthquake. If the mythic narrative is stripped away what remains here is a catalogue of the terrifying phenomena the Milesians and their successors set about to explain in terms of elemental processes eternally in motion: thunder, lightning, earthquakes, eclipses, and destructions of the world by water and fire. In effect the Milesian project of redefining divinity could not rest complete until it had included in its survey of the cosmos explanations, strictly in its own elemental terms, of the most terrifying phenomena the traditional gods were thought to control - on the control of which, indeed, their power and authority rested. The Olympian thunder had to be stolen.

Thunder, lightning, earthquakes, eclipses: these four topics are well-documented among the three Milesians. The topic we are most concerned with, however, world-destruction by water and fire, is first unambiguously attested only in Xenophanes and Heraclitus. Whether through accidents of preservation or because the agenda of these two neighbors of Miletos had different priorities, the very thinkers who attack Olympian theology most explicitly are also the ones we see foregrounding cataclysm as a central set of topics: by which elements, through what processes, and with what periodicities, could the world be repeatedly destroyed? In taking the measure of the original Milesian achievement, then, it becomes necessary to look in detail at the evidence that these topics were addressed by them too, whether peripherally or centrally. Though the evidence is more ambiguous in Anaximander and Anaximenes than in Xenophanes or Heraclitus, it is there to be found and a number of uniformitarian scholars have not failed to find it and to puzzle over it. A catastrophist reexamination of this evidence can profit from their beginnings. I shall argue, in particular, as have many earlier uniformitarian scholars, that the great fragment on cosmic retribution for which Anaximander is best known cannot be understood unless it is set in the largest possible context, namely, the successive destruction and regeneration of world-orders.

7. Dating and terminology

Since the topic of world-destruction in the Milesian agenda comes alive only when their work is juxtaposed to passages in Xenophanes and Heraclitus which explicitly mention it, I wish to keep these writers in view while moving through the central ideas of Anaximander and Anaximenes. Heraclitus, however, because of the complexity of his thought and the extensiveness of the remains of it available in his own words, deserves special treatment elsewhere; in this essay, therefore, I will be dealing comprehensively with Xenophanes and only obliquely with Heraclitus. The interweaving of Xenophanes in an exposition of the Milesians, moreover, provides a reminder of the continuity that binds together the Ionian beginnings with the Italian schools that follow soon upon them, since Xenophanes probably left the neighborhood of Miletos in his twenties and by his own account lived into his nineties - a life long enough for him to refer to both Thales and Pythagoras, to be attacked by Heraclitus, and to be supposed to be the teacher of Parmenides.[72] Indeed, the Xenophanean argument with which we will be most concerned proceeds from his observations of fossils in sites in both the eastern and western Mediterranean: Paros, Sicily, Malta.

The relative dating of Anaximander and Anaximenes presents few problems. Instead of being simply called Thales' pupil and put the conventional 40 years after him, Anaximander is also called his associate and his birth given at a non-arbitrary 14 years later than Thales'. He is also called a master of Pythagoras, a statement to be evaluated when we come to the latter. His maturity is associated in time with the reign of Polycrates, tyrant of Samos (c. 540-522), with whom Pythagoras' early years are also associated, but he is also said to have died the year of Cyrus' capture of Sardis, 546/5; the two datings are strictly incompatible but suffice for our purposes of securing a mid-Sixth Century time-frame. Anaximenes is dated by the ancient chronographers simply on the basis of his being an associate, presumably younger, of Anaximander.[73]

Anaximander was famous for having been the first to draw a map of the borders of land and sea in the inhabited world, and also a celestial sphere.[74] While Pherecydes of Syros (a master of Pythagoras who looms larger in his legend than does Anaximander) is said by Theopompus to have been "the first to write about nature and the gods" and in the *Suda* to have been "the first to bring out a book in prose", Anaximander is also credited by the *Suda* with a set of book-titles and the great fragment on cosmic retribution we shall discuss later is proof that it came from a prose treatise, for whose priority to Pherecydes Kahn has argued. [75] Diogenes reports that Anaximenes used "simple and economical Ionic diction", which also implies prose writings and stands in contrast to the "rather poetical language" in which the great fragment of Anaximander is said by its citer to have been couched.[76]

The evidence that Anaximander and Anaximenes wrote treatises, and wrote them in prose, encourages us to put under a microscope the terminology they reportedly used, in hopes that some at least of it can emerge as their own and not just the doxographers', and thus take us closer to the heart of their thought. This effort will have all the greater interest when juxtaposed to the study of Xenophanes' doctrines, most of which are quoted in his own verse and thus verifiably his own words.

The term at the heart of Anaximander's thought has indeed the ring of prose: *to apeiron* is an abstraction formed - like "the hot", "the wet", "the divine", or later "the good", "the true" and "the beautiful" - by prefixing the neuter singular definite article *to* to an adjective, in this case the adjective *apeiros*, which would have been familiar to Anaximander's readers (in various forms - *apeiron, apeiritos,* etc.) from Homer's and Hesiod's formulaic descriptions of earth and sea. The adjective seems to be formed by attaching a privative *a-* (which negates what follows it) to a stem consisting of the verbal root **per-*, which connotes forward motion across something to a limit at the other side. The verb *peraino* thus means "to traverse", the adverb *peran* "across", the noun *peirar* "the goal of a crossing". Earth and Sea are *apeiritos* for Homer and Hesiod because both are so immense that no mortal could ever traverse them from end to end, even though both in fact have *peirata*, utmost boundaries, which a goddess can reach and a poet inspired by the Muses can describe.[77] If in the hands of later philosophers like Aristotle *to apeiron* can mean "the infinite" and a whole set of different kinds of infinity can be discussed, we do better with an early thinker like Anaximander, transitional between the formulas of the poets and the terminology of the schools, to translate it as "the Boundless" or "the Indefinite" and to defer to individual contexts the evaluation of what specific kinds of limits Anaximander is claiming that his central principle lacks.

It is noteworthy that in contrasting Anaximenes' account of "the underlying nature" with Anaximander's Theophrastus says that the former too describes it with the adjective *apeiron* but adds that it is not, like Anaximander's principle, *aoriston*, "undefined"; for Anaximenes it is something defined or identifiable (*horismenen*), namely air. Diogenes, more simply, says that Anaximenes' principle was "air and the *apeiron*". In Xenophanes the adjective *apeiros* has more concrete uses which at first are startling. While we see the upper *peiras* (limit) of earth under our feet, underneath it continues *es apeiron* (indefinitely). Not only is the earth *apeiron* but there are *apeiroi* (innumerable) suns and moons, and the sun, which comes into being every day from tiny pieces of fire collected together, in fact moves onward *eis apeiron* (*ad infinitum*) though because of its distance from us it seems to move in a circle. Clearly for Xenophanes *apeiros* was an adjective with a number of specific cosmological uses rather than an abstraction in itself capable of explaining anything.

There is a final use of the adjective *apeiroi* ascribed by the doxographers to all three thinkers: in numerous passages each is said to have maintained that there are *apeiroi kosmoi*, innumerable world-orders. Scholars have debated whether the world-orders so characterized were deemed co-existent, successive or both, and the matter has to be thought through afresh for each thinker who is said to have used the phrase (a list that continues well beyond the Milesians, most notably to the atomists). Successive world-orders require that each in turn be generated and destroyed, and hence the concept entails the central theme of world-destruction which we are pursuing; I shall therefore defer the case-by-case evaluation of this phrase until other evidence for each thinker's concern with world-destruction has been amassed.

8. Surveys of Cosmos

Anaximander is known for a number of cosmological doctrines all of which suggest a mind in search of constant principles whose simplicity was satisfying to the intellect. He is said to have maintained that the earth rested in mid-air because "it is fitting that what is established at the center and equally related to the extremes not be borne up or down or to the sides". [78] The combined reading of several doxographers on his views of the diameters of the earth and the rings or wheels on which the sun, moon and stars were carried suggest that he posited a series of simple proportionate distances using the earth's diameter as a unit: the distance to the stars was 9, to the moon 18, to the sun 27. [79] There is also some (disputed) evidence that he introduced the notion of a celestial sphere to Greek astronomy, which, if true, suggests a fondness for geometrical simplicity in the universe comparable to the arithmetical simplicity of the above-mentioned ratios based on multiples of nine or three. [80]

Anaximander's cosmogony has been the subject of much disagreement among scholars in regard to its general purport and direction. At the beginning of the present *kosmos* he maintained that "something generative of hot and cold was separated off from the eternal". By a further series of separatings off the flame of the heavenly wheels was formed, "like bark around a tree", around the moist air surrounding the earth; the progresssive drying of the earth by the sun caused winds; these winds in turn cause rains, lightning, thunderbolts, "waterspouts" and "hurricanes", and are also responsible, obscurely, for the "turnings of the sun and moon", as though their annual and monthly motions in the heavens were affected by exhalations from the earth.[81] From the primal moisture of the earth the first animals were born, enclosed in barks (like the primal earth in the sphere of flame); at a certain point of evolution they came forth on to the drier part of the world.[82] As Ferrari has pointed out, there is thus a strong parallel between cosmogony and "anthropogony": in both instances Anaximander use the metaphor of bark to describe "a wet and cold substance surrounded by a hot and dry mantle which breaks at a determined moment and originates the present structure of the world".[83] Some scholars have interpreted this cosmogony as evidence that Anaximander was "the first uniformitarian" for believing that "the world arose from the same processes that maintain it"; this is a formulation I will discuss further at the end of this essay. Others have emphasized that since the direction of the present world's evolution is from moister to drier these same processes will eventually destroy it.[84]

In Anaximenes the desire for intellectual simplicity takes a turn more familiar to the modern scientific spirit, in that it seems to be seeking economy and generality in its hypotheses and to be doing so along more strictly materialist lines. [85] There seems to have been a kind of dialectic among the first three Milesians: if Thales posited a single material principle, water, and if Anaximander rejected it as too determinate and posited instead a materially indefinite *apeiron*, then one might say that Anaximenes, eager to take something from both predecessors, kept a single material principle but insisted that it be *apeiron* as well. Instead of water he posited air - infinite air - as the source of all things, but added something Thales' account of water as the source seems to have lacked, namely, terms for the *process* by which a single material principle could change into all other phenomena. This process was condensation and rarefaction: by becoming denser air turned into wind, cloud, water, earth, or stone; by becoming finer (*leptos*, which also means "light" or "subtle") it turned into fire. [86] (This latter term, *leptos*, has a great history ahead of it, since "lightness" or "fineness" will be invoked by Anaxagoras to describe the cosmic Mind and by the Epicureans likewise to account for the subtlety and mobility of the atoms that constitute the minds of men.)

Anaximenes' main achievement seems simply to have been to have added to a single material principle a pair of opposites which are neither material (e.g. water and fire) nor qualitative (e.g. the hot and the cold) but rather complementary operations of a concrete kind, observable everywhere and at all times. Whereas Anaximander's *apeiron* had to have "something productive of the hot and the cold," "separate off" from it at a cosmogonical starting-point, Anaximenes' condensation and rarefaction simply go on forever. Like Anaximander, Anaximenes can use his basic conceptions to account for a continuum of processes in all regions of the cosmos: the earth comes into being as a flat thing "compacted" from air, and being flat it rides on the air as a leaf does; the heavenly bodies come into being from earth as exhalations from it; though most bodies in the heavens are fiery, they carry around some which are made of earth and thus invisible, a proposition which seems to anticipate Anaxagoras in accounting for meteorites.[87]

Finally, when one turns to the details of cosmological speculation in Xenophanes, one is struck by the difficulty of connecting them to what seems to be his most driving idea, that of a completely non-anthropomorphic single god who does not move but who moves everything else. In speaking of neither the earth or the heavens does Xenophanes in fact invoke the action of this god to explain anything. He asserts that everything that comes to be is earth and water,[88] that sun, stars and even rainbows are really clouds,[89] and that there are innumerable suns and moons in different regions of the earth.[90] In an apparent rejection of all three Milesians' theories about the earth's shape and cause of stability he simply asserts that the earth continues to infinity under our feet.[91] The connection of this cosmology of infinites and his theological monism remains obscure, and one sympathizes with Aristotle's remark that as the first to postulate that everything was one Xenophanes made nothing clear but "looking off to the whole of heaven asserted that the One was God".[92]

9. Redefinition of divinity

Aristotle himself is our source for Anaximander's distinctive achievement in postulating a completely non-anthropomorphic principle which controls all things and is divine: "... of the *apeiron* there is no beginning... but it seems to be the beginning of other things, and to encompass everything and steer everything... This principle is the divine, for it is immortal and indestructible, as Anaximander asserts..."[93] Divinity in Anaximander's terms is thus both beginningless and endless, and must have been closely linked in his mind to the "eternal motion" (*aidios kinesis*) through which the opposites "separated off" from the *apeiron* when the present *kosmos* was formed.[94] Indeed, in speaking of this eternal motion Aristotle uses a phrase similar to the one he used for Anaximander's conception of divinity, calling it "immortal and indestructible",[95] and Hippolytus conflates the two passages in saying that Anaximander's *apeiron* was "immortal and unaging, and encompasses all the *kosmoi*".[96] Precisely how the *apeiron* "encompasses all" and "steers all" is nowhere spelled out, and has been much debated. If each *kosmos* is thought of as an ordered system with boundaries, then the simplest sense in which to thinking of the *apeiron* encompassing them is spatial. By being also beginningless and endless, it would thus be outside the limits of each *kosmos* in both time and space. As for its "steering" or "governing" (a term later to be used by Heraclitus, Parmenides, and Anaxagoras), several senses have been imagined for it, none of them spelled out by anything in Anaximander himself. Those who wish to see in him the first proponent of a nature whose order is immanent and "self-regulative", a "universe governed by law", have argued that the *apeiron* steers all things "by having initiated the world in such a way as to provide a continuing rule or law of change", this being the law of mutual retribution among the elements with which the great fragment concerns itself.[97] Others have argued that the "ordering of time" of which the fragment speaks intends time as an objective genitive in which the ordering is done by the *apeiron* itself, and "we are fortunate to have regular periodic phenomena because the Boundless steers them according to time.... Anaximander's physics dod not allow him to frame the idea of an immanent order of nature".[98]

Anaximenes is clearly a follower of Anaximander in his association of his principle, air, both with the divine and with eternal motion. Cicero says explicitly that Anaximenes determined that "air was God... measureless and infinite and always in motion".[99] According to Hippolytus he also said that "gods and things divine came into being" from air, which implies that, like Xenophanes, he had on his agenda a refutation of traditional understandings of the gods' origins and constitution.[100] At one point, in a passage which may use some of his own words, he seems to come close to Anaximander in connecting "encompassing" and "controlling": "As our soul, he says, being air, controls us, so do breath and air encompass the whole *kosmos*".[101] Though the scientific spirit may be uncomfortable with the anthropomorphism implicit in this man-world analogy, it is probable that Anaximenes intended the action of air in a human body as merely one instance of its omnipresent operation, and hence to strengthen the simplicity of his explanatory principle.

In Xenophanes we encounter the first explicit rejection of the Olympians' battles with Titans, Giants and Centaurs as fit themes for the songs men should sing at altar or symposium:

> First cheerful men must hymn the god
> with words that speak well and with pure stories...
> The man who as he drinks brings fine deeds to light...
> does not take as his theme the battle of Titans or Giants
> or Centaurs, those fictions of men of earlier days... [102]

Here as elsewhere Xenophanes appeals to a criterion of what is "fit" to say of "the god", by which theomachies are deemed unfit. The themes of motion and control in the Milesian accounts of the divine are each given crucial variations, and he seems to be as polemical against them as he is in his cosmological details, and as he is against traditional Olympian views of the gods. His denial of anthropomorphism we have in his own verse, and it could not be more explicit: there is "one God, greatest among gods and men, in no respect resembling mortals, either in body or in thought".[103] We also have in his own words the crucial doctrine that God is both unmoving and the cause of motion in everything else: "He remains always in the same place, in no way moving; nor does it befit him to betake himself in one direction at one time and in another at another, but without exertion, by the thought of his mind, he shakes everything".[104] Here the question, obscure in Anaximander, how the divine "steers" everything else, is addressed in remarkably concrete language, reminiscent of the Homeric Zeus shaking Olympos (both the vault of the heavens and the mountain on which the gods dwell) with a nod of his head.[105] Stirring as this first appearance of an unambiguous monotheism may be in Greek thought, it is not destined to be readily embraced by the Greek philosophical tradition: Parmenides' unmoving Being is separated from the world of motion as truth is separated from falsehood, and Aristotle's unmoved Mover is the cause of the motions of everything else only as the beloved is the cause of the motions of the lover. Like Anaxagoras' Mind, moreover, Xenophanes seems to invoke his God as little as possible to explain the actual workings of the rest of his cosmology, which exhibits all phenomenon taking rise from earth and water and gives no examples whatsoever of the divine mind "shaking" anything.

10. Elemental processes and world-destruction

It is possible to say that each of these thinkers was a uniformitarian, in the specific sense that each sought evidence of the continuous operation of a few simple elemental processes throughout the history of the present *kosmos* and in heaven and earth alike. It needs to be stressed all the more, therefore, that each in different ways gives evidence of the capacity to imagine that the same processes which brought the present world-order into being and are presently operative in it would also be capable of bringing to an end the balance on which it depends if it is to be truly an "order".[106] In their specific accounts of the elemental processes involved none of the three thinkers speaks explicitly of the destruction of the present *kosmos* as a whole; this, I will argue, is a notion to be looked for in other areas of the doxographers' account of their thought. Each does, however, speak of processes presently at work capable of leading to destruction on a large scale, and the language of the doxographers' account is therefore worth attending to closely.

Some thinkers, according to Aristotle in his *Meteorology*, thought that the earth, which was originally all moist, is being dried up by the sun, and that as a result the sea is shrinking and will end up by being entirely dry. A commentator says that Anaximander and Diogenes (a later theorist of air in the tradition of Anaximenes) were of this opinion, whereas Aristotle mentions only Democritus, who explicitly thought that the drying up of the sea would bring the world to an end.[107] Democritus was the author of a work entitled "The Great Year, or Astronomy", and Aristotle himself speaks of the Great Year in language glossed by Censorinus as follows: "The winter of this year is a supreme cataclysm, which we call the Deluge, and its summer is the

Conflagration, that is, the burning up of the world".[108] By the time of Aristotle the terms in which cataclysm was discussed had become highly schematic: two elemental opposites, water (the cold and the wet) versus fire (the hot and the dry), successively overwhelmed the world at fixed temporal cycles analogous to the fixed recurrence of summer and winter solstices. How far Anaximander himself had gone towards embracing all the elements of this schematism is unclear from these accounts of his observation that the sea was drying up and his projection that someday it would be completely dry. (It is a commonplace of modern scholarship to assume that Anaximander's projection was based on his observation of the gradual silting up of the area around the harbor of Miletos by the river Meander.) We should note, in any case, that the destruction of the present *kosmos*, defined specifically as an ordered relation of the heavenly bodies to an earth populated with living beings, follows necessarily from his theory of the way "the hot and the cold" proceeded in cosmogony. The hot and fiery sun and other heavenly bodies were separated off from the cold and moist earth; the sun's fire proceeded to produce exhalations from the earth which cause the motions of the heavenly bodies, and the sun's progressive drying of the earth caused living creatures to emerge from it. When the sun completely dries out the moisture of the earth there will be no more exhalations to cause the heavenly bodies to "turn", and it is also impossible to imagine animals continuing to live. The present *kosmos* will be destroyed in the sense that the orderly movements of heavenly bodies around the earth would cease, and the earth itself would be no longer populated. As Freudenthal has argued, if the Great Year was invoked by other ancient thinkers as the ultimate periodic cycle which no more destroyed the *kosmos* than did the lesser annual and daily periodicities, then Anaximander cannot have subscribed to it, since it implies that "once it came into existence, the world too, and not only the Boundless, will exist eternally". [109]

Anaximenes also speaks of alternations of the wet and the dry producing catastrophic effects, but in terms that do not explicitly amount to a world-destruction. The source is again Aristotle's *Meteorology*: "Anaximenes says that the earth, through being drenched and dried, breaks up, and that it undergoes seismic activity as a result of these peaks that are broken off and cave in. Therefore earthquakes happen in times both of drought and again of torrential rains; for in droughts, as has been stated, it dries up and caves in, and when it is made too moist by the waters it crumbles apart".[110] All we can say of this passage in isolation is that it forms part of the Milesian agenda of explaining terrifying phenomena such as earthquakes in terms of elemental processes which can be expected to recur periodically.

It is Xenophanes whose theory of two elements - earth and water, out of which he believed everthing else came to be - led him to conclude that a cycle between generation and destruction occurred as a specific result of the interactions he observed between them. Here we have not his own words but Hippolytus' detailed report:

> Xenophanes thinks that a mixture of the earth with the sea is occurring, and that over the course of time the earth is dissolved by the moist, and he claims that he has the following kinds of demonstrations: shells are found inland and on mountains, and he says that in the quarries in Syracuse the impression of a fish and of seaweed has been found, and the impression of a bay-leaf was found in Paros in the depth of the rock, and in Malta flat shapes of all kinds of marine objects. These, he says, came about when everything was covered with mud long ago, and the impression was dried in the mud. All mankind is destroyed whenever the earth, having been carried down into the sea, becomes mud; then there is another beginning of coming-to-be, and this foundation happens in all the *kosmoi*.[111]

The use of observations and inferences from all parts of the Mediterranean is impressively scientific. The cycle posited occurs between periods of cataclysmic blending of the two principle elements - when earth and water become mud everywhere all humans are destroyed - and fertile separation of them - as the mud dries out life forms evolve.

It is to be noted that in none of these three passages do we see an explicit connection established between the putative cycles to be inferred from the phenomena in question and the highest "principle" on which the thinker's world-view depends. Anaximander's *apeiron* somehow had within it something "generative of the hot and the cold" which separated out from it, but the passage in pseudo-Plutarch does not proceed the whole way from that cosmogony to the final phrase at which the sea is dried up. Anaximenes' account of earthquakes is entirely in terms of water and fire (torrents and droughts) and does not mention air. Xenophanes' cycles involving earth and water have nothing to do with the mind of the one unmoving God who "shakes" everything. To come closer to the relation of cycles of generation and destruction to the rest of their thinking we must return to what the doxographers say about their accounts of *kosmos* and *kosmoi*, the world or worlds which they say are generated and destroyed.

11. *Apeiroi Kosmoi*

Statements that the each of our three thinkers believed in innumerable worlds, *apeiroi kosmoi*, are provided by the doxographers, and in the case of Anaximander they exist in abundance.[112] Some scholars have proceeded by asking whether these innumerable worlds were thought of as being coexistent or successive, and have shown a desire to get rid of the whole topic by conjecturing that only the former is meant and that that view is really only an anachronistic retrojection onto the Milesians of views held by the atomists. Leucippus and Democritus posited infinite void and infinite atoms colliding with each other eternally, and proceeded to argue from these axioms that infinite *kosmoi* existed, constantly being generated by the combinations of these atoms in infinite void and constantly being destroyed when *kosmoi*, like atoms, collided with each other. These scholars have conjectured that when most doxographers found references to destruction of the *kosmos* or *kosmoi* in this or that Milesian they merely assumed he was like the atomists in assuming an infinity of coexistent *kosmoi*. [113]

The author on whom we depend for the Anaximander fragment, Simplicius, makes finer distinctions than the other doxographers, and goes some way towards defining what is really at issue. He distinguishes those who, like Anaximander and the Atomists, posit innumerable *kosmoi* and those who, like Anaximenes and Heraclitus, "say that the one *kosmos* comes into being and is destroyed"; the latter "affirm that there is always a *kosmos* but that it is not in fact always the same one, but rather becomes different at different times according to certain temporal periodicities". [114] We thus have clear evidence that more than one of the ancients maintained that there was an interminable series of world-destructions; when one world-order was destroyed it was suceeded by the coming-into-being of another, *ad infinitum*. The same general idea of recurrent world-destruction is inseparable from the *apeiroi kosmoi* which Simplicius attributes to Anaximander: "Those who posit *kosmoi* unlimited in number, such as the followers of Anaximander and Leukippus and Democritus and later those around Epicurus, suppose that they come into being and are destroyed indefinitely [*es apeiron*], some always coming into being while others are being destroyed, and they say that motion is eternal, for without motion there is no coming-into-being or destruction".[115] The possibility that later doxographers were confused in attributing *coexistent kosmoi* to Anaximander is thus an irrelevant side-issue, a question as to his original terminology which will probably remain irresoluble. What is unambiguous in the testimonia is that Anaximander believed in *successive* destruction of *kosmoi*, and without acknowledging this evidence we will be unable to situate his fragment in its proper context.

Given that modern scholars presuppose a large capacity for confusion on the part of the doxographers, and themselves claim a large authority to correct the mistakes of the ancients, one has to ask what is driving some of them to such insistence on the necessity to rewrite the ancient record. A glimpse into the uniformitarian assumptions underlying such scholars' arguments is to be had in the following sentences in Kirk, Raven and Schofield:

> If coexistent worlds might be suggested to some people... by the heavenly bodies, there is nothing whatever in 'the appearance of nature' to suggest *successive* worlds... as distinct from successive changes in the state of the one continuing world. These last are envisaged in the mythical catastrophes by fire and flood described in Plato's *Timaeus*, 22c-e, or in Deucalion's flood, and were to some extent suggested by natural phenomena [*reference is made to discussion of the passages just discussed on drying up of the sea in Anaximander and on destruction of humanity by mud in Xenophanes*]. But there was no reason to assume that the whole world was going to be destroyed, and that if destroyed it would be succeeded by another. It would be contrary to the whole mythical background of Greek thought and to the dictates of common sense to believe in a cycle of separate worlds; and their appearance in Anaximander would be most surprising.[116]

One should begin by setting aside the uniformitarian "appeal to common sense" here. Given the overwhelming preponderance of mythical cultures believing in cycles of separate worlds one should also be skeptical of any appeal to "the whole mythical background of Greek thought" to foreclose discussion. Ferrari, arguing against this kind of denial of world-destruction, has pointed out that even in Hesiod's account of the four generations of men in the *Works and Days* each generation (unlike in the myths of Deucalion or Noah) is "produced *ex novo* from the preceding, by the gods in general and Zeus in particular, after the total extinction or destruction of the preceding".[117] To be fair to scholars maintaining the position of Kirk, Raven and Schofield one must note their readiness to adduce other Greek instances of belief in cataclysmic destruction, both in the philosophers (Kahn mentions not only the passage in the *Timaeus* but also the frequent floods at *Laws* 677a and cycles of cosmic transformations at *Politicus* 269a, as well as Democritus and Aristotle on the Great Year) and in the myths of Phaethon and Deucalion.[118] At root the problem would seem to be that these uniformitarian scholars can acknowledge the existence of ancient accounts of cataclysms but can imagine no real human experience in which these accounts might be grounded; they therefore dutifully adduce the accounts, then bracket them and proceed as though they had no real weight in deciding the interpretation of ancient texts.

A fresh start in this controversy might be usefully made by reminding ourselves of the evolution of the word *kosmos* itself from the time of Thales to the time of the late doxographers. By the time of the latter it is equivalent to the Latin *mundus* and the English "world" or "world-system"; at the time of the former, as we have seen, it would still mean, more concretely, any pattern of well-ordered regularity such as one sees in contemplating the cyclical motions of the heavenly bodies around the earth or the mutual boundaries of sea and land. As Cornford points out, it was possible for the New Testament author of *2 Peter* to write of Noah's Flood as a cataclysmic destruction of "the then *kosmos*" which nevertheless left a habitable world of some kind to human survivors. [119] What we are specifically examining in the Milesian school and Xenophanes is the implied relation between periodic preponderances of one element over another on earth and the destruction of the well-ordered regularities of the motions of the bodies in the heavens. Such a relation is not available to the "common sense" of uniformitarians and they are not prepared to find evidence for it in "the appearance of nature". It is precisely, however, the object of catastrophist research.

Evidence that uniformitarian scholarship finds it difficult to register the early notion of *kosmos* still embedded in the late doxographers' phrases can be had in the phrases they spend the most time trying to make sense of. In introducing the passage in which Anaximander's fragment is embedded Simplicius, relying on Theophrastus, speaks of Anaximander saying that out of the *apeiron* came into being "all the heavens [*ouranoi*] and all the world-orders [*kosmoi*] in them", and the phrase has been deemed so unusual for a Peripatetic that it must have been taken by Theophrastus almost verbatim from the book of Anaximander.[120] There would be nothing unusual, in fact, for Anaximander, an observer of regularities in the heavens and a pupil and associate of Thales, to speak of "the heavens and the systems of regularity within them".

Similarly, Xenophanes has puzzled his modern interpreters by the final phrase, in the long passage from Hippolytos just quoted, stating that the new coming-to-be that occurs after each destruction of humanity by mud "happens in all the *kosmoi*". This has been glossed by Kirk, Raven and Schofield as "properly 'world-arrangements', i.e. of the earth's surface". But since Xenophanes also believed that there are many suns and moons according to regions, sections and zones of the earth, that in fact "there are innumerable suns and moons, and all things are made of earth", the notion of a "world-arrangement" cannot be limited to earth and must include the heavenly bodies as well.[121] Like Anaximander, Xenophanes is simply saying that patterns of order involving both heaven and earth come into being and are destroyed, and the Hippolytus passage should not be tortured into saying something more restricted.

To the discomfort of many uniformitarians, Anaximander, Anaximenes and Xenophanes all appear on doxographical lists of those who believed that there were *apeiroi kosmoi*; on one list, that of Stobaeus, all three appear together. As we have seen, the list given by our source for Anaximander's fragment, Simplicius, includes Anaximenes and goes out of its way to stress that the doctrine concerns the periodicities with which *successive kosmoi* replace each other: "All those who make the one *kosmos* generable and destructible - such as Anaximenes, Heraclitus, Diogenes, and later the Stoics - say that there is always a *kosmos* but that it is not always the same one; rather it becomes now one *kosmos* now another, according to certain periodicities of time". In the company of the eight passages in which Anaximander is said to have believed in *apeiroi kosmoi* Kahn places Aetius' list of all those who more specifically believed that the *kosmos* was subject to destruction, namely, Anaximander, Anaximenes, Anaxagoras, Archelaus, Diogenes, and Leucippus. Kahn is also honest enough to make his own list of all the seven ancient passages about Anaximander alone in which he is said to have maintained *kosmos* or *kosmoi* were destructible.[122]

An unprejudiced conclusion on the basis of this evidence is that all three of the thinkers with whom we are concerned believed both that there were innumerable *kosmoi* and that these *kosmoi* succeeded each other as one was destroyed and another in turn generated. Whether or not they also believed in an infinity of coexistent *kosmoi* does not affect my argument and need not be pursued further. What is essential is to understand what kinds of destruction they believed brought a *kosmos* to an end, and on what basis they also asserted that the succession of *kosmoi* was unending and indeed periodic. We have made a beginning of such understanding in our survey of the passages in which they spoke of elements moving through phases or cycles of preponderance leading to catastrophic destruction of the world inhabited by humans. But for the most fundamental philosophical principles on which the assumption of *periodicity* of destruction is based we must turn at last to Anaximander's great fragment itself.

12. The Fragment and its contexts

Much is made of "the extant fragment" of Anaximander quoted by Simplicius, both because of its vivid archaic metaphors and because, given testimony that Anaximander actually wrote a book, it is exciting to have a glimpse into it, no matter how brief. The passage of Simplicius in which the fragment is embedded is based on Theophrastus, who differs little from his master Aristotle in what he says about Anaximander's *apeiron*, and who presumably had Anaximander's book before him to quote from. Two other closely parallel passages on the *apeiron*, from Hippolytus and the pseudo-Plutarchan *Stromateis*, are also based on Theophrastus, and both also use phrases archaic and metaphorical enough to suggest that they too come ultimately from Anaximander's book.[123] Each of the three passages needs to be quoted and considered at a length sufficient for the context its author had in mind to emerge. This is, unfortunately, not often done. The same modern scholars who accuse ancient doxographers of distortion through selective quotation also build their own arguments out of selective quotation from the doxographers. These scholars also lament how the presuppositions of the philosophical schools in which a doxographer was trained - most often the Peripatetic - cause him to distort his early pre-Socratic source, but such distortion is also often evident in the uniformitarian "school" to which the scholars themselves belong, so thoroughgoingly they are not aware that any other alternative schools exist.

I shall therefore quote enough continuous text from each author to let his particular context for discussing Anaximander emerge. We can then try to understand what context Anaximander himself might have had in mind by seeing what contexts are common to the three passages in which his *apeiron* is adduced and relating that common context to the rest of his cosmogony.

The passage in the *Stromateis* moves from the *apeiron* itself to details of the cosmogony of the present *kosmos*. For the sake of bringing everything into consistent English, I will translate *kosmos* as "world-order" each time it appears in all three passages, the neuter substantive *to apeiron* as "the Boundless", and the plural adjective *apeiroi* as "boundless".

> Anaximander, the companion of Thales, said that the Boundless contained the whole cause of the coming into being and destruction of the totality of things; from it, he says, are separated off the heavens and in general all the boundless world-orders. He maintained that their destruction occurred, and much earlier their coming into being, from a boundless eternity, all of the world-orders recurring in cycles. He says that the earth is cylindrical in its shape, and that its depth is a third of its breadth. He says that what is from eternity generative of hot and cold was separated off at the coming into being of the present world-order, and that a certain sphere of fire was formed out of this around the air surrounding the earth, like bark around a tree; when it was broken off and shut up in certain circles, the sun and the moon and the stars were formed. He also says that in the beginning humans were generated out of animals of another species.[124]

Taken as a whole, this passage is discussing the *apeiron* to make two related points: that innumerable *kosmoi* come into being by separating off from the Boundless and are subsequently destroyed, and that the present *kosmos* (*toude tou kosmou,* the demonstrative *toude* referring to the "here and now") began when something generative of the hot and the cold separated off from the eternal and began processes which led both to the heavenly bodies we see around the earth and to the birth of human beings. As we have seen earlier, passages in other authors have been consistently combined with this one by scholars to indicate that the continuing of these processes will lead to the drying up of moisture on the earth and hence to the destruction of the present *kosmos.* Taken as a whole these processes would be an instance of a world-order coming into being and being destroyed, something which, in the language of the passage, "recurs in cycles".

The passage in Hippolytus has similar concerns:

> This man [Anaximander] said that the principle and element of the things that exist was the Boundless, having been the first to use this term "principle". In addition he said that the motion in which it comes about that the heavens come into being was eternal. He said that the principle of the things that exist was a certain nature of the Boundless, out of which come into being the heavens and the world-order in them, and that this principle was eternal and unaging, and that it surrounds all the world-orders. He speaks of time as though coming into being and existing and being destroyed were limited.[125]

Here too the topic is the coming into being *and destruction* of plural *kosmoi* according to some kind of scheme of temporal limitation. The *apeiron* surrounds all the world-orders, and the temporal limitations of their existence presumably have something to do with its eternal motion.

Finally, the complete passage in which the famous fragment is embedded is structured as a symmetrical account of the sources from which first the generation of the things that make up a *kosmos* and then their destruction occurs:

> Anaximander... said that the Boundless was the principle and element of the things that exist, having been the first to introduce this term "principle". He says that the principle is neither water not any of the other of the so-called elements, but some other nature which is boundless, out of which come into being the heavens and the world-orders in them. Out of those things out of which coming into being occurs for the things that exist, into those things again does the destruction of the things that exist come about according to necessity. For they make amends and reparation to each other for their offense, according to the ordering of time (speaking of them thus in rather poetical terms). It is clear that having observed the transformation into each other of the four elements he did not think any of them worthy to be made the substratum of the rest, but something else besides them.[126]

As in the other two passages, the general context is the complementary coming into being of plural *kosmoi* out of the *apeiron* and their destruction or reabsorption back into it. Destruction is causally linked to coming into being (*"For* they make amends") as Simplicius moves to what he is specifically interested in, the necessity and temporal ordering of the process by which everything that comes into being out of the elements is destroyed back into them.[127] For the more complex entities (from heavenly bodies to human beings) whose arrangement constitutes the order of any particular *kosmos*, coming into being is out of the same source into which these entities will later be destroyed. The process is subject to a temporal ordering, and by the time it is over amends have been made for the "injustice" committed by any one element in its preponderance over others.

The common context of all three passages is thus indisputably the coming into being of *kosmoi* out of the *apeiron* and their destruction back into it; the complementary nouns *genesis* and *phthora* occur in all three, and so does the plural *kosmoi*. Given the common dependence of all three passages on Theophrastus, and the evidence in each that he has before him Anaximander's own book and is at times using Anaximander's own phrasing, it is only fair to conclude, in the absence of decisive evidence to the contrary, that for Anaximander himself the coming into being and destruction of *kosmoi* was the theme which led him to posit an *apeiron* and to relate the existing beings of the present *kosmos* back to it. This he presumably did by speaking of "the hot and the cold", and very possibly other opposites so fundamental as to be deemed elements by later doxographers. Separation off of these opposites from the *apeiron* begins the process by which a *kosmos* evolves its more complex entities into an orderly arrangement, and that orderly arrangement comes to an end when these opposites are reabsorbed into the *apeiron* eternally surrounding them.

13. Time and Periodicity

What remain unclear in this consistent reading of the three passages are the phrases having to do with *time* in each. What did "pseudo-Plutarch" understand when he spoke of "all the *kosmoi* recurring in cycles" (*anakykloumenon panton auton*)? What was Hippolytus referring to when he said that Anaximander spoke "of time as though coming into being and existing and destruction were limited"? What is the "ordering of time" (*khronou taxin*) in Simplicius, and what are the "periodicities of time" with which he speaks of one *kosmos* being destroyed and replaced by another in Anaximenes and later thinkers?[128] Here uniformitarians are ready with answers which seem to them self-evident. If Anaximander spoke of "the hot and the cold", or "the wet and the dry", then surely his examples are to hand: in speaking of the ordering of time he must have meant the alternations of day and night, of summer and winter.[129] (The latter work particularly well for this thesis in the Mediterranean, with its rainy winter and dry summers, cold and wet alternating annually with hot and dry.) The truth of the matter, however, is that there is not a word in the testimonia about day and night or summer and winter; and as we have seen, the *apeiron* is being adduced to explain not these phenomena but rather the generation and destruction of whole *kosmoi*.

It is to the credit of a uniformitarian like Kahn that he does not limit his answers to this question to observable instances of the interaction of elemental opposites. He is ready to entertain the possibility "that Anaximander projected this pattern upon a still more majestic screen, and spoke (like Plato and his followers) of a Magnus Annus, in which the great astronomical cycles are to be accompanied by catastrophic transformations of the earth".[130] Kirk, Raven and Schofield, similarly, concede that Anaximander was familiar with the great legendary periods of fire and flood, in the ages of Phaethon and Deucalion; impressed by the recession of the sea from the Ionian coastline he might well have applied such periods to the whole history of the earth.[131] Where these scholars draw the line, however, is in allowing that such "great astronomical cycles" or "great legendary periods" were deemed by Anaximander to have ended up by destroying the whole *kosmos* within which they were occurring. They wish Anaximander to believe in an eternal *kosmos* "which takes the form of a rhythmically repeated cycle, executed by a system in dynamic equilibrium" whose periodicity is "the oldest formula of natural law". It is here that their uniformitarianism is leading them to distort the evidence in the same way that a Peripatetic believing in a single eternal *kosmos* is compelled to argue against the innumerable worlds of so many pre-Socratic predecessors. Fortunately, even without being explicitly catastrophist, more recent scholars such as Ferrari and Freudenthal have found the evidence for Anaximander's belief in an eventual destruction of the present *kosmos* by desiccation too compelling to let uniformitarian prejudices hold sway.[132]

Perhaps the ultimate cause for the distortions practiced by such distinguished scholars as Kirk and Kahn is that, in their admiration for Anaximander, they wish to claim him as spiritual founder for what they most esteem in modern science itself, namely, its insistence on a single universe governed by a single set of unchanging laws. In so doing they may be said to have committed the same error of projecting later beliefs onto a spiritual founder as modern scholars are so quick to suppose later doxographers did onto Thales. The history of early Greek thought from Thales to Democritus which I am attempting to reconstruct here from a catastrophist perspective need not expect to encounter in any one thinker a single giant leap from traumatized mythical belief to modern scientific explanation of all phenomena in terms of constant laws. If modern scholars wish to find a pre-Socratic insisting that there is in fact only one eternal *kosmos* with a single law (*logos*) underlying it they need only wait a couple of generations for Heraclitus. If they wish to find a completely mechanized and deanthropomorphized account of nature they need only wait a couple of generations more for Democritus.

If, like McKirahan, they wish to find a spiritual ancestor for uniformitarianism itself, they would do well to consider Stephen Jay Gould's account of the very different components which Lyell, its originator, combined into it. Gould distinguishes four kinds of uniformity given a common name by Lyell in a "hodgepodge of claims": uniformity of law, of process, of rate, of configuration. Uniformity of *law* holds that natural laws are constant in space and time. It is an "a priori claim of method that scientists must make in order to proceed with any analysis of the past",[133] and, as we have seen, scholars like Kahn are eager to claim that Anaximander was the first to assert this claim, while scholars like Freudenthal find in the Boundless vestiges of a divinity still acting from outside the natural world it encompasses and steering it by the periodicities it ordains. Uniformity of *process* insists that "processes now operating to mold the earth's surface should be invoked to explain the events of the past".[134] This is the sense in which McKirahan wishes to claim Anaximander as the first uniformitarian, as one who, like the geological school which came to prevail over catastrophism in the 19th century, "held that processes found today, such as erosion and volcanic activity, are responsible for the geological features of the earth";[135] in this sense Anaximander and modern uniformitarians are both making categorical assertions which block the way to catastrophist speculation. Uniformity of *rate* holds that "geological change is slow, gradual and steady, not cataclysmic or paroxysmal".[136] Whether Anaximander was this kind of uniformitarian depends entirely on how you imagine he meant the phrase *kata tou khronou taxin,* "according to the ordering of time"; this is nowhere spelled out. Finally, uniformity of configuration holds that "the earth has been fundamentally the same since its formation"; Lyell actually believed this, and in order to reconcile it with evidence of direction in the development of species he "supposed that the entire fossil record represents but one part of a 'great year' - a grand cycle that will occur again". [137] Anaximander emphatically did not believe in uniformity of configuration; he observed desiccation and extrapolated it ineluctably to the destruction of the present *kosmos.* His grand cycles, according to the *Stromateis*, involved whole *kosmoi.*

The first of Lyell's four uniformities, that of law, is still essential to science. The last of them, that of configuration, was never seriously pursued by science. Uniformity of process and rate have been disowned by science more and more explicitly over the last few decades of speculation about earth-impacts of asteroids and comets. Much of this speculation has tended, in fact, to find evidence for periodicities in the impact-caused mass extinctions the earth has undergone. For some astronomers the evidence of short-term periodicities (thousands of years rather than tens of thousands or millions) is strong enough to call for our trying to establish them precisely so as to make provisions against impact episodes in the foreseeable human future.[138]

As for the Milesians and their immediate successors - Xenophanes, Pythagoras, Heraclitus - the term uniformitarianism only unnecessarily confuses our evaluation of their achievement. They were at best proto-scientific. The process of deanthropomorphizing explanations of nature occurred in successive phases, and along the way many schematisms not based on testable observations were propounded. We should find it not dismaying, but only natural, if some of these thinkers' anthropomorphisms and schematisms still served the psychological function of allaying anxiety about the recurrence of catastrophic events in the sky capable of destroying the civilized world on the earth. Anaximander's divine Boundless, ordering the generation and destruction of *kosmoi* in the cycles by which it "encompassed" and "steered" all things, gives signs of still being devised to perform such psychological functions. If the destruction of the world by desiccation, and the generation of a new world later, are both subject to the governance of a divine being who arranges time according to ineluctable periodicities (both *kata to khreon* and *kata tou khronou taxin*, "according to necessity" and "according to [its] ordering of time"), then mankind need not concern itself with the propitiatory sacrifices to erratic gods which the traditional Olympian religion enjoined. The destructions will occur according to the periodicities an indefinable Being has established, and they may very well be in a very distant future, as the slowness of the recession of the coastline at Miletos might suggest.

Compared to the invocation of an indefinable Being who governed all things, Anaximenes' explanation of everything in terms of the condensation and rarefaction of a divine Air can be said to be constitute some progress towards "hard science". Since he is listed by Simplicius as one of those who believed that there always was a *kosmos* but that as one was destroyed it was succeeded by another, we must admit that he probably gave an account of the destruction of the present *kosmos* but that we know nothing about how he would have gone about it. In Xenophanes, on the other hand, there seems to be a split between scientific explanation - observation of fossils leading to conclusions about periodic destruction and rebirth of humanity from mud - and theological argument - there must be only one motionless God causing everything else. For moderns who find monotheism and science compatible Xenophanes will be a distinct sign of progress; for those who find science adequate to obviate all theories of divinity interacting with the world, he will be a retrograde step.

The Milesian proto-scientific accounts of cyclical processes caused only by elemental interaction were doing important work in 'explaining away' the myths of Phaethon and Deucalion, the Titanomachy and the battle of the gods at Troy, not to mention thunder and earthquakes, lightning and eclipses. They were enabling their subscribers to believe that the old stories were only primitive versions of a belief that previous world-destructions occurred, and so to distance themselves from the myths and the rituals based on them. But they were also presumably enabling their subscribers to believe that recurrences of world-destruction might occur only at vast distances of time in the future. (By the time - it remains unclear when exactly - the Great Year of 10,800 years has been introduced, each of its 360 "days" being a generation of 30 years, the winter of deluge and summer of conflagration have been located at secure temporal distances indeed.) It will constitute a sufficiently new analysis of these thinkers if we continue to take seriously both the terror with which the traditional story-telling about catastrophes was still charged and the speed with which memory of their true nature was being distorted from generation to generation. It is possible to 'explain away' the terrifying and still distort it. It is even possible to seem scientific in doing both - as uniformitarianism, with its "hodgepodge of claims", seemed for a century and a half.

Abbreviations

DK = Diels, H. & Kranz, H. 1951. *Die Fragmente der Vorsokratiker*, 6th ed. Berlin.

DL = Diogenes Laertius. 1972. *Lives of Eminent Philosophers*, trans. R. D. Hicks. Cambridge, Mass.

KRS = Kirk, G. S., Raven, J. E., & Schofield, M. 1983. *The Presocratic Philosophers:A Critical History with a Selection of Texts.*Cambridge.

Bibliography

Allen, T.W., 1912. *Homeri Opera, Tomus V.* Oxford.

Barnes, J., 1982. *The Presocratic Philosophers.* London.

Burkert, Walter. 1963. "Iranisches bei Anaximandros". *Rheinisches Museum,*106, pp. 97-134.

Cherniss, H. F. 1935. *Aristotle's Criticism of Presocratic Philosophy.* Baltimore.

Clube, V., and Napier, B., 1982. *The Cosmic Serpent: A Catastrophist View of Earth History.* New York.

Clube, V., and Napier, B., 1990. *The Cosmic Winter.* Oxford.

Cornford, F. M. 1934. "Innumerable Worlds in Presocratic Philosophy." *Classical Quarterly* XXVIII (1934), pp. 1-16.

DeGrazia, A., 1983. *Homo Schizo I: Human and Cultural Hologenesis.*Princeton.

DeGrazia, A. & Milton, E., 1984. *Solaria Binaria.* Princeton.

Dicks, D.R., 1959. "Thales." *Classical Quarterly* 9, pp. 294-309.

Diller, H., 1965. "Der vorphilosophische Gebrauch von *Kosmos* u. *Kosmein." Festchrift für Bruno Snell.* Munich.

Diller, H., 1966. "Solstices, Equinoxes, and the Presocratics". *Journal of Hellenic Studies* 86, pp. 26-40.

Ferrari, F., 1979. "Su Anassimandro B1." *La Parola del Passato* 184 (1979), pp. 118-126.

Freudenthal, G., 1986. "The Theory of the Opposites and an Ordered Universe: Physics and Metaphysics in Anaximander." *Phronesis* XXXI No. 3 (1986), pp. 198-228.

Furley, D., 1987. *The Greek Cosmologists. Vol. 1: The Formation of the Atomic Theory and its Earliest Critics.* Cambridge.

Gennaro, E., 1970. "La caduta speculativa di Taleti." Florence.

Gould, S.J., 1977. *Ever since Darwin: Reflections in Natural History.* Norton.

Hartner, W., 1969. "Eclipse Records and Thales' Prediction of a Solar Eclipse" *Centaurus* 14, pp. 60-71.

Heinsohn, G., 1988. *Die Sumerer gab es nicht.* Frankfurt/M.

Heinsohn, G., 1990. "Destruction Layers in Archaeological Sites: The Stratigraphy of Armageddon", in *Catastrophism 2000*, ed. M. Zysman & C. Whelton. Toronto.

Heinsohn, G., 1995. "The Restoration of Ancient History", paper presented at the Society for Historial Research, July 8, 1995.

Heinsohn, G. & Illig, H. 1990. *Wann lebten die Pharaonen?* Frankfurt/M.

Hölscher, U., 1968. *Anfängliches Fragen*. Göttingen.

Hoyle, F., 1993. *The Origin of the Universe and the Origin of Religion*. Wakefield.

Kahn, C., 1970. "Early Greek Astronomy". *Journal of Hellenic Studies* 90 (1970), pp. 99-116.

Kahn, C., 1985. *Anaximander and the Origins of Greek Cosmology*. Centrum Philadelphia, Philadelphia.

Kirk, G.S., 1955. "Some Problems in Anaximander." *Classical Quartlerly* 5 (1955), pp. 21-38.

Lesher, J. H., 1992. *Xenophanes of Colophon: Fragments: A Text and Translation with a Commentary*. Toronto.

Lewis, J.S., 1996. *Rain of Iron and Ice: The Very Real Threat of Comet and Asteroid Bombardment*. Reading, Mass.

McKirahan, R.D., Jr., 1994. *Philosophy before Socrates*. Indianapolis.

Mosshammer, A.A., 1976. "The Epoch of the Seven Sages". *California Studies in Classical Antiquity* 9, pp. 165-80.

-- 1981. "Thales' Eclipse." *Transactions of the American Philological Association* 111, pp. 145-55.

Mourelatos, A.P.D., 1988. "'X Is Really Y': Ionian Origins of a Thought Pattern", in *Ionian Philosophy*, ed. K. J. Boudouris. Athens.

Neugebauer, O., 1957. *The Exact Sciences in Antiquity*, 2nd ed. Providence.

Newton, R.R., 1970. *Ancient Astronomical Observations and the Accelerations of the Earth and Moon*. Baltimore.

Nietzsche, F., 1962. *Philosophy in the Tragic Age of the Greeks*, tr.,Marianne Cowan. South Bend.

Peiser, B., 1990. "Der Thales-Mythos oder 'How to believe six impossible Things before Breakfast". *Vorzeit-Frühzeit-Gegenwart* 2/3-90, pp. 85-99.

Peiser, B., 1990b. "Archilochos und Olympia." *Vorzeit-Frühzeit-Gegenwart* 5-90, pp. 20-37.

Peiser, B., 1993. *Das Dunkle Zeitalter Olympias: Kritische Untersuchungen der historischen, archäologischen und naturgeschichtlichen Probleme der griechiscen Achsenzeit am Beispiel der antiken Olympischen Spiele*. Frankfurt/M.

Reinhardt, K., 1916. *Parmenides und die Geschichte der griechischen Philosophie*. Bonn.

Robinson, T. M., 1987. *Heraclitus. Fragments: A Text and Translation with Commentary*. Toronto.

Seligman, P., 1962. *The Apeiron of Anaximander*. London.

Snell, B., 1944. "Die Nachrichten über die Lehren des Thales und die Anfänge der griechischen Philosophie- und Literaturgeschichte". *Philologus* xcvi pp. 170-82.

Snell, B., 1964. *Pindari Fragmenta*. Leipzig.

Talbott, D., 1980. *The Saturn Myth*. Garden City.

Velikovsky, I., 1950. *Worlds in Collision*. New York.

Vlastos, G., 1942. "Equality and Justice in Early Greek Cosmologies". *Classical Philology* 42.3 (July 1947), pp. 156-178.

Waerden, B.L. van der. 1954. *Science Awakening*, trans. Arnold Dresden. Groningen.

Wasserstein, A., 1955. "Thales' Determination of the Diameters of the Sun and Moon". *Journal of Hellenic Studies* lxxv, pp. 114-6.

References

1. For Mesopotamian chronology, see Heinsohn 1988, and for an updated summary in English Heinsohn 1995; for Egyptian, 1990; for Greek, Peiser 1993. In Heinsohn's chronology the last catastrophe before the 'Axial Age' occurs in the Seventh Century BCE (see Heinsohn 1990.243 and *passim*), and thus only a few generations before the major Sixth and Fifth Century figures to which I am limiting myself. Catastrophists who still adhere to traditional chronology, however, are also presumably concerned with securing a dating for the most recent global catastrophes before the Axial Age, and may well find their own reasons for dating it not many generations before the 6th century breakthroughs I am contemplating. (See for example, in this volume, in addition to the essay by Heinsohn, those by Peiser, MacKie, van Geel, and Wolfe.)
2. The guilty passage is *Metaphysics* A3, 983b6 = DK 11 A 12. The authoritative analysis of the distortions Aristotle has created by bringing his own doctrine of the four causes to earlier thinkers and finding them wanting in terms of it is Cherniss 1935.
3. See for instance KRS 89-91 & 93-94; McKirahan 1994.29-30.
4. DL 1.27 = DK 11 A 1. A comparable range of topics -- astronomical observations, water, soul, divinity -- is to be found in the summary of his doctrines in a scholiast on Plato (DK 11 A 3). The *Suda* (DK11 A 2) mentions astronomical matters and soul, but water and divinity have dropped out. Dicks 1959, insists on dividing all sources on Thales into those before and after 320 B.C.E and on treating only the former (Herodotus, Plato, Aristotle) as reliable (p. 294). His claim that only the later group of sources claims that Thales was "a pioneer in mathematics and astronomy" (p. 298) is belied by the fact that Herodotus claims that he predicted an eclipse, which I shall discuss at length shortly. The setting of 320 B.C.E as a dividing line among the sources is, moreover, tendentious, since Callimachus (whose status as librarian in Alexandria would have given him copious access to written sources about Thales extant in the fourth/third century BCE.) flourished not long after this date (ca. 305-240 BCE.) and mentions Thales precisely as a pioneer in astronomy (DL 1.23).
5. Aetius, DK 11 A 13b. The use of the word *kosmos* by both Diogenes and Aetius is, of course, no proof that this was the term actually used by Thales, and we have here at the outset a classical example of the pitfalls in the doxographical tradition, in which the terminology of later ages is constantly being retrojected to earlier ones. For the theory that the speculations Aetius attributes to Thales in fact stem from Aristotle's own interpretations of Thales, which were then incorporated in the doxographical tradition as Thales' own, see Snell 1944. Since only a generation or two later than Thales Heraclitus is found using the word *kosmos* (DK 22 B 30), its use by Thales is not implausible.
6. For uniformitarians, see Neugebauer 1957, Dicks 1959 and Newton 1970. For catastrophists see Peiser 1990. For a uniformitarian reply to Dicks' skepticism see Kahn 1970.99-10 1 and 115. That there is something remarkable about the crediting of the first accurate observations of the 365-day year, the fixing of solstices and equinoxes, and the prediction of eclipses to a man born as late as the -7th century was first pointed out by Velikovsky (1950.356-7).
7. For a non-catastrophist account of Thales' achievement as one of setting predictable processes over against primitive fears, see Gigon 1944.41-58.
8. This probably means that he defined it as a constellation in order to draw attention to its usefulness to sailors, since its smaller revolution "provides a more accurate fixed point than the Great Bear or Wain as a whole" (KRS 84).
9. DK 11 A 17 and A 3.
10. For an example of such sensitivity with respect to Thales see Gennaro 1970.
11. DK 11 A 1, A 2, A 17.
12. All sources are assembled by Diels under DK 11 A 5.
13. Peiser 1990, citing Dicks 1959, Neugebauer 1957, Newton 1970, Mosshammer 1981.
14. For a discussion of the imponderables in the dating of these battles themselves and the identification of the principal participants see Mosshammer 1981.
15. A slight variant for the canonization of the Seven Sages is 580,

the "fiftieth Olympiad". See discussion in KRS 76 and in Mosshammer 1976.

16. See above n. 5.
17. Dicks 1959, who cites the ancient sources assembled by Diels.
18. KRS 83 and 85. Herodotus (2.109) says that the Greeks learned of the gnomon, along with the celestial sphere and the twelve parts of the day, from the Babylonians.
19. DK 11 A 1 = DL 1.23-24, DK 11 A 3, DK 11 A 17; cf. the solstice-marker associated with Pherecydes in Seros, DL 1.119.
20. According to Heinsohn/Illig the pyramids may have been relatively new at the time of Thales' visit, and his calculation of their height would have made all the greater stir. For other instances in which their new chronology allows for speculation that influence is Greek on Egyptian, not Egyptian on Greek, see Heinsohn/Illig 1990 *passim.*
21. E.g. KRS 82, McKirahan 1994.24-5.
22. Peiser 1990.92. See also above n. 5.
23. Waerden 1954.86-7. Van der Waerden's impulses have been endorsed much more recently. McKirahan in his proposal of a Babylonian influence on Thales here states that the Babylonians "kept meticulous records from the mid-eighth century and so had a data base sufficient for such limited predictions. Adequate records for such predictions cannot be amassed in one person's lifetime", McKirahan 1994.25.
24. The specific cycle most often brought up by scholars in this discussion is the lunar cycle of 18 years 11 days (= 223 lunar months), which, once observed (after two or three recurrences?) would suffice only to indicate that an eclipse of some kind would was going to occur somewhere. For discussion of this cycle see Neugebauer 1957.142, who completely dismisses the possibility that Thales could have made use of it, and Dicks 1959.309 and Newton 1970.94-5, who though both skeptical of the specific prediction at the time of Herodotus' battle admit the possibility that Thales might have known of it. In the course of rejecting this cycle and giving his own formula van der Waerden once again undercuts the necessity he implies for borrowing from the Babylonians: "It is a fairy tale that, in making it [the eclipse prediction] Thales had used the 'Saros', the 18-year period which had been known to the Babylonians since about 400 B.C. I prefer to think that Thales, as well as the ancient Babylonians, start from the approximate relation: 51 draconitic lunar periods = 47 synodic months. According to this relation the possibility for the repetition of a lunar eclipse exists 47 months after a total lunar eclipse, while the chance of a solar eclipse occurs 23 1/2 months after a total lunar eclipse. Indeed, a considerable lunar eclipse could be seen 23 1/2 months before the eclipse of Thales", Waerden 1954.86-7.
25. Dicks 1959.295.
26. The possibility of such a connection, which stresses not Thales' specific ability to predict the eclipse at the battle mentioned by Herodotus but his general interest in eclipse cycles, is granted even by the most stringent critics of the "Thales' eclipse" tradition. "... he might possibly have heard of the 18-year cycle for lunar phenomena and might somehow have connected this with a solar eclipse so as to give rise to the story that he predicted it", (Dicks 1959.309). "The final version as we have it from Herodotus can best be understood as a literary assimilation of what were once entirely separate reports concerning the eclipse, the Lydo-Median War, and Thales' astronomical interests", (Mosshammer 1981.154) A complementary problem has been raised by Newton, namely, that knowledge of an eclipse cycle will necessarily lead one to predict eclipses by the day, not just the year, whereas Herodotus explicitly says that Thales had only predicted the year in which it would occur. Here again the assumption that a certain amount of garbling is going on in Herodotus need not lead to the conclusion that Thales had no interest in eclipses. The final word on the matter may well be Newton's: "[A commonly used] argument is that Thales did not make the prediction because cycles do not predict eclipses visible at a particular place with reliability. It seems to me that this argument misses the point of the legend. A legend that Thales predicted this eclipse, whether true or false, could not have existed if solar eclipse prediction had been reliable at the time. Because of their total reliability the eclipse predictions made in a modern ephemeris will never find a place in legend. Unreliable prediction was possible at this time, and the eclipse of Thales may commemorate one of the relatively rare successful predictions", 1970.94; cf. Kahn 1970.115, Mosshammer 1981.146, Peiser 1990.97.
27. DeGrazia 1983.
28. DL 1.34 = DK 11 A 1.
29. DK 22 B 105, referring to *Iliad* 18.251.
30. DK 11 A 13b.
31. The text is rendered problematic by scholarly disagreement about the words following *kosmos*: in Clement's version the fragment opens *kosmon ton auton hapanton, "kosmos*, which is the same for everyone"; in Plutarch's and Simplicius' it opens simply, *kosmon tonde*, "this *kosmos"*. See KRS 197-8 and Robinson 1987.98. For discussions of the evolution of the word *kosmos* see Diller 1965 and Kahn 1985.219-230.
32. See DL 9.8; Simplicius, *Commentary on Aristotle's Physics* 23.33; DK 22 B 31. Peiser 1990.88 analyses Heraclitus' fragments on Thales and Homer and poses only two alternatives: he meant by *astrologos* either someone who predicts men's fate by the constellations, or someone whose exact astronomical observations are pursued with mathematical rationality. But Heraclitus' own work stands in a middle position between these two alternatives: its claim that the universe moves in unvarying periodicities is buttressed by a few astronomical terms (the solstices) and perhaps also by a pseudo-mathematical speculation (that the Great Year was 10,800 years, each of its 360 "days" consisting of a human generation of 30 years - DK 22 A 13), but is also couched in terms of fate and destiny, though not governed by constellations. If we assume that Heraclitus was simply speaking of Thales as a predecessor of the same kind, then we will have to add that no language of fate or destiny is present in any of the testimonia on Thales we have. Peiser nevertheless, in the course of posing his alternatives, does ask a question in language which anticipates my central argument: "*War Thales mithin der erste Mensch, der sich durch exakte astronomische Beobachtungen vom allgemeinem Planetenkult seiner Landsleute, i.e. vom Glauben an das schicksalhafte Eingreifen der Gestirns- und Himmelgöttern, losgesagt hatte?*"
33. *Metaphysics* 1.5 986b10-25 = DK 21 A 30. Democritus, by contrast, and his immediate atomist predecessor Leucippus, affirmed that there were an infinite number of *kosmoi* (DK 68 A 40 and DK 67 A 1 = DL.9.31). As an author of treatises on such astronomical subjects as the planets and "celestial causes" (DK 68 A 33 = DL9.45-49) Democritus is likelier to have admired Thales, as Herodotus did, for specific astronomical predictions.
34. Kahn 1970.99-100 makes a similar point. Dicks shows some openness to such a distinction in language like the following: "The purpose of the present article is... to show that, whereas knowledge of the solstices does not presuppose any other than (relatively) simple observations, the concept of equinoxes is a much more sophisticated one, involving necessarily the complete picture of the spherical earth and the celestial sphere with equator and tropics and the ecliptic as a great circle traversed by the sun...", Dicks 1966.30. The phrases "knowledge of the solstices" and "concept of the equinoxes", however, betray an inability to push the distinction as far as it needs to be pushed; what Dicks really means by these phrases is not knowledge or concept but "ability to measure as moments in a given day". The concept of the equinoxes as two days in the year (as opposed to moments within days) is preliminarily available by extrapolation to anyone who has attempted to measure the solstices; it is simply the midpoint of equality through which day- and night-lengths must pass in their variation between the extremes of summer and winter solstice. The knowledge of *which* two days in the year in fact have equal lengths of day and night is more elusive, because of the slight eccentricity of the earth's orbit which makes for the "inequality of the seasons", which Thales himself is said to have discovered (see discussion below). As Kahn has shown in his rebuttal of Dicks (Kahn 1970.112-4), it is not nearly as difficult as Dicks makes out to determine which two days in the year are equinoctial, using the same gnomon Thales and Anaximander are said to have used to determine the solstices.
35. Here Dicks writes more wisely than he realizes: "The second of the two pitfalls mentioned above is a much less obvious one and much easier to fall into. It consists of a failure to recognize the tacit assumptions, based on the scientific theory of late antiquity or even (sometimes) of our own for which there is no evidence, but which are almost unconsciously made from our inability to dissociate our views on the thought of their early period from more modern concepts", Dicks 1966.28-9.

36. For professional astronomers invoking collisions with bodies smaller than planets, see for example Clube and Napier 1982 and 1990, Hoyle 1993, Lewis 1996; for speculators with training other than in astrophysics, see Velikovsky 1950, Talbott 1980. The collaboration of DeGrazia (a political scientist) and Milton (an astronomer) in *Solaria Binaria* (DeGrazia & Milton 1984), with its radical rethinking of the orbits and dating of all bodies in the present solar system, represents a notable exception to the above dichotomy.
37. A similar transition at Stonehenge from astronomy oriented towards impact episodes involving the ancestal comet of the Taurid meteor showers to astronomy oriented towards solar and lunar observations is hypothesized elsewhere in this volume in the essay by Duncan Steel.
38. The occurrence of the term in the texts of Homer and Hesiod as we have them is, of course, no absolute proof of their use prior to Thales, since these texts are diachronic compilations, as Dicks himself points out (1966.31 n. 27). Peiser's downdating of Homer and Hesiod, moreover, makes the poetic activity associated with their names more contemporary with Thales, and by no means necessarily several centuries earlier. In this respect Peiser even points out that while Heraclitus calls Homer an *astrologos*, he calls Thales the *first astrologos* (Peiser 1990.88).
39. KRS 83. These slight variations are due to the slight eccentricity of the earth's orbit, discussed in n. 34 above. For the various claims see DK 11 A 3 ("recognizing" the solstices), DL 1.23 = DK 11 A 1 (predicting the solstices), DL 1.24 = DK 11 A 1 (*parodos* and *periodos*).
40. DK 11 A 3 and DK 11 A 19. See discussion in Wasserstein 1955.
41. See discussion in KRS 83-4.
42. KRS 84. Sources are in DL 1.23 = DK 11 A 1 and DK 11 A 3a.
43. DK 11 A 13c. Again there are problems of possible anachronism. Since according to Aristotle, Thales said the earth rested on water, whereas his pupil Anaximander placed it immobile on the center with no support (DK11 A 12 and DK12 A 26), it is hard to see how Thales could have thought in terms of a complete starry sphere (see discussion in KRS 83-4), and the sphere itself sounds more Pythagorean than Thalean. On the larger difficulties inherent in the topic of the obliquity of the ecliptic, see Dicks 1966 and the reply in Kahn 1970.
44. DK 11 A 3 & A 17 (passage of the moon in front of the sun), DK 11 A 1, A 3, A 19 (regular orbits of the sun and moon).
45. *Politics* A 11, 1259a9 = DK 11 A 10. One might add that while Seneca tells us of Thales' theory that earthquakes were caused by fluctuations of the water the earth rested on (*Naturales Quaestiones* 3.14 = DK 11 A 15) Cicero speaks of Thales' "pupil" or "associate" Anaximander (who thought the earth rested on nothing but was simply "at the center") actually predicting an impending earthquake to the Spartans (*On Divination* 1.50.112 = DK 12 A 5a).
46. MCKirahan 1994.229. The original story is in Diogenes Laertius (DL 2.10 = DK 59 A 1), Pliny (DK 59 A 11), and Plutarch (DK 59 A 12). For a recent attempt by an astronomer to ground Anaxagoras' achievement in his observation of the annual recurrence of meteor showers, see Lewis 1996.13-14, 48, 145.
47. Compare the story in Plutarch's *Life of Pericles* (Ch.35) about Pericles explaining an eclipse to a steersman by passing his cloak in front of his face. Pericles was, of course, a friend and "pupil" of Anaxagoras, and the story is meant to illustrate the rationalism they held in common in explaining heavenly phenomena.
48. Pindar Fr. 52k in Snell 1964. A string of other phrases from poets on eclipses is offered by Plutarch, *On the Face of the Moon* 19. For the characteristics of "literary eclipses" see Newton 1970.99-100 and Peiser 1990.95-6.
49. Interestingly, Pindar is said to have indicated skepticism about the project of the *physiologoi*, the pre-Socratic philosophers who gave rational accounts on nature, by saying that they "plucked the fruit of wisdom unripe", Snell 1964 Fr. 209.
50. Scholars tend to reject these apothegms as false late attributions. In the case of this particular topic of "beginninglessness", however, it should be noted that "beginninglessness" was already a topic by the time of Thales' pupil or associate Anaximander (the *apeiron* is eternal, DK 12 A 11) and of the immediately respondent generation of Xenophanes (the gods are not born and there is no time when the gods did not exist, DK 21 A 12) and of Heraclitus (no god made this kosmos, but it always was, DK 22 B 30).
51. *Naturales Quaestiones* 3.14 = DK 11 A 15. The last point is made by McKirahan 1994.28-9.
52. *De Anima* A5, 411 a 7 = DK 11 A 22. The latter half of this sentence seems to have been alluded to by Plato as well in the context of souls being called gods (*Laws* 10, 899b).
53. E.g. KRS 97 n. 2.
54. DK 11 A 3, A 22. See discussion in KRS 96-7.
55. McKirahan 1994.31.
56. See for instance the discussion in McKirahan 1994.29-30based on KRS 92-94, which is in turn based on Hölscher 1968. The ancient view that Thales might have learned that water was the "source and birth of all things" is to be found in Plutarch *de Is. et Osir.* 34, 364d = DK 11 A 11. Plutarch suggests that in calling Okeanos the "birth of all things" Homer (*Il.* 14.201 & 246) also had learned from the Egyptians. Aristotle (*Meta.* A3 983b6= DK 11 A 12) is content to suggest that Thales got his idea from Homer, citing the same passage as Plutarch alludes to. Other ancient authors maintaining various Egyptian and other Near Eastern sources for Thales' learning are collected in DK 11 A 11. It needs to be stressed that moderns like Kirk, Raven and Schofield who adduce ancients like Plutarch to lend support to their speculation that Thales did not innovate but learned his central doctrines from Egyptians or Babylonians are merely making selective use of ancient sources; the vast majority of ancient testimonia insist on Thales as a true innovator. He is again and again called "the first" to do many things, and not just "the first Greek".
[57. McKirahan 1994.29.
58. One possible exception lies in the two testimonia juxtaposed by Diels in DK 11 A 23, Aetius' statement that Thales held that "God was the mind of the *kosmos*" and Cicero's statement that Thales' held that "water was the beginning of things, and God was the mind that fashioned everything from water." I concur with Kirk and Raven that these sentences are "anachronistic, and probably due to Stoic reinterpretation" (KRS 97 n. 1). Even if Thales said something like these two statements, however, they would indicate only a kind of deist clockmaker god, not one who later intervenes in his creation. The only other possible exception lies in one of the apothegms attributed to him by Diogenes (DL 1.35): "The most beautiful thing is the *kosmos*, for it is the work of God". This too is rejected as a late attribution by most scholars; and it too, even if genuine, speaks of only an original deist act of creation.
59. This point is made by McKirahan 1994.29.
60. Nietzsche 1962.39.
61. DK 11 A 19.
62. For a helpful recent discussion of different senses of the word "school" in this context see McKirahan 1994.77-8.
63. "All other cultures are put to shame by the marvellously idealized philosophical company represented by the ancient Greek masters Thales, Anaximander, Heraclitus, Parmenides, Anaxagoras, Empedocles, Democritus and Socrates. These men are monolithic. Their thinking and their character stand in a relationship characterized by strictest necessity. They are devoid of conventionality, for in their day there was no philosophic or academic professionalism. All of them, in magnificent solitude, were the only ones of their time whose lives were devoted to insight alone. They all possessed that virtuous energy of the ancients, herein excelling all men since, which led them to find their own individual form and to develop it through all its metamorphoses to its subtlest and greatest possibilities", Nietzsche 1962.31-2. This account applies well enough to Heraclitus, less well to others; in many cases it is simply impossible to tell whether it applies at all.
64. Kahn 1985.207; see his whole discussion, pp. 199-211.
65. Fine examples of this kind of philological interrogation of the doxographical tradition with reference to Anaximander can be found in Kahn 1985: for cosmos (*kosmos*) see pp. 219-230, for elements (*stoikheia*) pp. 120-126.
66. Xenophanes' attack: DK 21 B 11. Heraclitus' attack on Homer: DK 22 B 42; on both Hesiod and Xenophanes: DK 22 B 40.
67. DK 21 B 29.
68. See again Kahn 1985.126-133, which explains why Aristotle considers these two pairs more truly elements, since the traditional four "elements" could be derived from their four permutations: fire is the hot and the dry, air the hot and the wet, water the cold and the wet, earth the cold and the dry.
69. DK 13 A 5, A 7.

70. Catastrophic battles of Zeus:*Theogony* 685-722, 820- 880; different ages of men ordained by Zeus: *Works and Days* 127-173; Olympians joining fighting at Troy with devastating consequences on earth: *Iliad* 20. [71] The Trojan war itself is said elsewhere in the epic cycle, in a fragment from the lost *Cypria*, to have been devised by Zeus in order to rid the earth of population: see Allen 1912.118.
71. Peiser (1990a.95-6 and 1990b.35-6) has shown how two passages in Homer which speak of Zeus hiding the sun and darkening the sky in the midst of a battle (*Il.* 16.567ff. and *Od.* 20.354-7) may have influenced passages in Herodotus (1.74) and Archilochus (Fr. 74 D) commonly taken to be about solar eclipses. His argument is that since both of the later writers may have been simply assimilating a Homeric *topos* to their own narratives neither should be taken as eye-witnesses of an actual eclipse reliably datable by retrocalculation. In the catastrophic periods with which the *Iliad* and the *Odyssey* are concerned there are a number of other possible explanations for terrifying darkenings of the sky than solar eclipses. Peiser's argument is at least plausible, and it would not be inconsistent with it to argue that the possibility exists that the Sixth Century Milesians were (in a reversal of his scenario for Archilochus and Herodotus) assimilating their own experience of solar eclipses to a misunderstood Homeric prototype. In putting eclipses on their list of terrifying phenomena to be explained they would therefore be remaining consistent with the project of accounting for what was most terrifying in Homer.
72. See discussion in KRS 163-5.
73. For the specific sources, and the presuppositions to be kept in mind when evaluating these datings, see the discussion in KRS 101-2 and 143.
74. The map is discussed in Kahn 81-4, with reference to an extant Babylonian map which, being of "neo-Babylonian or Persian date", can at best be called contemporary with Anaximander and establishes no claim to Babylonian priority over Greek in map-making. The sphere is thought unlikely by KRS 104, but see Kahn 1970.105-110 and 1985.77-81.
75. See Kahn 1985.240, and KRS 51-2 and 102-3.
76. DL 2.3, and see KRS 144.
77. Etymology and epic passages are discussed in Kahn 1985.231-239.
78. Aristotle *De Caelo* 2.13 295b11-16 = DK 12 A 26; cf also DK 12 A 11. He is lauded by some for having anticipated Leibniz's Principle of Sufficient Reason, the earth's rest at the center being grounded here not in a physical reality like floating (Thales' earth floated on water, Anaximenes' on air) but rather in the lack of a sufficient reason for it to move in any direction. The connection to Leibniz is articulated and problematized by Barnes 1982.23-28; more recently it has been argued against by Freudenthal 1986.213 and Furley 1987.24-26.
79. For texts and discussion of the extent to which these ratios may be deemed "mystical" or "rational", see Kahn 1985.85-98 and KRS 134-7.
80. Texts discussed in Kahn 1985.60 and, at greater length in response to Dicks, in Kahn 1970.105-110, leading to the interesting formulation that "... in Greek astronomy, *the idea of a geometrical model for earth and heavens* played the same revolutionary role as the idea of proof in mathematics". See also discussion in Furley 1987.26-28, who suggests an association of the use of multiples of nine with "Hesiod's nine-day interval between heaven and earth, and earth and Tartarus", and also favors Burkert's association with "a knowledge of Iranian ideas, according to which the intervals measured by earth, stars, moon, sun represent steps in the ascent of the soul towards the light" (see Burkert 1963).
81. DK 12 A 10, 11, 23. The terms translated "waterspouts" and "hurricanes", *presteron* and *tuphonon*, may have catastrophic origins referring to more violent events than the presently oberservable processes designated by these words in English, e.g. electrical discharges or high winds generated during impact episodes.
82. DK 12 A 30.
83. Ferrari 1979.123-4, translation my own.
84. The phrase "the first uniformitarian", and the relevant passages, are in McKirahan 1994.36-38 and 41; see also KRS 138-140 for arguments that the drying up of the sea cannot have been seen as a violation of "cosmic stability". Ferrari 1979 and Freudenthal 1986 have both made convincing counter-arguments that the drying up of the present *kosmos* must lead to its reabsorption into the *apeiron*.
85. It is so praised by e.g. Barnes 1982.44-47.
86. DK 13 A 5, 7.
87. DK 13 A 6, 7, 14. See discussion in KRS.156, where the notion that the theory is meant to explain eclipses is discounted.
88. DK 21 B 29, DK 21 B 33. The numbers after the "B" in all Xenophanes citations in Diels-Kranz are the same as the fragment numbers in Lesher 1992.
89. DK 21A 32, 40, B 32.
90. DK 21 A 33. See Mourelatos 1988 for an analysis of the underlying pattern of thought in statements of "the sun is really cloud", or "X is really Y", type.
91. DK 21 B 28.
92. DK 21 A 30.
93. Aristotle *Physics* 203b 4-15 = DK 12 A 15.
94. DK 12 A 9.
95. DK 12 A 15.
96. DK 12 A 11.
97. "Self-regulative": Vlastos 1942.173. "Universe governed by law": Kahn 1985.166-193. "Continuing rule or law of change": KRS 116, after a discussion of other possible senses of "steering".
98. Ordering of time as objective genitive, with philological parallels: Ferrari 1979.126-7; the Boundless steering through periodic phenomena: Freudenthal 1986.209-210.
99. DK 13 A 10.
100. DK 13 A 7; see discussion in KRS 150-1.
101. DK 13 B 2.
102. Fr. 1.13-14 & 19-22. Both *euphemois muthois* and*katharoisi logoisi* are difficult to translate. As the contrast between *mythos* and *logos* came to mean more and more "stories others may tell but which I do not necessarily believe"; and *logos* itself, in the plural, can mean either "stories" or simply "words". *Euphemoi* as adjective describing *muthoi* could mean "keeping a pious silence"; literally it means "speaking well", but was often used to mean "show piety to a religious matter by keeping silence about it". Xenophanes verses may be assimilated into *topos* "Let war be far from the symposium", i.e. let men at drinking-parties neither pick fights with each other in their cups nor sing songs about the battles of others; the existence of this *topos*, however, does not diminish the force of his implied rejection of theomachy as a theme that speaks well and purely of "the god". Rather, it highlights the function of the symposium, and of the poetry appropriate to it, as a practice which civilizes by encouraging oblivion of the old catastrophic "myths". The oblivion is deepened by dismissing the myths as mere fictions of earlier generations (*plasmata ton proteron*).
103. DK 21 B 23.
104. DK 21 B 26 & 25.
105. *Il.* 1.530; the comparison is made by KRS 170.
106. See the distinction of four different senses of "uniformitarian" in Gould 1977.150-1, and discussion below.
107. DK 12 A 27; see discussion in KRS 139.
108. These passages are assembled and cited as possibly relevant to Anaximander by Kahn 1985.185.
109. Freudenthal 1986.219.
110. DK 13 A 21.
111. DK 21 A 33.
112. For Anaximander Kahn assembles and discusses eight such passages, 1985.46-53.
113. This is essentially the argument of KRS 1982.122-6, 151, 178, with references at 122 to the controversy on the matter involving Zeller, Burnet, and Cornford; Kirk originally made the argument alone in Kirk 1955.28-38. A similar argument appears in Kahn 1985.34-5 and 47-53. For an argument that Anaximander could have believed in infinite coexistent worlds on grounds internal to his own system, and not dependent on that of the atomists, see Freudenthal 1986.224-5.
114. List of those who maintain there is always a *kosmos* and that one changes into another according to periodicities: Simplicius *in Phys.* 1121.12 = DK 13 A 11.
115. Simplicius *in Phys.* 1121.5 = DK 12 A 17.
116. KRS 1982.122-3, with reference to further discussion on pp. 139-40.
117. Ferrari 1979.125.

118.Kahn on the relevance of the philosophers' account of world-destruction to Anaximander: Kahn 1985.184-6. Kirk, Raven and Schofield on the relevance of the myths of Deucalion and Phaethon: KRS 1982.122, 140, 178.

119.*ho tote kosmos hudati kataklustheis apoleto, 2 Peter* 3.6, in Cornford 1934.9. Cornford in fact argues two pages later that the earlier texts would have spoken more often of *ouranoi* than of *kosmoi*: "A close examination of the doxographic tradition shows that the further it gets from Aristotle and Theophrastus, the more often *kosmoi* is substituted for *ouranoi"*. By this reasoning it would follow that the original pre-Socratic texts spoke of the successive destructions of the heavens themselves.

120.Reinhardt 1916.175, discussed in Kahn 1985.49-50, who wishes the *ouranoi* to refer to "the various celestial rings which compose his structure for the visible heavens", and the *kosmoi* to be "some lower 'arrangements' of atmosphere or earth"; "the Boundless can thus surround both *ouranoi* and *kosmoi* within the framework of the one and only world system concerning which the testimonia for Anaximander give us any real information". Kahn later admits that this strained interpretation "is not the only one which can fit the textual evidence". While the later and less reliable variant of Hippolytus is discrepant in speaking not of "all the *kosmoi* in the heavens" but simply "the *kosmos* in the heavens", Hippolytus elsewhere states that Anaximander's *apeiron* "encompasses the *kosmoi*", using the same plural which Theophrastus uses.

121.KRS 1982.178, and passages from Aetius and Hippolytus cited by them on pp. 172-3.

122.Eight passages about Anaximander and *apeiroi kosmoi*: Kahn 1985.46-7; seven passages about Anaximander and destructible *kosmos* or *kosmoi*: Kahn 1985.185n3.

123.The phrase in Hippolytus is the Homeric "eternal and unaging"; the phrase in the *Stromateis* says that a sphere of fire was formed around the earth at the beginning of the present *kosmos* "like bark round a tree". The Anaximandrian origin of the Hippolytus phrase is suggested by Kahn 1985.43, of the *Stromateis* phrase by KRS 131.

124.DK 12 A 10. The words translated as "the totality of things" means literally "the All", *to pan.* David Furley (Furley 1987.1-9) has contrasted two conceptual schemes in antiquity which he calls the Closed World" and "the Infinite Universe"; the former tends to be designated as *to pan*, the latter is referred to a plurality of *kosmoi*.

125.DK 12 A 11. For the debate whether the parallel passages in Hippolytus and Simplicius mean that Anaximander was "the first to use this term 'principle' [*arkhe*]" or "the first to use this term *apeiron* [for the principle]", see Kahn 1985.29-32; neither reading would affect my arguments about the two passages.

126.DK 12 A 9.

127."[...] la fase di distruzione degli *onta* seque la stessa e inversa direttrice (*ex*!) della loro origine *poiché* gli *onta* 'pagano gli uni agli altri la pena e il risarcimento per l'ingiustizia commessa", Ferrari 1979.120-21. Ferrari here takes the *auta* as referring back to *tois ousi* in the preceding sentence, whereas Kahn insists (Kahn 1985.166-8) that the neuter plural of the *ex hon... eis tauta*, and also of the *auta*, must refer to the *stoikheia* of the preceding sentence. Kahn is surely right that the *ex hon* cannot be taken as a "generic plural" to refer to *to apeiron* itself, as most scholars persist in doing; whether the *auta* refers back to *tois ousi* or an implied *stoikheia* is a secondary issue not effecting arguments for reabsorption into the *apeiron* of what makes up a *kosmos*, the things that exist in it or the elements of which all other things in it are composed.

128.To the three passages in Anaximander Cornford argues that we should perhaps add a fourth. He finds the origins of the confusion of *kosmoi* with *ouranoi* in post-Epicurean doxography in the statement at the end of Aetius' *Peri Kosmou* (II.1.8) that Anaximander thought that his innumerable *kosmoi* "were set apart from each other by an equal amount" (*to ison autous apekhein allelon*); in this he was confusing Anaximander's notion that each of the three *ouranoi* in the sense of heavenly spheres of fire (for stars, moon and sun) were the same distance apart from each other (nine earth diameters) with Anaximander's notion that the *kosmoi* succeeded each other according to a periodicity: "Anaximander may have said that '*equal* intervals of *time* mark the becoming and perishing of the successive worlds". Cornford 1934.12.

129.E.g. Kahn 1985.184: "Anaximander must have seen this exchange transacted daily, in the alternation of light and darkness, of the fresh morning dew and the parching heat of noon.... or in the predominant cycle... that of the sun, since it is in step with this yearly movement that the seasons of heat and coolness, drought and rainfall succeed one another..."

130.Kahn 1985.184-6. In this discussion Kahn adduces not only all the catastrophist passages in Plato but also Xenophanes (on the destruction of humanity by mud) and Aristotle and Censorinus (on the "great year whose winter is deluge and whose summer is conflagration".)

131.KRS 140. Like Kahn, Kirk, Raven and Schofield here adduce both Xenophanes on mud and Aristotle on the Great Year.

131.The phrase "dynamic equilibrium" is taken by Kahn (1985.190) from Vlastos 1942.172, who in the same, highly influential article also explicitly argues against Anaximander's believing that "the universe's history has... been marked off into separate epochs" (Vlastos 1942.165). Freudenthal 1986, using the same medical texts on which Kahn had drawn for the quoted assertion, argues persuasively that the equilibrium of Anaximander's *kosmos* is not dynamic but static, and having been unbalanced heads for destruction: "The world perishes once it has reached the end of its natural lifespan (depending on the quantity of primeval moisture). Unlike cyclical (daily and yearly) meteorological phenomena, this is a distinctively irreversible process". Freudenthal 1986.222. As noted earlier, Freudenthal also argues (1986.210) against Kahn's "oldest formula of natural law" (1985.191). For a general overview of the debate in antiquity between adherents to a Closed World and an "Infinite Universe" see Furley 1987.1-9.

133.Gould 1977. 150.

134. Gould 1977.150.

135. McKirahan 1994.41.

136. Gould 1977.151.

137. Gould 1977.151-2.

138. E.g. Clube and Napier 1990, Hoyle 1993, Lewis 1996.

The 'Kultursturz' at the Bronze Age/ Iron Age Boundary

Irving Wolfe
Département d'études anglaises, Université de Montréal, C.P. 6128, Succursale Centre-Ville Montréal (Québec), Canada H3V 1B4

1. Introduction

Most of the papers in this volume present evidence for cosmically-induced catastrophes which occurred on Earth within the past 5000 years. There has been data from geology, archaeology, palaeontology, ancient history, meteorology, physics and astronomy, to name a few disciplines, all of which is indicative of the occurrence of large natural traumas to the Earth, especially in the third and second millennia BC. I will now ask you to consider a different sort of evidence for a natural catastrophe. Because it is historical, it may not seem as compelling as scientific evidence at first, but I will try to show that, when the items which I will present are viewed as a whole, as a general phenomenon occurring at one time across many cultures and thousands of miles, they will be seen to constitute a very large cultural shift in the first millennium BC. I will then argue that, from the magnitude of this cultural trauma, we may infer the occurrence of a severe physical trauma not long before. The majority of papers present physical evidence for catastrophe and have then inferred its effect on culture. My form of argumentation is the opposite side of the coin - I will present cultural evidence from which to infer the occurrence of a physical catastrophe.

The principle I will apply here is one I have been advocating for two decades. It is that *Nature produces Culture*, and more particularly that the natural cataclysms which our ancestors have collectively experienced have influenced and shaped the cultural artifacts created afterwards. To put it simply, cultures are what they have gone through. The past determines the present, and the cosmic past exerts the greatest influence. A culture, if properly interpreted, therefore becomes a mirror of what preceded it. That is the process which I shall try to prove for the 1st millennium by analysing in tandem several sets of cultural phenomena dated to the eighth, seventh and sixth centuries BC.

My analysis involves two principles adapted from physics and astronomy. The first is a cultural or social version of Newton's first law. It says that a cultural Age will continue in motion until it is acted on by a contrary force impelling it to change its nature and take a new path, only after which will it become a new Age. The implication is that, if there is no such drastic force acting, there is no reason for a cultural Age to change. It will simply continue on its original motion. For a cultural Age to be considered to have altered, the changes must be sufficiently pronounced to indicate a new order of things, (a new direction), and we must therefore guess that there must have been a cause drastic enough to provoke that new orientation. To put it in Newtonian terms, in order to cause such a re-action, we have to have an adequate action.

The second scientific dictum I will apply analogously to the cultural phenomena I have collected is the *Anthropic Principle*. This is a method of induction whereby one starts with what exists and reasons backward to what must have occurred in the past in order for that which exists to be present. For instance, in astrophysics, we have been told how fortunate our universe is, in that, at each critical juncture of its evolution, the very tiny window of opportunity that had to exist in order for our universe to evolve as it did, did (see Verschuur, this volume). This allowed us for example to retain our hydrogen, and then permitted the transformation of hydrogen to heavier atoms, and then the appearance of matter, and then the agglomeration of matter into clumps, leading to the formation of galaxies, the evolution of stellar dust clouds and gravity and finally to the appearance of planets and of life. Had not every one of these favourable conditions without exception presented itself at just the right time, we are told, the universe and its living entities as it is today could never have come into being. But it has, and that is how we can deduce its past from its present.

It is in terms of these two methods of approach to the physical past that I shall analyse the cultural past. That is to say, by studying cultural data, we will work backwards from what is, to try to reconstruct what should have happened in the past to bring us to the present condition, and, if the past is found to be significantly different, we will look for the drastic cause which resulted in the present. I would like to discuss today what may have been one such pivotal moment in our social history, the middle of the first millennium BC, which is hinted at by a contemporaneous set of rapid, violent and extensive cultural changes in many cultures across thousands of miles. I call it a *Kultursturz*, a term which is adapted from the meteorological term *Klimatsturz*, used by climate historians to refer to a weather trauma, a sudden, sharp, drastic alteration in large-scale climate leading to a new Age. That is a *Sturz* or physical crisis, and I see a *Kultursturz* as a trauma to an existing culture, creating an ideological crisis resulting in a large and decisive social revolution leading to a new cultural Age. Once we perceive it, we can then ask whether this *Kultursturz* or social crisis could indicate the occurrence of a large physical crisis or catastrophe not long before.

That is the question I will pose about the early first millennium BC, which we may rightly call the *Axial Age*. [14] I will present evidence for an enormous cultural transition at that time, after which things were no longer as they were before, and I will argue that its cause was drastic, and that it was cosmic rather than local or human. My cultural evidence will be not only what was ended (extinction) but also what was new (creation). I will present several cultural items, (new religions, new cosmologies, new athletic contests, new forms of government, new types of city), all of which first appeared during this period, which I think demonstrate a huge shift in outlook, virtually a spiritual transformation which seems to have occurred in many regions of the globe at the same time and in the same direction. Together, they coalesce into a composite picture of that era which shows that the changes were total and uniform, in that they were of the same particular sort, and may therefore be very significant.

2. An outburst of religions

I turn first to the religions of the mid-first millennium BC. This very short space of time has been called "the greatest of all religious eras," [6, p. 3] for during it no less than eight major religions appeared - Buddhism, Jainism, Upanishadic Hinduism, Confucianism, Taoism, Zoroastrianism, Pythagoreanism and Prophetic Judaism. Their appearance is dated even traditionally from 800 to 500 BC. [6] What we may safely surmise is that a "great age of religious consciousness and reformation" [6, p. 165] occurred from the Mediterranean across Asia to the Pacific Ocean sometime before the late 5th century BC. Of special interest to us is that there appears to be a common structure and history underlying each of these new faiths, to the point where they are almost variations on one theme. They all develop from an earlier predecessor religion, they all reform it in the same direction, and they all evolve the same new view of man's relation to the universe. It is therefore a collective ideological phenomenon of the highest significance, because it seems to indicate that large sections of humanity in different parts of the world almost simultaneously evolved similar new perceptions of the universe.

Buddhism will serve us as a paradigmatic example. Behind it lies the parent religion of classical Hinduism, which saw the eternal supreme god Brahman underlying the surface of constant change or *maya*. To the Hindu, the indestructible soul keeps being reborn into successive existences almost countlessly. If, however, each reincarnation is filled with good work or pursuit of truth or extreme devotion, the soul after many rebirths may eventually be released and join Brahman. If not, "life is a burden and rebirth a misfortune," [6, p. 25] and the gods must be continually propitiated with very complicated priestly rituals and sacrifices. Such was the religion of Hinduism as it came down in India from the second millennium BC.

Along comes Gautama, (the Buddha), some time in the 6th century BC, who rejects sacrifice and priesthood (and perhaps monotheism itself) and preaches that an end to reincarnation can be obtained much more quickly than Hinduism supposes, and by a different method - by private restraint and enlightenment. This is critical, for it involves the individual alone and his personal improvement, not the individual and deities who must be solicited. The goal is one's own *Nirvana*, or better *Parinirvana*, the final exit from rebirth, a solution totally different from the pessimistic Hindu portrait of almost incessant and painful rebirths except for a select few. Life, said Buddha, *is* pain when it is ruled by craving, but one who empties his mind of vain desire can end the pain and get "beyond rebirth." [33, p. 81] The Buddha agrees with standard Hinduism that punishment for sin (or *Karma*) is a reality, but he provides a system whereby the debits for ones *Karma* can be set aside relatively rapidly, after which the path is open "to final release [...] from all the taint of evil," [33] wherein one attains the end of rebirth by an "ultimate escape from the encumbrance of a material body." [31,p. 181] That escape is *Nirvana* - one is free of life and can exist in bliss.

What is significant is that this is essentially a process of *self*-salvation open to all persons of any sex or caste, which is very different from the Hindu belief that one's caste reflects one's *Karma* from previous lives and that salvation is granted only by the gods, and only to those at the top of the top caste, and mostly to men. The Buddhist message in contrast is universalist and merciful, unlike Hinduism, for all people may attain Buddhahood. *Nirvana* is a full and eternal personal identification with the deity, a holy "annihilation in the universal Brahman," [31] a state of "indescribable bliss" [33, p. 82] and "true beatitude" [33, p. 103] in which there "is no more *Karma*." It is open to anyone, and can be achieved in only a few life-times or even one, which is why some have called it a "short cut religion." [33, p. 106] The only criterion is individual worth, displayed by righteousness, mercy, asceticism, kindness and faith.

Here we have the distinctive pattern that characterises almost every new 1st-millennium religion. An older faith focused on worship, sacrifice and fear of heaven is rejected or modified in favour of an emphasis on interpersonal human values - righteousness, justice and mercy - and a quick salvation open to everyone. The torment of life as Hinduism saw it can be evaded by personal effort at improvement, by seeing, learning and then changing one's outlook, for the Buddhist standard of worthiness lies not in ritual but in individual and inter-human goodness. *Heaven is not involved*. This is the structure we will discover, with some variations, in most of the religions that appeared at once in the 1st millennium - there is a sudden shift from a theocentric vision of man's relation to heaven to a homocentric concept of man's relation to man. One's life and one's fate are not subject to divine caprice, but lie in ones own hands. Heaven is no longer feared, and does not require propitiation by worshippers but justice and mercy and morality between individuals. We shall have to ask what sort of change in the natural world might have provoked such a change in spiritual outlook.

Buddhism may be said to be a variant (if not a rival) of Hinduism, but within Hinduism itself there was a revolt of a similar kind, also dated traditionally some time between 800 to 600 BC. It is to be found in the writings called the *Upanishads*, which speak of a single ruling force (*Brahman*) which is benevolent (unlike the disinterested multifarious Hindu deities of the older religion) and strives for the well-being of humankind (i.e., it is not capricious or savage). In this reformed offspring of Hinduism, the end of rebirth can be attained through union with the Brahman, not through extensive ritual and worship as in the parent faith, but through ethics and correct philosophy. To discover how to behave, the Upanishadic believer must accept that there is but one supreme deity who rules the entire universe and is "absolute, impersonal, indescribable," [6, p. 25] and that all the different gods named in the earlier *Vedas* are simply variants of this one God. The goal of Upanishadic religion is union with this God, a "partaking of divinity and thus perhaps avoiding the repeated misery of rebirths," [6, p. 26] which is to be attained not through endless ritual but by recognition of the *Atman* or universal Self and devotion to it through "rules of social ethics," [6] which is to say *interpersonal* just behaviour. Upanishadic belief has been described as a "spirituality [...] invading Hindu thought," [6] and we see here the same shift from the theocentric to the homocentric that we had encountered in Buddhism, in that in the more ancient *Vedas* the *Brahman* had stood mostly for sacred wisdom uttered by the gods and requiring only obedience, whereas for the Upanishadic writers *Brahman* is a universal soul, not distant from humankind and therefore to be obsessively feared and propitiated, but a cosmic Self which is *in* everyone. It "existed before all else," it "is the source of all life and being," and it also "pervades the human soul." [6, p. 39] Everyone is therefore divine, or can be, if he understands that the Brahman wants only justice between people. We can therefore see that, apart from the Upanishadic stress upon monotheism, its major ethical precepts are almost identical to Buddhism.

A third religion of the same type to appear in India was Jainism. Its founder is said to be Mahavira, who flourished in the 6th century BC and preached that the eternal soul of each individual can be rescued from the grip of rebirth or *samsara* by good deeds. It is the aim of the soul to rise on the ladder of universal value by making atonement for and washing away its bad *karma*. This may take many rebirths, but, "through correct knowledge and conduct", [31,p. 186] which is through ethics and proper interpersonal values, the soul will at last "escape from the circle of births and deaths." [31] This pure condition is the Jainist equivalent of sainthood, when the individual is "finally beyond good and evil," [33, p. 55] and can get out of life. At that point the "soul will finally be freed from all *Karma*, bad and good, and will float like a bubble to the ceiling of the universe, there to abide in bliss untroubled". [33] It will have escaped not by ritual but by personal goodness. "This is isolation or *nirvana*," [33, p. 55] which he defines as a "going out," a release from the cycle of retribution. As with our previous examples, one's salvation is to be achieved "without resort to priest or sacrifice." [31, p. 185] It is a compact between the believer and heaven, and it depends upon correct seeing leading to correct behaviour. There *is* one supreme deity behind it all, but what heaven wants is not ritual but justice.

Turning to Iran, we find Zoroastrianism. Dated traditionally to some time in the 7th century, it is a magical and monotheist faith which envisages "a kingdom of justice that promised immortality and bliss." [31] It too reformed an earlier religion much like the Vedic in India, which had lots of gods and extensive polytheistic ceremonials. Zoroaster too spoke of one supreme god, Ahura Mazda, who is not an always-nearby oppressor of humankind needing to be placated by rites, but a distant deity who "makes rigid demands of righteousness. [...] This righteousness consists of truthfulness, kindness, benevolence, justice, devotion to God, and good works." [6, p. 162] To the Zoroastrian, these ethical actions "are more efficacious in the matter of salvation than [...] pious words, prayers, and religious ceremonies." [6, p. 162] Zoroaster's supreme God, just like every comparable deity in the new religions, is creator, lawgiver and judge, a "universal loving Father-God," [6, p. 169] and he rewards not ritual but good deeds. In the *Avesta* it is written that "The souls of the righteous [...] will dwell in eternal immortality [...] in his Kingdom." [6] Immortality can therefore be earned relatively easily by the Zoroastrian believer, who can become one with the divine by following the ethical rules which govern both gods and humans, for ethics is the working of the deity and the order of the universe. People therefore decide their own fate by choosing between good and evil. The good choose life, the evil select "not life", and those go to hell forever, while the righteous live immortally in heaven. In Zoroastrianism, God is wisdom and ethics, sacrifice is forbidden, ceremony is downplayed and all people can be saved. Heaven is not the enemy or severe critic of humankind, but the path to follow.

In Greece at this period we find Pythagoreanism and Orphism, which present a similar picture. Pythagoreanism is a purely ethical doctrine which believes that the soul reincarnates but can be purified by the "pursuit of personal holiness" [31, p. 216] so as to finally unite with the divine mathematical soul of the universe. Orphism is much like the faiths of India. It believes that there was a primordial battle between good and evil in which Zeus killed the Titans. Humans were formed from their ash and are therefore wicked, but they are also divine and can be saved. It is therefore "the mission of men," say the Orphists, "to rid themselves of the evil parts of their natures, leaving only the divine," [44, p. 436] but this process of self-purification was seen as long and slow, requiring "more than one lifetime." At the end, however, if the ethical precepts are followed, the body dies forever and the soul is liberated "to achieve its true life." [44] Within as short a time as only three virtuous lives, the Orphic believer could expect to be released from the 'wheel of birth' [...] and be allowed to spend eternity in Elysium." [44] It was all in the individual's own hands, and heaven was a willing participant.

In China, Confucianism and Taoism present a similar picture. To Confucius, who was much like Socrates, Heaven wants the *Tao* or correct way of life to be followed, and the universe, ruled by "Shang Ti, the Lord Above," [33, p. 95] rejects ritual and sacrifice in favour of righteousness. Education is ennobling, not birth, even the low-born can be saved and "virtue has its own reward." [33] We are told that his ideas were a "fundamental transformation" of the previous animistic religions, as was Taoism, where the physical and moral universe as a whole is Reality, and salvation is available to every believer, for "one is in Tao and survives death." [33, p. 98] By achieving the Taoist state of moral and perceptual purity, one may reach the Isles of the Blest, where the individual will then "never die." [33] This movement away from worship toward homocentricity concurs with every new religion we have analysed and even anticipates the early Christianity which followed some 600 years later.

The most powerful and most familiar instance of this shift in outlook is to be found in the Hebrew religion, in the sharp divergence which occurred between the parent Mosaic and the new Prophetic Judaism. As an example of the Old Religion, we turn to *Deuteronomy*, where the Deuteronomist warns

> "beware lest thou forget Jehovah, who brought thee forth out of the land of Egypt."

What must the admonished Israelite do?

> "Thou shalt fear Jehovah thy God, and him shalt thou serve [...] for Jehovah thy God [...] is a jealous God."

What is the punishment that awaits if the Israelites waver? It is nothing less than extinction. They must beware

> "[...] lest the anger of Jehovah thy God be kindled against thee, and he destroy thee from off the face of the earth." (Deuteronomy 6. 1-15)

The threat in these words is precise and menacing. The Israelites must "fear" Jehovah and "serve" Him with prayer and ritual, or else He will become "jealous" and give way to His "anger" and "destroy" the nation to the last person. This is a God who must be obeyed and propitiated. We need not concern ourselves that this text may have been fabricated in the first millennium, the very period we are looking at, and then fortuitously "discovered" and attributed backward to the second millennium. What is important for us is that it recapitulates the compact of Mosaic Judaism in its most severe and unyielding form precisely when it began to be attacked by the Prophetic reformers, just like many of the beleaguered Old Religions which were also undergoing confrontation by radical upstarts at that time. It is the same story.

Here in contrast is what the Prophets say.

> "I hate, I despise your feasts, and I will take no delight in your solemn assemblies. Yea, though you offer me your burnt-offerings and meal-offerings, I will not accept them; nor will I regard the peace-offerings of your fat beasts. Take then away from me the noise of thy songs; for I will not hear the melody of your viols. But let justice roll down as waters, and righteousness as a mighty stream." (Amos 5. 28-29)

Once more we find a de-emphasis of ritual, (of the relation between humans and the deity), and a stress upon justice (the relation of human to human). As Micah says

> " [...] shall I come before him with burnt-offerings, with calves

a year old? will Jehovah be pleased with thousands of rams, or with ten thousands of rivers of oil?"

No, says the Prophet, ceremony will not do, nor will sacrifice.

"He hath showed thee, O man, what is good; and what doth Jehovah require of thee but to do justly, and to love kindness, and to walk humbly with thy God!" (Micah 6.31)

Similarly in Hosea (6.6):

"For I desired mercy, and not sacrifice, and the knowledge of God more than burnt offerings."

It is found everywhere in the Prophets, and always it is the same message.

"To what purpose is the multitude of your sacrifices to me? saith the Lord [...] Cease to do evil; learn to do well; seek judgement, relieve the oppressed." (1 Isaiah)

The contrast between these two phases of Judaism could hardly be more complete. Not only is *The Lord* now the sole or supreme God, (whereas it has been argued that in the *Pentateuch* Jehovah is only one among several), but no longer is He pictured primarily as a God of Battle, the divine and invincible warrior who protects the nation of His choice with His mighty "outstretched arm" against the onslaughts of man or nature. Here God is ready to withdraw from central participation in human affairs and to allow the humans to determine their fates for themselves. It is therefore the same as with every new religion we have looked at: in the Prophets, heaven is only indirectly involved. It sets the rules, but humans determine the outcomes, and there is no atmosphere of fright or servitude, no stress upon ritual or sacrifice or fetish. We may say in general, therefore, that in all the new religions of the period a terror of irascible heaven has been replaced by a vision of peace between people on Earth, and by the belief that this holy condition can be achieved by willing and merciful human effort alone. This is what Heaven wants, but Heaven is not required. In a non-sacrilegious sense, for all practical purposes man has become the measure of things.

3. The cosmologies

Alongside the eruption of eight new religions around the 6th century, there was a complementary and equally sudden outburst of radical new philosophies, and they too tended in the same direction - from theocentric to geocentric. These new visions of the physical universe should more properly be called cosmologies, but they and their authors are included in the domain of philosophy in most studies of the period. They are however set off as "natural philosophy," which is the term for physical science in the classical world, and that is how we should consider them, as sciences. What is startlingly significant about them, (whatever we call them), is that they stand in total contrast to the Homeric or Olympian view which preceded them. In the older portrait, a congery of irascible gods indulge their passions heedlessly and irresponsibly Above and cause devastation Below on Earth, while in contrast this new breed of natural philosophers saw the world as regular, orderly and governed only by natural law. Such a change in outlook could hardly be more revolutionary, for the new world the pre-Socratics envisioned was not magical or vitalistic (under the sway of godly caprice) but materialist and naturalist, composed of definable objects and of immutable forces acting upon them.

These new thinkers are called the "pre-Socratic" philosophers. I do not think this is a very helpful description, however, since it refers mostly to the *time* in which they appeared, (before the "major" philosophers Socrates, Plato and Aristotle). If we were to choose a term which referred to the *place* of their ideas, they should more accurately be called *post-Homeric*, since they come after and diverge sharply from Homer but blend with some ease into Platonism. However that may be, it is said that they were the first in the ancient world to seek for orderly principles behind nature. I am not sure that this is true, but what is unique about the pre-Socratics for their time is not only that they sought order in nature, (as opposed to Homeric disorder), but their belief that order was not magical but natural. That is the distinction which makes them pre-Socratic (and post-Homeric).

It must be understood, therefore, in what way the pre-Socratic's quest was revolutionary. They embarked upon what was a genuinely scientific endeavour, to construct explanations of nature *as a unit* founded upon uniform laws and behaviour. They sought to explain the composition of natural objects, their properties and the processes they undergo, in terms of regular, orderly and systematic laws, without invoking the irregular intervention of any deities. Such a program was reductive but natural, in that it saw everything in the universe, heaven and earth, stars and humanity, as subject to one ubiquitous system of natural law. So far as is known, this was the first time that such an objective foundation for all of nature was postulated.

Different pre-Socratics then tried to identify the underlying master substance behind all the varied objects in nature, and the master set of unvarying rules which governed them. Whether it was water (Thales) or air (Anaximenes) or fire (Heraclitus) or the Infinite (Anaximander) or a cosmic Mind (Anaxagoras) or atoms (Democritus) or number (Phythagoras), and whether the process which formed the different natural objects out of the fundamental substance was rarefaction and condensation (Anaximenes) or a balance of tensions (Heraclitus) or the battle between Love and Strife (Empedocles) or random combinations (Democritus), there was no recourse to the intrusion of Homeric deities. They could not seem to agree on the specific nature of the fundamentals, for to the pre-Socratics the world was either constantly changing (which was observable by humankind) or was a stable immutable reality unable to be directly observed by people misled by sense perception, and its progress was random or teleological, but capricious gods played no role. Nature was its own ruler.

With that said, we must now attempt to explain this phenomenon in terms of a neo-catastrophist viewpoint. William Mullen claims that the Milesian cosmology of Thales, Anaximenes, Anaximander and Xenophanes was somehow a denial of catastrophe.

"The agenda [...] included explanations of phenomena most men deemed terrifying: thunder, lightning, earthquakes, eclipses, and periodic destruction of the kosmos itself. It set about to explain these phenomena in terms of the same elemental process [...].as it invoked to explain the orderly arrangement of the earth and the heavenly bodies." [Mullen, this volume]

The purpose, he says, was to negate a fear of heavenly disorder by showing that all events in the sky or on the earth, no matter how violent, were able to be included in a systematic set of processes restricted to invariant laws.

"In so doing, it implied the baselessness of the traditional Olympian religion which attributed lightning and earthquakes to whims of Zeus and Poseidon and world-destructions to battles of the sky gods." [Mullen]

Mullen believes that a large-scale destruction had occurred

sometime in the early 7th century, but that the skies had settled down to observable regularity fairly quickly, and by the mid-6th century humans wished to forget.

> "The ultimate Milesean agenda may therefore have been to liberate people from paralysing fear of the immediate recurrence of celestial disturbances in the recent past."

There were according to Mullen two objectives, the second an outcome of the first.

> "a) By insisting that world-destruction occurred only in vast cycles of time [...] the Milesian school was systematically distorting memories of recent disturbances, and its activities may be seen as part of a general pattern of oblivion and psychological distancing common to all cultures after the end of the Bronze Age catastrophes.
>
> b) [...] by insisting that these world-destructions occurred only as the result of unalterable elemental processes, it was also erecting a proto-scientific bulwark against apocalyptic thinking and behaviour."

Catastrophes of the past were denied, and consequently a fear of catastrophes in the future was avoided. Both ends of the memorial plenum (retrojection and projection) were protected. I have been arguing for two decades [45, 48] that most of the creations of our culture (idea systems, religions, cosmologies, sports, works of art) are unconsciously-directed denials of catastrophe, self-delusions designed to make us believe that cosmically-induced natural destructions did not occur and therefore will not. It is my belief that we are profoundly swayed consciously by these subtle fictions of ours to the point where we have become the unknowing victims of our unacknowledged past. We live in an unsuspected self-imposed blindness. In this instance, however, concerning the pre-Socratic cosmologies, I may have to differ somewhat from Mullen and from myself, for it seems to me that the coherent set of cultural phenomena which I am looking at in this article, (those of the Axial Age of the 6th century), may be open to a slightly more complex explanation.

What I mean is this: when we consider for example a soap opera of the late 2nd millennium *AD*, and we relate it to putative natural events of let us say the 2nd millennium *BC* as I have done, [46] then it is requisite that we invoke long-term processes of collective unconscious memory, fear, denial and hysterical rejection. If there is a connection, it must move along these sorts of channels. When, however, we are faced with the religions, cosmologies, athletic games, forms of government and ethos of urban organisation which burst onto the scene in the 6th century BC, we may have to deal with conscious as well as unconscious reactions, for the Axial phenomena appear to have emerged so closely after the natural cataclysms that it is not reasonable to rule out altogether the role of the world as it was then observed. It is to this extent (and to this extent only) that I diverge from Mullen. What happened *very soon after* the last catastrophe may have involved some deliberate acts.

To be specific, we have to explain why the outlook of the Greeks about the universe changed so radically and so abruptly in the 6th century. To say that it was the outcome of a different society (the *polis*) or a different economic system (commerce) or a different form of government (democracy), etc., etc., is merely to defer the question, for one would then have to ask *why* the polis or commerce or democracy arose. What is required is a more fundamental answer, one whose attributes would be likely to constitute on their own a definite break with the past, without recourse to anterior explanation-deferring secondary causes. My answer is that, if humankind *for the first time* began to look for and find objective order in the heavens during the pre-Socratic period, it did not happen because there was a sudden increase in Greek brainpower, or because of a sudden and unexplainable desire to perceive the heavens as they were rather than invent things that were not there in the sky. I think it occurred because *for the first time* the heavens displayed an objective order which could be observed. If there was a catastrophe around 700 BC, the skies might have settled down sufficiently by 600 BC or slightly later to permit recurrent observation of the year-length, the solstices and equinoxes and lunar cycles and even the Venus-Earth interlock, such that it would be apparent that *the heavens were orderly*. On the notion that Below always copies Above, human cosmology might have changed from the Olympian to the pre-Socratic, (from a chaotic sky filled with warring gods to a stable sky filled with planets and stars in orderly motion), because the sky itself changed in that manner. In this sense, there would have been no attempt by the pre-Socratics to deny celestial disturbance because *the sky was not disturbed*. The pre-Socratics merely followed what they saw.

Yet there is a sense in which pre-Socratic cosmology did undertake a "pattern of oblivion," as Mullen says, and that is in its systematic focus *only* on the orderly present state of the sky, without acknowledging any indications that the heavens above might not always have been orderly, even in the previous century. In my opinion, it is not likely that, only a few generations after the Bronze Age-ending catastrophes, they had already been fully forgotten. I think the process of forgetting-suppression-denial takes longer than that, and that stories of the cataclysms (so terrifying and memorable) were still part of conscious pre-Socratic culture. If this is correct, we would then have to describe the pre-Socratic cosmology as "psychological distancing" (Mullen), but of a *conscious* nature, whose purpose regarding earlier belief is plain. As one historian put it more profoundly than he knew, to describe the entire universe as

> "nature subject to law [...] insulated the universe [...] from the arbitrary acts of an unpredictable divine will." [31,p. 214]

Precisely. When the pre-Socratics insisted that the world was *entirely* intelligible, they removed the possibility of the arbitrary and unpredictable. Because they did this so soon after the catastrophe, we have to leave room for the possibility that it was done while the memories of the catastrophes were still present or partly present in the collective conscious, and that only afterwards did the process become purely unconscious.

In my opinion, this time scheme can help us to chart the path from apparent acknowledgement to clever denial of catastrophe which Greek cosmology took from the 6th to 4th centuries, (from the pre-Socratics to Aristotle). What we must perceive, as classicist Benjamin Farrington put it, is that the later cosmologies of Plato and Aristotle are "a radical departure from the first premise of the older school of natural philosophers." [17, p. 134] He calls them "a radical departure from the Ionian point of view." [17, p. 135] That is correct, for Aristotle and Plato *repudiate* the pre-Socratics, being as different from them as they were from Homer. Farrington describes the difference as religious.

> "[...] in the period of philosophy [...] from Thales to Aristotle [...] there is a double tradition, roughly defined as the religious and the scientific." [17, p. 137]

This cannot be correct, however, if Aristotle and Plato are "a radical departure." I do not think that these traditions co-existed. They were not a "double." I feel on the contrary that there was a *sequence*, and Farrington's distinction between religion and science allows us to perceive the subdivisions. To the scientific (6th and 5th centuries) belong the pre-Socratics, while to the religious (5th and 4th centuries) "belongs the theological astronomy of Pythagoras, Plato and Aristotle." [17, p. 138] The pre-Socratics (who came first) were in the first approximation secular, and only afterward was there a turn to the religious. This, I think, is a more accurate portrait of the "period...from Thales to Aristotle." It was not really a coherent period.

To trace more correctly from a neo-catastrophist point of view the evolution of Greek astronomy following the Axial Age, we will make use of Farrington's term "theological astronomy," which he used to describe Plato and Aristotle. It is more perspicacious than he suspects, for that is how I interpret the course which cosmology took in ancient Greece: it was a series of successive stages of burial and denial of catastrophe. First we have the Homeric picture of gods battling passionately and irresponsibly in the heavens. If it arose during the period of Axial upheaval, it probably reported what it saw (there is great disorder in the heavens, caused by the selfish "gods" who care not for humankind). That was countered perhaps a century and a half later by the pre-Socratic cosmology, which, if it arose after the skies had settled to *observable* regularity, may also have reported what it saw (there is *only order* in the heavens, and nature causes it). Here we might have the very fingerprint of the catastrophic divide in the Axial Age, so different are these views of heaven. Perhaps humankind did not feel sufficiently secure with this purely secular foundation of cosmic order, because it was replaced after less than two centuries by the Platonic and Aristotelian "theological" cosmologies (there is *ultimate* order in the heavens, despite local disorder, because a Divine Soul or Prime Mover causes it). Such a sequence, I feel, is a much truer description of the so-called "period... from Thales to Aristotle." It went from an apparently objective but really selective objectivity (the pre-Socratic emphasis only on the good side of nature) to the wishful reality of Plato and Aristotle (nature may be good and bad locally but is solely good ultimately).

The contribution which this "theological astronomy" made under the great illusionist Aristotle can therefore now be understood in its collective and self-delusive psychological sense. This portrait of the universe into which the traumatised shelter-seeking Greek cosmology finally evolved was the finest sort of wish-fulfilling pantheon, for two reasons. First, it made the universe orderly because *the deity which controls it is beneficent*. This meant that the universe is spiritually stable. Second, it extended this divinely-ordained cosmic order from the present to the entire past and the entire future of the universe. As Farrington (once again more astutely than he knew) put it

> "[...] from now on, the heavenly bodies, being divine and therefore immortal, cease to have a history." [17, p. 43]

Here is where collective denial of catastrophe has fully taken over, for the message of this *kindly* religious vision of the sky is simple: if the heavenly bodies have never changed, (if they do not "have a history"), then they do not have a violent past, *nor can they have a violent future*. They have always been what they are now and always will be, (just as in modern uniformitarianism). The bottom line of this paradigm is that the world is safe. Catastrophe has been made impossible. This line of increasing escape from cosmic fear is the motion that the so-called uniform and motionless "period...from Thales to Aristotle" followed. At all stages, it ran away from the cosmic past and hid itself in delusion. The result for our own age is evident: if each phase of Greek philosophy, up to and very much including Aristotle and Plato, was an escape from consciously facing the truth about the Axial catastrophe, then that is the message which the Greeks developed and the West inherited. It is only now in the late 20th century that we are beginning to face the whole of the catastrophic equation.

Perhaps the point may be well illustrated by an anecdote. Farrington, in his short little wartime book *Greek Science: Its Meaning for Us*, [17] points out that the Babylonians attributed all acts of natural creation and process to their god Marduk, whereas the pre-Socratics tried to "gather together into a coherent picture a number of observed facts *without letting Marduk in*." [17, p. 31] That was their compromise. In Plato and Aristotle, the catastrophic past still remains obscured in favour of a vision of eternal stability, but at a different price: Plato and Aristotle have *let Marduk back in*. With the pre-Socratics, astronomy had moved from the theological to the geocentric. In Plato and Aristotle, it had become theological again, but both denied catastrophe.

4. The great athletic contests

The third category of major cultural indicators of a possible physical catastrophe which might have occurred just before the Axial Age are the pan-Hellenic athletic competitions. There were four of these, (the Pythian, Isthmian, Nemean and Olympic Games), and all can be arguably dated as having emerged late in the 6th century. [34, 35] They are of course sporting events, but I will try to show that, when we interpret their *symbolic* function through neo-catastrophist eyes, (i.e., considering them as unsuspecting symbolic expressions of collective fears and feelings), we will see that they too, like the religions we have just looked at, imply that a cosmic menace to the Earth is over and gone. To wit, we shall discover that all of them celebrate the defeat of a giant monster by a giant hero, in which the champion represents the shining sky god who defeats the dark ominous dragon representing heavenly chaos. The hero then establishes a new order throughout the physical universe, it being the very order which has persisted to the present. What is being celebrated in all of these games is therefore the same thing - a cosmic victory over universal disorder and the emergence of the *present* stable sky.

It can be retorted of course that each great Game must be considered separately because it had its own distinctive hero and monster, but a closer look shows that they are all related as mere variations on one underlying theme. For instance, the Pythian Games at Delphi commemorated the victory of Apollo, son of Zeus, over Python, son of Ge, while the Nemean Games at Nemea honoured Herakles, son of Zeus, who strangled the Nemean Lion, offspring of Typhon and Echidna. At Corinth it was Poseidon (son of Cronos) in the Isthmian Games who was worshipped for his part in the defeat of the Titans, sons of Ge, while at Elis the Olympic Games celebrated the victory of Zeus, son of Cronos, over Typhon, son of Ge and father of Echidna. Ge however is the mother of Echidna and is also similar to Hera, in that, while Ge gave birth to the Nemean Lion and the Hydra, Hera is also said to have reared them and to be the mother of Typhon.

I have made reference to the confused and crazily interconnected genealogies of the different Greek monsters and heroes to show that Fontenrose [18] is correct that all of these myths are variants of one primal tale. The females as mothers of evil are alike (Hera bears Typhaeon, Ge Typhäous), the monsters are alike in parentage, attributes and attitudes to order (Typhon, Python, the Nemean Lion, the Titans and the Hydra all attack celestial order), and the heroes are also alike, as are the battles in which they engage and the outcomes. We may therefore guess that, however these original local combat stories may have been regularised in later redactions of Greek myth by making one participant the offspring or parent of the next, they are really parallel and analogous versions of one original source.

Fontenrose sees all of the great Greek combat myths "as essentially one." [18, p. 274] Whether it be major figures like Apollo and Zeus, or Herakles, Perseus or Kadmos, they are all variant "dragon slayers." [18, p. 275]

> "Herakles ranks with Zeus and Apollo as dragon fighter in the myth-making of the Greeks; any one of the three might be the champion in the great variety of tales that ring the changes on the combat theme." [18, p. 358]

The major items in the myth, as assembled by Fontenrose, give us the picture of a giant world-threatening but then happily world-saving battle in the sky. (The following selection follows Fontenroses numbers.)

> "1. The champion found not a single enemy [...] but two great enemies, male and female.
>
> 4. The combat-myth is a conflict between order and disorder, chaos and cosmos.
>
> 5. Chaos was [...] the habitation of monsters and demons, the land of the dead; it preceded the cosmos.
>
> 7. The dragon withheld the waters, which contained the materials of the ordered world; the hero had to kill him to release them.
>
> 16. [These are all stories] from which cosmic meanings have tended to disappear, though sometimes they cling to it in a disguised or obscure form." [18, pp. 465-466]

Fontenrose compares these Greek myths to the Canaanite religious tales "of Baal, Yam, and Mot," [18, p. 467] as well as to the Vedic myths of Indra, (Ibid.) and insists that in each instance every religious myths retells the same story - the hero's difficult but finally successful effort "to bring the cosmos into being". This is why he insists that

> "In the myths [...] we caught sight of the cosmic meanings of the combat." [18, p. 469]

It is a battle for the universe between totally different approaches to the universe.

> "[...] the hero-god champions the forces of creation, life, activity, and order. For creation is an active, life-asserting process that gives shape and form to unordered material." [18, p. 473]

This is what the hero represents, *change* which leads to improvement, change as a short period of violence and unlimitedness which then allows cosmic organisation and limit. (Cataclysm is therefore essential.) The dragon on the contrary opposes action and change.

> "But an order that resists all change [...] denies life and turns it into its opposite: it becomes a state of inactivity and death. The dragon enemy champions the forces of chaos, destruction, inaction, and death [...] It returns to chaos and death. "[18, p. 473]

What is important for our purposes is the agreement among many researchers that the Ur-story behind the combat myths is cosmic, for this is what allows us to propose that it may refer to a real historical event. We see this when, in each instance, the divine hero who is offspring of the major deities of the previous generation is opposed by an (equally divine) demonic dragon-enemy, and often a gigantic female, who attempt "to overthrow the gods and rule the universe." [44, p. 594] The combat which ensues is therefore a *battle for control of the physical world.* The hero, armed with his special divinely-sanctioned weapon, which is the only implement able to wound the almost-invulnerable monster, defends cosmic order and stability, while the monster seeks to destroy it. These are the true dimensions *and the stakes* of the contest, for, in the typical giant battle for control of the sky exemplified by Zeus and Typhon hurling mountains as missiles, the son of Chaos is defeated *forever.* In each version he is buried somewhere deep from where he will never emerge, or not for a long time. Significantly, after the monster is disposed of, the hero then sets about to restore order, but it is not the Old Order which he reconstitutes: it is a new heavenly structure with an important difference - it will hopefully last longer because it is less vulnerable to disequilibrium. The victory of the hero has therefore created long-term hope out of short-term violence, for it has saved heaven permanently from dissolution.

This great mythical battle, said one innocent classicist with unknowing truth, "is generally believed to have been a personification of terrifying natural phenomena...of cataclysmic proportions." [44, p. 595] He was, of course, referring strictly to terrestrial natural events, but we may go further and speculate that the original events mirrored in the myths were not merely *natural* but cosmically-induced, and that it is this *apparently world-saving* decisive moment in the skies which the Games embody and enshrine. "The struggle between god and snake," we are told, "figured importantly in various rituals performed at Delphic festivals of Apollo." [44, p. 508] We may extend this to every Apollonian festival, including the Pythian Games at Delphi, and to all the Games, as rituals honouring the violent but creative instance which brought the present Age into existence.

We can approach our point in a slightly different way by saying that, as we have already seen in our study of the new religions of the 1st millennium, the combat is a battle between symbolic generations. The Old Order, the Old Religion or the Old Nobility represent in the religious transition just what the dragon does in the combat stories - a clinging to a defunct, lifeless and oppressive status quo, which is a desire to impede the New Order desired by Heaven. It is malicious and counterproductive - in Fontenrose's phrase, it "champions [...] chaos, destruction, inaction, and death." The battle in the skies which the Games commemorate is therefore a contest between action or inaction, life or death, represented by the hero as Change or the dragon as Stasis. In the myth underlying each Game, it is change which is preferred, because change brings new life, it defeats death and eventually makes things better and more stable. That is the message behind all the combat myths, and, if a tumultuous cosmic event lies behind them, it would have to be one during which the "hero" as self-sacrificing offspring of the Old God does not shirk his duty, but *takes action* and defeats the Dragon of Disorder, (Joseph Campbell's "Hold Fast"), and then hands the world over to the New Time and the new gods, withdrawing to the distant skies. The significance of this master tale *for its time* is inescapable: it is the story of the origin of our present Age which was formed out of this black-versus-white battle in the sky. The Greek combat which the Games honoured is therefore *our cosmic patrimony.*

The common message underlying all the Games is therefore one of comfort and joy. It is that *the world is now orderly, i.e., it is safe from cosmic disturbance*, and Heaven will not afflict humankind from now on, or not for a very long time, and the world can celebrate. I would say that the Games served a therapeutic purpose, in that the athletes emulated the actions of the cosmic forces which had been seen on the screen of the sky, with the winners representing the original divine sky champion, able to defeat all enemies at cosmically symbolic feats like running or jumping or hurling weapons like the discus and javelin. [37] The Games therefore re-enacted in safe and re-assuring imitation the victory of stability over chaos in the sky. The Old Order, seen somehow to be feminine in its unstable evil, (women were forbidden at the Olympic games), deserved its destruction and was correspondingly brought to an end, replaced by a new masculine Order which will be stable forever. Death, in the form of a giant cosmic serpent, has been defeated, and it is life on Earth, life in the universe, Life itself, which has triumphed. That is the meaning of the Greek athletic ritual, and it would hardly have lasted for over 1000 years had it not been believed to be of vital importance.

There is evidence that this tendency toward celebratory and ritual re-enactment is found in many cultures. David Pankenier, for instance, outlines what he insists is the "primal conflict" in Chinese myth, which is the battle between the hero, known as the Yellow Emperor, and the cosmic dragon Chi You. (Pankenier, this volume.) We shall find that it is a mirror image of the combat in Greek myth.

> "Scenario: Sudden intrusion of chaos on peace and order. A battle to confront fiery disaster.
>
> Protagonists: Yellow Emperor, a dragon-like aspect; spirit father and astral double is the Pole Star > emblem of constancy and stability [...] plays a cosmogonic role in making the world safe and orderly." [Pankenier, this volume]

The Yellow Emperor, like Zeus and Apollo, "commands celestial spirits" and has a divinely-blessed weapon. He is "possessor of Heaven's Mandate" and his task is precisely to create change for the better by action: "initiates organised violence to achieve order." The enemy, meanwhile, is a double of Python and Typhon (and Marduk and Humbaba and Ahriman and Quetzalcoatl, and so on).

> "Chi You: most threatening miscreant and initiator of chaos [...] archetypal bestial warrior who kills indiscriminately; hideous aspect [...] wields weather weapons. [Pankenier]

Chi You rages in the sky but is ultimately defeated by the Yellow Emperor (YE) in the final eschatological battle, just as Marduk vanquished Tiamat or Theseus the Minotaur. He is then made harmless by being "transformed [...] into YE's votary" and afterward being placed in heaven by the Yellow Emperor as a comet. From the outcome we can see that the Chinese saw this battle as the event which established the present world order, and we are told that it is celebrated in the Qun Han New Year's ritual, which includes sporting events games and tests of strength (bull-grappling, cosmic kickball, horn-butting), all of which are "ritually reenacting primal combat." [Pankenier]

The similarity of this Chinese practice to the Greek Games and their common heavenly significance could hardly be closer. In both, a cosmic hero defeats a cosmic enemy forever, disposes of him and organises *the present order of the world*, and it is this act of cosmic salvation which is then annually re-enacted and celebrated to signal the disappearance of danger from the sky. The implication (for Chinese culture as for Greek) is clear - as long as this ritualistic *remembrance and re-occurrence* continues, world menace will not return. It is both sympathetic magic and ritual comforting. That is the ultimate purpose of these rituals and religions. They are devices by which we make ourselves feel good about the universe.

5. The Tyrants

It would be unlikely, of course, that the area of politics should be immune from the spiritual revolution which I am documenting. I turn to this topic now to show that the very same pattern occurred in the field of government. The major example comes from Greece, where a very significant political sequence is to be found in almost every region of the country and its colonies. It involves a change from an oligarchy to a tyranny to a democracy, it occurred all across Greece within the same very short space of time (late 7th to early 5th centuries), and it recapitulates the same set of upheavals that seemed to be implied in the religious, cosmological and athletic revolutions I have documented. In all of these major social areas, there was an evolution from an Old Order of servitude to a violent interregnum of confrontation to a final condition of civil peace and egalitarianism. We may therefore guess that this phenomenon too was a response to, a copy of, a flattering imitation of and a magical model for the future behaviour of, the heavens.

Let us look at Athens in the 7th and 6th centuries. It had been governed by the noble tribes, from among whom were selected the Archons who ruled the Council. Their administration may be epitomised in the "Draconian" laws of Draco, which were an intensification into aristocratic tyranny. He was replaced, however, by the much-different Solon, himself not a noble man, whose many reforms all went to help the middle and lower classes rather than the nobles. Solon eased the debt burden, redeemed slaves who had been sold, reduced the severe punishments for minor crimes, set up a public court of appeal against aristocratic decrees, established elections governed by universal suffrage, organised a Council of freemen to initiate legislation and re-evaluated the shaky currency. It was a wholesale victory of democracy over terror. Shortly after Solon's archonship ended, however, Peisistratus the tyrant took power, followed by his sons Hipparchus and Hippias, but they in turn were expelled by Cleisthenes, who finally completed the shift of power from the nobles to the people which Solon had begun a century before - he formed a ruling Council selected according to lots within a system of proportional representation. It seems to have taken two tries for the sequence to work itself out in Athens, but it did - an oligarchy was replaced rudely by a tyranny which led to democracy.

We have then a period of about one century during which this tremendous revolution in politics took place, and the eruption was echoed all over the Greek-speaking world: Thrasybulus the tyrant took power in Miletus, Polycrates in Samos, Theogenes in Megara, Periander in Corinth, Pheidon in Argos, Theron in Acragas, Gelon in Gela, Hieron in Syracuse, and so on. There was a veritable panorama of tyrants. That was not the end of the process, however, for a subsequent and equally revolutionary stage then ensued - by the middle of the fifth century, almost every tyrant in Sicily and Greece had been deposed and replaced by a democratic society. It is a phenomenon startling in its upheaval and its similarly from place to place.

We must be sure, however, that we appreciate just what this tyrant phenomenon represented. First, not every "tyrant" was cruel or tyrannical. The word as applied to these men did not carry the overtones of ruthless dictatorship that it later acquired. Second, many of the tyrants had considerable popular support. They were not gangsters ruling by terror. Third, they were essentially anti-aristocratic non-nobles who abolished class distinctions, broke the hold of the hereditary class and took power for the rising middle class from the aristocracy. It was a revolt of the citizens against the nobility that they represented, and in this we may see the same pattern that we saw earlier: an Old Order, associated with the Old Rulers and the Old View of Life, was violently supplanted by the intensive aggression of a New Force which defeated the Old Order by combat and reduced its power forever. This "epidemic of tyranny," as it has been called, was therefore a total revolution which destroyed a previous political Age.

The reign of the tyrants, however, was short-lived in every instance, just like its counterparts which overthrew the Old Religions or the Old Cosmologies, because it did not become an end-point, but functioned rather only as the medium of transition to the next stage. That next stage of course became the true New Order, which in this instance was the egalitarian, human-centred state of democracy which has persisted up to the present, conducted by humans for humans and not by slaves who serve the Deity and its aristocratic ministers. Just as the victory of Zeus (New Cosmic Force) over Typhon (Old Cosmic Force) paved the way for the pre-Socratic cosmology, or the revolutions of Buddha and the Biblical Prophets against the prevailing Old Religions of their time paved the way for the more homocentric New Religions of Buddhism and Prophetic Judaism, the victories of the tyrants over the Old Order of the nobles paved the way for the New Political Order, the homocentric democratic society which ensued everywhere.

6. New cities, new writing

Alongside the changes in religion, cosmology, government and large public ritual which we have noted, there was a corresponding alteration in the very nature of Greek cities, as one would expect in a severe ideological revolution. We know the new type of city as the Greek *polis*, and we are familiar with its political and artistic achievement, but it is important in terms of this study to understand in what particular ways the *polis* differed from Bronze-Age cities, for this will demonstrate in its way (as the new religions, new philosophies and new governments did in theirs) the drastic modifications in outlook which occurred in the Iron Age.

One must be careful not to over-generalise, but it can be safely said that the typical Bronze-Age city was most of all a temple centre, in that ultimately everything circled around and was only an adjunct to that religious building and its rites. All the residents were servants of that religious function, and even the monarch merely executed the desires of the deity. The Bronze-Age city was therefore a cult centre where society was ranged in a hierarchy run by the priests and the courtiers and established by one's relevance to the rituals, for everything was set up to serve the deity and his temporary incarnation, the ruler and his court. As a physical place, the City was the symbolic peak of the world and, as Eliade has persuasively argued, [15] the temple at its top was believed to be the gateway and connection to Heaven, the pipeline or umbilical cord up which supplication rose and down which blessings flowed if the rituals were well performed. The continued welfare of the Bronze-Age state therefore depended directly upon the maintenance of a friendly relationship with Heaven achieved by these rites, and it was to this that the Bronze-Age city was dedicated. In such a scheme, writing was an esoteric art and skill possessed only by the initiated few, and it was used almost exclusively to record for posterity the great truths about the world and the universe which had been revealed by the Deity.

The *polis*, in contrast, which emerged in the Iron Age, was a human community governed by human law, an egalitarian society of citizens in which the individual identified with the city as his community and dedicated himself to it for this reason, more than to any deity it served. [16, 32, 43] It was not ruled by the priests but by the citizens, and its aim was to aggrandise itself not as a religious site but as the centre of an agricultural, manufacturing and commercial economy. The land did not belong entirely to the temple or the monarch, but for the first time (as Gunnar Heinsohn has shown [23, 24]) was divided up among the citizens, who were independent and free and ruled themselves democratically through a legislative Assembly populated by equals. Such a City performed an earthly function for its participants, run by the elected representatives of the people, not a sacred function of rite and sacrifice run by priests, (the representatives of the deity), for it did not acquire its legitimacy from heaven but from the consent of the governed. Even when there was a single ruler, he was not seen as the intermediary with Heaven but as embodying the desires of the population. To use the well-known American phrase, the *polis* was a creation of, for and by the people. It is the same evolution which we have seen in all areas of Iron Age culture, and, if Below does echo Above, then we have to suspect that, in designing their cities after the Axial catastrophe, the people of the Iron Age no longer saw any need to keep on propitiating a menacing Heaven and began on the contrary to concentrate on themselves and their society as the prime centre of value.

Writing changed accordingly from an aristocratic and secret technique to a public skill which served the everyday concerns of the citizens. Cursive writing developed, which allowed quick and simple notation and suited the pen, and the materials used were papyrus and skins and wax and wood tablets. These were far more perishable than stone or clay incised writing, but they were not as bulky and therefore more portable and easy to store, all of which is consonant with a commercial rather than religious society. Even potsherds, shells and pieces of bone were used to write on, which are clearly indicative of the mundane use of writing. There was at the same time a rapid spread of literacy, so that writing was not confined mostly to the priests, and a simplification of the symbols, as writing became a "useful tool in the transaction of everyday business." [31, p. 146] So great was this change, this "democratization of learning" where writing and literature were no longer the "monopoly of a privileged clique," that it has been described by McNeill as "one of the major turning points in the history of civilization," a cultural shift which was "to alter fundamentally the cultural physiognomy of the Middle East" [31, p.146] in every region where it occurred. Writing changed from a theocentric preserve to a homocentric public possession, as if the people of the Iron Age felt free to neglect Heaven in favour of themselves and use writing for their own daily activities.

7. Dating the 1st millennium BC catastrophes

With the individual cultural items having been set forth, we can now put the evidence together to see where it leads. We can approach the data as a detective might a set of clues, to deduce what the items might be saying. With regard to dating, it appears to me that the combined social phenomenon which I have presented is coherent, and that it is datable to no earlier than the sixth or fifth centuries BC. What then are we to make of this distinct and revolutionary cultural outburst? Where is it to be integrated into a new and properly inclusive chronology of the Bronze and Iron Ages? If we accept the hypothesis that Culture follows Nature, then, in the light of what I have presented, we have no choice but to guess that something drastic happened in the skies not too long before the cultural upheaval, which leads us to ask of course when it occurred and what was its cause.

A significant amount of new evidence has appeared which points consistently to a very large and perhaps non-terrestrial cataclysm early in the 1st millennium BC, sometime around 800 to 700 BC. Palaeo-ecologists Bas van Geel *et al.* present evidence for an "abrupt climate change" in north-western Europe which he dates to what he calls "the Bronze Age/Iron Age transition," which he says occurred c. 850 to 760 BC. They have also added "palaeoecological, geological/geomorphological and archaeological evidence for a synchronous climatic change elsewhere in Europe and on other continents." [Van Geel *et al.*, this volume] Founding their deductions on the study of pollen and microfossils in organic-deposit strata like bogs, they argue for "world-wide dramatic changes in climate around 800 BC." [Van Geel *et al.*] A transition on this scale of degree and locale can hardly be attributed to a local or normal cause. Similarly, Lars Franzén and Thomas Larsson have studied sediment deposits in Sweden and Tunisia and discovered signs of very heavy flooding, severe frost and extreme sedimentation dated to about 1000 BC, which they say marked "the most abrupt transition from aridity to very humid conditions." [Franzén & Larsson, this volume] In the sediment, Franzén and Larsson found glassy spherules which could be of volcanic origin, and they speculate that they may have been formed by cometary impacts in ocean shelf sediments. They have also come across shocked quartz, (indicative of impact), and conclude that all of these signs point at a very turbulent Holocene with several sudden and dramatic climate events.

This too does not indicate a normal or strictly terrestrial cause for these events. In a similar manner, Euan MacKie has studied late-Bronze Age 'vitrified forts' and rock carvings in western Scotland, and his deductions converge on those of van Geels and Franzén, for he thinks his data could be evidence for earthquake-proofing and signs of unusual cometary activity just before the Iron Age. [MacKie, this volume]

To these complementary discoveries must be added three more items. The first is the archaeological research interpreted by Gunnar Heinsohn. [21, 22] Using indications of destruction layers in the strata of ancient civilisations, he has deduced that a large-scale catastrophe which terminated the Late Bronze Age and led to the Iron Age occurred about 730 to 700 BC. This downdating of the late catastrophic sequence by Heinsohn corresponds with the meteorological data collected by Charles Ginenthal in his just-published study of the disappearance of the mammoths [19]. Ginenthal has assembled evidence from the latest scientific literature which indicates abrupt changes in climate in the mid-2nd and then mid-1st millennia BC. These changes appear to have occurred quickly and were long-lasting, as illustrated by patterns of floral growth quite different from the present, (e.g., large trees far above the Arctic Circle). Third is the celestial interpretation of Egyptian history presented by classical historian Lynn Rose [38]. Using computer-aided astronomical retro-calculation to analyse Middle-Eastern observations of the heavens dated to the 8th or 7th centuries, Rose argues that they show a sky slightly but significantly different from today's. What can be said about the dating of the Axial catastrophe, therefore, is this: if the combined indications of van Geel, Franzén, MacKie, Heinsohn, Ginenthal, Mullen and Rose are in any way correct, they do point very powerfully to some sort of large cosmic disturbance beginning around 750 BC and ending perhaps at 700 BC which ended one historical period and started the next.

Taking as our working hypothesis the conjecture that a global natural trauma underlies the vast synchronous social transition I have traced, we then have to decide how long it took for that social revolution to come into being after the physical upheaval. I doubt that it would have taken more than a century and a half, long enough for the eco-system and the socio-system to recover, for observers to see and pronounce the regularity of cosmic order and then for the creators of culture to evolve the collective social artifacts of the Iron Age. These would have included items which embody the catastrophe (but only as myth), and items which deny or ignore the catastrophe in philosophy, religion, government and communal structures. Between them, these twin devices of remembering-in-pretence while forgetting-in-reality would safely bury any conscious awareness of cosmic menace, and this process has an established prototype. It is what an individual might do after a giant trauma, and, as the latest studies on collective trauma have solidly documented, [9] it is what a traumatised culture might also do. There is always seen an intense desire to escape from the pain of terror by any sort of comforting rite or illusion that can be devised, and this is always done as quickly as can be. Robert Jay Lifton, in his monumental *Death in Life: Survivors of Hiroshima* [28], describes responses that began only months after the atomic catastrophe. We must follow where the evidence leads, and the paths I have outlined lead to the same point. The moment we assign the Axial *cultural* revolution to the 6th and late 5th centuries, this makes the placement of the *physical* catastrophe in the 7th century (or later) imperative.

As an epilogue to this section on dating, I will make quick mention of one more item which, while it cannot be considered definitive, may in the future turn out to be relevant. Only time will tell. It is the theories of Julian Jaynes. He believes that our present mode of consciousness and our present form of dreaming came into being only at the beginning of the Iron Age, and Peiser "associates this sudden intellectual revolution during the so-called *Axial Age* with catastrophic events at the end of the Bronze Age." [36, p. 5] It is hard at the moment to know what to make of Jaynes, since he has not found much acceptance in psychology or cosmology, but his insistence that a previous condition of bicamerality or bilaterality in the human mind gave way to modern consciousness, and of "I-and-Thou" dreaming to modern dreams centred solely on the dreamer, and that all of this took place around the 7th or 8th centuries BC (which is when he dates the Axial upheaval), links him in some way to the neo-catastrophist endeavour.

8. The nature of the 1st millennium BC catastrophes

With regard to what happened, I said at the beginning that the data I would present as evidence for a catastrophe in the 1st millennium BC would not be scientific, but societal. (It would not be "hard," but indirect). If nevertheless Culture does follow Nature, then perhaps we can reconstruct something about the natural events from the cultural responses.

We see right off that there appear to be three stages in the story these data tell, which we might call Before, During and After whatever happened. *Before* is the previous Age, (the Bronze Age), represented by all that was rejected in later times - the aristocracy and priesthood and rituals and sacrifice and polytheism and temple centres and elaborate elitist texts and secret forms of writing and oligarchies and mythical tales of anthropomorphised petty uncaring gods. It is a world not concerned with ordinary people except as servants - one serves the nobles, the temple, the deities and their ceremonies and rites. All is subservience and propitiation, as can be gleaned from the available social evidence. Authority (as god or aristocrat or priest or scribe) was to be feared and mollified, or the consequence might be swift and arbitrary retribution.

To reconstruct *During*, which ended the Age Before, we turn to the story told in the combat myths which underlie the great athletic games. The tale depicts a terrible menace to the entire cosmos, a dark deadly evil dragon figure, divine like all heavenly bodies, who seeks to either subvert all order or to resurrect in hideous form the Old Regime (Python whom Apollo killed and replaced was the previous deity of the oracle at Delphi.) The Old Order, however, cannot be allowed to come back to life, for it led to the present world danger and might again. It therefore has to be permanently defeated, eliminated and replaced. This is the task of the champion of the New Order, also divine. He is the gleaming Light of the World, who battles the dark foe alone, the remaining "gods" having fled. After much to and fro, and lots of damage to the world and the hero, he is able at last to defeat the enemy, whom he imprisons in a nether place from where there can be no escape. It is then, when cosmic danger has been removed, (when the Old Order will never return), that the hero establishes a New Order which is far better because it is less volatile because it is more firmly founded. This last great battle in the sky therefore marks a point of *permanent* transition from the Old to the New, from instability to stability, from the decrepit to the young and strong, from the Past to the Future. It is *good*. That is the *During*.

The *After* is of course the New Age, the age of *humankind*, of democracy and the *polis* and cursive writing and commerce and materialist philosophy and religions of righteousness supervised by a kindly but distant monotheistic deity. This is quite opposite to the *Before* as I just described it, for here we have a set of purely homocentric concepts and structures in a *world which cares for people* and sets up *mundane* institutions for *human* betterment and progress. What is significant for our purposes now is to become aware that this portrait of current Earthly history is presented by our ancestors in two modes at once, the first astronomical, the second moral, with the pair said to be intrinsically related as effect and cause: the moral guides the physical. It is a cosmic tale with a moral structure.

To discern this, we note first that each part, the astronomical and the moral, is depicted as occurring in a sequence of three stages. The astronomical tells of an Old System that was intruded upon by some sort of highly energised body or mass of matter, which resulted in charged and tidal interactions, after which the intruder was ejected or diminished and a New System took over the order of the heavens. In the moral part, the aim in my opinion is to give this physical sequence a teleological and *benevolent* purpose. Here the three stages are recounted in such a way as to indicate that the terrible cosmic violence occurred only because the Old Order was about to collapse, and only the eruption of a cataclysmic interchange was able to create the New Order, which is better because it is more stable and will therefore endure longer. The message is therefore that the catastrophe *is not ultimately destructive but benevolent*. Despite the damage caused, it had a positive divine objective, which was to end a dangerous situation for Earth at all costs such that out of any regrettable chaos a new and better condition could emerge, (that condition, we are told, being what the benign Heaven wants). I.e. there had to be cosmic violence in order for there to be cosmic salvation. In my opinion, we have here a physical narrative of the heavens presented alongside a solacing moral overlay which was invented to put the physical in the kindest light. When this Before-During-and-After sequence is interpreted from a neo-catastrophist point of view, as evidence of a cosmic catastrophe followed by avoidance of the trauma, we perceive the real intent of post-Axial culture: it is to convey a theocentric message lying beneath and directing the homocentric surface. It is an attempt at the heavenly legitimation of the present (safe) World Order, and it is in this sense that the Culture of the Iron Age evolved from the Nature of the world during and after the Axial catastrophe. The first was a direct response to the trauma of the second.

There are several observations of a psychological nature which must be made about the moral interpretation which the survivors leading up to Aristotle put on these events, for they are what lead us to suspect that an unconscious memory of catastrophe persists in the human mind. To wit: the moral "spin" which the survivors gave to what was in reality a devastating occurrence was that it was all a very good thing in the long run. There *was* a great deal of violence and loss, certainly, but that violence was necessary, for only out of its liberation of energy could the new and better world, the homocentric and stable model, be created. The whole cosmic combat, therefore, no matter how destructive, was made into something ultimately good. It was presented as a moral clash between Good and Evil which *allowed change* as directed by heaven. One can then see how this attitude, that a divine force created disorder precisely to bring new order into being, might lead two centuries later to Plato's Divine World Soul, (which tolerates short-term disorder but always leads the universe to long-term perfection), and even to Aristotle's Prime Mover, (who allows occasional sub-lunar disturbance but keeps the super-lunar spheres singing in ordered harmony). It could also lead to the story of Osiris and, three centuries after that, to the Christian "spin" on the Fall of Man, whereby what at first appeared to be tragic became only a "Seeming Paradox," a "Fortunate Fall," as Adam's error has been described, because, had not Adam sinned, Jesus could not have sacrificed himself to *save humankind forever* from the Original Sin, just as, had not the Old Order in the sky degenerated and produced the Dragon, the Champion could not have established the New Order forever. Can it be that the origin of these moral reconstructions of physical destructiveness in ancient Egypt, Israel and Greece may lie in actual cosmic events not long before and how we first reacted to them?

9. A neo-catastrophist research programme

My concluding task in this paper is to assess the value of what I have presented and to try to define where it might fit into the burgeoning neo-catastrophist research agenda. We all accept that interactions between our Earth and bodies in the sky have caused cataclysms within the historical period. Of particular interest to me is the connection which many perceive between the occurrence of catastrophe and the emergence of new culture. Verschuur writes that the catastrophes "created the context within which life emerged and evolved." [Verschuur, this volume] This is very important, because it connects *catastrophe and civilisation*.

> "[...] we confront the stark fact that we live on a planet that has been shaped by violent and catastrophic events, catastrophic from the point of view of species and civilizations that might have been unfortunate enough to experience them, yet creative from the point of view of those that followed." [Verschuur, this volume]

Culture is being described as a creative response to the assaults of Nature. This leads Duncan Steel for example to propose that a pair of cosmic catastrophes in the 4th millennium provoked the construction of the Great Cursus and Stonehenge I. [Steel, this volume.] Mike Baillie similarly wonders whether abrupt climate changes in the second millennium BC might relate in some way to the start and the end of the Chinese Shang dynasty, [Baillie, this volume] which is supported by David Pankenier, who provides Chinese evidence bearing on the nature, extent and implications of disaster for Bronze Age civilisation in East Asia. [Pankenier, this volume.] Along the same lines, Bruce Masse believes that 20-30 at least locally catastrophic impacts likely occurred in various portions of the world during the past 6000 years, during which time occurred the major developments of human civilisation. [Masse, this volume.] These speculations have a common theme, that catastrophe leads to new civilisation: revolutions in Culture arise from the behaviour of Nature.

What is at issue, however, is the question of dating. I have been laying great stress on the 1st millennium BC because that is where my evidence leads. If we agree on the principle that cataclysms create culture, then the evidence for a massive cultural re-orientation mostly in the 6th century BC, one which involved a new religious outlook, a new cosmological outlook and a new social outlook, all in favour of humankind at the expense of the sky and its deities, has to call for a visible change in the heavens not long prior, and the gap between physical event and social result should not have been much over one century if the damage to the Earth was not extensive. It is this fundamental contrast which is the foundation of my argument that the Iron Age was different from the Bronze Age, so radically and abruptly different that we must look only to the skies for an explanation. Something quite drastic may have happened.

Here then is a project for the neo-catastrophist agenda, to sort out the catastrophic events of the 1st millennium BC and their dates and their causes and their effects. I am not saying that catastrophes occurred only in the 1st millennium, or that they were the most important, but that they cannot be overlooked. Anyone advocating a neo-catastrophist or specifically cometary model of cosmic behaviour to explain the punctuated appearance of revolutionary modifications in civilisation, belief and ritual in the Late Stone, Bronze and Iron Ages will have to be able to integrate this 1st-millennium data in combination into one coherent and plausible sequence. The event need not have been large in astronomical terms, for, as Clube and Napier point out, a catastrophe can be major in its human consequences even if it is relatively minor in its physical magnitude. [11] It would nevertheless have to be large enough to cause violent and sensational spectacles in the sky and correspondingly sensational effects on nature and culture.

Why is this important? I think it is because this moment of cosmic and then social revolution which I am arguing for in the 1st millennium BC was *the beginning of our present Age*, which, in its broadest dimensions, has not changed appreciably since that "Axial" moment. That moment is therefore perhaps the most critical point in the evolution and history of our present culture, for it was the time when our culture changed from cosmic to social, from religious to secular, from a vertical subjugation to Heaven to a horizontal equality between individuals. It was the very birth of our Age.

This is extremely significant, for, if we can understand what happened in the sky to cause that change, and what were the collective traumatic responses of our society to these terrifying cosmic events which we witnessed, it will explain us to ourselves. That is what is to be gained. Gerrit Verschuur writes that only now the mind of *Homo Sapiens* is becoming aware of its existence in a cosmic context. What we are aware of, he says, is truly terrifying because "we appreciate that we are perpetually poised on the edge of extinction". I consider this to be a great step forward, in contrast to what we have done consistently in the past, when, as I have shown throughout this article, we invented cosmological theories and social institutions structured to avoid any such awareness. A neo-catastrophist understanding of our place in the cosmos, no matter how frightening it might be, is much better because we as a species are inescapably the children of our collective past, and, if that past was heavily punctuated by catastrophe, we have to know this to understand where we came from and therefore *what we are*. To fail to do this would be to perpetuate the condition of self-imposed blindness to cosmic reality under which we all labour.

This condition can be perceived the moment we analyse the vision of nature which was finally adopted by the Greeks around 350 BC, for then the evolution of our hysterical terrified response to the fragility of heavenly stability becomes easy to chart. We merely postulate that a collective form of Post-Traumatic Stress Disorder (PTSD) took over, as most authorities on trauma would say, and at once a series of expectable steps in the avoidance of trauma is seen to have occurred. First there was a denial of the Old Order and its break-up as if it never existed, and one spoke only of the Safe New Order. Then the Safe New Order was enshrined not merely as the present condition but as divine law, as the eternal condition, which permitted a denial not only of past catastrophes but of future ones too, of *all* catastrophes, which *became forbidden*. They *could* not happen, ergo they had not, and the trauma had been totally suppressed. After that, we merely had to keep on believing it, which is what we have been doing for 2000 years from Aristotle to the Scholastics through Newton. [47] It is a typical post-trauma survivor reaction representing the triumph of desire. Neo-catastrophism is restoring a more rational view in its insistence that catastrophes have happened and may recur, and we must learn not to fear these indications of our subjection to our stellar environment, for they are our social genealogy and our reality. This is the sort of universe we live in, and we must accept it and learn to deal with it.

That is the hope I see in our mutual neo-catastrophist endeavour. Verschuur asks what will happen to humanity "now that we can no longer assume that civilization will go on forever" because we have learned about comet or asteroid impacts.

> "[...] once we become aware of this, what do we do next? [...] what is our species going to do with this knowledge?"

He wonders how catastrophically-knowledgeable humankind will handle the future. My answer is that we must first learn to handle the past. I therefore look at the question of neo-catastrophist awareness retrospectively rather than prospectively, for I feel that, the moment we fully understand the catastrophic nature of our past, *we can become free of the unconscious psychological pressure of it*. That is the major benefit to be gained, for then we will be able to prepare ourselves for the future unencumbered by the unsuspected hindrance of unconscious racial terror and its sad political and religious manifestations (as Victor Clube said) in "millenarianism, utopianism and uniformitarianism", not to mention avoidance of the truth and oppression of dissent. It is this intellectual freedom which we should aim for, and only a group as interdisciplinary and aware as ours is in a position to do this, to put social history and scientific evidence together and perhaps finally discern what happened in the last 6000 years, which would help relieve us of the burden of unsuspected self-delusion under which we have existed since Aristotle. I am confident that the results of neo-catastrophist research will allow us to re-consider and re-write human history in this light, and that this truth will set us free. That is the new programme. We have named the task, we see the work to be done, we are developing the tools and evolving the methods. Now we must go to it.

References

1. Adams, C.J. et al., eds., 1987, *The Encyclopedia of Religion,* Macmillan, New York.
2. Adkins, A.W.H. and P. White, eds.,1980, The Greek Polis, in *University of Chicago Readings in Western Civilization*, J.W. Beyer and J. Kishner, genl. eds., University of Chicago Press, Chicago.
3. Allen, R.E., ed., 1966, *Greek Philosophy: Thales to Aristotle,* Free Press, New York.
4. Andrewes, A., 1974, *The Greek Tyrants,* Holt and Rinehart.
5. Asher, D.J. and S.V.M. Clube, *An Extraterrestrial Influence During the Current Glacial-Interglacial.*
6. Ballou, R.O., ed., 1976, (1944), *The Portable World Bible,* Penguin Books, Harmondsworth, Middlesex, England.
7. Blakney, R.B., trans., 1983, (1955), *The Way of Life: Lao Tzu,* New American Library, New York.
8. Bouquet, A.C., 1941, *Comparative Religion,* Penguin Books, Harmondsworth, Middlesex, England.
9. Carruth, C., ed., 1995, *Trauma: Explorations in Memory,* The Johns Hopkins University Press, Baltimore.

10. Clube, S.V.M., 1992, The fundamental Role of Giant Comets in Earth History, in *Celestial Dynamics and Dynamical Astronomy,* 54, 179-193.
11. Clube, S.V.M. and Napier, B., 1982, *The Cosmic Serpent. A Catastrophist View of Earth History,* Universe Books, New York.
12. Copleston, F.S.J., 1962, (1946), *A History of Philosophy: Volume One, Greece and Rome,* Part 1, Image Books, Doubleday and Co., Garden City, New York.
13. Dicks, D.R., 1970, *Early Greek Astronomy to Aristotle,* Cornell University Press, Ithaca, New York.
14. Eisenstadt, S.N., ed., 1986, *The Origins and Diversity of Axial Age Civilizations,* State University of New York Press, Albany.
15. Eliade, M., 1974, (1958), *Patterns in Comparative Religion,* New American Library, New York.
16. Evans, J.A.S., ed., 1974, *Polis and Imperium.* Hakkert Publ. Co., Toronto.
17. Farrington, B., 1944, *Greek Science: Its Meaning for Us,* Penguin Books, Harmondsworth, Middlesex, England.
18. Fontenrose, J., 1980, (1959), *Python. A Study of Delphic Myth and its Origins,* University of California Press, Berkeley.
19. Ginenthal, C., 1997, *The Extinction of the Mammoth,* Special Edition, *The Velikovskian,* vol. III, Nos. 2 and 3, New York.
20. Harris, I., et. al., eds., 1992, *Contemporary Religion. A World Guide,* Longman, England.
21. Heinsohn, G., 1990, Destruction Layers in Archaeological Sites: the Stratigraphy of Armageddon?, in *Catastrophism 2000,* ed. C. Whelton and M. Zysman, Heretic Press, Toronto.
22. Heinsohn, G., 1988, *Die Sumerer gab es nicht,* Frankfurt/ Main.
23. Heinsohn, G., The Rise of Blood Sacrifice and Priest-Kingship in Mesopotamia: A 'Cosmic Decree'?, in *Religion,* 22, 309-336.
24. Heinsohn, G., November 1994, *The Restoration of Ancient History.* Paper presented at Portland, Oregon, November.
25. Hooke, S.H., 1963, *Middle Eastern Mythology.* Penguin Books, Harmondsworth, Middlesex, England.
26. Kauffman, W., 1959, 1960, *From Shakespeare to Existentialism,* Anchor Books, Doubleday and Co., Inc., Garden City, New York.
27. Kirk, G.S., Raven, G.E. and Schofield, M., 1983, *The Pre-Socratic Philosophers: A Critical History with a Selection of Texts,* Cambridge University Press, Cambridge.
28. Lifton, R.J., 1982, *Death in Life: Survivors of Hiroshima,* Random House, New York.
29. McGlew, J.F., 1993, *Tyranny and Political Culture in Ancient Greece,* Cornell University Press, Ithaca, New York.
30. McKirahan, R.D.,Jr., 1994, *Philosophy Before Socrates,* Hackett Publ. Co., Indianapolis.
31. McNeill, W.H., 1963, *The Rise of the West. A History of the Human Community,* The University of Chicago Press, Chicago.
32. Murray, O. and Price, S., eds., 1990, *The Greek City from Homer to Alexander,* Clarendon Press, Oxford.
33. Parrinder, G., 1964, *The World's Living Religions,* Pan Books, London.
34. Peiser, B.J., (1990), The Crime of Hippias, *STADION. International Journal of the History of Sport,* XVI, No. 1, 37-65.
35. Peiser, B.J., 1990, Greek History Begins in the Sixth Century BCE, *Aeon,* II, No. 3, 61-82.
36. Peiser, B.J., 1995, *Cultural Aspects of Neo-Catastrophism: What are the Implications for Archaeo-Astronomy?*
37. Peiser, B.J., 1995, The Divine Child in Ancient Greek Ritual and Myth, in *The Sports Historian* 15, 3-24.
38. Rose, L.E., October 1994, The Astronomical Evidence For Dating the End of the Middle Kingdom of Ancient Egypt to the Early Second Millenium: A Reassessment, *Journal of Near Eastern Studies,* 53, No. 4, 237-261.
39. Rose, L. and Vaughan, R.C., Analysis of the Babylonian Observations of Venus, *Kronos. A Journal of Interdisciplinary Synthesis,* Vol. II, No. 2, 3-26.
40. Smart, N., 1989, *The World's Religions.* Cambridge University Press, Cambridge.
41. Smart, N., 1996, *Dimensions of the Sacred. An Anatomy of the World's Beliefs,* University of California Press, Berkeley.
42. Smith, J.Z., ed., 1995, *The Harper Collins Dictionary of Religion,* Harper San Francisco, San Francisco.
43. Starr, C.G., 1986, *Individual and Community: The Rise of the Polis 800-500 B.C.,* Oxford University Press, New York.
44. Tripp, E., 1974, (1970), *The Meridian Handbook of Classical Mythology,* New American Library, New York.
45. Wolfe, I., 1978, 1979, Worlds in Collision and the Prince of Denmark, in *Society for Interdisciplinary Studies Review.* Part I, II, No. 4, Summer 1978, 37-52. Part II, Vol. III, No. 2, Spring 1979, 21-33.
46. Wolfe, I., 1982, Collective Amnesia and the Catastrophic Basis of Soap Opera, *Kronos. A Journal of Interdisciplinary Synthesis,* PartI, VII, No. 2, Winter 1982, 69-85. Part II, VII, No. 3, Spring 1982, 46-60.
47. Wolfe, I., 1994, A Catastrophic Interpretation of Western Cosmologies, in *Evidence that the Earth has Suffered Catastrophes of Cosmic Origin in Historical Times. Proceedings of the 1993 Cambridge Conference.* Society for Interdisciplinary Studies, London, England, p. 52-55.
48. Wolfe, I., 1994, A Catastrophist Reading of Religious Systems, *Chronology and Catastrophism Review. Journal of the Society for Interdisciplinary Studies,* XVI, p. 2-10.

The Problem of Historical Catastrophism

S.V.M. Clube

Oxford University, Department of Physics, Keeble Road, Oxford, UK & Armagh Observatory, College Hill, Armagh, UK

Summary

Astronomers at the dawn of civilisation perceived danger in the sky and society was notably unsettled [61]. Later, astronomers were to perceive order in the cosmos and society was to become notably less unsettled. The perception of danger never entirely disappeared however and there was considerable pressure on astronomers by the elite of society to uphold a model of the celestial environment in which celestial calm was very much the order of the day. This pressure still exists and leads to an elitist doctrine in which the biosphere's supposedly anti-apocalyptic environment is an essential feature of the cosmological setting. The latest version of this doctrine to acquire favour is largely the work of a single, dominant, intellectual influence, namely that of J H Oort [64] during the central years of the present century when there was a critical break in the general administration of astronomical affairs due to the Second World War. His unavoidable starting point, based upon the most accurate orbital measurements available for comets, was the cometary cloud around the Solar System and its critically valued dynamical isotropy. To preserve this condition against any serious perturbation, thereby avoiding any substantial changes in the near-Earth cometary flux during the course of geological and human history, there had to be very few stars moving slowly past the Sun. This condition was readily upheld if the primary (or mainstream) circulation of the Galactic disk was much the same as the observed (or so-called "standard") motion of spiral arm material in the solar vicinity. Oort essentially preserved the required condition therefore by ignoring two prominent asymmetries known to be present in the large-scale motion of the disk. The first of these asymmetries, orthogonal to the Solar-Galactocentric axis, meant disregarding a set of potentially very serious perturbations of the cometary cloud that would arise with a more probable mainstream circulation which included the Sun. The second of these asymmetries, along the Solar-Galactocentric axis, meant disregarding a potentially gross hubble parameter within the Galaxy in absolute conflict with the standard cosmological redshift. These deliberate evasions, never justified at the time nor since, have become the established basis of a highly distorted cosmic environment which is now systematically imposed upon the untutored public by uninformed specialists and commentators with a view to upholding the long-established anti-apocalyptic tradition. However these evasions are no longer valid since a substantial population of nearby slow-moving stars has been detected by the Hipparcos Astrometry Satellite and distant galaxies having purely cosmological redshifts evidently reveal gross concentrations of the hubble parameter (known as "quanta") in the vicinity of the intervening galaxies. By explaining rather than evading the prominent asymmetries therefore, the motion of the local spiral arm and the mainstream circulation of the disk are now clearly distinguished and it becomes more probable that successive, centrally injected, "grand design" spirals which condense into dark (or sub-stellar) bodies and undergo dynamical friction through the disk are intermittently replenishing the mainstream circulation. It follows in accordance with the geological and historical records that the cometary cloud perturbations giving rise to the near-Earth cometary flux are modulated by a vertical oscillation of the Sun which is predominantly under the influence of a "dark matter" mainstream circulation of the disk. It turns out in effect that the biosphere's anti-apocalyptic environment can no longer be preserved. Thus, by seeking to suppress, ridicule or eliminate the fear of cosmic disaster (instead of seeking to explain its origin, nature and dynamics), the purveyors of an elitist doctrine come to be merely the centrally placed advocates of a "modern enlightenment" which prolongs widespread ignorance as to the actual nature of the cosmic and terrestrial environments. Moreover by incorrectly claiming there is nothing to worry about in the sky, these advocates cultivate an intellectual and cultural climate of irrationality in which the socially most damaging versions of apocalypticism and anti-apocalypticism (eg fundamentalism and millenarianism), those involving an appeal to mystical, incorporeal or spiritual influences in Nature, continue to flourish. The resulting risk to civilization is nothing short of a scandal.

1. The anti-apocalyptic tradition

The space age evidence for apocalyptic threats has already put limits on the extent to which the principle of uniformitarianism can be admitted. The scholarly issue at stake therefore is nothing less than the survival of civilization in the face of such threats. However there remains a particularly vehement tradition within modern scholarship which not only seeks to eliminate apocalypticism from the course of classical, medieval and modern history but also, as a consequence, seeks to impose a highly distorted view of the cosmic environment upon the untutored public. This tradition, to do with comets, is only less vehement amongst the public at large on account of a general presumption, based on historical precedent, that the anti-apocalyptic tradition will nevertheless prevail. A 'scholarly' tradition under these circumstances which again ceases to prevail can only be regarded as a 'scholastic' tradition in due course. On the other hand, the public greatly values the freedoms which have come with the achievements of science and the current anti-apocalyptic tradition, in keeping with uniformitarianism, has been an increasingly established feature of science since its seventeenth century inception under the auspices of the Royal Society in England [69]. This ultimately means that our understanding of planets, stars and galaxies has broadly developed during the past three centuries without any reference to apocalypticism and that the public does not therefore fully appreciate the significance of comets. The reason for this distortion was better understood during the early stages of enlightenment when the anti-apocalyptic tradition was more clearly perceived as a matter of political expediency. Even the Society's most hallowed president Newton, for example, could only preserve his scientific reputation by publishing a posthumous apocalyptic tract [62]. The point here is that a revived anti-apocalyptic tradition did then prevail and was as much an imperial issue as a scholarly one, connecting the aspirations of an emergent Anglo-Saxon Enlightenment with those of Papist Rome and Hellenistic Greece. The tradition indeed can be traced back at least two millennia to Alexandria, the main Near-Eastern conduit through which the recorded knowledge of past apocalypticism first reached the West, and to a then highly distorted (ie Aristotelian/Ptolemaic) view of the cosmic environment which eventually needed all the powers of a medieval witchhunt and Inquisition to keep in place. In short, the history of Western civilization points to a continuous tradition of anti-apocalypticism and distorted cosmology whose advocates do become irrepressibly savage from time to time. This vehemence, it is clear, marks a ready determination on the part of bourgeois society to quell apocalypticism whenever it takes root. This mostly happens not only on account of the successive revelations which cause cosmic catastrophe to be anticipated (as the original Greek word implies) but also on account of the highly degenerate condition into which society is then commonly plunged as a result of its increasingly abject state of terror. But while it is perhaps easy to understand that bourgeois society might seek to impose a distorted view of the cosmic environment for the purposes of avoiding a degenerate condition of society whenever apocalypticism emerges, it is not so obvious in modern times that mankind can afford to allow such distortions to be introduced when there is a *practical* need to judge what should be done about the future state of the cosmic environment. It would appear that there is a particularly vehement tradition within modern scholarship which is no longer required!

The Greeks in fact were principally concerned to adapt eastern knowledge to the needs of empire and were the first to press for an essentially teleological account of apocalypticism which also carried the assurance of ecumenical (ie Christian) rather than nationalistic (ie Judaic) salvation. This priority is perhaps less certain nowadays since similar ecumenical principles were developed by the Zoroastrians [24] and the Greeks may have borrowed their ideas from another nation with imperial ambitions (the Persians). One way or another, the Greeks (ie the gnostics) ultimately took over a 2500 year old apocalyptic record originating from Chaldea and critically introduced incorporeal (ie spiritual) elements to what would otherwise have been a purely corporeal (ie material) account of the cosmos. These ingredients would eventually imply a tradition of anti-apocalypticism and distorted cosmology founded upon a supposedly partisan, or "providential", divinity. This "deus ex machina" ultimately served no deeper purpose than a lucky mascot or guardian angel but as a pervasive influence folded into the most obscure workings of the observable cosmos (see gnostic cosmology in particular [80]), the reassurance on offer through such sophistry evidently went beyond the everyday reach of bar-room critics and provided an inherently stronger social glue in times of urgent cosmological debate (note the acceptance of Christian cosmology in particular by the Roman administration). As such, the tradition would survive since it appeared to counter the otherwise terminal character of the highly degenerate condition engendered by apocalypse, basically removing core bourgeois societies at the heart of nations and empires from contention. Indeed, such a tradition would understandably be favoured by feudal "ancien regime" societies in the past. For the same reason, it would also be favoured by the narrow elites of academe, business, military and government at the heart of democracies today. The space age in other words, like past revelations, was bound to provoke a traditonally vehement response albeit the latter was also bound to be subdued to the extent that a suitably distorted cosmology remained adequately entrenched.

Historically of course these periods of complete breakdown in society, often involving extreme hedonistic tendencies, are also characterized by extreme anti-apocalyptic tendencies eg millenarianism. When these conditions arise, in the absence of a strong grip on reality, perfunctory government is often given to placing a wildly optimistic slant on the future state of the environment. Wise government on the other hand, retaining a strong grip on reality, does not reject apocalypticism per se. All the same, such government in the past has frequently resorted to the likes of prayer, sacrifice and moral exhortation in the presence of apocalyptic threats - apparently in the vain teleological belief such actions may have some direct influence on the supposed powers that give rise to cosmic catastrophe. Nowadays though, as long as teleology is expunged from science, the perceived obligation on hard-headed government is more to take such *practical* measures as are within its scope both to avert such catastrophe and to mitigate its effects. The main requirements evidently are (a) a proper understanding of the apocalyptic process, past and future; (b) a proper acknowledgement of its utter pertinence; and (c) such preparations as are essential to face up to the likely form future apocalypticism will take. Stages (b) and (c) are of course still barely with us at the present time and the dual purpose of this presentation, with (a) particularly in mind, is (i) to focus upon the perceived problem of cosmic catastrophism during the past 2000 years; and (ii) to outline its most probable explication in terms of the latest giant comet from "deep space" to penetrate and settle in the inner Solar System (ie cis-Jovian) environment. While this explication is clearly a matter for continuing scientific and historical research, it is perhaps salutory to keep in mind that the general level of late Roman civilization prior to the Dark Age was not recovered in Western Europe until the Enlightenment [71] - an interlude of a dozen centuries, at least. Indeed we should be under no illusions as to the capacity of the present cosmic environment (still very largely unseen) to inflict very significant changes on the condition of the biosphere (and hence human society) within centennial to millennial timescales. It follows that we need much better understanding in the wider community before the next bout of apocalyptic terror afflicts mankind.

2. The punctuation of uniformitarianism by comets

The water-bearing, life-supporting shell which surrounds our planet (the Earth) is now known to be suitably far from our star (the Sun) *and* suitably close to our Milky Way (the Galaxy). The state of the biosphere, in other words, appears to depend on two near-circular motions, one around the centre of our Solar System and the other around the centre of our Galactic disk. Indeed, since the biosphere probably derives from an intermittently replenished, original stock of moderately pristine cosmic material directly and indirectly delivered to the Earth by Oort cloud comets, there is growing circumstantial evidence that *both* near-circular motions are critical to the maintenance of a somewhat exceptional cosmic process which is also the evolution of life on Earth. Thus the former near-circular motion essentially controls the underlying "uniformitarian" condition of the biosphere while the latter near-circular motion essentially controls its superimposed "catastrophic" condition also known as apocalypticism. Together these imply a general state of "punctuated equilibrium", the occasional large-scale catastrophes due to cometary inputs corresponding to the mass speciations and extinctions which characterize biological evolution as a whole while the more frequent smaller-scale catastrophes are responsible for more local adaptations and eliminations of sub-species which in the case of recent human history, for example, can be considered to reflect the rise and fall of civilization, the first such rise corresponding to the emergence of *homo sapiens sapiens*. Any such working model of human biology and sociology is evidently devoid of any spiritual, teleological or magical connections and can therefore be discussed in fundamentally astrophysical and physical terms.

The larger of these near-circular motions, still essentially unsung in theories of biological evolution, is of course the one of concern here since it is the overall pattern of cometary deliveries to Earth which is expected to determine the overall pattern of "punctuated equilibrium". Broadly speaking, such cometary deliveries depend on the concentration of other stellar bodies close to the Sun which also participate in this near-circular motion. So far as the choice between apocalyptic and anti-apocalyptic scenarios is concerned therefore, it is the appropriately moving material in the solar neighbourhood and its interaction with the Oort cloud which tells us whether an appropriate flux of comets will be detected in cis-Jovian space. Until the emergence of radio-astronomy after the Second World War, there was a general suspicion that the very youngest stars in the solar neighbourhood (with ages ~ 0-50 myr) signified the greatest concentration of material close to the circular motion in question. These stars however were perceived as moving too quickly to have any serious long-term effect on the Oort cloud [64; 5, pp. 189 ff] and the tendency therefore was to discount the possibility of astronomical processes giving rise to "punctuated equilibrium".

An important asymmetry in the large-scale Galactic velocity field was then detected however (as between the first and fourth quadrants of galactic longitude [45]) which implied these very youngest stars belonged to the latest spiral arm system to penetrate the Galactic disk while the next youngest stars (with ages ~ 500-1000 myr), approximately co-moving with the Sun, were more obviously the ones we should associate with the near - circular motion in question [30]. Any such division of the nearby stars into physically distinct "stellar groups", as required by the most accurate proper motions then available, was not held in high regard by post-War astrophysicists and any such interpretation failed therefore to capture much attention. However the latest very accurate measurements of nearby stars using the Hipparcos Astrometry Satellite [14] appear to have strongly endorsed this interpretation. Both generations of stars are of course more youthful than the Solar System (age ~ 4500 myr) but are readily enough understood in the light of most of the evidence from other galaxies as reflecting extended bursts of star formation during the course of galactic history which are associated with the successive formation of spiral arm systems. Such bursts indeed are likely to be triggered every ~ 1000 myr or so and are perhaps at their most intense for an initial ~500 myr during the rapid arrestation of each successive spiral system (eg through the effects of dynamical friction [9]) as it settles into a broad annular circulation.

The type of astrophysical environment which is therefore implied for the Sun, and indeed for a proportion of stars generally, is then of profound significance for the Oort cloud. For, whereas the primordial star-forming material of spiral arms has long been regarded as the other visible components of galactic disks, namely gas and dust, an evolution of spiral arms which involves rather rapid settlement into a broad annular circulation would likely presuppose a largely unseen primordial population of sub-stellar bodies undergoing dynamical friction, otherwise referred to as "dark matter". Spiral arms and annular circulations comprising heavy concentrations of such dark matter would tend to suggest an evolutionary process in which primordial aggregates of such dark matter would combine to produce stars, gas and dust together, the more tenuous of these components becoming a part of the interstellar medium which returns to fuel the activity of galactic nuclei. This would then imply the Sun's vertical motion in the Galactic disk is most likely to be under the control of an annular concentration of dark matter aggregates close to the Galactic plane which are still in the process of evolving into low-mass stars. Not only would a configuration of material of this kind produce similar Galactic vertical and radial patterns in the visible distribution of Oort cloud comets, as observed [54], it would also lead us to expect a highly periodic history for the biosphere, as observed [22], in which the influx of Oort cloud comets would unquestionably play the dominant role.

Relatively short-lived spiral arms which are presumed to emanate repeatedly from the galactic nucleus (as jets) and which are then rather rapidly transformed into dark matter condensations prior to the the formation of stars are evidently a fundamental feature of the astronomical environment believed here to match the biosphere's apocalyptic history. But while this understanding reduces to one of comparing the geological and historical records in detail with a suitable astrophysical model describing the behaviour and evolution of Oort cloud comets, the galactic environment envisaged has important cosmological implications as well. These arise through the other important asymmetry in the Galactic velocity field that first came to be detected through radio astronomy. Thus it is through both these important asymmetries that we are aware of a possible "cosmological redshift" at the Sun's Galactocentric distance of ~33 kms^{-1} based on the apparent radial motions of the very youngest stars in the solar neighbourhood relative to the Galactic centre (~40 kms^{-1}) and relative to the Sun in near-circular orbit (~7 kms^{-1}). This discrepancy can be taken to imply a hubble parameter of ~4000 $kms^{-1}mpc^{-1}$ close to the Sun's potential but until recently, transmission arguments based upon the supposedly uniform flow of electromagnetic radiation through cosmic space have generally been taken to imply a sensibly constant hubble parameter throughout the cosmos of ~75 $kms^{-1}mpc^{-1}$. The apparent quantization of cosmological redshifts based on the Galactic centre [59] has fundamentally altered this perception however and a fluctuating hubble parameter with peaks at the potential wells of typical galaxies now seems very much more in accordance with the observational facts. The hubble parameter may therefore be produced by a *quadratic* doppler effect arising form the dominant field motions in an otherwise stationary universe along each line of sight [18] rather than by a supposed *linear* doppler effect due to a possibly underlying expansion of cosmic space [29].

While it is certainly the case at present that many astrophysicists assume the expanding universe is valid and are convinced as to the merits of "big bang" theory, it is also obviously the case that the currently discarded asymmetries in the Galactic velocity field represent a very serious distortion of the cosmic environment causing the likely apocalyptic role of comets to be repudiated without justification. The point to be recognized here then is that our understandings of comets, planets, stars and galaxies are not by any means so well established that we can say with certainty that an apocalyptic environment is excluded. On the other hand, alternative physical models can now be constructed and we can raise questions as to the likely predictions of these models in relation to the problem of historical catastrophism. A fundamental problem for modern scholarship therefore, taking together the various disciplines involved, is to make a judgement as to the adequacy of these various models. Already it is clear that Oort's evasion of two major asymmetries in the Galactic velocity field cannot be defended. The modern anti-apocalyptic doctrine cannot therefore be sustained. At the same time, the problem of historical catastrophism on the timescales of interest to civilization is not so much the frequency and intensity of celestial bombardments, drastic though these may be. Rather it is the disintegration and collapse of traumatized societies whether or not any anticipated celestial bombardment occurs. Indeed the very word most commonly used during the past two millennia to signify catastrophes of celestial origin - "apocalypsis" in Greek meaning "revelation" or "unveiling" - manifestly refers only to the evidence through which a predicted bombardment is widely expected. In effect it identifies advance astronomical knowledge as the source of a collective mental anguish which has been and still is critical to the survival of nations and empires Only by extension does it refer to the evidence of lost lives and environmental devastation that comes through the possible localized effects of celestial bombardment.

Such mental anguish is largely an affliction of societies whose collective grasp on reality is such as to readily induce an abandonment of rationalism and hence any thought of a realistic or organized response to the slings and arrows of outrageous fortune. Societies of this kind are often given to displays of uncontrollable emotion and mass hysteria whilst tending to dissociate themselves from many forms of ritualized behaviour. The official preservation of order and rationalism in society has therefore long been recognized as a hard-nosed responsibility of government and society requiring the utmost care and knowledge if apocalyptic traumatization is to be successfully controlled. These duties of care and knowledge have often devolved to ecclesiastical authorities in the past but such authorities lost a good deal of respect in late medieval European society when it was clear that cosmology was being falsified for the sake of preserving a preferred teleology. With the arrival of so-called "enlightenment" in Europe, there was at first a general expectation that the new untrammelled knowledge, to be effective, would need to be founded on physics to the exclusion of metaphysics. In other words, our understanding of Nature (and our reaction to it) could always be based upon a rational system of facts and explanatory processes which was open to public inspection and authentication (the very essence of scientific method), being thereby supposedly free of any irrational, incorporeal or supernatural elements. It was very soon realised however that such faith in the process of public inspection and authentication was to subscribe to a counsel of perfection and that any social organization was bound to include elements of metaphysics in public affairs in practice (see Section 6). Thus we now accept that our understanding of the world around us (and our reaction to it) is limited to a greater or lesser extent by certain conventions and paradigms which may, for example, cause us to downgrade historical facts at the expense of scientific facts, or vice versa, to the potential detriment of society generally. The modern tendency in fact is to write apocalypticism out of cosmology altogether with the result that neither government nor society has much awareness of the history of interaction between our planet and the cosmic environment. Indeed it is no longer widely appreciated that apocalypticism has always been the paramount perception since the dawn of civilization (ie since cuneiform records began) and that, while the Roman Catholic Church (in line with gnostic teaching generally) sought to impose an anti-apocalyptic outlook from about the second century AD onwards, its successful imposition was only spasmodic and not finally achieved until the European Enlightenment ca 1650 - 1850 AD.

As it happens, it was during this period that the balance of opinion in favour of apocalypticism rather than anti-apocalypticism was finally reversed. Thus we now recognise three critical interludes of revived apocalypticism and extreme social upheaval during the course of modern global history (ca 1650, ca 1790, ca 1850) when those who would emerge to fashion the anti-apocalyptic environment and its cosmological setting would find it necessary to invoke a material cosmos and yet keep mystical, incorporeal or spiritual influences in place.

The fundamental empirical fact to be recognized here nevertheless is a disintegration and collapse of society as a result of celestial traumatization which has been a seriously recurring problem throughout the course of history [23]. The traumatization commonly involves a perception that the "last days" are at hand or that the "end of the world" is coming. The association of this problem with prophecy based on revelation is of course well known [72]. However, in so far as the relevant astronomical knowledge is usually limited to a very small group of experts with relatively little historical knowledge, and vice versa, there has been and still is a very strong tendency to associate such prophecy more or less exclusively with comets [73]. This unfortunately results in a grossly simplified perception as to the general nature of revelation, causing the whole issue to become hopelessly clouded and seriously misunderstood in modern times. Thus, despite the major technological advances which have taken place during the course of the twentieth century, it can hardly be said that civilization is any the less at risk on account of the apocalyptic process.

3. The interruption of history by revelations

Fear to the point of absolute terror as a consequence of revelation, then, has been a feature of certain periods during the course of history when, as a result of specific observed phenomena in the sky, it was accepted that a cosmic catastrophe was imminent. Although these occasions did not necessarily materialise locally in the form that was ultimately feared, this was evidently not regarded as a serious reflection on the quality of the prediction since a longer term regularity in cosmic affairs was also widely perceived which meant that society as a whole would undoubtedly experience a calamitous cosmic bombardment at some stage in the possibly not so distant future. Thus historians of the Christian era whose ideas can now be traced back through Hellenism and Judaism to as early a source as Zoroaster during the second millennium BC [24], for example, would clearly characterise history by a sequence of lesser upheavals at intervals of a few centuries and by more serious upheavals at longer intervals of one or several millennia. This "millennial" perspective on the historical continuum was of course to be further modified in the hands of teleologically minded anti-apocalypticists who, as "millenarians", would be inclined to see their time or that following the next upheaval as a final, divinely ordained millennium. This overall perception of the physical environment has in fact been deeply ingrained in the general consciousness of Western civilization and has substantially weakened only during the course of the nineteenth and twentieth centuries AD. It does not follow of course that the past perception is therefore necessarily correct but no more is our present point of view validated by ignoring the past perception.

In accordance with this overall perception the underlying or essential reality of the evidently terrifying process described in the Christian *Book of Revelation* was never seriously in doubt. However the precise sequence of events relating to the successive upheavals has undergone several modifications during the course of history, no doubt reflecting the difficulty our ancestors had in determining the precise celestial process. One notes for example that the *Book of Revelation* is a product of the latter half of the first century AD being the accepted Christianized version of the previously existing Jewish *Book of Daniel*. The latter was originally composed during the Babylonian exile in the sixth century BC and underwent considerable revision during the Antiochan persecution in the second century BC before its adoption by the Christian movement at the turn of the millennium. This recurring interest in apocalypse evidently corresponds to new periods of revelation and a quickening sense of foreboding throughout the civilized nations of the Near East and the Mediterranean [68] which was eventually perceived as reaching some kind of nadir [75] during a new and rather serious period of revelation coupled with the fifth century decline and fall of the Roman Empire [7]. This period of revelation appeared sufficiently serious for a time thereafter to be taken to mark the start of a new world calendar (ie with 0 AM I = 500 AD), but the turn of events marking the subsequently perceived relationship between Rome's successor civilization in the West and its cosmic environment, after considerable dispute and a further "false start" (ie with 0 AM II = 600 AD), was not officially resolved until the adoption of the Christian calendar in 800 AD, supposedly with the ultimate Christian millennium by then well advanced [74].

Even then the pattern of events remained uncertain and with the arrival of further significant periods of revelation during the eleventh and fifteenth centuries AD, the pace of expectation quickened remarkably during the sixteenth and seventeenth centuries with increasingly serious attempts to establish when the final cosmic bombardment was due [76, 50, 37]. These efforts culminated with the supposedly definitive studies of revelation by Newton [8] only to be more or less immediately abandoned by the whiggish ascendancy in England; then by its counterparts in Anglo-Saxon America at the end of the eighteenth century AD and in the rest of Europe towards the middle of the nineteenth century AD (see Section 6 below). By this time, there was widespread appreciation of the fact that each period of revelation provoked a revival of historical knowledge and future prognostication based upon the records of earlier revelations. Thus the *Book of Daniel* took its lead in many respects from the *Book of Jubilees*, often attributed to Moses, and from the even earlier *First Book of Enoch* attributed to a partriarch around the beginning of the second millennium BC [24]. It follows that the *Book of Revelation* was widely seen as preparing the way for a major celestial bombardment some time beyond 100 AD in accordance with a past experience which involved both a significant revelation and a global catastrophe during the third millennium BC and a series of lesser revelations since which was chronicled by the Jewish Old Testament (covering some 2000 - 2500 years in modest detail) and the Christian New Testament (covering the subsequent 50 years in great detail). As it turned out, Christians were not the only ones by the fifth century AD to be convinced of an impending catastrophe and of the need to prepare for the final millennium [23]. Thus it was during this period and its lengthy aftermath that successive waves of invaders (barbarians and Vikings) penetrated the Roman Empire and settled in the so-called "deserted lands". These arose under conditions that were evidently alarming enough to be perceived as indicating the arrival of the forthcoming millennium [31]. These invaders, as often as not, were confronted only by an imperial administration whose appeal to Christian principle meant turning a blind eye to powerful squatters. By such means the Holy Roman Empire emerged, making up a patchwork of territories under the control of aristocracies naturally inclined to cooperate in the application of Christian principles. It has to be appreciated of course that while the ancient testaments provided a record of dramatic historical events (including revelations) by way of illustrating how the long interaction between cosmos and mankind affected a particular people, the main contemporary purpose of the testaments was to achieve a standard of social behaviour throughout Western Europe in the continued presence of such events that not only matched the intentions of a supposed cosmic authority but sustained the pre-eminence of an original Roman administration by appealing to its corresponding human authority and to its apostolic succession. It is reasonable to suppose these more secular aspects of "Christian policy" would have been of at least as much significance as the perceived teleological aspects of revelation. Thus the Holy Roman Empire to be originally moulded in association with "barbarian" aristocrats by 800 AD was to be penetrated and further moulded in association with "Viking" aristocrats, thereby laying the foundation for a deeply schismatic Christian church in the West whose Protestant and Catholic establishments increasingly polarised around doctrinal issues to do with the nature and purpose of revelations. Indeed there is little reason to doubt that Newton, for example, would have seen himself as but the latest in a long line of European scholars - still essentially within the general ambit of the Holy Roman Empire - who sought to give physical meaning to revelations both before and since the decline and fall of Rome itself (see Section 8).

To separate out issues of secular power and teleological catastrophism thus is to do not so much more than Seneca did for the Roman people when he drew upon the knowledge and experience of their Greek and Etruscan neighbours and presented his analysis of the situation around the same time as the author of the *Book of Revelation*. Seneca's great fame is that of a philosopher and adviser at the very heart of the Roman Empire during its ascendancy. He has been described as one who "disapproved of the world in which he lived [while playing] a leading role in it, as was appropriate for a tough Roman pragmatist" [25]. Whilst addressing the *Natural Questions* perceived to be of the greatest public significance, starting with "Lights in the Sky", he was to emphasise the two branches of knowledge, moral philosophy and cosmology, which in his opinion evidently mattered most to mankind: "one teaches us what ought to be done on Earth; the other what is done in heaven". It evidently followed that "the mind possesses the full and complete benefit of its human existence only when its spurns all evil, seeks the lofty and the deep, and enters the innermost secrets of Nature". There can be little doubt that the principal focus of Seneca's attention prompting such pronouncements was the longer term regularity (ie fixed intervals) that emerged in the observed incidence of a particular celestial phenomenon which corresponded to the unveiling of "advanced announcements of death by the universe". The celestial phenomenon involved is evidently the multiple incidence of "fiery shapes or so-called boards, balls, torches and blazes", various categories of what have later been classified as "blazing stars", "providences" or "fireballs" and which can now only be attributed to the hosts of defunct cometary debris which are intermittently encountered by our planet (see below). It is well known of course that the "fireball" phenomenon is not intrinsically dangerous and it is clear that Seneca looked upon these serious occurrences as involving rather modest "lightning bolts" which did us no harm. This however was not to deny their significance for the explicit warning provided by a regular series of such events was that of further events involving quite extreme fireballs which might sometimes bring about some good by causing damage to another party. Modest bolts did not therefore exclude the possibility of mightier bolts at the next occurrence in a revealed series of occurrences. Specifically relying upon the knowledge of the Etruscans, Seneca was evidently aware of far more calamitous bolts such as would "destroy whatever it strikes and, particularly, alter the state of private and public affairs". Revelation, it is clear, presented the possibility of large scale destruction but Seneca is at pains to provide relief for his oppressed fellow citizens by placing a positive slant on such happenings eg by setting an example through stoical resolve in the presence of such menace and by emphasising only the possibility of harm to others. This apparent tendency towards deviousness, to place a positive (or teleological) slant on catastrophe, to pour oil on what otherwise might be seen in wholly negative terms, to impose order on chaos, to override cosmological action with public morality, was evidently as much at home amidst the imperial ambitions of Augustan Rome as amidst those of Alexandrian Greece. The Greek approach to the problem however, as we shall see (Section 8), was altogether more subtle.

It is perhaps interesting to reflect that both Seneca and Newton attained academic distinction through their ability to place comets and fireballs in clearer perspective than their predecessors. Both furthered their careers and influence as thinkers and advisers at the heart of fledgling empires in their ascendancy. Both went on to write at length on the matter of lights in the sky and what they revealed. In both cases, the new glimmer of celestial understanding was to be set aside while the new parsimony which had expressed itself nationally in stoical and puritanical resolve was to be abandoned in favour of a liberal outlook and the new cosmically untroubled millennium.

4. Near-miss comets and their apocalyptic hosts

The facts of revelation tell us there are few grounds for sing the leaders of nations and empires in the past, together with their advisers (astronomers), were in the habit of regarding encounters with comets as either the only or the most likely celestial threat to civilization. Ordinary comets do of course happen to be the commonest independent bodies of interplanetary space to be readily seen from Earth as well as being the smallest independent bodies (at typical speeds of encounter with the Earth) capable of producing globally significant damage. It is certainly permissible therefore to suppose that perhaps one or even two cometary disasters of global extent may have chanced to be widely observed, experienced and remembered since the dawn of civilization ie during the past 5000 years. However active comets do not normally strike the Earth more than once per $\sim 10^7$ years and even defunct comets and asteroids, which were never readily seen in the past and which are still considered therefore to be irrelevant, do not normally strike the Earth more than once per $\sim 10^5$ years. It follows that it is really quite implausible to suppose that *direct* cometary encounters could have continued for millennia to be perceived as a common apocalyptic threat; or that *direct* asteroidal encounters will be so perceived today.

On the other hand, comets which travel in association with defunct or invisible sub-cometary debris and whose orbits are such as to bring about an occasional close passage by the Earth may give rise to lesser apocalyptic events whose effects are less likely to be global in extent. Under these circumstances, the range of phenomena known as blazing stars, providences or fireballs will sometimes be associated with cometary apparitions and it would be reasonable - without any knowledge of cometary orbits - to regard all comets as potential "signs" of impending disaster. Historical comets would then have sustained a public awareness of apocalypse just as their specific calculated influence nowadays tends to sustain the reverse! More interestingly though, any concentration or host of similar sub-cometary debris (with or without its parent comet) in suitably periodic orbits commensurable with the Earth - ie permitting the repetition *and* prediction of such encounters - would be justifiably regarded as an even more pressing apocalyptic threat than a random comet. Contrary therefore to a general impression which seems to have arisen in the aftermath of Halley and Newton's findings during the seventeenth century, although the perceived threat from comets was then considerably reduced, the sub-cometary apocalyptic threat still remained very much in place. Indeed, if the perception of cometary "signs was essentially rational, it is most likely to be explained in terms of a valid sub-cometary apocalyptic threat involving a host of debris.

In spite of these fairly obvious facts however, and notwithstanding the modern evaluation of encounter frequencies based on impact craters, some astronomers in modern times have continued to identify what might be called a significant low frequency/high energy "apocalyptic threat" as the one which now justifies the development of a full survey programme to discover all the potential near - Earth objects of cometary and asteroidal size ([57;cf 35]). It seems that they regard the nearest of these objects in future (say, during the next century) as the individual hazards requiring possible pre-emptive elimination on the grounds that their orbits would be too close for comfort! Neither governments nor public have been particularly impressed however by the suggested urgency of such a project since the presumed frequency of *actual* encounters of this kind is still low. Indeed it is clear that the primary factor of concern in relation to these near passages in the near future is not at all the likelihood of a direct cometary or asteroidal encounter but the likelihood of a sub-cometary or sub-asteroidal apocalyptic threat arising in conjunction with a relatively nearby "sign". By failing to take account of this matter of historical perception, therefore, these astronomers have failed to appreciate a basic reason for conducting the survey and the project has been seriously undersold. A similarly perverse consequence of this confusion over "low frequency" apocalyptic threats is the belief that the historical fear of comets [36] is trivial on account of its being based on an inaccurate or primitive determination of the hazard's frequency. There is no justification for this view which is evidently based on a profound ignorance of history. Indeed the comparative absence of any direct damage from comets has almost certainly been generally accepted since the beginnings of Sumerian civilization ca 5000 BP. This was probably well understood by Newton 300 years ago though not, as it happens, by his distinguished acolytes Halley and Whiston [15].

If then we credit the leaders of significant nations and their leaders in the past with the historical understanding that is realistically their due and if we accept the frequency with which nations and empires have been seriously traumatized in the past (see below), then it is clear beyond reasonable doubt that the sub-cometary apocalyptic threat must have been the one generally perceived. We do not have far to look in order to quantify this threat. We need only consider the independent bodies of interplanetary space at least 10 times smaller in linear size than ordinary comets and capable of causing biospheric damage on the scale of nations and empires (ie with the potential to produce multimegaton events); and those at least 100 times smaller than ordinary comets and capable of producing harmless atmospheric fireballs around the globe (ie with the potential to generate multikiloton events). *Neither* of these categories of interplanetary body is ever readily seen in space and so our understanding of such objects is very largely derived through the recorded encounters with our and other planets and their satellites. Based on the Tunguska and Shoemaker-Levy events this century for example, and the integral flux of such bodies recorded by lunar craters formed during the course of Earth history, the average global rate of such encounters causing damage on the scale of nations and empires is on the order of about one per century. This however is a simple long-term average which does not do justice to the known disintegration of comets and the actual swarms of debris that give rise to apocalypticism. If most of this debris originates from giant comets and we take account of the known variations in the cometary flux during galactic and human history [22] then the integral flux during *apocalyptic spells* is more on the order of about one per decade to one per year.

This rate is unlikely to be overestimated especially if we extend our perception of cometary hazards beyond simple explosive effects. First, it is often in the nature of cometary material to retain its (porous) physical structure whilst undergoing slow devolatilisation in space. Secondly, such interplanetary bodies having a mass sufficient to produce multimegaton events close to the ground can also be so fragile as to undergo premature disintegration during the final approach to Earth [33]. Under these circumstances there is something of a global threat to civilization since the incident dust resulting from such premature disintegration may in turn rapidly envelop the Earth, and a temporary veil (lasting years to decades) may then have a significant influence on the climate [19]. When the premature disintegration results in exceptionally fine particles, there is also the possibility of complex cometary chemicals surviving entry to give rise directly to virulent biospheric damage[42], albeit the study of such effects must still be regarded as being in its infancy.

If the bulk of these characteristic events is cometary in origin (necessarily reflecting the history of heirarchical disintegration which we associate with the largest cometary bodies settling in inner Solar System space), then it is easy to see why, after all, historical comets were feared. It follows directly from the process of heirarchical disintegration that the incoming celestial bodies to Earth must often be highly bunched in space; and especially so if, as the constitution of cometary dust indicates, they happen to be not particularly robust on account of recent devolatilisation. A near miss by an ordinary comet may therefore virtually guarantee a direct encounter with a significant sub-cometary body and produce damage on the scale of nations or empires. The probability of an encounter with a significant sub-cometary body is itself measurably enhanced when its orbital period is commensurable with the Earth's and the associated swarm of extreme sub-cometary fragments resulting also from disintegration is repeately encountered by the Earth so as to produce typical showers of ordinary atmospheric fireballs. In the latter instance of course we can also take it that the originating comet is either defunct or completely fragmented. Whichever is the case, encounters with such swarms of extreme sub-cometary debris are in fact a common feature of terrestrial history and occur in batches every other century or so [10, 17]. Herein lies the principal apocalyptic threat already discussed and the primary reason now for an observational programme to search out all the defunct comets likely to approach near-Earth space.

This issue can be viewed from another perspective. Astronomers now calculate on the basis of several hundred Earth-crossing asteroids and defunct comets which have been detected during the Space Age (mostly, in fact, within the last decade) that there are several thousand such bodies yet to be discovered. Up to 10 per cent or so of these objects on present reckoning may be derived from the latest giant comet to undergo heirarchical disintegration in inner Solar System space and are believed to be associated with the so-called Taurid meteoroid stream "complex" [19, 6, 70]. Originally identified this century as a clutch of short-period, low inclination, highly eccentric meteor streams extending over a sector of cis-Jovian space, whose orbital tracks directed to and from perihelion are concentrated within a couple of months centred on early November and late June respectively, the Taurid complex is now known to comprise a full range of sub-cometary or meteoroidal debris, most of which is defunct. The "giant comet" source has by now virtually disappeared since the heirarchical disintegration is a rapid process, cosmically speaking. Indeed the hundred or so objects which the stated 10 per cent or so represent cannot themselves be regarded as much of a threat to civilization since they strike the Earth at the rate of about once per $\sim 10^6$ years and are therefore likely to be removed by continuing disintegration before any such collision occurs! On the other hand, this hundred or so objects from a system undergoing heirarchical disintegration is bound to be associated with many more such objects which have themselves already undergone fragmentation and which continue to exist individually as coherent batches or hosts of sub-cometary debris. Earth "pseudo-encounters" with the unoccupied centres of these hosts which make up the bulk of the Taurid complex are thus likely to be occurring at the rate of about once per $\sim 10^5$ years. This rate is to be compared with the ordinary penetration of such hosts which seems to be occurring at the rate of about once per $\sim 10^2$ years. The implication is that any Taurid host with a source which either does or does not any longer exist and which passes within about 300,000 kilometres of the Earth, roughly three quarters of the distance to the Moon, is potentially an apocalyptic threat. This calculation is necessarily crude at the present time, due to the absence of any Spaceguard programme [57] to improve its precision, but the distance of nearest approach during recent close passages by near-Earth objects [13] does appear to indicate the "lunar sphere" may well be critical so far as the typical batches of encounters with individual hosts are concerned. In other words, it would not be surprising now if the discovery of a new asteroid in the Taurid complex which passes between the Earth and Moon also coincides with the revelation of a batch of encounters with its host of sub-cometary debris. Indeed if one encounter with such a batch were relatively close to the defunct cometary source, one might well expect to observe the latter's ghostly passage from horizon to horizon in association with a faintly luminous tail amidst the occasional blazes of light across the sky as otherwise invisible host members encounter the Earth's atmosphere and produce fireballs.

The characteristic feature of each such batch is an individual host of sub-cometary (or sub-asteroidal) debris in elliptical orbit around the Sun which is repeatedly and regularly intercepted by the Earth over a timescale of about a century, anything say between 50 and 150 years. Such interceptions are likely to be in accordance fairly frequently with a simple orbital commensurability between the host and the Earth eg at intervals of 13, 10 or 7 years, say, such as are associated with typical meteoroids with orbital periods of 31/4, 31/3, or 31/2 years respectively. Such patterns of encounter, limited by a near commensurability and by a traverse time due to relative orbital precession between the host and the Earth, will evidently result in repeated enhancements in the observed flux of fireballs and constitute a typical revelation. In such instances there is really no question as to the fact of possible encounters with larger meteoroids in the near future, only uncertainty as to the location, magnitude and frequency of devastation. The fear of apocalypse of course will be engendered by expert opinion which will naturally be encouraged to provide its most accurate assessment (necessarily uncertain!) of the realistic risk. Apocalyptic fear in other words is not ordinarily driven by charlatans and will inevitably continue until at least the peak density of the meteoroidal host has been traversed. The extreme sub-cometary bodies are not in themselves dangerous of course (cf Section 3) since they produce only relatively weak fireballs or clouds of dust and chemicals very high in the atmosphere. They nevertheless signify the presence of a proportion of ordinary sub-cometary bodies capable of producing very much more powerful fireballs closer to the surface of the Earth (ie at the level of multimegaton events) and high-level dust-veils severely influencing global climate.

Let us recapitulate here. By failing to recognise revelations and the apocalyptic record (Section 3) and by failing to consider their potentially rapid disintegration when reflecting on the effects of asteroids and defunct comets (Section 4), modern astronomers and historians have been able to construct a largely uneventful history for mankind which is in keeping with the perceived absence of any future trauma [57]. Government and society can take a good deal of comfort from this beguiling scheme, of course, but the disintegration of comets is also an established physical property of these objects [60] and it has been known for some thirty years that a recent giant comet is necessary to explain the extreme over-abundance of sub-cometary material in inner Solar System space [81]. It is clear that the full implications of comets to society and civilization can no longer be ignored.

5. Hosts of varying degree

Modern governments and their advisers (astronomers) are familiar enough with the situation that prevails during the normal course of events when there is no celestial threat. Thus it is well known that monastic and suicidal escapism will from time to time overtake millenarian sects and their like in a variety of disturbing forms. The escapism though commonly arises among highly rational and well organized societies which dissociate themselves from the mainstream on grounds which often seem at first to be acceptable within the normal standards of tolerance and freedom. Under these circumstances the mainstream authority is clearly expected to give only a suitably proportionate response to any deviation from civilized custom that arises and a mere tendency towards escapism is not of itself likely to be the signal for any decisive action. However the "knee-jerk" character of most of the official reaction which ensues when a deviation is no longer perceived to be acceptable can often be more disturbing than the deviation itself and a situation can therefore arise in which the supposedly principal stabilizing influence within society (its government) does in fact become its principal destabilizing influence. Even on a limited scale, when there is no celestial threat, this turn of events adequately illustrates the nature of the problem confronted by mankind [74]. Indeed governments and their advisers are evidently not very familiar with the documented occasions during the course of history when mass escapism very clearly arose and it is obvious therefore that there can be few grounds for confidence regarding the likely official response when such occasions arise again in future. The point here is that there have been frequent occasions in the past when it is clear that virtually the whole population has been concerned as to the accuracy of an apocalyptic prediction and has undergone widespread social disintegration - even to the extent that all semblance of political control was lost and overlordship soon passed to an alternative national or foreign authority [23]. It is very hard to believe an *uninformed* government will have any success in future confronting mass escapism when apocalypticism once again appears.

At the heart of these severe social breakdowns there clearly emerge deep theological concerns and it is a matter of historical fact that contemporary analysts often seek to alleviate the celestial traumatization of nations by propounding explanations of the revelations which offer some degree of emotional release from the danger (often known as "soothsaying" in the past). Two primary explanations seem to be dominant (see below). Both evidently start from the assumption that one is dealing with a manifestly obvious phenomenon. On the one hand it is claimed that the danger is selective and will not apply to "chosen people" (the doctrine of predestination); on the other hand it is claimed that the danger is more apparent than real or that the manifest phenomena are of a spiritual or disembodied character, and are not to be likened to the behaviour of ordinary matter (the doctrine of providence). The recourse to such desperate expedients, seemingly attractive to the intellectual class [23], is by no means guaranteed to succeed and the debate is usually left unresolved. Nevertheless the debates clearly arise in the context of apocalyptic host encounters and commonly degenerate into a state of civil unrest and revolution if not outright war. The resulting pattern of secular historiography implied by the successive batches of encounters with such hosts, admitting a step-function in the evolutionary response of civilization, evidently allows for the periods of eschatological concern which are known to punctuate the course of history [12]. Even at the best of times, of course, it is normal for nations to engage in war and we cannot necessarily expect that the correlation perceived here between apocalyptic/climatological events and their social consequences will be unalloyed by such consequences due to other factors. Nevertheless the tendency to civil unrest under specific, extreme, environmental conditions may will be so decisive as to raise the question whether these critical junctures in history when the reins of power are otherwise unaccountably relinquished in favour of new regimes (eg 1650, 1790 and 1850) are also associated with general depressions in global temperature (ie mini-ice ages), as expected. To explore this pattern of history further we need to clothe our astrophysical model with a bit more detail.

Much hangs, it would now appear, on the actual number of multimegaton airbursts or severe dust-veils which occur around the globe in association with typical meteoroidal hosts. For, let us say, ~100 such incidents in as many years (~75 over the ocean), almost every nation around the globe is likely to be directly or indirectly affected by a localized and/or global catastrophe tending to compound and exaggerate the local effects of traumatization through physical destruction and demographic loss eg by the enforced movement of whole peoples. Indeed we might expect that the civilizations associated with the largest empires under these circumstances will tend to collapse or, at best, metamorphose. On the other hand, for ~10 such incidents during a similar period of time (~7 over the ocean), large groups of nations around the globe will not be affected by localized catastrophes thereby limiting many peoples to the effects of traumatization alone, possibly allowing many nations to remain in a state of relative equilibrium - even to the extent of permitting large empires to consolidate once again. It follows then that a history of successive encounters with ordinary apocalyptic hosts which includes occasional encounters with more substantial hosts might impose a sequence of repeated national traumatizations interspersed with an occasional massive global collapse of civilization. This amounts to a pattern of history in which civilizations repeatedly undergo "Dark Age" declines widely separated in time whilst also displaying a subsequent tendency towards the regrouping of nations during the millennial aftermath of "Dark Ages" which leads to the fashioning of new empires and civilizations. Indeed, we might even consider a history of this kind to apply to the fifth century collapse of the Roman Empire and the subsequent "recovery" of Western civilization in its wake. If so, we might also recognize an earlier example of such collapse applied to the Sumerian Empire towards the end of the third millennium BC and the subsequent "recovery" of the Fertile Crescent in its wake. With this as our broad model, it is not without significance perhaps that cuneiform scholarship survived the first collapse ca 2000 BC to be the main bearer of astronomical wisdom until almost the time of Christ. Likewise we find Latin scholarship surviving the second collapse ca 400 - 600 AD to become the main bearer of astronomical wisdom in the West until the time of the Enlightenment. To the Enlightenment's uninitiated of course, this model has all the appearances of being a highly fanciful scenario. To some, it might even be unacceptable a priori because it makes a strong but highly depressing statement about the likely future state of civilization. Nevertheless it is clear that modern historians have never digested the full implications of past celestial signs and modern astrophysical insights concerning the general state of our celestial environment. Indeed, with notable exceptions [23], they have purposely turned a blind eye towards the apocalyptic record. It is a scenario that we shall need to keep in mind therefore as we continue our present review.

6. Modern hosts and the collapse of nations

Nations habitually jostle for position on the world-stage. Their individual purpose, it is apparent, is to maintain indefinite control over a section of the globe. In recognition of the various constraints under which each operates, it is customary for nations to act in concert amongst themselves so as to uphold an orderly state of dynamic equilibrium. Preserving the global status quo, in other words, is a favoured way of maintaining national control and hence national survival. The key to a modern nation's success in this continual struggle is evidently the government and the institutions through which the power of the people is effectively represented. Public opinion of course is, by its very nature, an elusive beast and while it is commonplace for modern governments to claim to represent people as a whole, such is the normal pace of events both government and people in practice usually defer to the well advantaged and to the well armed. As was perhaps more evident among ancien regimes in the past, therefore, it is aristocracies of a kind which generally hold the reins of national and international power. For such aristocracies, the ups and downs of nations through the balance of trade, armed aggression, natural disasters and so on are all part of the normal tapestry of events. Aristocracies indeed are sufficiently robust that even a global war may do relatively little to upset much of the status quo. Likewise, the status quo is hardly perturbed by a degree of substantial migration and mass genocide. During the twentieth century, for example, some 12 billion human lives have been lived and lost; and of these, some 80 million have been lost as a result of localized "end-time" revolutions largely provoked by fanatics [74]. Locally, this appalling statistic involves a demographic recession of, say, 10 - 50 per cent but globally it is less than 1 per cent. It is important therefore that we remain clear as to the level of catastrophe which is here discussed. The twentieth century by this and any other global reckoning is evidently somewhat uneventful! As a consequence it is those longer periods of upheaval around the globe to which mankind has intermittently succumbed during the course of history that aristocracies, and hence nations, have the greatest cause to fear: extended periods of famine, epidemic and war during which the demographic adjustments and losses globally increase above the customary background level (a few per cent) by at least an order of magnitude to, say, 10 - 20 per cent.

The three latest such periods, each lasting about 50 years, are centred around 1650, 1790 and 1850 [38] and closely match the three latest enhancements of the meteoroidal input to Earth [17]. Frequently identified as marking the starts of the English, French and German enlightenments respectively, these pivotal periods are of course historically conspicuous on account of the global upheaval and social revolution by which they are commonly characterised. It is a curiosity of the twentieth century however, possibly on account of its essentially uneventful character, that Western civilization now tends to spurn any connection between these social breakdowns and the celestial signs with which they are temporally associated. Admittedly some scholars have drawn attention to the climatological downturns at these epochs [49] which may of course be of celestial origin. Such proposals however remain the exception rather than the rule. In particular, historians fail to make anything of the greatly increased eschatological concern of these times, ie a very definite tendency on the part of those living to see critical aspects of their environment in a process of very rapid and terminal change [23]. Rather, in order to explain the upheavals, it is customary nowadays to invoke some kind of global breakdown in the diplomatic processes of representative government occasioned by widespread internal collapse, albeit any such so-called "whig interpretation of history" is now thought to be phenomenologically inconsistent and therefore seriously unsubstantiated (eg [16]:see also Section 8).

The whig interpretation of history, in particular its theoretical justification, derives from the highly respected philosophy of the well known English empiricists (namely Locke, Berkeley and Hume) which first emerged as the seventeenth century enhancement of the meteoroidal input to Earth went into decline. The acclaimed purpose of the new philosophy was "a grand onslaught on tradition, arbitrary government, and ecclesiastical authority in so far as all these things supported the old order which had been defeated once in 1640 and again in 1688" [52]. The victors of the English Civil War and Glorious Revolution saw this philosophy therefore as their justification in particular, a step back from the fundamentalist excesses and outworn authoritarianism of the Interregnum and a step forward to the millennialist dreams and "common sense" calculation of the Future. The foundation of the new philosophy was a general theory of knowledge based upon the empirical procedures of natural science rather than the divine ordinances of old. As is well known, this led almost at once to "the rise of a religious rationalism which dispensed with miracles and the whole paraphernalia of supernaturalism" (ie with revelations) "while continuing to believe in God" [52]. Upon such rationalism was then duly founded the new English Monarchy, the French Republic and the United States of America: new habits of thinking in which aristocracies and nations around the world would come to operate within a framework of political and economic liberalism.

This is not to say of course that these nations established a secure political and economic process based on such founding principles since, in practice, they proved suprisingly defective. Thus while it may be taken that an appeal to empirical procedures would introduce an element of precision into the measured relationship between apparently connected parameters, the physical connections would themselves be subject to choice and pragmatic judgement with extrapolated consequences that were not necessarily secure. A century or so after the English empiricists had embarked upon their line of enquiry, it was clear that the more or less exclusive dependence of such rationalism on natural science had run into severe theoretical difficulties; so much so that the whole line of thinking was deemed basically irrelevant by the influential school of German philosophers (namely Kant, Hegel and Marx). They essentially recognized that public affairs must in addition respect the empirical procedures of social science and hence rational metaphysics as much as rational physics, the unavoidable human tendency to base political action on supposition as much as on reality. Out of this were to grow the ideas of dialectical materialism and a crucial explanation of the great upheavals of history in terms of the natural instabilities that arise within and among nations through social and economic determinism. As is well known, the well nigh exclusive appeal to such determinisms is now commonplace in the service of modern history and modern politics, tending to displace any related intellectual discipline in which an appeal to physical determinism would also be involved. However, in arriving at their underlying principles, both the English empiricists and the German idealists have clearly relied upon a purely metaphysical interpretation of the recorded "miraculous" or "supernatural" phenomena of history. They have in effect taken a view as to the unphysical nature of "revelations" which modern astrophysicists are hardly able to endorse.

Here then is the basic reason why modern aristocracies (ie modern governments and their advisers) now largely overlook the evidence of meteoroidal inputs and the recurrence of eschatological concern during the great upheavals of recorded history. Indeed the original arguments on which the English empiricists and German idealists based their influential philosophies are now essentially forgotten. To recover these arguments, we need to recognise that it was the original cognoscenti of the Royal Society in London who first took a disparaging view of meteoroidal inputs and apocalyptic fears, blaming the public reaction of the contemporary English in particular [69]. These cognoscenti were clearly of a whiggish disposition and evidently did take such a view out of disrespect for the earlier intellectual support which was given fundamentalism rather than millennialism by the London Society of Astrologers [39]. The latter was disbanded at the end of the Interregnum and replaced by the Royal Society [73]. And in due course it did seem that this new view of things had been comprehensively endorsed when the new society's eventual president of distinction, Newton, failed to support his younger colleagues Halley and Whiston on the occasion of the publication of their ideas concerning a sustained *cometary* threat to Earth during the course of history [15].

Newton however was also a theological non-conformist and it is well known that he attached great significance to the known history of revelations (see below); thus he appears to have inclined more to the teleological or Calvinist view that the danger was predestined and selective rather than to the emerging mainstream view that it was more apparent than real. This should not be taken to imply an altogether blind faith associated with the typical puritanical outlook: rather it should be taken to imply a form of deism in which the cosmos naturally produces apocalyptic hosts to which the conduct of suitably guided societies needs to be adequately matched. In other words, Newton can be identified with those for whom eschatological concern and global upheaval were still primary considerations and it was the perceived business of society to survive apocalyptic terror as well as the perceived business of individuals belonging to society to make an earnest commitment or general covenant towards this end. Thus we are dealing with an era that still saw deism and teleology as two sides of the same coin, both operating in accordance with essentially the same natural law, a natural law moreover which applied to society as a whole and to which individuals exercising freedom of choice either conformed (good) or did not (evil). Later Newton was to publish his own ideas about the history of revelations and eschatological fear, essentially implicating an enhanced meteoroidal input of the kind observed during the course of his own lifetime [62, 8]. Broadly speaking these ideas were very closely aligned with those of late medieval scholarship in Europe [76] and hence with the Sumerian and Babylonian tradition stretching back to the very foundation of civilization [46]. Under these circumstances, to the extent that aristocratic governance is a natural condition of mankind (just as wolves hunt in packs!) and eschatology is a vital academic discipline associated with the activities of ancien regimes (from ~2500 BC-1800 AD), it is important that we reconsider the social imperatives once believed to arise in response to recognised celestial inputs. Indeed, all the more so if, as now seems likely, Newton at the end of his career was not properly understood by his younger contemporaries.

Although English empiricism eventually continued in the tradition of anti-apocalypticism, there can be little doubt as to its emergence in the knowledge of near-miss comets and apocalyptic hosts, causing nations to collapse. Thus, in likely awareness of revelations and their apocalyptic implications (Section 5) whilst reflecting on the attitude society should adopt (Section 6), the English empiricists evidently introduced a blatantly metaphysical interpretation of revelations which the remainder of Western civilisation proceeded to single out as advantageous during the course of two further periods of global upheaval around 1790 and 1850. It seems that a whiggish establishment which was initially expected to guide society in accordance with an apocalyptic fear which was historical and real (à la Locke: Section 7) was then privately persuaded that revelations were of no material substance in accordance with a teleological cosmology introduced by the "gnostic" Greeks (à la Berkeley: Section 8). We shall see that this persuasion was not altogether irrational.

7. The perception of "last times"

Any specification of these longer periods of upheaval around the globe (Section 6) in terms of a parochial (ie purely national) enlightenment can of course be very misleading. Thus we cannot be too enthusiastic about so-called English, French and German enlightenments when some scholars would perhaps justifiably connect the English and American upheavals of the seventeenth and eighteenth centuries with a general advance of predominantly Anglo-Saxon culture which French reactionaries did not at first acknowledge [27] but which was then admitted by most European nations during the upheaval of the nineteenth century. In point of fact, the successive enlightenments here are not as "progressive" as many commentators might now be inclined to claim since the post-revolutionary French enlightenment was in practice something of a wider European reaction against the English enlightenment which then failed to secure intellectual support. Thus the successive enlightenments perceived as revolutions were more in the nature of a whiggish reaction to apocalypticism as in England which later spread to America and then to Europe and to the rest of British Isles to become the orthodoxy of Western civilization based upon rather specific, pragmatic principles dictated by Berkeley rather than by Locke (see Section 8). Indeed, if we take an even more parochial view and regard these longer periods of upheaval as corresponding locally to the so-called English, Scottish and Irish enlightenments respectively, we can well see that these advances are probably not unconnected with the severity of the immediately preceding climatic downturns and hence possibly with the corresponding meteoroidal inputs. These were prominent enough within the British Isles to have implied a possible link between the social plights of each people in turn and the conspicuously adverse social conditions of the little ice-age, the highland clearances and the potato famine respectively [49]. There are problems evidently in handling such diversity among the parameters involved when seeking historical patterns in social behaviour but the rather similar correlations between astrophysical, geophysical, social and intellectual developments may seem well enough founded now to raise serious questions regarding the current linear model of historical evolution and to justify seeking to reinstate a cyclic model of history involving the successive incidence of apocalypse followed by renaissance [38]. Such a model does not necessarily imply renaissances which are progressive however and it is entirely possible that the apocalyptic experience provokes an increase or decrease in the authority or freedom with which society is controlled depending on the actual extent to which society believes the cosmos (or divinity) was involved.

With such a model, the intermittent and serious breakdown of society is of course the more readily understood and we can perhaps appreciate why our ancestors were so ready to concede mankind's innate propensity towards extreme anti-social behaviour in the presence of disturbing and terrifying conditions - the unavoidable disposition to "sin" - which could nevertheless be greatly alleviated by the creation of a rigidly structured society and by the early inculcation of suitable knowledge and social customs. The converse of course, attractive to Western civilization today, allows for a greater degree of individual freedom and toleration of individual habits within society whilst apparently failing to recognize the practical difficulties that then arise with the application of restraint in the presence of apocalyptic fear. Long accepted ideas about eschatology (the destiny of man after death) and kingship connected with the practices of omen astrology and the perception of cosmic catastrophe ("last times") which now seem strange did evidently arise however in association with ideas about group salvation (ie specifically favouring aristocracies) from prolonged cosmic terror [34]. Furthermore, accounts of the afterlife given in the ancient mythologies consistently reflect a situation of hopelessness and despair. These were not originally accounts of punishment in hell for misdeeds of the wicked, however, but of ordinary mortals transferred to a post-mortem environment, stories of inexplicable survival amidst equally inexplicable extinction [80]. The dreadful warning was that of a forthcoming state of the world, inevitably represented as a prolongation ad infinitum of tedium followed by terror, such as had already been experienced.

The form of governance known as "ancien regime", essentially recognising cosmic terror and serving aristocracies, can in fact be traced to its inception during the early stages of the Sumerian Empire and thereby to the very beginnings of systematic knowledge and civilization itself. As such it admitted the bodies of knowledge now referred to as omen astrology [77] and eschatology [12]. It is customary nowadays of course to dismiss these branches of knowledge as irrelevant but, contrary to a popular misconception, there are no serious grounds for supposing the essence of this secular knowledge is other than empirically based and subject to scientific interpretation in the modern sense of these terms [61]. In fact, the formal linkage of omen astrology, eschatology and "kingship" in the past was clearly fundamental to aristocracies with pretensions to hegemony. Territory submitted to rule by hegemony which is so extensive as to be vulnerable to the debilitating influence of apocalypse will commonly put the survival of a mere aristocracy and its possessions seriously at risk. It is no cause for surprise therefore that all aspects of social and intellectual existence in such a culture should have been geared to life beyond the apocalyptic events.

The discipline of eschatology moreover was no fanciful diversion: from its inception it dealt with a continuing apocalyptic process by which specific celestial agencies would inevitably reduce mankind and the terrestrial environment to states of mental and physical chaos. By classical times, it was very widely understood that the mightiest of these occurrences took the form of a "universal conflagration" or "flood" at long periodic intervals due to a circulation of celestial bodies which impinged upon the Earth. This was by no means all however since lesser eras during the course of history were also clearly involved, each with its associated "last times" characterized by celestial signs and frightening events as well as by wars, epidemics and famine. Comets on this account were just one category amongst a range of natural phenomena which might be heedlessly associated with the demarcation of critical periods in the course of natural history. Because of the relative frequency of these phenomena it is easy nowadays to imagine our ancestors have always had an exaggerated view therefore of the significance of their particular epoch. While this tendency cannot of course be discounted altogether, the inference is almost certainly misleading since it fails to take account of the primary apocalyptic agency (fireballs) through which the mental anguish generally develops. The point to be emphasised here perhaps is that the "uniformitarian" character of Nature is certainly defensible for a considerable range of phenomena for much of the time but because it is so deeply entrenched on this account, we must not fall into the error of supposing it must therefore apply to every celestial phenomenon. Indeed the clear impression that emerges from the strongly celestial orientation of much mythology and religion, not least during the classical era itself, is of a celestial agency in the past which did strengthen considerably at certain epochs to become traumatic in the extreme and which was then still an awesome prospect in the future (see also Section 5). In contrast any such extreme trauma seem to have been at a reduced level during the Enlightenment while the ancien regime has increasingly appeared to be an irrelevance and thus in a general state of abeyance.

Accordingly and until two hundred years ago, virtually all historiography was based upon a globally apocalyptic chronological scheme [12]. A world historical process was clearly envisaged in which ordinary natural processes would continue for the duration of an era until its inevitable termination in "last times". A very ancient pattern of such eras, now attributed to Zoroaster [24], is usually marked by the Creation (~5500 BC) and subsequently by successive Ages: the Golden (until ~3000 BC), the Silver (until ~2000 BC), the Bronze (until ~1000 BC) and the Iron (until ~[0] BC/AD). A more recent scheme originating with Eusebius and continuing into early modern times [62] involves a somewhat more elaborate division of time but recognizes also the passage of dominant civilizations: the Babylonians, the (unhistorical) Medes, the Persians, the Greeks and the Romans (~2000 BC-500 AD); followed by the emergence of the Holy Roman Empire (~500-1000 AD) and its renaissance as Western (European) civilization (~1000-1500 AD), subsequently undergoing reformation (~1500-1700 AD) and enlightenment (~1700-1800 AD) prior to its secular emancipation (~1800 AD - the present). One accepts of course that these transitions are to some extent still in general use but any suspicion that they correlate with environmental events has now largely been abandoned.

We should not be over-impressed by any apparent precision in these dates. Nevertheless the trend towards shorter eras as we approach the present almost certainly reflects an increased attention to detail regarding the correlation of celestial and terrestrial events. On the other hand, as we have seen, the periods of "last times" can endure for several generations and any chronological scheme involving single dates of demarcation is therefore necessarily imprecise. To gain some idea of its relative precision, we can refer to the most complete record of meteoroidal fireballs during the last ~2000 years which is that maintained by Chinese imperial astrologers [10]. This shows substantial enhancements roughly during the two centuries preceding and including the time of Christ (say ~150 BC-50 AD), the three centuries including the European Dark Ages (say ~350-650 AD), the two centuries that correspond to the bulk of the Crusades (say ~1000-1200 AD), the two centuries that precede and include the Reformation (say ~1350-1550 AD), the half-century that includes the English Civil War and Interregnum (say ~1625-1675 AD), the half-century that includes the American War of Independence and the French Revolution (say ~1765-1815 AD) and the few decades that includes the period of European Revolutions (say ~1830 -1860 AD).

Most of these enhancements are broadly concentrated in the northern hemisphere months of midsummer (end of June) and early winter (start of November) indicating a likely strong association with the Taurid stream of meteoroids [4]. This can reinforce the impression we are dealing with an incidence of "last times" which is due to hosts of debris ultimately originating from a single celestial agency. Whether the perception of a unified process was sustained in this manner is perhaps not clear but the idea of a celestial agency which intermittently exerts pressure in such a way as to apparently guide mankind to greater knowledge and understanding during successive stages of comparative calm but then also, at longer intervals, introduces a level of devastation that causes civilization to experience a particularly serious recession (cf the model of Section 5) was evidently commonplace. The details of apocalypticism indeed essentially upheld a picture of civilization in which the low-level evolution of mankind advances by a process of punctuated equilibrium in accordance with a random distribution of encounters between the Earth and Taurid hosts; but with the added feature of a longer term cycle as the greatest concentration of material within the Taurid stream apparently swings back and forth in the sky (see Section 9), regularly delivering mankind into the succesive depths of "Dark Age" every few thousand years or so while passing between the corresponding shallows of "enlightenment". This periodic timescale is of course very long in comparison with the human lifetime but it appears to have represented an additional yet certain threat in contrast to the uncertainty of random punctuations thus reinforcing the general incidence of mental anguish in the past. The important point to be emphasised here perhaps is that the incidence of "last times" in the past was never perceived to be a trifling matter, more an integral part of a supposedly real catastrophic evolutionary scheme which was generally understood to involve both "punctuations" and "cycles". Contrary to a current general impression therefore, the main precepts of the nineteenth century uniformitarianism-catastrophism debate relating to geological and biological evolution were already essentially in place, albeit in relation to much shorter timescales and seemingly more germane effects.

8. The disembodiment of revelations

The evolution of species is either "progressive" or "punctuational" depending whether the dominant environmental processes at play during the course of extinction and speciation are "uniformitarian" or "catastrophic". In principle the dominant environmental processes at play can be biological or physical, whence a catastrophe likely arises on account of an internal (biological) instability or an external (physical) perturbation. Likewise the evolution of civilization is either progressive or punctuational depending whether the dominant environmental process at play is uniformitarian or catastrophic. Catastrophes in this case however, as we have seen, refer to the rise and fall of "core societies" (aristocracies). Recent historians on the other hand have not generally looked to physical factors (eg climate, revelations etc) as the principal cause of such rise and fall with the result that the punctuational form of historiography that has been in place for more than two hundred years is one in which the longer periods of upheaval are mostly understood in terms of purely social factors in general acordance with the "whig interpretation of history". This widely accepted form of historiography clearly takes its lead from the English empiricists and German rationalists and, as we have already noted (Section 6), is currently undergoing critical re-evaluation [16]. An evidently fundamental issue however affecting the way this type of historiography came to be established is to be found in the role of biblical scholarship during the eighteenth century. Thus the Anglican biblical scholars of this period, many of them bishops and members of the whiggish ascendancy paying lip-service to the intellectual leadership of Locke and Newton, were heirs to a contemporary historical tradition which was made particularly manifest by the Protestant Bible, combining both the New and Old Testaments. The episodic character of history as expressed through the successive books of this Bible, indicating separate periods of interaction between the divine cosmos and mankind, was palpably reinforced as such through a perceived series of so-called "Christian evidences" cumulatively seen as providing "proof" of the fact in the manner of a euclidean text ([40, 41]cf [65]).

The Bible in other words was widely looked upon as a canon of evidential statements and prophetic utterances - propositions and theorems, one might suppose - which taken together essentially "proved" the given framework of historical knowledge. Interpolated between these statements and utterances of course was to be found the discussion and interpretation that would normally clarify the propositions and theorems in the context of the meaning and implications of the canon as a whole. For the most part, such discussion and interpretation are by way of subsequent commentary on the original canon and were recognized readily enough by their generally moral and literary style. The evidential statements and prophetic utterances on the other hand were clearly understood as the elements of an original, historically updated, document whose textual content plainly identified it as such and which was generally considered to be particularly well preserved.

This is not to say however that the meaning of the original document was considered to be immediately apparent. Rather it was accepted that the meaning may have drifted on account of repeated translation and redaction and that it was the business of biblical scholars to discern its true meaning. In other words, despite their substantially unaltered condition, the meaning of the propositions and theorems now had to be inferred as much from the context as from the words and was still a matter for scholarly debate. The whiggish scholars were thus very much aware that "whoever looks into the prophetical writings will find they are generally penned in a very exalted style.....and oftentimes in such images as cannot admit of literal interpretation" [66]. On the other hand the "exalted style" of the so-called "Christian evidences" was commonly seen as accurately reflecting the often dramatic or overpowering nature of the phenomena referred to, hence their usual interpretation as recorded instances of divine intervention in human affairs.

It would be quite out of character with the received theological wisdom of the time however to suppose these "revelations" of divine intervention were in any sense figurative or symbolic. Rather they were perceived as super-natural or extreme celestial events. In other words, there can be little doubt that the well preserved historical document subsumed by the Bible was widely perceived as a catalogue of very special celestial phenomena at certain epochs - such as astronomers are now familiar with in the case of the encyclopedic records of ancient Chinese observations of such phenomena - interwoven with a predictive or prophetic model linking together what would otherwise be regarded as distinctive but unconnected events. The modern astrophysicist has little difficulty in comprehending the general principles that underscore such a historical document. After all our modern picture of the universe is itself founded upon a predictive physical model designed to match or "prove" selected observational data in supposedly temporal sequence.

The analogy does not end there. For the eighteenth century biblical scholars and their predecessors, like modern astrophysicists, were plainly not unaware of the underlying, supposely more mundane, framework of natural phenomena upon which this category of more dramatic or supernatural events was superposed [43]. Thus a division of intellectual labour was evidently envisaged in which natural philosophers would largely supervise our understanding of more worldly phenomena and theologians at the top of the intellectual tree would supervise our understanding of divine intervention; that is, the whiggish ascendancy was no different from any other intellectual leadership and was ultimately in business to establish a definitive view of divine material (providence) and divine motion (its dynamic) along with an explanation of how these together were responsible for the special celestial events known as revelations. Soon after the mid-seventeenth century foundation of the Royal Society, as we have seen, Bishop Sprat had remarked upon the widespread concern in England as to the nature of "providences" and "prodigies" (their most extreme version) while others such as Baxter in England and Mather in New England [50] were advocating procedures whereby "providences" would be systematically observed and recorded over much of the globe. The frequency of meteoric "fireballs" due to the more substantial meteoroidal inputs (see Section 4) was in fact significantly enhanced above the usual background level during the seventeenth century and it can hardly be doubted therefore that these fireballs were the primary objects diverting contemporary attention towards thoughts of "last days" and a very natural association with earlier revelations as recorded in the principal canon of celestial knowledge. Normally of course the providential background is of no greater consequence to mankind than the incidence of lunar and solar eclipses or planetary conjunctions but if it is significantly increased then the occasional prodigy may inject vast amounts of dust into the atmosphere or generate a huge mutimegaton explosion, thereby becoming a very serious hazard to mankind. The well-preserved historical document - the Bible - we may be sure, was widely received as a very vital catalogue of extreme but very real providential events.

In fact, the science of celestial mechanics itself emerged from considerations like these relating to divine material and divine motion. Thus it is clearly no accident that Newton's *Principia* of the seventeenth century, which heralded the enquiries of the eighteenth century biblical scholars, was likewise structured like a euclidean text. It has been said that there was an unprecedented fusion of scientific reasoning and religious thought during the lifespan of Newton [8] but, while this may be undeniable, the prospect of an integrated scheme under one set of divine physical laws had been a major component of the Christian intellectual agenda since at least as far back as Aquinas [56]. What is perhaps not so fully appreciated these days, therefore, is the extent of the purely (astro)physical motivation behind much theological enquiry and of how much this represented a continuing desire to comprehend the flux of supposedly divine material, most especially its apocalyptic potential. Thus, by the seventeenth century, it was already recognized that providence would arrive during broad epochs (say 50-150 years) spasmodically distributed throughout the course of history (say at intervals of a few centuries) reaching back to the "deluge" and even to "creation" itself. If the historical analysis of such as Zeigler, Munster and Gifftheil [37], Ussher, Alsted, Brightman and Mede [76] and finally Newton [8] were to be upheld, the whole sequence of "biblical events" was stretched over some 6000 years and there was much to be learned about the future from a suitably careful examination of the past.

During these broad epochs, it is clear now that providence would arrive in characteristic but subordinate temporal patterns (ie at a particular month and a fixed, relatively small, number of years) such as would readily allow a rather straightforward prediction of the next influx together with such nervous anticipation of the "end of the world" as would be engendered. There was no absolute certainty of course that the "end of the world" would come but there was no mistaking the general pattern of downfall during history associated with these broad epochs. According to this pattern (Section 7), the great kingdoms of the past had succumbed in turn: first Babylonia and Media; then the Persians; later the Greeks; and finally the Romans. The latter of course is not identified in the Book of Revelations but by the sixteenth and seventeenth centuries, the European protestant movement had essentially added to the list of downfalls the so-called "great apostasy" when the Roman Empire governed from Constantinople failed to retain control of the West and power passed into the hands of the so-called "Antichrist" at Rome. Descendants of the massive population displaced into Western Europe during the great apostasy (Franks, Germans, Anglo-Saxons) seem therefore to have been the principal parties to perceive themselves as having been delivered (by the vikings?) into the hands of a huge ecclesiastical empire ruled by a despotic church from Rome. Their protestations evidently took hold alongside a perception of the environment which was most clearly presented in the Books of Daniel and Revelation, containing apocalpytic accounts of a most fearful and terrifying nature. These books, as we have already seen, are representative of a late Babylonian tradition before the time of Christ [77] which stretches back to the origin of omen astrology itself [61]. This is the very same tradition of course which led Keynes [46] half a century ago to categorize Newton as the "last of the Babylonians and Sumerians". There can be absolutely no doubt that the protestant whiggish ascendancy was heir to this tradition as well.

But whereas Newton, along with the principal puritanical movements of his time, would look back with considerable horror at the great apostasy and the original Christian sect identified with the apostolic succession which had penetrated and dominated the upper echelons of Roman society by the fourth century, the contemporary trend of whiggish opinion amongst English bishops was evidently towards condoning and reaffirming the critical fifth century decisions of the embryonic Holy Roman Church. The point here is that the Christian sect in question, whilst having gained in political clout as it moved away from the gnostic orthodoxy at the turn of the millennium in which Christ was primarily a cosmological phenomenon [80], was to use its position of dominance at the great Councils of Ephesus and Chalcedon to ascribe this cosmological phenomenon to a disembodied or spiritual Christ, precisely in accordance with the viewpoint now (re)adopted by the philosopher bishop, one Berkeley. It is well known of course nowadays that Newton subscribed (in private) to the so-called Arian heresy but the implications of this heresy with respect to the physical nature of the cosmological Christ (and hence revelations: see below) are now commonly overlooked by astrophysicists. Thus the basic point at issue amongst the English bishops was the precise interpretation to be given the "Christian evidences". Locke, on the one hand, inclined to the view that we must be dealing with quite exceptional natural events involving real material bodies and that "what is obscure to us may have been perfectly clear to those living at the same time in the past"; whereas Berkeley, on the other hand, argued that "the proper objects of vision constitute an universal language of the Author of Nature" and that "the language of Nature doth not vary in different ages or nations". Whatever was intended by the biblical records of revelation therefore, it was now supposed by the English bishops on the basis of the proposed immutability of Nature that the associated physical effects of revelation must always be essentially mundane, as the contemporary observation of Nature assuredly asserts. It follows that the dramatic quality of revelation must be understood to denote divine messages in the sky possessing only a spiritual rather than a material content while the processes of Nature must be regarded as essentially uniformitarian.

Such disembodied revelations were rather quickly approved by the whiggish ascendancy in England (to be followed by the rest of Western civilization in due course) with the result that physical revelations ceased to receive any intellectual attention and were essentially excluded altogether from the modern system of natural philosophy which came to be known as materialism. Non-physical revelations however, incorporated into a philosophical system known as hutchinsonianism [67, 41], were widely accepted and led to an essentially allegorical interpretation of the Bible [44] which was to become more or less axiomatic whilst also admitting a form of natural philosophy constrained by the principle of uniformitarianism [53, 26]. This far-reaching development was opposed during a significant revival of catastrophism between ~1790 and ~1850 in France [10, 48] and in Scotland [55] which saw the earliest scientific description of natural selection under the conditions of punctuated equilibrium but which then also saw this development rather quickly expunged from the scientific record [28]. This turn of events is historically significant since it demonstrates Darwin's particular concern, critical to the English enlightenment, which was to develop an account of natural selection compatible with uniformitarianism. The nineteenth century decline of catastrophism was essentially a repetition therefore of the early eighteenth century rejection of revelation as expressed through the differing views of Locke and Berkeley. During the twentieth century, Darwinian uniformitarianism became firmly established until the study of impact craters suggested a significant role for cosmic catastrophes underlying the course of evolution on geological timescales. During the same period, the physical study of fireballs has been reinstated but the findings nevertheless have tended to remain independent of the study of revelations with the result that three principal forms of natural philosophy relating to evolution on geological *and* historical timescales can now be discerned. These are (1) uniformitarianism-cum-hutchinsonianism (this essentially remains the mainstream view); (2) catastrophism-cum-hutchinsonianism [13, 58]; and (3) catastrophism [20, 22]. The differences of opinion rendered explicit here, though not intractable (see Section 9), remain a firmly established feature of modern enlightenment. The absence apparently of any satisfactory resolution to the problem is marked by a particularly vehement tradition within modern scholarship seeking to preserve anti-apocalypticism. Here it only remains to consider the physical theory of revelations that is permitted by their evidently material and historical nature.

The city of Alexandria, through its geographical location, eventually came to be singularly well placed for the scholars of imperial Greece undertaking comparative studies of the information emanating from the great civilizations of the Near East. Prominent amongst these scholars were the so-called "gnostics" who brought together the various common strands of Chaldean astrology, Zoroastrian dualism, Egyptian mythology, Jewish apocalypticism and Hellenistic speculation to construct a unified system of cosmology (and religion) which is now known to have been highly influential and extremely widespread across the temperate zone of three continents ([80],Chapter 3 in particular). The existence of this unified cosmology almost certainly betokens the spread of a unified (secular) culture as well [68], thus pointing to an early example of the kind of growth that we now associate with modern science and enlightenment at the heart of Western civilization. The underlying perception behind this gnostic or imperial cosmology was evidently that of the heavens which included an upper region (or celestial sphere: the stars), an intermediate region (or ecliptic plane: the planets) and a lower region (central to the sphere: including the Sun and Moon). The various visible components of these regions were clearly perceived as owing their physical existence and creation to a former unitary body, the cosmological Christ. Creation was not a once-and-for-all operation however but involved a process of heirarchical disintegration known as "emanation" which also led to the formation of other substantial celestial features, no longer extant. Foremost here was a circular embankment ("the flaming walls") in the form of a broad belt around the ecliptic, apparently just within or at the celestial sphere, which marked "the boundary" of the known universe. This boundary was also essentially continuous with a narrow bridge ("the rope of angels") which connected the circular embankment directly to the centre of the celestial sphere, thus joining the upper and lower regions of the heavens.

This boundary, including the bridge, appears to have been once visualized as a very luminous, material construct resulting from the emanations of the cosmological Christ which, however, was not entirely uniform in its general appearance. Thus it was made up of temporary shapes and evolving patterns, one such apparently including Christ itself in circulation between the heavens above and below. The lowest step of the bridge perceived as a ladder did eventually mark the most critical phase of emanation from the cosmological Christ. This critical phase was clearly taken to be an extremely dramatic event. It not only captured the attention of Earthlings below but was evidently considered to be the primary stimulus giving rise to the multiplicity of cosmic mythologies among the various nations of the known world. Indeed the event itself was a veritable "Pandora's Box" being the palpable focus for an outpouring of visible celestial entities whose existence thereafter appeared to capture the attention of witnesses all round the globe. Putting these mythologies together, the emanation was ultimately perceived by the gnostics to be in a direction essentially orthogonal to the line of the narrow bridge, projecting first on one side and then on the other to produce an identifiable "cross". This cross was never an isolated feature of the sky however (eg see Plato's "Timaeus": [51]) since it evidently connected with a more extended celestial structure. This structure, it would appear, corresponded to a closely bunched heirarchy of so-called "archons" (archaeons) and "aeons" - emasculated versions, apparently, of the archangels and angels who occupied the narrow bridge - whose rather similar orbital paths therefore intersected at a point on the bridge and whose distribution otherwise was such as to dominate the lower region of the heavens. The effect of the dramatic event therefore was to introduce new features in the sky which were a pale reflection of the earlier, more luminous, constructs and which furthermore did more obviously affect mankind. In fact these aeons were evidently regarded as hostile to the Earth and mankind and it was their emanations in particular which were subsequently associated with revelations and apocalyptic terror.

It eventually came to be well understood of course that revelations were equivocal in respect of the incidence of catastrophe and it was the Greeks apparently, as imperialists, who first sought to turn the incidence of potential non-catastrophe to public advantage by systematically painting revelations in wholly optimistic terms. The Greeks long before had taken the formal step of separating government and academe; and while it is clear the latter then took rationalism to even greater heights than before, the demands on government in the face of apocalyptic terror (cf the banishment of Anaxagoras: [82]) evidently meant a degree of sophistry was not out of place when presenting the realities of the cosmos. Organized religion and stage-managed miracles indeed became endemic amongst the cosmopolitan Greeks [32] and we cannot therefore be surprised at the cosmological phenomena which were embellished, ie. a fireball phenomenon which became providential rather than fickle; an original cosmological Christ whose body and initial emanations became spiritualized, good and divine; a fallen angel whose cross (regarded as Christ's burden) gave rise to the lower region of heaven as well as to mankind on Earth, both perceived as evolutionary stages of a defective material condition to be suffered by the entrained spirit; and a subsequent resurrection of the cosmological Christ perceived as an example to all other fallen spirits (ie all human souls) whose route back to the upper region of heaven needed to be made manifest. Nowadays of course we are more familiar with the teleological or moral philosophical aspects of the theological rationale that was invoked by way of explanation for the known cosmological facts; conversely we are rather less familiar with the interesting theoretical notion of a material condition which is some kind of negative or defective attribute of the (primary) spiritual condition, tending also to forget that we are now the beneficiaries of an intellectually hard won physical science in which only the (positive) material world exists. In fact, the significance of all the theoretical detail here must be the greater on account of its having been clearly related to the known cosmological facts. For we can hardly suppose that the Greek imperialists would expect their version of "gnosticism" to carry conviction with subordinate peoples unless the available cosmic mythology were very adequately explained. It has already been noted that the influence of gnosticism was wide: indeed it flourished for centuries alongside the rise of the Holy Roman Church. Subsequent understandings of apocalypticism would be expected therefore to match gnostic cosmology to a plausible or acceptable extent. It is hardly surprising that late medieval protestant heresy, inasmuch as it inclined to hard-nosed materialism, was a principal bearer of the original as opposed to the embellished gnostic cosmology which is an integral part of Christian theology.

To a later generation of imperialists in another place, however, who no longer knew which aspects of cosmology had been stage-managed and why some subtle aspects of gnosticism became heretical, it was a matter of simplicity, albeit of surprise, to save the surviving appearances by ultimately spiritualizing "the defective material condition" of the revelations that still remained: in other words, it was a matter of simplicity for English bishops, still rather ill-versed in the basics of Newtonian science, to disembody the most immediate parts of the observed cosmological environment for the sake of "a language of Nature" which "doth not vary in different ages or nations" and for the sake of an anti-apocalyptic environment which they sought to preserve.

9. Apocalypticism unravelled

We should not be too hard on English bishops! Subsequently this inclination to disembody the most immediate parts, if not all, of the observed cosmological environment has only been matched by academe's solicitude for secularisation: indeed it is something of a surprise nowadays to realise that this inclination can officially still be very much in place. Darwin and Huxley, as we have seen, were prominent nineteenth century influences who, by championing uniformitarianism-cum-hutchinsonianism, effectively made sure cosmological astrophysics remained disconnected from Earthbound physics right up to the Second World War. Before the Second World War, for example, Britain's most distinguished physicist (Rutherford) could easily set aside Britain's most distinguished astrophysicist (Eddington) as a mystic because he was all too obviously perceived as peddling a disembodied world. Thus did a founding architect of the knowledge that gave us "the bomb" deal with a founding architect of the knowledge that gave us "the expanding universe". After the Second World War however a mankind ridden with guilt disowned "the bomb" and as a consequence physicists were no longer held in high esteem. Britain's most influential astrophysicist (Woolley) could thus immediately position himself within the fastness of the cosmic arena and look across with disdain at the rocket-men on their new frontier and declare that [nearby] space was "bunk". America's most influential astrophysicists meanwhile were not so confident. Thus it was not until the 1960's they were able to persuade NASA to give up their fundamental in situ measurement of nearby space and their part in any 1986 rendezvous with Comet Halley in order to concentrate on the cosmic arena beyond. As a consequence, although the first rendezvous with a comet quickly firmed up the best (post-war) model for comets and the in situ measurement of nearby space rapidly established the particulate outflow of cis-Jovian space which laid bare the giant comet debris producing zodiacal dust, the perceived official position of astrophysicists in Western civilization is still one in which those who belong to "the cosmic arena" are expected to keep apocalypticism at bay.

It is easy to understand, therefore, why many astrophysicists were provoked by the writings of Velikovsky [78, 79] who, as a psychoanalyst, was justifiably very interested in the origins of apocalypticism but who unfortunately laid himself open to easy condemnation through his advocacy of an irrelevant and certainly inadequate "planetary" theory for some past comets. Normally, of course, astrophysicists would display some forbearance in the presence of theoretical inadequacies of this kind but, in this case, it was all too obvious that they had failed to address the historical evidence for apocalypticism and were merely intent on creating a cover for their own deficiencies. Whipple [81], on the other hand, did not address apocalypticism but did nevertheless introduce the first realistic theory of sub-cometary debris for nearby space involving a recent giant comet and this was widely approved by comet scientists [47]. Clube and Napier [21, 22] then developed this theory in the longer term galactic/geological context in such a way as to permit the issue of apocalypticism to be specifically addressed [1, 2, 17]. Very few astrophysicists now seem anxious to defend the English bishops although those hoping to eliminate apocalypticism from the course of classical, medieval and modern history [13, 58] still continue to express themselves rather vehemently.

The inner Solar System is dominated by apparently two distinct "bombardments" from the cometary cloud that accompanies the Sun in its orbit around the Galaxy [22]. One of these is essentially an "unevolved" bombardment which has undergone little or no interaction with the Sun and planets: it mostly comprises ordinary comets in orbits whose semi-major axes are relatively large and whose perihelia are isotropically distributed, implying a high mean ecliptic latitude (ie a population made up of so-called Halley-types [6]). The other is essentially an "evolved" bombardment which has undergone considerable physical and dynamical interaction with the Sun and the planets: it mostly comprises the sub-cometary debris of giant comets in orbits whose semi-major axes are relatively small and whose perihelia generally lie close to the ecliptic, implying a low mean ecliptic latitude (ie a population made up of so-called Encke-types). The dynamical histories of these bombardments are such that the Halley-types evidently sample the *direct* influx of comets perturbed from the so-called inner Oort cloud. This influx is periodic (~26 myr) in accordance with the cloud's probable penetration by impulsive perturbers in the form of low velocity dark matter from the nearby mainstream circulation of the Galactic disk. The Encke-types on the other hand appear to sample the *indirect* influx, comets perturbed from the inner Oort cloud which first settle in the outer Oort cloud and then preferentially transfer, in the case of objects that ultimately settle in the inner Solar System, to "sungrazing"/"Jupiter-avoiding" orbital configurations whose mean motion is resonant (eg 7 : 2 J; mean period = 3.39 years) and whose initial perihelion direction is in the general vicinity of Jupiter's. The perturbation flux in this instance reaching the Earth is likely to be triply periodic in accordance with the Sun's distance from the Galactic plane (~26 myr but 90° out of phase with the direct perturbation flux), the outer Oort cloud's perihelion distribution with respect to precession of the Jovian longitude of perihelion (~79 kyr) and the sungrazer nodal precession (~2.5 kyr) in relation to the Earth's orbit [2].

The physical characteristics of these direct and indirect perturbation fluxes are however fundamentally different. The Halley-types reaching the Earth are essentially dominated by the commonest inner Oort cloud comets, that is, by ordinary small comets with velocities close to the Solar System velocity of escape at the Earth's orbit. The Encke-types reaching the Earth are essentially dominated by the sub-cometary fragments originating from the rarest and largest inner Oort cloud comets, that is, by the meteoroid fragments of giant comets with velocities much closer to the velocity of circulation at the Earth's orbit. For a representative linear size ratio between small comets and meteoroid fragments of 10, the impact energy ratio is ~10,000 resulting in qualitatively different effects on the Earth. The *direct* perturbation flux essentially reaches the surface of the Earth and is responsible for craters, mass extinctions and major geological effects; the *indirect* perturbation flux, on the other hand, essentially reaches only the atmosphere producing impact fireballs and disintegration dust-veils, the latter primarily affecting the climate and the magnetic field. These general predictions are well illustrated by the differing mass extinction and geomagnetic reversal patterns of the late Phanerozoic respectively [22].

The current location of the Sun in the Milky Way, very close to the Galactic plane, implies the Earth is currently half way between the preceding and forthcoming mass extinction "bombardments" 26 myr apart but more or less within a major glaciation "bombardment". The latter bombardment is itself apparently cyclic with a sub-periodicity of ~79 kyr whose peaks roughly correspond to the most recent giant comets settling in inner Solar System space, the latest being responsible therefore for the last glacial starting ~60 kyr BP. The size and orbital configuration of giant comet sungrazers imply glacials that last ~50 kyr and the present overall picture is therefore consistent with a current giant comet which is likely to have reached the end of its physical and dynamical lifetime within (say) the last 10 - 5 kyr. The most recent glacial ~60-10 kyr BP is evidently in broad agreement therefore with this astronomical prediction and would lead us to expect a huge remnant "doughnut" of zodiacal dust (bounded by ecliptic latitude bands) where the sungrazing progenitor's original meteoroid stream has interacted with the asteroid belt and, in addition, another more substantial meteoroid stream resulting from a terminal encounter of the highly evolved progenitor with one of the inner planets (involving its complete tidal disintegration à la Shoemaker-Levy). Such a doughnut was first detected in modern times by the Infra Red Astronomical Satellite in 1983 and, although an original sungrazing meteoroid stream has yet to be discovered, is highly suggestive of the much earlier, more luminous, "boundary" configuration ("flaming walls" and "narrow bridge") which was such a prominent feature of gnostic cosmology. It follows that the "Taurid" meteoroid stream is very likely to have been formed as a result of a close encounter with an inner planet producing a complex of progenitor debris whose central core is also likely to be closely associated with the 7 : 2 J resonance. Substantial evidence now exists in fact for a general category of inner Solar System disruptions associated with the Taurid Complex involving devolatilised cometary bodies which provide the bulk of the fragile material reaching the Earth since ca 5000 BP [1, 2, 3, 17]. The general dynamical characteristics of the Taurid Complex are notably consistent with its formation some 5000 years ago in association with an initial luminous trail of the disrupted progenitor as the latter underwent gravitational deflection during its "terminal" close encounter with an inner planet (specifically in accordance with a cross at a low point of the narrow bridge). Apart from the relationship between this Complex and the generally random apocalyptic process discussed in the Sections above, the nodal precession of the Taurid progenitor is of particular significance since it results in an overriding long-term ~2.5 kyr "global cooling"/"global warming" cosmic dust cycle on Earth as well as the correlated "Dark Age"/"enlightenment" bombardment cycle particularly affecting mankind.

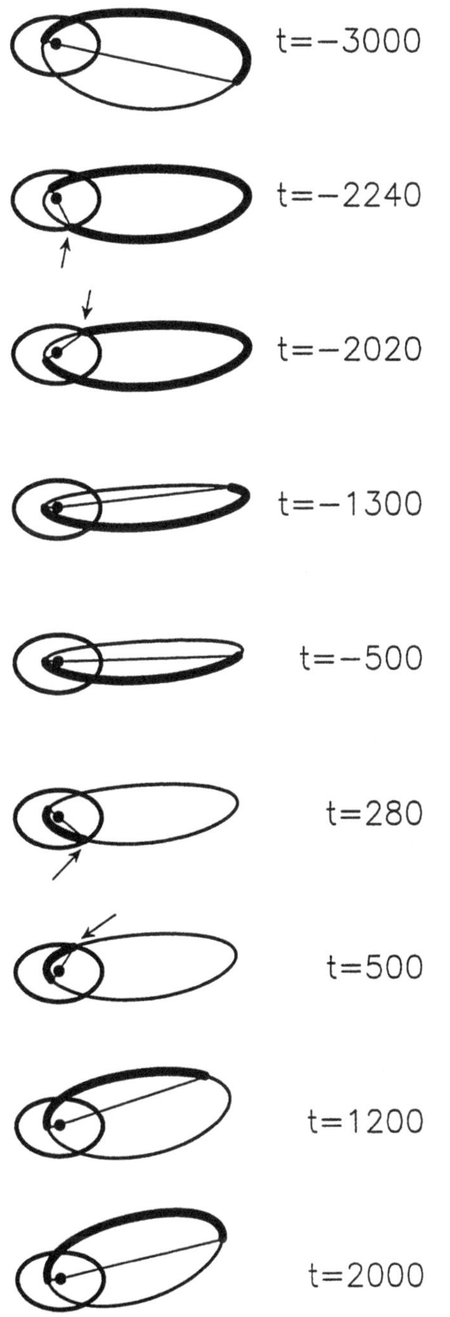

Figure 1: Illustration showing typical nodal precession for the core of the Taurid stream, in particular when the Earth's orbit is intersected (note the arrows). The parts of the orbit respectively above and below the ecliptic are shown as thick and thin lines, The timescale is t(AD). The Earth intersection epochs evidently occur ca 2200 - 2000 BC and ca 300 - 500 AD (implying a bombardment/dust cycle of around 2.5 kyr) while the corresponding perihelion positions are below and above the ecliptic

Strictly speaking the ~2.5 kyr dust/bombardment cycle is a ~5.0 kyr double cycle with alternating phases depending on the Taurid stream aspect with respect to the terrestrial polar axis (which is itself subject to a much slower precession) - see Figure 1. It is of interest therefore that the latest geophysical investigation of the pervasive "millennial-scale" cycle in the Holocene and Late Upper Pleistocene climate (ie since ~60 kyr BP) based on deep ocean sediment records of ice-rafting events in the North Atlantic [11] clearly indicates a dominant ~4.7 kyr cycle which is likely nevertheless, in view of the more detailed ~2.5 kyr cycle of the Holocene [63], to be a precessional double-cycle confirming the underlying role of the Taurid giant comet source. Of particular significance in this study is the additional existence of a more random "cycle" of 1.47 ± 0.50 kyr associated with sudden cooling events which evidently persist into the Holocene and which therefore bear witness to a continued interaction with the Earth involving unexpected but comparatively major apocalyptic events.

10. Conclusion

It has been shown here that the approximately centennial rise and fall of fireball streaming sometimes associated with Earth-approaching comets or asteroids is also the historical source of apocalyptic "signs". This streaming is a proxy for hazardous swarms of sub-cometary debris representing a higher flux to Earth than normally conceded of bodies in the mass range ~10^{12}-10^{15} g. Largely overlooked since early modern historical time (and even flatly proscribed by some authorities [13, 58]), this hazard appears most commonly to take the form of global climatic recessions, involving high-level dust, albeit low-level multi-megaton explosions associated with the most robust debris are by no means excluded. These recessions are a feature of the general flow of "Taurid" material to Earth recorded in polar ice-cores and ocean sediment-cores, now recognized as being responsible for a basic 5000 year double-cycle alternately producing global warming and global cooling. During the course of the Enlightenment, mankind has singularly failed to come to terms with this apparently centennial threat, having become strangely preoccupied during the Space Age with a very much less frequent threat (roughly a thousand times less frequent!) which is *directly* due to comets and asteroids. Whether or not mankind recognizes the appoximately centennial threat is tantamount to choosing between apocalyptic and anti-apocalyptic outlooks on the environment. This question, as I have shown, is of deep historical and political significance being intimately bound up with the origins of Christian doctrine and with the elitist desire to perpetuate anti-apocalypticism along with its appropriately distorted cosmological setting. In view of the intellectual and cultural climate of irrationality which arises thereby, it is a moot point whether mankind will meet the challenge posed by this question before the next bout of apocalyptic terror descends. Such a situation represents an intolerable risk to civilization.

11. Acknowledgements

I would like to thank Dr David Asher, Dr Alan Edwards and Dr Benny Peiser for critical comments on an earlier draft of this review; also the Society for Interdisciplinary Studies for the invitation to speak at Cambridge and the Leverhulme Trust for material support.

References

1. Asher, D. J. & Clube, S. V. M. 1993 *Q. J. R. Astron. Soc.* 34, 481
2. Asher, D. J. & Clube, S. V. M. 1998 *Cel. Mech.*, in press 3. Asher, D. J. & Izumi, K. 1998 *Mon Not . R. Astron. Soc* 297, 23
4. Astapovic, I. S. & Terenteva, A. K. 1968 *Physics and Dynamics of Meteors* eds. Kresak, L. & Millman, P. (Reidel, Dordrecht), 308
5. Bailey, M. E.,, Clube, S. V. M. & Napier, W. M. 1990 *The Origin of Comets* (Pergamon)
6. Bailey, M. E., Clube, S. V. M., Hahn, G., Napier, W.M. & Valsecchi, G.B. 1994 In: *Hazards due to Comets and Asteroids* ed. Gehrels, T. (University of Arizona Press, Tucson), 479
7. Barb, A. A. 1963 In: *The Conflict between Paganism and Christianity in the fourth Century* ed. Momigliano, A. (Clarendon Press, Oxford), 100
8. Barnett, S. J. 1997 In: *Prophecy* eds. Taithe, B. & Thornton, T. (Sutton Publishing), 101
9. Binney, J. & Tremaine, S. 1987 *Galactic Dynamics* (Princeton University Press)
10. Biot, E. 1807 Mem. *Acad. Sci. Inst. Nat. France* 10, 129
11. Bond, G., Showers, W., Cheseby, M., Lotti, R., Almasi, P., Menocal, P. de., Priore, P., Cullen, H., Hajdas, I., & Bonani, G. 1997 *Science* 278, 1257
12. Bultmann, R. 1957 *History and Eschatology* (Edinburgh University Press)
13. Chapman, C. R. 1998 *The Asteroid/Comet Impact Hazard* (draft)
14. Chereul, E., Creze, M. & Bienayme, O. 1998 *Astron & Astrophys*, in press
15. Christianson, G. E. 1984 *In the Presence of the Creator: Isaac Newton and his Times* (Free Press)
16. Clark, J. C. D. 1985 *English Society 1688 - 1832* (Cambridge University Press)
17. Clube, S. V. M. 1995 *Vistas in Astron.* 39, 673
18. Clube, S. V. M. 1994 *Frontiers of Fundamental Physics* eds Barone, M. & Selleri, F. (Plenum), 107
19. Clube, S. V. M. & Napier, W. M. 1984 *Mon.Not. R. Astron. Soc.* 211, 953
20. Clube, S. V. M. & Napier, W. M. 1986 In: *The Galaxy and the Solar System* eds. Smoluchowski R. et al (University of Arizona Press, Tucson), 260
21. Clube, S. V. M. & Napier, W. M. 1990 *The Cosmic Winter* (Basil Blackwell, Oxford)
22. Clube, S. V. M. & Napier, W. M. 1996 *Q. J. R. Astron. Soc.* 37, 617
23. Cohn, N. 1957 *The Pursuit of the Millennium* (Secker & Warburg, London)
24. Cohn, N. 1993 *Cosmos, Chaos and the World to come* (Yale University, New Haven & London)
25. Corcoran, T. H. 1971 (translator) *Seneca: Naturales Questiones* (Heinemann, London)
26. Darwin, C. 1872 *The Origin of Species* (John Murray, London)
27. Dawson, C. 1972 *The Gods of Revolution* (Sidgwick & Jackson)
28. Dempster, W. J. 1996 *Evolutionary Concepts in the Nineteenth Century* (Pentland Press)
29. Eddington, A. S. 1931 *Mon. Not. R. Astron. Soc.* 91, 412
30. Eggen, O. J. 1965 *Galactic Structure* eds Blaauw, A. & Schmidt, M. (University of Chicago Press), 111
31. Esmonde Cleary, A. S. 1989 *The Ending of Roman Britain* (Batsford, London)
32. Farrington, B. 1944 *Greek Science* (Penguin)
33. Fechtig, H. 1982 In: *Comets* ed. Wilkening, L. (University of Arizona Press, Tucson), 370
34. Frankfort, H. 1948 *Kingship and the Gods* (University of Chicago Press)
35. Gehrels, T. 1994 (editor) *Hazards due to Comets and Asteroids* (University of Arizona Press, Tucson), 370
36. Gibbon, E. 1776 *The Decline and Fall of the Roman Empire* ed. Low, D. M. (Guild Publishing, London)
37. Gibson, K. 1997 In: *Prophecy* eds. Taithe, B. & Thornton, T. (Sutton Publishing), 71
38. Goldstone, J. A. 1990 *Revolution and Rebellion in the Early Modern World* (University of California Press)
39. Hill, C. 1975 *Change and Continuity in Seventeenth Century England* (Weidenfeld & Nicolson)
40. Hitchin, N. W. 1995 *Anglican Theological Review* 77, 394
41. Hitchin, N. W. 1997 In: *Prophecy* eds. Taithe, B. & Thornton, T. (Sutton Publishing), 119
42. Hoyle, F. & Wickramasinghe, C. 1985 *Living Comets* (University College Cardiff Press)
43. Huxley, T. H. 1931 *Lectures and Essays* (Watts & Co, London)
44. Huxley, J. 1941 *Religion without Revelation* (Watts & Co, London)
45. Kerr, F. J., & Westerhout, G. 1965 *Galactic Structure* eds. Blaauw, A. & Schmidt, M. (University of Chicago Press), 167
46. Keynes, J. M. 1946 *The Royal Society Newton Tercentenary Celebrations* (Cambridge University Press), 27
47. Kresak, L. 1980 *Solid Particles in the Solar System* eds. Halliday, I. & McIntosh, B. A. (Reidel, Dordrecht), 211
48. Lamarck, J. B.1809 *Zoological Philosophy* trans. Elliot, H. E. (Macmillan, London)
49. Lamb, H. H. 1977 *Climate: Present, Past and Future* (Methuen, London)
50. Lamont, W. 1996 *Puritanism and Historical Controversy* (University College London Press)
51. Lee, H. D. P. 1971 (translator) Plato:*Timaeus and Critias* (Penguin, London)
52. Lewis, J. 1954 *Introduction to Philosophy* (Watts & Co, London)
53. Lyell, C. 1832 *The Principles of Geology* (John Murray, London)
54. Matese, J. & Whitmire, D. 1996 *Astrophys. J. Lett.*, preprint
55. Matthew, P. 1831 *Naval Timber and Arboriculture* (Longman et al, London)
56. McDermott, T. 1989 (translator) Aquinas: *Summa Theologiae* (Methuen, London)
57. Morrison, D. 1992 (chair) *The Spaceguard Survey: Report of the NASA International Near-Earth-Object Detection Workshop*
58. Morrison, D. 1997 *Skeptical Inquirer* 354, 22
59. Napier, W. M. 1994 *Frontiers of Fundamental Physics* eds Barone, M. & Selleri, F. (Plenum), 13
60. Napier, W. M. & Clube, S. V. M. 1997 *Rep. Prog. Phys.* 60, 293
61. Neugebauer, O. 1945 *Jour. Near East. Stud.* 4, 1
62. Newton, I. 1733 *Observations upon the Prophecies of Daniel and the Apocalypse of St John* (Darby, J. & Browne, T., London)
63. O' Brien, S. R., Mayewski, P. A., Meeker, L. D., Meese, D. A., Twickler, M. S. & Whitlow, S. I. 1995 *Science* 270, 1962
64. Oort, J. H. 1950 *Bull. Astron. Inst. Neth.* 11, 91
65. Paley, E. 1823 *An Account of the Life and Writings of William Paley, Works* (London)
66. Sherlock, T. 1724 *The Use and Intent of Prophecy, in the several Ages of the World* (Pemberton, J. & J., London)
67. Spearman, R. 1755 *An Abstract from the Works of John Hutchinson* (Withers, E., London)
68. Spengler, O. 1932 *The Decline of the West* ed. Hughes, H. S. (Oxford University Press)
69. Sprat, B. T. 1665 *History of the Royal Society ed. Cope,* J. I. & Jones, H. W. (St Louis 1959)
70. Steel, D. I., Asher, D. J., Napier, W. M., & Clube, S. V. M. 1994 In: *Hazards due to Comets and Asteroids* ed. Gehrels, T. (University of Arizona Press, Tucson), 463
71. Stone, L. & Fawtier Stone, J. C. 1984 *An Open Elite* (Oxford University Press)
72. Taithe, B. & Thornton, T. 1997 (editors) *Prophecy* (Sutton Publishing)

73. Thomas, K. 1971 *Religion and the Decline of Magic* (Weidenfeld & Nicholson)
74. Thompson, D. 1996 *The End of Time* (Sinclair-Stephenson)
75. Toynbee, A. J. 1945 *A Study in History* (Oxford University Press)
76. Trevor-Roper, H. 1987 *Catholics, Anglicans and Puritans* (Secker & Warburg, London)

77. van der Waerden, B. L. 1974 *Science Awakening II: the Birth of Astronomy* (Noordhof International Publishers, Leiden)
78. Velikovsky, I. 1950 *Worlds in Collision* (Victor Gollanz, London)
79. Velikovsky, I. 1982 *Mankind in Amnesia* (Sidgewick & Jackson, London)
80. Walker, B. 1983 *Gnosticism* (Crucible)
81. Whipple, F. L. 1967 *NASA SP* - 150, 409
82. Woolf, N. J. 1995 *Vistas in Astron*. 39, 699

Contributors

MARK E BAILEY is an astrophysicist and Director of the Armagh Observatory. His current research interests include the dynamical evolution of comets, asteroids and meteoroid streams; solar-system terrestrial interrelationships; and the implications of the cometary and meteoroid impact hazard. Asteroid (4050), discovered in 1976 by C.-I. Lagerkvist, was named Mebailey in March 1990 for his work on the dynamics and origin of comets. He is the co-author of *The Origin of Comets* (Pergamon, 1990).

MIKE BAILLIE is a professor in the Palaeoecology Centre, School of Geosciences, Queen's University, Belfast, N. Ireland. After reading physics at Queen's he moved into Archaeology and Palaeoecology taking a particular interest in chronology. His specialism is dendrochronology, and he has been involved in the construction of some of the first, long, oak chronologies. Using information from tree-ring records, he has attempted to identify abrupt environmental downturns in the past and to demonstrate their effects on past human populations. He publishes widely on these and related topics and is the author of *A Slice Through Time: dendrochronology and precision dating* (London: Routledge 1995).

VICTOR CLUBE is currently funded by the Leverhulme Trust in catastrophism/climate studies at Oxford University and Armagh Observatory (in collaboration with Bill Napier) having previously held a succession of career posts at Greenwich, Cape and Edinburgh Royal Observatories. These studies have the same bearing to astrophysics as Newton's historical/eschatological investigations had to dynamical astronomy. As such, they represent a still comparatively rare academic recognition of the highly erratic socio-political response to cosmic phenomena (mostly to do with giant comet debris) which commonly puts civilisation seriously at risk Publications include (as co-author) *The Cosmic Serpent*, *The Origins of Comets* and *The Cosmic Winter.*

MARIE-AGNÈS COURTY is a geologist and researcher at the French Centre for Scientific Research. Her scientific background is in Earth Sciences, Quaternary Geology and Prehistory. She is specialised in research on i) sedimentary formation processes of archaeological sites based on the use of soil micromorphology; ii) reconstruction of ancient paleogeography with a special focus on effects of geodynamical constrains on human settlement pattern and effects of natural factors on the preservation of past human landscapes; iii) palaeoclimatic reconstructions based on the study of soil archives. Her main research areas are North West Europe, North West India, the Middle East. Main publications: Courty, M.A., Goldberg, P. and Macphail, R.I. (1989). Soil, micromorphology and archaeology. Cambridge Manuals in Archaeology. Cambridge University Press. Courty, M.-A. and Fedoroff, N. (1985). Micromorphology of recent and buried soils in a semi-arid region of North-West India. Geoderma 35, 287-332. Fedoroff, N., Courty, M.A. and Thompson, M.L., (1990). Micromorphological evidence of palaeoenviromnental change in Pleistocene and Holocene paleosols. In: Soil Micro-morphology: a basic and applied science, L.A. Douglas (ed.), Elsevier, pp. 653-666. Courty, M.A.. 1994. Le cadre paléogéographique des occupations humaines dans le bassin du Haut Khabour (Syrie du nord-est). Premiers résultats. Paléorient, 20(1): 21-59, CNRS (ed.), Paris. GE, T., Courty, M.A., Wattez, J. and Matthews, W. (1993). Sedimentary formation processes of occupation surfaces. In: Formation processes in archaeological context, P. Goldberg, M. Petraglia and D.T. Nash (eds.). Monographs in World Archaeology, No 17: pp. 149-163. Madison, Wisconsin: Prehistory Press. Weiss, H., Courty, M.A.. Wetterstrom, W.. Meadow, R., Guichard, F.. Senior, L. and Curnow, A. (1993). The origin and collapse of Third Millenium North Mesopotamian Civilization. Science, 261, pp. 995-1004. Courty, M.A. Marlin, C., Dever, L., Tremblay P. and Vachier, P. (1994). Morphology, geochemistry and origin of calcitic pendents from the High Arctic (Spitsbergen). Geoderma, 61, pp. 71-102. Courty, M.-A. and Weiss, H. (1997). The scenario of environmental degradatation in the Tell Leilan region (N.E. Syria) during the late third millennium abrupt climate change. Proceedings of the NATO-ARW workshop on Third Millennium BC Abrupt Climate Change and Old World Social Collapse, N. Dalfes, G. Kukla and H. Weiss (eds), Berlin/Heidelberg: Springer Verlag.

LARS G. FRANZÉN is an associate professor in Physical Geography at the Department of Earth Sciences, Goeteborg University, Sweden. Initially studying peatlands from a nature-resource point of view, his present interest is peatlands as archives for atmospheric events during the Holocene, and their role as climate regulators during the entire Phanerozoic era. During the last 10 years he has also performed studies of the landscape and climate development in Tunisia during the Holocene. In the Tunisia study, playa deposits are used in a similar way as peat stratigraphies. During the field work in Tunisia several new archaeological sites were also found which could be associated with climate events.

GUNNAR HEINSOHN (*1943 in Gdynia/Poland) studied sociology, history, psychology, economics and religious studies at the *Freie Universität Berlin*. He gained a university diploma in sociology (1971) as well as doctorates (both *summa cum laude*) in the social sciences (1973) and in economics (1982). His publication list exceeds more than 440 titles including twenty monographies. Since 1984 he has been a tenured professor at the *Universität Bremen* where her serves as director of the *Raphael-Lemkin-Institut für Xenophobie und Genozidforschung* [Institute for xenophobia and genocide research] since 1993. His research focuses on the history and theory of civilization with special emphasis to (i) childhood/family/population, (ii) economy, (iii) religion, (iv) Genocide, and (v) chronology. Childhood/family/population was dealt with in (1) *Vorschulerziehung in der bürgerlichen Gesellschaft* [1991; Pre-school education in the modern age] (Frankfurt/M, 1974). (2) *Theorie des Kindergartens und der Spielpädagogik* [Theory of the kindergarten and the play-pedagogy] (Frankfurt/M, 1975; co-author: B.M.C. Knieper). (3) *Theorie des Familienrechts: Geschlechts-rollenaufhebung, Kindesvernachlässigung, Geburten-rückgang* [1974; Theory of family law: Diminishing sex role differences, child negligence and falling birth rates] (Frankfurt/M., 1976[2], co-author: R. Knieper). (4) *Menschenproduktion: Allgemeine Bevölkerungstheorie der Neuzeit* [Production of human beings: General population theory of Modern Times] (Frankfurt/M, 1979; co-authors: R. Knieper and O. Steiger). On the economy appeared (5) *Das Kibbuz-Modell - Bestandsaufnahme einer alternativen Wirtschafts- und Lebensform nach sieben Jahrzehnten* [The Kibbutz as a model: an alternative economy and life style after seven decades] (Frankfurt/M, 1982). (6) *Privateigentum, Patriarchat, Geldwirtschaft. Eine sozialtheoretische Rekonstruktion zur Antike* [Property, patriarchy and the monetary economy: A socio-theoretical reconstruction of antiquity] (Frankfurt/M, 1984). (7) *Eigentum, Zins und Geld: Ungelöste Rätsel der Wirtschaftswissenschaft* [Property, interest and money: Unsolved enigmas of economics] (Reinbek, 1996; co-author: O. Steiger). Religion became the subject of (8) *Was ist Antisemitismus? Der Ursprung von Monotheismus und Judenhaß. Warum Antizionismus?* [What is anti-Semitism? - The origin of monotheism and anti-Judaism. - Why anti-Zionism?] (Frankfurt/M, 1988), and (9) *Die Erschaffung der Götter: Das Opfer als Ursprung der Religion* [The formation of the gods: Sacrifice and the origin of religion] (Reinbek, 1997). Genocide was dealt with in (10) *Die Vernichtung der weisen Frauen: Beiträge zur Theorie und Geschichte von Bevölkerung und Kindheit* [The elimination of the wise women: Contributions to the theory and history of population and childhood] (München, 1994; Co-author: O. Steiger). (11) *Warum Auschwitz? Hitlers Plan und die Ratlosigkeit der Nachwelt* [Why Auschwitz? Hitlers plan and the perplexity of posteriority] (Reinbek, 1995). (12) *Was wollte Hitler? Auschwitz und die Lehre von den drei Weltzeitaltern* [What was Hitlers motive? Auschwitz and the doctrine of the three universal epochs] (Bremen, 1996). (13) *Ökologie und Völkermord* [Ecology and genocide] (Bremen, 1997). (14) *GULag und Auschwitz: Ein Wort zur Klärung der Differenz* [GULag and Auschwitz: a word on the difference] (Bremen, 1998). (15) *Lexikon der Völkermorde* [Encyclopedia of genocides] (Reinbek, 1998, in preparation). On chronology were published (16) *Nullpunkt Abraham. Abraham und die Chronologien Mesopotamiens und Ägyptens* [Point zero: Abraham and the chronologies of Mesopotamia and Egypt] (Basel, 1987). (17) *Wie alt ist das Menschengeschlecht?* [19911; How ancient is mankind?] (München, 1996). (18) *Die Sumerer gab es nicht: Von den Phantom-Imperien der Lehrbücher zur wirklichen Epochenabfolge in der Zivilisationswiege Südmesopotamien* [The Sumerians never existed: From the phantom empires of the textbooks to the real sequence of epochs in Southern Mesopotamias cradle of civilization

(Frankfurt/M, 1988). (19) *Wann lebten die Pharaonen? Archäologische und technologische Grundlagen für eine Neuschreibung der Geschichte Ägyptens und der übrigen Welt* [When did the pharaohs thrive? Archaeological and technological foundations for the re-writing of ancient history] (München, 1997; co-author: H. Illig). (20) *Wer herrschte im Industal? Die wiedergefundenen Imperien der Meder und der Perser* [19931; Who ruled in the Indus Valley? The rediscovery of the Medish and Akhaemenid Empires] (Gräfelfing, 1997). (21) *Assyrerkönige gleich Perserherrscher! Die Assyrienfunde bestätigen das Achämenidenreich* [1992; Assyrian kings as Persian rulers of Assyria: The archaeology of Assyria confirms the existence of the Akhaemenid Empire] (Gräfelfing, 1996).

EUAN MACKIE read Archaeological and Anthropological Tripos at St. John's College, Cambridge, in 1956-59. Ph.D at Glasgow Univ. 1963-1972. He was a member of the Cambridge Expedition to British Honduras 1959-60, responsible for the work of the archaeological section. The team uncovered some striking and unexpected evidence about the collapse of the Maya civilisation at this site. In October 1960, he was appointed Assistant Keeper of the Hunterian Museum, Glasgow. From 1985-1993 he served as the Museum's Deputy Director. His research areas include the Iron Age cultures of Atlantic (highland and island) Scotland and the monuments of the late Neolithic period. He has been mainly involved in excavating and studying brochs and allied structures, together with two 'vitrified forts', with the aim of dating them and placing them in a clear archaeological sequence. More recently the emphasis has been on trying to reconstruct the kind of society which produced these monumental drystone structures. Prehistoric standing stones, stone circles and rock carvings and have been an interest since about 1968. They present a different problem from that of the Iron Age; domestic sites and debris are largely absent and the evidence is mainly in the form of enigmatic ceremonial structures. The problem of reconstructing aspects of the ceremonial practices of the time has been pursued, particularly by excavating sites to check the calendrical hypothsis of A. Thom, itself based largely on statistical analysis. This procedure of testing ideas by excavation has proved rewarding though controversial. A book resulted from this work in 1977 [*The Megalith Builders*. Phaidon. 1977], and attempted a radical re-interpretation of the nature of late Neolithic society in Britain, [*Science and Society in Prehistoric Britain.* Paul Elek, 1977; see also the most recent publication: "Maeshowe and the winter solstice: ceremonial aspects of the Orkney Grooved ware culture." In: *Antiquity*, June 1997].

W. BRUCE MASSE is an environmental archaeologist with the U.S. Air Force and is affiliate graduate faculty at the University of Hawaii. He received his degrees in Anthropology from Stanford University (B.A.), the University of Arizona (M.A.), and Southern Illinois University at Carbondale (Ph.D.). His doctoral thesis on the archaeology and ecology of fishing in the Belau Islands of Micronesia was honored as runnerup for Outstanding Dissertation of the Year by the Society for American Archaeology. He has conducted extensive field research in Micronesia, Hawaii, and the American Southwest, and has published on a wide variety of topics including prehistoric irrigation and dry farming systems, zooarchaeology, settlement patterns, chronometrics, archaeoastronomy, cosmography, and public archaeology.

WILLIAM MULLEN is Professor of Classical Studies at Bard College, and is the author of Choreia: Pindar and Dance. In the 1970's he was an editor of Pensee and published several catastrophist papers there, including "A Reading of the Pyramid Texts" and "The Mesoamerican Record". In 1995 he presented "Worlds in Collision after Heinsohn" at the New York Velikovsky Centennial. He is now writing a book on Catastrophism and the 'Axial Age' (simultaneous breakthroughs in cultures from Greece to China).

BILL NAPIER is an astronomer at Armagh Observatory in Northern Ireland, where he is financed by the Leverhulme Trust to carry out research into 'Giant comets, catastrophes and civilisation'. After obtaining degrees in astronomy at Glasgow University, he spent the bulk of his career as an astronomer at the Royal Observatory, Edinburgh, where he gained an 'individual merit' grade, before moving briefly to Oxford and thence to Armagh. His main research interest is in the field of comets and catastrophes, and he has co-authored a number of recent and forthcoming books on these subjects. More recently he has also initiated a programme of testing claims made over the years by Arp, Tifft and others that some extragalactic redshifts cannot be explained in terms of currently understood physics.

AMOS NUR is the Wayne Loel Professor of Earth Sciences and Professor of Geophysics at Stanford University. Amos specializes in earthquake physics. For over twenty years, he has been investigating the temporal and spatial patterns of earthquakes throughout history to find clues useful for earthquake prediction. The longest and most complete record is in the Holy Land, where the Dead Sea seismic fault defines the Arabia - Africa plate boundary, as the San Andreas defines the N. American/Pacific plate boundary. Together with colleagues in archaeology, history, geology and geophysics at Stanford and Israel, Amos has organized an expedition to search for, and excavate and recover skeleton/s, artifacts, and Dead Sea scrolls buried 2000 years ago in the "Cave of Letters. in Israel's Judean desert by the devastating Dead Sea Earthquake of 31 BC. The first part of this expedition took place in March of this year. Amos is a winner of the Silver Apple Award for physical sciences at the National Educational Film Festival, 1991 for producing and directing a video on *Earthquakes in the Holy Land*. This has been shown extensively on a number of PBS stations around the country. Publications: Over 180 papers in refereed journals; and 3 books.

TREVOR PALMER graduated in biochemistry from the University of Cambridge and obtained a PhD from the University of London for research into inherited diseases. He is now Professor and Head of Department of Life Sciences, Dean of the Faculty of Science and Mathematics and Senior Dean at the Nottingham Trent University. He has been a Member of Council of the Society for Interdisciplinary Studies since 1986, and was Chairman from 1995 to 1998. His books include *Understanding Enzymes (1981,1985,1991,1994), Catastrophism, Neocatastrophism and Evolution* (1993), *Principles of Enzymology for Technological Applications* (1993) and *Controversy - Catastrophism and Evolution: The Ongoing Debate* (1998).

BENNY J PEISER was born in Israel and educated in West Germany. He is a historian and social anthropologist with particular research interest in neo-catastrophism and its implications for human and societal evolution. Benny has been researching, writing and lecturing about neo-catastrophism in the fields of ancient history, archaeology, cultural anthropology (and the origins of ancient combat sports) since 1987. He has presented and published numerous papers on the historical, intellectual and cultural implications of neo-catastrophism. He is the moderator of the electronic *Cambridge-Conference Network* (CCNet).

DUNCAN STEEL is director of Spaceguard Australia and Vice-President of The Spaceguard Foundation. Steel took B.Sc. and M.Sc. degrees at the University of London, studying at various colleges, and then moved to the University of Colorado where he worked on NASA's Pioneer Venus Orbiter program. He then moved to New Zealand, where he took his Ph.D. degree at the University of Canterbury for work on radar observations of meteors. He continued this work from 1985-96 at the University of Adelaide, Australia, spending 1987 as an ESA Fellow at the University of Lund,

Sweden. From 1990 until its termination in 1996 he directed the only southern hemisphere program for the discovery and tracking of near-Earth asteroids, based at the Anglo-Australian Observatory. In connection with this he has served on various international committees, including NASA's Spaceguard committee in 1991-92 and the International Astronomical Union's Working Group on Near-Earth Objects since 1991, and Working Group on the Prevention of Interplanetary Pollution since 1988. He is the author of over a hundred scientific papers, and the book Rogue Asteroids and Doomsday Comets" (Wiley, Chichester and New York, 1995). In addition, he has appeared in numerous TV and radio documentary programs concerned with the impact hazard.

DAVID PANKENIER (Ph.D. Stanford University, 1983) is Associate Professor of Chinese and Chair of the department of Modern Foreign Languages and Literatures at Lehigh University, Bethlehem, PA, USA. His research interests include ancient Chinese archaeology, astronomy, astrology, chronology, portentology, and the pre-imperial history of the foundational notion of heaven-sanctioned political legitimacy (Heaven's Mandate). This study has led to the identification and verification of several ancient records of rare planetary massings associated with dynastic transitions in the Chinese Bronze Age (2nd millennium BCE). A comprehensive discussion of these discoveries and their ideological implications may be found in "The Cosmo-Political Background of Heaven's Mandate", *Early China* 20 (1995):121-176.

BAS VAN GEEL is a paleo-ecologist in the research group Palynology and Paleo/Actuo-ecology at the University of Amsterdam, The Netherlands. He studies pollen, fungi, algae and other fossil objects from peat deposits, lake sediments and archaeological sites and reconstructs changes in vegetation and climate in the past. Climate change during the Bronze Age and especially the abrupt climate change at the Bronze Age/Iron Age transition was already one of the main topics in his Ph.D.- thesis (1976). Recently he applied the new strategy of 14C wiggle-match dating of peat deposits, which appeared to be of crucial importance for understanding the world-wide dramatic changes in climate around 800 BC. Changes in solar activity and related changes in cosmic rays, 14C-production, cloudiness and precipitation were very important factors for climate change in the past, and such changes in solar activity probably will also dominate climate change in the future. Bas van Geel is (co)author of more than 50 papers in international scientific journals and 8 chapters in books.

GERRIT L. VERSCHUUR is Adjunct Professor in the Physics Department, University of Memphis, Memphis, Tennessee. After obtaining his Ph. D. at the University of Manchester, where he worked at Jodrell Bank, he emigrated to the USA. As a professional radio astronomer he pioneered the measurement of the interstellar magnetic field in the context of broader studies of hydrogen gas between the stars. He has published over 70 papers in scientific journals, has published a similar number of articles on popular astronomy and has written seven books and co-authored or edited three more. His latest book is *Impact! The Threat of Comets and Asteroids* (Oxford University Press, 1996) grew out of his continued interest in the fact that life on earth exists and evolves in the context of cosmic phenomena. He is currently engaged in a number of commercial ventures based on devices he has invented in recent months.

IRVING WOLFE is professor agrégé, Département d'études anglaises, at the University of Montreal. (M.A. in English from McGill University, Canada; PhD in Drama from Bristol University, England; Speciality Shakespeare). In addition to his literary research, he has presented several dozen papers on catastrophism at conferences in England, Canada and the U.S., and has published papers in this field in several British and American journals. Has also contributed chapters on catastrophism to a number of anthologies. He is currently completing a book on Hitler and catastrophic terror and has prepared a screenplay for Hollywood of a mass shooting by Nazi soldiers. Professor Wolfe is concerned with the long-lasting effects of unconscious catastrophic memory on later collective human behaviour, and his work has therefore focused on such topics as catastrophism and creative art; catastrophism and religion; catastrophism and cosmologies; catastrophism and political extremism; catastrophism and epistemology; catastrophism and soap opera; catastrophism and Shakespeare; catastrophism and scientific repression; catastrophism and sport, and catastrophism and popular culture.

SOCIETY FOR INTERDISCIPLINARY STUDIES (SIS)

Despite the best efforts of scientists and historians, there is still much we do not know about the history of life, the Earth, its peoples and the Solar System and Universe they inhabit. Theories have been developed and researchers are working to refine them and fill in the details, but in many areas there is uncertainty.

The *Society for Interdisciplinary Studies* (SIS) is a UK-based organisation with worldwide membership, formed in response to a growing interest in the works of modern catastrophists who have stimulated controversy in the fields of cosmology, geology, and ancient history.

The SIS also provides a forum for independent and critical assessment of the works of modern as well as earlier catastrophists.

SIS membership brings together people from a wide variety of backgrounds and interests. Each year, members of the Society receive: Two issues of the Society's journal *Chronology and Catastrophism Review*, a twice-yearly high quality journal. Two issues of *SIS Internet Digest*, a twice-yearly publication summarising Web sites and news of interest to members.

For further information please write to Mrs. Val Pearce, SIS Membership Secretary, 10 Witley Green, Darley Heights, Stopsley, Bedfordshire LU2 8TR, UK

e-mail: SIS@knowledge.co.uk

SIS Homepage (on the www): http://www.knowledge.co.uk/xxx/cat/sis

www.ingramcontent.com/pod-product-compliance
Ingram Content Group UK Ltd.
Pitfield, Milton Keynes, MK11 3LW, UK
UKHW061846041026
15591UKWH00015B/942